T0224356

# Smooth Manifolds

Rajnikant Sinha

# Smooth Manifolds

Springer

Rajnikant Sinha
Department of Mathematics
Magadh University
Bodh Gaya, Bihar
India

ISBN 978-81-322-2955-1     ISBN 978-81-322-2104-3   (eBook)
DOI 10.1007/978-81-322-2104-3

Springer New Delhi Heidelberg New York Dordrecht London

Printed on acid-free paper

Springer is part of Springer Science+Business Media (www.springer.com)

# Preface

The theory of smooth manifolds is a frequently discussed area in modern mathematics among the broad class of audience. A number of popular books on this subject are available for general readers with little prerequisite in mathematics. On the other hand, numerous textbooks and monographs for advanced graduate students and research fellows are available as well. The intermediate books, allowing senior undergraduate students to enter the exciting field of smooth manifolds in a pedagogical way, are very few.

It is argued that new concepts involved in this field are too complicated to allow a simple introduction. Thus, the only way to master this theory is to be a time-consuming effort to accumulate the intelligible parts of advanced textbooks into a comprehensible collection of notes. No doubt, this approach has the advantage of making young students, with perseverance to go through such a learning process, a very well-trained future research worker. But unfortunately, this severely limits the number of students who ever really master the subject. This book modestly attempts to bridge the gap between a university curriculum and the more advanced books on *Smooth Manifolds*. Actually, the manuscript has evolved from my habit of writing elaborate proofs of different theorems over the past one decade.

This book intentionally supplies to the readers a high level of detail in arguments and derivations. More lengthy proofs of various theorems are, in general, outlined in such a way that they can be digested by an interested senior undergraduate student with little risk of ever getting lost. This book has been written in such a style that it invites the young students to fill in the gaps everywhere during its study. It is another matter that the author is there for the needy students with complete solutions. That is why, supplying a separate list of problems will only deviate us from the main aim of this book.

In my experience, some common problems for a young university student, trying to master mathematics, are the phrases in literatures like "it is clear that ...", "it is easy to see that ...", "it is straightforward", etc. If a student cannot supply the proof of such a "clearly," which most likely is the case, the common reaction under the time pressure of the studies is to accept the statement as true. And from this point on, throughout the rest part of the course, the deeper understanding of the subject is lost.

I have benefited from a number of advanced textbooks on the subject, and some unpublished works of my own and my distinguished friends. Some of these sources are mentioned in the bibliography. However, in an introductory book such as this, it is not possible to mention all literatures and all the people who have contributed to the understanding of this exciting field.

It is hoped that our readers (active and passive both) will find that I have substantially fulfilled the objective of bridging the gap between university curriculum and more advanced texts, and that they will enjoy the reading this book as much as I did while writing it.

Rajnikant Sinha

# Contents

# About the Author

Rajnikant Sinha is former professor of mathematics at Magadh University, Bodh Gaya, India. A passionate mathematician by heart, Prof. Sinha has published several interesting researches in international journals and contributed a book *Solutions to Weatherburn's Elementary Vector Analysis*. His research areas are topological vector spaces, differential geometry, and manifolds.

# Chapter 1
# Differentiable Manifolds

The pace of this chapter is rather slow for the simple reason that the definition of smooth manifold is itself forbidding for many unfamiliar readers. Although many young university students feel comfortable with the early parts of multivariable calculus, if one feels uncomfortable, then he should go first through some parts of Chap. 3. How verifications are done in various examples of smooth manifolds is a crucial thing to learn for a novice, so this aspect has been dealt here with some more detail.

## 1.1 Topological Manifolds

**Definition** Let $M$ be a Hausdorff topological space. Let $m$ be a positive integer. If, for every $x$ in $M$, there exists an open neighborhood $U$ of $x$ such that $U$ is homeomorphic to some open subset of Euclidean space $\mathbb{R}^m$, then we say that $M$ is an *m-dimensional topological manifold*.

**Definition** Let $M$ be an $m$-dimensional topological manifold. Let $x$ be an element of $M$. So there exists an open neighborhood $U$ of $x$ such that $U$ is homeomorphic to some open subset of Euclidean space $\mathbb{R}^m$. Hence, there exists a homeomorphism

$$\varphi_U : U \to \varphi_U(U)$$

such that $\varphi_U(U)$ is an open subset of $\mathbb{R}^m$. Here, the ordered pair $(U, \varphi_U)$ is called a *coordinate chart of M*. $(U, \varphi_U)$ is also simply denoted by $(U, \varphi)$.

**Definition** Let $M$ be an $m$-dimensional topological manifold. Let $(U, \varphi_U)$ be a coordinate chart of $M$. So

$$\varphi_U : U \to \varphi_U(U) \quad (\subset \mathbb{R}^m).$$

© Springer India 2014
R. Sinha, *Smooth Manifolds*, DOI 10.1007/978-81-322-2104-3_1

Hence, for every $x$ in $U$, $\varphi_U(x)$ is in $\mathbb{R}^m$. So there exist real numbers $u^1, \ldots, u^m$ such that

$$\varphi_U(x) = (u^1, \ldots, u^m).$$

Here, we say that $u^i (i = 1, \ldots, m)$ are the *local coordinates of the point $x$*. In short, we write

$$(\varphi_U(x))^i \equiv u^i.$$

**Lemma 1.1** *Let $M$ be an $m$-dimensional topological manifold. Let $(U, \varphi_U), (V, \varphi_V)$ be coordinate charts of $M$ such that $U \cap V$ is nonempty. Then,*

1. $\mathrm{dom}(\varphi_V \circ (\varphi_U)^{-1}) = \varphi_U(U \cap V)$,
2. $\mathrm{ran}(\varphi_V \circ (\varphi_U)^{-1}) = \varphi_V(U \cap V)$,
3. $\varphi_V \circ (\varphi_U)^{-1}$ *is a function,*
4. $\varphi_V \circ (\varphi_U)^{-1}$ *is 1–1,*
5. $\varphi_U(U \cap V), \varphi_V(U \cap V)$ *are open subsets of $\mathbb{R}^m$,*
6. $(\varphi_V \circ (\varphi_U)^{-1})^{-1} = \varphi_U \circ (\varphi_V)^{-1}$,
7. $(\varphi_V \circ (\varphi_U)^{-1}) : \varphi_U(U \cap V) \to \varphi_V(U \cap V)$, and $(\varphi_U \circ (\varphi_V)^{-1}) : \varphi_V(U \cap V) \to \varphi_U(U \cap V)$ *are homeomorphisms.*

*Proof of 1* Let us take any $x$ in $\mathrm{dom}(\varphi_V \circ (\varphi_U)^{-1})$. We want to show that $x$ is in $\varphi_U(U \cap V)$.

Since $x$ is in $\mathrm{dom}(\varphi_V \circ (\varphi_U)^{-1})$, by the definition of domain, there exists $y$ such that $(x, y)$ is in $\varphi_V \circ (\varphi_U)^{-1}$. Since $(x, y)$ is in $\varphi_V \circ (\varphi_U)^{-1}$, by the definition of composition, there exists $z$ such that $(x, z)$ is in $(\varphi_U)^{-1}$, and $(z, y)$ is in $\varphi_V$. Since $(x, z)$ is in $(\varphi_U)^{-1}$, $(z, x)$ is in $\varphi_U$. Since $(z, x)$ is in $\varphi_U$, and $\varphi_U : U \to \varphi_U(U)$, $\varphi_U(z) = x$, and $z$ is in $U$. Since $(z, y)$ is in $\varphi_V$, and $\varphi_V : V \to \varphi_V(V)$, $\varphi_V(z) = y$, and $z$ is in $V$. Here, $z$ is in $U$, and $z$ is in $V$, so $z$ is in $U \cap V$. Since $z$ is in $U \cap V$, and $x = \varphi_U(z)$, $x$ is in $\varphi_U(U \cap V)$. Thus, (see Fig. 1.1),

$$\mathrm{dom}\left(\varphi_V \circ (\varphi_U)^{-1}\right) \subset \varphi_U(U \cap V).$$

Next, let $x$ be in $\varphi_U(U \cap V)$. We want to prove that $x$ is in $\mathrm{dom}(\varphi_V \circ (\varphi_U)^{-1})$.

Since $x$ is in $\varphi_U(U \cap V)$, there exists $y$ in $U \cap V$ such that $\varphi_U(y) = x$. Since $y$ is in $U \cap V$, $y$ is in $U$, and $y$ is in $V$. Since $y$ is in $U$, and $\varphi_U : U \to \varphi_U(U)$, there exists $z$ such that $(y, z)$ is in $\varphi_U$. Since $y$ is in $V$, and $\varphi_V : V \to \varphi_V(V)$, there exists $w$ such that $(y, w)$ is in $\varphi_V$. Since $(y, z)$ is in $\varphi_U$, $(z, y)$ is in $(\varphi_U)^{-1}$. Since $(z, y)$ is in $(\varphi_U)^{-1}$, and

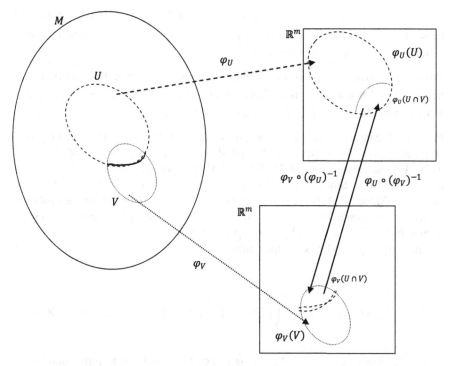

**Fig. 1.1** $m$-dimensional topological manifold

$(y, w)$ is in $\varphi_V$, $(z, w)$ is in $\varphi_V \circ (\varphi_U)^{-1}$, and hence, $z$ is in $\text{dom}(\varphi_V \circ (\varphi_U)^{-1})$. Since $(y, z)$ is in $\varphi_U$, and $\varphi_U$ is a function, $\varphi_U(y) = z$. Since $\varphi_U(y) = z$, and $\varphi_U(y) = x$, $x = z$. Since $x = z$, and $z$ is in $\text{dom}(\varphi_V \circ (\varphi_U)^{-1})$, $x$ is in $\text{dom}(\varphi_V \circ (\varphi_U)^{-1})$. Thus,

$$\varphi_U(U \cap V) \subset \text{dom}\Big(\varphi_V \circ (\varphi_U)^{-1}\Big).$$

$\square$

*Proof of 2* On using 1, we have $\text{dom}(\varphi_U \circ (\varphi_V)^{-1}) = \varphi_V(V \cap U)$. So

$$\begin{aligned}
\text{LHS} &= \text{ran}\Big(\varphi_V \circ (\varphi_U)^{-1}\Big) = \text{dom}\Big(\big(\varphi_V \circ (\varphi_U)^{-1}\big)^{-1}\Big) \\
&= \text{dom}\Big(\big((\varphi_U)^{-1}\big)^{-1} \circ (\varphi_V)^{-1}\Big) \\
&= \text{dom}\Big(\varphi_U \circ (\varphi_V)^{-1}\Big) \\
&= \varphi_V(V \cap U) \\
&= \varphi_V(U \cap V) = \text{RHS}.
\end{aligned}$$

$\square$

*Proof of 3* Since $(U, \varphi_U)$ is a coordinate chart of $M$, $\varphi_U$ is a 1–1 function, and hence, $(\varphi_U)^{-1}$ is a function. Since $(\varphi_U)^{-1}$ is a function, and $\varphi_V$ is a function, their composite $\varphi_V \circ (\varphi_U)^{-1}$ is a function.                               □

*Proof of 4* Since $(U, \varphi_U)$ is a coordinate chart of $M$, $\varphi_U$ is a 1–1 function, and hence, $(\varphi_U)^{-1}$ is a 1–1 function. Since $(V, \varphi_V)$ is a coordinate chart of $M$, $\varphi_V$ is a 1–1 function. Since $(\varphi_U)^{-1}$ is 1–1, and $\varphi_V$ is 1–1, $\varphi_V \circ (\varphi_U)^{-1}$ is 1–1.                               □

*Proof of 5* Since $(U, \varphi_U)$ is a coordinate chart of $M$, $\varphi_U : U \to \varphi_U(U)$ is a homeomorphism from open subset $U$ of $M$ onto open subset $\varphi_U(U)$ of $\mathbb{R}^m$. Similarly, $V$ is an open subset of $M$. Since $U, V$ are open subsets of $M$, $U \cap V$ is an open subset of $M$. Since $U, U \cap V$ are open subset of $M$, and $U \cap V \subset U$, $U \cap V$ is open in $U$. Since $U \cap V$ is open in $U$, and $\varphi_U : U \to \varphi_U(U)$ is a homeomorphism, $\varphi_U(U \cap V)$ is open in $\varphi_U(U)$. Since $\varphi_U(U \cap V)$ is open in $\varphi_U(U)$, and $\varphi_U(U)$ is an open subset of $\mathbb{R}^m$, $\varphi_U(U \cap V)$ is an open subset of $\mathbb{R}^m$.

Similarly, $\varphi_V(U \cap V)$ is an open subset of $\mathbb{R}^m$.                               □

*Proof of 6*

$$\text{LHS} = \left( \varphi_V \circ (\varphi_U)^{-1} \right)^{-1} = \left( \left( (\varphi_U)^{-1} \right)^{-1} \right) \circ (\varphi_V)^{-1} = \varphi_U \circ (\varphi_V)^{-1} = \text{RHS}.$$

□

*Proof of 7* Since $(\varphi_V \circ (\varphi_U)^{-1})^{-1} = \varphi_U \circ (\varphi_V)^{-1}$, it suffices to prove that

$$\left( \varphi_V \circ (\varphi_U)^{-1} \right) : \varphi_U(U \cap V) \to \varphi_V(U \cap V),$$

and

$$\left( \varphi_U \circ (\varphi_V)^{-1} \right) : \varphi_V(U \cap V) \to \varphi_U(U \cap V)$$

are continuous.

Since $(U, \varphi_U)$ is a coordinate chart of $M$, $\varphi_U$ is a homeomorphism, and hence, $(\varphi_U)^{-1}$ is continuous. Since $(V, \varphi_V)$ is a coordinate chart of $M$, $\varphi_V$ is a homeomorphism, and hence, $\varphi_V$ is continuous. Since $(\varphi_U)^{-1}, \varphi_V$ are continuous, their composite $\varphi_V \circ (\varphi_U)^{-1}$ is continuous.

Similarly, $\varphi_U \circ (\varphi_V)^{-1}$ is continuous. Hence, $\varphi_V \circ (\varphi_U)^{-1}$ is a homeomorphism. Similarly, $\varphi_U \circ (\varphi_V)^{-1}$ is a homeomorphism.                               □

## 1.2 Smooth Manifolds

**Definition** Let $M$ be an $m$-dimensional topological manifold. Let $(U, \varphi_U), (V, \varphi_V)$ be coordinate charts of $M$, and let $U \cap V$ be nonempty. So

$$\left(\varphi_V \circ (\varphi_U)^{-1}\right) : \varphi_U(U \cap V) \to \varphi_V(U \cap V)$$

is a homeomorphism from $\varphi_U(U \cap V)$ onto $\varphi_V(U \cap V)$, where $\varphi_U(U \cap V)$ and $\varphi_V(U \cap V)$ are open subsets of $\mathbb{R}^m$.

For every $(x^1, \ldots, x^m)$ in $\varphi_U(U \cap V)$, put

$$\left(\varphi_V \circ (\varphi_U)^{-1}\right)(x^1, \ldots, x^m) \equiv \left(f^1(x^1, \ldots, x^m), \ldots, f^m(x^1, \ldots, x^m)\right) = (y^1, \ldots, y^m),$$

where $(y^1, \ldots, y^m)$ is in $\varphi_V(U \cap V)$, and

$$\left(\varphi_U \circ (\varphi_V)^{-1}\right)(y^1, \ldots, y^m) \equiv \left(g^1(y^1, \ldots, y^m), \ldots, g^m(y^1, \ldots, y^m)\right) = (x^1, \ldots, x^m).$$

Since $\varphi_V \circ (\varphi_U)^{-1}$ is continuous, each $f^i : \varphi_U(U \cap V) \to \mathbb{R}$ is continuous. Similarly, each $g^i : \varphi_V(U \cap V) \to \mathbb{R}$ is continuous.

For fixed $i$, and for every $j = 1, \ldots, m$, if $(D_j f^i)(x^1, \ldots, x^m)$ exists at every point $(x^1, \ldots, x^m)$ of $\varphi_U(U \cap V)$, then we get functions $D_j f^i : \varphi_U(U \cap V) \to \mathbb{R}$. If every $D_j f^i : \varphi_U(U \cap V) \to \mathbb{R}$ is continuous, then we say that $f^i$ is $C^1$.

Similarly, by $g^i$ is $C^1$, we shall mean that each $D_j g^i : \varphi_V(U \cap V) \to \mathbb{R}$ $(j = 1, \ldots, m)$ is continuous. By $(U, \varphi_U)$ and $(V, \varphi_V)$ are $C^1$-compatible, we mean that

either $(U \cap V$ is an empty set),

or

$(f^i, g^i$ are $C^1$ for every $i = 1, \ldots, m$, whenever $U \cap V$ is a nonempty set).

For fixed $i$, and for every $j, k = 1, \ldots, m$, if $D_k(D_j f^i) (\equiv D_{kj} f^i)$ exists at every point of $\varphi_U(U \cap V)$, then we get functions $D_k(D_j f^i) : \varphi_U(U \cap V) \to \mathbb{R}$. If each $D_k(D_j f^i) : \varphi_U(U \cap V) \to \mathbb{R}$ is continuous, then we say that $f^i$ is $C^2$.

Similarly, by $g^i$ is $C^2$, we shall mean that each $D_k(D_j g^i) : \varphi_V(U \cap V) \to \mathbb{R}$ $(j, k = 1, \ldots, m)$ is continuous. By $(U, \varphi_U)$ and $(V, \varphi_V)$ are $C^2$-compatible, we mean that either $(U \cap V$ is an empty set),

or

$(f^i, g^i$ are $C^2$ for every $i = 1, \ldots, m$, whenever $U \cap V$ is a nonempty set).

Similar definitions for $(U, \varphi_U)$ and $(V, \varphi_V)$ are $C^3$-compatible can be supplied, etc.

It is clear that if $(U, \varphi_U)$ and $(V, \varphi_V)$ are $C^3$-compatible, then $(U, \varphi_U)$ and $(V, \varphi_V)$ are $C^2$-compatible, etc.

**Definition** Let $M$ be an $m$-dimensional topological manifold. Let $r$ be a positive integer. Let

$$\mathcal{A} \equiv \{(U, \varphi_U), (V, \varphi_V), (W, \varphi_W), \ldots\}$$

be a nonempty collection of coordinate charts of $M$. If

1. $\{U, V, W, \ldots\}$ is a cover of $M$, that is, $\cup\{U : (U, \varphi_U)$ is in $\mathcal{A}\} = M$,
2. all pairs of members in $\mathcal{A}$ are $C^r$-compatible,
3. $\mathcal{A}$ is *maximal* (in the sense that if $(\widetilde{U}, \varphi_{\widetilde{U}})$ is a coordinate chart of $M$, but not a member of $\mathcal{A}$, then there exists $(U, \varphi_U)$ in $\mathcal{A}$, such that $(\widetilde{U}, \varphi_{\widetilde{U}})$ and $(U, \varphi_U)$ are not $C^r$-compatible),

then we say that $\mathcal{A}$ *is a $C^r$-differentiable structure on $M$*, and the ordered pair $(M, \mathcal{A})$ is called a *$C^r$-differentiable manifold*. Here, members of $\mathcal{A}$ are called *admissible coordinate charts of $M$*.

**Theorem 1.2** *Let $M$ be an $m$-dimensional topological manifold. Let $r$ be a positive integer. Let*

$$\mathcal{A} \equiv \{(U, \varphi_U), (V, \varphi_V), (W, \varphi_W), \ldots\}$$

*be a nonempty collection of coordinate charts of $M$ satisfying*

1. *$\{U, V, W, \ldots\}$ is a cover of $M$, that is, $\cup\{U : (U, \varphi_U)$ is in $\mathcal{A}\} = M$,*
2. *all pairs of members in $\mathcal{A}$ are $C^r$-compatible.*

*Then, there exists a unique $C^r$-differentiable structure $\mathcal{B}$ on $M$ which contains $\mathcal{A}$.* (This theorem suggests that in constructing a differentiable manifold, it is enough to find a collection $\mathcal{A}$ of coordinate charts for which conditions 1 and 2 are satisfied.)

*Proof* **Existence** Let $\mathcal{B}$ be the collection of all coordinate charts $(U, \varphi_U)$ of $M$ such that $(U, \varphi_U)$ is $C^r$-compatible with every member of $\mathcal{A}$. First of all, we shall try to show that $\mathcal{B}$ contains $\mathcal{A}$. For this purpose, let us take any $(U, \varphi_U)$ in $\mathcal{A}$.

By the condition 2, $(U, \varphi_U)$ is $C^r$-compatible with all members of $\mathcal{A}$, so by the definition of $\mathcal{B}$, $(U, \varphi_U)$ is in $\mathcal{B}$. Hence, $\mathcal{A}$ is a subset of $\mathcal{B}$. Further since $\mathcal{A}$ is nonempty, and $\mathcal{A}$ is a subset of $\mathcal{B}$, $\mathcal{B}$ is nonempty.

Now we want to prove that $\mathcal{B}$ is a $C^r$-differentiable structure on $M$, that is,

1. $\cup\{U : (U, \varphi_U)$ is in $\mathcal{B}\} = M$,
2. all pairs of members of $\mathcal{B}$ are $C^r$-compatible.
3. $\mathcal{B}$ is maximal (in the sense that, if $(\widetilde{U}, \varphi_{\widetilde{U}})$ is a coordinate chart of $M$, but not a member of $\mathcal{B}$, then there exists $(U, \varphi_U)$ in $\mathcal{B}$ such that $(\widetilde{U}, \varphi_{\widetilde{U}})$ and $(U, \varphi_U)$ are not $C^r$-compatible).

For 1: Since $\mathcal{A}$ is contained in $\mathcal{B}$, $\cup\{U : (U, \varphi_U)$ is in $\mathcal{A}\} \subset \cup\{U : (U, \varphi_U)$ is in $\mathcal{B}\} \subset M$. Now, by the given condition 1, $\cup\{U : (U, \varphi_U)$ is in $\mathcal{B}\} = M$.
For 2: Let us take any $(U, \varphi_U)$ in $\mathcal{B}$, and $(V, \varphi_V)$ in $\mathcal{B}$. We have to prove that $(U, \varphi_U)$ and $(V, \varphi_V)$ are $C^r$-compatible.
Here, two cases arise: either $U \cap V$ is an empty set, or $U \cap V$ is a nonempty set. $\square$

Case I: when $U \cap V$ is an empty set. In this case, by the definition of $C^r$-compatible, $(U, \varphi_U)$ and $(V, \varphi_V)$ are $C^r$-compatible.
Case II: when $U \cap V$ is a nonempty set. For every $(x^1, \ldots, x^m)$ in $\varphi_U(U \cap V)$, put

$$\left(\varphi_V \circ (\varphi_U)^{-1}\right)(x^1, \ldots, x^m) \equiv (f^1(x^1, \ldots, x^m), \ldots, f^m(x^1, \ldots, x^m))$$
$$= (y^1, \ldots, y^m),$$

where $(y^1, \ldots, y^m)$ is in $\varphi_V(U \cap V)$, and

$$\left(\varphi_U \circ (\varphi_V)^{-1}\right)(y^1, \ldots, y^m) \equiv (g^1(y^1, \ldots, y^m), \ldots, g^m(y^1, \ldots, y^m))$$
$$= (x^1, \ldots, x^m).$$

We have to show that $f^i, g^i$ are $C^r$ for every $i = 1, \ldots, m$.
Since $U \cap V$ is nonempty, there exists an element $x$ in $U \cap V (\subset M)$. Since $x$ is in $M$, by condition 1, there exists $(W, \varphi_W)$ in $\mathcal{A}$ such that $x$ is in $W$. Since $(U, \varphi_U)$ is in $\mathcal{B}$, and $(W, \varphi_W)$ is in $\mathcal{A}$, by the definition of $\mathcal{B}$, $(U, \varphi_U)$ and $(W, \varphi_W)$ are $C^r$-compatible. Since $x$ is in $W$, and $x$ is in $U \cap V(\subset U)$, $x$ is in $W \cap U$, and hence, $U \cap W$ is a nonempty set. Since $(U, \varphi_U)$ and $(W, \varphi_W)$ are $C^r$-compatible, and $U \cap W$ is nonempty, by the definition of $C^r$-compatible, $f_1^i, g_1^i$ are $C^r$ for every $i = 1, \ldots, m$, where for every $(x^1, \ldots, x^m)$ in $\varphi_U(U \cap W)$,

$$\left(\varphi_W \circ (\varphi_U)^{-1}\right)(x^1, \ldots, x^m) \equiv (f_1^1(x^1, \ldots, x^m), \ldots, f_1^m(x^1, \ldots, x^m)) = (y^1, \ldots, y^m)$$
$$\in \varphi_W(U \cap W),$$

and

$$\left(\varphi_U \circ (\varphi_W)^{-1}\right)(y^1, \ldots, y^m) \equiv (g_1^1(y^1, \ldots, y^m), \ldots, g_1^m(y^1, \ldots, y^m)) = (x^1, \ldots, x^m).$$

Similarly, $f_2^i, g_2^i$ are $C^r$ for every $i = 1, \ldots, m$, where for every $(x^1, \ldots, x^m)$ in $\varphi_V(V \cap W)$,

$$\left(\varphi_W \circ (\varphi_V)^{-1}\right)(x^1, \ldots, x^m) \equiv \left(f_2^1(x^1, \ldots, x^m), \ldots, f_2^m(x^1, \ldots, x^m)\right) = (y^1, \ldots, y^m),$$
$$\in \varphi_W(V \cap W)$$

and

$$\left(\varphi_V \circ (\varphi_W)^{-1}\right)(y^1, \ldots, y^m) \equiv \left(g_2^1(y^1, \ldots, y^m), \ldots, g_2^m(y^1, \ldots, y^m)\right) = (x^1, \ldots, x^m).$$

Now for every $(x^1, \ldots, x^m)$ in $\varphi_U(U \cap V \cap W)$

$$\left(f^1(x^1, \ldots, x^m), \ldots, f^m(x^1, \ldots, x^m)\right)$$
$$= \left(\varphi_V \circ (\varphi_U)^{-1}\right)(x^1, \ldots, x^m)$$
$$= \left(\left(\varphi_V \circ (\varphi_W)^{-1}\right) \circ \left(\varphi_W \circ (\varphi_U)^{-1}\right)\right)(x^1, \ldots, x^m)$$
$$= \left(\varphi_V \circ (\varphi_W)^{-1}\right)\left(\left(\varphi_W \circ (\varphi_U)^{-1}\right)(x^1, \ldots, x^m)\right)$$
$$= \left(\varphi_V \circ (\varphi_W)^{-1}\right)\left(f_1^1(x^1, \ldots, x^m), \ldots, f_1^m(x^1, \ldots, x^m)\right)$$
$$= \left(g_2^1\left(f_1^1(x^1, \ldots, x^m), \ldots, f_1^m(x^1, \ldots, x^m)\right), \ldots, g_2^m\left(f_1^1(x^1, \ldots, x^m), \ldots, f_1^m(x^1, \ldots, x^m)\right)\right).$$

Hence,

$$f^1(x^1, \ldots, x^m) = g_2^1\left(f_1^1(x^1, \ldots, x^m), \ldots, f_1^m(x^1, \ldots, x^m)\right).$$

Since $g_2^1$, and each $f_1^k (k = 1, \ldots, m)$ are $C^r$,

$$\frac{\partial g_2^1\left(f_1^1(x^1, \ldots, x^m), \ldots, f_1^m(x^1, \ldots, x^m)\right)}{\partial x^1}$$
$$\left(= \sum_{k=1}^n \frac{\partial g_2^1\left(f_1^1(x^1, \ldots, x^m), \ldots, f_1^m(x^1, \ldots, x^m)\right)}{\partial f_1^k(x^1, \ldots, x^m)} \frac{\partial f_1^k(x^1, \ldots, x^m)}{\partial x^1}\right)$$

exists and is continuous. It follows that

$$\frac{\partial f^1(x^1, \ldots, x^m)}{\partial x^1} \left(= \frac{\partial g_2^1\left(f_1^1(x^1, \ldots, x^m), \ldots, f_1^m(x^1, \ldots, x^m)\right)}{\partial x^1}\right)$$

exists and is continuous. Similarly, $\frac{\partial f^1(x^1, \ldots, x^m)}{\partial x^2}$ exists and is continuous, etc.

So $f^1$ is $C^1$. Similarly, $f^1$ is $C^2$, etc. Hence, $f^1$ is $C^r$. Similarly, $f^2$ is $C^r$, etc. So each $f^i$ is $C^r$. Similarly, each $g^i$ is $C^r$. Thus, we have shown that, in all cases, $(U, \varphi_U)$ and $(V, \varphi_V)$ are $C^r$-compatible.

For 3: We claim that $\mathcal{B}$ is maximal. If not, otherwise, let $\mathcal{B}$ be not maximal. We have to arrive at a contradiction. Since $\mathcal{B}$ is not maximal, there exists a coordinate chart $(U, \varphi_U)$ of $M$ such that $(U, \varphi_U)$ is not a member of $\mathcal{B}$, and $(U, \varphi_U)$ is $C^r$-compatible with every member of $\mathcal{B}$. Since $(U, \varphi_U)$ is $C^r$-compatible with every member of $\mathcal{B}$, and $\mathcal{A}$ is contained in $\mathcal{B}$, $(U, \varphi_U)$ is $C^r$-compatible with every member of $\mathcal{A}$, and hence, by the definition of $\mathcal{B}$, $(U, \varphi_U)$ is in $\mathcal{B}$, a contradiction.

Thus, we have shown that $\mathcal{B}$ is a $C^r$-differentiable structure on $M$, which contains $\mathcal{A}$.

**Uniqueness** If not, otherwise, let $\mathcal{B}_1$ and $\mathcal{B}_2$ be two distinct $C^r$-differentiable structures on $M$, which contains $\mathcal{A}$. We have to arrive at a contradiction. Since $\mathcal{B}_1$ and $\mathcal{B}_2$ are two distinct $C^r$-differentiable structures on $M$, either (there exists a coordinate chart $(U, \varphi_U)$ of $M$ such that $(U, \varphi_U)$ is in $\mathcal{B}_1$, and $(U, \varphi_U)$ is not in $\mathcal{B}_2$) or (there exists a coordinate chart $(U, \varphi_U)$ of $M$ such that $(U, \varphi_U)$ is in $\mathcal{B}_2$, and $(U, \varphi_U)$ is not in $\mathcal{B}_1$).

Case I: when there exists a coordinate chart $(U, \varphi_U)$ of $M$ such that $(U, \varphi_U)$ is in $\mathcal{B}_1$, and $(U, \varphi_U)$ is not in $\mathcal{B}_2$. First of all, we shall show that $\mathcal{B}_1$ is contained in $\mathcal{B}$, where $\mathcal{B}$ is the differential structure constructed above. For this purpose, let us take any coordinate chart $(V, \varphi_V)$ of $M$ in $\mathcal{B}_1$. We have to show that $(V, \varphi_V)$ is in $\mathcal{B}$. Since $(V, \varphi_V)$ is in $\mathcal{B}_1$, and $\mathcal{B}_1$ is a $C^r$-differentiable structure on $M$ which contains $\mathcal{A}$, $(V, \varphi_V)$ is $C^r$-compatible with every member of $\mathcal{A}$, and hence, by the definition of $\mathcal{B}$, $(V, \varphi_V)$ is in $\mathcal{B}$. Thus, we have shown that $\mathcal{B}_1$ is contained in $\mathcal{B}$.

Similarly, $\mathcal{B}_2$ is contained in $\mathcal{B}$. Since $(U, \varphi_U)$ is not in $\mathcal{B}_2$, and $\mathcal{B}_2$ is $C^r$-differentiable structure on $M$, there exists $(V, \varphi_V)$ in $\mathcal{B}_2$, such that $(V, \varphi_V)$ and $(U, \varphi_U)$ are not $C^r$-compatible. Since $(V, \varphi_V)$ is in $\mathcal{B}_2$, and $\mathcal{B}_2$ is contained in $\mathcal{B}$, $(V, \varphi_V)$ is in $\mathcal{B}$. Since $(U, \varphi_U)$ is in $\mathcal{B}_1$, and $\mathcal{B}_1$ is contained in $\mathcal{B}$, $(U, \varphi_U)$ is in $\mathcal{B}$. Since $(U, \varphi_U), (V, \varphi_V)$ are in $\mathcal{B}$, and $\mathcal{B}$ is a $C^r$-differentiable structure on $M$, $(V, \varphi_V)$ and $(U, \varphi_U)$ are $C^r$-compatible, a contradiction.

Case II: when there exists a coordinate chart $(U, \varphi_U)$ of $M$ such that $(U, \varphi_U)$ is in $\mathcal{B}_2$, and $(U, \varphi_U)$ is not in $\mathcal{B}_1$. This case is similar to the case I. $\square$

**Theorem 1.3** *Let $M$ be an $m$-dimensional topological manifold. Let*

$$\mathcal{A} \equiv \{(U, \varphi_U), (V, \varphi_V), (W, \varphi_W), \ldots\}$$

*be a nonempty collection of coordinate charts of $M$ satisfying*

1. *$\{U, V, W, \ldots\}$ is a cover of $M$, that is, $\cup\{U : (U, \varphi_U)$ is in $\mathcal{A}\} = M$,*
2. *all pairs of members in $\mathcal{A}$ are $C^2$-compatible.*

*Then, there exists a unique $C^1$-differentiable structure $\mathcal{B}$ on $M$ which contains $\mathcal{A}$.*

*Proof* First of all, we shall show that $2'$: all pair of members in $\mathcal{A}$ are $C^1$-compatible. For this purpose, let us take any $(U, \varphi_U)$ and $(V, \varphi_V)$ in $\mathcal{A}$. Here, two cases arises: either $U \cap V$ is an empty set, or $U \cap V$ is a nonempty set. $\square$

Case I: when $U \cap V$ is an empty set. In this case, by the definition of $C^1$-compatible, $(U, \varphi_U)$ and $(V, \varphi_V)$ are $C^1$-compatible.

Case II: when $U \cap V$ is a nonempty set. For every $(x^1, \ldots, x^n)$ in $\varphi_U(U \cap V)$, put

$$\left(\varphi_V \circ (\varphi_U)^{-1}\right)(x^1, \ldots, x^m) \equiv \left(f^1(x^1, \ldots, x^m), \ldots, f^m(x^1, \ldots, x^m)\right) = (y^1, \ldots, y^m),$$

where $(y^1, \ldots, y^m)$ is in $\varphi_V(U \cap V)$, and

$$\left(\varphi_U \circ (\varphi_V)^{-1}\right)(y^1, \ldots, y^m) \equiv \left(g^1(y^1, \ldots, y^m), \ldots, g^m(y^1, \ldots, y^m)\right) = (x^1, \ldots, x^m).$$

It is enough to show that $f^i, g^i$ are $C^1$ for every $i = 1, \ldots, m$.

Since $(U, \varphi_U)$ and $(V, \varphi_V)$ in $\mathcal{A}$, by condition 2, $(U, \varphi_U)$ is $C^2$-compatible with $(V, \varphi_V)$. Since $(U, \varphi_U)$ is $C^2$-compatible with $(V, \varphi_V)$, and $U \cap V$ is nonempty, by the definition of $C^2$-compatible, $f^i, g^i$ are $C^2$ for every $i = 1, \ldots, m$. Since, for every $i = 1, \ldots, m$, $f^i, g^i$ are $C^2$, we have $f^i, g^i$ are $C^1$.

Now, on using conditions 1, 2', and Theorem 1.2, there exists a unique $C^1$-differentiable structure $\mathcal{B}$ on $M$ which contains $\mathcal{A}$.                                    □

**Note 1.4** As above, we can prove the following result:

Let $M$ be an $m$-dimensional topological manifold. Let $r, s$ be positive integers satisfying $0 < s \leq r$. Let

$$\mathcal{A} \equiv \{(U, \varphi_U), (V, \varphi_V), (W, \varphi_W), \ldots\}$$

be a nonempty collection of coordinate charts of $M$ satisfying

1. $\{U, V, W, \ldots\}$ is a cover of $M$, that is, $\cup\{U : (U, \varphi_U)$ is in $\mathcal{A}\} = M$,
2. all pairs of members in $\mathcal{A}$ are $C^r$-compatible.

Then, there exists a unique $C^s$-differentiable structure $\mathcal{B}$ on $M$ which contains $\mathcal{A}$.

**Definition** Let $M$ be an $m$-dimensional topological manifold. Let $\mathcal{A}$ be a nonempty collection of coordinate charts of $M$. If

1. $\cup\{U : (U, \varphi_U)$ is in $\mathcal{A}\} = M$,
2. all pairs of members in $\mathcal{A}$ are $C^\infty$-*compatible* (in the sense that every pair of members in $\mathcal{A}$ is $C^r$-compatible for every positive integer $r$),

then we say that $\mathcal{A}$ *is an atlas on* $M$.

**Definition** Let $M$ be an $m$-dimensional topological manifold. Let

$$\mathcal{A} \equiv \{(U, \varphi_U), (V, \varphi_V), (W, \varphi_W), \ldots\}$$

be a nonempty collection of coordinate charts of $M$. If

1. $\mathcal{A}$ is an atlas on $M$,
2. $\mathcal{A}$ *is maximal* (in the sense that if $(\widetilde{U}, \varphi_{\widetilde{U}})$ is a coordinate chart of $M$, but not a member of $\mathcal{A}$, then there exists $(U, \varphi_U)$ in $\mathcal{A}$, such that $(\widetilde{U}, \varphi_{\widetilde{U}})$ and $(U, \varphi_U)$ are not $C^\infty$-compatible),

then we say that $\mathcal{A}$ *is a $C^\infty$-differentiable structure on $M$*, and the pair $(M, \mathcal{A})$ is called a *$C^\infty$-differentiable manifold*. $C^\infty$-differentiable structure is also called *smooth structure*, and $C^\infty$-differentiable manifold is also called *smooth manifold*.

Here, members of $\mathcal{A}$ are called *admissible coordinate charts of M*.

**Note 1.5** We shall try to prove: if $\mathcal{A}$ is an atlas on an $m$-dimensional topological manifold $M$, then there exists a unique $C^\infty$-differentiable structure $\mathcal{B}$ on $M$ which contains $\mathcal{A}$.

**Existence** Let $\mathcal{B}$ be the collection of all coordinate charts $(U, \varphi_U)$ of $M$ such that $(U, \varphi_U)$ is $C^\infty$-compatible with every member of $\mathcal{A}$.

First of all, we shall try to show that $\mathcal{B}$ contains $\mathcal{A}$. For this purpose, let us take any $(U, \varphi_U)$ in $\mathcal{A}$. Since $\mathcal{A}$ is an atlas, and $(U, \varphi_U)$ is in $\mathcal{A}$, $(U, \varphi_U)$ is $C^\infty$-compatible with all members of $\mathcal{A}$, and hence, by the definition of $\mathcal{B}$, $(U, \varphi_U)$ is in $\mathcal{B}$. Hence, $\mathcal{A}$ is a subset of $\mathcal{B}$.

Further since $\mathcal{A}$ is nonempty, and $\mathcal{A}$ is a subset of $\mathcal{B}$, $\mathcal{B}$ is nonempty. Now we want to prove that $\mathcal{B}$ is a $C^\infty$-differentiable structure on $M$, that is,

1. $\cup\{U : (U, \varphi_U) \text{ is in } \mathcal{B}\} = M$,
2. all pairs of members of $\mathcal{B}$ are $C^\infty$-compatible.
3. $\mathcal{B}$ is maximal (in the sense that if $(\widetilde{U}, \varphi_{\widetilde{U}})$ a coordinate chart of $M$, but not a member of $\mathcal{B}$, then there exists $(U, \varphi_U)$ in $\mathcal{B}$ such that $(\widetilde{U}, \varphi_{\widetilde{U}})$ and $(U, \varphi_U)$ are not $C^\infty$-compatible.

For 1: Since $\mathcal{A}$ is contained in $\mathcal{B}$, $\cup\{U : (U, \varphi_U) \text{ is in } \mathcal{A}\} \subset \cup\{U : (U, \varphi_U)$ is in $\mathcal{B}\} \subset M$. Since $\mathcal{A}$ is an atlas, $\cup\{U : (U, \varphi_U) \text{ is in } \mathcal{A}\} = M$. Thus, $\cup\{U : (U, \varphi_U) \text{ is in } \mathcal{B}\} = M$.

For 2: Let us take any $(U, \varphi_U)$ in $\mathcal{B}$, and $(V, \varphi_V)$ in $\mathcal{B}$. We have to prove that $(U, \varphi_U)$ and $(V, \varphi_V)$ are $C^\infty$-compatible. Here, two cases arise: either $U \cap V$ is an empty set, or $U \cap V$ is a nonempty set.

Case I: when $U \cap V$ is an empty set. In this case, by the definition of $C^r$-compatible, $(U, \varphi_U)$ and $(V, \varphi_V)$ are $C^r$-compatible for every positive integer $r$, and hence, $(U, \varphi_U)$ and $(V, \varphi_V)$ are $C^\infty$-compatible.

Case II: when $U \cap V$ is a nonempty set. For every $(x^1, \ldots, x^m)$ in $\varphi_U(U \cap V)$, put

$$\left(\varphi_V \circ (\varphi_U)^{-1}\right)(x^1, \ldots, x^m) \equiv \left(f^1(x^1, \ldots, x^m), \ldots, f^m(x^1, \ldots, x^m)\right) = (y^1, \ldots, y^m),$$

where $(y^1, \ldots, y^m)$ is in $\varphi_V(U \cap V)$, and

$$\left(\varphi_U \circ (\varphi_V)^{-1}\right)(y^1, \ldots, y^m) \equiv (g^1(y^1, \ldots, y^m), \ldots, g^m(y^1, \ldots, y^m)) = (x^1, \ldots, x^m).$$

We have to show that $f^i, g^i$ are $C^\infty$ for every $i = 1, \ldots, m$.

Since $U \cap V$ is nonempty, there exists an element $x$ in $U \cap V$ ($\subset M$). Since $x$ is in $M$, and $\mathcal{A}$ is an atlas, there exists $(W, \varphi_W)$ in $\mathcal{A}$ such that $x$ is in $W$. Since $(U, \varphi_U)$ is in $\mathcal{B}$, and $(W, \varphi_W)$ is in $\mathcal{A}$, by the definition of $\mathcal{B}$, $(U, \varphi_U)$ and $(W, \varphi_W)$ are $C^\infty$-compatible. Since $x$ is in $W$, and $x$ is in $U \cap V (\subset U)$, $x$ is in $W \cap U$, and hence, $U \cap W$ is a nonempty set. Since $(U, \varphi_U)$ and $(W, \varphi_W)$ are $C^\infty$-compatible, and $U \cap W$ is nonempty, by the definition of $C^\infty$-compatible, $f_1^i, g_1^i$ are $C^\infty$ for every $i = 1, \ldots, m$, where for every $(x^1, \ldots, x^m)$ in $\varphi_U(U \cap W)$,

$$\left(\varphi_W \circ (\varphi_U)^{-1}\right)(x^1, \ldots, x^m) \equiv (f_1^1(x^1, \ldots, x^m), \ldots, f_1^m(x^1, \ldots, x^m))$$
$$= (y^1, \ldots, y^m) \in \varphi_W(U \cap W),$$

and

$$\left(\varphi_U \circ (\varphi_W)^{-1}\right)(y^1, \ldots, y^m) \equiv (g_1^1(y^1, \ldots, y^m), \ldots, g_1^m(y^1, \ldots, y^m)) = (x^1, \ldots, x^m).$$

Similarly, $f_2^i, g_2^i$ are $C^\infty$ for every $i = 1, \ldots, m$, where for every $(x^1, \ldots, x^m)$ in $\varphi_V(V \cap W)$,

$$\left(\varphi_W \circ (\varphi_V)^{-1}\right)(x^1, \ldots, x^m) \equiv (f_2^1(x^1, \ldots, x^m), \ldots, f_2^m(x^1, \ldots, x^m))$$
$$= (y^1, \ldots, y^m) \in \varphi_W(V \cap W),$$

and

$$\left(\varphi_V \circ (\varphi_W)^{-1}\right)(y^1, \ldots, y^m) \equiv (g_2^1(y^1, \ldots, y^m), \ldots, g_2^m(y^1, \ldots, y^m)) = (x^1, \ldots, x^m).$$

Now for every $(x^1, \ldots, x^m)$ in $\varphi_U(U \cap V \cap W)$

$$(f^1(x^1, \ldots, x^m), \ldots, f^m(x^1, \ldots, x^m))$$
$$= \left(\varphi_V \circ (\varphi_U)^{-1}\right)(x^1, \ldots, x^m)$$
$$= \left(\left(\varphi_V \circ (\varphi_W)^{-1}\right) \circ \left(\varphi_W \circ (\varphi_U)^{-1}\right)\right)(x^1, \ldots, x^m)$$
$$= \left(\varphi_V \circ (\varphi_W)^{-1}\right)\left(\left(\varphi_W \circ (\varphi_U)^{-1}\right)(x^1, \ldots, x^m)\right)$$
$$= \left(\varphi_V \circ (\varphi_W)^{-1}\right)(f_1^1(x^1, \ldots, x^m), \ldots, f_1^m(x^1, \ldots, x^m))$$
$$= (g_2^1(f_1^1(x^1, \ldots, x^m), \ldots, f_1^m(x^1, \ldots, x^m)), \ldots, g_2^m(f_1^1(x^1, \ldots, x^m), \ldots, f_1^m(x^1, \ldots, x^m))).$$

Hence,

$$f^1\left(x^1,\ldots,x^m\right) = g_2^1\left(f_1^1\left(x^1,\ldots,x^m\right),\ldots,f_1^m\left(x^1,\ldots,x^m\right)\right).$$

Since $g_2^1$, and each $f_1^k$ $(k = 1,\ldots,m)$ are $C^\infty$,

$$\frac{\partial g_2^1\left(f_1^1(x^1,\ldots,x^m),\ldots,f_1^m(x^1,\ldots,x^m)\right)}{\partial x^1}$$

$$\left(= \sum_{k=1}^n \frac{\partial g_2^1\left(f_1^1(x^1,\ldots,x^m),\ldots,f_1^m(x^1,\ldots,x^m)\right)}{\partial f_1^k(x^1,\ldots,x^m)} \frac{\partial f_1^k(x^1,\ldots,x^m)}{\partial x^1}\right)$$

exists and is continuous. It follows that

$$\frac{\partial f^1(x^1,\ldots,x^m)}{\partial x^1}\left(= \frac{\partial g_2^1\left(f_1^1(x^1,\ldots,x^m),\ldots,f_1^m(x^1,\ldots,x^m)\right)}{\partial x^1}\right)$$

exists and is continuous. Similarly, $\frac{\partial f^1(x^1,\ldots,x^m)}{\partial x^2}$ exists and is continuous, etc.

So $f^1$ is $C^1$. Similarly, $f^1$ is $C^2$, etc. Hence, $f^1$ is $C^\infty$. Similarly, $f^2$ is $C^\infty$, etc. So each $f^i$ is $C^\infty$. Similarly, each $g^i$ is $C^\infty$. Thus, we have shown that, in all cases, $(U, \varphi_U)$ and $(V, \varphi_V)$ are $C^\infty$-compatible.

For 3: We claim that $\mathcal{B}$ is maximal. If not, otherwise, let $\mathcal{B}$ be not maximal. We have to arrive at a contradiction. Since $\mathcal{B}$ is not maximal, there exists a coordinate chart $(U, \varphi_U)$ of $M$ such that $(U, \varphi_U)$ is not a member of $\mathcal{B}$, and $(U, \varphi_U)$ is $C^\infty$-compatible with every member of $\mathcal{B}$. Since $(U, \varphi_U)$ is $C^\infty$-compatible with every member of $\mathcal{B}$, and $\mathcal{A}$ is contained in $\mathcal{B}$, $(U, \varphi_U)$ is $C^\infty$-compatible with every member of $\mathcal{A}$, and hence, by the definition of $\mathcal{B}$, $(U, \varphi_U)$ is in $\mathcal{B}$, a contradiction. Thus, we have shown that $\mathcal{B}$ is a $C^\infty$-differentiable structure on $M$, which contains $\mathcal{A}$.

**Uniqueness** If not, otherwise, let $\mathcal{B}_1$ and $\mathcal{B}_2$ be two distinct $C^\infty$-differentiable structures on $M$, which contains $\mathcal{A}$. We have to arrive at a contradiction. Since $\mathcal{B}_1$ and $\mathcal{B}_2$ are two distinct $C^\infty$-differentiable structures on $M$, either

(there exists a coordinate chart $(U, \varphi_U)$ of $M$ such that $(U, \varphi_U)$ is in $\mathcal{B}_1$, and $(U, \varphi_U)$ is not in $\mathcal{B}_2$)

or

(there exists a coordinate chart $(U, \varphi_U)$ of $M$ such that $(U, \varphi_U)$ is in $\mathcal{B}_2$, and $(U, \varphi_U)$ is not in $\mathcal{B}_1$).

Case I: when there exists a coordinate chart $(U, \varphi_U)$ of $M$ such that $(U, \varphi_U)$ is in $\mathcal{B}_1$, and $(U, \varphi_U)$ is not in $\mathcal{B}_2$.

First of all, we shall show that $\mathcal{B}_1$ is contained in $\mathcal{B}$, where $\mathcal{B}$ is the differential structure constructed above. For this purpose, let us take any coordinate chart $(V, \varphi_V)$ of $M$ in $\mathcal{B}_1$. We have to show that $(V, \varphi_V)$ is in $\mathcal{B}$. Since $(V, \varphi_V)$ is in $\mathcal{B}_1$, and $\mathcal{B}_1$ is a $C^\infty$-differentiable structure on $M$ which contains $\mathcal{A}$, $(V, \varphi_V)$ is

$C^\infty$-compatible with every member of $\mathcal{A}$, and hence, by the definition of $\mathcal{B}$, $(V, \varphi_V)$ is in $\mathcal{B}$. Thus, we have shown that $\mathcal{B}_1$ is contained in $\mathcal{B}$.

Similarly, $\mathcal{B}_2$ is contained in $\mathcal{B}$. Since $(U, \varphi_U)$ is not in $\mathcal{B}_2$, and $\mathcal{B}_2$ is $C^\infty$-differentiable structure on $M$, there exists $(V, \varphi_V)$ in $\mathcal{B}_2$ such that $(V, \varphi_V)$ and $(U, \varphi_U)$ are not $C^\infty$-compatible. Since $(V, \varphi_V)$ is in $\mathcal{B}_2$, and $\mathcal{B}_2$ is contained in $\mathcal{B}$, $(V, \varphi_V)$ is in $\mathcal{B}$. Since $(U, \varphi_U)$ is in $\mathcal{B}_1$, and $\mathcal{B}_1$ is contained in $\mathcal{B}$, $(U, \varphi_U)$ is in $\mathcal{B}$. Since $(U, \varphi_U), (V, \varphi_V)$ are in $\mathcal{B}$, and $\mathcal{B}$ is a $C^\infty$-differentiable structure on $M$, $(V, \varphi_V)$ and $(U, \varphi_U)$ are $C^\infty$-compatible, a contradiction.

Case II: when there exists a coordinate chart $(U, \varphi_U)$ of $M$ such that $(U, \varphi_U)$ is in $\mathcal{B}_2$, and $(U, \varphi_U)$ is not in $\mathcal{B}_1$. This case is similar to the case I. Thus, in all cases, we get a contradiction.                                                $\square$

From now on, we shall assume without further mention that in our differentiable manifold $(M, \mathcal{A})$, $M$ is a second countable topological space.

## 1.3 Examples of Smooth Manifolds

*Example 1.6* (i) Let $G$ be a nonempty open subset of $\mathbb{R}^m$. Let us take $G$ for $M$.

Since $\mathbb{R}^m$ is a Hausdorff second countable topological space, $G$ with the induced topology is a Hausdorff second countable topological space. For every $x$ in $G$, $G$ is an open neighborhood of $x$. Since the identity mapping

$$\mathrm{Id}_G : G \to G$$

given by

$$\mathrm{Id}_G(x) = x$$

for every $x$ in $G$ is a homeomorphism from $G$ onto $G$, $G$ is an $m$-dimensional topological manifold. For $\mathcal{A}$, let us take the singleton set $\{(G, \mathrm{Id}_G)\}$. It is easy to observe that $\mathcal{A}$ satisfies the conditions 1 and 2 of Theorem 1.2 for every positive integer $r$. Hence, there exists a unique $C^\infty$-differentiable structure on $G$ which contains $\{(G, \mathrm{Id}_G)\}$. This $C^\infty$-differentiable structure on $G$ is called the *standard differentiable structure of $G$*. Thus, every nonempty open subset of $\mathbb{R}^m$ is an example of a smooth manifold.

(ii) Let $M$ be an $m$-dimensional smooth manifold, whose topology is $\mathcal{O}$, and differential structure is $\mathcal{A}$. Let $(U, \varphi_U) \in \mathcal{A}$.

Since $(U, \varphi_U) \in \mathcal{A}$, $U$ is an open subset of $M$. Let $\mathcal{O}_1$ be the induced topology over $U$, that is, $\mathcal{O}_1 = \{G : G \subset U, \text{ and } G \in \mathcal{O}\}$. Put $\mathcal{A}_1 \equiv \{(U, \varphi_U)\}$. Since $M$ is an $m$-dimensional smooth manifold, $M$ is a Hausdorff space. Since $M$ is a Hausdorff space, and $U$ is a subspace of $M$, $U$ with the induced topology is a Hausdorff space. Since $M$ is an $m$-dimensional smooth manifold, $M$ is a second countable space. Since $M$ is a second countable space, and $U$ is a subspace of $M$, $U$ with the induced topology is a second countable space. It is easy to observe that $\mathcal{A}$ satisfies the conditions 1 and 2 of Theorem 1.2 for every positive integer $r$. Hence, there exists a unique $C^\infty$-differentiable structure on $U$ which contains $\{(U, \varphi_U)\}$.

Thus for every admissible coordinate chart $(U, \varphi_U)$ of an $m$-dimensional smooth manifold $M$, $U$ is an $m$-dimensional smooth manifold.

*Example 1.7* For $m = 1, 2, 3, \ldots$, by the $m$-**dimensional unit sphere** $S^m$, we mean the set

$$\left\{ \left(x^1, \ldots x^{m+1}\right) : \left(x^1, \ldots x^{m+1}\right) \in \mathbb{R}^{m+1} \text{ and } \sqrt{(x^1)^2 + \cdots + (x^{m+1})^2} = 1 \right\}.$$

Since $\mathbb{R}^{m+1}$ is a Hausdorff second countable topological space, and $S^m$ is a subset of $\mathbb{R}^{m+1}$, $S^m$ with the induced topology is a Hausdorff second countable topological space. Here,

$$S^1 \equiv \left\{ \left(x^1, x^2\right) : \left(x^1, x^2\right) \in \mathbb{R}^2 \text{ and } \sqrt{(x^1)^2 + (x^2)^2} = 1 \right\},$$
$$S^2 \equiv \left\{ \left(x^1, x^2, x^3\right) : \left(x^1, x^2, x^3\right) \in \mathbb{R}^3 \text{ and } \sqrt{(x^1)^2 + (x^2)^2 + (x^3)^2} = 1 \right\}, \text{ etc.}$$

Now we shall try to prove that the *unit circle* $S^1$ is a 1-dimensional topological manifold. For this purpose, let us take any $(a^1, a^2)$ in $S^1$. We have to find an open neighborhood $G$ of $(a^1, a^2)$, and a homeomorphism $\varphi : G \to \varphi(G)$ such that $\varphi(G)$ is an open subset of $\mathbb{R}^1$. Since $(0, 0)$ does not lie in $S^1$, and $(a^1, a^2)$ is in $S^1$, both of $a^1, a^2$ cannot be 0. So two cases arise: either $a^1 \neq 0$ or $a^2 \neq 0$.

Case I: when $a^1 \neq 0$
Now two subcases arise: either $0 < a^1$ or $a^1 < 0$.
Subcase I: when $0 < a^1$. Let us take

$$\left\{ \left(x^1, x^2\right) : \left(x^1, x^2\right) \in S^1 \text{ and } 0 < x^1 \right\}$$

for $G_1$. Since $(a^1, a^2) \in S^1$ and $0 < a^1$, $(a^1, a^2) \in G_1$. Since

$$\left\{ \left(x^1, x^2\right) : \left(x^1, x^2\right) \in S^1 \text{ and } 0 < x^1 \right\} = \left\{ \left(x^1, x^2\right) : \left(x^1, x^2\right) \in \mathbb{R}^2 \text{ and } 0 < x^1 \right\} \cap S^1,$$

and $\left\{ \left(x^1, x^2\right) : \left(x^1, x^2\right) \in \mathbb{R}^2 \text{ and } 0 < x^1 \right\} (= (0, \infty) \times (-\infty, \infty))$ is open in $\mathbb{R}^2$, $\left\{ \left(x^1, x^2\right) : \left(x^1, x^2\right) \in S^1 \text{ and } 0 < x^1 \right\} (= G_1)$ is open in $S^1$ relative to the induced topology. Thus, $G_1$ is open in $S^1$. Since $(a^1, a^2) \in G_1$, and $G_1$ is open in $S^1$, $G_1$ is an open neighborhood of $(a^1, a^2)$. Since

$$G_1 = \left\{ (x^1, x^2) \; : \; (x^1, x^2) \in S^1 \text{ and } 0 < x^1 \right\}$$
$$= \left\{ (x^1, x^2) \; : \; \sqrt{(x^1)^2 + (x^2)^2} = 1 \text{ and } 0 < x^1 \right\},$$

it is clear that $\{ x^2 \; : \; (x^1, x^2) \in G_1 \}$ is equal to the open interval $(-1, 1)$.
Let us define a function

$$p_1 \; : \; G_1 \rightarrow \left\{ x^2 \; : \; (x^1, x^2) \in G \right\}$$

as follows: for every $(x^1, x^2)$ in $G_1$,

$$p_1(x^1, x^2) = x^2.$$

Clearly, $p_1$ is a 1–1 function from $G_1$ onto the open interval $(-1, 1)$. Further, since projection map is a continuous and open map,

$$p_1 \; : \; G_1 \rightarrow (-1, 1)$$

is a homeomorphism. Thus, we have shown that there exist an open neighborhood $G_1$ of $(a^1, a^2)$ and a homeomorphism $p_1 \; : \; G_1 \rightarrow (-1, 1)$, where $(-1, 1)$ is an open subset of $\mathbb{R}^1$.

Subcase II: when $a^1 < 0$. Proceeding as in subcase I, there exist an open neighborhood

$$G_2 \equiv \left\{ (x^1, x^2) \; : \; (x^1, x^2) \in S^1 \text{ and } x^1 < 0 \right\}$$

of $(a^1, a^2)$ and a homeomorphism

$$p_2 \; : \; G_2 \rightarrow (-1, 1)$$

defined as: for every $(x^1, x^2)$ in $G_2$,

$$p_2(x^1, x^2) = x^2.$$

Case II: when $a^2 \neq 0$
Now two subcases arise: either $0 < a^2$ or $a^2 < 0$.
Subcase I: when $0 < a^2$. Proceeding as above, there exist an open neighborhood

$$G_3 \equiv \left\{ (x^1, x^2) \; : \; (x^1, x^2) \in S^1 \text{ and } 0 < x^2 \right\}$$

of $(a^1, a^2)$ and a homeomorphism

$$p_3 \; : \; G_3 \rightarrow (-1, 1)$$

defined as: for every $(x^1, x^2)$ in $G_3$,

$$p_3(x^1, x^2) = x^1.$$

Subcase II: when $a^2 < 0$. Proceeding as above, there exist an open neighborhood

$$G_4 \equiv \left\{ (x^1, x^2) : (x^1, x^2) \in S^1 \text{ and } x^2 < 0 \right\}$$

of $(a^1, a^2)$ and a homeomorphism

$$p_4 : G_4 \to (-1, 1)$$

defined as: for every $(x^1, x^2)$ in $G_4$,

$$p_4(x^1, x^2) = x^1.$$

Thus, we see that in all cases, there exist an open neighborhood $G$ of $(a^1, a^2)$, and a homeomorphism $\varphi : G \to \varphi(G)$, where $\varphi(G)$ is an open subset of $\mathbb{R}^1$.

Hence, by the definition of topological manifold, $S^1$ is a 1-dimensional topological manifold. Here, we get four coordinate charts $(G_1, p_1), (G_2, p_2), (G_3, p_3), (G_4, p_4)$. Let us take

$$\{(G_1, p_1), (G_2, p_2), (G_3, p_3), (G_4, p_4)\}$$

for $\mathcal{A}$. We want to show that $\mathcal{A}$ has the following two properties:

1. $\cup\{U : (U, \varphi_U) \text{ is in } \mathcal{A}\} = S^1$,
2. all pairs of members in $\mathcal{A}$ are $C^\infty$-compatible.

For 1: Here,

$$\cup\{U : (U, \varphi_U); \text{ is in } \mathcal{A}\}$$
$$= G_1 \cup G_2 \cup G_3 \cup G_4$$
$$= \left\{ (x^1, x^2) : (x^1, x^2) \in S^1 \text{ and } 0 < x^1 \right\} \cup \left\{ (x^1, x^2) : (x^1, x^2) \in S^1 \text{ and } x^1 < 0 \right\}$$
$$\cup \left\{ (x^1, x^2) : (x^1, x^2) \in S^1 \text{ and } 0 < x^2 \right\} \cup \left\{ (x^1, x^2) : (x^1, x^2) \in S^1 \text{ and } x^2 < 0 \right\}$$
$$= S^1.$$

For 2: Let us take $(G_1, p_1), (G_2, p_2)$ as a pair of members of $\mathcal{A}$. Since

$$G_1 \cap G_2 = \left( \left\{ (x^1, x^2) : (x^1, x^2) \in S^1 \text{ and } 0 < x^1 \right\} \cap \left\{ (x^1, x^2) : (x^1, x^2) \in S^1 \text{ and } x^1 < 0 \right\} \right)$$

is an empty set, by the definition of $C^r$-compatible, $(G_1, p_1), (G_2, p_2)$ are $C^r$-compatible for every positive integer $r$. Hence, $(G_1, p_1), (G_2, p_2)$ are $C^\infty$-compatible.

Next, let us take $(G_1, p_1), (G_3, p_3)$ as a pair of members of $\mathcal{A}$. Here,

$$G_1 \cap G_3 = \{(x^1, x^2) : (x^1, x^2) \in S^1 \text{ and } 0 < x^1\} \cap \{(x^1, x^2) : (x^1, x^2) \in S^1 \text{ and } 0 < x^2\}$$
$$= \{(x^1, x^2) : (x^1, x^2) \in S^1, 0 < x^1 \text{ and } 0 < x^2\}.$$

We want to prove that

$$\left(p_1 \circ \left(p_3^{-1}\right)\right) : p_3(G_1 \cap G_3) \to p_1(G_1 \cap G_3)$$

is $C^1$.

By Lemma 1.1, $p_1 \circ (p_3^{-1})$ is a homeomorphism from open subset $p_3(G_1 \cap G_3)$ of $\mathbb{R}^1$ onto open subset $p_1(G_1 \cap G_3)$ of $\mathbb{R}^1$. Now let us take any $p_3(x)$ in $p_3(G_1 \cap G_3)$, where $x$ is in $G_1 \cap G_3$. Since $x$ is in $G_1 \cap G_3$, and $G_1 \cap G_3 = \{(x^1, x^2) : (x^1, x^2) \in S^1, 0 < x^1 \text{ and } 0 < x^2\}$, there exist real numbers $x^1, x^2$ such that $x = (x^1, x^2) \in S^1$, $0 < x^1$ and $0 < x^2$. Since

$$(x^1, x^2) \in S^1 = \left\{ (x^1, x^2) : (x^1, x^2) \in \mathbb{R}^2 \text{ and } \sqrt{(x^1)^2 + (x^2)^2} = 1 \right\},$$

and $0 < x^2$, $x^2 = \sqrt{1 - (x^1)^2}$. Since $0 < x^1$, $0 < x^2$, and $(x^1)^2 + (x^2)^2 = 1$, it follows that $0 < x^1 < 1$, and $0 < x^2 < 1$. Hence, $t \equiv p_3(x) = p_3(x^1, x^2) = x^1 \in (0, 1)$. Now it is clear that $p_3(G_1 \cap G_3)$ is equal to the open interval $(0, 1)$.
Further,

$$\left(p_1 \circ (p_3^{-1})\right)(t) = \left(p_1 \circ (p_3^{-1})\right)(p_3(x)) = p_1\left((p_3^{-1})(p_3(x))\right)$$
$$= p_1(x) = p_1(x^1, x^2) = x^2 = \sqrt{1 - (x^1)^2}$$
$$= \sqrt{1 - (p_3(x^1, x^2))^2} = \sqrt{1 - (p_3(x))^2} = \sqrt{1 - t^2}$$

or,

$$\left(p_1 \circ (p_3^{-1})\right)(t) = \sqrt{1 - (t)^2}$$

for every $t$ in $p_3(G_1 \cap G_3)(=(0, 1))$. Therefore,

$$\left(p_1 \circ (p_3^{-1})\right)'(t) = \frac{1}{2\sqrt{1 - t^2}}(2t) = \frac{t}{\sqrt{1 - t^2}}$$

or,

$$\left(p_1 \circ (p_3^{-1})\right)'(t) = \frac{t}{\sqrt{1 - t^2}} \quad (t \in (0, 1)).$$

It follows that $t \mapsto (p_1 \circ (p_3^{-1}))(t)$ is a differentiable function on the open subset $p_3(G_1 \cap G_3)$ of $\mathbb{R}^1$. Also, since

$$\left(p_1 \circ \left(p_3^{-1}\right)\right)'(t) = \frac{t}{\sqrt{1 - t^2}}.$$

and $t \mapsto \frac{t}{\sqrt{1-t^2}}$ is a continuous function,

$$\left(p_1 \circ \left(p_3^{-1}\right)\right) : p_3(G_1 \cap G_3) \to p_1(G_1 \cap G_3)$$

is $C^1$. Similarly, we can show that

$$\left(p_1 \circ \left(p_3^{-1}\right)\right) : p_3(G_1 \cap G_3) \to p_1(G_1 \cap G_3)$$

is $C^2$, etc. Thus,

$$\left(p_1 \circ \left(p_3^{-1}\right)\right) : p_3(G_1 \cap G_3) \to p_1(G_1 \cap G_3)$$

is $C^r$ for every positive integer $r$. Hence,

$$\left(p_1 \circ \left(p_3^{-1}\right)\right) : p_3(G_1 \cap G_3) \to p_1(G_1 \cap G_3)$$

is $C^\infty$. Thus, we have shown that the pair $(G_1, p_1), (G_3, p_3)$ are $C^\infty$-compatible.

Similarly, we can show that all other pairs of members in $\mathcal{A}$ are $C^\infty$-compatible. This proves 2. Hence, by Theorem 1.2, there exists a unique $C^\infty$-differentiable structure $\mathcal{B}$ on $S^1$ which contains $\mathcal{A}$. Thus, the ordered pair $(S^1, \mathcal{B})$ is an example of a smooth manifold of dimension 1.

Similarly, we can construct a $C^\infty$-differentiable structure on $S^2$, etc. In short, the unit sphere $S^m$ is a smooth manifold of dimension $m$.

*Example 1.8* Let us define a relation $\sim$ over $\mathbb{R}^2 - \{(0,0)\}$ as follows:
For every $(x^1, x^2), (y^1, y^2)$ in $\mathbb{R}^2 - \{(0,0)\}$, by

$$\left(x^1, x^2\right) \sim \left(y^1, y^2\right)$$

we shall mean there exists a real number $t$ such that

$$\left(x^1, x^2\right) = t\left(y^1, y^2\right).$$

We shall try to show that $\sim$ is an equivalence relation over $\mathbb{R}^2 - \{(0,0)\}$, that is,

1. $(x^1, x^2) \sim (x^1, x^2)$ for every $(x^1, x^2)$ in $\mathbb{R}^2 - \{(0,0)\}$,
2. if $(x^1, x^2) \sim (y^1, y^2)$, then $(y^1, y^2) \sim (x^1, x^2)$,
3. if $(x^1, x^2) \sim (y^1, y^2)$, and $(y^1, y^2) \sim (z^1, z^2)$ then $(x^1, x^2) \sim (z^1, z^2)$.

For 1: Since $(x^1, x^2) = 1(x^1, x^2)$, $(x^1, x^2) \sim (x^1, x^2)$.

For 2: Since $(x^1, x^2) \sim (y^1, y^2)$, there exists a real number $t$ such that $(x^1, x^2) = t(y^1, y^2)$. Since $(x^1, x^2), (y^1, y^2)$ are in $\mathbb{R}^2 - \{(0,0)\}$, $(x^1, x^2)$ and $(y^1, y^2)$ are nonzero. Since $(x^1, x^2)$ and $(y^1, y^2)$ are nonzero, and $(x^1, x^2) = t(y^1, y^2)$, $t$ is nonzero, and hence, $(y^1, y^2) = \frac{1}{t}(x^1, x^2)$. Since $(y^1, y^2) = \frac{1}{t}(x^1, x^2)$, $(y^1, y^2) \sim (x^1, x^2)$.

For 3: Since $(x^1, x^2) \sim (y^1, y^2)$, there exists a real number $t$ such that $(x^1, x^2) = t(y^1, y^2)$. Since $(y^1, y^2) \sim (z^1, z^2)$, there exists a real number $s$ such that $(y^1, y^2) = s(z^1, z^2)$. Since $(x^1, x^2) = t(y^1, y^2)$, and $(y^1, y^2) = s(z^1, z^2)$, $(x^1, x^2) = t(s(z^1, z^2)) = (st)(z^1, z^2)$, and hence, $(x^1, x^2) \sim (z^1, z^2)$.

Thus, we have shown that $\sim$ is an equivalence relation over $\mathbb{R}^2 - \{(0,0)\}$. If $(x^1, x^2) \neq (0,0)$, then the equivalence class of $(x^1, x^2)$ is given by

$$[(x^1, x^2)] \equiv \{(y^1, y^2) : (y^1, y^2) \sim (x^1, x^2)\}$$
$$= \{(y^1, y^2) : (y^1, y^2) = t(x^1, x^2) \text{ for some real } t\}$$

or,

$$[(x^1, x^2)] = \{t(x^1, x^2) : t \neq 0\}.$$

We shall denote the quotient space

$$\left(\mathbb{R}^2 - \{(0,0)\}\right)/\sim$$

of all equivalence classes by $P^1$. Thus,

$$P^1 \equiv \{[(x^1, x^2)] : (x^1, x^2) \neq (0,0)\}$$

or,

$$P^1 = \{\{t(x^1, x^2) : t \neq 0\} : (x^1, x^2) \neq (0,0)\}.$$

The topology of $\mathbb{R}^2 - \{(0,0)\}$ is the induced topology of the standard topology of $\mathbb{R}^2$, and the topology of the quotient space $(\mathbb{R}^2 - \{(0,0)\})/\sim (= P^1)$ is the quotient topology. Thus, a subset $G$ of $P^1$ is open means the set

$$\cup\{[(x^1, x^2)] : [(x^1, x^2)] \in G\}$$

is open in $\mathbb{R}^2 - \{(0,0)\}$, that is,

$$\cup\{\{t(x^1, x^2) : t \neq 0\} : [(x^1, x^2)] \in G\}$$

is open in $\mathbb{R}^2 - \{(0,0)\}$, that is, either

$$\cup\{\{t(x^1,x^2) : t \neq 0\} : [(x^1,x^2)] \in G\}$$

or

$$\cup\{\{t(x^1,x^2) : t \neq 0\} : [(x^1,x^2)] \in G\} \cup \{(0,0)\}$$

is open in $\mathbb{R}^2$.

Since $\mathbb{R}^2$ is a Hausdorff space, its subspace $\mathbb{R}^2 - \{(0,0)\}$ is a Hausdorff space, and hence, its quotient space $(\mathbb{R}^2 - \{(0,0)\})/\sim (= P^1)$ is a Hausdorff space. Since $\mathbb{R}^2$ is a second countable space, its subspace $\mathbb{R}^2 - \{(0,0)\}$ is a second countable space, and hence, its quotient space $(\mathbb{R}^2 - \{(0,0)\})/\sim (= P^1)$ is a second countable space. Put

$$U_1 \equiv \{[(x^1,x^2)] : x^1 \neq 0\}.$$

We want to show that $U_1$ is open in $P^2$.
For this purpose, we must show that

$$\cup\{\{t(x^1,x^2) : t \neq 0\} : [(x^1,x^2)] \in U_1\}$$

is open in $\mathbb{R}^2 - \{(0,0)\}$.
Since (see Fig. 1.2),

$$\cup\{\{t(x^1,x^2) : t \neq 0\} : [(x^1,x^2)] \in U_1\} = \cup\{\{t(x^1,x^2) : t \neq 0\} : x^1 \neq 0\}$$
$$= \mathbb{R}^2 - \{(0,x^2) : x^2 \text{ is any real}\},$$

and $\mathbb{R}^2 - \{(0,x^2) : x^2 \text{ is any real}\}$ is open in $\mathbb{R}^2$, $\cup\{\{t(x^1,x^2) : t \neq 0\} : [(x^1,x^2)] \in U_1\}$ is open in $\mathbb{R}^2 - \{(0,0)\}$. Thus, we have shown that $U_1$ is open in $P^1$.

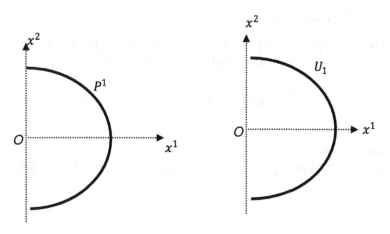

**Fig. 1.2** Hausdorff space

Now let us define a function

$$\varphi_1 : U_1 \to \mathbb{R}$$

as follows: for every $[(x^1, x^2)]$ in $U_1$

$$\varphi_1([(x^1, x^2)]) \equiv \frac{x^2}{x^1}.$$

We shall try to show that

1. $\varphi_1$ is well-defined,
2. $\varphi_1$ is 1–1,
3. $\varphi_1$ is continuous,
4. $\varphi_1(U_1)$ is open in $\mathbb{R}$,
5. $\varphi_1^{-1} : \varphi_1(U_1) \to U_1$ is continuous.

For 1: Let $[(x^1, x^2)] = [(y^1, y^2)]$, where $[(x^1, x^2)], [(y^1, y^2)]$ are in $U_1$. Since $[(x^1, x^2)], [(y^1, y^2)]$ are in $U_1$, $(x^1, x^2), (y^1, y^2)$ are nonzero members of $\mathbb{R}^2$, and $x^1, y^1$ are nonzero. Since $x^1, y^1$ are nonzero, $\frac{x^2}{x^1}, \frac{y^2}{y^1}$ are real numbers.
We must prove that

$$\frac{x^2}{x^1} = \frac{y^2}{y^1}.$$

Since $[(x^1, x^2)] = [(y^1, y^2)]$, $(x^1, x^2) \sim (y^1, y^2)$. Further since $(x^1, x^2), (y^1, y^2)$ are nonzero members of $\mathbb{R}^2$, there exists a nonzero real number $t$ such that

$$(x^1, x^2) = t(y^1, y^2) = (ty^1, ty^2).$$

So $x^1 = ty^1$, and $x^2 = ty^2$. Since $x^1 = ty^1, x^2 = ty^2$, and $x^1, y^1, t$ are nonzero, $\frac{x^2}{x^1} = \frac{ty^2}{ty^1} = \frac{y^2}{y^1}$. This proves 1.

For 2: Let $\varphi_1([(x^1, x^2)]) = \varphi_1([(y^1, y^2)])$. We have to prove that $[(x^1, x^2)] = [(y^1, y^2)]$, that is, there exists a real number $t$ such that $(x^1, x^2) = t(y^1, y^2)$. Here,

$$\frac{x^2}{x^1} = \varphi_1([(x^1, x^2)]) = \varphi_1([(y^1, y^2)]) = \frac{y^2}{y^1}.$$

Case I: when $x^2 \neq 0$
Since $\frac{x^2}{x^1} = \frac{y^2}{y^1}$, and $x^2 \neq 0, y^2 \neq 0$, and hence, $x^1 = \frac{x^2}{y^2} y^1$. So $(x^1, x^2) = \frac{x^2}{y^2}(y^1, y^2)$.

Case II: when $x^2 = 0$

Since $\frac{x^2}{x^1} = \frac{y^2}{y^1}$, and $x^2 = 0$, $y^2 = 0$. So $(x^1, x^2) = (x^1, 0) = \frac{x^1}{y^1}(y^1, 0) = \frac{x^1}{y^1}(y^1, y^2)$.
Thus, we see that in all cases, there exists a real number $t$ such that $(x^1, x^2) = t(y^1, y^2)$. This proves 2.

For 3. Let us note that the equivalence class $[(x^1, x^2)]$ of $(x^1, x^2)$ is simply the straight line joining the origin $O$ and $(x^1, x^2)$, but origin $O$ deleted. Further, the slope $(= \tan\theta)$ of this line is the $\varphi_1$-image of $[(x^1, x^2)]$. Since $\tan\theta$ is a continuous function over $(-\frac{\pi}{2}, \frac{\pi}{2})$, it is clear that

$$\varphi_1 : U_1 \to \mathbb{R}$$

is a continuous function.

For 4. It is clear that

$$\varphi_1(U_1) = (-\infty, +\infty) = \mathbb{R},$$

and $\mathbb{R}$ is an open set, so $\varphi_1(U_1)$ is an open set of real numbers.

For 5. In 4, we have seen that $\varphi_1(U_1) = \mathbb{R}$, so it remains to prove that $\varphi_1^{-1} : \mathbb{R} \to U_1$ is continuous.

Since, for every $m$ in $\mathbb{R}$, $[(1, m)]$ is in $U_1$, and $\varphi_1([(1, m)]) = \frac{m}{1} = m$,

$$\varphi_1^{-1}(m) = [(1, m)].$$

Since $m \mapsto (1, m)$ is continuous, and $(x^1, x^2) \mapsto [(x^1, x^2)]$ is continuous, their composite $m \mapsto [(1, m)]$ is continuous, that is, $\varphi_1^{-1}$ is a continuous function. This proves 5.

Similarly, we can show that

$$U_2 \equiv \{[(x^1, x^2)] : x^2 \neq 0\}$$

is an open subset of $P^1$, and the function

$$\varphi_2 : U_2 \to \mathbb{R}$$

defined as follows: for every $[(x^1, x^2)]$ in $U_2$

$$\varphi_2([(x^1, x^2)]) \equiv \frac{x^1}{x^2},$$

is a homeomorphism from $U_2$ onto $\mathbb{R}$. Thus, the open subset $U_2$ of $P^2$ is homeomorphic to $\mathbb{R}$.

Since $P^1$ is a Hausdorff space, and $\{U_1, U_2\}$ is an open cover of $P^1$, $P^1$ is a 1-dimensional topological manifold. Here, we get two coordinate charts $(U_1, \varphi_1)$, $(U_2, \varphi_2)$. Let us take

$$\{(U_1, \varphi_1), (U_2, \varphi_2)\}.$$

for $\mathcal{A}$. We want to show that $\mathcal{A}$ has the following two properties:

1. $U_1 \cup U_2 = P^1$,
2. all pairs of members in $\mathcal{A}$ are $C^r$-compatible, for every $r = 1, 2, 3, \ldots$.

For 1: It is clear from the construction of $U_1$ and $U_2$.

For 2: Let us take $(U_1, \varphi_1), (U_2, \varphi_2)$ as a pair of members in $\mathcal{A}$. Here,

$$U_1 \cap U_2 = \{[(x^1, x^2)] \ : \ x^1 \neq 0, \text{ and } x^2 \neq 0\}.$$

Further

$$\varphi_2 \circ (\varphi_1)^{-1} : \ \varphi_1(U_1 \cap U_2) \to \varphi_2(U_1 \cap U_2)$$

is a continuous function. Also, for every $m$ in $\varphi_1(U_1 \cap U_2)$, there exists $[(x^1, x^2)]$ in $U_1 \cap U_2$ such that

$$m = \varphi_1([(x^1, x^2)]).$$

Since $[(x^1, x^2)]$ is in $U_1 \cap U_2$, $x^1 \neq 0$ and $x^2 \neq 0$. Now $m = \varphi_1([(x^1, x^2)]) = \frac{x^2}{x^1}$, and hence,

$$\left(\varphi_2 \circ (\varphi_1)^{-1}\right)(m) = \varphi_2\left((\varphi_1)^{-1}(m)\right) = \varphi_2\left((\varphi_1)^{-1}(\varphi_1([(x^1, x^2)]))\right)$$
$$= \varphi_2([(x^1, x^2)]) = \frac{x^1}{x^2} = \frac{1}{m}.$$

Now since $\varphi_1(U_1 \cap U_2) = \mathbb{R} - \{0\} = \varphi_2(U_1 \cap U_2)$, $(\varphi_2 \circ (\varphi_1)^{-1}) : m \mapsto \frac{1}{m}$ is $C^r$ for every $r = 1, 2, 3, \ldots$. This proves 2. Hence, by Theorem 1.2, there exists a unique $C^\infty$-differentiable structure $\mathcal{B}$ on $P^1$ which contains $\mathcal{A}$. Thus, the ordered pair $(P^1, \mathcal{B})$ is an example of a smooth manifold of dimension 1. Similarly, we can construct a $C^\infty$-differentiable structure on $P^2$, etc.

In short, we say that the *projective space* $P^m$ is a smooth manifold of dimension $m$.

*Example 1.9* Let $M$ be a 2-dimensional smooth manifold with differentiable structure $\mathcal{A}$, and let $N$ be a 3-dimensional smooth manifold with differentiable structure $\mathcal{B}$.

Now we want to show that the Cartesian product $M \times N$, with the product topology, is a $(2 + 3)$-dimensional topological manifold. For this purpose, we must prove that

1. $M \times N$, with the product topology, is a Hausdorff topological space,
2. for every $(x, y)$ in $M \times N$, there exists an open neighborhood of $(x, y)$ in $M \times N$ which is homeomorphic to some open subset of the Euclidean space $\mathbb{R}^{2+3}$ (see Fig 1.3).

For 1: Since $M$ is a smooth manifold, the topology of $M$ is Hausdorff. Similarly the topology of $N$ is Hausdorff. Since the topologies of $M$ and $N$ are Hausdorff, their product topology is Hausdorff. This proves 1.

For 2: Let us take any $(x, y)$ in $M \times N$, where $x$ is in $M$, and $y$ is in $N$. Since $x$ is in $M$, and $M$ is a 2-dimensional smooth manifold with differentiable structure $\mathcal{A}$, there exists a coordinate chart $(U, \varphi_U) \in \mathcal{A}$ such that $x$ is in $U$. Similarly, there exists a coordinate chart $(V, \psi_V) \in \mathcal{B}$ such that $y$ is in $V$. Since $(U, \varphi_U) \in \mathcal{A}$, and $\mathcal{A}$ is a differentiable structure on $M$, $\varphi_U$ is a homeomorphism from the open subset $U$ of $M$ onto the open subset $\varphi_U(U)$ of $\mathbb{R}^2$. Similarly, $\psi_V$ is a homeomorphism from the open subset $V$ of $N$ onto the open subset $\psi_V(V)$ of $\mathbb{R}^3$.

Since $x$ is in $U$, and $U$ is open in $M$, $U$ is an open neighborhood of $x$. Similarly, $V$ is an open neighborhood of $y$. Since $U$ is an open neighborhood of $x$, and $V$ is an open neighborhood of $y$, the Cartesian product $U \times V$ is an open neighborhood of $(x, y)$ in $M \times N$. Since $\varphi_U(U)$ is open in $\mathbb{R}^2$, and $\psi_V(V)$ is open in $\mathbb{R}^3$, their Cartesian product $(\varphi_U(U)) \times (\psi_V(V))$ is open in $\mathbb{R}^2 \times \mathbb{R}^3 (= \mathbb{R}^{2+3})$.
Now let us define a function

$$(\varphi_U \times \psi_V) : U \times V \to ((\varphi_U(U)) \times (\psi_V(V)))$$

as follows: for every $(x, y)$ in $U \times V$,

$$(\varphi_U \times \psi_V)(x, y) \equiv (\varphi_U(x), \psi_V(y)).$$

It remains to prove that $(\varphi_U \times \psi_V)$ is a homeomorphism, that is,

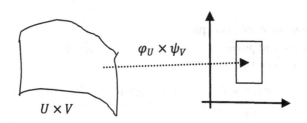

$U \times V$        $\varphi_U \times \psi_V$

**Fig. 1.3** Hausdorff topological space

1. $(\varphi_U \times \psi_V)$ is $1-1$,
2. $(\varphi_U \times \psi_V)$ is onto,
3. $(\varphi_U \times \psi_V)$ is continuous,
4. $(\varphi_U \times \psi_V)^{-1}$ is continuous.

For 1: Let $(\varphi_U \times \psi_V)(x, y) = (\varphi_U \times \psi_V)(z, w)$, where $x, z$ are in $U$, and $y, w$ are in $V$. We have to show that $(x, y) = (z, w)$, that is, $x = z$, and $y = w$. Since

$$(\varphi_U(x), \psi_V(y)) = (\varphi_U \times \psi_V)(x, y) = (\varphi_U \times \psi_V)(z, w) = (\varphi_U(z), \psi_V(w)),$$

$$\varphi_U(x) = \varphi_U(z), \text{ and } \psi_V(y) = \psi_V(w).$$

Since $\varphi_U$ is a homeomorphism, $\varphi_U$ is $1$-$1$. Since $\varphi_U$ is $1$-$1$, and $\varphi_U(x) = \varphi_U(z)$, $x = z$. Similarly $y = w$. This proves 1.

For 2: We have to prove that

$$(\varphi_U \times \psi_V) : U \times V \to ((\varphi_U(U)) \times (\psi_V(V)))$$

is onto. For this purpose, let us take any $(s, t)$ in $(\varphi_U(U)) \times (\psi_V(V))$. Now we have to find a member of $U \times V$, whose $(\varphi_U \times \psi_V)$-image is $(s, t)$. Since $(s, t)$ is in $(\varphi_U(U)) \times (\psi_V(V))$, $s$ is in $\varphi_U(U)$, and $t$ is in $\psi_V(V)$. Since $s$ is in $\varphi_U(U)$, there exists an element $x$ in $U$ such that $s = \varphi_U(x)$. Similarly, there exists an element $y$ in $V$ such that $t = \psi_V(y)$. Since $x$ is in $U$, and $y$ is in $V$, $(x, y)$ is in $U \times V$. Further

$$(\varphi_U \times \psi_V)(x, y) = (\varphi_U(x), \psi_V(y)) = (s, t).$$

This proves 2.

For 3: Since $\varphi_U$ is a homeomorphism, $\varphi_U$ is a continuous function, and hence, $x \mapsto \varphi_U(x)$ is continuous. Since, in the product topology, $(x, y) \mapsto x$ is continuous, and $x \mapsto \varphi_U(x)$ is continuous, their composite $(x, y) \mapsto \varphi_U(x)$ is continuous. Similarly, $(x, y) \mapsto \psi_V(y)$ is continuous. Since $(x, y) \mapsto \varphi_U(x)$ is continuous, and $(x, y) \mapsto \psi_V(y)$ is continuous, with respect to product topology, $(x, y) \mapsto (\varphi_U(x), \psi_V(y))(= (\varphi_U \times \psi_V)(x, y))$ is continuous, and hence,

$$(\varphi_U \times \psi_V) : U \times V \to ((\varphi_U(U)) \times (\psi_V(V)))$$

is a continuous function. This proves 3.

For 4: We have to prove that $(\varphi_U \times \psi_V)^{-1} : ((\varphi_U(U)) \times (\psi_V(V))) \to U \times V$ is continuous, that is,

$$(\varphi_U \times \psi_V) : U \times V \to ((\varphi_U(U)) \times (\psi_V(V)))$$

is an open mapping. For this purpose, let us take any open subset $G \times H$ of $U \times V$, where $G$ is open in $U$, and $H$ is open in $V$. We have to prove that $(\varphi_U \times \psi_V)(G \times H)$ is open in $(\varphi_U(U)) \times (\psi_V(V))$. By the definition of $(\varphi_U \times \psi_V)$,

$$(\varphi_U \times \psi_V)(G \times H) = (\varphi_U(G) \times \psi_V(H)).$$

Since $\varphi_U : U \to (\varphi_U(U))$ is a homeomorphism, and $G$ is open in $U$, $\varphi_U(G)$ is an open subset of $\varphi_U(U)$. Similarly, $\psi_V(H)$ is an open subset of $\psi_V(V)$. Since $\varphi_U(G)$ is an open subset of $\varphi_U(U)$, and $\psi_V(H)$ is an open subset of $\psi_V(V)$, their Cartesian product $\varphi_U(G) \times \psi_V(H)(= (\varphi_U \times \psi_V)(G \times H))$ is an open subset of the product space $\varphi_U(U) \times \psi_V(V)$. This proves $\underline{4}$.
Thus,

$$(\varphi_U \times \psi_V) : U \times V \to ((\varphi_U(U)) \times (\psi_V(V)))$$

is a homeomorphism. This proves 2.
Hence, $M \times N$, with the product topology, is a $(2 + 3)$-dimensional topological manifold. Since $M$ is a 2-dimensional smooth manifold, $M$ is a second countable space. Similarly, $N$ is a second countable space. Since $M, N$ are second countable spaces, their product $M \times N$ is a second countable space. Here, for every $(U, \varphi_U) \in \mathcal{A}$, and for every $(V, \psi_V) \in \mathcal{B}$, we get the coordinate chart $(U \times V, (\varphi_U \times \psi_V))$ of $M \times N$. Let us denote the collection

$$\{(U \times V, (\varphi_U \times \psi_V)) : (U, \varphi_U) \in \mathcal{A}, (V, \psi_V) \in \mathcal{B}\}$$

of coordinate charts of $M \times N$ by $\mathcal{C}$.
We want to show that $\mathcal{C}$ has the following two properties:

**1.** $\cup\{U \times V : (U \times V, (\varphi_U \times \psi_V)) \in \mathcal{C}\} = M \times N$,
**2.** all pairs of members in $\mathcal{C}$ are $C^r$-compatible, for every $r = 1, 2, 3, \ldots$.

For **1**: This is clear from the above discussion.
For **2**: Let us take any $(U \times V, (\varphi_U \times \psi_V)), (\hat{U} \times \hat{V}, (\varphi_{\hat{U}} \times \psi_{\hat{V}}))$ as a pair of members in $\mathcal{C}$. Here,

$$(U \times V) \cap (\hat{U} \times \hat{V}) = (U \cap \hat{U}) \times (V \cap \hat{V}).$$

Now we want to prove that the mapping

$$\left( (\varphi_{\hat{U}} \times \psi_{\hat{V}}) \circ (\varphi_U \times \psi_V)^{-1} \right) : (\varphi_U \times \psi_V)((U \times V) \cap (\hat{U} \times \hat{V}))$$
$$\to (\varphi_{\hat{U}} \times \psi_{\hat{V}})((U \times V) \cap (\hat{U} \times \hat{V}))$$

is $C^1$, that is,

$$\left( (\varphi_{\hat{U}} \times \psi_{\hat{V}}) \circ (\varphi_U \times \psi_V)^{-1} \right) : (\varphi_U \times \psi_V)((U \cap \hat{U}) \times (V \cap \hat{V}))$$
$$\to (\varphi_{\hat{U}} \times \psi_{\hat{V}})((U \cap \hat{U}) \times (V \cap \hat{V}))$$

is $C^1$. Here, $(U \times V, (\varphi_U \times \psi_V))$ is in $\mathcal{C}$, so $(U, \varphi_U) \in \mathcal{A}$ and $(V, \psi_V) \in \mathcal{B}$. Similarly, $(\hat{U}, \varphi_{\hat{U}}) \in \mathcal{A}$ and $(\hat{V}, \psi_{\hat{V}}) \in \mathcal{B}$. Since $(U, \varphi_U) \in \mathcal{A}$, $(\hat{U}, \varphi_{\hat{U}}) \in \mathcal{A}$, and $\mathcal{A}$ is a differentiable structure, $(U, \varphi_U)$ and $(\hat{U}, \varphi_{\hat{U}})$ are $C^r$-compatible for every positive integer $r$. Since $(U, \varphi_U)$ and $(\hat{U}, \varphi_{\hat{U}})$ are $C^r$-compatible for every positive integer $r$, $(U, \varphi_U)$ and $(\hat{U}, \varphi_{\hat{U}})$ are $C^1$-compatible. Since $(U, \varphi_U)$ and $(\hat{U}, \varphi_{\hat{U}})$ are $C^1$-compatible, $f^1, f^2, g^1, g^2$ are $C^1$, where, for every $(x^1, x^2)$ in $\varphi_U(U \cap \hat{U})(\subset \mathbb{R}^2)$,

$$\left( \varphi_{\hat{U}} \circ (\varphi_U)^{-1} \right)(x^1, x^2) \equiv (f^1(x^1, x^2), f^2(x^1, x^2)) = (y^1, y^2) \in \varphi_{\hat{U}}(U \cap \hat{U}),$$

and

$$\left( \varphi_U \circ (\varphi_{\hat{U}})^{-1} \right)(y^1, y^2) \equiv (g^1(y^1, y^2), g^2(y^1, y^2)) = (x^1, x^2).$$

Similarly, $h^1, h^2, h^3, k^1, k^2, k^3$ are $C^1$, where for every $(x^1, x^2, x^3)$ in $\psi_V(V \cap \hat{V})(\subset \mathbb{R}^3)$,

$$\left( \psi_{\hat{V}} \circ (\psi_V)^{-1} \right)(x^1, x^2, x^3) \equiv (h^1(x^1, x^2, x^3), h^2(x^1, x^2, x^3), h^3(x^1, x^2, x^3))$$
$$= (y^1, y^2, y^3) \in \psi_{\hat{V}}(V \cap \hat{V}),$$

and

$$\left( \psi_V \circ (\psi_{\hat{V}})^{-1} \right)(y^1, y^2, y^3) \equiv (k^1(y^1, y^2, y^3), k^2(y^1, y^2, y^3), k^3(y^1, y^2, y^3))$$
$$= (x^1, x^2, x^3).$$

Let us take any $(x^1, x^2, x^3, x^4, x^5)$ in $(\varphi_U \times \psi_V)((U \cap \hat{U}) \times (V \cap \hat{V}))(\subset \mathbb{R}^{2+3})$. So there exist $x$ in $U \cap \hat{U}$ and $x^*$ in $V \cap \hat{V}$ such that

$$(\varphi_U(x), \psi_V(x^*)) = (\varphi_U \times \psi_V)(x, x^*) = (x^1, x^2, x^3, x^4, x^5).$$

Hence,

$$(\varphi_U \times \psi_V)^{-1}(x^1, x^2, x^3, x^4, x^5) = (x, x^*),$$

or

$$\varphi_U(x) = (x^1, x^2),$$

and

$$\psi_V(x^*) = (x^3, x^4, x^5).$$

Now

$$\left( (\varphi_{\hat{U}} \times \psi_{\hat{V}}) \circ (\varphi_U \times \psi_V)^{-1} \right) (x^1, x^2, x^3, x^4, x^5)$$

$$= (\varphi_{\hat{U}} \times \psi_{\hat{V}}) \left( (\varphi_U \times \psi_V)^{-1} (x^1, x^2, x^3, x^4, x^5) \right)$$

$$= (\varphi_{\hat{U}} \times \psi_{\hat{V}})(x, x^*)$$

$$= \left( \left( \varphi_{\hat{U}} \circ \left( (\varphi_U)^{-1} \circ \varphi_U \right) \right)(x), \left( \psi_{\hat{V}} \circ \left( (\psi_V)^{-1} \circ \psi_V \right) \right)(x^*) \right)$$

$$= \left( \left( \left( \varphi_{\hat{U}} \circ (\varphi_U)^{-1} \right) \circ \varphi_U \right)(x), \left( \left( \psi_{\hat{V}} \circ (\psi_V)^{-1} \right) \circ \psi_V \right)(x^*) \right)$$

$$= \left( \left( \varphi_{\hat{U}} \circ (\varphi_U)^{-1} \right)(\varphi_U(x)), \left( \psi_{\hat{V}} \circ (\psi_V)^{-1} \right)(\psi_V(x^*)) \right)$$

$$= \left( \left( \varphi_{\hat{U}} \circ (\varphi_U)^{-1} \right)(x^1, x^2), \left( \psi_{\hat{V}} \circ (\psi_V)^{-1} \right)(x^3, x^4, x^5) \right)$$

$$= \left( \left( \varphi_{\hat{U}} \circ (\varphi_U)^{-1} \right)(x^1, x^2), \left( \psi_{\hat{V}} \circ (\psi_V)^{-1} \right)(x^3, x^4, x^5) \right)$$

$$= ((f^1(x^1, x^2), f^2(x^1, x^2)), (h^1(x^3, x^4, x^5), h^2(x^3, x^4, x^5), h^3(x^3, x^4, x^5)))$$

$$= (f^1(x^1, x^2), f^2(x^1, x^2), h^1(x^3, x^4, x^5), h^2(x^3, x^4, x^5), h^3(x^3, x^4, x^5)).$$

Thus, for every $(x^1, x^2, x^3, x^4, x^5)$ in $(\varphi_U \times \psi_V)((U \cap \hat{U}) \times (V \cap \hat{V}))(\subset \mathbb{R}^{2+3})$,

$$\left( (\varphi_{\hat{U}} \times \psi_{\hat{V}}) \circ (\varphi_U \times \psi_V)^{-1} \right)(x^1, x^2, x^3, x^4, x^5)$$

$$= (f^1(x^1, x^2), f^2(x^1, x^2), h^1(x^3, x^4, x^5), h^2(x^3, x^4, x^5), h^3(x^3, x^4, x^5)).$$

Now we want to prove that $F^1 : (x^1, x^2, x^3, x^4, x^5) \mapsto f^1(x^1, x^2)$ is $C^1$. Here, $F^1 = f^1 \circ \alpha$, where $\alpha : \mathbb{R}^5 \to \mathbb{R}^2$ is defined by $\alpha(x^1, x^2, x^3, x^4, x^5) \equiv (x^1, x^2)$. So for $x = (x^1, x^2, x^3, x^4, x^5)$,

$$(DF^1)(x) = (D(f^1 \circ \alpha))(x) = (Df^1)(\alpha(x)) \circ ((D\alpha)(x))$$

or,

$$(DF^1)(x) = \left[ \left( \frac{\partial f^1}{\partial x^1} \right)(\alpha(x)) \quad \left( \frac{\partial f^1}{\partial x^2} \right)(\alpha(x)) \right] \begin{bmatrix} 1 & 0 & 0 & 0 & 0 \\ 0 & 1 & 0 & 0 & 0 \end{bmatrix}$$

$$= \left[ \left( \frac{\partial f^1}{\partial x^1} \right)(\alpha(x)) \quad \left( \frac{\partial f^1}{\partial x^2} \right)(\alpha(x)) \quad 0 \quad 0 \quad 0 \right].$$

Hence,

$$DF^1 = \left[ (D_1 f^1) \circ \alpha \quad (D_2 f^1) \circ \alpha \quad 0 \quad 0 \quad 0 \right].$$

Since $f^1$ is $C^1$, $(D_1 f^1)$ is continuous. Since $(D_1 f^1)$ is continuous, and $\alpha$ is continuous, their composite $(D_1 f^1) \circ \alpha$ is continuous. Similarly, $(D_2 f^1) \circ \alpha$ is continuous. Since $(D_1 f^1) \circ \alpha$, $(D_2 f^1) \circ \alpha$ are continuous, and

$$DF^1 = \left[ (D_1 f^1) \circ \alpha \quad (D_2 f^1) \circ \alpha \quad 0 \quad 0 \quad 0 \right],$$

$F^1$ is $C^1$. Thus, we have shown that $(x^1, x^2, x^3, x^4, x^5) \mapsto f^1(x^1, x^2)$ is $C^1$.

Similarly, the functions $(x^1, x^2, x^3, x^4, x^5) \mapsto f^2(x^1, x^2)$, $(x^1, x^2, x^3, x^4, x^5) \mapsto h^1(x^3, x^4, x^5)$,

$$(x^1, x^2, x^3, x^4, x^5) \mapsto h^2(x^3, x^4, x^5), (x^1, x^2, x^3, x^4, x^5) \mapsto h^3(x^3, x^4, x^5)$$

are $C^1$. Since $(x^1, x^2, x^3, x^4, x^5) \mapsto f^1(x^1, x^2)$, $(x^1, x^2, x^3, x^4, x^5) \mapsto f^2(x^1, x^2)$,

$$(x^1, x^2, x^3, x^4, x^5) \mapsto h^1(x^3, x^4, x^5), (x^1, x^2, x^3, x^4, x^5) \mapsto h^2(x^3, x^4, x^5),$$
$$(x^1, x^2, x^3, x^4, x^5) \mapsto h^3(x^3, x^4, x^5)$$

are $C^1$, and

$$\left( (\varphi_{\hat{U}} \times \psi_{\hat{V}}) \circ (\varphi_U \times \psi_V)^{-1} \right) (x^1, x^2, x^3, x^4, x^5)$$
$$= \left( f^1(x^1, x^2), f^2(x^1, x^2), h^1(x^3, x^4, x^5), h^2(x^3, x^4, x^5), h^3(x^3, x^4, x^5) \right),$$

$$\left( (\varphi_{\hat{U}} \times \psi_{\hat{V}}) \circ (\varphi_U \times \psi_V)^{-1} \right) : (\varphi_U \times \psi_V)((U \cap \hat{U}) \times (V \cap \hat{V}))$$
$$\to (\varphi_{\hat{U}} \times \psi_{\hat{V}})((U \cap \hat{U}) \times (V \cap \hat{V}))$$

is $C^1$.

Similarly, we can prove that $(\varphi_{\hat{U}} \times \psi_{\hat{V}}) \circ (\varphi_U \times \psi_V)^{-1}$ is $C^2$, etc. Hence, the pair $(U \times V, (\varphi_U \times \psi_V)), (\hat{U} \times \hat{V}, (\varphi_{\hat{U}} \times \psi_{\hat{V}}))$ of members in $C$ are $C^r$-compatible, for every $r = 1, 2, 3, \ldots$. This proves **2**.

Hence, by Theorem 1.2, there exists a unique $C^\infty$-differentiable structure $\mathcal{D}$ on $M \times N$ which contains $C$. Thus, the ordered pair $(M \times N, \mathcal{D})$ is an example of a smooth manifold of dimension $(2 + 3)$. Similarly, if $M$ is an $m$-dimensional smooth manifold, and $N$ is an $n$-dimensional smooth manifold, then the Cartesian product $M \times N$, with the product topology, becomes an $(m + n)$-dimensional smooth manifold. In short, we say that the *product manifold $M \times N$* is an $(m + n)$-dimensional smooth manifold.

# Chapter 2
# Tangent Spaces

Above dimension three, Euclidean space loses its quality of being visible. So, in order to investigate higher-dimensional smooth manifolds, we need to generalize the concept of tangent line in the case of curves in two-dimensional Euclidean space, and tangent plane in the case of surfaces in three-dimensional Euclidean space, using abstract tools of algebra, and analysis. This generalized concept has been named the tangent space. Among many other concepts that facilitate the enquiry into the properties of higher-dimensional smooth manifolds as we shall see later, the notion of tangent space is foremost. It is possible to introduce tangent space in various ways. Because each method has its own merit, we shall introduce it in multiple ways in this chapter for deeper understanding. Remember that in the process of generalization, unlike curves and surfaces, smooth manifolds are devoid of any ambient space. It was in this sense a challenging task for early mathematicians.

## 2.1 Smooth Functions

**Definition** Let $M$ be an $m$-dimensional smooth manifold. Let $f : M \to \mathbb{R}$ be any function. Let $p$ be an element of $M$. If, for every admissible coordinate chart $(U, \varphi_U)$ of $M$ satisfying $p \in U$, $(f \circ (\varphi_U)^{-1}) : \varphi_U(U) \to \mathbb{R}$ is $C^\infty$ at the point $\varphi_U(p)$ in $\mathbb{R}^m$, then we say that $f$ is $C^\infty$ at $p$ in $M$.

Observe that, if $f$ is $C^\infty$ at $p$ in $M$, then $f$ is continuous at $p$.

Reason: Since $M$ is an $m$-dimensional smooth manifold, and $p$ is an element of $M$, there exists an admissible coordinate chart $(U, \varphi_U)$ of $M$ satisfying $p \in U$. Since $f$ is $C^\infty$ at $p$ in $M$, $(f \circ (\varphi_U)^{-1}) : \varphi_U(U) \to \mathbb{R}$ is $C^\infty$ at the point $\varphi_U(p)$ in $\mathbb{R}^m$. Since $(f \circ (\varphi_U)^{-1}) : \varphi_U(U) \to \mathbb{R}$ is $C^\infty$ at the point $\varphi_U(p)$ in $\mathbb{R}^m$, $(f \circ (\varphi_U)^{-1}) : \varphi_U(U) \to \mathbb{R}$ is continuous at the point $\varphi_U(p)$. Since $(U, \varphi_U)$ is an admissible coordinate chart of $M$, $\varphi_U : U \to \varphi_U(U)$ is continuous. Since $\varphi_U : U \to \varphi_U(U)$ is continuous, and $p \in U$, $\varphi_U : U \to \varphi_U(U)$ is continuous at $p$. Since $\varphi_U : U \to \varphi_U(U)$ is continuous at $p$, and $(f \circ (\varphi_U)^{-1}) : \varphi_U(U) \to \mathbb{R}$ is continuous at the point $\varphi_U(p)$,

© Springer India 2014
R. Sinha, *Smooth Manifolds*, DOI 10.1007/978-81-322-2104-3_2

their composite function $\left(f \circ (\varphi_U)^{-1}\right) \circ \varphi_U \left(= f \circ ((\varphi_U)^{-1} \circ \varphi_U) = f\right)$ is continuous at $p$. Thus, $f$ is continuous at $p$.

**Theorem 2.1** *Let $M$ be an $m$-dimensional smooth manifold. Let $f : M \to \mathbb{R}$ be any function. Let $p$ be an element of $M$. If there exists an admissible coordinate chart $(U, \varphi_U)$ of $M$ satisfying $p \in U$ such that $(f \circ (\varphi_U)^{-1}) : \varphi_U(U) \to \mathbb{R}$ is $C^\infty$ at the point $\varphi_U(p)$ in $\mathbb{R}^m$, then $f$ is $C^\infty$ at $p$ in $M$.*

*Proof* Let us take any admissible coordinate chart $(V, \varphi_V)$ of $M$ satisfying $p \in V$. We have to show that $(f \circ (\varphi_V)^{-1}) : \varphi_V(V) \to \mathbb{R}$ is $C^\infty$ at the point $\varphi_V(p)$ in $\mathbb{R}^m$. Since

$$\left(f \circ (\varphi_V)^{-1}\right) = \left(f \circ \left((\varphi_U)^{-1} \circ \varphi_U\right)\right) \circ (\varphi_V)^{-1} = \left(\left(f \circ (\varphi_U)^{-1}\right) \circ \varphi_U\right) \circ (\varphi_V)^{-1}$$

$$= \left(f \circ (\varphi_U)^{-1}\right) \circ \left(\varphi_U \circ (\varphi_V)^{-1}\right),$$

$(\varphi_U \circ (\varphi_V)^{-1})$ is $C^\infty$ at the point $\varphi_V(p)$ in $\mathbb{R}^m$, and $(f \circ (\varphi_U)^{-1})$ is $C^\infty$ at the point $\varphi_U(p)$ in $\mathbb{R}^m$, their composite $(f \circ (\varphi_U)^{-1}) \circ (\varphi_U \circ (\varphi_V)^{-1})(= (f \circ (\varphi_V)^{-1}))$ is $C^\infty$ at the point $\varphi_V(p)$ in $\mathbb{R}^m$. Thus, we have shown that $(f \circ (\varphi_V)^{-1}) : \varphi_V(V) \to \mathbb{R}$ is $C^\infty$ at the point $\varphi_V(p)$ in $\mathbb{R}^m$. $\qquad\square$

**Definition** Let $M$ be an $m$-dimensional smooth manifold. Let $f : M \to \mathbb{R}$ be any function. By $f$ is $C^\infty$ on $M$ (or $f$ is *smooth* on $M$), we mean that $f$ is $C^\infty$ at every point $p$ in $M$. The set of all smooth functions $f : M \to \mathbb{R}$ on $M$ is denoted by $C^\infty(M)$.

Observe that, if $f$ is $C^\infty$ on $M$, then $f$ is a continuous function.

**Definition** Let $M$ be an $m$-dimensional smooth manifold, and $N$ be an $n$-dimensional smooth manifold. Let $f : M \to N$ be any continuous function. Let $p$ be an element of $M$. If for every admissible coordinate chart $(U, \varphi_U)$ of $M$ satisfying $p \in U$, and for every admissible coordinate chart $(V, \psi_V)$ of $N$ satisfying $f(p) \in V$, each of the $n$ component functions of the mapping

$$\psi_V \circ \left(f \circ (\varphi_U)^{-1}\right) : \varphi_U(U \cap f^{-1}(V)) \to \psi_V(V)$$

is $C^\infty$ at the point $\varphi_U(p)$ in $\mathbb{R}^m$ (in short, the mapping

$$\psi_V \circ \left(f \circ (\varphi_U)^{-1}\right) : \varphi_U(U \cap f^{-1}(V)) \to \psi_V(V)$$

is $C^\infty$ at the point $\varphi_U(p)$ in $\mathbb{R}^m$); then, we say that $f$ is $C^\infty$ at $p$ in $M$.

As above we observe that if $f : M \to N$ is $C^\infty$ at $p$ in $M$, then $f$ is continuous at $p$.

**Theorem 2.2** *Let $M$ be an $m$-dimensional smooth manifold, and $N$ be an $n$-dimensional smooth manifold. Let $f : M \to N$ be any continuous function. Let $p$*

*be an element of M. If there exist an admissible coordinate chart $(U, \varphi_U)$ of M satisfying $p \in U$, and an admissible coordinate chart $(V, \psi_V)$ of N satisfying $f(p) \in V$ such that each of the n component functions of the mapping*

$$\psi_V \circ \left( f \circ (\varphi_U)^{-1} \right) : \varphi_U(U \cap f^{-1}(V)) \to \psi_V(V)$$

*is $C^\infty$ at the point $\varphi_U(p)$ in $\mathbb{R}^m$; then, f is $C^\infty$ at p in M.*

*Proof* Let us take any admissible coordinate chart $(\widehat{U}, \varphi_{\widehat{U}})$ of M satisfying $p \in \widehat{U}$, and any admissible coordinate chart $(\widehat{V}, \psi_{\widehat{V}})$ of N satisfying $f(p) \in \widehat{V}$. We have to show that

$$\psi_{\widehat{V}} \circ \left( f \circ \left( \varphi_{\widehat{U}} \right)^{-1} \right) : \varphi_{\widehat{U}} \left( \widehat{U} \cap f^{-1} \left( \widehat{V} \right) \right) \to \psi_{\widehat{V}} \left( \widehat{V} \right)$$

is $C^\infty$ at the point $\varphi_{\widehat{U}}(p)$ in $\mathbb{R}^m$. Since

$$\begin{aligned}
\psi_{\widehat{V}} \circ \left( f \circ \left( \varphi_{\widehat{U}} \right)^{-1} \right) &= \psi_{\widehat{V}} \circ \left( \left( f \circ \left( (\varphi_U)^{-1} \circ \varphi_U \right) \right) \circ \left( \varphi_{\widehat{U}} \right)^{-1} \right) \\
&= \psi_{\widehat{V}} \circ \left( \left( f \circ (\varphi_U)^{-1} \right) \circ \left( \varphi_U \circ \left( \varphi_{\widehat{U}} \right)^{-1} \right) \right) \\
&= \psi_{\widehat{V}} \circ \left( \left( \left( \left( (\psi_V)^{-1} \circ \psi_V \right) \circ f \right) \circ (\varphi_U)^{-1} \right) \circ \left( \varphi_U \circ \left( \varphi_{\widehat{U}} \right)^{-1} \right) \right) \\
&= \left( \psi_{\widehat{V}} \circ (\psi_V)^{-1} \right) \circ \left( \left( (\psi_V \circ f) \circ (\varphi_U)^{-1} \right) \circ \left( \varphi_U \circ \left( \varphi_{\widehat{U}} \right)^{-1} \right) \right),
\end{aligned}$$

$(\varphi_U \circ (\varphi_U)^{-1})$ is $C^\infty$ at the point $\varphi_{\widehat{U}}(p)$ in $\mathbb{R}^m$, $\psi_V \circ (f \circ (\varphi_U)^{-1})$ is $C^\infty$ at the point $\varphi_U(p)$ in $\mathbb{R}^m$, and $(\psi_{\widehat{V}} \circ (\psi_V)^{-1})$ is $C^\infty$ at the point $\psi_V(f(p))$ in $\mathbb{R}^n$, their composite $(\psi_{\widehat{V}} \circ (\psi_V)^{-1}) \circ (((\psi_V \circ f) \circ (\varphi_U)^{-1}) \circ (\varphi_U \circ (\varphi_{\widehat{U}})^{-1}))$ $(= \psi_{\widehat{V}} \circ (f \circ (\varphi_{\widehat{U}})^{-1}))$ is $C^\infty$ at the point $\varphi_{\widehat{U}}(p)$ in $\mathbb{R}^m$.

Thus, we have shown that $\psi_{\widehat{V}} \circ (f \circ (\varphi_{\widehat{U}})^{-1}) : \varphi_{\widehat{U}}(\widehat{U} \cap f^{-1}(\widehat{V})) \to \psi_{\widehat{V}}(\widehat{V})$ is $C^\infty$ at the point $\varphi_{\widehat{U}}(p)$ in $\mathbb{R}^m$.  □

**Lemma 2.3** *Let M be an m-dimensional smooth manifold, N be an n-dimensional smooth manifold, and K be a k-dimensional smooth manifold. Let p be an element of M. Let $f : M \to N$ be $C^\infty$ at p in M, and $g : N \to K$ be $C^\infty$ at $f(p)$ in N. Then, $g \circ f$ is $C^\infty$ at p in M.*

*Proof* We have to prove that $g \circ f$ is $C^\infty$ at p in M. By Theorem 2.2, we must find an admissible coordinate chart $(U, \varphi_U)$ of M satisfying $p \in U$, and an admissible coordinate chart $(W, \chi_w)$ of K satisfying $(g \circ f)(p) \in W$ such that the mapping

$$\chi_w \circ \left( (g \circ f) \circ (\varphi_U)^{-1} \right) : \varphi_U \Big( U \cap (g \circ f)^{-1}(W) \Big) \to \chi_w(W)$$

is $C^\infty$ at the point $\varphi_U(p)$ in $\mathbb{R}^m$. Since $p$ is in $M$, and $M$ is an $m$-dimensional smooth manifold, there exists an admissible coordinate chart $(U, \varphi_U)$ of $M$ satisfying $p \in U$. Since $f : M \to N$, and $p$ is in $M$, $f(p)$ is in $N$. Since $f(p)$ is in $N$, and $N$ is an $n$-dimensional smooth manifold, there exists an admissible coordinate chart $(V, \psi_V)$ of $N$ satisfying $f(p) \in V$. Since $f : M \to N$ is $C^\infty$ at $p$ in $M$, $(U, \varphi_U)$ is an admissible coordinate chart of $M$ satisfying $p \in U$, and $(V, \psi_V)$ is an admissible coordinate chart of $N$ satisfying $f(p) \in V$, the mapping

$$\psi_V \circ \left( f \circ (\varphi_U)^{-1} \right) : \varphi_U(U \cap f^{-1}(V)) \to \psi_V(V)$$

is $C^\infty$ at the point $\varphi_U(p)$ in $\mathbb{R}^m$.

Similarly, there exists an admissible coordinate chart $(W, \chi_w)$ of $K$ satisfying $g(f(p)) \in W$ such that the mapping

$$\chi_w \circ \left( g \circ (\psi_V)^{-1} \right) : \psi_V(V \cap g^{-1}(W)) \to \chi_w(W)$$

is $C^\infty$ at the point $\psi_V(f(p))$ in $\mathbb{R}^n$. Since $\psi_V \circ (f \circ (\varphi_U)^{-1})$ is $C^\infty$ at the point $\varphi_U(p)$ in $\mathbb{R}^m$, and $\chi_w \circ (g \circ (\psi_V)^{-1})$ is $C^\infty$ at the point $\psi_V(f(p))$ in $\mathbb{R}^n$, their composite $(\chi_w \circ (g \circ (\psi_V)^{-1})) \circ (\psi_V \circ (f \circ (\varphi_U)^{-1})) (= \chi_w \circ ((g \circ f) \circ (\varphi_U)^{-1}))$ is $C^\infty$ at the point $\varphi_U(p)$ in $\mathbb{R}^m$. □

**Definition** Let $M$ be an $m$-dimensional smooth manifold, and $N$ be an $n$-dimensional smooth manifold. Let $p$ be an element of $M$, and let $G$ be an open neighborhood of $p$ in $M$. Let $f : G \to N$ be any continuous function. If for every admissible coordinate chart $(U, \varphi_U)$ of $M$ satisfying $p \in U$, and for every admissible coordinate chart $(V, \psi_V)$ of $N$ satisfying $f(p) \in V$, each of the $n$ component functions of the mapping

$$\psi_V \circ \left( f \circ (\varphi_U)^{-1} \right) : \varphi_U(U \cap G \cap f^{-1}(V)) \to \psi_V(V)$$

is $C^\infty$ at the point $\varphi_U(p)$ in $\mathbb{R}^m$ (in short, the mapping

$$\psi_V \circ \left( f \circ (\varphi_U)^{-1} \right) : \varphi_U(U \cap G \cap f^{-1}(V)) \to \psi_V(V)$$

is $C^\infty$ at the point $\varphi_U(p)$ in $\mathbb{R}^m$); then, we say that $f$ is $C^\infty$ at $p$ in $M$.

**Definition** Let $M$ be an $m$-dimensional smooth manifold, and $N$ be an $n$-dimensional smooth manifold. Let $G$ be a nonempty open subset of $M$. Let $f : G \to N$ be any continuous function. By $f$ is a *smooth map from $G$ to $N$*, we mean that $f$ is $C^\infty$ at every point $p$ in $G$.

**Definition** Let $M$ and $N$ be $m$-dimensional smooth manifolds. Let $f : M \to N$ be a 1–1, onto mapping. If $f$ is a smooth map from $M$ to $N$, and $f^{-1}$ is a smooth map from $N$ to $M$, then we say that $f$ is a *diffeomorphism from M onto N*. We observe that if $f$ is a diffeomorphism from $M$ onto $N$, then $f$ is a homeomorphism from $M$ onto $N$.

**Definition** Let $M$ be an $m$-dimensional smooth manifold, and $N$ be an $m$-dimensional smooth manifold. If there exists a function $f : M \to N$ such that $f$ is a diffeomorphism from $M$ onto $N$, then we say that the *manifolds M and N are isomorphic* (or, *diffeomorphic*).

**Definition** Let $M$ be an $m$-dimensional smooth manifold. Let $a$ and $b$ be any real numbers such that $a < b$. Let $\gamma$ be a continuous function from the open interval $(a, b)$ to $M$. Recall that the open interval $(a, b)$ is a 1-dimensional smooth manifold. Hence, it is meaningful to say that $\gamma$ is a smooth map from $(a, b)$ to $M$. If $\gamma$ is a smooth map from $(a, b)$ to $M$, then we say that $\gamma$ *is a parametrized curve in the manifold M*.

**Definition** Let $M$ be an $m$-dimensional smooth manifold. Let $p \in M$. Let $\gamma$ be a parametrized curve in the manifold $M$. If there exists a real number $\varepsilon > 0$ such that $\gamma$ is defined on the open interval $(-\varepsilon, \varepsilon)$, and $\gamma(0) = p$, then we say that $\gamma$ *is a parametrized curve in M through p*. The set of all parametrized curves in $M$ through $p$ is denoted by $\Gamma_p(M)$ (or, simply $\Gamma_p$).

*Example 2.4* In the Example 1.9, let us define

$$\pi_1 : M \times N \to M$$

as follows: for every $(x, y)$ in $M \times N$,

$$\pi_1(x, y) = x.$$

We want to prove that $\pi_1$ is a smooth function from the product manifold $M \times N$ onto $M$. For this purpose, let us take any $(p, q)$ in $M \times N$. Now, there exists a coordinate chart $(U \times V, (\varphi_U \times \psi_V))$ of the product manifold $M \times N$ such that $(U, \varphi_U) \in \mathcal{A}, (V, \psi_V) \in \mathcal{B}$, and $(p, q)$ is in $U \times V$. Hence, $(U, \varphi_U)$ is a coordinate chart of $M$ such that $p(= \pi_1(p, q))$ is in $U$. We want to show that

$$\varphi_U \circ \pi_1 \circ (\varphi_U \times \psi_V)^{-1} : (\varphi_U \times \psi_V)(U \times V) \to \varphi_U(U)$$

is $C^\infty$ at the point $(\varphi_U \times \psi_V)(p, q)(= (\varphi_U(p), \psi_V(q)))$ in $\mathbb{R}^{2+3}$. Let us take any $(x^1, x^2, x^3, x^4, x^5)$ in $(\varphi_U \times \psi_V)(U \times V)$. So, there exist $u$ in $U$, and $v$ in $V$ such that

$$(\varphi_U(u), \psi_V(v)) = (\varphi_U \times \psi_V)(u, v) = (x^1, x^2, x^3, x^4, x^5).$$

Hence,

$$(\varphi_U \times \psi_V)^{-1}(x^1, x^2, x^3, x^4, x^5) = (u, v),$$

or

$$\varphi_U(u) = \left(x^1, x^2\right), \ \psi_V(v) = \left(x^3, x^4, x^5\right).$$

So

$$\begin{aligned}
\left(\varphi_U \circ \pi_1 \circ (\varphi_U \times \psi_V)^{-1}\right)\left(x^1, x^2, x^3, x^4, x^5\right) &= (\varphi_U \circ \pi_1)\left((\varphi_U \times \psi_V)^{-1}\left(x^1, x^2, x^3, x^4, x^5\right)\right) \\
&= (\varphi_U \circ \pi_1)(u, v) = \varphi_U(\pi_1(u, v)) = \varphi_U(u) \\
&= \left(x^1, x^2\right).
\end{aligned}$$

Hence, $\left(\varphi_U \circ \pi_1 \circ (\varphi_U \times \psi_V)^{-1}\right) : (x^1, x^2, x^3, x^4, x^5) \mapsto (x^1, x^2)$, which is trivially $C^\infty$.

So, by Theorem 2.2, the *natural projection* $\pi_1$ of the product manifold $M \times N$ is a smooth map. Similarly, the natural projection $\pi_2$ of the product manifold $M \times N$ is a smooth map.

**Definition** Let $M$ be an $m$-dimensional smooth manifold. Let $p$ be an element of $M$. By $C_p^\infty(M)$ (or, simply $C_p^\infty$), we mean the collection of all real-valued functions $f$ whose $\mathrm{dom} f$ is an open neighborhood of $p$ in $M$, and for every admissible coordinate chart $(U, \varphi_U)$ of $M$ satisfying $p \in U$,

$$\left(f \circ (\varphi_U)^{-1}\right) : \varphi_U((\mathrm{dom} f) \cap U) \to \mathbb{R}$$

is $C^\infty$ at the point $\varphi_U(p)$ in $\mathbb{R}^m$. Observe that if $f$ is in $C_p^\infty(M)$, then $f$ is continuous on some open neighborhood of $p$.

**Definition** Let $M$ be an $m$-dimensional smooth manifold. Let $f : M \to \mathbb{R}^k$ be any function. By $f$ is smooth, we mean for every $p$ in $M$ there exists an admissible coordinate chart $(U, \varphi_U)$ of $M$ satisfying $p \in U$ such that the composite function $f \circ (\varphi_U)^{-1} : \varphi_U(U) \to \mathbb{R}^k$ is smooth in the sense of ordinary calculus. Remember, $\varphi_U(U)$ is an open subset of $\mathbb{R}^m$ containing $\varphi_U(p)$.

**Note 2.5** Let $G$ be a nonempty open subset of $\mathbb{R}^m$. We know that $G$ is an $m$-dimensional smooth manifold with the standard differentiable structure containing $\{(G, \mathrm{Id}_G)\}$. Let $f : G \to \mathbb{R}^k$ be any smooth function in the sense just defined. We want to show that $f : G \to \mathbb{R}^k$ is a smooth function in the sense of ordinary calculus. Since $G$ is nonempty, there exists $a$ in $G$. Since $a$ is in $G$, and $f : G \to \mathbb{R}^k$ be any smooth function in the sense just defined, there exists an admissible coordinate chart $(U, \varphi_U)$ of $G$ satisfying $a \in U$ such that the composite function $f \circ (\varphi_U)^{-1} : \varphi_U(U) \to \mathbb{R}^k$ is smooth in the sense of ordinary calculus. Since $(U, \varphi_U)$ is in $\{(G, \mathrm{Id}_G)\}$, $U = G$, and $\varphi_U = \mathrm{Id}_G$. Since $U = G$, and $\varphi_U = \mathrm{Id}_G$, $f \circ (\varphi_U)^{-1} =$

$f \circ (\mathrm{Id}_G)^{-1} = f \circ \mathrm{Id}_G = f$, and $\varphi_U(U) = \mathrm{Id}_G(U) = U = G$. Since $\varphi_U(U) = G$, and $f \circ (\varphi_U)^{-1} = f$, and $f \circ (\varphi_U)^{-1} : \varphi_U(U) \to \mathbb{R}^k$ is smooth in the sense of ordinary calculus, $f : G \to \mathbb{R}^k$ is smooth in the sense of ordinary calculus.

Conversely, let $f : G \to \mathbb{R}^k$ be smooth in the sense of ordinary calculus. We want to show that $f : G \to \mathbb{R}^k$ is a smooth function in the sense just defined. Let us take any $a$ in $G$. We have to find an admissible coordinate chart $(U, \varphi_U)$ of $G$ satisfying $a \in U$ such that the composite function $f \circ (\varphi_U)^{-1} : \varphi_U(U) \to \mathbb{R}^k$ is smooth in the sense of ordinary calculus. Let us take $(G, \mathrm{Id}_G)$ for $(U, \varphi_U)$. Hence, $\varphi_U(U) = \mathrm{Id}_G(U) = U = G$, and $f \circ (\varphi_U)^{-1} = f \circ (\mathrm{Id}_G)^{-1} = f \circ \mathrm{Id}_G = f$. Since $\varphi_U(U) = G, f \circ (\varphi_U)^{-1} = f$, and $f : G \to \mathbb{R}^k$ is smooth in the sense of ordinary calculus, $f \circ (\varphi_U)^{-1} : \varphi_U(U) \to \mathbb{R}^k$ is smooth in the sense of ordinary calculus.

Thus, we have shown that $f : G \to \mathbb{R}^k$ is smooth in the sense of ordinary calculus if and only if $f : G \to \mathbb{R}^k$ is a smooth function in the sense just defined.

**Note 2.6** Let $M$ be an $m$-dimensional smooth manifold. Let $f : M \to \mathbb{R}^k$ be any smooth function. Similar to the proof of Theorem 2.1, it can be shown that for every $p$ in $M$, and for every admissible coordinate chart $(U, \varphi_U)$ of $M$ satisfying $p \in U$, the composite function $f \circ (\varphi_U)^{-1} : \varphi_U(U) \to \mathbb{R}^k$ is smooth in the sense of ordinary calculus.

## 2.2 Algebra of Smooth Functions

**Note 2.7** Let $M$ be an $m$-dimensional smooth manifold. By $C^\infty(M)$, we mean the collection of all smooth functions $f : M \to \mathbb{R}$. For every $f, g$ in $C^\infty(M)$, we define $(f + g) : M \to \mathbb{R}$ as follows: for every $x$ in $M$,

$$(f + g)(x) \equiv f(x) + g(x).$$

For every $f$ in $C^\infty(M)$, and for every real $t$, we define $(tf) : M \to \mathbb{R}$ as follows: for every $x$ in $M$,

$$(tf)(x) \equiv t(f(x)).$$

It is clear that the set $C^\infty(M)$, together with vector addition, and scalar multiplication defined as above, constitutes a real linear space. In short, $C^\infty(M)$ is a real linear space.

For every $f, g$ in $C^\infty(M)$, we define $(f \cdot g) : M \to \mathbb{R}$ as follows: for every $x$ in $M$,

$$(f \cdot g)(x) \equiv (f(x))(g(x)).$$

It is easy to see that

1. for every $f, g, h$ in $C^\infty(M)$, $(f \cdot g) \cdot h = f \cdot (g \cdot h)$,
2. for every $f, g$ in $C^\infty(M)$, $f \cdot g = g \cdot f$,
3. for every $f, g, h$ in $C^\infty(M)$, $(f + g) \cdot h = f \cdot h + g \cdot h$,
4. for every real $t$, and for every $f, g$ in $C^\infty(M)$, $t(f \cdot g) = (tf) \cdot g$.

In short, $C^\infty(M)$ is an algebra.

**Note 2.8** Let $M$ be an $m$-dimensional smooth manifold. Let $p \in M$. Let $(U, \varphi_U)$ be an admissible coordinate chart of $M$ satisfying $p \in U$. Let $\gamma$ be in $\Gamma_p(M)$. Since $\gamma$ is in $\Gamma_p(M)$, there exists a real number $\delta > 0$ such that $\gamma$ is a smooth map from $(-\delta, \delta)$ to $M$, and $\gamma(0) = p$. Since $\gamma$ is a smooth map from $(-\delta, \delta)$ to $M$, and $0$ is in $(-\delta, \delta)$, $\gamma$ is continuous at $0$. Since $\gamma$ is continuous at $0$, $\gamma(0) = p$, and $U$ is an open neighborhood of $p$, $\gamma^{-1}(U)$ is an open neighborhood of $0$, and hence, $(-\delta, \delta) \cap \gamma^{-1}(U)$ is an open neighborhood of $0$. Also since $\mathrm{dom}(\varphi_U) = U$, and $\mathrm{dom}(\gamma) = (-\delta, \delta)$, $\mathrm{dom}(\varphi_U \circ \gamma) = (-\delta, \delta) \cap \gamma^{-1}(U)$. Since $(\varphi_U \circ \gamma) : ((-\delta, \delta) \cap \gamma^{-1}(U)) \to \mathbb{R}^m$, $(-\delta, \delta) \cap \gamma^{-1}(U)$ is an open neighborhood of $0$, $\varphi_U$ is a smooth, and $\gamma$ is a smooth map, $(\varphi_U \circ \gamma)'(0)$ exists and is a member of $\mathbb{R}^m$.

**Definition** Let $M$ be an $m$-dimensional smooth manifold. Let $p \in M$. Let $\gamma, \gamma_1$ be in $\Gamma_p(M)$. By $\gamma \approx \gamma_1$, we shall mean that for every admissible coordinate chart $(U, \varphi_U)$ of $M$ satisfying $p \in U$, $(\varphi_U \circ \gamma)'(0) = (\varphi_U \circ \gamma_1)'(0)$.

**Lemma 2.9** *Let $M$ be an $m$-dimensional smooth manifold. Let $p \in M$. Let $\gamma, \gamma_1$ be in $\Gamma_p(M)$. If there exists an admissible coordinate chart $(U, \varphi_U)$ of $M$ satisfying $p \in U$, and $(\varphi_U \circ \gamma)'(0) = (\varphi_U \circ \gamma_1)'(0)$, then $\gamma \approx \gamma_1$.*

*Proof* Let us take any admissible coordinate chart $(V, \psi_V)$ of $M$ satisfying $p \in V$. We have to prove that $(\psi_V \circ \gamma)'(0) = (\psi_V \circ \gamma_1)'(0)$.

$$
\begin{aligned}
\mathrm{LHS} = (\psi_V \circ \gamma)'(0) &= \left( \psi_V \circ \left( (\varphi_U)^{-1} \circ \varphi_U \right) \circ \gamma \right)'(0) \\
&= \left( \left( \psi_V \circ (\varphi_U)^{-1} \right) \circ (\varphi_U \circ \gamma) \right)'(0) \\
&= \left( \left( \left( \psi_V \circ (\varphi_U)^{-1} \right) \right)'((\varphi_U \circ \gamma)(0)) \right)((\varphi_U \circ \gamma)'(0)) \\
&= \left( \left( \left( \psi_V \circ (\varphi_U)^{-1} \right) \right)'((\varphi_U \circ \gamma)(0)) \right)((\varphi_U \circ \gamma_1)'(0)) \\
&= \left( \left( \left( \psi_V \circ (\varphi_U)^{-1} \right) \right)'(\varphi_U(\gamma(0))) \right)((\varphi_U \circ \gamma_1)'(0)) \\
&= \left( \left( \left( \psi_V \circ (\varphi_U)^{-1} \right) \right)'(\varphi_U(p)) \right)((\varphi_U \circ \gamma_1)'(0)) \\
&= \left( \left( \left( \psi_V \circ (\varphi_U)^{-1} \right) \right)'(\varphi_U(\gamma_1(0))) \right)((\varphi_U \circ \gamma_1)'(0)) \\
&= \left( \left( \left( \psi_V \circ (\varphi_U)^{-1} \right) \right)'((\varphi_U \circ \gamma_1)(0)) \right)((\varphi_U \circ \gamma_1)'(0)),
\end{aligned}
$$

and

$$\text{RHS} = (\psi_V \circ \gamma_1)'(0) = \left(\psi_V \circ \left((\varphi_U)^{-1} \circ \varphi_U\right) \circ \gamma_1\right)'(0)$$

$$= \left(\left(\psi_V \circ (\varphi_U)^{-1}\right) \circ (\varphi_U \circ \gamma_1)\right)'(0)$$

$$= \left(\left(\left(\psi_V \circ (\varphi_U)^{-1}\right)\right)'((\varphi_U \circ \gamma_1)(0))\right)\left((\varphi_U \circ \gamma_1)'(0)\right).$$

Hence, LHS = RHS. □

**Lemma 2.10** *Let M be an m-dimensional smooth manifold. Let $p \in M$. Then, $\approx$ is an equivalence relation over $\Gamma_p(M)$.*

*Proof* Here, we must prove

1. for every $\gamma$ in $\Gamma_p(M)$, $\gamma \approx \gamma$,
2. if $\gamma \approx \gamma_1$, then $\gamma_1 \approx \gamma$,
3. if $\gamma \approx \gamma_1$ and $\gamma_1 \approx \gamma_2$, then $\gamma \approx \gamma_2$.

For 1: Since $p \in M$, and $M$ is an $m$-dimensional smooth manifold, there exists an admissible coordinate chart $(U, \varphi_U)$ of $M$ satisfying $p \in U$. Since $(\varphi_U \circ \gamma)'(0) = (\varphi_U \circ \gamma)'(0)$, by Lemma 2.9, $\gamma \approx \gamma$.

For 2: Let $\gamma \approx \gamma_1$. We have to prove that $\gamma_1 \approx \gamma$. For this purpose, let us take any admissible coordinate chart $(U, \varphi_U)$ of $M$ satisfying $p \in U$. We must show that $(\varphi_U \circ \gamma_1)'(0) = (\varphi_U \circ \gamma)'(0)$. Since $\gamma \approx \gamma_1$, $(\varphi_U \circ \gamma)'(0) = (\varphi_U \circ \gamma_1)'(0)$, and hence, $(\varphi_U \circ \gamma_1)'(0) = (\varphi_U \circ \gamma)'(0)$.

For 3: Let $\gamma \approx \gamma_1$ and $\gamma_1 \approx \gamma_2$. We have to prove that $\gamma \approx \gamma_2$. For this purpose, let us take any admissible coordinate chart $(U, \varphi_U)$ of $M$ satisfying $p \in U$. We must show that $(\varphi_U \circ \gamma)'(0) = (\varphi_U \circ \gamma_2)'(0)$. Since $\gamma \approx \gamma_1$, $(\varphi_U \circ \gamma)'(0) = (\varphi_U \circ \gamma_1)'(0)$. Since $\gamma_1 \approx \gamma_2$, $(\varphi_U \circ \gamma_1)'(0) = (\varphi_U \circ \gamma_2)'(0)$. Since $(\varphi_U \circ \gamma)'(0) = (\varphi_U \circ \gamma_1)'(0)$, and $(\varphi_U \circ \gamma_1)'(0) = (\varphi_U \circ \gamma_2)'(0)$, $(\varphi_U \circ \gamma)'(0) = (\varphi_U \circ \gamma_2)'(0)$. □

**Note 2.11** Let $M$ be an $m$-dimensional smooth manifold. Let $p \in M$. By the Lemma 2.10, the quotient set $\Gamma_p(M)/\approx$ is the collection of all equivalence classes $[\![\gamma]\!]$, where $\gamma$ is in $\Gamma_p(M)$. Thus,

$$\Gamma_p(M)/\approx \equiv \left\{[\![\gamma]\!] : \gamma \in \Gamma_p(M)\right\}$$

where

$$[\![\gamma]\!] \equiv \left\{\gamma_1 \in \Gamma_p(M) : \gamma_1 \approx \gamma\right\}.$$

Intuitively, the Lemma 2.10 says that in $\Gamma_p(M)$ we will not distinguish between $\gamma_1$ and $\gamma$ whenever $\gamma_1 \approx \gamma$.

**Definition** Let $M$ be an $m$-dimensional smooth manifold. Let $p \in M$. Let $(U, \varphi_U)$ be any admissible coordinate chart of $M$ satisfying $p \in U$. Let $\gamma, \gamma_1$ be in $\Gamma_p(M)$. If $[\![\gamma]\!] = \gamma_1$, then $\gamma \approx \gamma_1$, and hence, $(\varphi_U \circ \gamma)'(0) = (\varphi_U \circ \gamma_1)'(0)$. This shows that $[\![\gamma]\!] \mapsto (\varphi_U \circ \gamma)'(0)$ is a well-defined function from $\Gamma_p(M)/\approx$ to $\mathbb{R}^m$. We shall denote this function by $\varphi_{U*}$. Thus,

$$\varphi_{U*} : (\Gamma_p(M)/\approx) \to \mathbb{R}^m$$

is the function defined as follows: for every $[\![\gamma]\!]$ in $\Gamma_p(M)/\approx$ where $\gamma$ is in $\Gamma_p(M)$,

$$\varphi_{U*}([\![\gamma]\!]) \equiv (\varphi_U \circ \gamma)'(0).$$

**Lemma 2.12** *Let $M$ be an $m$-dimensional smooth manifold. Let $p \in M$. Let $(U, \varphi_U)$ be any admissible coordinate chart of $M$ satisfying $p \in U$. Then, the function*

$$\varphi_{U*} : (\Gamma_p(M)/\approx) \to \mathbb{R}^m$$

*is 1–1 and onto.*

*Proof* 1–1: Let $\varphi_{U*}([\![\gamma]\!]) = \varphi_{U*}([\![\gamma_1]\!])$, where $\gamma, \gamma_1$ are in $\Gamma_p(M)$. We have to prove that $[\![\gamma]\!] = [\![\gamma_1]\!]$, that is, $\gamma \approx \gamma_1$. Since $(\varphi_U \circ \gamma)'(0) = \varphi_{U*}([\![\gamma]\!]) = \varphi_{U*}([\![\gamma_1]\!]) = (\varphi_U \circ \gamma_1)'(0)$, $(\varphi_U \circ \gamma)'(0) = (\varphi_U \circ \gamma_1)'(0)$, and hence, by Lemma 2.9, $\gamma \approx \gamma_1$.

Onto: Let us take any $v$ in $\mathbb{R}^m$. We have to find $\gamma$ in $\Gamma_p(M)$ such that $\varphi_{U*}([\![\gamma]\!]) = v$; that is, $(\varphi_U \circ \gamma)'(0) = v$; that is, $\lim_{t \to 0} \frac{(\varphi_U \circ \gamma)(t) - (\varphi_U \circ \gamma)(0)}{t} = v$; that is, $\lim_{t \to 0} \frac{\varphi_U(\gamma(t)) - \varphi_U(\gamma(0))}{t} = v$; that is, $\lim_{t \to 0} \frac{\varphi_U(\gamma(t)) - \varphi_U(p)}{t} = v$. Let us define a function $\gamma_1 : (-1, 1) \to \mathbb{R}^m$ as follows: For every $t$ in the open interval $(-1, 1)$,

$$\gamma_1(t) \equiv tv + \varphi_U(p).$$

Put

$$\gamma \equiv (\varphi_U)^{-1} \circ \gamma_1.$$

We shall try to prove that

1. $\gamma(0) = p$,
2. 0 is an interior point of the domain of $\gamma$,
3. $\gamma$ is a smooth map from $(-\delta, \delta)$ to $M$ for some $\delta > 0$,
4. $\lim_{t \to 0} \frac{\varphi_U(\gamma(t)) - \varphi_U(p)}{t} = v$.

For 1: Here,

$$\text{LHS} = \gamma(0) = \left((\varphi_U)^{-1} \circ \gamma_1\right)(0) = (\varphi_U)^{-1}(\gamma_1(0)) = (\varphi_U)^{-1}(0v + \varphi_U(p))$$
$$= (\varphi_U)^{-1}(\varphi_U(p)) = p = \text{RHS}.$$

This proves 1.

For 2: By the definition of $\gamma_1$, the function $\gamma_1 : (-1, 1) \to \mathbb{R}^m$ is continuous. Since $\gamma_1 : (-1, 1) \to \mathbb{R}^m$ is continuous, and 0 is in the open interval $(-1, 1)$, $\gamma_1$ is continuous at 0. Since $(U, \varphi_U)$ is a coordinate chart of $M$, $(\varphi_U)^{-1} : \varphi_U(U) \to U$ is a 1-1, onto, continuous function, and $\varphi_U(U)$ is an open subset of $\mathbb{R}^m$. Since $p \in U$, $\varphi_U(p) \in \varphi_U(U)$. Since $\gamma_1(0) = 0v + \varphi_U(p) = \varphi_U(p) \in \varphi_U(U)$, and $(\varphi_U)^{-1} : \varphi_U(U) \to U$ is continuous, $(\varphi_U)^{-1} : \varphi_U(U) \to U$ is continuous at the point $\gamma_1(0)$. Since $\gamma_1(0) \in \varphi_U(U)$, and $\varphi_U(U)$ is an open subset of $\mathbb{R}^m$, $\varphi_U(U)$ is an open neighborhood of $\gamma_1(0)$. Since $\gamma_1 : (-1, 1) \to \mathbb{R}^m$ is continuous, $\varphi_U(U)$ is an open neighborhood of $\gamma_1(0)$, there exists $\delta > 0$ such that $\delta < 1$ and, for every $t$ in $(-\delta, \delta)$, we have $\gamma_1(t) \in \varphi_U(U)$, and hence, $\gamma(t) = ((\varphi_U)^{-1} \circ \gamma_1)(t) = (\varphi_U)^{-1}(\gamma_1(t)) \in U$. Since $\gamma(t) \in U$ for every $t$ in $(-\delta, \delta)$, it follows that $(-\delta, \delta)$ is a subset of the dom$(\gamma)$. Hence, 0 is an interior point of the domain of the function $\gamma$. This proves 2.

For 3: In 2, we have seen that $\gamma$ is defined over $(-\delta, \delta)$. Now, it remains to be proved that $\gamma$ is a smooth map from $(-\delta, \delta)$ to $M$; that is, $\gamma$ is $C^\infty$ at every point $t$ in $(-\delta, \delta)$. Here, $\gamma$ is defined over $(-\delta, \delta)$, so $\gamma(t)$ is meaningful for every $t$ in $(-\delta, \delta)$. Since $\gamma(t)$ is meaningful for every $t$ in $(-\delta, \delta)$, and

$$\gamma(t) = \left((\varphi_U)^{-1} \circ \gamma_1\right)(t) = (\varphi_U)^{-1}(\gamma_1(t)) = (\varphi_U)^{-1}(tv + \varphi_U(p)),$$

$(\varphi_U)^{-1}(tv + \varphi_U(p))$ is meaningful for every $t$ in $(-\delta, \delta)$. Since $(U, \varphi_U)$ is a coordinate chart of $M$, $(\varphi_U)^{-1} : \varphi_U(U) \to U$. Since $(\varphi_U)^{-1} : \varphi_U(U) \to U$, and $(\varphi_U)^{-1}(tv + \varphi_U(p))$ is meaningful for every $t$ in $(-\delta, \delta)$, $(\varphi_U)^{-1}(tv + \varphi_U(p))$ is in $U$ for every $t$ in $(-\delta, \delta)$. Since $(\varphi_U)^{-1}(tv + \varphi_U(p))$ is in $U$ for every $t$ in $(-\delta, \delta)$, and $\gamma(t) = (\varphi_U)^{-1}(tv + \varphi_U(p))$, $\gamma(t)$ is in $U$ for every $t$ in $(-\delta, \delta)$. Thus, $\gamma$ maps $(-\delta, \delta)$ to $U$.

Now, it remains to be proved that $\gamma$ is a smooth map from $(-\delta, \delta)$ to $M$; that is, $\gamma$ is $C^\infty$ at every point $t$ in $(-\delta, \delta)$. For this purpose, let us fix any $t$ in $(-\delta, \delta)$. Since $\gamma_1$ is $C^\infty$ at $t$, $\gamma = (\varphi_U)^{-1} \circ \gamma_1$, $\gamma$ maps $(-\delta, \delta)$ to $U$, and $t$ is in $(-\delta, \delta)$, $\varphi_U \circ \gamma$ is $C^\infty$ at $t$. Since $\varphi_U \circ \gamma$ is $C^\infty$ at $t$, and $(U, \varphi_U)$ is an admissible coordinate chart of $M$ satisfying $\gamma(t) \in U$, by Theorem 2.2, $\gamma$ is $C^\infty$ at $t$ in $(-\delta, \delta)$. This proves 3. From 1, 2, and 3, we find that $\gamma$ is a parametrized curve in $M$ through $p$; that is, $\gamma$ is in $\Gamma_p(M)$.

4. LHS $= \lim_{t \to 0} \dfrac{\varphi_U(\gamma(t)) - \varphi_U(p)}{t} = \lim_{t \to 0} \dfrac{\varphi_U(((\varphi_U)^{-1} \circ \gamma_1)(t)) - \varphi_U(p)}{t} = \lim_{t \to 0} \dfrac{\gamma_1(t) - \varphi_U(p)}{t}$

$= \lim_{t \to 0} \dfrac{(tv + \varphi_U(p)) - \varphi_U(p)}{t} = v = $ RHS.

This proves that $\varphi_{U*} : (\Gamma_p(M)/\approx) \to \mathbb{R}^m$ is onto.  $\square$

**Note 2.13** Let $M$ be an $m$-dimensional smooth manifold. Let $p \in M$. Let $(U, \varphi_U)$ be any admissible coordinate chart of $M$ satisfying $p \in U$.

Since $\varphi_{U*} : (\Gamma_p(M)/\approx) \to \mathbb{R}^m$ is 1–1 onto, $(\varphi_{U*})^{-1} : \mathbb{R}^m \to (\Gamma_p(M)/\approx)$ exists, and is 1–1 onto. This fact allows us to define a binary operation $+$ over $\Gamma_p(M)/\approx$ as follows:

For every $[\![\gamma]\!], [\![\gamma_1]\!]$ in $\Gamma_p(M)/\approx$ where $\gamma, \gamma_1$ are in $\Gamma_p(M)$, by $[\![\gamma]\!] + [\![\gamma_1]\!]$, we mean

$$(\varphi_{U*})^{-1}(\varphi_{U*}([\![\gamma]\!]) + \varphi_{U*}([\![\gamma_1]\!])).$$

Next, for every $[\![\gamma]\!]$ in $\Gamma_p(M)/\approx$ where $\gamma$ is in $\Gamma_p(M)$, and for every real $t$, by $t[\![\gamma]\!]$, we mean

$$(\varphi_{U*})^{-1}(t(\varphi_{U*}(\gamma))).$$

Hence, $\varphi_{U*}([\![\gamma]\!] + [\![\gamma_1]\!]) = \varphi_{U*}([\![\gamma]\!]) + \varphi_{U*}([\![\gamma_1]\!])$, and $\varphi_{U*}(t[\![\gamma]\!]) = t(\varphi_{U*}([\![\gamma]\!]))$. Now, we want to verify that $\Gamma_p(M)/\approx$ is a real linear space:

1. $+$ is associative: Take any $[\![\gamma]\!], [\![\gamma_1]\!], [\![\gamma_2]\!]$ in $\Gamma_p(M)/\approx$ where $\gamma, \gamma_1, \gamma_2$ are in $\Gamma_p(M)$. We have to show that $([\![\gamma]\!] + [\![\gamma_1]\!]) + [\![\gamma_2]\!] = [\![\gamma]\!] + ([\![\gamma_1]\!] + [\![\gamma_2]\!])$.

$$\text{LHS} = ([\![\gamma]\!] + [\![\gamma_1]\!]) + [\![\gamma_2]\!] = \left((\varphi_{U*})^{-1}(\varphi_{U*}([\![\gamma]\!]) + \varphi_{U*}([\![\gamma_1]\!]))\right) + [\![\gamma_2]\!]$$

$$= (\varphi_{U*})^{-1}\left(\varphi_{U*}\left((\varphi_{U*})^{-1}(\varphi_{U*}([\![\gamma]\!]) + \varphi_{U*}([\![\gamma_1]\!]))\right) + \varphi_{U*}([\![\gamma_2]\!])\right)$$

$$= (\varphi_{U*})^{-1}((\varphi_{U*}([\![\gamma]\!]) + \varphi_{U*}([\![\gamma_1]\!])) + \varphi_{U*}([\![\gamma_2]\!]))$$

$$= (\varphi_{U*})^{-1}(\varphi_{U*}([\![\gamma]\!]) + (\varphi_{U*}([\![\gamma_1]\!]) + \varphi_{U*}([\![\gamma_2]\!]))),$$

$$\text{RHS} = [\![\gamma]\!] + ([\![\gamma_1]\!] + [\![\gamma_2]\!]) = [\![\gamma]\!] + \left((\varphi_{U*})^{-1}(\varphi_{U*}([\![\gamma_1]\!]) + \varphi_{U*}([\![\gamma_2]\!]))\right)$$

$$= (\varphi_{U*})^{-1}\left(\varphi_{U*}([\![\gamma]\!]) + \varphi_{U*}\left((\varphi_{U*})^{-1}(\varphi_{U*}([\![\gamma_1]\!]) + \varphi_{U*}([\![\gamma_2]\!]))\right)\right)$$

$$= (\varphi_{U*})^{-1}(\varphi_{U*}([\![\gamma]\!]) + (\varphi_{U*}([\![\gamma_1]\!]) + \varphi_{U*}([\![\gamma_2]\!]))).$$

Hence, LHS = RHS.

2. Existence of zero element: Here $\varphi_{U*} : (\Gamma_p(M)/\approx) \to \mathbb{R}^m$ is 1–1 onto, and $0 \in \mathbb{R}^m$, so $(\varphi_{U*})^{-1}(0)$ is in $(\Gamma_p(M)/\approx)$. We shall try to prove that $(\varphi_{U*})^{-1}(0)$

serves the purpose of zero element. For this purpose, let us take any $[\![\gamma]\!]$ in $\Gamma_p(M)/\approx$ where $\gamma$ is in $\Gamma_p(M)$. We have to show that

$$(\varphi_{U*})^{-1}(0) + [\![\gamma]\!] = [\![\gamma]\!].$$

$$
\begin{aligned}
\text{LHS} &= (\varphi_{U*})^{-1}(0) + [\![\gamma]\!] \\
&= (\varphi_{U*})^{-1}\Big(\varphi_{U*}\big((\varphi_{U*})^{-1}(0)\big) + \varphi_{U*}([\![\gamma]\!])\Big) \\
&= (\varphi_{U*})^{-1}(0 + \varphi_{U*}([\![\gamma]\!])) \\
&= (\varphi_{U*})^{-1}(\varphi_{U*}([\![\gamma]\!])) = [\![\gamma]\!] = \text{RHS}.
\end{aligned}
$$

3. Existence of negative element: Let us take any $[\![\gamma]\!]$ in $\Gamma_p(M)/\approx$ where $\gamma$ is in $\Gamma_p(M)$. Since $[\![\gamma]\!]$ is in $\Gamma_p(M)/\approx$, and $\varphi_{U*} : (\Gamma_p(M)/\approx) \to \mathbb{R}^m$, $\varphi_{U*}([\![\gamma]\!])$ is in $\mathbb{R}^m$, and hence, $-(\varphi_{U*}([\![\gamma]\!]))$ is in $\mathbb{R}^m$. So $(\varphi_{U*})^{-1}(-(\varphi_{U*}([\![\gamma]\!])))$ is in $(\Gamma_p(M)/\approx)$. We have to show that

$$(\varphi_{U*})^{-1}(-(\varphi_{U*}([\![\gamma]\!]))) + [\![\gamma]\!] = (\varphi_{U*})^{-1}(0).$$

$$
\begin{aligned}
\text{LHS} &= (\varphi_{U*})^{-1}(-(\varphi_{U*}([\![\gamma]\!]))) + [\![\gamma]\!] \\
&= (\varphi_{U*})^{-1}\Big(\varphi_{U*}\big((\varphi_{U*})^{-1}(-(\varphi_{U*}([\![\gamma]\!])))\big) + \varphi_{U*}([\![\gamma]\!])\Big) \\
&= (\varphi_{U*})^{-1}((-(\varphi_{U*}([\![\gamma]\!]))) + \varphi_{U*}([\![\gamma]\!])) = (\varphi_{U*})^{-1}(0) = \text{RHS}.
\end{aligned}
$$

4. $+$ is commutative: Take any $[\![\gamma]\!], [\![\gamma_1]\!]$ in $\Gamma_p(M)/\approx$ where $\gamma, \gamma_1$ are in $\Gamma_p(M)$. We have to show that $[\![\gamma]\!] + [\![\gamma_1]\!] = [\![\gamma_1]\!] + [\![\gamma]\!]$.

$$
\begin{aligned}
\text{LHS} &= [\![\gamma]\!] + [\![\gamma_1]\!] = (\varphi_{U*})^{-1}(\varphi_{U*}([\![\gamma]\!]) + \varphi_{U*}([\![\gamma_1]\!])) \\
&= (\varphi_{U*})^{-1}(\varphi_{U*}([\![\gamma_1]\!]) + \varphi_{U*}([\![\gamma]\!])) \\
&= [\![\gamma_1]\!] + [\![\gamma]\!] = \text{RHS}.
\end{aligned}
$$

5. Take any $[\![\gamma]\!], [\![\gamma_1]\!]$ in $\Gamma_p(M)/\approx$ where $\gamma, \gamma_1$ are in $\Gamma_p(M)$. Take any real number $t$. We have to prove that

$$t([\![\gamma]\!] + [\![\gamma_1]\!]) = t[\![\gamma]\!] + t[\![\gamma_1]\!].$$

$$
\begin{aligned}
\text{LHS} &= t([\![\gamma]\!] + [\![\gamma_1]\!]) = (\varphi_{U*})^{-1}(t(\varphi_{U*}([\![\gamma]\!] + [\![\gamma_1]\!]))) \\
&= (\varphi_{U*})^{-1}\Big(t\Big(\varphi_{U*}\big((\varphi_{U*})^{-1}(\varphi_{U*}([\![\gamma]\!]) + \varphi_{U*}([\![\gamma_1]\!]))\big)\Big)\Big) \\
&= (\varphi_{U*})^{-1}(t(\varphi_{U*}([\![\gamma]\!]) + \varphi_{U*}([\![\gamma_1]\!]))), \\
\text{RHS} &= t[\![\gamma]\!] + t[\![\gamma]\!]_1 = (\varphi_{U*})^{-1}(t(\varphi_{U*}([\![\gamma]\!]))) + (\varphi_{U*})^{-1}\big(t(\varphi_{U*}([\![\gamma]\!]_1))\big).
\end{aligned}
$$

So it remains to be proved that

$$(\varphi_{U*})^{-1}(t(\varphi_{U*}([\![\gamma]\!]) + \varphi_{U*}([\![\gamma_1]\!]))) = (\varphi_{U*})^{-1}(t(\varphi_{U*}([\![\gamma]\!])))$$
$$+ (\varphi_{U*})^{-1}(t(\varphi_{U*}([\![\gamma_1]\!])))$$

that is,

$$t(\varphi_{U*}([\![\gamma]\!]) + \varphi_{U*}([\![\gamma_1]\!])) = \varphi_{U*}\Big((\varphi_{U*})^{-1}(t(\varphi_{U*}([\![\gamma]\!]))) + (\varphi_{U*})^{-1}(t(\varphi_{U*}([\![\gamma_1]\!])))\Big).$$

$$\text{RHS} = \varphi_{U*}\Big((\varphi_{U*})^{-1}(t(\varphi_{U*}([\![\gamma]\!]))) + (\varphi_{U*})^{-1}(t(\varphi_{U*}([\![\gamma_1]\!])))\Big)$$
$$= \varphi_{U*}\Big((\varphi_{U*})^{-1}(t(\varphi_{U*}([\![\gamma]\!])))\Big) + \varphi_{U*}\Big((\varphi_{U*})^{-1}(t(\varphi_{U*}([\![\gamma_1]\!])))\Big)$$
$$= t(\varphi_{U*}([\![\gamma]\!])) + t(\varphi_{U*}([\![\gamma_1]\!]))$$
$$= t(\varphi_{U*}([\![\gamma]\!]) + \varphi_{U*}([\![\gamma_1]\!])) = \text{LHS}.$$

6. Take any $[\![\gamma]\!]$ in $\Gamma_p(M)/\approx$ where $\gamma$ is in $\Gamma_p(M)$. Take any real numbers $s, t$. We have to prove that

$$(s + t)[\![\gamma]\!] = s[\![\gamma]\!] + t[\![\gamma]\!].$$

$$\text{LHS} = (s + t)[\![\gamma]\!] = (\varphi_{U*})^{-1}((s + t)(\varphi_{U*}([\![\gamma]\!])))$$
$$= (\varphi_{U*})^{-1}(s(\varphi_{U*}([\![\gamma]\!])) + t(\varphi_{U*}([\![\gamma]\!]))),$$

$$\text{RHS} = s[\![\gamma]\!] + t[\![\gamma]\!] = (\varphi_{U*})^{-1}(\varphi_{U*}(s[\![\gamma]\!]) + \varphi_{U*}(t[\![\gamma]\!])).$$

So it remains to be proved that

$$(\varphi_{U*})^{-1}(s(\varphi_{U*}([\![\gamma]\!])) + t(\varphi_{U*}([\![\gamma]\!]))) = (\varphi_{U*})^{-1}(\varphi_{U*}(s[\![\gamma]\!]) + \varphi_{U*}(t[\![\gamma]\!])),$$

that is,

$$s(\varphi_{U*}([\![\gamma]\!])) + t(\varphi_{U*}([\![\gamma]\!])) = \varphi_{U*}(s[\![\gamma]\!]) + \varphi_{U*}(t[\![\gamma]\!]).$$

$$\text{RHS} = \varphi_{U*}(s[\![\gamma]\!]) + \varphi_{U*}(t[\![\gamma]\!]) = s(\varphi_{U*}([\![\gamma]\!])) + t(\varphi_{U*}([\![\gamma]\!])) = \text{LHS}.$$

7. Take any $[\![\gamma]\!]$ in $\Gamma_p(M)/\approx$ where $\gamma$ is in $\Gamma_p(M)$. Take any real numbers $s, t$. We have to prove that

$$(st)[\![\gamma]\!] = s(t[\![\gamma]\!]).$$

$$\text{LHS} = (st)[\![\gamma]\!] = (\varphi_{U*})^{-1}((st)(\varphi_{U*}([\![\gamma]\!]))),$$

$$\text{RHS} = s(t[\![\gamma]\!]) = (\varphi_{U*})^{-1}(s(\varphi_{U*}(t[\![\gamma]\!]))) = (\varphi_{U*})^{-1}(s(t(\varphi_{U*}([\![\gamma]\!]))))$$
$$= (\varphi_{U*})^{-1}((st)(\varphi_{U*}([\![\gamma]\!]))).$$

Hence, LHS = RHS.

8. Take any $[\![\gamma]\!]$ in $\Gamma_p(M)/\approx$ where $\gamma$ is in $\Gamma_p(M)$. We have to prove that

$$1[\![\gamma]\!] = [\![\gamma]\!].$$

$$\text{LHS} = 1[\![\gamma]\!] = (\varphi_{U*})^{-1}(1(\varphi_{U*}([\![\gamma]\!]))) = (\varphi_{U*})^{-1}(\varphi_{U*}([\![\gamma]\!])) = [\![\gamma]\!] = \text{RHS.} \quad \square$$

Thus, we have shown that $\Gamma_p(M)/\approx$ is a real linear space.

Since $\Gamma_p(M)/\approx$ is a real linear space, $\varphi_{U*} : (\Gamma_p(M)/\approx) \to \mathbb{R}^m$ is 1–1, onto,

$$\varphi_{U*}([\![\gamma]\!] + [\![\gamma_1]\!]) = \varphi_{U*}([\![\gamma]\!]) + \varphi_{U*}([\![\gamma_1]\!]), \text{ and } \varphi_{U*}(t[\![\gamma]\!]) = t(\varphi_{U*}([\![\gamma]\!])),$$

so $\varphi_{U*}$ is an isomorphism between real linear space $\Gamma_p(M)/\approx$ and the real linear space $\mathbb{R}^m$. Further since the dimension of $\mathbb{R}^m$ is $m$, $\Gamma_p(M)/\approx$ is a real linear space of dimension $m$.

Conclusion: Let $M$ be an $m$-dimensional smooth manifold. Let $p$ be an element of $M$. Then, $\Gamma_p(M)/\approx$ is isomorphic to $\mathbb{R}^m$ for some vector addition, and scalar multiplication over $\Gamma_p(M)/\approx$.

**Definition** Let $M$ be an $m$-dimensional smooth manifold. Let $p$ be an element of $M$. The set $\Gamma_p(M)/\approx$ is denoted by $T_pM$ or $T_p(M)$ (see Fig. 2.1).

**Note 2.14** Let $M$ be a 2-dimensional smooth manifold. Let $p \in M$. Let $(U, \varphi_U)$ be any admissible coordinate chart of $M$ satisfying $p \in U$. Let us define functions

$$u^1 : U \to \mathbb{R}, \ u^2 : U \to \mathbb{R}$$

as follows: for every $x$ in $U$,

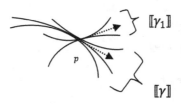

**Fig. 2.1** Dimensional smooth manifold

$$\varphi_U(x) \equiv \left(u^1(x), u^2(x)\right).$$

(In short, we say that $u^1, u^2$ are the component functions of $\varphi_U$.) We shall try to prove that $u^1$ is in $C_p^\infty(M)$. By the definition of $C_p^\infty(M)$, it remains to be proved that for every admissible coordinate chart $(V, \varphi_V)$ of $M$ satisfying $p \in V$,

$$\left(u^1 \circ (\varphi_V)^{-1}\right) : \varphi_V(U \cap V) \to \mathbb{R}$$

is $C^\infty$ at the point $\varphi_V(p)$ in $\mathbb{R}^2$.

Since $M$ is a 2-dimensional smooth manifold, and $(U, \varphi_U), (V, \varphi_V)$ are admissible coordinate charts of $M$ satisfying $p \in U \cap V$, by the definition of smooth manifold,

$$\left(\varphi_U \circ (\varphi_V)^{-1}\right) : \varphi_V(U \cap V) \to \mathbb{R}$$

is $C^\infty$ at the point $\varphi_V(p)$ in $\mathbb{R}^2$.

For every $(y^1, y^2)$ in $\varphi_V(U \cap V)$, we have $(\varphi_V)^{-1}(y^1, y^2) \in U \cap V \subset U$, and hence,

$$
\begin{aligned}
\left(\varphi_U \circ (\varphi_V)^{-1}\right)(y^1, y^2) &= \varphi_U\left((\varphi_V)^{-1}(y^1, y^2)\right) \\
&= \left(u^1\left((\varphi_V)^{-1}(y^1, y^2)\right), u^2\left((\varphi_V)^{-1}(y^1, y^2)\right)\right) \\
&= \left(\left(u^1 \circ (\varphi_V)^{-1}\right)(y^1, y^2), \left(u^2 \circ (\varphi_V)^{-1}\right)(y^1, y^2)\right)
\end{aligned}
$$

So $(u^1 \circ (\varphi_V)^{-1}), (u^2 \circ (\varphi_V)^{-1})$ are the component functions of $(\varphi_U \circ (\varphi_V)^{-1})$. Next, since $(\varphi_U \circ (\varphi_V)^{-1})$ is $C^\infty$ at the point $\varphi_V(p)$ in $\mathbb{R}^2$, its component functions $(u^1 \circ (\varphi_V)^{-1}), (u^2 \circ (\varphi_V)^{-1})$ are $C^\infty$ at the point $\varphi_V(p)$ in $\mathbb{R}^2$. This proves that $u^1$ is in $C_p^\infty(M)$. Similarly, $u^2$ is in $C_p^\infty(M)$.

**Theorem 2.15** Let M be an m-dimensional smooth manifold. Let p be an element of M.

Let $\sim$ be a relation over $C_p^\infty(M)$ defined as follows: for every f and g in $C_p^\infty(M)$, by $f \sim g$, we shall mean that there exists an open neighborhood H of p such that $f(x) = g(x)$ for every x in H. Then, $\sim$ is an equivalence relation over $C_p^\infty(M)$.

*Proof* Here, we must prove

1. for every f in $C_p^\infty(M), f \sim f$,
2. if $f \sim g$, then $g \sim f$,
3. if $f \sim g$, and $g \sim h$, then $f \sim h$.

For 1. Let us take any $f$ in $C_p^\infty(M)$. So, by the definition of $C_p^\infty(M)$, $\mathrm{dom} f$ is an open neighborhood of $p$ in $M$. Since $\mathrm{dom} f$ is an open neighborhood of $p$ in $M$, and $f(x) = f(x)$ for every $x$ in $\mathrm{dom} f$, by the definition of $\sim$, $f \sim f$. This proves 1.

For 2. Let $f \sim g$ where $f$ and $g$ are in $C_p^\infty(M)$. So, by the definition of $\sim$, there exists an open neighborhood $H$ of $p$ such that $f(x) = g(x)$ for every $x$ in $H$. Hence, $g(x) = f(x)$ for every $x$ in $H$, where $H$ is an open neighborhood of $p$. Therefore, by the definition of $\sim$, $g \sim f$, this proves 2.

For 3. Let $f \sim g$, and $g \sim h$, where $f, g, h$ are in $C_p^\infty(M)$. Since $f \sim g$, by the definition of $\sim$, there exists an open neighborhood $H$ of $p$ such that $f(x) = g(x)$ for every $x$ in $H$. Similarly, there exists an open neighborhood $K$ of $p$ such that $g(x) = h(x)$ for every $x$ in $K$. Since $H, K$ are open neighborhoods of $p$, $H \cap K$ is an open neighborhood of $p$. Also we have $f(x) = h(x)$ for every $H \cap K$. Hence, by the definition of $\sim$, $f \sim h$. This proves 3.

Hence, $\sim$ is an equivalence relation over $C_p^\infty(M)$.                       $\square$

**Definition** Let $M$ be an $m$-dimensional smooth manifold. Let $p$ be an element of $M$.

From Theorem 2.15, the relation $\sim$ is an equivalence relation over $C_p^\infty(M)$. The quotient set $C_p^\infty(M)/\sim$, of all equivalence classes is denoted by $\mathcal{F}_p(M)$ (or, simply $\mathcal{F}_p$). Thus,

$$\mathcal{F}_p(M) \equiv \left\{ [f] : f \in C_p^\infty(M) \right\}$$

where

$$[f] \equiv \left\{ g : g \in C_p^\infty(M) \text{ and } g \sim f \right\}.$$

Intuitively, Theorem 2.15 says that in $C_p^\infty(M)$ we will not distinguish between $f$ and $g$ whenever $f \sim g$.

**Definition** Let $M$ be an $m$-dimensional smooth manifold. Let $p$ be an element of $M$. Let $f, g$ be in $C_p^\infty(M)$. By the *sum* $f + g$, we mean the function whose domain is $(\mathrm{dom} f) \cap (\mathrm{dom} g)$, and for every $x$ in $(\mathrm{dom} f) \cap (\mathrm{dom} g)$,

$$(f + g)(x) \equiv f(x) + g(x).$$

By the *product* $f \cdot g$, we mean the function whose domain is $(\mathrm{dom} f) \cap (\mathrm{dom} g)$, and for every $x$ in $(\mathrm{dom} f) \cap (\mathrm{dom} g)$,

$$(f \cdot g)(x) \equiv f(x) \cdot g(x).$$

For any real $t$, by the *scalar multiple* $tf$, we mean the function whose domain is $\mathrm{dom} f$, and for every $x$ in $\mathrm{dom} f$,

$$(tf)(x) \equiv t(f(x)).$$

**Theorem 2.16** *Let $M$ be an $m$-dimensional smooth manifold. Let $p$ be an element of $M$. If $f, g$ are in $C_p^\infty(M)$, then $f + g$ is in $C_p^\infty(M)$.*

*Proof* Here, let $f, g$ be in $C_p^\infty(M)$. We must prove:

1. $\mathrm{dom}(f + g)$ is an open neighborhood of $p$ in $M$,
2. for every admissible coordinate chart $(U, \varphi_U)$ of $M$ satisfying $p \in U$,

$$\left((f + g) \circ (\varphi_U)^{-1}\right) : \varphi_U((\mathrm{dom}(f + g)) \cap U) \to \mathbb{R}$$

is $C^\infty$ at the point $\varphi_U(p)$ in $\mathbb{R}^m$.

For 1. Since $f, g$ are in $C_p^\infty(M)$, $\mathrm{dom} f$ and $\mathrm{dom} g$ are open neighborhoods of $p$ in $M$, and hence, their intersection $(\mathrm{dom} f) \cap (\mathrm{dom} g)$ is an open neighborhood of $p$ in $M$. Since $(\mathrm{dom} f) \cap (\mathrm{dom} g)$ is an open neighborhood of $p$ in $M$, and domain of $f + g$ is $(\mathrm{dom} f) \cap (\mathrm{dom} g)$, $\mathrm{dom}(f + g)$ is an open neighborhood of $p$ in $M$. This proves 1.

For 2. Let us take any admissible coordinate chart $(U, \varphi_U)$ of $M$ satisfying $p \in U$. Since $f$ is in $C_p^\infty(M)$, by the definition of $C_p^\infty(M)$,

$$\left(f \circ (\varphi_U)^{-1}\right) : \varphi_U((\mathrm{dom} f) \cap U) \to \mathbb{R}$$

is $C^\infty$ at the point $\varphi_U(p)$ in $\mathbb{R}^m$. Similarly,

$$\left(g \circ (\varphi_U)^{-1}\right) : \varphi_U((\mathrm{dom} g) \cap U) \to \mathbb{R}$$

is $C^\infty$ at the point $\varphi_U(p)$ in $\mathbb{R}^m$. Further, since

$$\left(f \circ (\varphi_U)^{-1}\right) : \varphi_U((\mathrm{dom} f) \cap U) \to \mathbb{R}$$

and

$$\left(g \circ (\varphi_U)^{-1}\right) : \varphi_U((\mathrm{dom} g) \cap U) \to \mathbb{R},$$

their sum is the function

$$\left(f \circ (\varphi_U)^{-1}\right) + \left(g \circ (\varphi_U)^{-1}\right) : (\varphi_U((\mathrm{dom} f) \cap U)) \cap (\varphi_U((\mathrm{dom} g) \cap U)) \to \mathbb{R},$$

and is $C^\infty$ at the point $\varphi_U(p)$ in $\mathbb{R}^m$. Now, since

$$\text{dom}\left(\left(f \circ (\varphi_U)^{-1}\right) + \left(g \circ (\varphi_U)^{-1}\right)\right) = (\varphi_U((\text{dom} f) \cap U)) \cap (\varphi_U((\text{dom} g) \cap U))$$
$$= \varphi_U(((\text{dom} f) \cap U) \cap ((\text{dom} g) \cap U))$$
$$= \varphi_U(((\text{dom} f) \cap (\text{dom} g)) \cap U)$$
$$= \varphi_U((\text{dom}(f + g)) \cap U)$$
$$= \text{dom}\left((f + g) \circ (\varphi_U)^{-1}\right)$$

and, for every $x$ in $\text{dom}\left((f + g) \circ (\varphi_U)^{-1}\right)$,

$$\left(\left(f \circ (\varphi_U)^{-1}\right) + \left(g \circ (\varphi_U)^{-1}\right)\right)(x) = \left(\left(f \circ (\varphi_U)^{-1}\right)\right)(x) + \left(\left(g \circ (\varphi_U)^{-1}\right)\right)(x)$$
$$= f\left(\left((\varphi_U)^{-1}\right)(x)\right) + g\left(\left((\varphi_U)^{-1}\right)(x)\right)$$
$$= (f + g)\left(\left((\varphi_U)^{-1}\right)(x)\right)$$
$$= \left((f + g) \circ (\varphi_U)^{-1}\right)(x)$$

so,

$$\left(f \circ (\varphi_U)^{-1}\right) + \left(g \circ (\varphi_U)^{-1}\right) = (f + g) \circ (\varphi_U)^{-1}.$$

Since $(f \circ (\varphi_U)^{-1}) + (g \circ (\varphi_U)^{-1}) = (f + g) \circ (\varphi_U)^{-1}$, and $(f \circ (\varphi_U)^{-1}) + (g \circ (\varphi_U)^{-1})$ is $C^\infty$ at the point $\varphi_U(p)$ in $\mathbb{R}^m$, $(f + g) \circ \varphi_U)^{-1}$ is $C^\infty$ at the point $\varphi_U(p)$ in $\mathbb{R}^m$. This proves 2. Hence, $f + g$ is in $C_p^\infty(M)$. $\square$

**Theorem 2.17** *Let $M$ be an $m$-dimensional smooth manifold. Let $p$ be an element of $M$. If $f, g$ are in $C_p^\infty(M)$, then $f \cdot g$ is in $C_p^\infty(M)$.*

*Proof* Its proof is quite similar to the proof of Theorem 2.16. $\square$

**Theorem 2.18** *Let $M$ be an $m$-dimensional smooth manifold. Let $p$ be an element of $M$. If $t$ is a real, and $f$ is in $C_p^\infty(M)$, then $tf$ is in $C_p^\infty(M)$.*

*Proof* Its proof is similar to the proof of Theorem 2.16. $\square$

**Theorem 2.19** *Let $M$ be an $m$-dimensional smooth manifold. Let $p$ be an element of $M$. If $f, g, h, k$ are in $C_p^\infty(M)$, $f \sim g$, and $h \sim k$, then $(f + h) \sim (g + k)$.*

*Proof* Let us take any $f, g, h, k$ in $C_p^\infty(M)$, and let $f \sim g$, $h \sim k$. Since $f, g$ are in $C_p^\infty(M)$, by Theorem 2.16, $f + h$ is in $C_p^\infty(M)$. Similarly $g + k$ is in $C_p^\infty(M)$. We have to prove that $(f + h) \sim (g + k)$. So, by the definition of $\sim$, we must find an open neighborhood $H$ of $p$ such that $(f + h)(x) = (g + k)(x)$ for every $x$ in $H$. Since $f \sim g$, there exists an open neighborhood $H_1$ of $p$ such that $f(x) = g(x)$ for every $x$ in $H_1$. Since $h \sim k$, there exists an open neighborhood $H_2$ of $p$ such that $h(x) = k(x)$ for every $x$ in $H_2$. Since $H_1, H_2$ are open neighborhoods of $p$, $H_1 \cap H_2$ is an open neighborhoods of $p$. Also, for every $x$ in $H_1 \cap H_2$, $(f + h)(x) = f(x) + h(x) = g(x) + k(x) = (g + k)(x)$. □

**Theorem 2.20** *Let M be an m-dimensional smooth manifold. Let p be an element of M. If $f, g, h, k$ are in $C_p^\infty(M), f \sim g$, and $h \sim k$, then $(f \cdot h) \sim (g \cdot k)$.*

*Proof* Its proof is quite similar to the proof of Theorem 2.19. □

**Theorem 2.21** *Let M be an m-dimensional smooth manifold. Let p be an element of M. If $f, g$ are in $C_p^\infty(M), f \sim g$, and t is a real, then $(tf) \sim (tg)$.*

*Proof* Let us take any $f, g$ in $C_p^\infty(M)$, and let $f \sim g$. Since $f, g$ are in $C_p^\infty(M)$, by Theorem 2.18, $tf$ is in $C_p^\infty(M)$. Similarly, $tg$ is in $C_p^\infty(M)$. We have to prove that $(tf) \sim (tg)$. So, by the definition of $\sim$, we must find an open neighborhood $H$ of $p$ such that $(tf)(x) = (tg)(x)$ for every $x$ in $H$. Since $f \sim g$, there exists an open neighborhood $H$ of $p$ such that $f(x) = g(x)$ for every $x$ in $H$. So, for every $x$ in $H$, $(tf)(x) = t(f(x)) = t(g(x)) = (tg)(x)$. □

**Note 2.22** Theorems 2.19, 2.20, and 2.21 give guarantee that the following definitions are legitimate.

**Definition** For every $f, g$ in $C_p^\infty(M)$, and for every real $t$,

$$[f] + [g] \equiv [f + g],$$
$$t[f] \equiv [tf],$$
$$[f][g] \equiv [f \cdot g].$$

**Theorem 2.23** *Let M be an m-dimensional smooth manifold. Let p be an element of M. The quotient set $C_p^\infty(M)/\sim$, together with vector addition, and scalar multiplication is defined as follows: for every $f, g$ in $C_p^\infty(M)$, and for every real t,*

$$[f] + [g] \equiv [f + g], \quad t[f] \equiv [tf],$$

*constitute a real linear space. In short, $\mathcal{F}_p(M)$ is a real linear space.*

Let us define multiplication operation over $C_p^\infty(M)/\sim$ as follows: for every $f, g$ in $C_p^\infty(M)$,

$$[f][g] \equiv [f \cdot g].$$

Then,

1. for every $f, g, h$ in $C_p^\infty(M)$, $([f][g])[h] = [f]([g][h])$,
2. for every $f, g$ in $C_p^\infty(M)$, $[f][g] = [g][f]$,
3. for every $f, g, h$ in $C_p^\infty(M)$, $([f] + [g])[h] = [f][h] + [g][h]$,
4. for every real $t$, and for every $f, g$ in $C_p^\infty(M)$, $t([f][g]) = (t[f])[g]$.

In short, $\mathcal{F}_p(M)$ is an algebra.

*Proof* The conditions for linear space remain to be verified.

1. $+$ is associative: Let us take any $f, g, h$ in $C_p^\infty(M)$. We have to prove that

$$([f] + [g]) + [h] = [f] + ([g] + [h]).$$

LHS $= ([f] + [g]) + [h] = [f + g] + [h] = [(f + g) + h] = [f + (g + h)]$
$= [f] + [g + h] = [f] + ([g] + [h]) = $ RHS.

2. Existence of zero element: Let us define the constant function $0 : M \to \mathbb{R}$ as follows: For every $x$ in $M$, $0(x) \equiv 0$. We want to prove that $0$ is in $C_p^\infty(M)$. For this purpose, we must prove

   1. The domain of the function $0$ is an open neighborhood of $p$ in $M$,
   2. for every admissible coordinate chart $(U, \varphi_U)$ of $M$ satisfying $p \in U$,

$$\left(0 \circ (\varphi_U)^{-1}\right) : \varphi_U((\text{dom } 0) \cap U) \to \mathbb{R}$$

is $C^\infty$ at the point $\varphi_U(p)$ in $\mathbb{R}^m$.

   For 1: Here, the domain of the function $0$ is $M$ which is an open neighborhood of $p$ in $M$. This proves 1.
   For 2: Let us take any admissible coordinate chart $(U, \varphi_U)$ of $M$ satisfying $p \in U$. Now, by the definition of the function $0$, $(0 \circ (\varphi_U)^{-1})$ is the constant function $0$ defined on the open subset $\varphi_U(U)$ of $\mathbb{R}^m$, which is known to be $C^\infty$ at the point $\varphi_U(p)$ in $\mathbb{R}^m$. This proves 2.
   Thus, we have shown that $0$ is in $C_p^\infty(M)$. Now, it remains to be showed that $[0] + [f] = [f] = [f] + [0]$ for every $f$ in $C_p^\infty(M)$. Here,

$$\text{LHS} = [0] + [f] = [0 + f] = [f] = [f + 0] = [f] + [0] = \text{RHS}.$$

This proves 2.

3. Existence of negative element: Let us take any $f$ in $C_p^\infty(M)$. So, by the definition of $C_p^\infty(M)$, $f$ is a real-valued function $f$ whose $\mathrm{dom}f$ is an open neighborhood of $p$ in $M$, and for every admissible coordinate chart $(U, \varphi_U)$ of $M$ satisfying $p \in U$,

$$\left(f \circ (\varphi_U)^{-1}\right) : \varphi_U((\mathrm{dom}f) \cap U) \to \mathbb{R}$$

is $C^\infty$ at the point $\varphi_U(p)$ in $\mathbb{R}^m$.

Now, let us define a function $(-f) : (\mathrm{dom}f) \to \mathbb{R}$ as follows:

For every $x$ in $\mathrm{dom}f$, $(-f)(x) \equiv -(f(x))$. We want to prove that $(-f)$ is in $C_p^\infty(M)$. For this purpose, we must prove

1. The domain of the function $(-f)$ is an open neighborhood of $p$ in $M$,
2. For every admissible coordinate chart $(U, \varphi_U)$ of $M$ satisfying $p \in U$,

$$\left((-f) \circ (\varphi_U)^{-1}\right) : \varphi_U((\mathrm{dom}(-f)) \cap U) \to \mathbb{R}$$

is $C^\infty$ at the point $\varphi_U(p)$ in $\mathbb{R}^m$.

For 1: Since $f$ is in $C_p^\infty(M)$, by the definition of $C_p^\infty(M)$, $\mathrm{dom}f$ is an open neighborhood of $p$ in $M$. Since $\mathrm{dom}f$ is an open neighborhood of $p$ in $M$, and, by the definition of $(-f)$, $\mathrm{dom}(-f) = \mathrm{dom}f$, $\mathrm{dom}(-f)$ is an open neighborhood of $p$ in $M$. This proves 1.

For 2: Let us take an admissible coordinate chart $(U, \varphi_U)$ of $M$ satisfying $p \in U$. Now, by the definition of the function $(-f)$, $(-f) \circ (\varphi_U)^{-1} = -(f \circ (\varphi_U)^{-1})$. Since $(f \circ (\varphi_U)^{-1})$ is $C^\infty$ at the point $\varphi_U(p)$ in $\mathbb{R}^m$, $-(f \circ (\varphi_U)^{-1})$ is $C^\infty$ at the point $\varphi_U(p)$ in $\mathbb{R}^m$. Since $-(f \circ (\varphi_U)^{-1})$ is $C^\infty$ at the point $\varphi_U(p)$ in $\mathbb{R}^m$, and $(-f) \circ (\varphi_U)^{-1} = -(f \circ (\varphi_U)^{-1})$, $(-f) \circ (\varphi_U)^{-1}$ is $C^\infty$ at the point $\varphi_U(p)$ in $\mathbb{R}^m$. This proves 2.

Thus, we have shown that $(-f)$ is in $C_p^\infty(M)$. Now, it remains to be showed that $[(-f)] + [f] = [0] = [f] + [(-f)]$. Here,

$$\text{LHS} = [(-f)] + [f] = [(-f) + f] = [0] = [f + (-f)] = [f] + [(-f)] = \text{RHS}.$$

This proves 3.

4. $+$ is commutative: Let us take any $f, g$ in $C_p^\infty(M)$. We have to prove that

$$[f] + [g] = [g] + [f].$$

Here

$$\text{LHS} = [f] + [g] = [f + g] = [g + f] = [g] + [f] = \text{RHS}.$$

This proves 4.

5. For every real $s, t$, and $f$ in $C_p^\infty(M)$, $(s + t)[f] = s[f] + t[f]$ :
Here

$$\text{LHS} = (s + t)[f] = [(s + t)f] = [(sf) + (tf)] = [(sf)] + [(tf)] = s[f] + t[f]$$
$$= \text{RHS}.$$

6. For every real $s, t$, and $f$ in $C_p^\infty(M)$, $(st)[f] = s(t[f])$ :
Here,

$$\text{LHS} = (st)[f] = [(st)f] = [s(tf)] = s[(tf)] = s(t[f]) = \text{RHS}.$$

7. For every real $t$, and for every $f, g$ in $C_p^\infty(M)$, $t([f] + [g]) = t[f] + t[g]$ :
Here,

$$\text{LHS} = t([f] + [g]) = t[(f + g)] = [t(f + g)] = [(tf) + (tg)]$$
$$= [(tf)] + [(tg)] = t[f] + t[g] = \text{RHS}.$$

8. For every $f$ in $C_p^\infty(M)$, $1[f] = [f]$ :
Here,

$$\text{LHS} = 1[f] = [1f] = [f] = \text{RHS}.$$

Thus, we have shown that quotient set $C_p^\infty(M)/\sim$ is a real linear space. Now, we want to prove that $\mathcal{F}_p(M)$ is an algebra.

1. Multiplication is associative:

$$\text{LHS} = ([f][g])[h] = [f \cdot g] + [h] = [(f \cdot g) \cdot h] = [f \cdot (g \cdot h)]$$
$$= [f][g \cdot h] = [f]([g][h]) = \text{RHS}.$$

2. Multiplication is commutative:

$$\text{LHS} = [f][g] = [f \cdot g] = [g \cdot f] = [g][f] = \text{RHS}.$$

3. Multiplication distributes over +:

$$\text{LHS} = ([f] + [g])[h] = [f + g][h] = [(f + g) \cdot h] = [(f \cdot h) + (g \cdot h)]$$
$$= [f \cdot h] + [g \cdot h] = [f][h] + [g][h] = \text{RHS}.$$

4. $\text{LHS} = t([f][g]) = t[f \cdot g] = [t(f \cdot g)] = [(tf) \cdot g] = [tf][g] = (t[f])[g] = \text{RHS}.$  $\qquad\square$

## 2.3 Smooth Germs on Smooth Manifolds

**Definition** Let $M$ be an $m$-dimensional smooth manifold. Let $p$ be an element of $M$. Under the vector addition, and scalar multiplication as defined in Theorem 2.23, $\mathcal{F}_p(M)$ is a real linear space. The members of $\mathcal{F}_p(M)$ are called $C^\infty$-germs at $p$ on $M$. Thus, if $f$ is in $C_p^\infty(M)$, then the equivalence class $[f]$ is in $\mathcal{F}_p(M)$, and hence, $[f]$ is a $C^\infty$-germ at $p$ on $M$.

**Theorem 2.24** *Let $M$ be an $m$-dimensional smooth manifold. Let $p$ be an element of $M$. Let $\gamma$ be in $\Gamma_p(M)$. Let $f$ be in $C_p^\infty(M)$. Then,*

1. *0 is an interior point of the domain of the real-valued function $f \circ \gamma$,*
2. $\lim_{t \to 0} \frac{(f \circ \gamma)(0+t) - (f \circ \gamma)(0)}{t}$ *exists,*
3. *if $f \sim g$, then*

$$\lim_{t \to 0} \frac{(f \circ \gamma)(0+t) - (f \circ \gamma)(0)}{t} = \lim_{t \to 0} \frac{(g \circ \gamma)(0+t) - (g \circ \gamma)(0)}{t}.$$

*Proof*

1:  Since $\gamma$ is in $\Gamma_p(M)$, by the definition of $\Gamma_p(M)$, there exists a real number $\delta > 0$ such that $\gamma : (-\delta, \delta) \to M$, $\gamma(0) = p$, and $\gamma$ is a smooth map from $(-\delta, \delta)$ to $M$. Since $\gamma$ is a smooth map from $(-\delta, \delta)$ to $M$, $\gamma$ is a continuous map.

Since $f$ is in $C_p^\infty(M)$, $f : (\mathrm{dom} f) \to \mathbb{R}$ and $\mathrm{dom} f$ is an open neighborhood of $p$ in $M$. Since $\gamma(0) = p$, and $p$ is in $\mathrm{dom} f$, 0 is an element of the domain of $f \circ \gamma$. Since $f$ is in $C_p^\infty(M)$, $f$ is continuous on some open neighborhood $U$ of $p$. Since $\gamma : (-\delta, \delta) \to M$ is a continuous map, $\gamma(0) = p$, and $U$ is an open neighborhood of $p$, there exists $\varepsilon > 0$ such that $\varepsilon < \delta$ and, for every $t$ in $(-\varepsilon, \varepsilon)$, we have $\gamma(t) \in U$ $(\subset (\mathrm{dom} f))$, and hence, $(f \circ \gamma)(t) = f(\gamma(t)) \in \mathbb{R}$. It follows that $(-\varepsilon, \varepsilon)$ is a subset of $\mathrm{dom}(f \circ \gamma)$. Hence, 0 is an interior point of the domain of the function $f \circ \gamma$. This proves 1.

2:  Since 0 is an interior point of the domain of the real-valued function $f \circ \gamma$, it is meaningful to write

$$\lim_{t \to 0} \frac{(f \circ \gamma)(0+t) - (f \circ \gamma)(0)}{t},$$

provided it exists. Now we shall try to show that $\lim_{t \to 0} \frac{(f \circ \gamma)(0+t) - (f \circ \gamma)(0)}{t}$ exists.

Since $p \in M$, and $M$ is an $m$-dimensional smooth manifold, there exists an admissible coordinate chart $(U, \varphi_U)$ of $M$ satisfying $p \in U$. Since $f$ is in $C_p^\infty(M)$, and $(U, \varphi_U)$ is an admissible coordinate chart of $M$ satisfying $p \in U$,

$$\left( f \circ (\varphi_U)^{-1} \right) : \varphi_U((\mathrm{dom} f) \cap U) \to \mathbb{R}$$

is $C^\infty$ at the point $\varphi_U(p)$ in $\mathbb{R}^m$. Observe that the domain of $(f \circ (\varphi_U)^{-1})$; that is, $\varphi_U((\mathrm{dom} f) \cap U)$ is an open neighborhood of $\varphi_U(p)$ $(= \varphi_U(\gamma(0)) = (\varphi_U \circ \gamma)(0))$ in $\mathbb{R}^m$. Since $\gamma$ is a smooth map from $(-\delta, \delta)$ to $M$, and $0 \in (-\delta, \delta)$, $\gamma$ is $C^\infty$ at 0 in $\mathbb{R}$. Since $\gamma$ is $C^\infty$ at 0 in $\mathbb{R}$, $\gamma(0) = p$, and $(U, \varphi_U)$ is an admissible coordinate chart of $M$ satisfying $p \in U$, $(\varphi_U \circ \gamma) : (-\delta, \delta) \to \mathbb{R}^m$ is $C^\infty$ at the point 0 in $\mathbb{R}$. Since

$$(\varphi_U \circ \gamma) : (-\delta, \delta) \to \mathbb{R}^m$$

is $C^\infty$ at the point 0,

$$\left( f \circ (\varphi_U)^{-1} \right) : \varphi_U((\mathrm{dom} f) \cap U) \to \mathbb{R}$$

is $C^\infty$ at the point $(\varphi_U \circ \gamma)(0)$ in $\mathbb{R}^m$, and the domain of $(f \circ (\varphi_U)^{-1})$ is an open neighborhood of $(\varphi_U \circ \gamma)(0)$ in $\mathbb{R}^m$, the composite function $(f \circ (\varphi_U)^{-1}) \circ (\varphi_U \circ \gamma)(= f \circ \gamma)$ is $C^\infty$ at the point 0 in $\mathbb{R}$. Hence,

$$\left. \frac{\mathrm{d}(f \circ \gamma)(t)}{\mathrm{d}t} \right|_{t=0}$$

that is,

$$\lim_{t \to 0} \frac{(f \circ \gamma)(0 + t) - (f \circ \gamma)(0)}{t}$$

exists. This proves 2.

3:   Let $f \sim g$.

We have to prove that

$$\lim_{t \to 0} \frac{(f \circ \gamma)(0 + t) - (f \circ \gamma)(0)}{t} = \lim_{t \to 0} \frac{(g \circ \gamma)(0 + t) - (g \circ \gamma)(0)}{t}.$$

Since $f \sim g$, there exists an open neighborhood $H$ of $p(= \gamma(0))$ such that $f(x) = g(x)$ for every $x$ in $H$. Since $f(x) = g(x)$ for every $x$ in $H$, and $\gamma(0)$ is in $H$,

$$(f \circ \gamma)(0) = f(\gamma(0)) = g(\gamma(0)) = (g \circ \gamma)(0).$$

Since $\gamma : (-\delta, \delta) \to M$ is a continuous map, and $H$ is an open neighborhood of $\gamma(0)$, there exists $\varepsilon_1 > 0$ such that $\varepsilon_1 < \delta$, and for every $t$ in the open interval $(-\varepsilon_1, \varepsilon_1)$, $\gamma(t)$ is in $H$. Since 0 is an interior point of the domain of the function $f \circ \gamma$, there exists $\varepsilon_2 > 0$ such that $\varepsilon_2 < \delta$, and the open interval $(-\varepsilon_2, \varepsilon_2)$ is contained in the domain of the function $f \circ \gamma$. Similarly, there exists $\varepsilon_3 > 0$ such that $\varepsilon_3 < \delta$, and the open interval $(-\varepsilon_3, \varepsilon_3)$ is contained in the domain of the function $g \circ \gamma$. Put

$$\varepsilon \equiv \min\{\varepsilon_1, \varepsilon_2, \varepsilon_3\}.$$

Clearly, $\varepsilon > 0$. If $t$ is in the open interval $(-\varepsilon, \varepsilon)$, then $t$ is in the open interval $(-\varepsilon_2, \varepsilon_2)$, and hence, $(f \circ \gamma)(t)$ is meaningful. Similarly, if $t$ is in the open interval $(-\varepsilon, \varepsilon)$, then $(g \circ \gamma)(t)$ is meaningful. If $t$ is in the open interval $(-\varepsilon, \varepsilon)$, then $t$ is in the open interval $(-\varepsilon_1, \varepsilon_1)$, and hence, $\gamma(t)$ is in $H$. Since, for every $t$ in $(-\varepsilon, \varepsilon)$, $\gamma(t)$ is in $H$, and since $f(x) = g(x)$ for every $x$ in $H$, for every $t$ in $(-\varepsilon, \varepsilon)$,

$$(f \circ \gamma)(t) = f(\gamma(t)) = g(\gamma(t)) = (g \circ \gamma)(t).$$

Hence,

$$\lim_{t \to 0} \frac{(f \circ \gamma)(0 + t) - (f \circ \gamma)(0)}{t} =$$

$$\lim_{t \to 0} \frac{(f \circ \gamma)(t) - (f \circ \gamma)(0)}{t} = \lim_{\substack{t \to 0, \\ t \text{ in } (-\varepsilon, \varepsilon)}} \frac{(f \circ \gamma)(t) - (f \circ \gamma)(0)}{t}$$

$$= \lim_{\substack{t \to 0, \\ t \text{ in } (-\varepsilon, \varepsilon)}} \frac{(g \circ \gamma)(t) - (g \circ \gamma)(0)}{t}$$

$$= \lim_{\substack{t \to 0, \\ t \text{ in } (-\varepsilon, \varepsilon)}} \frac{(g \circ \gamma)(0 + t) - (g \circ \gamma)(0)}{t}$$

$$= \lim_{t \to 0} \frac{(g \circ \gamma)(0 + t) - (g \circ \gamma)(0)}{t}.$$

This proves 3.                                                                    □

**Note 2.25** Let $M$ be an $m$-dimensional smooth manifold. Let $p$ be an element of $M$. Let $\gamma$ be in $\Gamma_p(M)$. Let $[f]$ be in $\mathcal{F}_p(M)$ where $f$ is in $C_p^\infty(M)$. Let $[g]$ be in $\mathcal{F}_p(M)$ where $g$ is in $C_p^\infty(M)$. If $[f] = [g]$, then $f \sim g$, and hence,

$$\lim_{t\to 0}\frac{(f\circ \gamma)(0+t)-(f\circ \gamma)(0)}{t}=\lim_{t\to 0}\frac{(g\circ \gamma)(0+t)-(g\circ \gamma)(0)}{t}.$$

So the following definition is well defined.

**Definition** Let $M$ be an $m$-dimensional smooth manifold. Let $p$ be an element of $M$. Let $\gamma$ be in $\Gamma_p(M)$. Let $[f]$ be in $\mathcal{F}_p(M)$ where $f$ is in $C_p^\infty(M)$. Then,

$$\lim_{t\to 0}\frac{(f\circ \gamma)(0+t)-(f\circ \gamma)(0)}{t}$$

exists, and is denoted by $\ll \gamma,[f]\gg$. Thus,

$$\ll \gamma,[f]\gg \equiv \lim_{t\to 0}\frac{(f\circ \gamma)(0+t)-(f\circ \gamma)(0)}{t}$$

or,

$$\ll \gamma,[f]\gg \equiv \frac{\mathrm{d}(f\circ \gamma)(t)}{\mathrm{d}t}\bigg|_{t=0}.$$

**Theorem 2.26** *Let $M$ be an m-dimensional smooth manifold. Let $p$ be an element of $M$. Let $\gamma$ be in $\Gamma_p(M)$. Let $[f]$ be in $\mathcal{F}_p(M)$ where $f$ is in $C_p^\infty(M)$. Let $[g]$ be in $\mathcal{F}_p(M)$ where $g$ is in $C_p^\infty(M)$. Then,*

1. $\ll \gamma,[f]+[g]\gg = \ll \gamma,[f]\gg + \ll \gamma,[g]\gg$,
2. $\ll \gamma,\alpha[f]\gg = \alpha\ll \gamma,[f]\gg$ *for every real $\alpha$.*

*In short, we say that $\ll,\gg$ is linear in the second variable.*

*Proof*

1. Here

$$\mathrm{LHS}=\ll \gamma,[f]+[g]\gg = \ll \gamma,[f+g]\gg = \lim_{t\to 0}\frac{((f+g)\circ \gamma)(0+t)-((f+g)\circ \gamma)(0)}{t}$$

$$=\lim_{t\to 0}\frac{(f+g)(\gamma(t))-(f+g)(\gamma(0))}{t}=\lim_{t\to 0}\frac{(f(\gamma(t))+g(\gamma(t)))-(f(\gamma(0))+g(\gamma(0)))}{t}$$

$$=\lim_{t\to 0}\frac{(f(\gamma(t))-f(\gamma(0)))+(g(\gamma(t))-g(\gamma(0)))}{t}$$

$$=\lim_{t\to 0}\left(\frac{(f(\gamma(t))-f(\gamma(0)))}{t}+\frac{(g(\gamma(t))-g(\gamma(0)))}{t}\right)$$

$$=\lim_{t\to 0}\frac{(f(\gamma(t))-f(\gamma(0)))}{t}+\lim_{t\to 0}\frac{(g(\gamma(t))-g(\gamma(0)))}{t}$$

$$=\lim_{t\to 0}\frac{(f(\gamma(t+0))-f(\gamma(0)))}{t}+\lim_{t\to 0}\frac{(g(\gamma(t+0))-g(\gamma(0)))}{t}$$

$$=\ll \gamma,[f]\gg + \ll \gamma,[g]\gg = \mathrm{RHS}.$$

2. Here

$$\text{LHS} = \ll \gamma, \alpha[f] \gg = \ll \gamma, [\alpha f] \gg = \lim_{t \to 0} \frac{((\alpha f) \circ \gamma)(0+t) - ((\alpha f) \circ \gamma)(0)}{t}$$

$$= \lim_{t \to 0} \frac{(\alpha f)(\gamma(t)) - (\alpha f)(\gamma(0))}{t} = \lim_{t \to 0} \frac{\alpha(f(\gamma(t))) - \alpha(f(\gamma(0)))}{t}$$

$$= \lim_{t \to 0} \alpha\left(\frac{f(\gamma(t)) - f(\gamma(0))}{t}\right) = \alpha\left(\lim_{t \to 0} \frac{f(\gamma(t)) - f(\gamma(0))}{t}\right)$$

$$= \alpha\left(\lim_{t \to 0} \frac{f(\gamma(t+0)) - f(\gamma(0))}{t}\right) = \alpha \ll \gamma, [f] \gg = \text{RHS}. \qquad \square$$

**Note 2.27** Let $M$ be an $m$-dimensional smooth manifold. Let $p \in M$. Let $[[\gamma]], [[\gamma_1]]$ be in $T_pM(= \Gamma_p(M)/\approx)$ where $\gamma, \gamma_1$ are in $\Gamma_p(M)$. Let $[f], [f_1]$ be in $\mathcal{F}_p(M)$ where $f, f_1$ are in $C_p^\infty(M)$. Let $[[\gamma]] = [[\gamma_1]]$, and $[f] = [f_1]$. We shall try to show that

$$\left.\frac{d(f \circ \gamma)(t)}{dt}\right|_{t=0} = \left.\frac{d(f_1 \circ \gamma_1)(t)}{dt}\right|_{t=0}.$$

Since $p \in M$, and $M$ is an $m$-dimensional smooth manifold, there exists an admissible coordinate chart $(U, \varphi_U)$ of $M$ satisfying $p \in U$. Now since $[[\gamma]] = [[\gamma_1]]$, $\gamma \approx \gamma_1$, and hence, $(\varphi_U \circ \gamma)'(0) = (\varphi_U \circ \gamma_1)'(0)$. Since $[f] = [f_1], f \sim g$, and hence, there exists an open neighborhood $H$ of $p(= \gamma(0))$ such that $f(x) = g(x)$ for every $x$ in $H$.

$$\text{LHS} = \left.\frac{d(f \circ \gamma)(t)}{dt}\right|_{t=0} = \left.\frac{d\left(f \circ \left((\varphi_U)^{-1} \circ \varphi_U\right) \circ \gamma\right)(t)}{dt}\right|_{t=0}$$

$$= \left.\frac{d\left(\left(f \circ (\varphi_U)^{-1}\right) \circ (\varphi_U \circ \gamma)\right)(t)}{dt}\right|_{t=0}$$

$$= \left(\left(f \circ (\varphi_U)^{-1}\right)'((\varphi_U \circ \gamma)(0))\right)((\varphi_U \circ \gamma)'(0))$$

$$= \left(\left(f \circ (\varphi_U)^{-1}\right)'((\varphi_U \circ \gamma)(0))\right)((\varphi_U \circ \gamma_1)'(0))$$

$$= \left(\left(f \circ (\varphi_U)^{-1}\right)'((\varphi_U(\gamma(0))))\right)((\varphi_U \circ \gamma_1)'(0))$$

$$= \left(\left(f \circ (\varphi_U)^{-1}\right)'((\varphi_U(p)))\right)((\varphi_U \circ \gamma_1)'(0))$$

$$= \left(\left(f \circ (\varphi_U)^{-1}\right)'((\varphi_U(\gamma_1(0))))\right)((\varphi_U \circ \gamma_1)'(0))$$

$$= \left(\left(f \circ (\varphi_U)^{-1}\right)'((\varphi_U \circ \gamma_1)(0))\right)((\varphi_U \circ \gamma_1)'(0))$$

$$= \left.\frac{d\left(\left(f \circ (\varphi_U)^{-1}\right) \circ (\varphi_U \circ \gamma_1)\right)(t)}{dt}\right|_{t=0} = \left.\frac{d(f \circ \gamma_1)(t)}{dt}\right|_{t=0} = \text{RHS}.$$

Hence, the following definition is legitimate:

**Definition** Let $M$ be an $m$-dimensional smooth manifold. Let $p \in M$. Let $[\![\gamma]\!]$ be in $T_pM$ where $\gamma$ is in $\Gamma_p(M)$. Let $[f]$ be in $\mathcal{F}_p(M)$ where $f$ is in $C_p^\infty(M)$. By $[\![\gamma]\!][f]$, we shall mean $\frac{d(f \circ \gamma)(t)}{dt}\Big|_{t=0}(= \ll\gamma, [f]\gg)$.

**Lemma 2.28** *Let $M$ be an $m$-dimensional smooth manifold. Let $p \in M$. Let $[\![\gamma]\!]$ be in $T_pM$ where $\gamma$ is in $\Gamma_p(M)$. Then*

1. $[\![\gamma]\!]([f] + [f_1]) = [\![\gamma]\!][f] + [\![\gamma]\!][f_1]$ *for every* $[f], [f_1]$ *in* $\mathcal{F}_p(M)$ *where* $f, f_1$ *are in* $C_p^\infty(M)$,
2. $[\![\gamma]\!](t[f]) = t([\![\gamma]\!][f])$ *for every* $[f]$ *in* $\mathcal{F}_p(M)$ *where* $f$ *is in* $C_p^\infty(M)$, *and for every real* $t$,
3. $[\![\gamma]\!]([f][f_1]) = ([\![\gamma]\!][f])(f_1(p)) + (f(p))([\![\gamma]\!][f_1])$ *for every* $[f], [f_1]$ *in* $\mathcal{F}_p(M)$ *where* $f, f_1$ *are in* $C_p^\infty(M)$.

*Proof*

1: LHS $= [\![\gamma]\!]([f] + [f_1]) = [\![\gamma]\!]([f + f_1]) = \ll\gamma, [f + f_1]\gg = \ll\gamma + [f_1]\gg$
$= \ll\gamma, [f]\gg + \ll\gamma, [f_1]\gg = \gamma[f] + \gamma[f_1] = $ RHS.

2: LHS $= [\![\gamma]\!](t[f]) = [\![\gamma]\!]([tf]) = \ll\gamma, t[f]\gg = t\ll\gamma, [f]\gg = t([\![\gamma]\!][f]) = $ RHS.

3: LHS $= [\![\gamma]\!]([f][f_1]) = [\![\gamma]\!]([f \cdot f_1]) = \frac{d((f \cdot f_1) \circ \gamma)(t)}{dt}\Big|_{t=0} = \frac{d((f \cdot f_1)(\gamma(t)))}{dt}\Big|_{t=0}$

$= \frac{d((f(\gamma(t)) \cdot f_1(\gamma(t))))}{dt}\Big|_{t=0} = \frac{d((((f \circ \gamma)(t)) \cdot ((f_1 \circ \gamma)(t))))}{dt}\Big|_{t=0}$

$= \frac{d((f \circ \gamma)(t))}{dt}\Big|_{t=0}((f_1 \circ \gamma)(0)) + ((f \circ \gamma)(0))\frac{d((f_1 \circ \gamma)(t))}{dt}\Big|_{t=0}$

$= ([\![\gamma]\!][f])((f_1 \circ \gamma)(0)) + ((f \circ \gamma)(0))([\![\gamma]\!][f])$

$= ([\![\gamma]\!][f])((f_1(\gamma(0)))) + (f(\gamma(0)))([\![\gamma]\!][f_1])$

$= ([\![\gamma]\!][f])((f_1(p))) + (f(p))([\![\gamma]\!][f_1]) = $ RHS. $\qquad\square$

## 2.4 Derivations

**Definition** Let $M$ be an $m$-dimensional smooth manifold. Let $p$ be an element of $M$. Let $D : \mathcal{F}_p(M) \to \mathbb{R}$ be any function. If

1. $D([f] + [f_1]) = D[f] + D[f_1]$ for every $[f], [f_1]$ in $\mathcal{F}_p(M)$ where $f, f_1$ are in $C_p^\infty(M)$,
2. $D(t[f]) = t(D[f])$ for every $[f]$ in $\mathcal{F}_p(M)$ where $f$ is in $C_p^\infty(M)$, and for every real $t$,
3. $D([f][f_1]) = (D[f])(f_1(p)) + (f(p))(D[f_1])$ for every $[f], [f_1]$ in $\mathcal{F}_p(M)$ where $f, f_1$ are in $C_p^\infty(M)$,

then we say that $D$ is a *derivation at* $p$. Here, the collection of all derivations at $p$ is denoted by $\mathcal{D}_p(M)$.

*Example 2.29* Let $M$ be an $m$-dimensional smooth manifold. Let $p \in M$. Let $[\![\gamma]\!]$ be in $T_pM$ where $\gamma$ is in $\Gamma_p(M)$.

By the Lemma 2.28, the mapping $[f] \mapsto [\![\gamma]\!][f]$ is an example of a derivation at $p$. We shall denote this derivation by $D_{[\gamma]}$. Thus,

$$D_{[\gamma]}([f]) \equiv [\![\gamma]\!][f]$$

for every $[f]$ in $\mathcal{F}_p(M)$ where $f$ is in $C_p^\infty(M)$. Also

$$\{D_\gamma : [\![\gamma]\!] \in T_pM\} \subset \mathcal{D}_p(M).$$

**Note 2.30** Let $M$ be an $m$-dimensional smooth manifold. Let $p$ be an element of $M$. For every $D, D_1$ in $\mathcal{D}_p(M)$, we define $D + D_1 : \mathcal{F}_p(M) \to \mathbb{R}$ as follows: For every $[f]$ in $\mathcal{F}_p(M)$ where $f$ is in $C_p^\infty(M)$, $(D + D_1)([f]) \equiv D[f] + D_1[f]$. We shall try to show that $D + D_1$ is in $\mathcal{D}_p(M)$, that is,

1. $(D + D_1)([f] + [f_1]) = (D + D_1)[f] + (D + D_1)[f_1]$ for every $[f], [f_1]$ in $\mathcal{F}_p(M)$ where $f, f_1$ are in $C_p^\infty(M)$,
2. $(D + D_1)(t[f]) = t((D + D_1)[f])$ for every $[f]$ in $\mathcal{F}_p(M)$ where $f$ is in $C_p^\infty(M)$, and for every real $t$,
3. $(D + D_1)([f][f_1]) = ((D + D_1)[f])(f_1(p)) + (f(p))((D + D_1)[f_1])$ for every $[f], [f_1]$ in $\mathcal{F}_p(M)$ where $f, f_1$ are in $C_p^\infty(M)$.

For 1: LHS $= (D + D_1)([f] + [f_1]) = D([f] + [f_1]) + D_1([f] + [f_1])$
$= (D[f] + D[f_1]) + (D_1[f] + D_1[f_1])$
$= (D[f] + D_1[f]) + (D[f_1] + D_1[f_1])$
$= (D + D_1)[f] + (D + D_1)[f_1] =$ RHS.

For 2: LHS $= (D + D_1)(t[f]) = D(t[f]) + D_1(t[f])$
$= t(D[f]) + t(D_1[f]) = t(D[f] + D_1[f])$
$= t((D + D_1)[f]) =$ RHS.

For 3: LHS $= (D + D_1)([f][f_1]) = D([f][f_1]) + D_1([f][f_1])$
$= ((D[f])(f_1(p)) + (f(p))(D[f_1])) + ((D_1[f])(f_1(p)) + (f(p))(D_1[f_1]))$
$= ((D[f]) + D_1[f])(f_1(p)) + (f(p))(D[f_1] + D_1[f_1])$
$= ((D + D_1)[f])(f_1(p)) + (f(p))((D + D_1)[f_1]) =$ RHS.

This proves that $D + D_1$ is in $\mathcal{D}_p(M)$.

For every $D$ in $\mathcal{D}_p(M)$, and for every real $t$, we define $tD : \mathcal{F}_p(M) \to \mathbb{R}$ as follows: For every $[f]$ in $\mathcal{F}_p(M)$ where $f$ is in $C_p^\infty(M)$, $(tD)([f]) \equiv t(D[f])$. We shall try to show that $tD$ is in $\mathcal{D}_p(M)$, that is,

1. $(tD)([f] + [f_1]) = (tD)[f] + (tD)[f_1]$ for every $[f], [f_1]$ in $\mathcal{F}_p(M)$ where $f, f_1$ are in $C_p^\infty(M)$,
2. $(tD)(s[f]) = s((tD)[f])$ for every $[f]$ in $\mathcal{F}_p(M)$ where $f$ is in $C_p^\infty(M)$, and for every real $s$,

3. $(tD)([f][f_1]) = ((tD)[f])(f_1(p)) + (f(p))((tD)[f_1])$ for every $[f], [f_1]$ in $\mathcal{F}_p(M)$ where $f, f_1$ are in $C_p^\infty(M)$.

For 1: LHS $= (tD)([f] + [f_1]) = t(D([f] + [f_1]))$
$$= t(D[f] + D[f_1]) = t(D[f]) + t(D[f_1]) = (tD)[f] + (tD)[f_1] = \text{RHS}.$$
For 2: LHS $= (tD)(s[f]) = t(D(s[f])) = t(s(D[f]))$
$$= s(t(D[f])) = s((tD)[f]) = \text{RHS}.$$
For 3: LHS $= (tD)([f][f_1]) = t(D([f][f_1])) = t((D[f])(f_1(p)) + (f(p))(D[f_1]))$
$$= t(D[f])(f_1(p)) + (f(p))t(D[f_1])$$
$$= ((tD)[f])(f_1(p)) + (f(p))((tD)[f_1]) = \text{RHS}.$$

This proves that $tD$ is in $\mathcal{D}_p(M)$.

Now, we shall try to verify that $\mathcal{D}_p(M)$ is a real linear space.

1. $+$ is associative: Let us take any $D, D_1, D_2$ in $\mathcal{D}_p(M)$. We have to prove that $(D + D_1) + D_2 = D + (D_1 + D_2)$, that is, for every $[f]$ in $\mathcal{F}_p(M)$ where $f$ is in $C_p^\infty(M)$, $((D + D_1) + D_2)[f] = (D + (D_1 + D_2))[f]$.

LHS $= ((D + D_1) + D_2)[f] = (D + D_1)[f] + D_2[f] = (D[f] + D_1[f]) + D_2[f]$
$$D[f] + (D_1[f] + D_2[f]) = D[f] + (D_1 + D_2)[f] = (D + (D_1 + D_2))[f] = \text{RHS}.$$

2. Existence of zero element: Let us define the constant function $0 : \mathcal{F}_p(M) \to \mathbb{R}$ as follows: for every $[f]$ in $\mathcal{F}_p(M)$, $0[f] \equiv 0$. We want to prove that $0$ is in $\mathcal{D}_p(M)$. For this purpose, we must prove

   1. $0([f] + [f_1]) = 0[f] + 0[f_1]$ for every $[f], [f_1]$ in $\mathcal{F}_p(M)$ where $f, f_1$ are in $C_p^\infty(M)$,
   2. $0(t[f]) = t(0[f])$ for every $[f]$ in $\mathcal{F}_p(M)$ where $f$ is in $C_p^\infty(M)$, and for every real $t$,
   3. $0([f][f_1]) = (0[f])(f_1(p)) + (f(p))(0[f_1])$, for every $[f], [f_1]$ in $\mathcal{F}_p(M)$ where $f, f_1$ are in $C_p^\infty(M)$.

For 1: LHS $= 0([f] + [f_1]) = 0 = 0 + 0 = 0[f] + 0[f_1] = \text{RHS}.$
For 2: LHS $= 0(t[f]) = 0 = t \cdot 0 = t(0[f]) = \text{RHS}.$
For 3: LHS $= 0([f][f_1]) = 0([f \cdot f_1]) = 0 = 0(f_1(p)) + (f(p))0 = (0[f])(f_1(p)) + (f(p))(0[f_1]) = \text{RHS}.$

Thus, we have shown that $0$ is in $\mathcal{D}_p(M)$. Now, it remains to be showed that $0 + D = D$ for every $D$ in $\mathcal{D}_p(M)$, that is, for every $[f]$ in $\mathcal{F}_p(M)$ where $f$ is in $C_p^\infty(M)$, $(0 + D)[f] = D[f]$.

LHS $= (0 + D)[f] = 0[f] + D[f] = 0 + D[f] = D[f] = \text{RHS}.$

Hence, $0$ serves the purpose of zero element in $\mathcal{D}_p(M)$.

3. Existence of negative element: Let us take any $D$ in $\mathcal{D}_p(M)$. Now, let us define $(-D) : \mathcal{F}_p(M) \to \mathbb{R}$ as follows: For every $[f]$ in $\mathcal{F}_p(M)$, $(-D)[f] \equiv -(D[f])$. We want to prove that $(-D)$ is in $\mathcal{D}_p(M)$. For this purpose, we must prove

1. $(-D)([f] + [f_1]) = (-D)[f] + (-D)[f_1]$ for every $[f], [f_1]$ in $\mathcal{F}_p(M)$ where $f$, $f_1$ are in $C_p^\infty(M)$,
2. $(-D)(t[f]) = t((-D)[f])$ for every $[f]$ in $\mathcal{F}_p(M)$ where $f$ is in $C_p^\infty(M)$, and for every real $t$,
3. $(-D)([f][f_1]) = ((-D)[f])(f_1(p)) + (f(p))((-D)[f_1])$, for every $[f][f_1]$ in $\mathcal{F}_p(M)$ where $f, f_1$ are in $C_p^\infty(M)$

For 1: LHS $= (-D)([f] + [f_1]) = -(D([f] + [f_1])) = -(D[f] + D[f_1])$
$= -(D[f]) + (-(D[f_1])) = (-D)[f] + (-D)[f_1] = $ RHS.
This proves 1.
For 2: LHS $= (-D)(t[f]) = -(D(t[f])) = -(t(D[f]))$
$= t(-(D[f])) = t((-D)[f]) = $ RHS.
For 3: LHS $= (-D)([f][f_1]) = -(D([f][f_1])) = -((D[f])(f_1(p)) + (f(p))(D[f_1]))$
$= (-(D[f]))(f_1(p)) + (f(p))(-(D[f_1]))$
$= ((-D)[f])(f_1(p)) + (f(p))((-D)[f_1]) = $ RHS.

Thus, we have shown that $(-D)$ is in $\mathcal{D}_p(M)$. Now, it remains to be showed that $(-D) + D = 0$, that is, for every $[f]$ in $\mathcal{F}_p(M)$ where $f$ is in $C_p^\infty(M)$, $((-D) + D)[f] = 0[f]$.

LHS $= ((-D) + D)[f] = (-D)[f] + D[f] = -(D[f]) + D[f] = 0 = 0[f] = $ RHS.

Hence, $(-D)$ serves the purpose of negative element of $D$ in $\mathcal{D}_p(M)$.

4. $+$ is commutative: Let us take any $D, D_1$ in $\mathcal{D}_p(M)$. We have to prove that $D + D_1 = D_1 + D$, that is, for every $[f]$ in $\mathcal{F}_p(M)$ where $f$ is in $C_p^\infty(M)$, $(D + D_1)[f] = (D_1 + D)[f]$.

LHS $= (D + D_1)[f] = D[f] + D_1[f] = D_1[f] + D[f] = (D_1 + D)[f] = $ RHS.

This proves 4.
5. For every real $s, t$, and $D$ in $\mathcal{D}_p(M)$, we have to prove that $(s + t)D = sD + tD$, that is, for every $[f]$ in $\mathcal{F}_p(M)$ where $f$ is in $C_p^\infty(M)$, $((s + t)D)[f] = (sD + tD)[f]$. Here

LHS $= ((s + t)D)[f] = (s + t)(D[f]) = s(D[f]) + t(D[f]) = (sD)[f] + (tD)[f]$
$= (sD + tD)[f] = $ RHS.

6. For every real $s, t$, and $D$ in $\mathcal{D}_p(M)$, we have to prove that $(st)D = s(tD)$, that is, for every $[f]$ in $\mathcal{F}_p(M)$ where $f$ is in $C_p^\infty(M)$, $((st)D)[f] = (s(tD))[f]$. Here,

LHS $= ((st)D)[f] = (st)(D[f]) = s(t(D[f])) = s((tD)[f]) = (s(tD))[f] = $ RHS.

7. For every real $t$, and for every $D, D_1$ in $\mathcal{D}_p(M)$, we have to prove that $t(D + D_1) = tD + tD_1$, that is, for every $[f]$ in $\mathcal{F}_p(M)$ where $f$ is in $C_p^\infty(M)$, $(t(D + D_1))[f] = (tD + tD_1)[f]$. Here

$$\text{LHS} = (t(D + D_1))[f] = t((D + D_1)[f]) = t(D[f] + D_1[f]) = t(D[f]) + t(D_1[f])$$
$$= (tD)[f] + (tD_1)[f] = (tD + tD_1)[f] = \text{RHS}.$$

8. For every $D$ in $\mathcal{D}_p(M)$, we have to show that $1D = D$, that is, for every $[f]$ in $\mathcal{F}_p(M)$ where $f$ is in $C_p^\infty(M)$, $(1D)[f] = D[f]$. Here,

$$\text{LHS} = (1D)[f] = 1(D[f]) = D[f] = \text{RHS}.$$

Thus, we have shown that $\mathcal{D}_p(M)$ is a real linear space.

**Lemma 2.31** *Let $M$ be an m-dimensional smooth manifold. Let $p$ be an element of $M$. Let $D : \mathcal{F}_p(M) \to \mathbb{R}$ be a derivation at $p$. Let $[f]$ be in $\mathcal{F}_p(M)$ where $f$ is in $C_p^\infty(M)$. If $f$ is a constant function, then $D[f] = 0$.*

*Proof* Since $f$ is a constant function, there exists a real number $c$ such that $f(x) = c$ for every $x$ in the domain of $f$. Since

$$D(1) = D(1 \cdot 1) = (D(1))(1(p)) + (1(p))(D(1)) = (D(1))1 + 1(D(1))$$
$$= 2(D(1)),$$

$D(1) = 0$. Now,

$$\text{LHS} = D[f] = D(c) = D(c1) = c(D(1)) = c \cdot 0 = 0 = \text{RHS}. \qquad \square$$

**Note 2.32** Let $M$ be a 2-dimensional smooth manifold. Let $p \in M$. Let $(U, \varphi_U)$ be any admissible coordinate chart of $M$ satisfying $p \in U$. Let $u^1, u^2$ be the component functions of $\varphi_U$. So

$$u^1 : U \to \mathbb{R}, \ u^2 : U \to \mathbb{R},$$

and for every $x$ in $U$,

$$\varphi_U(x) = (u^1(x), u^2(x)).$$

Since $u^1$ is in $C_p^\infty(M)$, $[u^1]$ is in $\mathcal{F}_p(M)$. Similarly, $[u^2]$ is in $\mathcal{F}_p(M)$.

Further, let us observe that

$$\left(D_1\left(u^1 \circ (\varphi_U)^{-1}\right)\right)(\varphi_U(p))$$

$$= \left(D_1\left(u^1 \circ (\varphi_U)^{-1}\right)\right)(u^1(p), u^2(p))$$

$$= \lim_{t \to 0} \frac{\left(u^1 \circ (\varphi_U)^{-1}\right)(u^1(p) + t, u^2(p)) - \left(u^1 \circ (\varphi_U)^{-1}\right)(u^1(p), u^2(p))}{t}$$

$$= \lim_{t \to 0} \frac{\left(u^1 \circ (\varphi_U)^{-1}\right)((u^1(p), u^2(p)) + (t, 0)) - \left(u^1 \circ (\varphi_U)^{-1}\right)(u^1(p), u^2(p))}{t}$$

$$= \lim_{t \to 0} \frac{\left(u^1 \circ (\varphi_U)^{-1}\right)(\varphi_U(p) + (t, 0)) - \left(u^1 \circ (\varphi_U)^{-1}\right)(\varphi_U(p))}{t}$$

$$= \lim_{t \to 0} \frac{\left(u^1 \circ (\varphi_U)^{-1}\right)(\varphi_U(p) + (t, 0)) - u^1(p)}{t}$$

$$= \lim_{t \to 0} \frac{u^1\left((\varphi_U)^{-1}(\varphi_U(p) + (t, 0))\right) - u^1(p)}{t}$$

$$= \lim_{t \to 0} \frac{u^1(q) - u^1(p)}{t}$$

where $q \equiv (\varphi_U)^{-1}(\varphi_U(p) + (t, 0))$, that is,

$$\left(u^1(q), u^2(q)\right) = \varphi_U(q) = \varphi_U(p) + (t, 0) = \left(u^1(p), u^2(p)\right) + (t, 0)$$
$$= \left(u^1(p) + t, u^2(p)\right).$$

Hence, $u^1(q) = u^1(p) + t$. This shows that

$$\left(D_1\left(u^1 \circ (\varphi_U)^{-1}\right)\right)(\varphi_U(p)) = \lim_{t \to 0} \frac{u^1(q) - u^1(p)}{t} = \lim_{t \to 0} \frac{(u^1(p) + t) - u^1(p)}{t} = 1.$$

Thus,

$$\left(D_1\left(u^1 \circ (\varphi_U)^{-1}\right)\right)(\varphi_U(p)) = 1.$$

Next

$$\left(D_2\left(u^1 \circ (\varphi_U)^{-1}\right)\right)(\varphi_U(p))$$

$$= \left(D_2\left(u^1 \circ (\varphi_U)^{-1}\right)\right)(u^1(p), u^2(p))$$

$$= \lim_{t \to 0} \frac{\left(u^1 \circ (\varphi_U)^{-1}\right)(u^1(p), u^2(p) + t) - \left(u^1 \circ (\varphi_U)^{-1}\right)(u^1(p), u^2(p))}{t}$$

$$= \lim_{t \to 0} \frac{\left(u^1 \circ (\varphi_U)^{-1}\right)((u^1(p), u^2(p)) + (0, t)) - \left(u^1 \circ (\varphi_U)^{-1}\right)(u^1(p), u^2(p))}{t}$$

$$= \lim_{t \to 0} \frac{\left(u^1 \circ (\varphi_U)^{-1}\right)(\varphi_U(p) + (0, t)) - \left(u^1 \circ (\varphi_U)^{-1}\right)(\varphi_U(p))}{t}$$

$$= \lim_{t \to 0} \frac{\left(u^1 \circ (\varphi_U)^{-1}\right)(\varphi_U(p) + (0, t)) - u^1(p)}{t}$$

$$= \lim_{t \to 0} \frac{u^1\left((\varphi_U)^{-1}(\varphi_U(p) + (0, t))\right) - u^1(p)}{t}$$

$$= \lim_{t \to 0} \frac{u^1(q) - u^1(p)}{t}$$

where $q \equiv (\varphi_U)^{-1}(\varphi_U(p) + (0, t))$, that is,

$$\begin{aligned}(u^1(q), u^2(q)) = \varphi_U(q) &= \varphi_U(p) + (0, t) = (u^1(p), u^2(p)) + (0, t) \\ &= (u^1(p), u^2(p) + t).\end{aligned}$$

Hence, $u^1(q) = u^1(p)$. This shows that

$$\left(D_2\left(u^1 \circ (\varphi_U)^{-1}\right)\right)(\varphi_U(p)) = \lim_{t \to 0} \frac{u^1(p) - u^1(p)}{t} = 0.$$

Thus,

$$\left(D_2\left(u^1 \circ (\varphi_U)^{-1}\right)\right)(\varphi_U(p)) = 0.$$

Similarly, $(D_2(u^2 \circ (\varphi_U)^{-1}))(\varphi_U(p)) = 1$, and $(D_1(u^2 \circ (\varphi_U)^{-1}))(\varphi_U(p)) = 0$.

**Definition** Let $M$ be a 2-dimensional smooth manifold. Let $p$ be an element of $M$. Let $(U, \varphi_U)$ be any admissible coordinate chart of $M$ satisfying $p \in U$. Let $u^1, u^2$ be the component functions of $\varphi_U$. The function $\frac{\partial}{\partial u^1}\big|_p : \mathcal{F}_p(M) \to \mathbb{R}$ is defined as follows: For every $[f]$ in $\mathcal{F}_p(M)$,

$$\left(\frac{\partial}{\partial u^1}\bigg|_p\right)[f] \equiv \left(D_1\left(f \circ (\varphi_U)^{-1}\right)\right)(\varphi_U(p)).$$

The function $\frac{\partial}{\partial u^2}\big|_p : \mathcal{F}_p(M) \to \mathbb{R}$ is defined as follows: For every $[f]$ in $\mathcal{F}_p(M)$,

$$\left(\frac{\partial}{\partial u^2}\bigg|_p\right)[f] \equiv \left(D_2\left(f \circ (\varphi_U)^{-1}\right)\right)(\varphi_U(p)).$$

From the Note 2.32,

$$\left(\frac{\partial}{\partial u^i}\bigg|_p\right)[u^j] = \begin{cases} 1 & \text{if} \quad i = j \\ 0 & \text{if} \quad i \neq j. \end{cases}$$

**Note 2.33** We shall prove that $\frac{\partial}{\partial u^1}\big|_p$ is in $\mathcal{D}_p(M)$. For this purpose, we must prove

1. $(\frac{\partial}{\partial u^1}\big|_p)([f] + [f_1]) = (\frac{\partial}{\partial u^1}\big|_p)[f] + (\frac{\partial}{\partial u^1}\big|_p)[f_1]$ for every $[f], [f_1]$ in $\mathcal{F}_p(M)$ where $f, f_1$ are in $C_p^\infty(M)$,
2. $(\frac{\partial}{\partial u^1}\big|_p)(t[f]) = t((\frac{\partial}{\partial u^1}\big|_p)[f])$ for every $[f]$ in $\mathcal{F}_p(M)$ where $f$ is in $C_p^\infty(M)$, and for every real $t$,
3. $(\frac{\partial}{\partial u^1}\big|_p)([f][f_1]) = ((\frac{\partial}{\partial u^1}\big|_p)[f])(f_1(p)) + (f(p))((\frac{\partial}{\partial u^1}\big|_p)[f_1])$, for every $[f], [f_1]$ in $\mathcal{F}_p(M)$ where $f, f_1$ are in $C_p^\infty(M)$

For 1: LHS $= \left(\frac{\partial}{\partial u^1}\bigg|_p\right)([f] + [f_1]) = \left(\frac{\partial}{\partial u^1}\bigg|_p\right)([f + f_1]) = \left(D_1\left((f + f_1) \circ (\varphi_U)^{-1}\right)\right)(\varphi_U(p))$

$= \left(D_1\left(\left(f \circ (\varphi_U)^{-1}\right) + \left(f_1 \circ (\varphi_U)^{-1}\right)\right)\right)(\varphi_U(p)) = \left(D_1\left(f \circ (\varphi_U)^{-1}\right)\right)(\varphi_U(p))$

$+ \left(D_1\left(f_1 \circ (\varphi_U)^{-1}\right)\right)(\varphi_U(p)) = \left(\frac{\partial}{\partial u^1}\bigg|_p\right)[f] + \left(\frac{\partial}{\partial u^1}\bigg|_p\right)[f_1] = $ RHS.

For 2: LHS $= \left(\frac{\partial}{\partial u^1}\bigg|_p\right)(t[f]) = \left(D_1\left((tf) \circ (\varphi_U)^{-1}\right)\right)(\varphi_U(p)) = \left(D_1\left(t\left(f \circ (\varphi_U)^{-1}\right)\right)\right)(\varphi_U(p))$

$= t\left(\left(D_1\left(f \circ (\varphi_U)^{-1}\right)\right)(\varphi_U(p))\right) = t\left(\left(\frac{\partial}{\partial u^1}\bigg|_p\right)[f]\right) = $ RHS.

For 3: LHS $= \left(\frac{\partial}{\partial u^1}\bigg|_p\right)([f][f_1]) = \left(\frac{\partial}{\partial u^1}\bigg|_p\right)([f \cdot f_1]) = \left(D_1\left((f \cdot f_1) \circ (\varphi_U)^{-1}\right)\right)(\varphi_U(p))$

$= \left(D_1\left(\left(\left(f \circ (\varphi_U)^{-1}\right) \cdot \left(f_1 \circ (\varphi_U)^{-1}\right)\right)\right)\right)(\varphi_U(p))$

$= \left(\left(D_1\left(f \circ (\varphi_U)^{-1}\right)\right)(\varphi_U(p))\right)\left(\left(f_1 \circ (\varphi_U)^{-1}\right)(\varphi_U(p))\right)$

$+ \left(\left(f \circ (\varphi_U)^{-1}\right)(\varphi_U(p))\right)\left(\left(D_1\left(f_1 \circ (\varphi_U)^{-1}\right)\right)(\varphi_U(p))\right)$

$= \left(\left(D_1\left(f \circ (\varphi_U)^{-1}\right)\right)(\varphi_U(p))\right)(f_1(p)) + (f(p))\left(\left(D_1\left(f_1 \circ (\varphi_U)^{-1}\right)\right)(\varphi_U(p))\right)$

$= \left(\left(\frac{\partial}{\partial u^1}\bigg|_p\right)[f]\right)(f_1(p)) + (f(p))\left(\left(\frac{\partial}{\partial u^1}\bigg|_p\right)[f_1]\right) = $ RHS.

This proves that $\frac{\partial}{\partial u^1}|_p$ is in $\mathcal{D}_p(M)$. Similarly, $\frac{\partial}{\partial u^2}|_p$ is in $\mathcal{D}_p(M)$.

**Note 2.34** We shall prove that $\frac{\partial}{\partial u^1}|_p, \frac{\partial}{\partial u^2}|_p$ are linearly independent. For this purpose, let $t_1(\frac{\partial}{\partial u^1}|_p) + t_2(\frac{\partial}{\partial u^2}|_p) = 0$. We have to show that $t_1 = t_2 = 0$. Since $t_1(\frac{\partial}{\partial u^1}|_p) + t_2(\frac{\partial}{\partial u^2}|_p) = 0$, so

$$0 = 0[u^1] = \left(t_1\left(\frac{\partial}{\partial u^1}\bigg|_p\right) + t_2\left(\frac{\partial}{\partial u^2}\bigg|_p\right)\right)[u^1] = t_1\left(\left(\frac{\partial}{\partial u^1}\bigg|_p\right)[u^1]\right) + t_2\left(\left(\frac{\partial}{\partial u^2}\bigg|_p\right)[u^1]\right)$$

$$= t_1(1) + t_2(0) = t_1.$$

Thus, $t_1 = 0$. Similarly, $t_2 = 0$. Thus, we have shown that $\frac{\partial}{\partial u^1}|_p, \frac{\partial}{\partial u^2}|_p$ are linearly independent.

**Lemma 2.35** *Let $M$ be a 2-dimensional smooth manifold. Let $p$ be an element of $M$. Let $(U, \varphi_U)$ be any admissible coordinate chart of $M$ satisfying $p \in U$. Let us define functions*

$$u^1 : U \to \mathbb{R}, \ u^2 : U \to \mathbb{R}$$

*as follows: For every $x$ in $U$,*

$$\varphi_U(x) \equiv \left(u^1(x), u^2(x)\right).$$

*If $\varphi_U(p) = (0,0)$, then $\{\frac{\partial}{\partial u^1}|_p, \frac{\partial}{\partial u^2}|_p\}$ is a basis of the real linear space $\mathcal{D}_p(M)$.*

*Proof* From the above discussion, it remains to be proved that $\{\frac{\partial}{\partial u^1}|_p, \frac{\partial}{\partial u^2}|_p\}$ generates the whole space $\mathcal{D}_p(M)$. For this purpose, let us take any $D$ in $\mathcal{D}_p(M)$. We have to find real numbers $t_1, t_2$ such that $D = t_1(\frac{\partial}{\partial u^1}|_p) + t_2(\frac{\partial}{\partial u^2}|_p)$, that is, for every $[f]$ in $\mathcal{F}_p(M)$ where $f$ is in $C_p^\infty(M)$, we have

$$D[f] = \left(t_1\left(\frac{\partial}{\partial u^1}\bigg|_p\right) + t_2\left(\frac{\partial}{\partial u^2}\bigg|_p\right)\right)[f].$$

Let us take any $[f]$ in $\mathcal{F}_p(M)$ where $f$ is in $C_p^\infty(M)$. It suffices to prove that

$$D[f] = \left((D[u^1])\left(\frac{\partial}{\partial u^1}\bigg|_p\right) + (D[u^2])\left(\frac{\partial}{\partial u^2}\bigg|_p\right)\right)[f],$$

that is,

$$D[f] = (D[u^1])\left(\left(\frac{\partial}{\partial u^1}\bigg|_p\right)[f]\right) + (D[u^2])\left(\left(\frac{\partial}{\partial u^2}\bigg|_p\right)[f]\right).$$

Let us note that for every $\varphi_U(q)(= (u^1(q), u^2(q)))$ in $\varphi_U(U)$, where $q$ is in $U$,

$$
\begin{aligned}
f(q) - f(p) &= f(q) - f\left((\varphi_U)^{-1}(0,0)\right) \\
&= f\left((\varphi_U)^{-1}\left(u^1(q), u^2(q)\right)\right) - f\left((\varphi_U)^{-1}(0,0)\right) \\
&= \left(f \circ (\varphi_U)^{-1}\right)\left(u^1(q), u^2(q)\right) - \left(f \circ (\varphi_U)^{-1}\right)(0,0) \\
&= \left.\left(\left(f \circ (\varphi_U)^{-1}\right)\left((u^1(q))t, (u^2(q))t\right)\right)\right|_{t=0}^{t=1} \\
&= \int_0^1 \left(\frac{\mathrm{d}}{\mathrm{d}t}\left(f \circ (\varphi_U)^{-1}\right)\left((u^1(q))t, (u^2(q))t\right)\right)\mathrm{d}t \\
&= \int_0^1 \left(\left(\left(D_1\left(f \circ (\varphi_U)^{-1}\right)\right)\left((u^1(q))t, (u^2(q))t\right)\right)\left(\frac{\mathrm{d}}{\mathrm{d}t}\left((u^1(q))t\right)\right)\right. \\
&\qquad \left. + \left(\left(D_2\left(f \circ (\varphi_U)^{-1}\right)\right)\left((u^1(q))t, (u^2(q))t\right)\right)\left(\frac{\mathrm{d}}{\mathrm{d}t}\left((u^2(q))t\right)\right)\right)\mathrm{d}t \\
&= \int_0^1 \left(\left(\left(D_1\left(f \circ (\varphi_U)^{-1}\right)\right)\left((u^1(q))t, (u^2(q))t\right)\right)\left(u^1(q)\right)\right. \\
&\qquad \left. + \left(\left(D_2\left(f \circ (\varphi_U)^{-1}\right)\right)\left((u^1(q))t, (u^2(q))t\right)\right)\left(u^2(q)\right)\right)\mathrm{d}t \\
&= (u^1(q)) \int_0^1 \left(\left(D_1\left(f \circ (\varphi_U)^{-1}\right)\right)\left((u^1(q))t, (u^2(q))t\right)\right)\mathrm{d}t \\
&\quad + (u^2(q)) \int_0^1 \left(\left(D_2\left(f \circ (\varphi_U)^{-1}\right)\right)\left((u^1(q))t, (u^2(q))t\right)\right)\mathrm{d}t \\
&= (u^1(q)) \int_0^1 \left(\left(D_1\left(f \circ (\varphi_U)^{-1}\right)\right)\left(t(u^1(q), u^2(q))\right)\right)\mathrm{d}t \\
&\quad + (u^2(q)) \int_0^1 \left(\left(D_2\left(f \circ (\varphi_U)^{-1}\right)\right)\left(t(u^1(q), u^2(q))\right)\right)\mathrm{d}t \\
&= (u^1(q)) \int_0^1 \left(\left(D_1\left(f \circ (\varphi_U)^{-1}\right)\right)\left(t(\varphi_U(q))\right)\right)\mathrm{d}t \\
&\quad + (u^2(q)) \int_0^1 \left(\left(D_2\left(f \circ (\varphi_U)^{-1}\right)\right)\left(t(\varphi_U(q))\right)\right)\mathrm{d}t \\
&= (u^1(q)) \int_0^1 \left(\left(\left.\frac{\partial}{\partial u^1}\right|_{(\varphi_U)^{-1}(t(\varphi_U(q)))}\right)[f]\right)\mathrm{d}t \\
&\quad + (u^2(q)) \int_0^1 \left(\left(\left.\frac{\partial}{\partial u^2}\right|_{(\varphi_U)^{-1}(t(\varphi_U(q)))}\right)[f]\right)\mathrm{d}t \\
&= (u^1(q))(g_1(q)) + (u^2(q))(g_2(q)),
\end{aligned}
$$

where

$$g_1(q) \equiv \int\limits_0^1 \left( \left( \frac{\partial}{\partial u^1}\bigg|_{(\varphi_U)^{-1}(t(\varphi_U(q)))} \right)[f] \right) dt,$$

and

$$g_2(q) \equiv \int\limits_0^1 \left( \left( \frac{\partial}{\partial u^2}\bigg|_{(\varphi_U)^{-1}(t(\varphi_U(q)))} \right)[f] \right) dt$$

for every $q$ in $U$. Hence,

$$f(q) = \left(u^1(q)\right)(g_1(q)) + \left(u^2(q)\right)(g_2(q)) + f(p) = \left(u^1 g_1\right)(q) + \left(u^2 g_2\right)(q) + f(p)$$
$$= \left(u^1 g_1 + u^2 g_2 + f(p)\right)(q)$$

for every $q$ in $U$. So

$$f = u^1 g_1 + u^2 g_2 + f(p).$$

Next

$$D[f] = D[f(p) + u^1 g_1 + u^2 g_2] = D[f(p)] + D[u^1 g_1] + D[u^2 g_2]$$
$$= 0 + D[u^1 g_1] + D[u^2 g_2] = D[u^1 g_1] + D[u^2 g_2]$$
$$= \left((D[u^1])(g_1(p)) + (u^1(p))(D[g_1])\right) + \left((D[u^2])(g_2(p)) + (u^2(p))(D[g_2])\right)$$
$$= \left((D[u^1])(g_1(p)) + (0)(D[g_1])\right) + \left((D[u^2])(g_2(p)) + (0)(D[g_2])\right)$$
$$= (D[u^1])(g_1(p)) + (D[u^2])(g_2(p))$$
$$= (D[u^1])\left( \int\limits_0^1 \left( \left( \frac{\partial}{\partial u^1}\bigg|_{(\varphi_U)^{-1}(t(\varphi_U(p)))} \right)[f] \right) dt \right)$$
$$+ (D[u^2])\left( \int\limits_0^1 \left( \left( \frac{\partial}{\partial u^2}\bigg|_{(\varphi_U)^{-1}(t(\varphi_U(p)))} \right)[f] \right) dt \right)$$
$$= (D[u^1])\left( \int\limits_0^1 \left( \left( \frac{\partial}{\partial u^1}\bigg|_{(\varphi_U)^{-1}(t(0,0))} \right)[f] \right) dt \right)$$
$$+ (D[u^2])\left( \int\limits_0^1 \left( \left( \frac{\partial}{\partial u^2}\bigg|_{(\varphi_U)^{-1}(t(0,0))} \right)[f] \right) dt \right)$$

$$= (D[u^1]) \left( \int_0^1 \left( \left( \left. \frac{\partial}{\partial u^1} \right|_{(\varphi_U)^{-1}(0,0)} \right) [f] \right) dt \right)$$

$$+ (D[u^2]) \left( \int_0^1 \left( \left( \left. \frac{\partial}{\partial u^2} \right|_{(\varphi_U)^{-1}(0,0)} \right) [f] \right) dt \right)$$

$$= (D[u^1]) \left( \int_0^1 \left( \left( \left. \frac{\partial}{\partial u^1} \right|_p \right) [f] \right) dt \right) + (D[u^2]) \left( \int_0^1 \left( \left( \left. \frac{\partial}{\partial u^2} \right|_p \right) [f] \right) dt \right)$$

$$= (D[u^1]) \left( \left( \left( \left. \frac{\partial}{\partial u^1} \right|_p \right) [f] \right) \int_0^1 1 dt \right) + (D[u^2]) \left( \left( \left( \left. \frac{\partial}{\partial u^2} \right|_p \right) [f] \right) \int_0^1 1 dt \right)$$

$$= (D[u^1]) \left( \left( \left. \frac{\partial}{\partial u^1} \right|_p \right) [f] \right) 1 + (D[u^2]) \left( \left( \left. \frac{\partial}{\partial u^2} \right|_p \right) [f] \right) 1$$

$$= (D[u^1]) \left( \left( \left. \frac{\partial}{\partial u^1} \right|_p \right) [f] \right) + (D[u^2]) \left( \left( \left. \frac{\partial}{\partial u^2} \right|_p \right) [f] \right) = \text{RHS}. \qquad \square$$

**Lemma 2.36** *Let $M$ be a 2-dimensional smooth manifold. Let $p$ be an element of $M$. Then, $\mathcal{D}_p(M)$ is a 2-dimensional real linear space.*

*Proof* Since $M$ is a 2-dimensional smooth manifold, and $p$ is in $M$, there exists an admissible coordinate chart $(U, \varphi_U)$ of $M$ satisfying $p \in U$. Here, we can find admissible coordinate chart $(U, \psi_U)$ of $M$ satisfying $\psi_U(x) = \varphi_U(x) - \varphi_U(p)$ for every $x$ in $U$. Hence, $\psi_U(p) = 0$. Let us define functions

$$u^1 : U \to \mathbb{R}, \quad u^2 : U \to \mathbb{R}$$

as follows: For every $x$ in $U$,

$$\left( u^1(x), u^2(x) \right) \equiv \psi_U(x).$$

Then, by Lemma 2.35, $\{\left. \frac{\partial}{\partial u^1} \right|_p, \left. \frac{\partial}{\partial u^2} \right|_p\}$ is a basis of the real linear space $\mathcal{D}_p(M)$. Hence, the dimension of $\mathcal{D}_p(M)$ is 2. $\qquad \square$

As above, we can prove the following

**Lemma 2.37** *Let $M$ be an m-dimensional smooth manifold. Let $p$ be an element of $M$. Then, $\mathcal{D}_p(M)$ is an m-dimensional real linear space. Also, if $(U, \varphi_U)$ is any admissible coordinate chart of $M$ satisfying $p \in U$, then $\{\left. \frac{\partial}{\partial u^1} \right|_p, \ldots, \left. \frac{\partial}{\partial u^m} \right|_p\}$ is a basis of $\mathcal{D}_p(M)$, where $u^1, \ldots, u^m$ are the component functions of $\varphi_U$.*

**Theorem 2.38** *Let $M$ be an $m$-dimensional smooth manifold. Let $p$ be an element of $M$. Put*

$$\mathcal{H}_p(M) \equiv \{[f] : [f] \in \mathcal{F}_p(M), \text{ and } \ll\gamma, [f]\gg = 0 \quad \text{for every } \gamma \text{ in } \Gamma_p(M)\}.$$

Then, $\mathcal{H}_p(M)$ is a linear subspace of the real linear space $\mathcal{F}_p(M)$. In short, we say that $\mathcal{H}_p$ is a linear subspace of $\mathcal{F}_p$.

*Proof* Let us try to prove that $\mathcal{H}_p(M)$ is nonempty. Let us recall that $[0] \in \mathcal{F}_p(M)$, where 0 denotes the constant function zero defined on $M$. Since, for every $\gamma$ in $\Gamma_p(M)$,

$$\ll\gamma, [0]\gg = \lim_{t\to 0}\frac{0(\gamma(t+0)) - 0(\gamma(0))}{t} = \lim_{t\to 0}\frac{0-0}{t} = \lim_{t\to 0} 0 = 0,$$

by the definition of $\mathcal{H}_p(M)$, $[0] \in \mathcal{H}_p(M)$. Hence, $\mathcal{H}_p(M)$ is nonempty.

Now, it remains to be proved that

1. If $[f] \in \mathcal{H}_p(M), [g] \in \mathcal{H}_p(M)$, then $[f] + [g] \in \mathcal{H}_p(M)$,
2. For every real $\alpha$, and for every $[f] \in \mathcal{H}_p(M)$, $\alpha[f] \in \mathcal{H}_p(M)$.

For 1: Let $[f] \in \mathcal{H}_p(M), [g] \in \mathcal{H}_p(M)$. Since $[f] \in \mathcal{H}_p(M), [g] \in \mathcal{H}_p(M)$, and $\mathcal{H}_p(M)$ is contained in $\mathcal{F}_p(M)$, $[f] \in \mathcal{F}_p(M), [g] \in \mathcal{F}_p(M)$. Since $[f] \in \mathcal{F}_p(M), [g] \in \mathcal{F}_p(M)$, and by Theorem 2.23, $\mathcal{F}_p(M)$ is a real linear space, $[f] + [g] \in \mathcal{F}_p(M)$.

Now, let us take any $\gamma$ in $\Gamma_p(M)$. Since $[f] \in \mathcal{H}_p(M)$, by the definition of $\mathcal{H}_p(M)$, $\ll\gamma, [f]\gg = 0$. Similarly, $\ll\gamma, [g]\gg = 0$. Now, by Theorem 2.26,

$$\ll\gamma, [f] + [g]\gg = \ll\gamma, [f]\gg + \ll\gamma, [g]\gg = 0 + 0 = 0.$$

So, by the definition of $\mathcal{H}_p(M)$, $[f] + [g] \in \mathcal{H}_p(M)$. This proves 1.

For 2: Let $\alpha$ be any real number, and let $[f] \in \mathcal{H}_p(M)$. Since $[f] \in \mathcal{H}_p(M)$, and $\mathcal{H}_p(M)$ is contained in $\mathcal{F}_p(M)$, $[f] \in \mathcal{F}_p(M)$. Since $\alpha$ is a real number, $[f] \in \mathcal{F}_p(M)$, and by Theorem 2.23, $\mathcal{F}_p(M)$ is a real linear space, $\alpha[f] \in \mathcal{F}_p(M)$.

Now, let us take any $\gamma$ in $\Gamma_p(M)$. Since $[f] \in \mathcal{H}_p(M)$, by the definition of $\mathcal{H}_p(M)$, $\ll\gamma, [f]\gg = 0$. Now, by Theorem 2.26,

$$\ll\gamma, \alpha[f]\gg = \alpha\ll\gamma, [f]\gg = \alpha(0) = 0.$$

So, by the definition of $\mathcal{H}_p(M)$, $\alpha[f] \in \mathcal{H}_p(M)$. This proves 2. Hence, $\mathcal{H}_p(M)$ is a linear subspace of the real linear space $\mathcal{F}_p(M)$.                                  $\square$

## 2.5  Cotangent Spaces

**Definition** Let $M$ be an $m$-dimensional smooth manifold. Let $p$ be an element of $M$. The set

$$\{[f] : [f] \in \mathcal{F}_p(M), \text{ and } \ll\gamma, [f]\gg \, = 0 \text{ for every } \gamma \text{ in } \Gamma_p(M)\}$$

is denoted by $\mathcal{H}_p(M)$ (or, simply $\mathcal{H}_p$). From Theorem 2.38, $\mathcal{H}_p$ is a linear subspace of the real linear space $\mathcal{F}_p$. So it is meaningful to write the quotient space $\frac{\mathcal{F}_p}{\mathcal{H}_p}$. We know that

$$\frac{\mathcal{F}_p}{\mathcal{H}_p} \equiv \{[f] + \mathcal{H}_p : [f] \in \mathcal{F}_p\}.$$

Here, vector addition, and scalar multiplication over $\frac{\mathcal{F}_p}{\mathcal{H}_p}$ are defined as follows: For every $[f], [g] \in \mathcal{F}_p$, and for every real $t$,

$$([f] + \mathcal{H}_p) + ([g] + \mathcal{H}_p) \equiv ([f] + [g]) + \mathcal{H}_p = [f + g] + \mathcal{H}_p,$$
$$t([f] + \mathcal{H}_p) \equiv (t[f]) + \mathcal{H}_p = [tf] + \mathcal{H}_p.$$

The quotient space $\frac{\mathcal{F}_p}{\mathcal{H}_p}$ is denoted by $T_p^*(M)$ (or, simply $T_p^*$) and is called the *cotangent space* of $M$ at $p$. Since $\frac{\mathcal{F}_p}{\mathcal{H}_p}$ is a real linear space, the cotangent space $T_p^*$ of $M$ at $p$ is a real linear space. Intuitively, in $T_p^*$, we will not distinguish between $[f]$ and $[g]$ whenever $[f] - [g]$ is in $\mathcal{H}_p$.

**Definition** Let $M$ be an $m$-dimensional smooth manifold. Let $p$ be an element of $M$. Let $[f] \in \mathcal{F}_p$. $[f] + \mathcal{H}_p$ is denoted by $\tilde{[f]}$ (or, $[f]^\sim$) or $(df)_p$, and is called the *cotangent vector* on $M$ at $p$ determined by the term $[f]$. Thus, we can write

$$T_p^* = \{\tilde{[f]} : [f] \in \mathcal{F}_p\}$$

or,

$$T_p^* = \left\{(df)_p : [f] \in \mathcal{F}_p\right\}.$$

From the above discussion, we get the following formulae:

$$\begin{cases} [f]^\sim + [g]^\sim = [f + g]^\sim \\ t[f]^\sim = [tf]^\sim \end{cases}$$

or,

$$\begin{cases} (df)_p + (dg)_p = (d(f+g))_p \\ t\big((df)_p\big) = (d(tf))_p. \end{cases}$$

**Theorem 2.39** *Let M be an m-dimensional smooth manifold. Let p be an element of M. Let [f] be in $\mathcal{F}_p(M)$ where f is in $C_p^\infty(M)$. Let $(U, \varphi_U)$ be an admissible coordinate chart of M satisfying $p \in U$. Then, $[f] \in \mathcal{H}_p(M)$ if and only if*

$$\left(D_1\big(f \circ (\varphi_U)^{-1}\big)\right)(\varphi_U(p)) = \left(D_2\big(f \circ (\varphi_U)^{-1}\big)\right)(\varphi_U(p)) = \cdots$$
$$= \left(D_m\big(f \circ (\varphi_U)^{-1}\big)\right)(\varphi_U(p)) = 0.$$

*In other words, $[f] \in \mathcal{H}_p(M)$ if and only if $(\frac{\partial}{\partial u^1}|_p)[f] = (\frac{\partial}{\partial u^2}|_p)[f] = \cdots = (\frac{\partial}{\partial u^m}|_p)[f] = 0$.*

*Proof* (if part): Let

$$\left(D_1\big(f \circ (\varphi_U)^{-1}\big)\right)(\varphi_U(p)) = \cdots = \left(D_m\big(f \circ (\varphi_U)^{-1}\big)\right)(\varphi_U(p)) = 0.$$

We have to show that $[f] \in \mathcal{H}_p(M)$, that is, $\ll\gamma, [f]\gg = 0$ for every $\gamma$ in $\Gamma_p(M)$. For this purpose, let us take any $\gamma$ in $\Gamma_p(M)$. Since $\ll\gamma, [f]\gg = \frac{d(f \circ \gamma)(t)}{dt}\big|_{t=0}$, we must prove that

$$\frac{d(f \circ \gamma)(t)}{dt}\bigg|_{t=0} = 0.$$

Since $\gamma$ is in $\Gamma_p(M)$, by the definition of $\Gamma_p(M)$, $\gamma$ is a parametrized curve in $M$ through $p$, and hence, there exists a real number $\delta > 0$ such that $\gamma$ is defined on the open interval $(-\delta, \delta)$, $\gamma(0) = p$, and $\gamma$ is a smooth map from $(-\delta, \delta)$ to $M$. As in the proof of Theorem 2.24, we can show that 0 is an interior point of the domain of the mapping $\varphi_U \circ \gamma$. Since 0 is an interior point of the domain of the mapping $\varphi_U \circ \gamma$, there exists $\varepsilon > 0$ such that $\varepsilon < \delta$, and each of the $m$ component functions of the mapping

$$(\varphi_U \circ \gamma) : (-\varepsilon, \varepsilon) \to \mathbb{R}^m$$

is $C^\infty$ at 0 in $(-\varepsilon, \varepsilon)$, and hence, each of the $m$ functions $F_1 : (-\varepsilon, \varepsilon) \to \mathbb{R}, \ldots, F_m : (-\varepsilon, \varepsilon) \to \mathbb{R}$ is $C^\infty$ at 0 in $(-\varepsilon, \varepsilon)$ where

$$(F_1(t), \ldots, F_m(t)) \equiv (\varphi_U \circ \gamma)(t)$$

for every $t$ in $(-\varepsilon, \varepsilon)$. Since $f$ is in $C_p^\infty(M)$, by the definition of $C_p^\infty(M)$, dom$f$ is an open neighborhood of $p$ in $M$, and

$$\left(f \circ (\varphi_U)^{-1}\right) : \varphi_U((\mathrm{dom} f) \cap U) \to \mathbb{R}$$

is $C^\infty$ at the point $\varphi_U(p)$ $(= \varphi_U(\gamma(0)) = (\varphi_U \circ \gamma)(0))$ in $\mathbb{R}^m$, and hence, the partial derivatives $(D_1(f \circ (\varphi_U)^{-1}))((\varphi_U \circ \gamma)(0)), \ldots, (D_m(f \circ (\varphi_U)^{-1}))((\varphi_U \circ \gamma)(0))$ exist. Now,

$$\mathrm{LHS} = \left.\frac{d(f \circ \gamma)(t)}{dt}\right|_{t=0} = \left.\frac{d\left(\left(f \circ (\varphi_U)^{-1}\right) \circ (\varphi_U \circ \gamma)\right)(t)}{dt}\right|_{t=0}$$

$$= \left[ \left(D_1\left(f \circ (\varphi_U)^{-1}\right)\right)((\varphi_U \circ \gamma)(0)) \quad \cdots \quad \left(D_m\left(f \circ (\varphi_U)^{-1}\right)\right)((\varphi_U \circ \gamma)(0)) \right] \begin{bmatrix} \dfrac{dF_1(t)}{dt} \\ \vdots \\ \dfrac{dF_m(t)}{dt} \end{bmatrix}_{t=0}$$

$$= \left[ \left(D_1\left(f \circ (\varphi_U)^{-1}\right)\right)(\varphi_U(p)) \quad \cdots \quad \left(D_m\left(f \circ (\varphi_U)^{-1}\right)\right)(\varphi_U(p)) \right] \begin{bmatrix} \dfrac{dF_1(t)}{dt} \\ \vdots \\ \dfrac{dF_m(t)}{dt} \end{bmatrix}_{t=0}$$

$$= [0 \quad \cdots \quad 0] \begin{bmatrix} \dfrac{dF_1(t)}{dt} \\ \vdots \\ \dfrac{dF_m(t)}{dt} \end{bmatrix}_{t=0} = \left[0 \cdot \left(\dfrac{dF_1(t)}{dt}\right) + \ldots + 0 \cdot \left(\dfrac{dF_m(t)}{dt}\right)\right]_{t=0} = 0 = \mathrm{RHS}.$$

$\square$

*Proof* (only if part): Let $[f] \in \mathcal{H}_p(M)$. We have to show that

$$\left(D_1\left(f \circ (\varphi_U)^{-1}\right)\right)(\varphi_U(p)) = \cdots = \left(D_m\left(f \circ (\varphi_U)^{-1}\right)\right)(\varphi_U(p)) = 0.$$

If not, otherwise, let $(D_1(f \circ (\varphi_U)^{-1}))(\varphi_U(p))$ (for simplicity) be nonzero. We have to arrive at a contradiction. Let us define a function $\gamma_1 : (-1, 1) \to \mathbb{R}^m$ as follows: For every $t$ in the open interval $(-1, 1)$,

$$\gamma_1(t) \equiv (t, 0, \ldots, 0) + \varphi_U(p).$$

Put

$$\gamma \equiv (\varphi_U)^{-1} \circ \gamma_1.$$

We shall try to prove that

1. $\gamma(0) = p$,
2. $0$ is an interior point of the domain of $\gamma$,
3. $\gamma$ is a smooth map from $(-\delta, \delta)$ to $M$ for some $\delta > 0$.

For 1: Here,

$$\text{LHS} = \gamma(0) = \left( (\varphi_U)^{-1} \circ \gamma_1 \right)(0) = (\varphi_U)^{-1}(\gamma_1(0)) = (\varphi_U)^{-1}((0,0,\ldots,0) + \varphi_U(p))$$
$$= (\varphi_U)^{-1}(\varphi_U(p)) = p = \text{RHS}.$$

This proves 1.

For 2: By the definition of $\gamma_1$, the function $\gamma_1 : (-1,1) \to \mathbb{R}^m$ is continuous. Since $\gamma_1 : (-1,1) \to \mathbb{R}^m$ is continuous, and 0 is in the open interval $(-1,1)$, $\gamma_1$ is continuous at 0.

Since $(U, \varphi_U)$ is a coordinate chart of $M$, $(\varphi_U)^{-1} : \varphi_U(U) \to U$ is a 1–1, onto, continuous function, and $\varphi_U(U)$ is an open subset of $\mathbb{R}^m$. Since $p \in U$, $\varphi_U(p) \in \varphi_U(U)$. Since $\gamma_1(0) = (0,0,\ldots,0) + \varphi_U(p) = \varphi_U(p) \in \varphi_U(U)$, and $(\varphi_U)^{-1} : \varphi_U(U) \to U$ is continuous, $(\varphi_U)^{-1} : \varphi_U(U) \to U$ is continuous at the point $\gamma_1(0)$. Since $\gamma_1(0) \in \varphi_U(U)$, and $\varphi_U(U)$ is an open subset of $\mathbb{R}^m$, $\varphi_U(U)$ is an open neighborhood of $\gamma_1(0)$. Since $\gamma_1 : (-1,1) \to \mathbb{R}^m$ is continuous, $\varphi_U(U)$ is an open neighborhood of $\gamma_1(0)$, there exists $\delta > 0$ such that $\delta < 1$ and, for every $t$ in $(-\delta, \delta)$ we have $\gamma_1(t) \in \varphi_U(U)$, and hence, $\gamma(t) = ((\varphi_U)^{-1} \circ \gamma_1)(t) = (\varphi_U)^{-1}(\gamma_1(t)) \in U$. Since $\gamma(t) \in U$ for every $t$ in $(-\delta, \delta)$, it follows that $(-\delta, \delta)$ is a subset of the $\text{dom}(\gamma)$. Hence, 0 is an interior point of the domain of the function $\gamma$. This proves 2.

For 3: In 2, we have seen that $\gamma$ is defined over $(-\delta, \delta)$. Now, it remains to be proved that $\gamma$ is a smooth map from $(-\delta, \delta)$ to $M$, that is, $\gamma$ is $C^\infty$ at every point $t$ in $(-\delta, \delta)$.

Here, $\gamma$ is defined over $(-\delta, \delta)$, so $\gamma(t)$ is meaningful for every $t$ in $(-\delta, \delta)$. Since $\gamma(t)$ is meaningful for every $t$ in $(-\delta, \delta)$, and

$$\gamma(t) = \left( (\varphi_U)^{-1} \circ \gamma_1 \right)(t) = (\varphi_U)^{-1}(\gamma_1(t)) = (\varphi_U)^{-1}((t,0,\ldots,0) + \varphi_U(p)),$$

$(\varphi_U)^{-1}((t,0,\ldots,0) + \varphi_U(p))$ is meaningful for every $t$ in $(-\delta, \delta)$. Since $(U, \varphi_U)$ is a coordinate chart of $M$, $(\varphi_U)^{-1} : \varphi_U(U) \to U$. Since $(\varphi_U)^{-1} : \varphi_U(U) \to U$, and $(\varphi_U)^{-1}((t,0,\ldots,0) + \varphi_U(p))$ is meaningful for every $t$ in $(-\delta, \delta)$, $(\varphi_U)^{-1}((t,0,\ldots,0) + \varphi_U(p))$ is in $U$ for every $t$ in $(-\delta, \delta)$. Since $(\varphi_U)^{-1}((t,0,\ldots,0) + \varphi_U(p))$ is in $U$ for every $t$ in $(-\delta, \delta)$, and $\gamma(t) = (\varphi_U)^{-1}((t,0,\ldots,0) + \varphi_U(p))$, $\gamma(t)$ is in $U$ for every $t$ in $(-\delta, \delta)$. Thus, $\gamma$ maps $(-\delta, \delta)$ to $U$.

Now, it remains to be proved that $\gamma$ is a smooth map from $(-\delta, \delta)$ to $M$, that is, $\gamma$ is $C^\infty$ at every point $t$ in $(-\delta, \delta)$. For this purpose, let us fix any $t$ in $(-\delta, \delta)$. From the definition of $\gamma_1$, $\gamma_1$ is $C^\infty$ at $t$. Since $t$ is in $(-\delta, \delta)$, and $\gamma$ maps $(-\delta, \delta)$ to $U$, $\gamma(t)$ is in $U$. Since $\gamma_1$ is $C^\infty$ at $t$, $\gamma = (\varphi_U)^{-1} \circ \gamma_1$, $\gamma$ maps $(-\delta, \delta)$ to $U$, and $t$ is in $(-\delta, \delta)$, $\varphi_U \circ \gamma$ is $C^\infty$ at $t$. Since $\varphi_U \circ \gamma$ is $C^\infty$ at $t$, and $(U, \varphi_U)$ is an admissible coordinate chart of $M$ satisfying $\gamma(t) \in U$, by Theorem 2.2, $\gamma$ is $C^\infty$ at $t$ in $(-\delta, \delta)$. This proves 3.

From 1, 2, and 3, we find that $\gamma$ is a parametrized curve in $M$ through $p$, that is, $\gamma$ is in $\Gamma_p(M)$. Since $[f] \in \mathcal{H}_p(M)$, and $\gamma$ is in $\Gamma_p(M)$, by the definition of $\mathcal{H}_p(M)$, $\ll \gamma, [f] \gg = 0$. Put

$$\varphi_U(p) \equiv (a_1, \ldots, a_m).$$

Next, let us define $m$ functions $F_1 : (-\delta, \delta) \to \mathbb{R}$, ..., $F_m : (-\delta, \delta) \to \mathbb{R}$ such that for every $t$ in $(-\delta, \delta)$,

$$\begin{aligned}(F_1(t), \ldots, F_m(t)) &\equiv (\varphi_U \circ \gamma)(t) = \gamma_1(t) = (t, 0, \ldots, 0) + \varphi_U(p) \\ &= (t, 0, \ldots, 0) + (a_1, \ldots, a_m) = (t + a_1, a_2, \ldots, a_m).\end{aligned}$$

Hence,

$$0 = \ll \gamma, [f] \gg = \left. \frac{\mathrm{d}(f \circ \gamma)(t)}{\mathrm{d}t} \right|_{t=0} = \left. \frac{\mathrm{d}\left( \left( f \circ (\varphi_U)^{-1} \right) \circ (\varphi_U \circ \gamma) \right)(t)}{\mathrm{d}t} \right|_{t=0}$$

$$= \left[ \left( D_1 \left( f \circ (\varphi_U)^{-1} \right) \right)((\varphi_U \circ \gamma)(0)) \quad \cdots \quad \left( D_m \left( f \circ (\varphi_U)^{-1} \right) \right)((\varphi_U \circ \gamma)(0)) \right] \left. \begin{bmatrix} \dfrac{\mathrm{d}(t + a_1)}{\mathrm{d}t} \\ \dfrac{\mathrm{d}(a_2)}{\mathrm{d}t} \\ \vdots \\ \dfrac{\mathrm{d}(a_m)}{\mathrm{d}t} \end{bmatrix} \right|_{t=0}$$

$$= \left[ \left( D_1 \left( f \circ (\varphi_U)^{-1} \right) \right)(\varphi_U(p)) \quad \cdots \quad \left( D_m \left( f \circ (\varphi_U)^{-1} \right) \right)(\varphi_U(p)) \right] \left. \begin{bmatrix} 1 \\ 0 \\ \vdots \\ 0 \end{bmatrix} \right|_{t=0}$$

$$= \left( D_1 \left( f \circ (\varphi_U)^{-1} \right) \right)(\varphi_U(p)) \cdot 1 + 0 + \cdots + 0 = \left( D_1 \left( f \circ (\varphi_U)^{-1} \right) \right)(\varphi_U(p)) \neq 0,$$

which is a contradiction.                                                                                  □

**Theorem 2.40** *Let $M$ be a $m$-dimensional smooth manifold. Let $p$ be an element of $M$. Let $f^1, f^2 \in C_p^\infty(M)$. Let $G$ be an open neighborhood of $(f^1(p), f^2(p))$ in $\mathbb{R}^2$. Let $F : G \to \mathbb{R}$ be a smooth function. Then, there exists $f$ in $C_p^\infty(M)$ such that for every $x$ in $\mathrm{dom} f$,*

$$f(x) = F(f^1(x), f^2(x)).$$

*Proof* For this purpose, by the definition of $C_p^\infty(M)$, we must find a function $f$ such that

1. $\mathrm{dom} f$ is an open neighborhood of $p$ in $M$,
2. $f(x) = F(f^1(x), f^2(x))$ for every $x$ in $\mathrm{dom} f$, and
3. for every admissible coordinate chart $(U, \varphi_U)$ of $M$ satisfying $p \in U$,

$$\left(f \circ (\varphi_U)^{-1}\right) : \varphi_U((\mathrm{dom} f) \cap U) \to \mathbb{R}$$

is $C^\infty$ at the point $\varphi_U(p)$ in $\mathbb{R}^m$.

Since $f^1 \in C_p^\infty(M)$, by the definition of $C_p^\infty(M)$, $f^1$ is a real-valued function whose $\mathrm{dom}(f^1)$ is an open neighborhood of $p$ in $M$. Similarly, $f^2$ is a real-valued function whose $\mathrm{dom}(f^2)$ is an open neighborhood of $p$ in $M$. Since $\mathrm{dom}(f^1)$ is an open neighborhood of $p$ in $M$, and $\mathrm{dom}(f^2)$ is an open neighborhood of $p$ in $M$, their intersection $\mathrm{dom}(f^1) \cap \mathrm{dom}(f^2)$ is an open neighborhood of $p$ in $M$. Put

$$V_1 \equiv \mathrm{dom}(f^1) \cap \mathrm{dom}(f^2).$$

Let us define a function

$$g : V_1 \to \mathbb{R}^2$$

as follows: For every $x$ in $V_1$

$$g(x) \equiv \left(f^1(x), f^2(x)\right).$$

Since $f^1 \in C_p^\infty(M)$, $f^1$ is continuous at $p$. Similarly, $f^2$ is continuous at $p$. Since $f^1, f^2$ are continuous at $p$, and $g(x) = (f^1(x), f^2(x))$ for every $x$ in $V_1$, $g : V_1 \to \mathbb{R}^2$ is continuous at $p$. Since $g : V_1 \to \mathbb{R}^2$ is continuous at $p$, and $G$ is an open neighborhood of $(f^1(p), f^2(p))(= g(p))$ in $\mathbb{R}^2$, there exists an open neighborhood $V(\subset V_1 = \mathrm{dom}(f^1) \cap \mathrm{dom}(f^2))$ of $p$ such that $g(V)$ is contained in $G$. Since $g(V)$ is contained in $G$, and $F : G \to \mathbb{R}$, for every $x$ in $V$, $F(g(x))(= F(f^1(x), f^2(x)))$ is a real number.

Now, let us define a function

$$f : V \to \mathbb{R}$$

as follows: For every $x$ in $V$,

$$f(x) \equiv F\left(f^1(x), f^2(x)\right).$$

Clearly, the conditions 1 and 2 are satisfied.

For 3, let us take an admissible coordinate chart $(U, \varphi_U)$ of $M$ satisfying $p \in U$. We have to show that the function

$$\left(f \circ (\varphi_U)^{-1}\right) : \varphi_U(V \cap U) \to \mathbb{R}$$

is $C^\infty$ at the point $\varphi_U(p)$ in $\mathbb{R}^m$. Here, for every $\varphi_U(x)$ in $\varphi_U(V \cap U)$, where $x$ is in $V \cap U$, we have

$$\left(f \circ (\varphi_U)^{-1}\right)(\varphi_U(x))$$
$$= \left(\left(f \circ (\varphi_U)^{-1}\right) \circ \varphi_U\right)(x) = \left(f \circ \left((\varphi_U)^{-1} \circ \varphi_U\right)\right)(x) = f(x)$$
$$= F(f^1(x), f^2(x)) = F\left(f^1\left(\left((\varphi_U)^{-1} \circ \varphi_U\right)(x)\right), f^2\left(\left((\varphi_U)^{-1} \circ \varphi_U\right)(x)\right)\right)$$
$$= F\left(\left(\left(f^1 \circ \left((\varphi_U)^{-1} \circ \varphi_U\right)\right)(x)\right), \left(\left(f^2 \circ \left((\varphi_U)^{-1} \circ \varphi_U\right)\right)(x)\right)\right)$$
$$= F\left(\left(\left(f^1 \circ (\varphi_U)^{-1}\right) \circ \varphi_U\right)(x), \left(\left(f^2 \circ (\varphi_U)^{-1}\right) \circ \varphi_U\right)(x)\right)$$
$$= F\left(\left(f^1 \circ (\varphi_U)^{-1}\right)(\varphi_U(x)), \left(f^2 \circ (\varphi_U)^{-1}\right)(\varphi_U(x))\right).$$

Since $f^1 \in C_p^\infty(M)$, and $(U, \varphi_U)$ is an admissible coordinate chart $(U, \varphi_U)$ of $M$ satisfying $p \in U$, by the definition of $C_p^\infty(M)$,

$$\left(f^1 \circ (\varphi_U)^{-1}\right) : \varphi_U\left(\mathrm{dom}(f^1) \cap U\right) \to \mathbb{R}$$

is $C^\infty$ at the point $\varphi_U(p)$ in $\mathbb{R}^m$. Since $(\mathrm{dom}f) \cap U = V \cap U \subset \mathrm{dom}(f^1)$ $\cap \mathrm{dom}(f^2) \cap U \subset \mathrm{dom}(f^1) \cap U$, $\varphi_U((\mathrm{dom}f) \cap U) \subset \varphi_U((\mathrm{dom}(f^1)) \cap U)$. Since $\varphi_U$ $((\mathrm{dom}f) \cap U) \subset \varphi_U((\mathrm{dom}(f^1)) \cap U)$, and $(f^1 \circ (\varphi_U)^{-1}) : \varphi_U(\mathrm{dom}(f^1) \cap U) \to \mathbb{R}$ is $C^\infty$ at the point $\varphi_U(p)$ in $\mathbb{R}^m$, the restriction of $f^1 \circ (\varphi_U)^{-1}$ to $\varphi_U((\mathrm{dom}\, f) \cap U)$ is $C^\infty$ at the point $\varphi_U(p)$ in $\mathbb{R}^m$. Therefore, for every $x$ in $(\mathrm{dom}\, f) \cap U$, $\varphi_U(x) \mapsto \left(f^1 \circ (\varphi_U)^{-1}\right)(\varphi_U(x))$ is $C^\infty$ at the point $\varphi_U(p)$ in $\mathbb{R}^m$. Similarly, for every $x$ in $(\mathrm{dom}\, f) \cap U$, $\varphi_U(x) \mapsto \left(f^2 \circ (\varphi_U)^{-1}\right)(\varphi_U(x))$ is $C^\infty$ at the point $\varphi_U(p)$ in $\mathbb{R}^m$. It follows that, for every $x$ in $(\mathrm{dom}\, f) \cap U$, $\varphi_U(x) \mapsto \left(\left(f^1 \circ (\varphi_U)^{-1}\right)(\varphi_U(x)),\right.$ $\left(f^2 \circ (\varphi_U)^{-1}\right)(\varphi_U(x)))$ is $C^\infty$ at the point $\varphi_U(p)$ in $\mathbb{R}^m$. Since, for every $x$ in $(\mathrm{dom}f) \cap U$, $\varphi_U(x) \mapsto ((f^1 \circ (\varphi_U)^{-1})(\varphi_U(x)), (f^2 \circ (\varphi_U)^{-1})(\varphi_U(x)))$ is $C^\infty$ at the point $\varphi_U(p)$ in $\mathbb{R}^m$, $G$ is an open neighborhood of $(f^1(p), f^2(p))$ in $\mathbb{R}^2$, and $F : G \to \mathbb{R}$ is a smooth function, for every $x$ in $(\mathrm{dom}f) \cap U (= V \cap U)$, the function $\varphi_U(x) \mapsto F((f^1 \circ (\varphi_U)^{-1})(\varphi_U(x)), (f^2 \circ (\varphi_U)^{-1})(\varphi_U(x))) (= (f \circ (\varphi_U)^{-1})(\varphi_U(x)))$ is $C^\infty$ at the point $\varphi_U(p)$ in $\mathbb{R}^m$, and hence,

$$\left(f \circ (\varphi_U)^{-1}\right) : \varphi_U(V \cap U) \to \mathbb{R}$$

is $C^\infty$ at the point $\varphi_U(p)$ in $\mathbb{R}^m$. This proves 3.                  $\square$

**Theorem 2.41** *Let $M$ be an $m$-dimensional smooth manifold. Let $p$ be an element of $M$. Let $f^1, f^2 \in C_p^\infty(M)$. Let $G$ be an open neighborhood of $(f^1(p), f^2(p))$ in $\mathbb{R}^2$. Let $F : G \to \mathbb{R}$ be a smooth function. Let $f, g$ be in $C_p^\infty(M)$ such that for every $x$ in* dom $f$,

$$f(x) = F(f^1(x), f^2(x))$$

*and, for every $x$ in* dom $g$,

$$g(x) = F(f^1(x), f^2(x)).$$

  *Then, $f \sim g$.*

*Proof* By Theorem 2.40, the existence of $f, g$ are guaranteed. Since $f$ is in $C_p^\infty(M)$, dom $f$ is an open neighborhood of $p$ in $M$. Similarly, dom $g$ is an open neighborhood of $p$ in $M$. Since dom $f$ is an open neighborhood of $p$ in $M$, and dom $g$ is an open neighborhood of $p$ in $M$, their intersection (dom $f$) $\cap$ (dom $g$) is an open neighborhood of $p$ in $M$. Next, let us take any $x$ in (dom $f$) $\cap$(dom $g$). It remains to be showed that $f(x) = g(x)$. Since $x$ is in (dom $f$) $\cap$(dom $g$), $x$ is in dom $f$. Since $x$ is in dom $f$, $f(x) = F(f^1(x), f^2(x))$. Similarly, $g(x) = F(f^1(x), f^2(x))$. Since $f(x) = F(f^1(x), f^2(x))$ and $g(x) = F(f^1(x), f^2(x))$, $f(x) = g(x)$.                  $\square$

**Theorem 2.42** *Let $M$ be an $m$-dimensional smooth manifold. Let $p$ be an element of $M$. Let $f^1, f^2 \in C_p^\infty(M)$. Let $G$ be an open neighborhood of $(f^1(p), f^2(p))$ in $\mathbb{R}^2$. Let $F : G \to \mathbb{R}$ be a smooth function. Then, there exists a unique $[f]$ in $\mathcal{F}_p(M)$, where $f$ is in $C_p^\infty(M)$, such that for every $x$ in* dom $f$,

$$f(x) = F(f^1(x), f^2(x)).$$

  *Also*

$$[f]^\sim = \left((D_1 F)(f^1(p), f^2(p))\right)[f^1]^\sim + \left((D_2 F)(f^1(p), f^2(p))\right)[f^2]^\sim.$$

*Proof* **Existence**: By Theorem 2.40, there exists $f$ in $C_p^\infty(M)$ such that for every $x$ in dom $f$,

$$f(x) = F(f^1(x), f^2(x)).$$

Further, since $f$ is in $C_p^\infty(M)$, the $C^\infty$-germ $[f]$ is in $\mathcal{F}_p(M)$. This proves the existence part of the theorem.

**Uniqueness**: For this purpose, let $[f], [g]$ be in $\mathcal{F}_p(M)$, where $f, g$ are in $C_p^\infty(M)$, such that for every $x$ in $\mathrm{dom} f$,

$$f(x) = F(f^1(x), f^2(x))$$

and, for every $x$ in $\mathrm{dom} g$,

$$g(x) = F(f^1(x), f^2(x)).$$

We have to prove that $[f] = [g]$.

By Theorem 2.41, $f \sim g$. Since $f, g$ are in $C_p^\infty(M)$, $f \sim g$, and $\sim$ is an equivalence relation over $C_p^\infty(M)$, the equivalence class $[f]$ determined by $f$, and the equivalence class $[g]$ determined by $g$ are equal, that is, $[f] = [g]$. This proves the uniqueness part of the theorem.                                                                                    $\square$

Since $[f] \in \mathcal{F}_p(M)$, the cotangent vector $[f]^\sim$ (that is, $[f] + \mathcal{H}_p$) on $M$ at $p$ determined by the germ $[f]$ is a member of the quotient space $\frac{\mathcal{F}_p}{\mathcal{H}_p}$ (that is, the cotangent space $T_p^*(M)$ of $M$ at $p$).

Next, since $f^1 \in C_p^\infty(M)$, the equivalence class $[f^1]$ determined by $f^1$ is a member of $\mathcal{F}_p(M)$, and hence, the cotangent vector $[f^1]^\sim$ on $M$ at $p$ determined by the term $[f^1]$ is a member of the quotient space $\frac{\mathcal{F}_p}{\mathcal{H}_p}$. Similarly, the cotangent vector $[f^2]^\sim$ on $M$ at $p$ determined by the term $[f^2]$ is a member of the quotient space $\frac{\mathcal{F}_p}{\mathcal{H}_p}$.

Since $[f^1]^\sim, [f^2]^\sim$ are in $\frac{\mathcal{F}_p}{\mathcal{H}_p}$, and $\frac{\mathcal{F}_p}{\mathcal{H}_p}$ is a real linear space, the linear combination $((D_1 F)(f^1(p), f^2(p)))[f^1]^\sim + ((D_2 F)(f^1(p), f^2(p)))[f^2]^\sim$ of $[f^1]^\sim, [f^2]^\sim$ is in $\frac{\mathcal{F}_p}{\mathcal{H}_p}$.

Finally, we have to show that

$$[f]^\sim = ((D_1 F)(f^1(p), f^2(p)))[f^1]^\sim + ((D_2 F)(f^1(p), f^2(p)))[f^2]^\sim.$$

Since

$$\begin{aligned} & ((D_1 F)(f^1(p), f^2(p)))[f^1]^\sim + ((D_2 F)(f^1(p), f^2(p)))[f^2]^\sim \\ &= [((D_1 F)(f^1(p), f^2(p)))f^1]^\sim + [((D_2 F)(f^1(p), f^2(p)))f^2]^\sim \\ &= [((D_1 F)(f^1(p), f^2(p)))f^1 + ((D_2 F)(f^1(p), f^2(p)))f^2]^\sim \end{aligned}$$

we have to show that

$$[f]^\sim = [((D_1 F)(f^1(p), f^2(p)))f^1 + ((D_2 F)(f^1(p), f^2(p)))f^2]^\sim,$$

that is,

$$f + \mathcal{H}_p = \left( \left( (D_1F)\left(f^1(p), f^2(p)\right)\right)f^1 + \left((D_2F)\left(f^1(p), f^2(p)\right)\right)f^2 \right) + \mathcal{H}_p,$$

that is,

$$\left( \left((D_1F)\left(f^1(p), f^2(p)\right)\right)f^1 + \left((D_2F)\left(f^1(p), f^2(p)\right)\right)f^2 \right) - f \in \mathcal{H}_p,$$

that is, for every $\gamma$ in $\Gamma_p(M)$,

$$\ll \gamma, \left[ \left( \left((D_1F)\left(f^1(p), f^2(p)\right)\right)f^1 + \left((D_2F)\left(f^1(p), f^2(p)\right)\right)f^2 \right) - f \right] \gg = 0,$$

that is, for every $\gamma$ in $\Gamma_p(M)$,

$$\ll \gamma, \left((D_1F)\left(f^1(p), f^2(p)\right)\right)\left[f^1\right] + \left((D_2F)\left(f^1(p), f^2(p)\right)\right)\left[f^2\right] - [f] \gg = 0,$$

that is, for every $\gamma$ in $\Gamma_p(M)$,

$$\left((D_1F)\left(f^1(p), f^2(p)\right)\right)\ll \gamma, \left[f^1\right] \gg + \left((D_2F)\left(f^1(p), f^2(p)\right)\right)\ll \gamma, \left[f^2\right] \gg$$
$$- \ll \gamma, [f] \gg = 0,$$

that is, for every $\gamma$ in $\Gamma_p(M)$,

$$\left((D_1F)\left(f^1(p), f^2(p)\right)\right)\ll \gamma, \left[f^1\right] \gg + \left((D_2F)\left(f^1(p), f^2(p)\right)\right)\ll \gamma, \left[f^2\right] \gg$$
$$= \ll \gamma, [f] \gg.$$

For this purpose, let us take any $\gamma$ in $\Gamma_p(M)$. Since $\gamma$ is in $\Gamma_p(M)$, there exists a real number $\delta > 0$ such that $\gamma$ is defined on the open interval $(-\delta, \delta)$, and $\gamma(0) = p$. We have to prove that

$$\left((D_1F)\left(f^1(p), f^2(p)\right)\right)\ll \gamma, \left[f^1\right] \gg + \left((D_2F)\left(f^1(p), f^2(p)\right)\right)\ll \gamma, \left[f^2\right] \gg$$
$$= \ll \gamma, [f] \gg.$$

$$\text{LHS} = \left((D_1F)\left(f^1(p), f^2(p)\right)\right)\ll \gamma, \left[f^1\right] \gg + \left((D_2F)\left(f^1(p), f^2(p)\right)\right)\ll \gamma, \left[f^2\right] \gg$$

$$= \left((D_1F)\left(f^1(p), f^2(p)\right)\right)\frac{\mathrm{d}(f^1 \circ \gamma)(t)}{\mathrm{d}t}\bigg|_{t=0} + \left((D_2F)\left(f^1(p), f^2(p)\right)\right)\frac{\mathrm{d}(f^2 \circ \gamma)(t)}{\mathrm{d}t}\bigg|_{t=0}$$

$$= \left((D_1F)\left(f^1(p), f^2(p)\right)\right)\frac{\mathrm{d}(f^1(\gamma(t)))}{\mathrm{d}t}\bigg|_{t=0} + \left((D_2F)\left(f^1(p), f^2(p)\right)\right)\frac{\mathrm{d}(f^2(\gamma(t)))}{\mathrm{d}t}\bigg|_{t=0}.$$

$$\text{RHS} = \ll \gamma, [f] \gg = \frac{\mathrm{d}(f \circ \gamma)(t)}{\mathrm{d}t}\bigg|_{t=0} = \frac{\mathrm{d}(f(\gamma(t)))}{\mathrm{d}t}\bigg|_{t=0} = \frac{\mathrm{d}(F(f^1(\gamma(t)), f^2(\gamma(t))))}{\mathrm{d}t}\bigg|_{t=0}$$

$$= \frac{\mathrm{d}(F((f^1 \circ \gamma)(t), (f^2 \circ \gamma)(t)))}{\mathrm{d}t}\bigg|_{t=0}$$

$$= \left((D_1F)\left((f^1 \circ \gamma)(0), (f^2 \circ \gamma)(0)\right)\right)\frac{d(f^1 \circ \gamma)(t)}{dt}\bigg|_{t=0}$$

$$+ \left((D_2F)\left((f^1 \circ \gamma)(0), (f^2 \circ \gamma)(0)\right)\right)\frac{d(f^2 \circ \gamma)(t)}{dt}\bigg|_{t=0}$$

$$= \left((D_1F)\left(f^1(\gamma(0)), f^2(\gamma(0))\right)\right)\frac{d(f^1(\gamma(t)))}{dt}\bigg|_{t=0}$$

$$+ \left((D_2F)\left(f^1(\gamma(0)), f^2(\gamma(0))\right)\right)\frac{d(f^2(\gamma(t)))}{dt}\bigg|_{t=0}$$

$$= \left((D_1F)\left(f^1(p), f^2(p)\right)\right)\frac{d(f^1(\gamma(t)))}{dt}\bigg|_{t=0} + \left((D_2F)\left(f^1(p), f^2(p)\right)\right)\frac{d(f^2(\gamma(t)))}{dt}\bigg|_{t=0}.$$

Hence, LHS = RHS. $\square$

**Note 2.43** Since $[f]^\sim$ is also denoted by $(df)_p$, the formula in Theorem 2.17 can be written as:

$$(df)_p = \left((D_1F)\left(f^1(p), f^2(p)\right)\right)(df^1)_p + \left((D_2F)\left(f^1(p), f^2(p)\right)\right)(df^2)_p.$$

Since $(D_1F)(f^1(p), f^2(p))$ is classically written as $\frac{\partial F}{\partial x^1}(f^1(p), f^2(p))$, etc., we can write:

$$\left(dF(f^1, f^2)\right)_p = (df)_p = \left(\frac{\partial F}{\partial x^1}(f^1(p), f^2(p))\right)(df^1)_p + \left(\frac{\partial F}{\partial x^1}(f^1(p), f^2(p))\right)(df^2)_p.$$

**Note 2.44** Since the multiplication operation of real numbers is a smooth function from $\mathbb{R}^2$ to $\mathbb{R}$, in Theorem 2.42, we can take multiplication operation in place of $F$. So $\frac{\partial F}{\partial x^1}(f^1(p), f^2(p))$ becomes $f^2(p)$, and $\frac{\partial F}{\partial x^2}(f^1(p), f^2(p))$ becomes $f^1(p)$. Hence, we get

$$\left(d(f^1 \cdot f^2)\right)_p = (df)_p = \left(\frac{\partial F}{\partial x^1}(f^1(p), f^2(p))\right)(df^1)_p + \left(\frac{\partial F}{\partial x^2}(f^1(p), f^2(p))\right)(df^2)_p$$

$$= (f^2(p))(df^1)_p + (f^1(p))(df^2)_p$$

or,

$$\left(d(f^1 \cdot f^2)\right)_p = (f^2(p))(df^1)_p + (f^1(p))(df^2)_p.$$

Similarly, for every $f^1, f^2, f^3$ in $C_p^\infty(M)$, we have

$$\left(d(f^1 \cdot f^2 \cdot f^3)\right)_p = (f^2(p))(f^3(p))(df^1)_p + (f^3(p))(f^1(p))(df^2)_p + (f^1(p))(f^2(p))(df^3)_p,$$

etc. If we recollect our previous results, we get the following theorem.

**Theorem 2.45** *Let M be an m-dimensional smooth manifold. Let p be an element of M. Let $f, g \in C_p^\infty(M)$. Then,*

1. $(d(f + g))_p = (df)_p + (dg)_p$,
2. $(d(tf))_p = t((df)_p)$, *for every real t,*
3. $(d(f \cdot g))_p = (g(p))(df)_p + (f(p))(dg)_p$.

*Proof* Proofs have already been supplied.                                   □

**Theorem 2.46** *Let M be an m-dimensional smooth manifold. Let p be an element of M. Let $f^1, f^2, f^3 \in C_p^\infty(M)$. Let G be an open neighborhood of $(f^1(p), f^2(p), f^3(p))$ in $\mathbb{R}^3$. Let $F : G \to \mathbb{R}$ be a smooth function. Then, there exists a unique $[f]$ in $\mathcal{F}_p(M)$, where f is in $C_p^\infty(M)$, such that for every x in $\mathrm{dom} f$,*

$$f(x) = F(f^1(x), f^2(x), f^3(x)).$$

*Also*

$$[f]^\sim = ((D_1F)(f^1(p), f^2(p), f^3(p)))[f^1]^\sim + ((D_2F)(f^1(p), f^2(p), f^3(p)))[f^2]^\sim$$
$$+ ((D_3F)(f^1(p), f^2(p), f^3(p)))[f^3]^\sim.$$

*Proof* Its proof is quite similar to the proof of Theorem 2.42, etc.         □

**Theorem 2.47** *Let M be a 2-dimensional smooth manifold. Let p be an element of M. Let $(U, \varphi_U)$ be an admissible coordinate chart of M satisfying $p \in U$. Let $u^1, u^2$ be the component functions of $\varphi_U$. Then,*

1. $u^1, u^2$ *are in* $C_p^\infty(M)$,
2. $[u^1]^\sim, [u^2]^\sim$ *are linearly independent in the real linear space* $T_p^*(M)$,
3. *for every $[f]^\sim$ in $T_p^*(M)$, where f is in $C_p^\infty(M)$,*

$$[f]^\sim = \left(\left(D_1\left(f \circ (\varphi_U)^{-1}\right)\right)(u^1(p), u^2(p))\right)[u^1]^\sim + \left(\left(D_2\left(f \circ (\varphi_U)^{-1}\right)\right)(u^1(p), u^2(p))\right)[u^2]^\sim.$$

*In other words,*

$$(df)_p = \left(\left(\frac{\partial}{\partial u^1}\bigg|_p\right)[f]\right)((du^1)_p) + \left(\left(\frac{\partial}{\partial u^2}\bigg|_p\right)[f]\right)((du^2)_p).$$

4. $\{[u^1]^\sim, [u^2]^\sim\}$ is a basis of the real linear space $T_p^*(M)$,
5. $\dim(T_p^*(M)) = 2$.

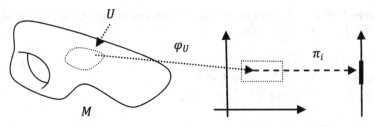

$$u^i \equiv \pi_i \circ \varphi_U$$

**Fig. 2.2** Cotangent space

*Proof*

1. This fact has been shown earlier.
2. In 1, we have seen that $u^1$ is in $C_p^\infty(M)$. Since $u^1$ is in $C_p^\infty(M)$, the $C^\infty$-germ $[f]$ is in $\mathcal{F}_p(M)$, and hence, $[u^1]^\sim (= [u^1] + \mathcal{H}_p)$ is in the cotangent space $T_p^*$ of $M$ at $p$. Similarly, $[u^2]^\sim (= [u^2] + \mathcal{H}_p)$ is in the cotangent space $T_p^*$ of $M$ at $p$ (see Fig. 2.2).

We have to show that $[u^1]^\sim, [u^2]^\sim$ are linearly independent. For this purpose, let $a_1[u^1]^\sim + a_2[u^2]^\sim = 0$, where $a_1, a_2$ are real numbers. We have to show that $a_1 = 0$, and $a_2 = 0$. Since $0 = a_1[u^1]^\sim + a_2[u^2]^\sim = [a_1u^1]^\sim + [a_2u^2]^\sim = [a_1u^1 + a_2u^2]^\sim$, $[a_1u^1 + a_2u^2]$ is in $\mathcal{H}_p$, and hence, by the definition of $\mathcal{H}_p$, $\ll \gamma, [a_1u^1 + a_2u^2] \gg = 0$ for every $\gamma$ in $\Gamma_p(M)$. Therefore, for every $\gamma$ in $\Gamma_p(M)$,

$$0 = \ll \gamma, [a_1u^1 + a_2u^2] \gg = \ll \gamma, [a_1u^1] + [a_2u^2] \gg = \ll \gamma, a_1[u^1] + a_2[u^2] \gg$$
$$= a_1 \ll \gamma, [u^1] \gg + a_2 \ll \gamma, [u^2] \gg = a_1 \frac{d(u^1 \circ \gamma)(t)}{dt}\Big|_{t=0} + a_2 \frac{d(u^2 \circ \gamma)(t)}{dt}\Big|_{t=0},$$

that is,

$$a_1 \frac{d(u^1 \circ \gamma)(t)}{dt}\Big|_{t=0} + a_2 \frac{d(u^2 \circ \gamma)(t)}{dt}\Big|_{t=0} = 0 \ldots (*)$$

for every $\gamma$ in $\Gamma_p(M)$.

Let us define a function

$$\gamma_1 : \mathbb{R} \to \mathbb{R}^2$$

as follows: For every $t$ in $\mathbb{R}$,

$$\gamma_1(t) \equiv \left( u^1(p) + t, u^2(p) \right).$$

Clearly, $\gamma_1$ is a smooth function.

Here, $\gamma_1$ is continuous, $\gamma_1(0) = (u^1(p) + 0, u^2(p)) = (u^1(p), u^2(p)) = \varphi_U(p)$, and $\varphi_U(U)$ is an open neighborhood of $\varphi_U(p)$, so there exists $\varepsilon > 0$ such that for every $t$ in the open interval $(-\varepsilon, \varepsilon)$, we have $\gamma_1(t)$ is in $\varphi_U(U)$, and hence, $(\varphi_U)^{-1}(\gamma_1(t))$ is in $U$.

Now, let us define a function

$$\gamma : (-\varepsilon, \varepsilon) \to M$$

as follows: For every $t$ in $(-\varepsilon, \varepsilon)$,

$$\gamma(t) \equiv (\varphi_U)^{-1}(\gamma_1(t)) = \left((\varphi_U)^{-1} \circ \gamma_1\right)(t).$$

We shall try to see that $\gamma$ is in $\Gamma_p(M)$. By the definition of $\Gamma_p(M)$, we must prove that

1. $\gamma(0) = p$,
2. $\gamma$ is a parametrized curve in the manifold $M$, that is, $\gamma : (-\varepsilon, \varepsilon) \to M$ is a smooth map from $(-\varepsilon, \varepsilon)$ to $M$, that is, $\gamma$ is $C^\infty$ at every point $p$ of $(-\varepsilon, \varepsilon)$.

For 1: Here, LHS $= \gamma(0) = (\varphi_U)^{-1}(\gamma_1(0)) = (\varphi_U)^{-1}(u^1(p) + 0, u^2(p)) = (\varphi_U)^{-1}(u^1(p), u^2(p)) = (\varphi_U)^{-1}(\varphi_U(p)) = p = $ RHS.

For 2: For this purpose, let us take any $t_0$ in $(-\varepsilon, \varepsilon)$. We have to show that $\gamma : (-\varepsilon, \varepsilon) \to M$ is $C^\infty$ at $t_0$.

We want to apply Theorem 2.2. Since $t_0$ is in $(-\varepsilon, \varepsilon)$, by the definition of $\gamma$, $\gamma(t_0)(= (\varphi_U)^{-1}(\gamma_1(t_0)))$ is in $U$, and hence, $(U, \varphi_U)$ is an admissible coordinate chart of $M$ satisfying $\gamma(t_0) \in U$. Since $\varphi_U \circ \gamma = \varphi_U \circ ((\varphi_U)^{-1} \circ \gamma_1) = (\varphi_U \circ (\varphi_U)^{-1}) \circ \gamma_1 = \gamma_1$, and $\gamma_1$ is a smooth function, $\varphi_U \circ \gamma$ is a smooth function. Since $\gamma : (-\varepsilon, \varepsilon) \to M$, $t_0$ is in $(-\varepsilon, \varepsilon)$, $(U, \varphi_U)$ is an admissible coordinate chart of $M$ satisfying $\gamma(t_0) \in U$, and $\varphi_U \circ \gamma$ is a smooth function so, by Theorem 2.2, $\gamma$ is $C^\infty$ at $t_0$ in $(-\varepsilon, \varepsilon)$. This proves 2. Thus, we have shown that $\gamma$ is in $\Gamma_p(M)$. Hence, from (*),

$$0 = a_1 \frac{d\left(u^1 \circ \left((\varphi_U)^{-1} \circ \gamma_1\right)\right)(t)}{dt}\Bigg|_{t=0} + a_2 \frac{d\left(u^2 \circ \left((\varphi_U)^{-1} \circ \gamma_1\right)\right)(t)}{dt}\Bigg|_{t=0}$$

$$= a_1 \frac{d\left(\left(u^1 \circ (\varphi_U)^{-1}\right) \circ \gamma_1\right)(t)}{dt}\Bigg|_{t=0} + a_2 \frac{d\left(\left(u^2 \circ (\varphi_U)^{-1}\right) \circ \gamma_1\right)(t)}{dt}\Bigg|_{t=0}$$

$$= a_1 \left[\left(D_1\left(u^1 \circ (\varphi_U)^{-1}\right)\right)(\gamma_1(0)) \quad \left(D_2\left(u^1 \circ (\varphi_U)^{-1}\right)\right)(\gamma_1(0))\right] \begin{bmatrix} \frac{d}{dt}(u^1(p) + t) \\ \frac{d}{dt}(u^2(p)) \end{bmatrix}_{t=0}$$

$$+ a_2 \left[ \left( D_1 \left( u^2 \circ (\varphi_U)^{-1} \right) \right) (\gamma_1(0)) \quad \left( D_2 \left( u^2 \circ (\varphi_U)^{-1} \right) \right) (\gamma_1(0)) \right] \left[ \begin{matrix} \frac{d}{dt}(u^1(p) + t) \\ \frac{d}{dt}(u^2(p)) \end{matrix} \right]_{t=0}$$

$$= a_1 \left[ \left( D_1 \left( u^1 \circ (\varphi_U)^{-1} \right) \right) (\gamma_1(0)) \quad \left( D_2 \left( u^1 \circ (\varphi_U)^{-1} \right) \right) (\gamma_1(0)) \right] \left[ \begin{matrix} 1 \\ 0 \end{matrix} \right]$$

$$+ a_2 \left[ \left( D_1 \left( u^2 \circ (\varphi_U)^{-1} \right) \right) (\gamma_1(0)) \quad \left( D_2 \left( u^2 \circ (\varphi_U)^{-1} \right) \right) (\gamma_1(0)) \right] \left[ \begin{matrix} 1 \\ 0 \end{matrix} \right]$$

$$= a_1 \left( \left( D_1 \left( u^1 \circ (\varphi_U)^{-1} \right) \right) (\gamma_1(0)) \right) + a_2 \left( \left( D_1 \left( u^2 \circ (\varphi_U)^{-1} \right) \right) (\gamma_1(0)) \right)$$

$$= a_1 \left( \left( D_1 \left( u^1 \circ (\varphi_U)^{-1} \right) \right) (\varphi_U(p)) \right) + a_2 \left( \left( D_1 \left( u^2 \circ (\varphi_U)^{-1} \right) \right) (\varphi_U(p)) \right) = a_1(1) + a_2(0) = a_1.$$

Thus, $a_1 = 0$. Similarly, $a_2 = 0$. Thus, we have shown that $[u^1]^\sim, [u^2]^\sim$ are linearly independent. This proves 2.

3. Let us take any $[f]^\sim$, where $f$ is in $C_p^\infty(M)$.

We have to find reals $a_1, a_2$ such that $[f]^\sim = a_1[u^1]^\sim + a_2[u^2]^\sim$. Since $f$ is in $C_p^\infty(M)$ so, by the definition of $C_p^\infty(M)$, $f$ is a real-valued function whose $\mathrm{dom} f$ is an open neighborhood of $p$ in $M$, and

$$\left( f \circ (\varphi_U)^{-1} \right) : \varphi_U((\mathrm{dom} f) \cap U) \to \mathbb{R}$$

is $C^\infty$ at the point $\varphi_U(p)$ in $\mathbb{R}^2$. Here, $\varphi_U((\mathrm{dom} f) \cap U)$ is an open neighborhood of $\varphi_U(p)(= (u^1(p), u^2(p)))$. Since $(f \circ (\varphi_U)^{-1}) : \varphi_U((\mathrm{dom} f) \cap U) \to \mathbb{R}$ is $C^\infty$ at the point $\varphi_U(p)$ in $\mathbb{R}^2$, for some open neighborhood $G(\subset \varphi_U((\mathrm{dom} f) \cap U) \subset \mathbb{R}^2)$ of $\varphi_U(p)$, $(f \circ (\varphi_U)^{-1})$ is smooth on $G$. For every $x$ in $(\mathrm{dom} f) \cap U$, we have

$$f(x) = \left( f \circ (\varphi_U)^{-1} \right) (\varphi_U(x)) = \left( f \circ (\varphi_U)^{-1} \right) (u^1(x), u^2(x)).$$

Now, we want to apply Theorem 2.42. Since, for every $x$ in $(\mathrm{dom} f) \cap U$,

$$f(x) = \left( f \circ (\varphi_U)^{-1} \right) (u^1(x), u^2(x)), \quad u^1, u^2 \in C_p^\infty(M),$$

$G$ is an open neighborhood of $(u^1(p), u^2(p))(= \varphi_U(p))$ in $\mathbb{R}^2$, and $(f \circ (\varphi_U)^{-1}) : G \to \mathbb{R}$ is a smooth function so, by Theorem 2.42,

$$[f]^\sim = \left( \left( D_1 \left( f \circ (\varphi_U)^{-1} \right) \right) (u^1(p), u^2(p)) \right) [u^1]^\sim + \left( \left( D_2 \left( f \circ (\varphi_U)^{-1} \right) \right) (u^1(p), u^2(p)) \right) [u^2]^\sim.$$

This proves 3.

4. From 2, 3, we find that $\{ [u^1]^\sim, [u^2]^\sim \}$ is a basis of the real linear space $T_p^*(M)$. This proves 4.

5. From 4, $\{[u^1]^\sim, [u^2]^\sim\}$ is a basis of the real linear space $T_p^*(M)$, and the number of elements in the basis is 2, so the dimension of $T_p^*(M)$ is 2. This proves 5.   □

As above, we can prove the following theorem.

**Theorem 2.48** *Let $M$ be an $m$-dimensional smooth manifold. Let $p$ be an element of $M$. Let $(U, \varphi_U)$ be an admissible coordinate chart of $M$ satisfying $p \in U$. Let us define $m$ functions*

$$u^1 : U \to \mathbb{R}, \ldots, u^m : U \to \mathbb{R}$$

*as follows: For every $x$ in $U$,*

$$\varphi_U(x) \equiv \left(u^1(x), \ldots, u^m(x)\right).$$

*Then,*

1. $u^1, \ldots, u^m$ are in $C_p^\infty(M)$.
2. $[u^1]^\sim, \ldots, [u^m]^\sim$ are linearly independent in the real linear space $T_p^*(M)$.
3. For every $[f]^\sim$ in $T_p^*(M)$, where $f$ is in $C_p^\infty(M)$,

$$[f]^\sim = \left(\left(D_1\left(f \circ (\varphi_U)^{-1}\right)\right)\left(u^1(p), \ldots, u^m(p)\right)\right)[u^1]^\sim + \cdots$$
$$+ \left(\left(D_m\left(f \circ (\varphi_U)^{-1}\right)\right)\left(u^1(p), \ldots, u^m(p)\right)\right)[u^m]^\sim.$$

*In other words,*

$$(df)_p = \left(\left(\left.\frac{\partial}{\partial u^1}\right|_p\right)[f]\right)\left((du^1)_p\right) + \cdots + \left(\left(\left.\frac{\partial}{\partial u^m}\right|_p\right)[f]\right)\left((du^m)_p\right).$$

4. $\{[u^1]^\sim, \ldots, [u^m]^\sim\}$ is a basis of the real linear space $T_p^*(M)$.
5. $\dim(T_p^*(M)) = m$.

*Proof* Its proof is quite similar to the proof of Theorem 2.47.   □

**Definition** Let $M$ be an $m$-dimensional smooth manifold. Let $p$ be an element of $M$. Let $(U, \varphi_U)$ be an admissible coordinate chart of $M$ satisfying $p \in U$.

The basis $\{[u^1]^\sim, \ldots, [u^m]^\sim\}$ of the cotangent space $T_p^*(M)$ of $M$ at $p$, as defined in Theorem 2.48, is called the *natural basis* of $T_p^*(M)$ determined by $(U, \varphi_U)$.

**Note 2.49** Let $M$ be an $m$-dimensional smooth manifold. Let $p$ be an element of $M$. Let $\gamma$ be in $\Gamma_p(M)$. Let $[f]^\sim = [g]^\sim$, where $f, g$ are in $C_p^\infty(M)$. We shall show that $\ll \gamma, [f] \gg = \ll \gamma, [g] \gg$. Since $[f]^\sim = [g]^\sim$, $[f - g] = [f] - [g] \in \mathcal{H}_p$, and hence,

$$0 = \ll \gamma, [f] - [g] \gg = \ll \gamma, [f] \gg - \ll \gamma, [g] \gg$$

or,

$$\ll \gamma, [f] \gg = \ll \gamma, [g] \gg.$$

This shows that the following definition is unambiguous.

**Definition** Let $M$ be an $m$-dimensional smooth manifold. Let $p$ be an element of $M$. Let $\gamma$ be in $\Gamma_p(M)$. Let $[f]^\sim$ be in $T_p^*(M)$, where $f$ is in $C_p^\infty(M)$. By $\ll \gamma, [f]^\sim \gg$, we mean $\ll \gamma, [f] \gg$.

**Theorem 2.50** *Let $M$ be an $m$-dimensional smooth manifold. Let $p$ be an element of $M$. Let us define a relation $\sim$ over $\Gamma_p(M)$ as follows: for every $\gamma, \gamma'$ in $\Gamma_p(M)$, by $\gamma \sim \gamma'$, we shall mean that for every $[f]^\sim$ in $T_p^*(M)$, $\ll \gamma, [f]^\sim \gg = \ll \gamma', [f]^\sim \gg$. Then, $\sim$ is an equivalence relation over $\Gamma_p(M)$.*

*Proof* Here, we must prove:

1. $\gamma \sim \gamma$ for every $\gamma$ in $\Gamma_p(M)$,
2. if $\gamma \sim \gamma'$ then $\gamma' \sim \gamma$,
3. if $\gamma \sim \gamma'$ and $\gamma' \sim \gamma''$, then $\gamma \sim \gamma''$.

For 1: Let us take any $\gamma$ in $\Gamma_p(M)$. We have to prove that $\gamma \sim \gamma$. For this purpose, let us take any $[f]^\sim$ in $T_p^*(M)$, where $f$ is in $C_p^\infty(M)$. Since $\ll \gamma, [f]^\sim \gg = \ll \gamma, [f]^\sim \gg$, by the definition of $\sim$, $\gamma \sim \gamma$. This proves 1.

For 2: Let $\gamma \sim \gamma'$. We have to prove that $\gamma' \sim \gamma$, that is, $\ll \gamma', [f]^\sim \gg = \ll \gamma, [f]^\sim \gg$ for every $[f]^\sim$ in $T_p^*(M)$, where $f$ is in $C_p^\infty(M)$. For this purpose, let us take any $[f]^\sim$ in $T_p^*(M)$, where $f$ is in $C_p^\infty(M)$. Since $\gamma \sim \gamma'$, by the definition of $\sim$, $\ll \gamma, [f]^\sim \gg = \ll \gamma', [f]^\sim \gg$, and hence, $\ll \gamma', [f]^\sim \gg = \ll \gamma, [f]^\sim \gg$. This proves 2.

For 3: Let $\gamma \sim \gamma'$, and $\gamma' \sim \gamma''$. We have to prove that $\gamma \sim \gamma''$, that is, $\ll \gamma, [f]^\sim \gg = \ll \gamma'', [f]^\sim \gg$ for every $[f]^\sim$ in $T_p^*(M)$, where $f$ is in $C_p^\infty(M)$. For this purpose, let us take any $[f]^\sim$ in $T_p^*(M)$, where $f$ is in $C_p^\infty(M)$. Since $\gamma \sim \gamma'$, by the definition of $\sim$, $\ll \gamma, [f]^\sim \gg = \ll \gamma', [f]^\sim \gg$. Since $\gamma' \sim \gamma''$, by the definition of $\sim$, $\ll \gamma', [f]^\sim \gg = \ll \gamma'', [f]^\sim \gg$. Since $\ll \gamma, [f]^\sim \gg = \ll \gamma', [f]^\sim \gg$ and $\ll \gamma', [f]^\sim \gg = \ll \gamma'', [f]^\sim \gg$, $\ll \gamma, [f]^\sim \gg = \ll \gamma'', [f]^\sim \gg$. This proves 3.

Thus, we have shown that $\sim$ is an equivalence relation over $\Gamma_p(M)$.  $\square$

**Definition** Let $M$ be an $m$-dimensional smooth manifold. Let $p$ be an element of $M$. We have seen that the relation $\sim$, as defined in Theorem 2.50, is an equivalence relation over $\Gamma_p(M)$. For any $\gamma$ in $\Gamma_p(M)$, the equivalence class of $\gamma$ is denoted by $[\gamma]$. Thus, for any $\gamma$ in $\Gamma_p(M)$,

$$[\gamma] \equiv \{ \gamma' : \gamma' \in \Gamma_p(M) \text{ and } \gamma' \sim \gamma \}.$$

We shall denote the quotient set $\Gamma_p(M)/\sim$ by $\Gamma_p^*(M)$ (or, simply, $\Gamma_p^*$). Thus,

$$\Gamma_p^*(M) \equiv \{[\gamma] : \gamma \in \Gamma_p(M)\}.$$

Intuitively, in $\Gamma_p^*(M)$, we will not distinguish between $\gamma'$ and $\gamma$ whenever $\gamma' \sim \gamma$.

**Note 2.51** Let $M$ be an $m$-dimensional smooth manifold. Let $p$ be an element of $M$. Let $[f]^\sim$ be in $T_p^*(M)$, where $f$ is in $C_p^\infty(M)$. Let $[\gamma] = [\gamma']$, where $\gamma, \gamma'$ are in $\Gamma_p(M)$. We shall show that $\ll \gamma, [f]^\sim \gg = \ll \gamma', [f]^\sim \gg$. Since $[\gamma] = [\gamma']$, $\gamma \sim \gamma'$, and hence, by Theorem 2.22,

$$\ll \gamma, [f]^\sim \gg = \ll \gamma', [f]^\sim \gg.$$

This shows that the following definition is unambiguous.

**Definition** Let $M$ be an $m$-dimensional smooth manifold. Let $p$ be an element of $M$. Let $[f]^\sim$ be in $T_p^*(M)$, where $f$ is in $C_p^\infty(M)$. Let $\gamma$ be in $\Gamma_p(M)$. By $\langle [\gamma], [f]^\sim \rangle$, we mean $\ll \gamma, [f]^\sim \gg$. Thus,

$$\langle [\gamma], [f]^\sim \rangle = \ll \gamma, [f]^\sim \gg = \ll \gamma, [f] \gg = \frac{d(f \circ \gamma)(t)}{dt}\bigg|_{t=0}$$
$$= \lim_{t \to 0} \frac{f(\gamma(0+t)) - f(\gamma(0))}{t},$$

which is a real number.

**Theorem 2.52** *Let $M$ be a 2-dimensional smooth manifold. Let $p$ be an element of $M$. Let $(U, \varphi_U)$ be an admissible coordinate chart of $M$ satisfying $p \in U$. Let $\gamma$ be in $\Gamma_p(M)$. Let us define 2 functions*

$$u^1 : U \to \mathbb{R}, \quad u^2 : U \to \mathbb{R}$$

*as follows: For every $x$ in $U$,*

$$\varphi_U(x) \equiv \left(u^1(x), u^2(x)\right).$$

*(In short, $u^1, u^2$ are the component functions of $\varphi_U$.) Then, for every $[f]^\sim$ in $T_p^*(M)$, where $f$ is in $C_p^\infty(M)$,*

$$\langle [\gamma], [f]^\sim \rangle = \left(\left(D_1\left(f \circ (\varphi_U)^{-1}\right)\right)(u^1(p), u^2(p))\right)\left(\frac{d}{dt}(u^1 \circ \gamma)(t)\right)\bigg|_{t=0}$$
$$+ \left(\left(D_2\left(f \circ (\varphi_U)^{-1}\right)\right)(u^1(p), u^2(p))\right)\left(\frac{d}{dt}(u^2 \circ \gamma)(t)\right)\bigg|_{t=0}.$$

*In other words,*

$$\langle [\gamma], [f]^{\sim} \rangle = \left( \left( \frac{\partial}{\partial u^1} \Big|_p \right) [f] \right) \left( \frac{d}{dt} (u^1 \circ \gamma)(t) \right) \Big|_{t=0} + \left( \left( \frac{\partial}{\partial u^2} \Big|_p \right) [f] \right) \left( \frac{d}{dt} (u^2 \circ \gamma)(t) \right) \Big|_{t=0},$$

or

$$\langle [\gamma], [f]^{\sim} \rangle = \left( \left( \frac{\partial}{\partial u^1} \Big|_p \right) [f] \right) \left( \left( \frac{d}{dt} \Big|_{t=0} \right) (u^1 \circ \gamma)(t) \right)$$
$$+ \left( \left( \frac{\partial}{\partial u^2} \Big|_p \right) [f] \right) \left( \left( \frac{d}{dt} \Big|_{t=0} \right) (u^2 \circ \gamma)(t) \right).$$

*Proof* Since $\gamma$ is in $\Gamma_p(M)$, by the definition of $\Gamma_p(M)$, there exists a real number $\delta > 0$ such that $\gamma$ is defined on the open interval $(-\delta, \delta)$, $\gamma(0) = p$, and $\gamma$ is a smooth map from $(-\delta, \delta)$ to $M$. Also, $\gamma$ is continuous at 0. Since $\gamma$ is continuous at 0, $U$ is an open neighborhood of $p(=\gamma(0))$, there exists $\varepsilon > 0$ such that $\varepsilon < \delta$, and for every $t$ in the open interval $(-\varepsilon, \varepsilon)$ we have $\gamma(t) \in U$. So, for every $t$ in $(-\varepsilon, \varepsilon)$,

$$(\varphi_U \circ \gamma)(t) = \varphi_U(\gamma(t)) = \left( u^1(\gamma(t)), u^2(\gamma(t)) \right) = \left( (u^1 \circ \gamma)(t), (u^2 \circ \gamma)(t) \right).$$

Thus, in the open neighborhood $(-\varepsilon, \varepsilon)$ of 0, the component functions of $\varphi_U \circ \gamma$ are $u^1 \circ \gamma$, and $u^2 \circ \gamma$. Also,

$$(\varphi_U \circ \gamma)(0) = \varphi_U(\gamma(0)) = \varphi_U(p) = \left( u^1(p), u^2(p) \right).$$

Now

$$\text{LHS} = \langle [\gamma], [f]^{\sim} \rangle = \ll \gamma, [f]^{\sim} \gg = \ll \gamma, [f] \gg = \frac{d(f \circ \gamma)(t)}{dt} \Big|_{t=0}$$

$$= \frac{d \left( f \circ \left( (\varphi_U)^{-1} \circ \varphi_U \right) \circ \gamma \right)(t)}{dt} \Big|_{t=0}$$

$$= \frac{d \left( \left( f \circ (\varphi_U)^{-1} \right) \circ (\varphi_U \circ \gamma) \right)(t)}{dt} \Big|_{t=0}$$

$$= \left[ \left( D_1 \left( f \circ (\varphi_U)^{-1} \right) \right) ((\varphi_U \circ \gamma)(0)) \quad \left( D_2 \left( f \circ (\varphi_U)^{-1} \right) \right) (\varphi_U \circ \gamma)(0) \right]$$
$$\times \begin{bmatrix} \frac{d}{dt} (u^1 \circ \gamma)(t) \\ \frac{d}{dt} (u^2 \circ \gamma)(t) \end{bmatrix} \Big|_{t=0}$$

$$= \left[ \left( D_1 \left( f \circ (\varphi_U)^{-1} \right) \right) (u^1(p), u^2(p)) \quad \left( D_2 \left( f \circ (\varphi_U)^{-1} \right) \right) (u^1(p), u^2(p)) \right]$$

$$\times \left[ \begin{array}{c} \frac{d}{dt} (u^1 \circ \gamma)(t) \\ \frac{d}{dt} (u^2 \circ \gamma)(t) \end{array} \right] \Bigg|_{t=0}$$

$$= \left( \left( D_1 \left( f \circ (\varphi_U)^{-1} \right) \right) (u^1(p), u^2(p)) \right) \left( \frac{d}{dt} (u^1 \circ \gamma)(t) \right) \Bigg|_{t=0}$$

$$+ \left( \left( D_2 \left( f \circ (\varphi_U)^{-1} \right) \right) (u^1(p), u^2(p)) \right) \left( \frac{d}{dt} (u^2 \circ \gamma)(t) \right) \Bigg|_{t=0} = \text{RHS.}$$

$\square$

## 2.6 Tangent Space as a Dual Space

**Note 2.53** Here, we will digress slightly from the current topic. Let $V$ be any real linear space. Let $V^*$ be the collection of all linear functionals of $V$. We know that $V^*$ is also a real linear space under pointwise vector addition, and pointwise scalar multiplication operations. $V^*$ is called the dual space of $V$. Let $n$ be the dimension of $V$. Let $\{e^1, \ldots, e^n\}$ be a basis of $V$. Let us define a function $e_1^* : V \to \mathbb{R}$ as follows: For every real $t_1, \ldots, t_n$

$$e_1^* \left( t_1 e^1 + \cdots + t_n e^n \right) = t_1.$$

Clearly, $e_1^*$ is in $V^*$. Similarly, we define function $e_2^* : V \to \mathbb{R}$ as follows: For every real $t_1, \ldots, t_n$

$$e_2^* \left( t_1 e^1 + \cdots + t_n e^n \right) = t_2,$$

etc. We shall try to show that $\{e_1^*, \ldots, e_n^*\}$ constitutes a basis of $V^*$. For this purpose, we must prove

1. $e_1^*, \ldots, e_n^*$ are linearly independent, that is, $t_1 e_1^* + \cdots + t_n e_1^* = 0$ implies $t_1 = 0, \ldots, t_n = 0$.
2. $\{e_1^*, \ldots, e_n^*\}$ generates $V^*$, that is, for every $f$ in $V^*$, there exist reals $t_1, \ldots, t_n$ such that $t_1 e_1^* + \cdots + t_n e_n^* = f$.

For 1: Let $t_1 e_1^* + \cdots + t_n e_1^* = 0$. So,

$$0 = 0(e^1) = \left( t_1 e_1^* + \cdots + t_n e_1^* \right) (e^1) = t_1 \cdot \left( e_1^*(e^1) \right) + \cdots + t_n \cdot \left( e_n^*(e^1) \right)$$
$$= t_1 \cdot 1 + t_2 \cdot 0 + \cdots + t_n \cdot 0 = t_1.$$

Hence, $t_1 = 0$. Similarly, $t_2 = 0, \ldots, t_n = 0$.

For 2: Let us take any $f$ in $V^*$. We shall show that $(f(e^1))e_1^* + \cdots + (f(e^n))e_n^* = f$. For this purpose, it is enough to prove that

$$((f(e^1))e_1^* + \cdots + (f(e^n))e_n^*)(e_1) = f(e^1), \ldots, ((f(e^1))e_1^* + \cdots + (f(e^n))e_n^*)(e^n) = f(e^n).$$

Here,

$$\text{LHS} = ((f(e^1))e_1^* + \cdots + (f(e^n))e_n^*)(e_1) = (f(e^1))e_1^*(e_1) + \cdots + (f(e^n))e_n^*(e_1)$$
$$= (f(e^1)) \cdot 1 + (f(e^2)) \cdot 0 + \cdots + (f(e^n)) \cdot 0 = f(e^1) = \text{RHS},$$

etc. Thus, we have shown that $\{e_1^*, \ldots, e_n^*\}$ constitutes a basis of $V^*$. The basis $\{e_1^*, \ldots, e_n^*\}$ of the dual space $V^*$ is called the dual basis of the basis $\{e^1, \ldots, e^n\}$. Observe that

1. $e_j^*(e^i) = \delta_j^i$.
2. $\dim(V^*) = n = \dim(V)$.

Now, let us consider the mapping $v : V \to V^*$ defined as follows: For every real $t_1, \ldots, t_n$

$$v(t_1 e^1 + \cdots + t_n e^n) = t_1 e_1^* + \cdots + t_n e_n^*.$$

We shall show that $v$ is an isomorphism from $V$ onto $V^*$. For this purpose, we must prove

1. $v$ is 1–1
2. $v$ is onto.
3. $v(sx + ty) = s(v(x)) + t(v(y))$, for every real $s, t$, and for every $x, y$ in $V$.

For 1: Let $v(x) = v(y)$. We have to prove that $x = y$. Let $x = s_1 e^1 + \cdots + s_n e^n$, and $y = t_1 e^1 + \cdots + t_n e^n$. Now, we have

$$s_1 e_1^* + \cdots + s_n e_n^* = v(s_1 e^1 + \cdots + s_n e^n) = v(x) = v(y) = v(t_1 e^1 + \cdots + t_n e^n)$$
$$= t_1 e_1^* + \cdots + t_n e_n^*,$$

and hence, $s_1 = t_1, \ldots, s_n = t_n$. Since $s_1 = t_1, \ldots, s_n = t_n$, $x = s_1 e^1 + \cdots + s_n e^n = t_1 e^1 + \cdots + t_n e^n = y$. This proves 1.

For 2: For this purpose, let us take any $f$ in $V^*$. We have to find an element in $V$ whose $v$-image is $f$. Since $f$ is in $V^*$, and $\{e_1^*, \ldots, e_n^*\}$ constitutes a basis of $V^*$, there exist reals $t_1, \ldots, t_n$ such that

$$t_1 e_1^* + \cdots + t_n e_n^* = f.$$

Now

$$v\left(t_1 e^1 + \cdots + t_n e^n\right) = t_1 e_1^* + \cdots + t_n e_n^* = f,$$

where $t_1 e^1 + \cdots + t_n e^n$ is in $V$. This proves 2.

For 3: Let us take any $x \equiv s_1 e^1 + \cdots + s_n e^n$, and $y \equiv t_1 e^1 + \cdots + t_n e^n$. Let us take any reals $s, t$. Now

$$\begin{aligned}
\text{LHS} &= v(sx + ty) = v\left(s\left(s_1 e^1 + \cdots + s_n e^n\right) + t\left(t_1 e^1 + \cdots + t_n e^n\right)\right) \\
&= v\left((ss_1 + tt_1)e^1 + \cdots + (ss_n + tt_n)e^n\right) = (ss_1 + tt_1)e_1^* + \cdots + (ss_n + tt_n)e_n^* \\
&= s\left(s_1 e_1^* + \cdots + s_n e_n^*\right) + t\left(t_1 e_1^* + \cdots + t_n e_n^*\right) = s\left(v\left(s_1 e^1 + \cdots + s_n e^n\right)\right) \\
&\quad + t\left(v\left(t_1 e^1 + \cdots + t_n e^n\right)\right) \\
&= s(v(x)) + t(v(y)) = \text{RHS}.
\end{aligned}$$

Thus, we have shown that $v$ is an isomorphism from $V$ onto $V^*$. Hence, $V$ and $V^*$ are isomorphic real linear spaces.

**Definition** Let $M$ be an $m$-dimensional smooth manifold. Let $p$ be an element of $M$. We will denote the dual space of the real linear space $T_p^*(M)$ by $\mathcal{T}_p(M)$. Thus, $\mathcal{T}_p(M)$ is the collection of all linear functionals defined on $T_p^*(M)$. It is known that $\mathcal{T}_p(M)$ is a real linear space. Now, let $(U, \varphi_U)$ be an admissible coordinate chart of $M$ satisfying $p \in U$.

Let us define $m$ functions

$$u^1 : U \to \mathbb{R}, \ldots, u^m : U \to \mathbb{R}$$

as follows: For every $x$ in $U$,

$$\varphi_U(x) \equiv \left(u^1(x), \ldots, u^m(x)\right).$$

By Theorem 2.48, $\{[u^1]^\sim, \ldots, [u^m]^\sim\}$ is a basis of the real linear space $T_p^*(M)$. Let us define the function $[u_1]^* : T_p^*(M) \to \mathbb{R}$ as follows: For every real $t_1, \ldots, t_m$,

$$[u_1]^* \left(t_1 [u^1]^\sim + \cdots + t_m [u^m]^\sim\right) \equiv t_1.$$

It is easy to see that $[u_1]^*$ is in $\mathcal{T}_p(M)$.

As above, we define $[u_2]^* : T_p^*(M) \to \mathbb{R}$ as follows: For every real $t_1, \ldots, t_m$,

$$[u_2]^* \left( t_1 [u^1]^{\sim} + \cdots + t_m [u^m]^{\sim} \right) \equiv t_2,$$

etc. We know that $\{[u_1]^*, \ldots, [u_m]^*\}$ constitutes a basis of the dual space $\mathcal{T}_p(M)$ of $T_p^*(M)$. $\{[u_1]^*, \ldots, [u_m]^*\}$ is called the *dual basis of the basis* $\{[u^1]^{\sim}, \ldots, [u^m]^{\sim}\}$ Clearly,

$$[u_i]^* \left( [u^j]^{\sim} \right) = \delta_i^j,$$

where $\delta_i^j$ denotes the Kronecker delta symbol. In other words,

$$[u_i]^* \left( (du^j)_p \right) = \delta_i^j.$$

Also,

$$\dim \left( \mathcal{T}_p(M) \right) = \dim \left( T_p^*(M) \right) = m.$$

**Theorem 2.54** *Let M be a 2-dimensional smooth manifold. Let p be an element of M. Let $\gamma$ be in $\Gamma_p(M)$. Let us define a function*

$$[\gamma]^*: T_p^*(M) \to \mathbb{R},$$

*as follows: For every $[f]^{\sim}$ in $T_p^*(M)$, where f is in $C_p^\infty(M)$,*

$$[\gamma]^*([f]^{\sim}) \equiv \langle [\gamma], [f]^{\sim} \rangle.$$

*Then, $[\gamma]^*$ is a linear functional on the real linear space $T_p^*(M)$, that is, $\{[\gamma]^* : \gamma \in \Gamma_p(M)\}$ is a subset of the dual space $\mathcal{T}_p(M)$ of $T_p^*(M)$.*

**Proof** Here, we must prove:
For every real $\alpha, \beta$, and for every $[f]^{\sim}, [g]^{\sim}$ in $T_p^*(M)$, where $f, g$ are in $C_p^\infty(M)$,

$$[\gamma]^*(\alpha[f]^{\sim} + \beta[g]^{\sim}) = \alpha([\gamma]^*([f]^{\sim})) + \beta([\gamma]^*([g]^{\sim})).$$

LHS $= [\gamma]^*(\alpha[f]^{\sim} + \beta[g]^{\sim}) = [\gamma]^*([\alpha f]^{\sim} + [\beta g]^{\sim}) = [\gamma]^*([\alpha f + \beta g]^{\sim}) = \langle [\gamma], [\alpha f + \beta g]^{\sim} \rangle$
$= \ll \gamma, [\alpha f + \beta g] \gg = \ll \gamma, [\alpha f + \beta g] \gg = \ll \gamma, [\alpha f] + [\beta g] \gg$
$= \alpha \ll \gamma, [f] \gg + \beta \ll \gamma, [g] \gg = \alpha \ll \gamma, [f]^{\sim} \gg + \beta \ll \gamma, [g]^{\sim} \gg = \langle \alpha[\gamma], [f]^{\sim} \rangle + \beta \langle [\gamma], [g]^{\sim} \rangle$
$= \alpha([\gamma]^*([f]^{\sim})) + \beta([\gamma]^*([g]^{\sim})) =$ RHS.                                     $\square$

**Lemma 2.55** *Let M be a 2-dimensional smooth manifold. Let p be an element of M. Let $(U, \varphi_U)$ be an admissible coordinate chart $(U, \varphi_U)$ of M such that p is in U.*

Let $u^1, u^2$ be the component functions of $\varphi_U$. Let $T$ be a *linear functional of $T_p^*(M)$.* Let

$$\gamma_1 : \mathbb{R} \to \mathbb{R}^2$$

*be the function defined as follows: For every $t$ in $\mathbb{R}$,*

$$\gamma_1(t) \equiv \left( (T([u^1]^\sim))t, (T([u^2]^\sim))t \right) + \varphi_U(p).$$

*Then*

1. $(\varphi_U)^{-1} \circ \gamma_1$ is in $\Gamma_p(M)$
2. for every $f$ in $C_p^\infty(M)$,

$$T([f]^\sim) = \left\langle \left[ (\varphi_U)^{-1} \circ \gamma_1 \right], [f]^\sim \right\rangle.$$

*Proof*

1: Clearly, $\gamma_1$ is a smooth function. Here, $\gamma_1$ is continuous, $\gamma_1(0) = ((T([u^1]^\sim)) \cdot 0, (T([u^2]^\sim)) \cdot 0) + \varphi_U(p) = \varphi_U(p)$, and $\varphi_U(U)$ is an open neighborhood of $\varphi_U(p)$, so there exists $\varepsilon > 0$ such that for every $t$ in the open interval $(-\varepsilon, \varepsilon)$, we have $\gamma_1(t)$ in $\varphi_U(U)$, and hence, $(\varphi_U)^{-1}(\gamma_1(t))$ is in $U$. Now, let us define a function

$$\gamma : (-\varepsilon, \varepsilon) \to M$$

as follows: For every $t$ in $(-\varepsilon, \varepsilon)$,

$$\gamma(t) \equiv (\varphi_U)^{-1}(\gamma_1(t)) = \left( (\varphi_U)^{-1} \circ \gamma_1 \right)(t).$$

So,

$$\gamma = (\varphi_U)^{-1} \circ \gamma_1.$$

We shall try to see that $\gamma$ is in $\Gamma_p(M)$. By the definition of $\Gamma_p(M)$, we must prove that

1. $\gamma(0) = p$,
2. $\gamma$ is a parametrized curve in the manifold $M$, that is, $\gamma : (-\varepsilon, \varepsilon) \to M$ is a smooth map from $(-\varepsilon, \varepsilon)$ to $M$, that is, $\gamma$ is $C^\infty$ at every point of $(-\varepsilon, \varepsilon)$.

For 1: Here LHS $= \gamma(0) = (\varphi_U)^{-1}(\gamma_1(0)) = (\varphi_U)^{-1}(\varphi_U(p)) = p =$ RHS. This proves 1.

For 2: For this purpose, let us take any $t_0$ in $(-\varepsilon, \varepsilon)$. We have to show that $\gamma : (-\varepsilon, \varepsilon) \to M$ is $C^\infty$ at $t_0$. We want to apply Theorem 2.2. Since $t_0$ is in $(-\varepsilon, \varepsilon)$, by the definition of $\gamma$, $\gamma(t_0)(= (\varphi_U)^{-1}(\gamma_1(t_0)))$ is in $U$, and hence, $(U, \varphi_U)$ is an admissible coordinate chart of $M$ satisfying $\gamma(t_0) \in U$. Since $\varphi_U \circ \gamma = \varphi_U \circ ((\varphi_U)^{-1} \circ \gamma_1) = (\varphi_U \circ (\varphi_U)^{-1}) \circ \gamma_1 = \gamma_1$, and $\gamma_1$ is a smooth function, $\varphi_U \circ \gamma$ is a smooth function. Since $\gamma : (-\varepsilon, \varepsilon) \to M$, $t_0$ is in $(-\varepsilon, \varepsilon)$, $(U, \varphi_U)$ is an admissible coordinate chart of $M$ satisfying $\gamma(t_0) \in U$, and $\varphi_U \circ \gamma$ is a smooth function, by Theorem 2.2, $\gamma$ is $C^\infty$ at $t_0$ in $(-\varepsilon, \varepsilon)$. This proves 2. Thus, we have shown that $\gamma$ is in $\Gamma_p(M)$. This proves 1.

2: Since $[f]^\sim$ is in $T_p^*(M)$ so, by Theorem 2.47,

$$[f]^\sim = \left(\left(D_1\left(f \circ (\varphi_U)^{-1}\right)\right)(u^1(p), u^2(p))\right)[u^1]^\sim + \left(\left(D_2\left(f \circ (\varphi_U)^{-1}\right)\right)(u^1(p), u^2(p))\right)[u^2]^\sim,$$

and hence,

$$\begin{aligned}
\text{LHS} &= T([f]^\sim) = T\left(\left((D_1\left(f \circ (\varphi_U)^{-1}\right))(u^1(p), u^2(p))\right)[u^1]^\sim \right.\\
&\quad \left. + \left((D_2\left(f \circ (\varphi_U)^{-1}\right))(u^1(p), u^2(p))\right)[u^2]^\sim\right) \\
&= \left((D_1\left(f \circ (\varphi_U)^{-1}\right))(u^1(p), u^2(p))\right)\left(T([u^1]^\sim)\right) \\
&\quad + \left((D_2\left(f \circ (\varphi_U)^{-1}\right))(u^1(p), u^2(p))\right)\left(T([u^2]^\sim)\right).
\end{aligned}$$

Next, by Theorem 2.52,

$$\begin{aligned}
\text{RHS} &= \left\langle \left[(\varphi_U)^{-1} \circ \gamma_1\right], [f]^\sim \right\rangle = \langle [\gamma], [f]^\sim \rangle \\
&= \left.\left(\left(D_1\left(f \circ (\varphi_U)^{-1}\right)\right)(u^1(p), u^2(p))\right)\left(\frac{d}{dt}(u^1 \circ \gamma)(t)\right)\right|_{t=0} \\
&\quad + \left.\left(\left(D_2\left(f \circ (\varphi_U)^{-1}\right)\right)(u^1(p), u^2(p))\right)\left(\frac{d}{dt}(u^2 \circ \gamma)(t)\right)\right|_{t=0}.
\end{aligned}$$

Now, it remains to be proved that

$$\left.\frac{d\left(u^1 \circ \left((\varphi_U)^{-1} \circ \gamma_1\right)\right)(t)}{dt}\right|_{t=0} = T([u^1]^\sim),$$

and

$$\left.\frac{\mathrm{d}\left(u^2 \circ \left((\varphi_U)^{-1} \circ \gamma_1\right)\right)(t)}{\mathrm{d}t}\right|_{t=0} = \left(T\left([u^2]^\sim\right)\right).$$

Here,

$$\mathrm{LHS} = \left.\frac{\mathrm{d}\left(u^1 \circ \left((\varphi_U)^{-1} \circ \gamma_1\right)\right)(t)}{\mathrm{d}t}\right|_{t=0} = \left.\frac{\mathrm{d}\left(\left(u^1 \circ (\varphi_U)^{-1}\right) \circ \gamma_1\right)(t)}{\mathrm{d}t}\right|_{t=0}$$

$$= \left[\left(D_1\left(u^1 \circ (\varphi_U)^{-1}\right)\right)(\gamma_1(0)) \quad \left(D_2\left(u^1 \circ (\varphi_U)^{-1}\right)\right)(\gamma_1(0))\right]\left[\begin{matrix}\frac{\mathrm{d}}{\mathrm{d}t}((T([u^1]^\sim))t + u^1(p)) \\ \frac{\mathrm{d}}{\mathrm{d}t}((T([u^2]^\sim))t + u^2(p))\end{matrix}\right]_{t=0}$$

$$= \left[\left(D_1\left(u^1 \circ (\varphi_U)^{-1}\right)\right)(\gamma_1(0)) \quad \left(D_2\left(u^1 \circ (\varphi_U)^{-1}\right)\right)(\gamma_1(0))\right]\left[\begin{matrix}T([u^1]^\sim) \\ T([u^2]^\sim)\end{matrix}\right]$$

$$= \left(\left(D_1\left(u^1 \circ (\varphi_U)^{-1}\right)\right)(\varphi_U(p))\right)\left(T([u^1]^\sim)\right) + \left(\left(D_2\left(u^1 \circ (\varphi_U)^{-1}\right)\right)(\varphi_U(p))\right)\left(T([u^2]^\sim)\right)$$

$$= (1)\left(T([u^1]^\sim)\right) + (0)\left(T([u^2]^\sim)\right) = \left(T([u^1]^\sim)\right) = \mathrm{RHS}.$$

Similarly,

$$\left.\frac{\mathrm{d}\left(u^2 \circ \left((\varphi_U)^{-1} \circ \gamma_1\right)\right)(t)}{\mathrm{d}t}\right|_{t=0} = \left(T\left([u^2]^\sim\right)\right). \qquad \square$$

**Theorem 2.56** *Let $M$ be a 2-dimensional smooth manifold. Let $p$ be an element of $M$. Let $T$ be a linear functional on $T_p^*(M)$. Then, there exists $\gamma$ in $\Gamma_p(M)$ such that for every $[f]^\sim$ in $T_p^*(M)$,*

$$T([f]^\sim) = \langle [\gamma], [f]^\sim \rangle.$$

*Proof* Since $M$ is a 2-dimensional smooth manifold, and $p$ is in $M$, there exists an admissible coordinate chart $(U, \varphi_U)$ of $M$ such that $p$ is in $U$. Let us define 2 functions

$$u^1 : U \to \mathbb{R}, \quad u^2 : U \to \mathbb{R}$$

as follows: For every $x$ in $U$,

$$\varphi_U(x) \equiv \left(u^1(x), u^2(x)\right).$$

Let

$$\gamma_1 : \mathbb{R} \to \mathbb{R}^2$$

be the function defined as follows: for every $t$ in $\mathbb{R}$,

$$\gamma_1(t) \equiv ((T([u^1]^{\sim}))t, (T([u^2]^{\sim}))t) + \varphi_U(p).$$

Then, by Lemma 2.55, $(\varphi_U)^{-1} \circ \gamma_1$ is in $\Gamma_p(M)$ and for every $[f]^{\sim}$ in $T_p^*(M)$,

$$T([f]^{\sim}) = \langle \left[(\varphi_U)^{-1} \circ \gamma_1\right], [f]^{\sim} \rangle. \qquad \square$$

**Theorem 2.57** *Let $M$ be a 2-dimensional smooth manifold. Let $p$ be an element of $M$. For every $\gamma$ in $\Gamma_p(M)$ we define a function*

$$[\gamma]^*: T_p^*(M) \rightarrow \mathbb{R},$$

*as follows: For every $[f]^{\sim}$ in $T_p^*(M)$, where $f$ is in $C_p^{\infty}(M)$,*

$$[\gamma]^*([f]^{\sim}) \equiv \langle [\gamma], [f]^{\sim} \rangle.$$

*Then, the mapping $[\gamma] \mapsto [\gamma]^*$ is a 1–1 mapping from $\{[\gamma] : \gamma \in \Gamma_p(M)\}$ onto $T_p(M)$.*

*Proof* Here, we must prove

1. $[\gamma] \mapsto [\gamma]^*$ is a well-defined mapping from $\{[\gamma] : \gamma \in \Gamma_p(M)\}$ to $T_p(M)$,
2. if $[\gamma]^* = [\gamma']^*$, where $\gamma, \gamma' \in \Gamma_p(M)$, then $[\gamma] = [\gamma']$,
3. for every $T$ in $T_p(M)$, there exists $\gamma$ in $\Gamma_p(M)$ such that $[\gamma]^* = T$.

For 1: Let $[\gamma] = [\gamma']$, that is, $\gamma \sim \gamma'$. We have to show that $[\gamma]^* = [\gamma']^*$, that is, for every $[f]^{\sim}$ in $T_p^*(M)$, $[\gamma]^*([f]^{\sim}) = [\gamma']^*([f]^{\sim})$. For this purpose, let us take any $[f]^{\sim}$ in $T_p^*(M)$. Since $\gamma \sim \gamma'$, and $[f]^{\sim}$ is in $T_p^*(M)$, by the definition of $\sim$, $\ll \gamma, [f]^{\sim} \gg = \ll \gamma', [f]^{\sim} \gg$. Now

$$\text{LHS} = [\gamma]^*([f]^{\sim}) = \langle [\gamma], [f]^{\sim} \rangle = \ll \gamma, [f]^{\sim} \gg = \ll \gamma', [f]^{\sim} \gg = \langle [\gamma'], [f]^{\sim} \rangle$$
$$= [\gamma']^*([f]^{\sim}) = \text{RHS}.$$

So $[\gamma] \mapsto [\gamma]^*$ is a well-defined mapping.

Next, by Theorem 2.54, $[\gamma]^*$ is in $T_p(M)$ for every $\gamma$ in $\Gamma_p(M)$. This proves 1.

For 2: Let $[\gamma]^* = [\gamma']^*$, where $\gamma, \gamma' \in \Gamma_p(M)$. We have to prove that $[\gamma] = [\gamma']$, that is, $\gamma \sim \gamma'$, that is, for every $[f]^{\sim}$ in $T_p^*(M)$, $\ll \gamma, [f]^{\sim} \gg = \ll \gamma', [f]^{\sim} \gg$. For this purpose, let us take any $[f]^{\sim}$ in $T_p^*(M)$. We have to show that

$$\ll \gamma, [f]^{\sim} \gg = \ll \gamma', [f]^{\sim} \gg.$$

Since $[\gamma]^* = [\gamma']^*$, $[\gamma]^* : T_p^*(M) \to \mathbb{R}$, $[\gamma']^* : T_p^*(M) \to \mathbb{R}$, and $[f]^\sim$ is in $T_p^*(M)$,

$$
\begin{aligned}
\text{LHS} &= \ll \gamma, [f]^\sim \gg = \langle [\gamma], [f]^\sim \rangle = [\gamma]^*([f]^\sim) = [\gamma']^*([f]^\sim) = \langle [\gamma'], [f]^\sim \rangle \\
&= \ll \gamma', [f]^\sim \gg = \text{RHS}.
\end{aligned}
$$

For 3: Let us take any $T$ in $\mathcal{T}_p(M)$. Since $T$ is in $\mathcal{T}_p(M)$, and $\mathcal{T}_p(M)$ is dual space of the real linear space $T_p^*(M)$, $T$ is a linear functional on $T_p^*(M)$, and hence, by Theorem 2.56, there exists $\gamma$ in $\Gamma_p(M)$ such that for every $[f]^\sim$ in $T_p^*(M)$,

$$
T([f]^\sim) = \langle [\gamma], [f]^\sim \rangle.
$$

Since, for every $[f]^\sim$ in $T_p^*(M)$, we have $T([f]^\sim) = \langle [\gamma], [f]^\sim \rangle = [\gamma]^*([f]^\sim)$, $T = [\gamma]^*$. This proves 3.                                                                           $\square$

Similarly, we can prove the following theorem.

**Theorem 2.58** *Let M be an m-dimensional smooth manifold. Let p be an element of M. For every $\gamma$ in $\Gamma_p(M)$, we define a function*

$$
[\gamma]^* : T_p^*(M) \to \mathbb{R},
$$

*as follows: For every $[f]^\sim$ in $T_p^*(M)$, where f is in $C_p^\infty(M)$,*

$$
[\gamma]^*([f]^\sim) \equiv \langle [\gamma], [f]^\sim \rangle.
$$

Then, the mapping $[\gamma] \mapsto [\gamma]^*$ is a 1–1 mapping from $\{[\gamma] : \gamma \in \Gamma_p(M)\}$ onto $\mathcal{T}_p(M)$.

*Proof* Its proof is quite similar to the proof of Theorem 2.57.                        $\square$

**Definition** Let $M$ be an $m$-dimensional smooth manifold. Let $p$ be an element of $M$. Let $\eta$ be the mapping $[\gamma] \mapsto [\gamma]^*$ from $\{[\gamma] : \gamma \in \Gamma_p(M)\}$ to $\mathcal{T}_p(M)$, as defined in Theorem 2.58. By Theorem 2.58, $\eta$ is a 1–1 mapping from $\{[\gamma] : \gamma \in \Gamma_p(M)\}$ onto $\mathcal{T}_p(M)$, and hence, $\eta^{-1}$ is a 1–1 mapping from $\mathcal{T}_p(M)$ onto $\{[\gamma] : \gamma \in \Gamma_p(M)\}$.

Define *vector addition*, and *scalar multiplication* over the set $\{[\gamma] : \gamma \in \Gamma_p(M)\}$ as follows: For every $\gamma, \gamma'$ in $\Gamma_p(M)$, and for every real $t$,

$$
[\gamma] + [\gamma'] = \eta^{-1}(\eta([\gamma]) + \eta([\gamma'])),
$$

and

$$t[\gamma] = \eta^{-1}(t(\eta([\gamma]))).$$

Since $\mathcal{T}_p(M)$ is a real linear space, and $\eta$ is a 1–1 mapping from $\{[\gamma] : \gamma \in \Gamma_p(M)\}$ onto $\mathcal{T}_p(M)$, $\{[\gamma] : \gamma \in \Gamma_p(M)\}$ is also a real linear space. Clearly, $\eta : [\gamma] \mapsto [\gamma]^*$ is an isomorphism from real linear space $\{[\gamma] : \gamma \in \Gamma_p(M)\}$ onto real linear space $\mathcal{T}_p(M)$. Thus, we have the following formulae:

$$\eta([\gamma] + [\gamma']) = \eta([\gamma]) + \eta([\gamma'])$$

and

$$\eta(t[\gamma]) = t(\eta([\gamma])).$$

In other words,

$$([\gamma] + [\gamma'])^* = ([\gamma])^* + ([\gamma'])^*$$

and

$$(t[\gamma])^* = t([\gamma])^*,$$

because $\eta$ is the mapping $[\gamma] \mapsto [\gamma]^*$ from $\{[\gamma] : \gamma \in \Gamma_p(M)\}$ to $\mathcal{T}_p(M)$, as defined in Theorem 2.58.

Since $\eta : [\gamma] \mapsto [\gamma]^*$ is an isomorphism from real linear space $\{[\gamma] : \gamma \in \Gamma_p(M)\}$ onto real linear space $\mathcal{T}_p(M)$, $\eta^{-1} : [\gamma]^* \mapsto [\gamma]$ is also an isomorphism. Also,

$$\dim\{[\gamma] : \gamma \in \Gamma_p(M)\} = \dim(\mathcal{T}_p(M)).$$

Further, since $\dim(\mathcal{T}_p(M)) = \dim(T_p^*(M)) = $ (the dimension of the manifold), so

$$\dim\{[\gamma] : \gamma \in \Gamma_p(M)\} = \text{(the dimension of the manifold)}.$$

It follows that the real linear space $\{[\gamma] : \gamma \in \Gamma_p(M)\}$ is isomorphic to $\mathbb{R}^m$.

**Theorem 2.59** *Let M be a 2-dimensional smooth manifold. Let p be an element of M. Let $(U, \varphi_U)$ be an admissible coordinate chart of M satisfying $p \in U$. Let $\gamma : (-\delta, \delta) \to M$, and $\gamma' : (-\delta', \delta') \to M$ be in $\Gamma_p(M)$, where $\delta > 0$, and $\delta' > 0$. Then, $[\gamma] = [\gamma']$ if and only if*

$$\left.\frac{\mathrm{d}(\varphi_U \circ \gamma)(t)}{\mathrm{d}t}\right|_{t=0} = \left.\frac{\mathrm{d}(\varphi_U \circ \gamma')(t)}{\mathrm{d}t}\right|_{t=0}.$$

*In other words,  $\gamma \sim \gamma'$  if and only if  $\gamma \approx \gamma'$ , that is,  $[\gamma] = \gamma$ . And hence,*

$$\{[\gamma] : \gamma \in \Gamma_p(M)\} = T_pM.$$

*Proof of "only if" part:* Let  $[\gamma] = [\gamma']$ . Since  $[\gamma] = [\gamma']$ ,  $\gamma \sim \gamma'$ . Let us define 2 functions

$$u^1 : U \to \mathbb{R}, \quad u^2 : U \to \mathbb{R}$$

as follows: For every  $x$  in  $U$ ,

$$\varphi_U(x) \equiv \left(u^1(x), u^2(x)\right).$$

By Theorem 2.48,  $[u^1]^{\sim}, [u^2]^{\sim}$  are in  $T_p^*(M)$ . Since  $\gamma \sim \gamma'$ , and  $[u^1]^{\sim}, [u^2]^{\sim}$  are in  $T_p^*(M)$ , by the definition of  $\sim$ ,  $\ll \gamma, [u^1]^{\sim} \gg = \ll \gamma', [u^1]^{\sim} \gg$ , and  $\ll \gamma, [u^2]^{\sim} \gg = \ll \gamma', [u^2]^{\sim} \gg$ . Now

$$\begin{aligned}
\text{LHS} &= \left.\frac{d(\varphi_U \circ \gamma)(t)}{dt}\right|_{t=0} = \left.\frac{d(\varphi_U(\gamma(t)))}{dt}\right|_{t=0} = \left.\frac{d(u^1(\gamma(t)), u^2(\gamma(t)))}{dt}\right|_{t=0} \\
&= \left.\left(\frac{d(u^1(\gamma(t)))}{dt}, \frac{d(u^2(\gamma(t)))}{dt}\right)\right|_{t=0} = \left.\left(\frac{d((u^1 \circ \gamma)(t))}{dt}, \frac{d((u^2 \circ \gamma)(t))}{dt}\right)\right|_{t=0} \\
&= \left(\left.\frac{d((u^1 \circ \gamma)(t))}{dt}\right|_{t=0}, \left.\frac{d((u^2 \circ \gamma)(t))}{dt}\right|_{t=0}\right) = \left(\ll \gamma, [u^1]^{\sim} \gg, \ll \gamma, [u^2]^{\sim} \gg\right) \\
&= \left(\ll \gamma', [u^1]^{\sim} \gg, \ll \gamma', [u^2]^{\sim} \gg\right) = \left(\left.\frac{d((u^1 \circ \gamma')(t))}{dt}\right|_{t=0}, \left.\frac{d((u^2 \circ \gamma')(t))}{dt}\right|_{t=0}\right) \\
&= \left.\left(\frac{d((u^1 \circ \gamma')(t))}{dt}, \frac{d((u^2 \circ \gamma')(t))}{dt}\right)\right|_{t=0} = \left.\left(\frac{d(u^1(\gamma'(t)))}{dt}, \frac{d(u^2(\gamma'(t)))}{dt}\right)\right|_{t=0} \\
&= \left.\frac{d(u^1(\gamma'(t)), u^2(\gamma'(t)))}{dt}\right|_{t=0} = \left.\frac{d(\varphi_U(\gamma'(t)))}{dt}\right|_{t=0} = \left.\frac{d(\varphi_U \circ \gamma')(t)}{dt}\right|_{t=0} = \text{RHS}.
\end{aligned}$$

*Proof of "if" part* Let

$$\left.\frac{d(\varphi_U \circ \gamma)(t)}{dt}\right|_{t=0} = \left.\frac{d(\varphi_U \circ \gamma')(t)}{dt}\right|_{t=0}.$$

We have to show that  $[\gamma] = [\gamma']$ , that is,  $\gamma \sim \gamma'$ , that is, for every  $[f]^{\sim}$  in  $T_p^*(M)$ ,  $\ll \gamma, [f]^{\sim} \gg = \ll \gamma', [f]^{\sim} \gg$ . For this purpose, let us take any  $[f]^{\sim}$  in  $T_p^*(M)$ . We have to show that  $\ll \gamma, [f]^{\sim} \gg = \ll \gamma', [f]^{\sim} \gg$ . Now, since

$$\left(\ll\gamma,\left[u^1\right]^\sim\gg,\ll\gamma,\left[u^2\right]^\sim\gg\right) = \left(\left.\frac{d((u^1\circ\gamma)(t))}{dt}\right|_{t=0},\left.\frac{d((u^2\circ\gamma)(t))}{dt}\right|_{t=0}\right)$$

$$= \left.\left(\frac{d((u^1\circ\gamma)(t))}{dt},\frac{d((u^2\circ\gamma)(t))}{dt}\right)\right|_{t=0}$$

$$= \left.\left(\frac{d(u^1(\gamma(t)))}{dt},\frac{d(u^2(\gamma(t)))}{dt}\right)\right|_{t=0} = \left.\frac{d(u^1(\gamma(t)),u^2(\gamma(t)))}{dt}\right|_{t=0}$$

$$= \left.\frac{d(\varphi_U(\gamma(t)))}{dt}\right|_{t=0}$$

$$= \left.\frac{d(\varphi_U\circ\gamma)(t)}{dt}\right|_{t=0} = \left.\frac{d(\varphi_U\circ\gamma')(t)}{dt}\right|_{t=0}$$

$$= \left.\frac{d(\varphi_U(\gamma'(t)))}{dt}\right|_{t=0} = \left.\frac{d(u^1(\gamma'(t)),u^2(\gamma'(t)))}{dt}\right|_{t=0}$$

$$= \left.\left(\frac{d(u^1(\gamma'(t)))}{dt},\frac{d(u^2(\gamma'(t)))}{dt}\right)\right|_{t=0}$$

$$= \left.\left(\frac{d((u^1\circ\gamma')(t))}{dt},\frac{d((u^2\circ\gamma')(t))}{dt}\right)\right|_{t=0}$$

$$= \left(\left.\frac{d((u^1\circ\gamma')(t))}{dt}\right|_{t=0},\left.\frac{d((u^2\circ\gamma')(t))}{dt}\right|_{t=0}\right)$$

$$= \left(\ll\gamma',\left[u^1\right]^\sim\gg,\ll\gamma',\left[u^2\right]^\sim\gg\right),$$

so,

$$\ll\gamma,\left[u^1\right]^\sim\gg = \ll\gamma',\left[u^1\right]^\sim\gg \quad\text{and}\quad \ll\gamma,\left[u^2\right]^\sim\gg = \ll\gamma',\left[u^2\right]^\sim\gg.$$

Since $[f]^\sim$ in $T_p^*(M)$, by Theorem 2.48,

$$[f]^\sim = \left(\left(D_1\left(f\circ(\varphi_U)^{-1}\right)\right)\left(u^1(p),u^2(p)\right)\right)\left[u^1\right]^\sim + \left(\left(D_2\left(f\circ(\varphi_U)^{-1}\right)\right)\left(u^1(p),u^2(p)\right)\right)\left[u^2\right]^\sim.$$

Now,

$$\text{LHS} = \ll\gamma,[f]^\sim\gg$$

$$= \ll\gamma,\left(\left(D_1\left(f\circ(\varphi_U)^{-1}\right)\right)\left(u^1(p),u^2(p)\right)\right)\left[u^1\right]^\sim + \left(\left(D_2\left(f\circ(\varphi_U)^{-1}\right)\right)\left(u^1(p),u^2(p)\right)\right)\left[u^2\right]^\sim\gg$$

$$= \ll\gamma,\left(\left(\left(D_1\left(f\circ(\varphi_U)^{-1}\right)\right)\left(u^1(p),u^2(p)\right)\right)\left[u^1\right] + \left(\left(D_2\left(f\circ(\varphi_U)^{-1}\right)\right)\left(u^1(p),u^2(p)\right)\right)\left[u^2\right]\right)^\sim\gg$$

$$= \ll\gamma,\left(\left(\left(D_1\left(f\circ(\varphi_U)^{-1}\right)\right)\left(u^1(p),u^2(p)\right)\right)\left[u^1\right] + \left(\left(D_2\left(f\circ(\varphi_U)^{-1}\right)\right)\left(u^1(p),u^2(p)\right)\right)\left[u^2\right]\right)\gg$$

$$= \left(\left(D_1\left(f\circ(\varphi_U)^{-1}\right)\right)\left(u^1(p),u^2(p)\right)\right)\ll\gamma,\left[u^1\right]\gg + \left(\left(D_2\left(f\circ(\varphi_U)^{-1}\right)\right)\left(u^1(p),u^2(p)\right)\right)\ll\gamma,\left[u^2\right]\gg$$

$$= \left(\left(D_1\left(f\circ(\varphi_U)^{-1}\right)\right)\left(u^1(p),u^2(p)\right)\right)\ll\gamma',\left[u^1\right]\gg + \left(\left(D_2\left(f\circ(\varphi_U)^{-1}\right)\right)\left(u^1(p),u^2(p)\right)\right)\ll\gamma',\left[u^2\right]\gg$$

$$= \ll\gamma',\left(\left(\left(D_1\left(f\circ(\varphi_U)^{-1}\right)\right)\left(u^1(p),u^2(p)\right)\right)\left[u^1\right] + \left(\left(D_2\left(f\circ(\varphi_U)^{-1}\right)\right)\left(u^1(p),u^2(p)\right)\right)\left[u^2\right]\right)\gg$$

$$= \ll\gamma',\left(\left(\left(D_1\left(f\circ(\varphi_U)^{-1}\right)\right)\left(u^1(p),u^2(p)\right)\right)\left[u^1\right] + \left(\left(D_2\left(f\circ(\varphi_U)^{-1}\right)\right)\left(u^1(p),u^2(p)\right)\right)\left[u^2\right]\right)^\sim\gg$$

$$= \ll\gamma',\left(\left(D_1\left(f\circ(\varphi_U)^{-1}\right)\right)\left(u^1(p),u^2(p)\right)\right)\left[u^1\right]^\sim + \left(\left(D_2\left(f\circ(\varphi_U)^{-1}\right)\right)\left(u^1(p),u^2(p)\right)\right)\left[u^2\right]^\sim\gg$$

$$= \ll\gamma',[f]^\sim\gg = \text{RHS}.$$

$\square$

**Note 2.60** Let $M$ be an $m$-dimensional smooth manifold. Let $p$ be an element of $M$. Since $\{[\gamma] : \gamma \in \Gamma_p(M)\} = T_pM$, and $\{[\gamma] : \gamma \in \Gamma_p(M)\}$ is a real linear space, $T_pM$ is a real linear space. Since $\eta : [\gamma] \mapsto [\gamma]^*$ is an isomorphism from real linear space $\{[\gamma] : \gamma \in \Gamma_p(M)\}$ onto real linear space $\mathcal{T}_p(M)$, $\{[\gamma] : \gamma \in \Gamma_p(M)\}$ and $\mathcal{T}_p(M)$ are isomorphic, and hence, $T_pM$ and $\mathcal{T}_p(M)$ are isomorphic.

**Definition** Let $M$ be an $m$-dimensional smooth manifold. Let $p$ be an element of $M$. The real linear space $T_pM$ is called the *tangent space of $M$ at $p$*, and the elements of $T_pM$ are called *tangent vectors at $p$*.

**Theorem 2.61** *Let $M$ be a 2-dimensional smooth manifold. Let $p$ be an element of $M$. Then, the mapping $([\gamma], [f]^\sim) \mapsto \langle [\gamma], [f]^\sim \rangle$ from the real linear space $T_p(M) \times T_p^*(M)$ to $\mathbb{R}$ is bilinear, that is, linear in each variable.*

*Proof* Let $(U, \varphi_U)$ be an admissible coordinate chart of $M$ satisfying $p \in U$. We have to show that

1. For every $[\gamma]$ in $\{[\gamma] : \gamma \in \Gamma_p(M)\}$, for every $[f]^\sim, [g]^\sim$ in $T_p^*(M)$, and for every real $s, t$,

$$\langle [\gamma], s[f]^\sim + t[g]^\sim \rangle = s\langle [\gamma], [f]^\sim \rangle + t\langle [\gamma], [g]^\sim \rangle.$$

2. For every $\gamma, \gamma'$ in $\Gamma_p(M)$, for every $[f]^\sim$ in $T_p^*(M)$,

$$\langle [\gamma] + [\gamma'], [f]^\sim \rangle = \langle [\gamma], [f]^\sim \rangle + \langle [\gamma'], [f]^\sim \rangle.$$

3. For every $[f]^\sim$ in $T_p^*(M)$, for every $\gamma$ in $\Gamma_p(M)$, and for every real $s$,

$$\langle s[\gamma], [f]^\sim \rangle = s\langle [\gamma], [f]^\sim \rangle.$$

For 1: LHS $= \langle [\gamma], s[f]^\sim + t[g]^\sim \rangle = \langle [\gamma], [sf]^\sim + [tg]^\sim \rangle$
$= \langle [\gamma], [sf + tg]^\sim \rangle = \ll \gamma, [sf + tg] \gg$
$= s \ll \gamma, [f] \gg + t \ll \gamma, [g] \gg = s \ll \gamma, [f]^\sim \gg + t \ll \gamma, [g]^\sim \gg$
$= s\langle [\gamma], [f]^\sim \rangle + t\langle [\gamma], [g]^\sim \rangle$
$=$ RHS.

For 2: Let us take any $\gamma, \gamma'$ in $\Gamma_p(M)$. Let us take any $[f]^\sim$ in $T_p^*(M)$. We have to show that $\langle [\gamma] + [\gamma'], [f]^\sim \rangle = \langle [\gamma], [f]^\sim \rangle + \langle [\gamma'], [f]^\sim \rangle$.

LHS $= \langle [\gamma] + [\gamma'], [f]^\sim \rangle = \langle [\gamma''], [f]^\sim \rangle$ where $[\gamma''] \equiv [\gamma] + [\gamma']$ for some $\gamma''$ in $\Gamma_p(M)$. Since $[\gamma''] \equiv [\gamma] + [\gamma']$, $([\gamma''])^* = ([\gamma] + [\gamma'])^* = ([\gamma])^* + ([\gamma'])^*$, and hence,

$$\langle [\gamma''], [f]^\sim \rangle = ([\gamma''])^*([f]^\sim) = (([\gamma])^* + ([\gamma'])^*)([f]^\sim)$$
$$= (([\gamma])^*)([f]^\sim) + (([\gamma'])^*)([f]^\sim) = \langle [\gamma], [f]^\sim \rangle + \langle [\gamma'], [f]^\sim \rangle.$$

Thus, we have shown that $\langle [\gamma] + [\gamma'], [f]^{\sim} \rangle = \langle [\gamma], [f]^{\sim} \rangle + [\gamma'], [f]^{\sim}$. This proves 2.

For 3: Let us take any $[f]^{\sim}$ in $T_p^*(M)$. Let us take any $\gamma$ in $\Gamma_p(M)$, and any real $s$. We have to show that $\langle s[\gamma], [f]^{\sim} \rangle = s\langle [\gamma], [f]^{\sim} \rangle$.

LHS $= \langle s[\gamma], [f]^{\sim} \rangle = \langle [\gamma'], [f]^{\sim} \rangle$ where $[\gamma'] \equiv s[\gamma]$ for some $\gamma'$ in $\Gamma_p(M)$. Since $[\gamma'] = s[\gamma]$, $([\gamma'])^* = (s[\gamma])^* = s(([\gamma])^*)$, and hence,

$$\langle [\gamma'], [f]^{\sim} \rangle = ([\gamma'])^* ([f]^{\sim}) = (s(([\gamma])^*))([f]^{\sim}) = s((([\gamma])^*)([f]^{\sim}))$$
$$= s\langle [\gamma], [f]^{\sim} \rangle.$$

Thus, we have shown that $\langle s[\gamma], [f]^{\sim} \rangle = s\langle [\gamma], [f]^{\sim} \rangle$. This proves 3.  □

**Theorem 2.62** *Let $M$ be an $m$-dimensional smooth manifold. Let $p$ be an element of $M$. Then, the mapping $([\gamma], [f]^{\sim}) \mapsto \langle [\gamma], [f]^{\sim} \rangle$ from the real linear space $T_p(M) \times T_p^*(M)$ to $\mathbb{R}$ is bilinear, that is, linear in each variable.*

*Proof* Its proof is quite similar to that of Theorem 2.61, etc.  □

**Definition** Let $M$ be a 2-dimensional smooth manifold. Let $p$ be an element of $M$. Let $(U, \varphi_U)$ be an admissible coordinate chart of $M$ such that $p$ is in $U$. Let $u^1, u^2$ be the component functions of $\varphi_U$. Let us define the function $\mathcal{U}_1 : T_p^*(M) \to \mathbb{R}$ as follows: For every real $t_1, t_2$,

$$\mathcal{U}_1\left(t_1[u^1]^{\sim} + t_2[u^2]^{\sim}\right) \equiv t_1.$$

Clearly, $\mathcal{U}_1$ is a linear functional on $T_p^*(M)$, and hence, $\mathcal{U}_1$ is in $\mathcal{T}_p(M)$. Let us define the function $\mathcal{U}_2 : T_p^*(M) \to \mathbb{R}$ as follows: For every real $t_1, t_2$,

$$\mathcal{U}_2\left(t_1[u^1]^{\sim} + t_2[u^2]^{\sim}\right) \equiv t_2.$$

Clearly, $\mathcal{U}_2$ is a linear functional on $T_p^*(M)$, and hence, $\mathcal{U}_2$ is in $\mathcal{T}_p(M)$.

Since $\eta : [\gamma] \mapsto [\gamma]^*$ is an isomorphism from real linear space $\{[\gamma] : \gamma \in \Gamma_p(M)\}$ onto real linear space $\mathcal{T}_p(M)$, and $\{[\gamma] : \gamma \in \Gamma_p(M)\} = T_p(M)$, $\eta : [\gamma] \mapsto [\gamma]^*$ is an isomorphism from tangent space $T_p(M)$ onto real linear space $\mathcal{T}_p(M)$, and hence, there exists a unique $[\lambda_1]$ in $T_p(M)$ such that $[\lambda_1]^* = \mathcal{U}_1$.

Similarly, there exists a unique $[\lambda_2]$ in $T_p(M)$ such that $[\lambda_2]^* = \mathcal{U}_2$. We know that $\{\mathcal{U}_1, \mathcal{U}_2\}$ is a basis of the dual space $\mathcal{T}_p(M)$ of $T_p^*(M)$. Further, since $[\lambda_1]^* = \mathcal{U}_1$, and $[\lambda_2]^* = \mathcal{U}_2$, $\{[\lambda_1]^*, [\lambda_2]^*\}$ is a basis of $\mathcal{T}_p(M)$. Also,

$$\langle [\lambda_i], [u^j]^{\sim} \rangle = [\lambda_i]^* ([u^j]^{\sim}) = \delta_i^j.$$

Since $\{[\lambda_1]^*, [\lambda_2]^*\}$ is a basis of $\mathcal{T}_p(M)$, and $\eta : [\gamma] \mapsto [\gamma]^*$ is an isomorphism from tangent space $T_p(M)$ onto $\mathcal{T}_p(M)$, $\{[\lambda_1], [\lambda_2]\}$ is a basis of $T_p(M)$.

Here, we say that $\{[\lambda_1], [\lambda_2]\}$ is the *natural basis of the tangent space* $T_p(M)$ under local coordinate system $(u^i)$. Similar definitions can be supplied for a 3-dimensional smooth manifold, etc.

In continuation with the above discussion for 2-dimensional smooth manifold $M$, let us note that for every $[f]^\sim$ in $T_p^*(M)$,

$$
\begin{aligned}
\langle [\lambda_1], (df)_p \rangle &= \langle [\lambda_1], [f]^\sim \rangle \\
&= \langle [\lambda_1], \left( \left( D_1 \left( f \circ (\varphi_U)^{-1} \right) \right) (u^1(p), u^2(p)) \right) [u^1]^\sim \\
&\quad + \left( \left( D_2 \left( f \circ (\varphi_U)^{-1} \right) \right) (u^1(p), u^2(p)) \right) [u^2]^\sim \rangle \\
&= \left( \left( D_1 \left( f \circ (\varphi_U)^{-1} \right) \right) (u^1(p), u^2(p)) \right) \langle [\lambda_1], [u^1]^\sim \rangle \\
&\quad + \left( \left( D_2 \left( f \circ (\varphi_U)^{-1} \right) \right) (u^1(p), u^2(p)) \right) \langle [\lambda_1], [u^2]^\sim \rangle \\
&= \left( \left( D_1 \left( f \circ (\varphi_U)^{-1} \right) \right) (u^1(p), u^2(p)) \right) (1) \\
&\quad + \left( \left( D_2 \left( f \circ (\varphi_U)^{-1} \right) \right) (u^1(p), u^2(p)) \right) (0) \\
&= \left( D_1 \left( f \circ (\varphi_U)^{-1} \right) \right) (u^1(p), u^2(p)).
\end{aligned}
$$

Thus, for every $[f]^\sim$ in $T_p^*(M)$,

$$
\langle [\lambda_1], (df)_p \rangle = \left( D_1 \left( f \circ (\varphi_U)^{-1} \right) \right) (u^1(p), u^2(p)) = \left( \frac{\partial}{\partial u^1} \Big|_p \right) [f].
$$

So

$$
\langle [\lambda_1], (df)_p \rangle = \left( \frac{\partial}{\partial u^1} \Big|_p \right) [f].
$$

Similarly, we have

$$
\langle [\lambda_2], (df)_p \rangle = \left( \frac{\partial}{\partial u^2} \Big|_p \right) [f].
$$

So, for every $i = 1, 2$, and for every $[f]$ in $\mathcal{F}_p(M)$,

$$
\langle [\lambda_i], [f]^\sim \rangle = \left( \frac{\partial}{\partial u^i} \Big|_p \right) [f].
$$

Hence, when there is no confusion, we denote $[\lambda_i]$ by $\frac{\partial}{\partial u^i}\big|_p$.

Further, since $\{[\lambda_1], [\lambda_2]\}$ is the natural basis of the tangent space $T_p(M)$ under local coordinate system $(u^i)$, $\{\frac{\partial}{\partial u^1}\big|_p, \frac{\partial}{\partial u^2}\big|_p\}$ is the natural basis of the tangent space $T_p(M)$ under local coordinate system $(u^i)$. Thus, for every $[f]$ in $\mathcal{F}_p(M)$,

$$\left\langle \frac{\partial}{\partial u^i}\bigg|_p , (df)_p \right\rangle = \left( \frac{\partial}{\partial u^i}\bigg|_p \right)[f].$$

Also, we have

$$\left\langle \frac{\partial}{\partial u^i}\bigg|_p , (du^j)_p \right\rangle = \delta_i^j.$$

**Theorem 2.63** *Let M be a 2-dimensional smooth manifold. Let p be an element of M. Let $(U, \varphi_U)$ be an admissible coordinate chart of M satisfying $p \in U$. Let $\gamma$ be in $\Gamma_p(M)$. Let $u^1$, $u^2$ be the component functions of $\varphi_U$. Then,*

$$\left( \frac{d}{dt}(u^1 \circ \gamma)(t) \right)\bigg|_{t=0}, \quad \left( \frac{d}{dt}(u^2 \circ \gamma)(t) \right)\bigg|_{t=0}$$

*are the components of the tangent vector $[\gamma]$ at p, with respect to the natural basis*

$$\left\{ \frac{\partial}{\partial u^1}\bigg|_p , \frac{\partial}{\partial u^2}\bigg|_p \right\}$$

*of the tangent space $T_p(M)$ under local coordinate system $(u^i)$.*

*Proof* Let us take any $[f]^\sim$ in $T_p^*(M)$, where $f$ is in $C_p^\infty(M)$. Since

$$
\begin{aligned}
[\gamma]^*([f]^\sim) = \langle [\gamma], [f]^\sim \rangle &= \left( \left( D_1\left( f \circ (\varphi_U)^{-1} \right) \right)(u^1(p), u^2(p)) \right)\left( \frac{d}{dt}(u^1 \circ \gamma)(t) \right)\bigg|_{t=0} \\
&\quad + \left( \left( D_2\left( f \circ (\varphi_U)^{-1} \right) \right)(u^1(p), u^2(p)) \right)\left( \frac{d}{dt}(u^2 \circ \gamma)(t) \right)\bigg|_{t=0} \\
&= \langle [\lambda_1], (df)_p \rangle \left( \frac{d}{dt}(u^1 \circ \gamma)(t) \right)\bigg|_{t=0} + \langle [\lambda_2], (df)_p \rangle \left( \frac{d}{dt}(u^2 \circ \gamma)(t) \right)\bigg|_{t=0} \\
&= \left( \left( \frac{d}{dt}(u^1 \circ \gamma)(t) \right)\bigg|_{t=0} \right)\langle [\lambda_1], [f]^\sim \rangle + \left( \left( \frac{d}{dt}(u^2 \circ \gamma)(t) \right)\bigg|_{t=0} \right)\langle [\lambda_2], [f]^\sim \rangle \\
&= \left( \left( \frac{d}{dt}(u^1 \circ \gamma)(t) \right)\bigg|_{t=0} \right)([\lambda_1]^*([f]^\sim)) + \left( \left( \frac{d}{dt}(u^2 \circ \gamma)(t) \right)\bigg|_{t=0} \right)([\lambda_2]^*([f]^\sim)) \\
&= \left( \left( \left( \frac{d}{dt}(u^1 \circ \gamma)(t) \right)\bigg|_{t=0} \right)[\lambda_1]^* \right)([f]^\sim) + \left( \left( \left( \frac{d}{dt}(u^2 \circ \gamma)(t) \right)\bigg|_{t=0} \right)[\lambda_2]^* \right)([f]^\sim) \\
&= \left( \left( \left( \frac{d}{dt}(u^1 \circ \gamma)(t) \right)\bigg|_{t=0} \right)[\lambda_1] \right)^*([f]^\sim) + \left( \left( \left( \frac{d}{dt}(u^2 \circ \gamma)(t) \right)\bigg|_{t=0} \right)[\lambda_2] \right)^*([f]^\sim) \\
&= \left( \left( \left( \left( \frac{d}{dt}(u^1 \circ \gamma)(t) \right)\bigg|_{t=0} \right)[\lambda_1] \right)^* + \left( \left( \left( \frac{d}{dt}(u^2 \circ \gamma)(t) \right)\bigg|_{t=0} \right)[\lambda_2] \right)^* \right)([f]^\sim) \\
&= \left( \left( \left( \frac{d}{dt}(u^1 \circ \gamma)(t) \right)\bigg|_{t=0} \right)[\lambda_1] + \left( \left( \frac{d}{dt}(u^2 \circ \gamma)(t) \right)\bigg|_{t=0} \right)[\lambda_2] \right)^*([f]^\sim),
\end{aligned}
$$

so

$$[\gamma]^* = \left( \left( \left( \frac{d}{dt}(u^1 \circ \gamma)(t) \right) \Big|_{t=0} \right)[\lambda_1] + \left( \left( \frac{d}{dt}(u^2 \circ \gamma)(t) \right) \Big|_{t=0} \right)[\lambda_2] \right)^*.$$

Further, since

$$\eta([\gamma]) = [\gamma]^* = \left( \left( \left( \frac{d}{dt}(u^1 \circ \gamma)(t) \right) \Big|_{t=0} \right)[\lambda_1] + \left( \left( \frac{d}{dt}(u^2 \circ \gamma)(t) \right) \Big|_{t=0} \right)[\lambda_2] \right)^*$$

$$= \eta \left( \left[ \left( \left( \frac{d}{dt}(u^1 \circ \gamma)(t) \right) \Big|_{t=0} \right)[\lambda_1] + \left( \left( \frac{d}{dt}(u^2 \circ \gamma)(t) \right) \Big|_{t=0} \right)[\lambda_2] \right] \right),$$

and $\eta$ is a 1–1 mapping from $T_p(M)$ onto $\mathcal{T}_p(M)$,

$$[\gamma] = \left( \left( \frac{d}{dt}(u^1 \circ \gamma)(t) \right) \Big|_{t=0} \right)[\lambda_1] + \left( \left( \frac{d}{dt}(u^2 \circ \gamma)(t) \right) \Big|_{t=0} \right)[\lambda_2]$$

$$= \left( \left( \frac{d}{dt}(u^1 \circ \gamma)(t) \right) \Big|_{t=0} \right)\left( \frac{\partial}{\partial u^1} \Big|_p \right) + \left( \left( \frac{d}{dt}(u^2 \circ \gamma)(t) \right) \Big|_{t=0} \right)\left( \frac{\partial}{\partial u^2} \Big|_p \right).$$

Further, since $\{[\lambda_1], [\lambda_2]\}$ is a basis of the tangent space $T_p(M)$, that is,

$$\left\{ \left( \frac{\partial}{\partial u^1} \Big|_p \right), \left( \frac{\partial}{\partial u^2} \Big|_p \right) \right\}$$

is a basis of the tangent space $T_p(M)$,

$$\left( \frac{d}{dt}(u^1 \circ \gamma)(t) \right) \Big|_{t=0}, \left( \frac{d}{dt}(u^2 \circ \gamma)(t) \right) \Big|_{t=0}$$

are the components of the tangent vector $[\gamma]$ at $p$, with respect to the natural basis. $\square$

**Note 2.64** As above, for $m$-dimensional smooth manifold $M$, we get the formula:

$$[\gamma] = \left( \left( \frac{d}{dt}(u^1 \circ \gamma)(t) \right) \Big|_{t=0} \right)\left( \frac{\partial}{\partial u^1} \Big|_p \right) + \cdots + \left( \left( \frac{d}{dt}(u^m \circ \gamma)(t) \right) \Big|_{t=0} \right)\left( \frac{\partial}{\partial u^m} \Big|_p \right).$$

Further, since $\{[\lambda_1], \ldots, [\lambda_m]\}$ is a basis of the tangent space $T_p(M)$, that is,

$$\left\{ \left( \frac{\partial}{\partial u^1} \Big|_p \right), \ldots, \left( \frac{\partial}{\partial u^m} \Big|_p \right) \right\}$$

is a basis of the tangent space $T_p(M)$,

$$\left.\left(\frac{d}{dt}(u^1 \circ \gamma)(t)\right)\right|_{t=0}, \ldots, \left.\left(\frac{d}{dt}(u^m \circ \gamma)(t)\right)\right|_{t=0}$$

are the components of the tangent vector $[\gamma]$ at $p$, with respect to the natural basis.

**Note 2.65** When there is no confusion, we generally avoid writing the lower index $p$ of the tangent vectors, cotangent vectors, etc.

**Theorem 2.66** *Let $M$ be an $m$-dimensional smooth manifold. Let $p$ be an element of $M$. Let $(df)_p, (dg)_p$ be in $T_p^*$, where $f, g$ are in $C_p^\infty$. Let $X$ be in $T_pM$. Let $s, t$ be any real numbers. Then,*

1. $\langle X, (d(sf + tg))_p \rangle = s \cdot \langle X, (df)_p \rangle + t \cdot \langle X, (dg)_p \rangle,$
2. $\langle X, (d(f \cdot g))_p \rangle = f(p) \cdot \langle X, (dg)_p \rangle + g(p) \cdot \langle X, (df)_p \rangle.$

*Proof*

1: By Theorem 2.45, $\langle X, (d(sf + tg))_p \rangle = \langle X, s \cdot ((df)_p) + t \cdot ((dg)_p) \rangle$. Further, since the mapping $([\gamma], [f]^\sim) \mapsto \langle [\gamma], [f]^\sim \rangle$ from the real linear space $T_p \times T_p^*$ to $\mathbb{R}$ is linear in the second variable, $\langle X, s \cdot ((df)_p) + t \cdot ((dg)_p) \rangle = s \cdot \langle X, (df)_p + t \cdot X, (dg)_p \rangle$, and hence,

$$\langle X, (d(sf + tg))_p \rangle = s \cdot \langle X, (df)_p + t \cdot X, (dg)_p \rangle.$$

This proves 1.

2: By Theorem 2.45, $\langle X, (d(f \cdot g))_p \rangle = \langle X, (g(p))(df)_p + (f(p))(dg)_p \rangle$. Further, since the mapping $([\gamma], [f]^\sim) \mapsto \langle [\gamma], [f]^\sim \rangle$ from the real linear space $T_p \times T_p^*$ to $\mathbb{R}$ is linear in the second variable, $\langle X, (g(p))(df)_p + (f(p))(dg)_p \rangle = (g(p)) \cdot \langle X, (df)_p \rangle + (f(p)) \cdot \langle X, (dg)_p \rangle$, and hence,

$$\langle X, (d(f \cdot g))_p \rangle = (g(p)) \cdot \langle X, (df)_p \rangle + (f(p)) \cdot \langle X, (dg)_p \rangle.$$

This proves 2.                                                                    □

## 2.7 Contravariant Vectors and Covariant Vectors

**Definition** Let $M$ be an $m$-dimensional smooth manifold. Let $p$ be an element of $M$. Let $(df)_p$ be in $T_p^*$, where $f$ is in $C_p^\infty$. Let $X$ be in $T_pM$. By $Xf$, we mean $\langle X, (df)_p \rangle$, and is called the *directional derivative of the function $f$ along the tangent vector $X$*.

Now, the Theorem 2.66 can be restated as follows:

**Theorem 2.67** *Let M be an m-dimensional smooth manifold. Let p be an element of M. Let X be in $T_pM$. Then, for every $f, g$ in $C_p^\infty$, and for every real $s, t$,*

1. $X(sf + tg) = s \cdot (Xf) + t \cdot (Xg)$,
2. $X(f \cdot g) = f(p) \cdot X(f) + g(p) \cdot X(g)$.

**Theorem 2.68** *Let M be a 2-dimensional smooth manifold. Let p be an element of M. Let $[\gamma]$ be in $T_pM$, where $\gamma$ is in $\gamma \in \Gamma_p(M)$. Let $(U, \varphi_U), (U', \varphi'_U)$ be two admissible coordinate charts of M satisfying $p \in U$, and $p \in U'$. Let $u^1, u^2$ be the component functions of $\varphi_U$. Let $u'^1, u'^2$ be the component functions of $\varphi'_U$. Then,*

$$\left(\frac{d}{dt}(u'^1 \circ \gamma)(t)\right)\bigg|_{t=0} = \left(\frac{\partial u'^1}{\partial u^1}\right)\left(\frac{d}{dt}(u^1 \circ \gamma)(t)\right)\bigg|_{t=0} + \left(\frac{\partial u'^1}{\partial u^2}\right)\left(\frac{d}{dt}(u^2 \circ \gamma)(t)\right)\bigg|_{t=0},$$

*where*

$$\frac{\partial u'^1}{\partial u^1} \equiv \left(D_1\left(u'^1 \circ (\varphi_U)^{-1}\right)\right)(u^1(p), u^2(p)) = \left(\frac{\partial}{\partial u^1}\bigg|_p\right)[u'^1],$$

*and*

$$\frac{\partial u'^1}{\partial u^2} \equiv \left(D_2\left(u'^1 \circ (\varphi_U)^{-1}\right)\right)(u^1(p), u^2(p)) = \left(\frac{\partial}{\partial u^2}\bigg|_p\right)[u'^1].$$

*Proof* Here,

$$\left(\frac{d}{dt}(u^1 \circ \gamma)(t)\right)\bigg|_{t=0}, \quad \left(\frac{d}{dt}(u^2 \circ \gamma)(t)\right)\bigg|_{t=0}$$

are the components of the tangent vector $[\gamma]$ at $p$, with respect to the natural basis

$$\left\{\frac{\partial}{\partial u^1}\bigg|_p, \frac{\partial}{\partial u^2}\bigg|_p\right\}$$

of the tangent space $T_pM$ under local coordinate system $(u^i)$. Also,

$$\left(\frac{d}{dt}(u'^1 \circ \gamma)(t)\right)\bigg|_{t=0}, \quad \left(\frac{d}{dt}(u'^2 \circ \gamma)(t)\right)\bigg|_{t=0}$$

are the components of the tangent vector $[\gamma]$ at $p$, with respect to the natural basis

$$\left\{ \frac{\partial}{\partial u'^1}\Big|_p, \frac{\partial}{\partial u'^2}\Big|_p \right\}$$

of the tangent space $T_pM$ under local coordinate system $(u'^i)$. Since $\gamma$ is in $\Gamma_p(M)$, $\gamma(0) = p$. We have to prove that

$$\left( \frac{d}{dt}(u'^1 \circ \gamma)(t) \right)\Big|_{t=0} = \left( \frac{\partial u'^1}{\partial u^1} \right)\left( \frac{d}{dt}(u^1 \circ \gamma)(t) \right)\Big|_{t=0} + \left( \frac{\partial u'^1}{\partial u^2} \right)\left( \frac{d}{dt}(u^2 \circ \gamma)(t) \right)\Big|_{t=0}.$$

$$\begin{aligned}
\text{LHS} &= \left( \frac{d}{dt}(u'^1 \circ \gamma)(t) \right)\Big|_{t=0} = \left( \frac{d}{dt}\left((\varphi'_U \circ \gamma)^1\right)(t) \right)\Big|_{t=0} \\
&= \left( \frac{d}{dt}\left(\left(\left(\varphi'_U \circ \left((\varphi_U)^{-1} \circ \varphi_U\right) \circ \gamma\right)\right)^1\right)(t) \right)\Big|_{t=0} \\
&= \left( \frac{d}{dt}\left(\left(\left(\varphi'_U \circ (\varphi_U)^{-1}\right) \circ (\varphi_U \circ \gamma)^1\right)\right)(t) \right)\Big|_{t=0} \\
&= \left( \frac{d}{dt}\left(\left(\varphi'_U \circ (\varphi_U)^{-1}\right)^1 \circ (\varphi_U \circ \gamma)\right)(t) \right)\Big|_{t=0} \\
&= \left[ \left(D_1\left(\left(\varphi'_U \circ (\varphi_U)^{-1}\right)^1\right)\right)((\varphi_U \circ \gamma)(0)) \quad \left(D_2\left(\left(\varphi'_U \circ (\varphi_U)^{-1}\right)^1\right)\right)((\varphi_U \circ \gamma)(0)) \right] \\
&\quad \times \begin{bmatrix} \left(\frac{d}{dt}\left((\varphi_U \circ \gamma)^1\right)(t)\right)\Big|_{t=0} \\ \left(\frac{d}{dt}\left((\varphi_U \circ \gamma)^2\right)(t)\right)\Big|_{t=0} \end{bmatrix} \\
&= \left[ \left(D_1\left(\left(\varphi'_U \circ (\varphi_U)^{-1}\right)^1\right)\right)(\varphi_U(\gamma(0))) \quad \left(D_2\left(\left(\varphi'_U \circ (\varphi_U)^{-1}\right)^1\right)\right)(\varphi_U(\gamma(0))) \right] \\
&\quad \times \begin{bmatrix} \left(\frac{d}{dt}\left((\varphi_U \circ \gamma)^1\right)(t)\right)\Big|_{t=0} \\ \left(\frac{d}{dt}\left((\varphi_U \circ \gamma)^2\right)(t)\right)\Big|_{t=0} \end{bmatrix} \\
&= \left[ \left(D_1\left(\left(\varphi'_U \circ (\varphi_U)^{-1}\right)^1\right)\right)(\varphi_U(p)) \quad \left(D_2\left(\left(\varphi'_U \circ (\varphi_U)^{-1}\right)^1\right)\right)(\varphi_U(p)) \right] \\
&\quad \times \begin{bmatrix} \left(\frac{d}{dt}\left((\varphi_U \circ \gamma)^1\right)(t)\right)\Big|_{t=0} \\ \left(\frac{d}{dt}\left((\varphi_U \circ \gamma)^2\right)(t)\right)\Big|_{t=0} \end{bmatrix} \\
&= \left[ \left(D_1\left(\left(\varphi'_U \circ (\varphi_U)^{-1}\right)^1\right)\right)(\varphi_U(p)) \quad \left(D_2\left(\left(\varphi'_U \circ (\varphi_U)^{-1}\right)^1\right)\right)(\varphi_U(p)) \right] \\
&\quad \times \begin{bmatrix} \left(\frac{d}{dt}\left(\left((\varphi_U)^1\right) \circ \gamma\right)(t)\right)\Big|_{t=0} \\ \left(\frac{d}{dt}\left(\left((\varphi_U)^2\right) \circ \gamma\right)(t)\right)\Big|_{t=0} \end{bmatrix}
\end{aligned}$$

$$= \left[ \left( D_1\left( \left( \varphi'_U \circ (\varphi_U)^{-1} \right)^1 \right) \right)(\varphi_U(p)) \quad \left( D_2\left( \left( \varphi'_U \circ (\varphi_U)^{-1} \right)^1 \right) \right)(\varphi_U(p)) \right]$$

$$\times \begin{bmatrix} \left( \frac{\mathrm{d}}{\mathrm{d}t}\left( \left((\varphi_U)^1\right)(\gamma(t)) \right) \right)\Big|_{t=0} \\ \left( \frac{\mathrm{d}}{\mathrm{d}t}\left( \left((\varphi_U)^2\right)(\gamma(t)) \right) \right)\Big|_{t=0} \end{bmatrix}$$

$$= \left[ \left( D_1\left( \left( \varphi'_U \circ (\varphi_U)^{-1} \right)^1 \right) \right)(\varphi_U(p)) \quad \left( D_2\left( \left( \varphi'_U \circ (\varphi_U)^{-1} \right)^1 \right) \right)(\varphi_U(p)) \right]$$

$$\times \begin{bmatrix} \left( \frac{\mathrm{d}}{\mathrm{d}t}\left((u^1)(\gamma(t))\right) \right)\Big|_{t=0} \\ \left( \frac{\mathrm{d}}{\mathrm{d}t}\left((u^2)(\gamma(t))\right) \right)\Big|_{t=0} \end{bmatrix}$$

$$= \left[ \left( D_1\left( \left( \varphi'_U \circ (\varphi_U)^{-1} \right)^1 \right) \right)(\varphi_U(p)) \quad \left( D_2\left( \left( \varphi'_U \circ (\varphi_U)^{-1} \right)^1 \right) \right)(\varphi_U(p)) \right]$$

$$\times \begin{bmatrix} \left( \frac{\mathrm{d}}{\mathrm{d}t}(u^1 \circ \gamma)(t) \right)\Big|_{t=0} \\ \left( \frac{\mathrm{d}}{\mathrm{d}t}(u^2 \circ \gamma)(t) \right)\Big|_{t=0} \end{bmatrix}$$

$$= \left[ \left( D_1\left( \left( \varphi'_U \circ (\varphi_U)^{-1} \right)^1 \right) \right)(u^1(p), u^2(p)) \quad \left( D_2\left( \left( \varphi'_U \circ (\varphi_U)^{-1} \right)^1 \right) \right)(u^1(p), u^2(p)) \right]$$

$$\times \begin{bmatrix} \left( \frac{\mathrm{d}}{\mathrm{d}t}(u^1 \circ \gamma)(t) \right)\Big|_{t=0} \\ \left( \frac{\mathrm{d}}{\mathrm{d}t}(u^2 \circ \gamma)(t) \right)\Big|_{t=0} \end{bmatrix}$$

$$= \left( \left( D_1\left( \left( \varphi'_U \circ (\varphi_U)^{-1} \right)^1 \right) \right)(u^1(p), u^2(p)) \right)\left( \frac{\mathrm{d}}{\mathrm{d}t}(u^1 \circ \gamma)(t) \right)\Big|_{t=0}$$
$$+ \left( \left( D_2\left( \left( \varphi'_U \circ (\varphi_U)^{-1} \right)^1 \right) \right)(u^1(p), u^2(p)) \right)\left( \frac{\mathrm{d}}{\mathrm{d}t}(u^2 \circ \gamma)(t) \right)\Big|_{t=0} .$$

$$\text{RHS} = \left( \frac{\partial u'^1}{\partial u^1} \right)\left( \frac{\mathrm{d}}{\mathrm{d}t}(u^1 \circ \gamma)(t) \right)\Big|_{t=0} + \left( \frac{\partial u'^1}{\partial u^2} \right)\left( \frac{\mathrm{d}}{\mathrm{d}t}(u^2 \circ \gamma)(t) \right)\Big|_{t=0}$$

$$= \left( \left( D_1\left( u'^1 \circ (\varphi_U)^{-1} \right) \right)(u^1(p), u^2(p)) \right)\left( \frac{\mathrm{d}}{\mathrm{d}t}(u^1 \circ \gamma)(t) \right)\Big|_{t=0}$$
$$+ \left( \left( D_2\left( u'^1 \circ (\varphi_U)^{-1} \right) \right)(u^1(p), u^2(p)) \right)\left( \frac{\mathrm{d}}{\mathrm{d}t}(u^2 \circ \gamma)(t) \right)\Big|_{t=0}$$

$$= \left( \left( D_1\left( \left( \varphi'_U \circ (\varphi_U)^{-1} \right)^1 \right) \right)(u^1(p), u^2(p)) \right)\left( \frac{\mathrm{d}}{\mathrm{d}t}(u^1 \circ \gamma)(t) \right)\Big|_{t=0}$$
$$+ \left( \left( D_2\left( \left( \varphi'_U \circ (\varphi_U)^{-1} \right)^1 \right) \right)(u^1(p), u^2(p)) \right)\left( \frac{\mathrm{d}}{\mathrm{d}t}(u^2 \circ \gamma)(t) \right)\Big|_{t=0} .$$

Hence, LHS = RHS.                                                                                           $\square$

**Note 2.69** Similarly, under the same conditions as in theorem 34, we get

$$\left(\frac{d}{dt}\left(u'^2 \circ \gamma\right)(t)\right)\bigg|_{t=0} = \left(\frac{\partial u'^2}{\partial u^1}\right)\left(\frac{d}{dt}\left(u^1 \circ \gamma\right)(t)\right)\bigg|_{t=0} + \left(\frac{\partial u'^2}{\partial u^2}\right)\left(\frac{d}{dt}\left(u^2 \circ \gamma\right)(t)\right)\bigg|_{t=0}.$$

In short, we write

$$\left(\frac{d}{dt}\left(u'^i \circ \gamma\right)(t)\right)\bigg|_{t=0} = \left(\frac{\partial u'^i}{\partial u^1}\right)\left(\frac{d}{dt}\left(u^1 \circ \gamma\right)(t)\right)\bigg|_{t=0} + \left(\frac{\partial u'^i}{\partial u^2}\right)\left(\frac{d}{dt}\left(u^2 \circ \gamma\right)(t)\right)\bigg|_{t=0},$$

or

$$\left(\frac{d}{dt}\left(u'^i \circ \gamma\right)(t)\right)\bigg|_{t=0} = \sum_{j=1}^{2}\left(\frac{\partial u'^i}{\partial u^j}\right)\left(\frac{d}{dt}\left(u^j \circ \gamma\right)(t)\right)\bigg|_{t=0},$$

and we say that *tangent vectors are contravariant vectors*.

Similar results can be obtained for 3-dimensional smooth manifold, etc.

**Theorem 2.70** *Let M be a 2-dimensional smooth manifold. Let p be an element of M. Let* $(df)_p$ *be in* $T_p^*$, *where f is in* $C_p^\infty$. *Let* $(U, \varphi_U), (U', \varphi'_U)$ *be two admissible coordinate charts of M satisfying* $p \in U$, $p \in U'$. *Let* $u^1, u^2$ *be the component functions of* $\varphi_U$. *Let* $u'^1, u'^2$ *be the component functions of* $\varphi'_U$. *Then,*

$$\frac{\partial f}{\partial u'^1} = \left(\frac{\partial u^1}{\partial u'^1}\right)\frac{\partial f}{\partial u^1} + \left(\frac{\partial u^2}{\partial u'^1}\right)\frac{\partial f}{\partial u^2},$$

*where*

$$\frac{\partial f}{\partial u'^1} \equiv \left(D_1\left(f \circ (\varphi'_U)^{-1}\right)\right)\left(u'^1(p), u'^2(p)\right) = \left(\frac{\partial}{\partial u'^1}\bigg|_p\right)[f],$$

$$\frac{\partial f}{\partial u^1} \equiv \left(D_1\left(f \circ (\varphi_U)^{-1}\right)\right)\left(u^1(p), u^2(p)\right) = \left(\frac{\partial}{\partial u^1}\bigg|_p\right)[f],$$

$$\frac{\partial f}{\partial u^2} \equiv \left(D_2\left(f \circ (\varphi_U)^{-1}\right)\right)\left(u^1(p), u^2(p)\right) = \left(\frac{\partial}{\partial u^1}\bigg|_p\right)[f].$$

*Proof* By Theorem 2.48,

$$\left(\frac{\partial f}{\partial u^1}\right)_p \left(\equiv \left(D_1\left(f \circ (\varphi_U)^{-1}\right)\right)\left(u^1(p), u^2(p)\right)\right), \left(\frac{\partial f}{\partial u^2}\right)_p \left(\equiv \left(D_2\left(f \circ (\varphi_U)^{-1}\right)\right)\left(u^1(p), u^2(p)\right)\right)$$

are the components of the cotangent vector $(df)_p$ at $p$, with respect to the natural basis $\{(du^1)_p, (du^2)_p\}$ of the cotangent space $T_p^*$ under local coordinate system $(u^i)$. Also,

$$\left(\frac{\partial f}{\partial u'^1}\right)_p \left(\equiv \left(D_1\left(f \circ (\varphi'_U)^{-1}\right)\right)(u'^1(p), u'^2(p))\right), \left(\frac{\partial f}{\partial u'^2}\right)_p \left(\equiv \left(D_2\left(f \circ (\varphi'_U)^{-1}\right)\right)(u'^1(p), u'^2(p))\right)$$

are the components of the cotangent vector $(df)_p$ at $p$, with respect to the natural basis $\{(du'^1)_p, (du'^2)_p\}$ of the cotangent space $T_p^*$ under local coordinate system $(u'^i)$. We have to prove that

$$\frac{\partial f}{\partial u'^1} = \left(\frac{\partial u^1}{\partial u'^1}\right)\frac{\partial f}{\partial u^1} + \left(\frac{\partial u^2}{\partial u'^1}\right)\frac{\partial f}{\partial u^2}.$$

$$\text{LHS} = \frac{\partial f}{\partial u'^1} = \left(D_1\left(f \circ (\varphi'_U)^{-1}\right)\right)(u'^1(p), u'^2(p))$$

$$= \left(D_1\left(f \circ \left((\varphi_U)^{-1} \circ \varphi_U\right) \circ (\varphi'_U)^{-1}\right)\right)(u'^1(p), u'^2(p))$$

$$= \left(D_1\left(\left(f \circ (\varphi_U)^{-1}\right) \circ \left(\varphi_U \circ (\varphi'_U)^{-1}\right)\right)\right)(u'^1(p), u'^2(p))$$

$$= \text{1st column of the matrix product}$$

$$\left[\left(D_1\left(f \circ (\varphi_U)^{-1}\right)\right)\left(\left(\left(\varphi_U \circ (\varphi'_U)^{-1}\right)\right)(u'^1(p), u'^2(p))\right)\right.$$
$$\left.\left(D_2\left(f \circ (\varphi_U)^{-1}\right)\right)\left(\left(\left(\varphi_U \circ (\varphi'_U)^{-1}\right)\right)(u'^1(p), u'^2(p))\right)\right]_{1 \times 2}$$

$$\times \left[\begin{array}{cc} \left(\left(D_1\left(\left(\varphi_U \circ (\varphi'_U)^{-1}\right)^1\right)\right)(u'^1(p), u'^2(p))\right) & \left(\left(D_2\left(\left(\varphi_U \circ (\varphi'_U)^{-1}\right)^1\right)\right)(u'^1(p), u'^2(p))\right) \\ \left(\left(D_1\left(\left(\varphi_U \circ (\varphi'_U)^{-1}\right)^2\right)\right)(u'^1(p), u'^2(p))\right) & \left(\left(D_2\left(\left(\varphi_U \circ (\varphi'_U)^{-1}\right)^2\right)\right)(u'^1(p), u'^2(p))\right) \end{array}\right]$$

$$= \left(\left(D_1\left(f \circ (\varphi_U)^{-1}\right)\right)\left(\left(\left(\varphi_U \circ (\varphi'_U)^{-1}\right)\right)(u'^1(p), u'^2(p))\right)\right)$$
$$\left(\left(D_1\left(\left(\varphi_U \circ (\varphi'_U)^{-1}\right)^1\right)\right)(u'^1(p), u'^2(p))\right)$$
$$+ \left(\left(D_2\left(f \circ (\varphi_U)^{-1}\right)\right)\left(\left(\left(\varphi_U \circ (\varphi'_U)^{-1}\right)\right)(u'^1(p), u'^2(p))\right)\right)$$
$$\left(\left(D_1\left(\left(\varphi_U \circ (\varphi'_U)^{-1}\right)^2\right)\right)(u'^1(p), u'^2(p))\right)$$

$$= \left(\left(D_1\left(f \circ (\varphi_U)^{-1}\right)\right)\left(\left(\left(\varphi_U \circ (\varphi'_U)^{-1}\right)\right)(\varphi'_U(p))\right)\right)\left(\left(D_1\left(\left(\varphi_U \circ (\varphi'_U)^{-1}\right)^1\right)\right)(\varphi'_U(p))\right)$$
$$+ \left(\left(D_2\left(f \circ (\varphi_U)^{-1}\right)\right)\left(\left(\left(\varphi_U \circ (\varphi'_U)^{-1}\right)\right)(\varphi'_U(p))\right)\right)\left(\left(D_1\left(\left(\varphi_U \circ (\varphi'_U)^{-1}\right)^2\right)\right)(\varphi'_U(p))\right)$$

$$= \left(\left(D_1\left(f \circ (\varphi_U)^{-1}\right)\right)\left(\left(\left(\varphi_U \circ (\varphi'_U)^{-1}\right) \circ \varphi'_U\right)(p)\right)\right)\left(\left(D_1\left(\left(\varphi_U \circ (\varphi'_U)^{-1}\right)^1\right)\right)(\varphi'_U(p))\right)$$
$$+ \left(\left(D_2\left(f \circ (\varphi_U)^{-1}\right)\right)\left(\left(\left(\varphi_U \circ (\varphi'_U)^{-1}\right) \circ \varphi'_U\right)(p)\right)\right)\left(\left(D_1\left(\left(\varphi_U \circ (\varphi'_U)^{-1}\right)^2\right)\right)(\varphi'_U(p))\right)$$

$$
= \left( \left( D_1 \left( f \circ (\varphi_U)^{-1} \right) \right) \left( \left( \varphi_U \circ \left( (\varphi_U')^{-1} \circ \varphi_U' \right) \right)(p) \right) \right) \left( \left( D_1 \left( \left( \varphi_U \circ (\varphi_U')^{-1} \right)^1 \right) \right)(\varphi_U'(p)) \right)
$$

$$
+ \left( \left( D_2 \left( f \circ (\varphi_U)^{-1} \right) \right) \left( \left( \varphi_U \circ \left( (\varphi_U')^{-1} \circ \varphi_U' \right) \right)(p) \right) \right) \left( \left( D_1 \left( \left( \varphi_U \circ (\varphi_U')^{-1} \right)^2 \right) \right)(\varphi_U'(p)) \right)
$$

$$
= \left( \left( D_1 \left( f \circ (\varphi_U)^{-1} \right) \right)(\varphi_U(p)) \right) \left( \left( D_1 \left( \left( \varphi_U \circ (\varphi_U')^{-1} \right)^1 \right) \right)(\varphi_U'(p)) \right)
$$

$$
+ \left( \left( D_2 \left( f \circ (\varphi_U)^{-1} \right) \right)(\varphi_U(p)) \right) \left( \left( D_1 \left( \left( \varphi_U \circ (\varphi_U')^{-1} \right)^2 \right) \right)(\varphi_U'(p)) \right)
$$

$$
= \left( \left( D_1 \left( \left( \varphi_U \circ (\varphi_U')^{-1} \right)^1 \right) \right)(\varphi_U'(p)) \right) \left( \left( D_1 \left( f \circ (\varphi_U)^{-1} \right) \right)(\varphi_U(p)) \right)
$$

$$
+ \left( \left( D_1 \left( \left( \varphi_U \circ (\varphi_U')^{-1} \right)^2 \right) \right)(\varphi_U'(p)) \right) \left( \left( D_2 \left( f \circ (\varphi_U)^{-1} \right) \right)(\varphi_U(p)) \right).
$$

$$
\text{RHS} = \left( \frac{\partial u^1}{\partial u'^1} \right) \frac{\partial f}{\partial u^1} + \left( \frac{\partial u^2}{\partial u'^1} \right) \frac{\partial f}{\partial u^2}
$$

$$
= \left( \left( D_1 \left( u^1 \circ (\varphi_U')^{-1} \right) \right) \left( u'^1(p), u'^2(p) \right) \right) \frac{\partial f}{\partial u^1}
$$

$$
+ \left( \left( D_1 \left( u^2 \circ (\varphi_U')^{-1} \right) \right) \left( u'^1(p), u'^2(p) \right) \right) \frac{\partial f}{\partial u^2}
$$

$$
= \left( \left( D_1 \left( \left( \varphi_U \circ (\varphi_U')^{-1} \right)^1 \right) \right) \left( u'^1(p), u'^2(p) \right) \right) \frac{\partial f}{\partial u^1}
$$

$$
+ \left( \left( D_1 \left( \left( \varphi_U \circ (\varphi_U')^{-1} \right)^2 \right) \right) \left( u'^1(p), u'^2(p) \right) \right) \frac{\partial f}{\partial u^2}
$$

$$
= \left( \left( D_1 \left( \left( \varphi_U \circ (\varphi_U')^{-1} \right)^1 \right) \right) \left( u'^1(p), u'^2(p) \right) \right)
$$

$$
\times \left( \left( D_1 \left( f \circ (\varphi_U)^{-1} \right) \right) \left( u^1(p), u^2(p) \right) \right)
$$

$$
+ \left( \left( D_1 \left( \left( \varphi_U \circ (\varphi_U')^{-1} \right)^2 \right) \right) \left( u'^1(p), u'^2(p) \right) \right)
$$

$$
\times \left( \left( D_2 \left( f \circ (\varphi_U)^{-1} \right) \right) \left( u^1(p), u^2(p) \right) \right)
$$

$$
= \left( \left( D_1 \left( \left( \varphi_U \circ (\varphi_U')^{-1} \right)^1 \right) \right)(\varphi_U'(p)) \right) \left( \left( D_1 \left( f \circ (\varphi_U)^{-1} \right) \right)(\varphi_U(p)) \right)
$$

$$
+ \left( \left( D_1 \left( \left( \varphi_U \circ (\varphi_U')^{-1} \right)^2 \right) \right)(\varphi_U'(p)) \right) \left( \left( D_2 \left( f \circ (\varphi_U)^{-1} \right) \right)(\varphi_U(p)) \right).
$$

So, LHS = RHS.                                                                              $\square$

**Note 2.71** Similarly, under the same conditions as in Theorem 2.70, we get

$$\frac{\partial f}{\partial u'^2} = \left(\frac{\partial u^1}{\partial u'^2}\right)\frac{\partial f}{\partial u^1} + \left(\frac{\partial u^2}{\partial u'^2}\right)\frac{\partial f}{\partial u^2}.$$

In short, we write

$$\frac{\partial f}{\partial u'^i} = \sum_{j=1}^{2}\left(\frac{\partial u^j}{\partial u'^i}\right)\frac{\partial f}{\partial u^j},$$

and we say that *cotangent vectors are covariant vectors.*

Similar results can be obtained for 3-dimensional smooth manifold, etc.

**Note 2.72** Let $M$ be an $m$-dimensional smooth manifold, and $N$ be an $n$-dimensional smooth manifold. Let $p$ be an element of $M$, and let $F : M \to N$ be any smooth map. Then, for every $f$ in $C_{F(p)}^{\infty}(N)$, $f \circ F$ is in $C_p^{\infty}(M)$.

*Reason:* Since $f$ is in $C_{F(p)}^{\infty}(N)$, $f$ is a real-valued function such that $\mathrm{dom}f$ is an open neighborhood of $F(p)$ in $N$. Since $F : M \to N$ is a smooth map, and $p$ is in $M$, $F : M \to N$ is continuous at $p$. Since $F : M \to N$ is continuous at $p$, and $\mathrm{dom}f$ is an open neighborhood of $F(p)$ in $N$, $F^{-1}(\mathrm{dom}f)$ is an open neighborhood of $p$ in $M$. Since $\mathrm{dom}(f \circ F) = F^{-1}(\mathrm{dom}f)$, and $F^{-1}(\mathrm{dom}f)$ is an open neighborhood of $p$ in $M$, $\mathrm{dom}(f \circ F)$ is an open neighborhood of $p$ in $M$. Now, it remains to be proved that $f \circ F$ is $C^{\infty}$ at $p$.

Since $F : M \to N$ is a smooth map, and $p$ is in $M$, $F$ is $C^{\infty}$ at $p$. Since $f$ is in $C_{F(p)}^{\infty}(N)$, $f$ is a real-valued function such that $\mathrm{dom}\,f$ is an open neighborhood of $F(p)$ in $N$, and $f$ is $C^{\infty}$ at $F(p)$ in $N$. Since the $F$-image of $\mathrm{dom}(f \circ F)$ is contained in $\mathrm{dom}f$, $F$ is $C^{\infty}$ at $p$, and $f$ is $C^{\infty}$ at $F(p)$, the composite function $f \circ F$ is $C^{\infty}$ at $p$.

Hence, $f \circ F$ is in $C_p^{\infty}(M)$.

**Theorem 2.73** *Let $M$ be any $m$-dimensional smooth manifold, and let $N$ be any $n$-dimensional smooth manifold. Let $F : M \to N$ be any smooth map, and let $p$ be in $M$. Let*

$$F^* : T_{F(p)}^* \to T_p^*$$

*be the function defined as follows: For every $(df)_{F(p)}$ in $T_{F(p)}^*$, where $f$ is in $C_{F(p)}^{\infty}(N)$,*

$$F^*\left((df)_{F(p)}\right) = (d(f \circ F))_p.$$

*Then, $F^*$ is linear.*

*Proof* For every $f$ in $C_{F(p)}^\infty(N)$, from the above note, $f \circ F$ is in $C_p^\infty(M)$, and hence, $(d(f \circ F))_p$ is in $T_p^*$. Since, for every $f$ in $C_{F(p)}^\infty(N)$, $F^*((df)_{F(p)})$ $(\equiv (d(f \circ F))_p)$ is in $T_p^*$, $F^* : T_{F(p)}^* \to T_p^*$ is a function. Next, let us take any $f, g$ in $C_{F(p)}^\infty(N)$, and any real $s, t$. We have to show that

$$F^*\Big((d(sf + tg))_{F(p)}\Big) = s\Big(F^*\big((df)_{F(p)}\big)\Big) + t\Big(F^*\big((dg)_{F(p)}\big)\Big).$$

On using Theorem 2.45, we have

$$\text{LHS} = F^*\Big((d(sf + tg))_{F(p)}\Big) = (d((sf + tg) \circ F))_p = (d(((sf) \circ F + (tg) \circ F)))_p$$

$$= (d((s(f \circ F) + t(g \circ F))))_p = s\Big((d(f \circ F))_p\Big) + t\Big((d(g \circ F))_p\Big)$$

$$= s\Big(F^*\big((df)_{F(p)}\big)\Big) + t\Big(F^*\big((dg)_{F(p)}\big)\Big) = \text{RHS}. \qquad \square$$

**Note 2.74** Let $M$ be any $m$-dimensional smooth manifold, and $N$ be any $n$-dimensional smooth manifold. Let $p$ be in $M$. Let $F : M \to N$ be any smooth map. Fix any $\gamma$ in $T_p(M)$, where $\gamma$ is in $\Gamma_p(M)$. As in Theorem 2.73, $F^* : T_{F(p)}^* \to T_p^*$. So for every $(df)_{F(p)}$ in $T_{F(p)}^*$ where $f$ is in $C_{F(p)}^\infty(N)$, $F^*((df)_{F(p)})$ is in $T_p^*$. Since $F^*((df)_{F(p)})$ is in $T_p^*$ and, $[\gamma]$ is in $T_p(M)$, $\langle [\gamma], F^*((df)_{F(p)}) \rangle$ is a real number. We shall try to show that the mapping $(df)_{F(p)} \mapsto \langle [\gamma], F^*((df)_{F(p)}) \rangle$ from real linear space $T_{F(p)}^*$ to $\mathbb{R}$ is linear, that is, for every $f, g$ in $C_{F(p)}^\infty(N)$, and for every real $s, t$,

$$\langle [\gamma], F^*\Big(s\big((df)_{F(p)}\big) + t\big((dg)_{F(p)}\big)\Big) \rangle = s\langle [\gamma], F^*\big((df)_{F(p)}\big) \rangle$$
$$+ t\langle [\gamma], F^*\big((dg)_{F(p)}\big) \rangle.$$

By Theorem 2.73,

$$\text{LHS} = \langle [\gamma], F^*\Big(s\big((df)_{F(p)}\big) + t\big((dg)_{F(p)}\big)\Big) \rangle$$
$$= \langle [\gamma], s\Big(F^*\big((df)_{F(p)}\big)\Big) + t\Big(F^*\big((dg)_{F(p)}\big)\Big) \rangle.$$

Further, by Theorem 2.62,

$$\langle [\gamma], s\Big(F^*\big((df)_{F(p)}\big)\Big) + t\Big(F^*\big((dg)_{F(p)}\big)\Big) \rangle$$
$$= s\langle [\gamma], F^*\big((df)_{F(p)}\big) \rangle + t\langle [\gamma], F^*\big((dg)_{F(p)}\big) \rangle = \text{RHS}$$

so, LHS = RHS. So, $(df)_{F(p)} \mapsto \langle [\gamma], F^*((df)_{F(p)}) \rangle$ is a linear functional on the real linear space $T^*_{F(p)}$, and hence, the mapping $(df)_{F(p)} \mapsto \langle [\gamma], F^*((df)_{F(p)}) \rangle$ is a member of $\mathcal{T}_{F(p)}$. Let us denote the mapping $(df)_{F(p)} \mapsto \langle [\gamma], F^*((df)_{F(p)}) \rangle$ by $\widehat{F}([\gamma])$. Thus, for every $[\gamma]$ in $T_p(M)$, $\widehat{F}([\gamma])$ is in $\mathcal{T}_{F(p)}$. Since $\widehat{F}([\gamma])$ is in $\mathcal{T}_{F(p)}$, there exists a unique $F_*([\gamma])$ in $T_{F(p)}(N)$ such that $(F_*([\gamma]))^* = \widehat{F}([\gamma])$, that is, for every $(df)_{F(p)}$ in $T^*_{F(p)}$ where $f$ is in $C^\infty_{F(p)}(N)$,

$$(F_*([\gamma]))^* \left( (df)_{F(p)} \right) = \langle [\gamma], F^* \left( (df)_{F(p)} \right) \rangle.$$

Since for every $[\gamma]$ in $T_p(M)$, $F_*([\gamma])$ is a unique member of $T_{F(p)}(N)$, $F_* : T_p(M) \to T_{F(p)}(N)$.

## 2.8  Tangent Maps

**Definition** Let $M$ be any $m$-dimensional smooth manifold, and $N$ be any $n$-dimensional smooth manifold. Let $p$ be in $M$. Let $F : M \to N$ be any smooth map. Let $p$ be in $M$. The mapping $F_* : T_p(M) \to T_{F(p)}(N)$, as defined in the Note 2.74, is called the *tangent map induced by F*.

We have seen that, for every $(df)_{F(p)}$ in $T^*_{F(p)}$ where $f$ is in $C^\infty_{F(p)}(N)$, and for every $[\gamma]$ in $T_p(M)$ where $\gamma$ is in $\Gamma_p(M)$,

$$(F_*([\gamma]))^* \left( (df)_{F(p)} \right) = \langle [\gamma], F^* \left( (df)_{F(p)} \right) \rangle.$$

**Theorem 2.75** *Let $M$ be any $m$-dimensional smooth manifold, and let $N$ be any $n$-dimensional smooth manifold. Let $F : M \to N$ be any smooth map, and let $p$ be in $M$. Then, the tangent map $F_* : T_p(M) \to T_{F(p)}(N)$ induced by $F$ is linear.*

**Proof** Let us take any $[\gamma_1], [\gamma_2]$ in $T_p(M)$, and any real $s, t$. We have to show that

$$F_*(s[\gamma_1] + t[\gamma_2]) = s(F_*([\gamma_1])) + t(F_*([\gamma_2])).$$

We first try to prove that, for every $(df)_{F(p)}$ in $T^*_{F(p)}$,

$$(F_*(s[\gamma_1] + t[\gamma_2]))^* \left( (df)_{F(p)} \right) = (s(F_*([\gamma_1])) + t(F_*([\gamma_2])))^* \left( (df)_{F(p)} \right).$$

$$\text{LHS} = (F_*(s[\gamma_1] + t[\gamma_2]))^* \Big((df)_{F(p)}\Big) = \langle s[\gamma_1] + t[\gamma_2], F^*\Big((df)_{F(p)}\Big)\rangle$$

$$= s\langle[\gamma_1], F^*\Big((df)_{F(p)}\Big)\rangle + t\langle[\gamma_2], F^*\Big((df)_{F(p)}\Big)\rangle$$

$$= s\Big((F_*([\gamma_1]))^*\Big((df)_{F(p)}\Big)\Big) + t\Big((F_*([\gamma_2]))^*\Big((df)_{F(p)}\Big)\Big)$$

$$= (s((F_*([\gamma_1]))^*))\Big((df)_{F(p)}\Big) + (t((F_*([\gamma_2]))^*))\Big((df)_{F(p)}\Big)$$

$$= (s((F_*([\gamma_1]))^*) + t((F_*([\gamma_2]))^*))\Big((df)_{F(p)}\Big)$$

$$= (s(F_*([\gamma_1])) + t(F_*([\gamma_2])))^*\Big((df)_{F(p)}\Big) = \text{RHS}.$$

Hence, LHS = RHS. Thus, we have shown that, for every $(df)_{F(p)}$ in $T^*_{F(p)}$,

$$(F_*(s[\gamma_1] + t[\gamma_2]))^* \Big((df)_{F(p)}\Big) = (s(F_*([\gamma_1])) + t(F_*([\gamma_2])))^* \Big((df)_{F(p)}\Big).$$

Hence,

$$(F_*(s[\gamma_1] + t[\gamma_2]))^* = (s(F_*([\gamma_1])) + t(F_*([\gamma_2])))^*.$$

Now, by the definition of $\eta$,

$$\eta(F_*(s[\gamma_1] + t[\gamma_2])) = (F_*(s[\gamma_1] + t[\gamma_2]))^* = (s(F_*([\gamma_1])) + t(F_*([\gamma_2])))^*$$
$$= \eta(s(F_*([\gamma_1])) + t(F_*([\gamma_2])))$$

so,

$$\eta(F_*(s[\gamma_1] + t[\gamma_2])) = \eta(s(F_*([\gamma_1])) + t(F_*([\gamma_2]))).$$

Further, since $\eta$ is 1–1,

$$F_*(s[\gamma_1] + t[\gamma_2]) = s(F_*([\gamma_1])) + t(F_*([\gamma_2])). \qquad \square$$

**Note 2.76** Let $M$ be any 2-dimensional smooth manifold, and $N$ be any 3-dimensional smooth manifold. Let $p$ be in $M$. Let $F : M \to N$ be any smooth map. Let $(U, \varphi_U)$ be an admissible coordinate chart of $M$ satisfying $p \in U$, and let $(V, \psi_V)$ be an admissible coordinate chart of $N$ satisfying $F(p) \in V$. Let $u^1, u^2$ be the component functions of $\varphi_U$, and let $v^1, v^2, v^3$ be the component functions of $\psi_V$.

Since $(V, \psi_V)$ is an admissible coordinate chart of $N$ satisfying $F(p) \in V$, and $v^1, v^2, v^3$ are the component functions of $\psi_V$, $\{(dv^1)_{F(p)}, (dv^2)_{F(p)}, (dv^3)_{F(p)}\}$ is the natural basis of the cotangent space $T^*_{F(p)}(N)$ determined by $(V, \psi_V)$. Since $(U, \varphi_U)$ is an admissible coordinate chart of $M$ satisfying $p \in U$, and $u^1, u^2$ are the component functions of $\varphi_U$, $\{(du^1)_p, (du^2)_p\}$ is the natural basis of the cotangent space

$T_p^*(M)$ determined by $(U, \varphi_U)$. Since $F^* : T_{F(p)}^* \to T_p^*$, and $(dv^1)_{F(p)}, (dv^2)_{F(p)},$ $(dv^3)_{F(p)}$ are in $T_{F(p)}^*$, $F^*((dv^1)_{F(p)}), F^*((dv^2)_{F(p)}), F^*((dv^3)_{F(p)})$ are in $T_p^*$. Also, by the definition of $F^*$,

$$F^*\left((dv^1)_{F(p)}\right) = (d(v^1 \circ F))_p.$$

Next, by Theorem 2.48,

$$
\begin{aligned}
(d(v^1 \circ F))_p &= \left(\left(D_1\left((v^1 \circ F) \circ (\varphi_U)^{-1}\right)\right)(u^1(p), u^2(p))\right)(d(u^1))_p \\
&\quad + \left(\left(D_2\left((v^1 \circ F) \circ (\varphi_U)^{-1}\right)\right)(u^1(p), u^2(p))\right)(d(u^2))_p \\
&= \left(\left(D_1\left(v^1 \circ \left(F \circ (\varphi_U)^{-1}\right)\right)\right)(u^1(p), u^2(p))\right)(d(u^1))_p \\
&\quad + \left(\left(D_2\left(v^1 \circ \left(F \circ (\varphi_U)^{-1}\right)\right)\right)(u^1(p), u^2(p))\right)(d(u^2))_p \\
&= \left(\left(D_1\left(\left(\psi_V \circ \left(F \circ (\varphi_U)^{-1}\right)\right)^1\right)\right)(u^1(p), u^2(p))\right)(d(u^1))_p \\
&\quad + \left(\left(D_2\left(\left(\psi_V \circ \left(F \circ (\varphi_U)^{-1}\right)\right)^1\right)\right)(u^1(p), u^2(p))\right)(d(u^2))_p \\
&= \sum_{i=1}^{2} \left(\left(D_i\left(\left(\psi_V \circ \left(F \circ (\varphi_U)^{-1}\right)\right)^1\right)\right)(u^1(p), u^2(p))\right)(d(u^i))_p.
\end{aligned}
$$

Thus,

$$F^*\left((dv^1)_{F(p)}\right) = \sum_{i=1}^{2} \left(\left(D_i\left(\left(\psi_V \circ \left(F \circ (\varphi_U)^{-1}\right)\right)^1\right)\right)(u^1(p), u^2(p))\right)(d(u^i))_p.$$

Similarly,

$$F^*\left((dv^2)_{F(p)}\right) = \sum_{i=1}^{2} \left(\left(D_i\left(\left(\psi_V \circ \left(F \circ (\varphi_U)^{-1}\right)\right)^2\right)\right)(u^1(p), u^2(p))\right)(d(u^i))_p,$$

and

$$F^*\left((dv^3)_{F(p)}\right) = \sum_{i=1}^{2} \left(\left(D_i\left(\left(\psi_V \circ \left(F \circ (\varphi_U)^{-1}\right)\right)^3\right)\right)(u^1(p), u^2(p))\right)(d(u^i))_p.$$

In short, we write

$$F^*\left((dv^\alpha)_{F(p)}\right) = \sum_{i=1}^{2}\left(\left(D_i\left(\left(\psi_V \circ \left(F \circ (\varphi_U)^{-1}\right)\right)^\alpha\right)\right)(u^1(p), u^2(p))\right)(d(u^i))_p$$

or

$$F^*\left((dv^\alpha)_{F(p)}\right) = \sum_{i=1}^{2}\left(\frac{\partial F^\alpha}{\partial u^i}\right)_p (d(u^i))_p,$$

where

$$\left(\frac{\partial F^\alpha}{\partial u^i}\right)_p \equiv \left(D_i\left(\left(\psi_V \circ \left(F \circ (\varphi_U)^{-1}\right)\right)^\alpha\right)\right)(u^1(p), u^2(p)).$$

If there is no confusion, we simply write

$$F^*(dv^\alpha) = \sum_{i=1}^{2}\frac{\partial F^\alpha}{\partial u^i}du^i.$$

If $M$ is an $m$-dimensional smooth manifold, and $N$ is an $n$-dimensional smooth manifold, then as above, we get the formula: For every $\alpha = 1, \ldots, n$,

$$F^*(dv^\alpha) = \sum_{i=1}^{m}\frac{\partial F^\alpha}{\partial u^i}du^i.$$

**Note 2.77** Let $M$ be any $m$-dimensional smooth manifold, and $N$ be any $n$-dimensional smooth manifold. Let $p$ be in $M$. Let $F : M \to N$ be any smooth map. Let $(U, \varphi_U)$ be an admissible coordinate chart of $M$ satisfying $p \in U$, and let $(V, \psi_V)$ be an admissible coordinate chart of $N$ satisfying $F(p) \in V$. Let $u^1, \ldots, u^m$ be the component functions of $\varphi_U$, and let $v^1, \ldots, v^n$ be the component functions of $\psi_V$.

For $i = 1, \ldots, m$, and for $\alpha = 1, \ldots, n$, $\frac{\partial}{\partial u^i}\big|_p$ is in $T_p(M)$, and $(dv^\alpha)_{F(p)}$ is in $T^*_{F(p)}$. So, by the definition of $F_*$,

$$\left(F_*\left(\frac{\partial}{\partial u^i}\bigg|_p\right)\right)^*\left((dv^\alpha)_{F(p)}\right) = \left\langle \frac{\partial}{\partial u^i}\bigg|_p, F^*\left((dv^\alpha)_{F(p)}\right)\right\rangle$$

$$= \left\langle \frac{\partial}{\partial u^i}\bigg|_p, \sum_{j=1}^{m}\left(\frac{\partial F^\alpha}{\partial u^j}\right)_p (d(u^j))_p\right\rangle$$

$$= \sum_{j=1}^{m} \left(\frac{\partial F^{\alpha}}{\partial u^{j}}\right)_{p} \left\langle \frac{\partial}{\partial u^{i}}\bigg|_{p}, (d(u^{j}))_{p} \right\rangle$$

$$= \sum_{j=1}^{m} \left(\frac{\partial F^{\alpha}}{\partial u^{j}}\right)_{p} \delta_{i}^{j} = \left(\frac{\partial F^{\alpha}}{\partial u^{i}}\right)_{p}.$$

Thus,

$$\left(F_{*}\left(\frac{\partial}{\partial u^{i}}\bigg|_{p}\right)\right)^{*} \left((dv^{\alpha})_{F(p)}\right) = \left(\frac{\partial F^{\alpha}}{\partial u^{i}}\right)_{p}.$$

We want to prove that

$$\left(F_{*}\left(\frac{\partial}{\partial u^{i}}\bigg|_{p}\right)\right)^{*} = \sum_{\beta=1}^{n} \left(\frac{\partial F^{\beta}}{\partial u^{i}}\right)_{p} \left(\left(\frac{\partial}{\partial v^{\beta}}\bigg|_{F(p)}\right)^{*}\right),$$

that is, for every $(df)_{F(p)}$ in $T_{F(p)}^{*}$, where $f$ is in $C_{F(p)}^{\infty}(N)$,

$$\left(F_{*}\left(\frac{\partial}{\partial u^{i}}\bigg|_{p}\right)\right)^{*} \left((df)_{F(p)}\right) = \left(\sum_{\beta=1}^{n} \left(\frac{\partial F^{\beta}}{\partial u^{i}}\right)_{p} \left(\left(\frac{\partial}{\partial v^{\beta}}\bigg|_{F(p)}\right)^{*}\right)\right) \left((df)_{F(p)}\right).$$

Here,

$$\text{LHS} = \left(F_{*}\left(\frac{\partial}{\partial u^{i}}\bigg|_{p}\right)\right)^{*} \left((df)_{F(p)}\right) = \left\langle \frac{\partial}{\partial u^{i}}\bigg|_{p}, F^{*}\left((df)_{F(p)}\right) \right\rangle$$

$$= \left\langle \frac{\partial}{\partial i}\bigg|_{p}, F^{*}\left(\sum_{\alpha=1}^{n} \left(\left(D_{\alpha}\left(f \circ (\psi_{V})^{-1}\right)\right)(v^{1}(p), \ldots, v^{n}(p))\right)(dv^{\alpha})_{F(p)}\right) \right\rangle$$

$$= \left\langle \frac{\partial}{\partial u^{i}}\bigg|_{p}, \sum_{\alpha=1}^{n} \left(\left(D_{\alpha}\left(f \circ (\psi_{V})^{-1}\right)\right)(v^{1}(p), \ldots, v^{n}(p))\right) F^{*}\left((dv^{\alpha})_{F(p)}\right) \right\rangle$$

$$= \sum_{\alpha=1}^{n} \left(\left(D_{\alpha}\left(f \circ (\psi_{V})^{-1}\right)\right)(v^{1}(p), \ldots, v^{n}(p))\right) \left\langle \frac{\partial}{\partial u^{i}}\bigg|_{p}, F^{*}\left((dv^{\alpha})_{F(p)}\right) \right\rangle$$

$$= \sum_{\alpha=1}^{n} \left(\left(D_{\alpha}\left(f \circ (\psi_{V})^{-1}\right)\right)(v^{1}(p), \ldots, v^{n}(p))\right) \left(\left(F_{*}\left(\frac{\partial}{\partial u^{i}}\bigg|_{p}\right)\right)^{*} \left((dv^{\alpha})_{F(p)}\right)\right)$$

$$= \sum_{\alpha=1}^{n} \left(\left(D_{\alpha}\left(f \circ (\psi_{V})^{-1}\right)\right)(v^{1}(p), \ldots, v^{n}(p))\right) \left(\frac{\partial F^{\alpha}}{\partial u^{i}}\right)_{p},$$

and

$$\text{RHS} = \left( \sum_{\beta=1}^{n} \left( \frac{\partial F^{\beta}}{\partial u^i} \right)_p \left( \left( \frac{\partial}{\partial v^{\beta}} \Big|_{F(p)} \right) \right)^* \right) \left( (df)_{F(p)} \right)$$

$$= \sum_{\beta=1}^{n} \left( \frac{\partial F^{\beta}}{\partial u^i} \right)_p \left( \left( \frac{\partial}{\partial v^{\beta}} \Big|_{F(p)} \right)^* \left( (df)_{F(p)} \right) \right)$$

$$= \sum_{\beta=1}^{n} \left( \frac{\partial F^{\beta}}{\partial u^i} \right)_p \left( \left( \frac{\partial}{\partial v^{\beta}} \Big|_{F(p)} \right)^* \right.$$
$$\left. \left( \sum_{\alpha=1}^{n} \left( \left( D_{\alpha}\left( f \circ (\psi_V)^{-1} \right) \right) \left( v^1(p), \ldots, v^n(p) \right) \right) (dv^{\alpha})_{F(p)} \right) \right)$$

$$= \sum_{\beta=1}^{n} \left( \frac{\partial F^{\beta}}{\partial u^i} \right)_p \left( \sum_{\alpha=1}^{n} \left( \left( D_{\alpha}\left( f \circ (\psi_V)^{-1} \right) \right) \left( v^1(p), \ldots, v^n(p) \right) \right) \right.$$
$$\left. \left( \left( \frac{\partial}{\partial v^{\beta}} \Big|_{F(p)} \right)^* \left( (dv^{\alpha})_{F(p)} \right) \right) \right)$$

$$= \sum_{\beta=1}^{n} \left( \frac{\partial F^{\beta}}{\partial u^i} \right)_p \left( \sum_{\alpha=1}^{n} \left( \left( D_{\alpha}\left( f \circ (\psi_V)^{-1} \right) \right) \left( v^1(p), \ldots, v^n(p) \right) \right) \left\langle \frac{\partial}{\partial v^{\beta}} \Big|_{F(p)}, (dv^{\alpha})_{F(p)} \right\rangle \right)$$

$$= \sum_{\beta=1}^{n} \left( \frac{\partial F^{\beta}}{\partial u^i} \right)_p \left( \sum_{\alpha=1}^{n} \left( \left( D_{\alpha}\left( f \circ (\psi_V)^{-1} \right) \right) \left( v^1(p), \ldots, v^n(p) \right) \right) \delta^{\alpha}_{\beta} \right)$$

$$= \sum_{\beta=1}^{n} \left( \frac{\partial F^{\beta}}{\partial u^i} \right)_p \left( \left( D_{\beta}\left( f \circ (\psi_V)^{-1} \right) \right) \left( v^1(p), \ldots, v^n(p) \right) \right)$$

$$= \sum_{\alpha=1}^{n} \left( \frac{\partial F^{\alpha}}{\partial u^i} \right)_p \left( \left( D_{\alpha}\left( f \circ (\psi_V)^{-1} \right) \right) \left( v^1(p), \ldots, v^n(p) \right) \right).$$

Hence, LHS = RHS. Thus, we have shown that

$$\left( F_*\left( \frac{\partial}{\partial u^i} \Big|_p \right) \right)^* = \sum_{\beta=1}^{n} \left( \frac{\partial F^{\beta}}{\partial u^i} \right)_p \left( \left( \frac{\partial}{\partial v^{\beta}} \Big|_{F(p)} \right)^* \right).$$

Since, by the definition of $\eta$,

$$\eta\left( F_*\left( \frac{\partial}{\partial u^i} \Big|_p \right) \right) = \left( F_*\left( \frac{\partial}{\partial u^i} \Big|_p \right) \right)^* = \sum_{\beta=1}^{n} \left( \frac{\partial F^{\beta}}{\partial u^i} \right)_p \left( \left( \frac{\partial}{\partial v^{\beta}} \Big|_{F(p)} \right)^* \right)$$

$$= \sum_{\beta=1}^{n} \left( \frac{\partial F^{\beta}}{\partial u^i} \right)_p \left( \eta\left( \frac{\partial}{\partial v^{\beta}} \Big|_{F(p)} \right) \right) = \eta\left( \sum_{\beta=1}^{n} \left( \frac{\partial F^{\beta}}{\partial u^i} \right)_p \left( \frac{\partial}{\partial v^{\beta}} \Big|_{F(p)} \right) \right)$$

so,

$$\eta\left(F_*\left(\frac{\partial}{\partial u^i}\bigg|_p\right)\right) = \eta\left(\sum_{\beta=1}^n \left(\frac{\partial F^\beta}{\partial u^i}\right)_p \left(\frac{\partial}{\partial v^\beta}\bigg|_{F(p)}\right)\right).$$

Further, since $\eta$ is 1–1,

$$F_*\left(\frac{\partial}{\partial u^i}\bigg|_p\right) = \sum_{\beta=1}^n \left(\frac{\partial F^\beta}{\partial u^i}\right)_p \left(\frac{\partial}{\partial v^\beta}\bigg|_{F(p)}\right) = \sum_{\alpha=1}^n \left(\frac{\partial F^\alpha}{\partial u^i}\right)_p \left(\frac{\partial}{\partial v^\alpha}\bigg|_{F(p)}\right)$$

In short, we write

$$F_*\left(\frac{\partial}{\partial u^i}\right) = \sum_{\alpha=1}^n \frac{\partial F^\alpha}{\partial u^i} \frac{\partial}{\partial v^\alpha}.$$

**Note 2.78** Let $M$ and $N$ be any 3-dimensional smooth manifolds. Let $p$ be in $M$. Let $F : M \to N$ be any smooth map. Let $(U, \varphi_U)$ be an admissible coordinate chart of $M$ satisfying $p \in U$, and let $(V, \psi_V)$ be an admissible coordinate chart of $N$ satisfying $F(p) \in V$. Let $u^1, u^2, u^3$ be the component functions of $\varphi_U$, and let $v^1, v^2, v^3$ be the component functions of $\psi_V$. Here,

$$\left\{\frac{\partial}{\partial u^1}\bigg|_p, \frac{\partial}{\partial u^2}\bigg|_p, \frac{\partial}{\partial u^3}\bigg|_p\right\}$$

is the natural basis of the tangent space $T_p(M)$ under local coordinate system $(u^i)$, and

$$\left\{\frac{\partial}{\partial v^1}\bigg|_{F(p)}, \frac{\partial}{\partial v^2}\bigg|_{F(p)}, \frac{\partial}{\partial v^3}\bigg|_{F(p)}\right\}$$

is the natural basis of the tangent space $T_{F(p)}(N)$ under local coordinate system $(v^i)$. Also, by Theorem 2.75, the tangent map $F_*$ is a linear transformation from linear space $T_p(M)$ to linear space $T_{F(p)}(N)$. From the above note,

$$F_*\left(\frac{\partial}{\partial u^i}\bigg|_p\right) = \sum_{j=1}^3 \left(\frac{\partial F^j}{\partial u^i}\right)_p \left(\frac{\partial}{\partial v^j}\bigg|_{F(p)}\right).$$

So the matrix representation of $F_*$ under the natural bases

$$\left\{\frac{\partial}{\partial u^1}\bigg|_p, \frac{\partial}{\partial u^2}\bigg|_p, \frac{\partial}{\partial u^3}\bigg|_p\right\} \quad \text{and} \quad \left\{\frac{\partial}{\partial v^1}\bigg|_{F(p)}, \frac{\partial}{\partial v^2}\bigg|_{F(p)}, \frac{\partial}{\partial v^3}\bigg|_{F(p)}\right\}$$

is

$$\begin{bmatrix} \left(\frac{\partial F^1}{\partial u^1}\right)_p & \left(\frac{\partial F^1}{\partial u^2}\right)_p & \left(\frac{\partial F^1}{\partial u^3}\right)_p \\[2mm] \left(\frac{\partial F^2}{\partial u^1}\right)_p & \left(\frac{\partial F^2}{\partial u^2}\right)_p & \left(\frac{\partial F^2}{\partial u^3}\right)_p \\[2mm] \left(\frac{\partial F^3}{\partial u^1}\right)_p & \left(\frac{\partial F^3}{\partial u^2}\right)_p & \left(\frac{\partial F^3}{\partial u^3}\right)_p \end{bmatrix}_{3\times 3} ,$$

where $\left(\frac{\partial F^j}{\partial u^i}\right)_p$ stands for $(D_i((\psi_V \circ (F \circ (\varphi_U)^{-1}))^j))(u^1(p), u^2(p), u^2(p))$.

Further, the linear transformation $F_* : T_p(M) \to T_{F(p)}(N)$ is an isomorphism if and only if

$$\det \begin{bmatrix} \left(\frac{\partial F^1}{\partial u^1}\right)_p & \left(\frac{\partial F^1}{\partial u^2}\right)_p & \left(\frac{\partial F^1}{\partial u^3}\right)_p \\[2mm] \left(\frac{\partial F^2}{\partial u^1}\right)_p & \left(\frac{\partial F^2}{\partial u^2}\right)_p & \left(\frac{\partial F^2}{\partial u^3}\right)_p \\[2mm] \left(\frac{\partial F^3}{\partial u^1}\right)_p & \left(\frac{\partial F^3}{\partial u^2}\right)_p & \left(\frac{\partial F^3}{\partial u^3}\right)_p \end{bmatrix} \neq 0.$$

**Note 2.79** Let $E$ be an open subset of $\mathbb{R}^3$. Let $F : E \to \mathbb{R}^3$. Let $p$ be in $E$. By Example 1.6, $E$ is a smooth manifold with the standard differentiable structure of $E$. Here, differentiable structure on $E$ contains $\{(E, \mathrm{Id}_E)\}$. Similarly, $\mathbb{R}^3$ is a smooth manifold with the standard differentiable structure of $\mathbb{R}^3$. Here, differentiable structure on $\mathbb{R}^3$ contains $\{(\mathbb{R}^3, \mathrm{Id}_{\mathbb{R}^3})\}$. Now, let $F : E \to \mathbb{R}^3$ be a smooth function. From the Note 2.78, the tangent map $F_*$ induced by $F$ is an isomorphism if and only if

$$\det \begin{bmatrix} \left(\frac{\partial F^1}{\partial u^1}\right)_p & \left(\frac{\partial F^1}{\partial u^2}\right)_p & \left(\frac{\partial F^1}{\partial u^3}\right)_p \\[2mm] \left(\frac{\partial F^2}{\partial u^1}\right)_p & \left(\frac{\partial F^2}{\partial u^2}\right)_p & \left(\frac{\partial F^2}{\partial u^3}\right)_p \\[2mm] \left(\frac{\partial F^3}{\partial u^1}\right)_p & \left(\frac{\partial F^3}{\partial u^2}\right)_p & \left(\frac{\partial F^3}{\partial u^3}\right)_p \end{bmatrix} \neq 0,$$

where

$$\left(\frac{\partial F^j}{\partial u^i}\right)_p = \left(D_i\left(\left(i_{\mathbb{R}^3} \circ \left(F \circ (i_E)^{-1}\right)\right)^j\right)\right)(i_E(p)) = (D_i F^j)(i_E(p)) = (D_i F^j)(p).$$

Thus, $F_*$ is an isomorphism if and only if

$$\det \begin{bmatrix} (D_1 F^1)(p) & (D_2 F^1)(p) & (D_3 F^1)(p) \\ (D_1 F^2)(p) & (D_2 F^2)(p) & (D_3 F^2)(p) \\ (D_1 F^3)(p) & (D_2 F^3)(p) & (D_3 F^3)(p) \end{bmatrix} \neq 0.$$

# Chapter 3
# Multivariable Differential Calculus

In the very definition of smooth manifold, there is a statement like "...for every point $p$ in the topological space $M$ there exist an open neighborhood $U$ of $p$, an open subset $V$ of Euclidean space $\mathbb{R}^n$, and a homeomorphism $\varphi$ from $U$ onto $V$ ...". This phrase clearly indicates that homeomorphism $\varphi$ can act as a messenger in carrying out the well-developed multivariable differential calculus of $\mathbb{R}^n$ into the realm of unornamented topological space $M$. So the role played by multivariable differential calculus in the development of the theory of smooth manifolds is paramount. Many theorems of multivariable differential calculus are migrated into manifolds like their generalizations. Although most students are familiar with multivariable differential calculus from their early courses, but the exact form of theorems we need here are generally missing from their collection of theorems. So this is somewhat long chapter on multivariable differential calculus that is pertinent here for a good understanding of smooth manifolds.

## 3.1 Linear Transformations

**Definition** The set of all linear transformations from $\mathbb{R}^n$ to $\mathbb{R}^m$ is denoted by $L(\mathbb{R}^n, \mathbb{R}^m)$. By $L(\mathbb{R}^n)$, we mean $L(\mathbb{R}^n, \mathbb{R}^n)$. Thus, $L(\mathbb{R}^n)$ denotes the set of all linear transformations from $\mathbb{R}^n$ to $\mathbb{R}^n$.

We know that $L(\mathbb{R}^n, \mathbb{R}^m)$ is a real linear space under pointwise vector addition and pointwise scalar multiplication.

**Definition** Let $A$ be in $L(\mathbb{R}^n, \mathbb{R}^m)$ and $B$ be in $L(\mathbb{R}^m, \mathbb{R}^k)$. By the *product BA*, we mean the mapping

$$BA : \mathbb{R}^n \to \mathbb{R}^k$$

defined as follows: For every $x$ in $\mathbb{R}^n$,

$$(BA)(x) \equiv B(A(x)).$$

© Springer India 2014
R. Sinha, *Smooth Manifolds*, DOI 10.1007/978-81-322-2104-3_3

It is easy to see that $BA$ is a linear transformation from $\mathbb{R}^n$ to $\mathbb{R}^k$, and hence, $BA$ is in $L(\mathbb{R}^n, \mathbb{R}^k)$.

**Note 3.1** Let $A$ be in $L(\mathbb{R}^3, \mathbb{R}^2)$. Let $(t_1, t_2, t_3)$ be any member of $\mathbb{R}^3$ satisfying $|(t_1, t_2, t_3)| \leq 1$. Since $|t_1| = \sqrt{(t_1)^2} \leq \sqrt{(t_1)^2 + (t_2)^2 + (t_3)^2} = |(t_1, t_2, t_3)| \leq 1$, so $|t_1| \leq 1$. Similarly, $|t_2| \leq 1$, and $|t_3| \leq 1$. Further, since $A$ is a linear transformation from $\mathbb{R}^3$ to $\mathbb{R}^2$,

$$
\begin{aligned}
|A((t_1, t_2, t_3))| &= |A(t_1(1,0,0) + t_2(0,1,0) + t_3(0,0,1))| \\
&= |t_1(A((1,0,0))) + t_2(A((0,1,0))) + t_3(A((0,0,1)))| \\
&\leq |t_1(A((1,0,0)))| + |t_2(A((0,1,0)))| + |t_3(A((0,0,1)))| \\
&= |t_1||A((1,0,0))| + |t_2||A((0,1,0))| + |t_3||A((0,0,1))| \\
&\leq 1 \cdot |A((1,0,0))| + 1 \cdot |A((0,1,0))| + 1 \cdot |A((0,0,1))| \\
&= |A((1,0,0))| + |A((0,1,0))| + |A((0,0,1))|.
\end{aligned}
$$

Thus, we have seen that for every $(t_1, t_2, t_3)$ in $\mathbb{R}^3$ satisfying $|(t_1, t_2, t_3)| \leq 1$, we have

$$|A((t_1, t_2, t_3))| \leq (|A((1,0,0))| + |A((0,1,0))| + |A((0,0,1))|).$$

So, $(|A((1,0,0))| + |A((0,1,0))| + |A((0,0,1))|)$ is an upper bound of the set

$$\left\{ |A((t_1, t_2, t_3))| : (t_1, t_2, t_3) \text{ is in } \mathbb{R}^3, \text{and } |(t_1, t_2, t_3)| \leq 1 \right\}.$$

Thus, $\{ |A((t_1, t_2, t_3))| : (t_1, t_2, t_3) \text{ is in } \mathbb{R}^3, \text{and } |(t_1, t_2, t_3)| \leq 1 \}$ is a nonempty bounded above set of real numbers, and hence,

$$\sup\left\{ |A((t_1, t_2, t_3))| : (t_1, t_2, t_3) \text{ is in } \mathbb{R}^3, \text{and } |(t_1, t_2, t_3)| \leq 1 \right\}$$

exists. Further, we have seen that

$$
\begin{aligned}
&\sup\left\{ |A((t_1, t_2, t_3))| : (t_1, t_2, t_3) \text{ is in } \mathbb{R}^3, \text{and } |(t_1, t_2, t_3)| \leq 1 \right\} \\
&\leq (|A((1,0,0))| + |A((0,1,0))| + |A((0,0,1))|).
\end{aligned}
$$

Similarly, if $A$ is in $L(\mathbb{R}^n, \mathbb{R}^m)$, then $\sup\{|A(x)| : x \text{ is in } \mathbb{R}^n, \text{and } |x| \leq 1\}$ exists, and

$$\sup\{|A(x)| : x \text{ is in } \mathbb{R}^n, \text{and } |x| \leq 1\} \leq (|A((1,\ldots,0))| + \cdots + |A((0,\ldots,1))|).$$

**Definition** Let $A$ be in $L(\mathbb{R}^n, \mathbb{R}^m)$. By $\|A\|$, we mean the real number

$$\sup\{|A(x)| : x \text{ is in } \mathbb{R}^n, \text{and } |x| \leq 1\}.$$

$\|A\|$ is called the *norm of A*.

From the above discussion, we find that for every $A$ in $L(\mathbb{R}^n, \mathbb{R}^m)$,

$$\|A\| \leq (|A((1,\ldots,0))| + \cdots + |A((0,\ldots,1))|).$$

**Theorem 3.2**

1. *If $A$ is in $L(\mathbb{R}^n, \mathbb{R}^m)$, then for every $x$ in $\mathbb{R}^n$,*

$$|A(x)| \leq \|A\||x|.$$

2. *If $A$ is in $L(\mathbb{R}^n, \mathbb{R}^m)$, then the mapping*

$$A : \mathbb{R}^n \to \mathbb{R}^m$$

*is uniformly continuous.*
3. *If $A$, $B$ are in $L(\mathbb{R}^n, \mathbb{R}^m)$, then*

$$\|A + B\| \leq \|A\| + \|B\|.$$

4. *If $A$ is in $L(\mathbb{R}^n, \mathbb{R}^m)$, and $t$ is a real number, then*

$$\|tA\| = |t|\|A\|.$$

5. *If, for every $A$, $B$ in $L(\mathbb{R}^n, \mathbb{R}^m)$,*

$$d(A, B) \equiv \|A - B\|,$$

*then $L(\mathbb{R}^n, \mathbb{R}^m)$ is a metric space with the metric d.*
6. *The mapping $A \mapsto \|A\|$ from $L(\mathbb{R}^n, \mathbb{R}^m)$ to $\mathbb{R}$ is continuous.*
7. *If $A$ is in $L(\mathbb{R}^n, \mathbb{R}^m)$, and $B$ is in $L(\mathbb{R}^m, \mathbb{R}^k)$, then*

$$\|BA\| \leq \|B\|\|A\|.$$

*Proof*

1: Let $A$ be in $L(\mathbb{R}^n, \mathbb{R}^m)$. Let $x$ be in $\mathbb{R}^n$.

Case I: when $x = 0$

$$|A(x)| = |A(0)| = |0| = 0 \leq 0 = \|A\|0 = \|A\||0| = \|A\||x|.$$

Case II: when $x \neq 0$. Since $x \neq 0$, $|x|$ is a nonzero real number, and hence, $\frac{1}{|x|}$ is a real number. Also, $|\frac{1}{|x|}x| = |\frac{1}{|x|}||x| = \frac{1}{|x|}|x| = 1 \leq 1$. It follows that $|A(\frac{1}{|x|}x)|$ is a member of the set $\{|A(y)| : y$ is in $\mathbb{R}^n$, and $|y| \leq 1\}$, and hence, from the definition of $\|A\|$, $|A(\frac{1}{|x|}x)| \leq \|A\|$. Now, since

$$\frac{1}{|x|}|A(x)| = \left|\frac{1}{|x|}\right||A(x)| = \left|\frac{1}{|x|}(A(x))\right| = \left|A\left(\frac{1}{|x|}x\right)\right| \leq \|A\|,$$

$$|A(x)| \leq \|A\||x|.$$

Thus, we see that in all cases, $|A(x)| \leq \|A\||x|$. This proves 1.

2: Let us take $\varepsilon > 0$.

Case I: when $\|A\| \neq 0$. For any $x, y$ in $\mathbb{R}^n$ satisfying $|x - y| < \frac{\varepsilon}{\|A\|}$, we have

$$|A(x) - A(y)| = |A(x - y)| \leq \|A\||x - y| < \|A\|\left(\frac{\varepsilon}{\|A\|}\right) = \varepsilon.$$

Case II: when $\|A\| = 0$. For any $x, y$ in $\mathbb{R}^n$ satisfying $|x - y| < 1$, we have

$$|A(x) - A(y)| = |A(x - y)| \leq \|A\||x - y| = 0 \cdot |x - y| = 0 < \varepsilon.$$

Thus, we see that in all cases, for every $\varepsilon > 0$, there exists $\delta > 0$ such that for every $x, y$ in $\mathbb{R}^n$ satisfying $|x - y| < \delta$, we have $|A(x) - A(y)| < \varepsilon$. So, by the definition of uniform continuity, $A$ is uniformly continuous. This proves 2.

3: Let $A, B$ be in $L(\mathbb{R}^n, \mathbb{R}^m)$. Let us take any $x$ in $\mathbb{R}^n$ satisfying $|x| \leq 1$. Now,

$$\begin{aligned}|(A + B)(x)| &= |A(x) + B(x)| \leq |A(x)| + |B(x)| \leq \|A\||x| + |B(x)| \leq \|A\||x| \\ &\quad + \|B\||x| \\ &= (\|A\| + \|B\|)|x| \leq (\|A\| + \|B\|) \cdot 1 = \|A\| + \|B\|.\end{aligned}$$

So, $\|A\| + \|B\|$ is an upper bound of the set $\{|(A + B)(x)| : x$ is in $\mathbb{R}^n$, and $|x| \leq 1\}$, and hence,

$$\|A + B\| = \sup\{|(A + B)(x)| : x \text{ is in } \mathbb{R}^n, \text{ and } |x| \leq 1\} \leq \|A\| + \|B\|.$$

This proves 3.

4: Let $A$ be in $L(\mathbb{R}^n, \mathbb{R}^m)$, and let $t$ be a real number. Let us take any $x$ in $\mathbb{R}^n$ satisfying $|x| \leq 1$. Now,

$$|(tA)(x)| = |t(A(x))| = |t||A(x)| \leq |t|(\|A\||x|) = (|t|\|A\|)|x| \leq (|t|\|A\|) \cdot 1$$
$$= |t|\|A\|.$$

So, $|t|\|A\|$ is an upper bound of the set $\{|(tA)(x)| : x \text{ is in } \mathbb{R}^n, \text{ and } |x| \leq 1\}$, and hence,

$$\|tA\| = \sup\{|(tA)(x)| : x \text{ is in } \mathbb{R}^n, \text{ and } |x| \leq 1\} \leq |t|\|A\|.$$

So, for every real $t$,

$$\|tA\| \leq |t|\|A\| \cdots (*).$$

Case I: when $t \neq 0$. Since $t \neq 0$, $\frac{1}{t}$ is a real number, and hence, by $(*)$,

$$\|A\| = \left\|\frac{1}{t}(tA)\right\| \leq \left|\frac{1}{t}\right|\|tA\| = \frac{1}{|t|}\|tA\|.$$

Thus, $|t|\|A\| \leq \|tA\|$. Hence, by $(*)$,

$$\|tA\| = |t|\|A\|.$$

Case II: when $t = 0$. Here,

$$\|tA\| = \|0A\| = \|0\| = \sup\{|0(x)| : x \text{ is in } \mathbb{R}^n, \text{ and } |x| \leq 1\} = \sup\{|0|\}$$
$$= \sup\{0\} = 0 = 0 \cdot \|A\|$$
$$= |0|\|A\| = |t|\|A\|.$$

Thus, we see that in all cases,

$$\|tA\| = |t|\|A\|.$$

This proves 4.

5: Here, we must prove

i. For every $A$ in $L(\mathbb{R}^n, \mathbb{R}^m)$,

$$\|A - A\| = 0.$$

ii. For every $A$, $B$ in $L(\mathbb{R}^n, \mathbb{R}^m)$, if $\|A - B\| = 0$, then $A = B$.

iii. For every $A$, $B$ in $L(\mathbb{R}^n, \mathbb{R}^m)$,

$$\|A - B\| = \|B - A\|.$$

iv. for every $A$, $B$, $C$ in $L(\mathbb{R}^n, \mathbb{R}^m)$,

$$\|A - B\| \leq \|A - C\| + \|C - B\|.$$

For   i:   $\|A - A\| = \|0\| = \sup\{|0(x)| : x \text{ is in } \mathbb{R}^n, \text{ and } |x| \leq 1\} = \sup\{|0|\} = \sup\{0\} = 0$.

For ii: Let $\|A - B\| = 0$. Let us take any $x$ in $\mathbb{R}^n$. We have to prove that $A(x) = B(x)$. Since $0 \leq |(A - B)(x)| \leq \|A - B\| |x| = 0 \cdot |x| = 0$, $|(A - B)(x)| = 0$, and hence, $A(x) - B(x) = (A - B)(x) = 0$. Thus, $A(x) = B(x)$.

For iii: By 4,

$$\begin{aligned}\text{LHS} &= \|A - B\| = \|(-1)(B - A)\| = |-1| \|(B - A)\| = 1 \cdot \|B - A\| \\ &= \|B - A\| = \text{RHS}.\end{aligned}$$

For iv: By 3,

$$\|A - B\| = \|(A - C) + (C - B)\| \leq \|A - C\| + \|C - B\|.$$

This completes the proof of iv.

6: Let us take any $A$ in $L(\mathbb{R}^n, \mathbb{R}^m)$. We have to prove that the mapping $A \mapsto \|A\|$ is continuous. For this purpose, let us take any $\varepsilon > 0$. Let us take any $B$ in $L(\mathbb{R}^n, \mathbb{R}^m)$ satisfying $\|B - A\| < \varepsilon$. We have to prove that $\big|\|B\| - \|A\|\big| < \varepsilon$. Since $\|B\| = \|(B - A) + A\| \leq \|B - A\| + \|A\|$, $\|B\| - \|A\| \leq \|B - A\|$. Similarly, $-(\|B\| - \|A\|) = \|A\| - \|B\| \leq \|A - B\| = \|B - A\|$. Hence,

$$\big|\|B\| - \|A\|\big| = \max\{\|B\| - \|A\|, -(\|B\| - \|A\|)\} \leq \|B - A\| < \varepsilon.$$

This proves 6.

7: Let us take any $A$ in $L(\mathbb{R}^n, \mathbb{R}^m)$ and $B$ in $L(\mathbb{R}^m, \mathbb{R}^k)$. Let us take any $x$ in $\mathbb{R}^n$ satisfying $|x| \leq 1$. Since

$$\begin{aligned}|(BA)(x)| &= |B(A(x))| \leq \|B\| |A(x)| \leq \|B\| \cdot (\|A\| |x|) \\ &= (\|B\| \|A\|) |x| \leq \|B\| \|A\| \cdot 1 = \|B\| \|A\|,\end{aligned}$$

$$\|BA\| = \sup\{|(BA)(x)| : x \text{ is in } \mathbb{R}^n, \text{ and } |x| \leq 1\} \leq \|B\| \|A\|.$$

This proves 7.                                                                   $\square$

**Theorem 3.3** *Let $\Omega$ be the set of all invertible (i.e., 1–1 and onto) members of* $L(\mathbb{R}^n)$.

1. *If $A$ is in $\Omega$, then $\|A^{-1}\|$ is nonzero.*
2. *If $A$ is in $\Omega$, then the open sphere $S_{(1/\|A^{-1}\|)}(A)$ is contained in $\Omega$, that is, if $A$ is in $\Omega$, $B$ is in $L(\mathbb{R}^n)$, and $\|B - A\| < \frac{1}{\|A^{-1}\|}$, then $B$ is in $\Omega$.*
3. *$\Omega$ is an open subset of $L(\mathbb{R}^n)$.*
4. *The mapping $A \mapsto A^{-1}$ from $\Omega$ to $\Omega$ is continuous.*

*Proof*

1: Let us take any $A$ in $\Omega$. We claim that $\|A^{-1}\|$ is nonzero. If not, otherwise, let $\|A^{-1}\| = 0$. We have to arrive at a contradiction. Since $A$ is in $\Omega$, by the definition of $\Omega$, $A^{-1}$ exists and $A^{-1}$ is in $\Omega$. Also, $AA^{-1} = I$, where $I$ denotes the identity transformation from $\mathbb{R}^n$ to $\mathbb{R}^n$. For every $x$ in $\mathbb{R}^n$, $0 \le |A^{-1}(x)| \le \|A^{-1}\||x| = 0 \cdot |x| = 0$, so $|A^{-1}(x)| = 0$, and hence, $A^{-1}(x) = 0$. Therefore,

$$(1,0,\ldots,0) = I((1,0,\ldots,0)) = (AA^{-1})((1,0,\ldots,0))$$
$$= A\big(A^{-1}((1,0,\ldots,0))\big)$$
$$= A((0,0,\ldots,0)) = (0,0,\ldots,0),$$

and hence, $1 = 0$, a contradiction. So, our claim is true, that is, $\|A^{-1}\|$ is nonzero. This proves 1.

2: Let $A$ be in $\Omega$, and let $B$ be in $L(\mathbb{R}^n)$. Since $A$ is in $\Omega$, by 1, $\|A^{-1}\|$ is nonzero, and hence, $\frac{1}{\|A^{-1}\|}$ is a real number. Also, by the definition of norm $\|\|$, $\frac{1}{\|A^{-1}\|}$ is a positive real number. Let

$$\|B - A\| < \frac{1}{\|A^{-1}\|}.$$

We have to show that $B$ is in $\Omega$, that is, $B : \mathbb{R}^n \to \mathbb{R}^n$ is 1–1 and onto.
1–1ness: Let $B(x) = B(y)$. We have to prove that $x = y$, that is, $x - y = 0$. If not, otherwise, let $x - y \ne 0$. We have to arrive at a contradiction. Since $x - y \ne 0$, $0 < |x - y|$. Since $B(x) = B(y)$, and $B$ is a linear transformation, $0 = B(x) - B(y) = B(x - y)$. Hence,

$$|x - y| = |(A^{-1}A)(x - y)| = |(A^{-1})(A(x - y))| \le \|A^{-1}\||A(x - y)|$$
$$= \|A^{-1}\||A(x - y) - 0|$$
$$= \|A^{-1}\||A(x - y) - B(x - y)|$$
$$= \|A^{-1}\||(A - B)(x - y)| \le \|A^{-1}\|(\|A - B\||x - y|)$$
$$= (\|A^{-1}\|\|A - B\|)|x - y|$$
$$= (\|A^{-1}\|\|B - A\|)|x - y| < \left(\|A^{-1}\|\frac{1}{\|A^{-1}\|}\right)|x - y| = |x - y|.$$

Thus, $|x - y| < |x - y|$, which gives a contradiction. This proves that $B$ is 1–1. Ontoness: Let us take any $y$ in $\mathbb{R}^n$. Since $B : \mathbb{R}^n \to \mathbb{R}^n$ is 1–1, $B$ $((1, 0, \ldots, 0)), \ldots, B((0, \ldots, 0, 1))$ are $n$ distinct elements of $\mathbb{R}^n$. Now, we shall show that $B((1, 0, \ldots, 0)), \ldots, B((0, \ldots, 0, 1))$ are linearly independent. For this purpose, let $t_1(B((1, 0, \ldots, 0))) + \cdots + t_n(B((0, \ldots, 0, 1))) = 0$. We have to show that $t_1 = 0, \ldots, t_n = 0$. Here,

$$
\begin{aligned}
B((0, \ldots, 0)) = (0, \ldots, 0) = 0 &= t_1(B((1, 0, \ldots, 0))) + \cdots \\
&\quad + t_n(B((0, \ldots, 0, 1))) \\
&= B(t_1(1, 0, \ldots, 0)) + \cdots + B(t_n(0, \ldots, 0, 1)) \\
&= B((t_1, 0, \ldots, 0)) + \cdots + B((0, \ldots, 0, t_n)) \\
&= B((t_1, 0, \ldots, 0) + \cdots + (0, \ldots, 0, t_n)) = B((t_1, \ldots, t_n)),
\end{aligned}
$$

and $B$ is 1–1, so $(0, \ldots, 0) = (t_1, \ldots, t_n)$, and hence, $t_1 = 0, \ldots, t_n = 0$. Thus, $B$ $((1, 0, \ldots, 0)), \ldots, B((0, \ldots, 0, 1))$ constitute a basis of the $n$-dimensional real linear space $\mathbb{R}^n$. Since $B((1, 0, \ldots, 0)), \ldots, B((0, \ldots, 0, 1))$ constitute a basis of space $\mathbb{R}^n$, and $y$ is in $\mathbb{R}^n$, there exist real numbers $s_1, \ldots, s_n$ such that

$$
\begin{aligned}
y &= s_1(B((1, 0, \ldots, 0))) + \cdots + s_n(B((0, \ldots, 0, 1))) \\
&= B(s_1(1, 0, \ldots, 0)) + \cdots + B(s_n(0, \ldots, 0, 1)) \\
&= B((s_1, \ldots, s_n)).
\end{aligned}
$$

Thus, $B((s_1, \ldots, s_n)) = y$, where $(s_1, \ldots, s_n)$ is in $\mathbb{R}^n$. This proves that $B$ is onto. This proves 2.

3: Let us take any $A$ in $\Omega$. Since $A$ is in $\Omega$, by 1, $\|A^{-1}\|$ is nonzero. Now by 2, the open sphere $\{B : B \in L(\mathbb{R}^n), \text{ and } \|B - A\| < \frac{1}{\|A^{-1}\|}\}$ is contained in $\Omega$. Hence, by the definition of open set, $\Omega$ is an open subset of $L(\mathbb{R}^n)$. This proves 3.

4: Let us take any $A$ in $\Omega$. We have to prove that the mapping $B \mapsto B^{-1}$ is continuous at $A$. Since $A$ is in $\Omega$, by 1, $\|A^{-1}\|$ is nonzero. If $B$ is in $L(\mathbb{R}^n)$ satisfying $\|B - A\| < \frac{1}{\|A^{-1}\|}$, then by 2, $B$ is in $\Omega$, and hence, $B^{-1}$ exists. Now, it is enough to prove that $\lim_{B \to A} \|B^{-1} - A^{-1}\|$ exists, and its value is 0. Here,

$$
\begin{aligned}
\|B^{-1} - A^{-1}\| &= \|(A^{-1}A)B^{-1} - A^{-1}(BB^{-1})\| = \|A^{-1}(AB^{-1}) - A^{-1}(BB^{-1})\| \\
&= \|A^{-1}(AB^{-1} - BB^{-1})\| = \|A^{-1}(A - B)B^{-1}\| \le \|A^{-1}\|\|A - B\|\|B^{-1}\| \\
&= (\|A^{-1}\|\|A - B\|) \cdot \sup\{|(B^{-1})(x)| : x \text{ is in } \mathbb{R}^n, \text{ and } |x| \le 1\}
\end{aligned}
$$

or,

$$
\begin{aligned}
\|B^{-1} - A^{-1}\| &\le (\|A^{-1}\|\|A - B\|) \\
&\quad \cdot \sup\{|(B^{-1})(x)| : x \text{ is in } \mathbb{R}^n, \text{ and } |x| \le 1\} \cdots (*).
\end{aligned}
$$

Observe that for any $x$ in $\mathbb{R}^n$ satisfying $|x| \leq 1$, we have

$$\left|B^{-1}(x)\right| = \left|\left((A^{-1})A\right)\left(B^{-1}(x)\right)\right| = \left|(A^{-1})\left(A\left(B^{-1}(x)\right)\right)\right| \leq \left\|A^{-1}\right\|\left|A\left(B^{-1}(x)\right)\right|$$
$$= \left\|A^{-1}\right\|\left|(A-B)\left(B^{-1}(x)\right) + x\right| \leq \left\|A^{-1}\right\|\left(\left|(A-B)\left(B^{-1}(x)\right)\right| + |x|\right)$$
$$\leq \left\|A^{-1}\right\|\left(\|A-B\|\left|B^{-1}(x)\right| + |x|\right) \leq \left\|A^{-1}\right\|\left(\|B-A\|\left|B^{-1}(x)\right| + 1\right)$$

so,

$$\left|B^{-1}(x)\right|\left(1 - \left\|A^{-1}\right\|\|B-A\|\right) \leq \left\|A^{-1}\right\|.$$

Further, since $\|B-A\| < \frac{1}{\|A^{-1}\|}$,

$$\left|B^{-1}(x)\right| \leq \frac{\left\|A^{-1}\right\|}{1 - \left\|A^{-1}\right\|\|B-A\|}.$$

Hence,

$$\sup\left\{\left|(B^{-1})(x)\right| : x \text{ is in } \mathbb{R}^n \quad \text{and} \quad |x| \leq 1\right\} \leq \frac{\left\|A^{-1}\right\|}{1 - \left\|A^{-1}\right\|\|B-A\|}.$$

Now, from (*),

$$\left\|B^{-1} - A^{-1}\right\| \leq \left(\left\|A^{-1}\right\|\|B-A\|\right) \cdot \frac{\left\|A^{-1}\right\|}{1 - \left\|A^{-1}\right\|\|B-A\|} = \left\|A^{-1}\right\|\frac{\|B-A\|\left\|A^{-1}\right\|}{1 - \left\|A^{-1}\right\|\|B-A\|}$$
$$= \left\|A^{-1}\right\|\left(-1 + \frac{1}{1 - \left\|A^{-1}\right\|\|B-A\|}\right).$$

Thus, if $\|B-A\| < \frac{1}{\|A^{-1}\|}$, then

$$0 \leq \left\|B^{-1} - A^{-1}\right\| \leq \left\|A^{-1}\right\|\left(-1 + \frac{1}{1 - \left\|A^{-1}\right\|\|B-A\|}\right).$$

By Theorem 3.2, the mapping $B \mapsto \|B\|$ from $\Omega$ to $\mathbb{R}$ is continuous, so $\lim_{B \to A} \|B-A\| = 0$. Also,

$$\lim_{B \to A} \left\|A^{-1}\right\|\left(-1 + \frac{1}{1 - \left\|A^{-1}\right\|\|B-A\|}\right) = \left\|A^{-1}\right\|\left(-1 + \frac{1}{1 - \left\|A^{-1}\right\| \cdot 0}\right)$$
$$= \left\|A^{-1}\right\|(-1 + 1) = 0.$$

Since, for every $B$ satisfying $\|B - A\| < \frac{1}{\|A^{-1}\|}$, we have

$$0 \leq \|B^{-1} - A^{-1}\| \leq \|A^{-1}\| \left( -1 + \frac{1}{1 - \|A^{-1}\|\|B - A\|} \right),$$

and

$$\lim_{B \to A} \|A^{-1}\| \left( -1 + \frac{1}{1 - \|A^{-1}\|\|B - A\|} \right) = 0$$

so, $\lim_{B \to A} \|B^{-1} - A^{-1}\|$ exists, and its value is 0. This shows that the mapping $B \mapsto B^{-1}$ from $\Omega$ to $\Omega$ is continuous at $A$. This proves 4. $\qquad \square$

**Definition** Let $\{e_1, e_2, e_3\}$ be any basis of real linear space $\mathbb{R}^3$, and let $\{f_1, f_2\}$ be any basis of real linear space $\mathbb{R}^2$. Let $A$ be in $L(\mathbb{R}^3, \mathbb{R}^2)$. Let

$$\begin{aligned}
A(e_1) &\equiv a_{11}f_1 + a_{21}f_2, \\
A(e_2) &\equiv a_{12}f_1 + a_{22}f_2, \\
A(e_3) &\equiv a_{13}f_1 + a_{23}f_2.
\end{aligned}$$

The matrix

$$\begin{bmatrix} a_{11} & a_{12} & a_{13} \\ a_{21} & a_{22} & a_{23} \end{bmatrix}_{2 \times 3}$$

is denoted by $[A]$.

Similar definition can be supplied for $A$ in $L(\mathbb{R}^n, \mathbb{R}^m)$.

**Theorem 3.4** *Let $\mathcal{M}$ be the collection of all $2 \times 3$ matrices with real entries. Let $\{e_1, e_2, e_3\}$ be a basis of real linear space $\mathbb{R}^3$, and let $\{f_1, f_2\}$ be a basis of real linear space $\mathbb{R}^2$. Then, the mapping $A \mapsto [A]$ from $L(\mathbb{R}^3, \mathbb{R}^2)$ to $\mathcal{M}$ is 1–1 and onto.*

*Proof* 1–1ness: Let $[A] = [B]$, where

$$[A] \equiv \begin{bmatrix} a_{11} & a_{12} & a_{13} \\ a_{21} & a_{22} & a_{23} \end{bmatrix}_{2 \times 3} \quad \text{and} \quad [B] \equiv \begin{bmatrix} b_{11} & b_{12} & b_{13} \\ b_{21} & b_{22} & b_{23} \end{bmatrix}_{2 \times 3}.$$

We have to prove that the mappings $A : \mathbb{R}^3 \to \mathbb{R}^2$ and $B : \mathbb{R}^3 \to \mathbb{R}^2$ are equal, that is, for every $t_1 e_1 + t_2 e_2 + t_3 e_3$ in $\mathbb{R}^3$,

$$A(t_1 e_1 + t_2 e_2 + t_3 e_3) = B(t_1 e_1 + t_2 e_2 + t_3 e_3).$$

For this purpose, let us take any $t_1 e_1 + t_2 e_2 + t_3 e_3$ in $\mathbb{R}^3$. Since $[A] = [B]$, $a_{ij} = b_{ij}$ for every $i = 1, 2,$ and for every $j = 1, 2, 3$. Further,

$$[A] = \begin{bmatrix} a_{11} & a_{12} & a_{13} \\ a_{21} & a_{22} & a_{23} \end{bmatrix}_{2\times3} \quad \text{and} \quad [B] = \begin{bmatrix} b_{11} & b_{12} & b_{13} \\ b_{21} & b_{22} & b_{23} \end{bmatrix}_{2\times3}$$

so $A(e_1) = a_{11}f_1 + a_{21}f_2 = b_{11}f_1 + b_{21}f_2 = B(e_1)$. Similarly, $A(e_2) = B(e_2)$ and $A(e_3) = B(e_3)$.

So,

$$\begin{aligned} \text{LHS} &= A(t_1e_1 + t_2e_2 + t_3e_3) = t_1(A(e_1)) + t_2(A(e_2)) + t_3(A(e_3)) \\ &= t_1(B(e_1)) + t_2(B(e_2))t_3(B(e_3)) = B(t_1e_1 + t_2e_2 + t_3e_3) = \text{RHS}. \end{aligned}$$

This proves that $A \mapsto [A]$ is 1–1.

Ontoness: Let us take any $2 \times 3$ matrix

$$\begin{bmatrix} a_{11} & a_{12} & a_{13} \\ a_{21} & a_{22} & a_{23} \end{bmatrix}_{2\times3}.$$

Let us define a function $A : \mathbb{R}^3 \to \mathbb{R}^2$ as follows: For every $t_1e_1 + t_2e_2 + t_3e_3$ in $\mathbb{R}^3$,

$$A(t_1e_1 + t_2e_2 + t_3e_3) \equiv t_1(a_{11}f_1 + a_{21}f_2) + t_2(a_{12}f_1 + a_{22}f_2) + t_3(a_{13}f_1 + a_{23}f_2).$$

We shall show that $A$ is in $L(\mathbb{R}^3, \mathbb{R}^2)$. For this purpose, let us take any $x \equiv s_1e_1 + s_2e_2 + s_3e_3$, $y \equiv t_1e_1 + t_2e_2 + t_3e_3$, and any real number $t$. We have to prove that

1. $A(x + y) = A(x) + A(y)$,
2. $A(tx) = t(A(x))$.

For 1: 
$$\begin{aligned} \text{LHS} &= A(x + y) = A((s_1e_1 + s_2e_2 + s_3e_3) + (t_1e_1 + t_2e_2 + t_3e_3)) \\ &= A((s_1 + t_1)e_1 + (s_2 + t_2)e_2 + (s_3 + t_3)e_3) \\ &= (s_1 + t_1)(a_{11}f_1 + a_{21}f_2) + (s_2 + t_2)(a_{12}f_1 + a_{22}f_2) \\ &\quad + (s_3 + t_3)(a_{13}f_1 + a_{23}f_2). \end{aligned}$$

and

$$\begin{aligned} \text{RHS} &= A(x) + A(y) = A(s_1e_1 + s_2e_2 + s_3e_3) + A(t_1e_1 + t_2e_2 + t_3e_3) \\ &= s_1(a_{11}f_1 + a_{21}f_2) + s_2(a_{12}f_1 + a_{22}f_2) + s_3(a_{13}f_1 + a_{23}f_2) \\ &\quad + t_1(a_{11}f_1 + a_{21}f_2) + t_2(a_{12}f_1 + a_{22}f_2) + t_3(a_{13}f_1 + a_{23}f_2) \\ &= (s_1 + t_1)(a_{11}f_1 + a_{21}f_2) + (s_2 + t_2)(a_{12}f_1 + a_{22}f_2) \\ &\quad + (s_3 + t_3)(a_{13}f_1 + a_{23}f_2). \end{aligned}$$

So, LHS = RHS. This proves 1.

For 2:  LHS $= A(tx)$

$= A(t(s_1e_1 + s_2e_2 + s_3e_3)) = A((ts_1)e_1 + (ts_2)e_2 + (ts_3)e_3)$

$= (ts_1)(a_{11}f_1 + a_{21}f_2) + (ts_2)(a_{12}f_1 + a_{22}f_2) + (ts_3)(a_{13}f_1 + a_{23}f_2)$

$= (ts_1a_{11} + ts_2a_{12} + ts_3a_{13})f_1 + (ts_1a_{21} + ts_2a_{22} + ts_3a_{23})f_2$

$= t(s_1a_{11} + s_2a_{12} + s_3a_{13})f_1 + t(s_1a_{21} + s_2a_{22} + s_3a_{23})f_2$

$= ts_1(a_{11}f_1 + a_{21}f_2) + ts_2(a_{12}f_1 + a_{22}f_2) + ts_3(a_{13}f_1 + a_{23}f_2)$

$= t(s_1(a_{11}f_1 + a_{21}f_2) + s_2(a_{12}f_1 + a_{22}f_2) + s_3(a_{13}f_1 + a_{23}f_2)).$

and

RHS $= t(A(x)) = t(A(s_1e_1 + s_2e_2 + s_3e_3))$

$= t(s_1(a_{11}f_1 + a_{21}f_2) + s_2(a_{12}f_1 + a_{22}f_2) + s_3(a_{13}f_1 + a_{23}f_2)).$

So, LHS = RHS. This proves 2.

Finally, it remains to prove that

$$[A] = \begin{bmatrix} a_{11} & a_{12} & a_{13} \\ a_{21} & a_{22} & a_{23} \end{bmatrix}.$$

Since

$A(e_1) = A(1e_1 + 0e_2 + 0e_3)$

$= 1(a_{11}f_1 + a_{21}f_2) + 0(a_{12}f_1 + a_{22}f_2) + 0(a_{13}f_1 + a_{23}f_2),$

$$A(e_1) = a_{11}f_1 + a_{21}f_2.$$

Similarly,

$$A(e_2) = a_{12}f_1 + a_{22}f_2,$$
$$A(e_3) = a_{13}f_1 + a_{23}f_2.$$

Hence,

$$[A] = \begin{bmatrix} a_{11} & a_{12} & a_{13} \\ a_{21} & a_{22} & a_{23} \end{bmatrix}. \qquad \square$$

**Note 3.5** The result similar to Theorem 3.4 can be proved as above for $L(\mathbb{R}^n, \mathbb{R}^m)$.

**Theorem 3.6** *Let* $\{e_1, e_2, e_3, e_4\}$ *be any basis of real linear space* $\mathbb{R}^4$, *let* $\{f_1, f_2, f_3\}$ *be any basis of real linear space* $\mathbb{R}^3$, *and let* $\{g_1, g_2\}$ *be any basis of real linear space* $\mathbb{R}^2$. *Let A be in* $L(\mathbb{R}^4, \mathbb{R}^3)$, *and let B be in* $L(\mathbb{R}^3, \mathbb{R}^2)$. *Then,*

$$[BA] = [B][A].$$

*Proof* Here, $A$ is in $L(\mathbb{R}^4, \mathbb{R}^3)$ and $B$ is in $L(\mathbb{R}^3, \mathbb{R}^2)$, so their product $BA$ (i.e., the composite $B \circ A$) is in $L(\mathbb{R}^4, \mathbb{R}^2)$, and hence, LHS$(= [BA])$ is a $2 \times 4$ matrix. Since $A$ is in $L(\mathbb{R}^4, \mathbb{R}^3)$, $[A]$ is a $3 \times 4$ matrix. Since $B$ is in $L(\mathbb{R}^3, \mathbb{R}^2)$, $[B]$ is a $2 \times 3$ matrix. Since $[B]$ is a $2 \times 3$ matrix, and $[A]$ is a $3 \times 4$ matrix, RHS$(= [B][A])$ is a $2 \times 4$ matrix. So, matrices in LHS and RHS have the same order. Let

$$[A] = \begin{bmatrix} a_{11} & a_{12} & a_{13} & a_{14} \\ a_{21} & a_{22} & a_{23} & a_{24} \\ a_{31} & a_{32} & a_{33} & a_{34} \end{bmatrix}_{3 \times 4}$$

and

$$[B] = \begin{bmatrix} b_{11} & b_{12} & b_{13} \\ b_{21} & b_{22} & b_{23} \end{bmatrix}_{2 \times 3}.$$

Hence,

$$(BA)(e_1) = B(A(e_1)) = B(a_{11}f_1 + a_{21}f_2 + a_{31}f_3) = a_{11}(B(f_1)) + a_{21}(B(f_2)) + a_{31}(B(f_3))$$
$$= a_{11}(b_{11}g_1 + b_{21}g_2) + a_{21}(b_{12}g_1 + b_{22}g_2) + a_{31}(b_{13}g_1 + b_{23}g_2)$$
$$= (a_{11}b_{11} + a_{21}b_{12} + a_{31}b_{13})g_1 + (a_{11}b_{21} + a_{21}b_{22} + a_{31}b_{23})g_2.$$

So the first column of $[BA]$ is

$$a_{11}b_{11} + a_{21}b_{12} + a_{31}b_{13}$$
$$a_{11}b_{21} + a_{21}b_{22} + a_{31}b_{23}$$

Now,

$$[B][A] = \begin{bmatrix} b_{11} & b_{12} & b_{13} \\ b_{21} & b_{22} & b_{23} \end{bmatrix}_{2 \times 3} \begin{bmatrix} a_{11} & a_{12} & a_{13} & a_{14} \\ a_{21} & a_{22} & a_{23} & a_{24} \\ a_{31} & a_{32} & a_{33} & a_{34} \end{bmatrix}_{3 \times 4}$$
$$= \begin{bmatrix} b_{11}a_{11} + b_{12}a_{21} + b_{13}a_{31} & \cdots & \cdots & \cdots \\ b_{21}a_{11} + b_{22}a_{21} + b_{23}a_{31} & \cdots & \cdots & \cdots \end{bmatrix}_{2 \times 4}$$
$$= \begin{bmatrix} a_{11}b_{11} + a_{21}b_{12} + a_{31}b_{13} & \cdots & \cdots & \cdots \\ a_{11}b_{21} + a_{21}b_{22} + a_{31}b_{23} & \cdots & \cdots & \cdots \end{bmatrix}_{2 \times 4}$$

Hence, first column of $[BA]$ is the same as the first column of $[B][A]$. Similarly, second column of $[BA]$ is the same as the second column of $[B][A]$, third column of $[BA]$ is the same as the third column of $[B][A]$, and fourth column of $[BA]$ is the same as the fourth column of $[B][A]$. This shows that $[BA] = [B][A]$.    $\square$

**Note 3.7** The result similar to Theorem 3.6 can be proved as above for $A$ in $L(\mathbb{R}^n, \mathbb{R}^m)$ and $B$ in $L(\mathbb{R}^m, \mathbb{R}^p)$.

**Note 3.8** Let $A$ be in $L(\mathbb{R}^3, \mathbb{R}^2)$. Let

$$[A] = \begin{bmatrix} a_{11} & a_{12} & a_{13} \\ a_{21} & a_{22} & a_{23} \end{bmatrix}_{2 \times 3}$$

relative to the basis $\{(1, 0, 0), (0, 1, 0), (0, 0, 1)\}$ of $\mathbb{R}^3$ and the basis $\{(1, 0), (0, 1)\}$ of $\mathbb{R}^2$. Let $(t_1, t_2, t_3)$ be any member of $\mathbb{R}^3$ satisfying $|(t_1, t_2, t_3)| \leq 1$. Since $A$ is a linear transformation from $\mathbb{R}^3$ to $\mathbb{R}^2$,

$$
\begin{aligned}
A((t_1, t_2, t_3)) &= A(t_1(1, 0, 0) + t_2(0, 1, 0) + t_3(0, 0, 1)) \\
&= t_1(A((1, 0, 0))) + t_2(A((0, 1, 0))) + t_3(A((0, 0, 1))) \\
&= t_1(a_{11}, a_{21}) + t_2(a_{12}, a_{22}) + t_3(a_{13}, a_{23}) \\
&= (t_1 a_{11} + t_2 a_{12} + t_3 a_{13}, t_1 a_{21} + t_2 a_{22} + t_3 a_{23}) \\
&= ((t_1, t_2, t_3) \cdot (a_{11}, a_{12}, a_{13}), (t_1, t_2, t_3) \cdot (a_{21}, a_{22}, a_{23})),
\end{aligned}
$$

and hence,

$$
\begin{aligned}
|A((t_1, t_2, t_3))| &= |((t_1, t_2, t_3) \cdot (a_{11}, a_{12}, a_{13}), (t_1, t_2, t_3) \cdot (a_{21}, a_{22}, a_{23}))| \\
&= \sqrt{|(t_1, t_2, t_3) \cdot (a_{11}, a_{12}, a_{13})|^2 + |(t_1, t_2, t_3) \cdot (a_{21}, a_{22}, a_{23})|^2} \\
&\leq \sqrt{(|(t_1, t_2, t_3)||(a_{11}, a_{12}, a_{13})|)^2 + (|(t_1, t_2, t_3)||(a_{21}, a_{22}, a_{23})|)^2} \\
&\leq \sqrt{(1|(a_{11}, a_{12}, a_{13})|)^2 + (1|(a_{21}, a_{22}, a_{23})|)^2} \\
&= \sqrt{|(a_{11}, a_{12}, a_{13})|^2 + |(a_{21}, a_{22}, a_{23})|^2} \\
&= \sqrt{(a_{11})^2 + (a_{12})^2 + (a_{13})^2 + (a_{21})^2 + (a_{22})^2 + (a_{23})^2}.
\end{aligned}
$$

This shows that

$$
\begin{aligned}
\|A\| &= \sup\{|A((t_1, t_2, t_3))| : (t_1, t_2, t_3) \text{ is in } \mathbb{R}^n, \text{ and } |(t_1, t_2, t_3)| \leq 1\} \\
&\leq \sqrt{(a_{11})^2 + (a_{12})^2 + (a_{13})^2 + (a_{21})^2 + (a_{22})^2 + (a_{23})^2} = \sqrt{\sum_{j=1}^{3}\left(\sum_{i=1}^{2}(a_{ij})^2\right)}.
\end{aligned}
$$

Conclusion: If $A$ is in $L(\mathbb{R}^3, \mathbb{R}^2)$ and

$$[A] = \begin{bmatrix} a_{11} & a_{12} & a_{13} \\ a_{21} & a_{22} & a_{23} \end{bmatrix}_{2 \times 3}$$

relative to the basis $\{(1, 0, 0), (0, 1, 0), (0, 0, 1)\}$ of $\mathbb{R}^3$ and the basis $\{(1, 0), (0, 1)\}$ of $\mathbb{R}^2$, then

$$\|A\| \leq \sqrt{\sum_{j=1}^{3}\left(\sum_{i=1}^{2}(a_{ij})^2\right)}.$$

The result similar to this conclusion can be proved as above for $\mathbb{R}^n$ in place of $\mathbb{R}^3$ and $\mathbb{R}^m$ in place of $\mathbb{R}^2$.

**Theorem 3.9** *Let $X$ be a metric space. Let $\tau$ be the 1–1, onto function $A \mapsto [A]$ of Theorem 3.4, relative to the basis $\{(1, 0, 0), (0, 1, 0), (0, 0, 1)\}$ of $\mathbb{R}^3$ and the basis $\{(1, 0), (0, 1)\}$ of $\mathbb{R}^2$. If, for $i = 1, 2,$ and $j = 1, 2, 3,$ each $a_{ij} : X \to \mathbb{R}$ is continuous, then the mapping*

$$x \mapsto \tau^{-1}\left(\begin{bmatrix} a_{11}(x) & a_{12}(x) & a_{13}(x) \\ a_{21}(x) & a_{22}(x) & a_{23}(x) \end{bmatrix}_{2\times3}\right)$$

*from $X$ to $L(\mathbb{R}^3, \mathbb{R}^2)$ is also continuous.*

*Proof* Let us observe that for every $x$, $y$ in $X$,

$$\left\| \tau^{-1}\left(\begin{bmatrix} a_{11}(x) & a_{12}(x) & a_{13}(x) \\ a_{21}(x) & a_{22}(x) & a_{23}(x) \end{bmatrix}\right) - \tau^{-1}\left(\begin{bmatrix} a_{11}(y) & a_{12}(y) & a_{13}(y) \\ a_{21}(y) & a_{22}(y) & a_{23}(y) \end{bmatrix}\right) \right\|$$
$$= \|A_x - A_y\|,$$

where

$$A_x \equiv \tau^{-1}\left(\begin{bmatrix} a_{11}(x) & a_{12}(x) & a_{13}(x) \\ a_{21}(x) & a_{22}(x) & a_{23}(x) \end{bmatrix}\right) \in L(\mathbb{R}^3, \mathbb{R}^2),$$

$$A_y \equiv \tau^{-1}\left(\begin{bmatrix} a_{11}(y) & a_{12}(y) & a_{13}(y) \\ a_{21}(y) & a_{22}(y) & a_{23}(y) \end{bmatrix}\right) \in L(\mathbb{R}^3, \mathbb{R}^2).$$

Since

$$A_x = \tau^{-1}\left(\begin{bmatrix} a_{11}(x) & a_{12}(x) & a_{13}(x) \\ a_{21}(x) & a_{22}(x) & a_{23}(x) \end{bmatrix}\right),$$

$$[A_x] = \tau(A_x) = \begin{bmatrix} a_{11}(x) & a_{12}(x) & a_{13}(x) \\ a_{21}(x) & a_{22}(x) & a_{23}(x) \end{bmatrix},$$

and hence,

$$A_x((1,0,0)) = (a_{11}(x))(1,0) + (a_{21}(x))(0,1) = (a_{11}(x), a_{21}(x)),$$
$$A_x((0,1,0)) = (a_{12}(x))(1,0) + (a_{22}(x))(0,1) = (a_{12}(x), a_{22}(x)),$$
$$A_x((0,0,1)) = (a_{13}(x))(1,0) + (a_{23}(x))(0,1) = (a_{13}(x), a_{23}(x)).$$

Similarly,

$$A_y((1,0,0)) = (a_{11}(y), a_{21}(y)),$$
$$A_y((0,1,0)) = (a_{12}(y), a_{22}(y)),$$
$$A_y((0,0,1)) = (a_{13}(y), a_{23}(y)).$$

Since $A_x$ is in $L(\mathbb{R}^3, \mathbb{R}^2)$, and $A_y$ is in $L(\mathbb{R}^3, \mathbb{R}^2)$, $A_x - A_y$ is in $L(\mathbb{R}^3, \mathbb{R}^2)$, and hence, by Note 3.1,

$$
\begin{aligned}
\|A_x - A_y\| &\leq \left|(A_x - A_y)((1,0,0))\right| + \left|(A_x - A_y)((0,1,0))\right| \\
&\quad + \left|(A_x - A_y)((0,0,1))\right| \\
&= \left|(A_x((1,0,0)) - A_y((1,0,0)))\right| + \left|(A_x((0,1,0)) - A_y((0,1,0)))\right| \\
&\quad + \left|(A_x((0,0,1)) - A_y((0,0,1)))\right| \\
&= \left|((a_{11}(x), a_{21}(x)) - (a_{11}(y), a_{21}(y)))\right| \\
&\quad + \left|((a_{12}(x), a_{22}(x)) - (a_{12}(y), a_{22}(y)))\right| \\
&\quad + \left|((a_{13}(x), a_{23}(x)) - (a_{13}(y), a_{23}(y)))\right| \\
&= \left|(a_{11}(x) - a_{11}(y), a_{21}(x) - a_{21}(y))\right| \\
&\quad + \left|(a_{12}(x) - a_{12}(y), a_{22}(x) - a_{22}(y))\right| \\
&\quad + \left|(a_{13}(x) - a_{13}(y), a_{23}(x) - a_{23}(y))\right| \\
&\leq (|a_{11}(x) - a_{11}(y)| + |a_{21}(x) - a_{21}(y)|) \\
&\quad + (|a_{12}(x) - a_{12}(y)| + |a_{22}(x) - a_{22}(y)|) \\
&\quad + (|a_{13}(x) - a_{13}(y)| + |a_{23}(x) - a_{23}(y)|).
\end{aligned}
$$

Thus, for every $x$, $y$ in $X$,

$$
\begin{aligned}
&\left\| \tau^{-1}\left(\begin{bmatrix} a_{11}(x) & a_{12}(x) & a_{13}(x) \\ a_{21}(x) & a_{22}(x) & a_{23}(x) \end{bmatrix}\right) - \tau^{-1}\left(\begin{bmatrix} a_{11}(y) & a_{12}(y) & a_{13}(y) \\ a_{21}(y) & a_{22}(y) & \dot{a}_{23}(y) \end{bmatrix}\right) \right\| \\
&\leq (|a_{11}(x) - a_{11}(y)| + |a_{21}(x) - a_{21}(y)|) \\
&\quad + (|a_{12}(x) - a_{12}(y)| + |a_{22}(x) - a_{22}(y)|) \\
&\quad + (|a_{13}(x) - a_{13}(y)| + |a_{23}(x) - a_{23}(y)|) \cdots (*).
\end{aligned}
$$

Now, let us fix any $x$ in $X$.
It is enough to prove that the function

$$x \mapsto \tau^{-1}\left(\begin{bmatrix} a_{11}(x) & a_{12}(x) & a_{13}(x) \\ a_{21}(x) & a_{22}(x) & a_{23}(x) \end{bmatrix}_{2 \times 3}\right)$$

from $X$ to $L(\mathbb{R}^3, \mathbb{R}^2)$ is continuous at $x$. For this purpose, let us take any $\varepsilon > 0$. Since each $a_{ij} : X \to \mathbb{R}$ is continuous at $x$, there exists $\delta > 0$ such that for every $y$ in the open neighborhood $B_\delta(x)$ of $x$, we have $|a_{ij}(x) - a_{ij}(y)| < \frac{\varepsilon}{6}$ (for every $i = 1$, 2, and $j = 1, 2, 3$). Hence, for every $y$ in the open neighborhood $B_\delta(x)$ of $x$, we have from (*),

$$\left\| \tau^{-1}\left(\begin{bmatrix} a_{11}(x) & a_{12}(x) & a_{13}(x) \\ a_{21}(x) & a_{22}(x) & a_{23}(x) \end{bmatrix}\right) - \tau^{-1}\left(\begin{bmatrix} a_{11}(y) & a_{12}(y) & a_{13}(y) \\ a_{21}(y) & a_{22}(y) & a_{23}(y) \end{bmatrix}\right) \right\|$$
$$\leq (|a_{11}(x) - a_{11}(y)| + |a_{21}(x) - a_{21}(y)|) + (|a_{12}(x) - a_{12}(y)| + |a_{22}(x) - a_{22}(y)|)$$
$$+ (|a_{13}(x) - a_{13}(y)| + |a_{23}(x) - a_{23}(y)|) < \frac{\varepsilon}{6} + \frac{\varepsilon}{6} + \frac{\varepsilon}{6} + \frac{\varepsilon}{6} + \frac{\varepsilon}{6} + \frac{\varepsilon}{6} = \varepsilon. \qquad \square$$

**Note 3.10** The result similar to Theorem 3.9 can be proved as above for $\mathbb{R}^n$ in place of $\mathbb{R}^3$ and $\mathbb{R}^m$ in place of $\mathbb{R}^2$.

## 3.2 Differentiation

**Theorem 3.11** Let $E$ be an open subset of $\mathbb{R}^3$. Let $f : E \to \mathbb{R}^2$. Let $f_1 : E \to \mathbb{R}$, $f_2 : E \to \mathbb{R}$ be the component functions of $f$, that is, $f(h) = (f_1(h), f_2(h))$ for every $h$ in $E$. Let $x$ be in $E$.

If there exist $A$ and $B$ in $L(\mathbb{R}^3, \mathbb{R}^2)$ such that

$$\lim_{h \to 0} \left( \frac{1}{|h|} ((f_1(x+h), f_2(x+h)) - (f_1(x), f_2(x)) - A(h)) \right) = 0,$$

and

$$\lim_{h \to 0} \left( \frac{1}{|h|} ((f_1(x+h), f_2(x+h)) - (f_1(x), f_2(x)) - B(h)) \right) = 0,$$

then $A = B$.

*Proof* We have to prove that $A = B$, that is, for every $h$ in $\mathbb{R}^3$, $A(h) - B(h) = 0$.
For this purpose, let us take any $h^* \equiv (h_1, h_2, h_3)$ in $\mathbb{R}^3$.
It is sufficient to prove that $|(A - B)(h^*)| = 0$.

Case I: when $h^* = 0$

$$\text{LHS} = |(A - B)(h^*)| = |(A - B)(0)| = |0| = 0 = \text{RHS}.$$

Case II: when $h^* \neq 0$

Since $\lim_{h \to 0}(\frac{1}{|h|}((f_1(x+h), f_2(x+h)) - (f_1(x), f_2(x)) - A(h))) = 0$, $t \mapsto th^*$ is a continuous function, and $h^* \neq 0$, and hence,

$$\lim_{t \to 0}\left(\frac{1}{|th^*|}((f_1(x+th^*), f_2(x+th^*)) - (f_1(x), f_2(x)) - A(th^*))\right) = 0.$$

Similarly,

$$\lim_{t \to 0}\left(\frac{1}{|th^*|}((f_1(x+th^*), f_2(x+th^*)) - (f_1(x), f_2(x)) - B(th^*))\right) = 0.$$

Hence,

$$\lim_{t \to 0}\left(\frac{1}{|th^*|}((f_1(x+th^*), f_2(x+th^*)) - (f_1(x), f_2(x)) - A(th^*))\right.$$
$$\left. - \frac{1}{|th^*|}((f_1(x+th^*), f_2(x+th^*)) - (f_1(x), f_2(x)) - B(th^*))\right)$$

exists, and its value is $0 \,(= 0 - 0)$. Further, since $|h^*| > 0$,

$$0 = \lim_{t \to 0}\left(\frac{1}{|th^*|}((f_1(x+th^*), f_2(x+th^*)) - (f_1(x), f_2(x)) - A(th^*))\right.$$
$$\left. - \frac{1}{|th^*|}((f_1(x+th^*), f_2(x+th^*)) - (f_1(x), f_2(x)) - B(th^*))\right)$$
$$= \lim_{t \to 0}\left(\frac{1}{|th^*|}(B(th^*) - A(th^*))\right)$$
$$= \lim_{t \to 0}\left(\frac{1}{|th^*|}(t(B(h^*)) - t(A(h^*)))\right) = \lim_{t \to 0}\left(\frac{1}{|th^*|}t(B(h^*) - A(h^*))\right)$$
$$= \lim_{t \to 0}\left(\frac{1}{|th^*|}t((B - A)(h^*))\right) = \lim_{t \to 0}\left(\frac{1}{|t||h^*|}t((B - A)(h^*))\right).$$

Since $\lim_{t \to 0}(\frac{1}{|t||h^*|}t((B - A)(h^*))) = 0$,

$$0 = \lim_{t \to 0}\left|\frac{1}{|t||h^*|}t((B - A)(h^*))\right| = \lim_{t \to 0}\frac{1}{|h^*|}|(B - A)(h^*)| = \frac{1}{|h^*|}|(B - A)(h^*)|.$$

Since $\frac{1}{|h^*|}|(B - A)(h^*)| = 0$, LHS $= |(A - B)(h^*)| = |(B - A)(h^*)| = 0 = $ RHS.
Thus, we have shown that in all cases, LHS = RHS.                                          $\square$

**Note 3.12** The result similar to Theorem 3.11 can be proved as above for $\mathbb{R}^n$ in place of $\mathbb{R}^3$ and $\mathbb{R}^m$ in place of $\mathbb{R}^2$. According to Theorem 3.11, if there exists $A$ in $L(\mathbb{R}^3, \mathbb{R}^2)$ such that

$$\lim_{h \to 0} \left( \frac{1}{|h|} ((f_1(x+h), f_2(x+h)) - (f_1(x), f_2(x)) - A(h)) \right) = 0,$$

then such an $A$ is unique.

**Definition** Let $E$ be an open subset of $\mathbb{R}^3$. Let $f : E \to \mathbb{R}^2$. Let $f_1 : E \to \mathbb{R}$, $f_2 : E \to \mathbb{R}$ be the component functions of $f$, that is, $f(h) = (f_1(h), f_2(h))$ for every $h$ in $E$. Let $x$ be in $E$.

If there exists $A$ in $L(\mathbb{R}^3, \mathbb{R}^2)$ such that

$$\lim_{h \to 0} \left( \frac{1}{|h|} ((f_1(x+h), f_2(x+h)) - (f_1(x), f_2(x)) - A(h)) \right) = 0,$$

then we say that $f$ is *differentiable at x*, and we write

$$f'(x) = A.$$

Here, $f'(x)$ is called the *differential of f at x*, or the *total derivative of f at x*. It is also denoted by $Df(x)$.

In short, $f'(x) = A$ means $\lim_{h \to 0} (\frac{1}{|h|} (f(x+h) - f(x) - A(h))) = 0$.

If $f$ is differentiable at every point of $E$, then we say that $f$ is *differentiable on E*. Similar definition can be supplied for $\mathbb{R}^n$ in place of $\mathbb{R}^3$ and $\mathbb{R}^m$ in place of $\mathbb{R}^2$.

**Theorem 3.13** *Let E be an open subset of $\mathbb{R}^n$. Let $f : E \to \mathbb{R}^m$. Let x be in E. Let A be in $L(\mathbb{R}^n, \mathbb{R}^m)$.*

*$f'(x) = A$ if and only if there exists a function $r : \{h : x + h \in E\} \to \mathbb{R}^m$ such that*

(i)   $r(0) = 0$,
(ii)  $f(x+h) = f(x) + A(h) + |h|(r(h))$ for every $h$ in $\{h : x + h \in E\}$,
(iii) *$r$ is continuous at 0.*

*Proof of "if" part*
Suppose that there exists a function $r : \{h : x + h \in E\} \to \mathbb{R}^m$ such that

(i)   $r(0) = 0$,
(ii)  $f(x + h) = f(x) + A(h) + |h|(r(h))$ for every $h$ in $\{h : x + h \in E\}$,
(iii) $r$ is continuous at 0.

We have to show that $f'(x) = A$, that is,

$$\lim_{h \to 0} \left( \frac{1}{|h|} (f(x+h) - f(x) - A(h)) \right) = 0.$$

Since $r$ is continuous at 0,

$$\text{RHS} = 0 = r(0) = \lim_{h \to 0} r(h) = \lim_{h \to 0} \left( \frac{1}{|h|} \left( f(x+h) - f(x) - A(h) \right) \right) = \text{LHS}.$$

*Proof of "only if" part*
Let $f'(x) = A$, that is,

$$\lim_{h \to 0} \left( \frac{1}{|h|} \left( f(x+h) - f(x) - A(h) \right) \right) = 0 \cdots (*).$$

Let us define $r : \{h : x+h \in E\} \to \mathbb{R}^m$ as follows: For every $h$ in $\{h : x+h \in E\}$,

$$r(h) \equiv \begin{cases} \frac{1}{|h|} \left( f(x+h) - f(x) - A(h) \right), & \text{if } h \neq 0 \\ 0, & \text{if } h = 0 \end{cases}.$$

Clearly, $r(0) = 0$. Also, it is clear that for every $h$ in $\{h : x+h \in E\}$, $f(x+h) = f(x) + A(h) + |h|(r(h))$. By (*),

$$\lim_{h \to 0} r(h) = \lim_{h \to 0} \left( \frac{1}{|h|} \left( f(x+h) - f(x) - A(h) \right) \right) = 0 = r(0),$$

and hence, $r$ is continuous at 0.                                                       □

**Theorem 3.14** *Let $E$ be an open subset of $\mathbb{R}^3$. Let $f : E \to \mathbb{R}^2$. Let $f_1 : E \to \mathbb{R}$, $f_2 : E \to \mathbb{R}$ be the component functions of $f$, that is, $f(h) = (f_1(h), f_2(h))$ for every $h$ in $E$. Let $x$ be in $E$. Let $A \in L(\mathbb{R}^3, \mathbb{R}^2)$, where*

$$[A] = \begin{bmatrix} a_{11} & a_{12} & a_{13} \\ a_{21} & a_{22} & a_{23} \end{bmatrix}_{2 \times 3}$$

*relative to the basis $\{(1, 0, 0), (0, 1, 0), (0, 0, 1)\}$ of $\mathbb{R}^3$ and the basis $\{(1, 0), (0, 1)\}$ of $\mathbb{R}^2$. $f'(x) = A$ if and only if*

$$\lim_{(h_1, h_2, h_3) \equiv h \to 0} \left( \frac{1}{|h|} \left( f_1(x+h) - f_1(x) - (h_1, h_2, h_3) \cdot (a_{11}, a_{12}, a_{13}) \right) \right) = 0,$$

$$\lim_{(h_1, h_2, h_3) \equiv h \to 0} \left( \frac{1}{|h|} \left( f_2(x+h) - f_2(x) - (h_1, h_2, h_3) \cdot (a_{21}, a_{22}, a_{23}) \right) \right) = 0.$$

*Proof*
$f'(x) = A$ means

$$(0,0) = 0 = \lim_{h \to 0} \left( \frac{1}{|h|} (f(x+h) - f(x) - A(h)) \right)$$

$$= \lim_{(h_1,h_2,h_3) \equiv h \to 0} \left( \frac{1}{|h|} ((f_1(x+h), f_2(x+h)) - (f_1(x), f_2(x)) - A(h)) \right).$$

Since

$$\begin{aligned}
A(h) &= A((h_1, h_2, h_3)) \\
&= h_1(A((1,0,0))) + h_2(A((0,1,0))) + h_3(A((0,0,1))) \\
&= h_1(a_{11}, a_{21}) + h_2(a_{12}, a_{22}) + h_3(a_{13}, a_{23}) \\
&= (h_1 a_{11} + h_2 a_{12} + h_3 a_{13}, h_1 a_{21} + h_2 a_{22} + h_3 a_{23}) \\
&= ((h_1, h_2, h_3) \cdot (a_{11}, a_{12}, a_{13}), (h_1, h_2, h_3) \cdot (a_{21}, a_{22}, a_{23})),
\end{aligned}$$

$f'(x) = A$ if and only if

$$(0,0) = \lim_{(h_1,h_2,h_3) \equiv h \to 0} \left( \frac{1}{|h|} ((f_1(x+h), f_2(x+h)) - (f_1(x), f_2(x))) \right.$$

$$\left. - ((h_1, h_2, h_3) \cdot (a_{11}, a_{12}, a_{13}), (h_1, h_2, h_3) \cdot (a_{21}, a_{22}, a_{23})) \right)$$

$$= \lim_{(h_1,h_2,h_3) \equiv h \to 0} \left( \frac{1}{|h|} (f_1(x+h) - f_1(x) - (h_1, h_2, h_3) \cdot (a_{11}, a_{12}, a_{13})), \right.$$

$$\left. \frac{1}{|h|} (f_2(x+h) - f_2(x) - (h_1, h_2, h_3) \cdot (a_{21}, a_{22}, a_{23})) \right)$$

$$= \left( \lim_{(h_1,h_2,h_3) \equiv h \to 0} \frac{1}{|h|} (f_1(x+h) - f_1(x) - (h_1, h_2, h_3) \cdot (a_{11}, a_{12}, a_{13})), \right.$$

$$\left. \lim_{(h_1,h_2,h_3) \equiv h \to 0} \frac{1}{|h|} (f_2(x+h) - f_2(x) - (h_1, h_2, h_3) \cdot (a_{21}, a_{22}, a_{23})) \right).$$

Thus, $f'(x) = A$ if and only if

$$\lim_{(h_1,h_2,h_3) \equiv h \to 0} \frac{1}{|h|} (f_1(x+h) - f_1(x) - (h_1, h_2, h_3) \cdot (a_{11}, a_{12}, a_{13})) = 0,$$

$$\lim_{(h_1,h_2,h_3) \equiv h \to 0} \frac{1}{|h|} (f_2(x+h) - f_2(x) - (h_1, h_2, h_3) \cdot (a_{21}, a_{22}, a_{23})) = 0. \qquad \square$$

**Note 3.15** The result similar to Theorem 3.14 can be proved as above for $\mathbb{R}^n$ in place of $\mathbb{R}^3$ and $\mathbb{R}^m$ in place of $\mathbb{R}^2$.

**Theorem 3.16** *Let E be an open subset of* $\mathbb{R}^n$. *Let* $f : E \to \mathbb{R}^m$. *Let x be in E.*
*If f is differentiable at x, then f is continuous at x.*

*Proof* Let $f$ be differentiable at $x$.

We have to prove that $f$ is continuous at $x$, that is, $\lim_{h \to 0}(f(x+h) - f(x)) = 0$.
For this purpose, let us take any $\varepsilon > 0$.

Since $f$ is differentiable at $x$, there exists a linear transformation $A$ from $\mathbb{R}^n$ to $\mathbb{R}^m$
such that $0 = \lim_{h \to 0}(\frac{1}{|h|}(f(x+h) - f(x) - A(h)))$, and hence, there exists $\delta > 0$
such that for every $h$ in $\{h : x + h \in E\}$ satisfying $0 < |h| < \delta$, we have

$$\frac{1}{|h|}|f(x+h) - f(x) - A(h)| = \left|\frac{1}{|h|}(f(x+h) - f(x) - A(h)) - 0\right| < \frac{\varepsilon}{2}.$$

Case I: when $\|A\| \neq 0$

Now, if $h$ is in $\{h : x + h \in E\}$ satisfying $0 < |h| < \min\{1, \frac{\varepsilon}{2\|A\|}, \delta\}$, then

$$|(f(x+h) - f(x)) - 0| = \left||h|\left(\frac{1}{|h|}(f(x+h) - f(x) - A(h))\right) + A(h)\right|$$

$$\leq |h|\left|\frac{1}{|h|}(f(x+h) - f(x) - A(h))\right| + |A(h)|$$

$$\leq |h|\left|\frac{1}{|h|}(f(x+h) - f(x) - A(h))\right| + \|A\|\|h\|$$

$$\leq 1 \cdot \left|\frac{1}{|h|}(f(x+h) - f(x) - A(h))\right| + \|A\|\|h\|$$

$$< 1 \cdot \frac{\varepsilon}{2} + \|A\| \cdot \frac{\varepsilon}{2\|A\|} = \varepsilon.$$

So, $f$ is continuous at $x$.

Case II: when $\|A\| = 0$

Here, $\|A\| = 0$, so $A = 0$. Now, if $h$ is in $\{h : x + h \in E\}$ satisfying
$0 < |h| < \min\{1, \delta\}$, then

$$|(f(x+h) - f(x)) - 0| = \left||h|\left(\frac{1}{|h|}(f(x+h) - f(x) - 0(h))\right)\right|$$

$$= |h|\left(\frac{1}{|h|}|(f(x+h) - f(x) - 0(h))|\right)$$

$$\leq 1 \cdot \left(\frac{1}{|h|}|(f(x+h) - f(x) - A(h))|\right) < 1 \cdot \frac{\varepsilon}{2} < \varepsilon.$$

So, $f$ is continuous at $x$.

Thus, we see that in all cases, $f$ is continuous at $x$.                                      $\square$

**Note 3.17** If $A \in L(\mathbb{R}^n, \mathbb{R}^m)$, then $A$ is differentiable on $\mathbb{R}^n$, and for every $x$ in $\mathbb{R}^n$, $A'(x) = A$.

Reason: Let us take any $x$ in $\mathbb{R}^n$. Since

$$\lim_{h \to 0} \left( \frac{1}{|h|} (A(x+h) - A(x) - A(h)) \right)$$
$$= \lim_{h \to 0} \left( \frac{1}{|h|} ((A(x) + A(h)) - A(x) - A(h)) \right)$$
$$= \lim_{h \to 0} (0) = 0, \ A'(x) = A.$$

Further, since $A$ is differentiable on $\mathbb{R}^n$, by Theorem 3.16, $A$ is continuous on $\mathbb{R}^n$.

**Theorem 3.18** *Let $E$ be an open subset of $\mathbb{R}^n$. Let $x$ be in $E$. Let $f : E \to \mathbb{R}^m$ be differentiable at $x$. Let $G$ be an open subset of $\mathbb{R}^m$, and $G$ contains the range of $f$. Let $g : G \to \mathbb{R}^p$ be differentiable at $f(x)$.*

*Then,*

1. *the composite function $(g \circ f) : E \to \mathbb{R}^p$ is differentiable at $x$,*
2. $(g \circ f)'(x) = (g'(f(x)))(f'(x))$.

*Proof* Here, $f : E \to \mathbb{R}^m$ is differentiable at $x$, so by Theorem 3.13, there exists a function $r : \{h : x + h \in E\} \to \mathbb{R}^m$ such that

  (i) $r(0) = 0$,
  (ii) $f(x+h) = f(x) + (f'(x))(h) + |h|(r(h))$ for every $h$ in $\{h : x + h \in E\}$,
  (iii) $r$ is continuous at 0.

Here, $g : G \to \mathbb{R}^p$ is differentiable at $f(x)$, so by Theorem 3.17, there exists a function $s : \{k : f(x) + k \in G\} \to \mathbb{R}^p$ such that

  (i') $s(0) = 0$,
  (ii') $g(f(x) + k) = g(f(x)) + (g'(f(x)))(k) + |k|(s(k))$ for every $k$ in $\{k : f(x) + k \in G\}$,
  (iii') $s$ is continuous at 0.

We have to show that the composite function $(g \circ f) : E \to \mathbb{R}^p$ is differentiable at $x$, and $(g \circ f)'(x) = (g'(f(x)))(f'(x))$. By Theorem 3.13, it is enough to find a function $t : \{h : x + h \in E\} \to \mathbb{R}^p$ such that

  (I) $t(0) = 0$,
  (II) $(g \circ f)(x+h) = (g \circ f)(x) + ((g'(f(x)))(f'(x)))(h) + |h|(t(h))$ for every $h$ in $\{h : x + h \in E\}$,
  (III) $t$ is continuous at 0.

Here, for every $h$ in $\{h : x + h \in E\}$,

$$
\begin{aligned}
(g \circ f)(x + h) &= g(f(x + h)) \\
&= g(f(x) + (f'(x))(h) + |h|(r(h))) \\
&= g(f(x) + ((f'(x))(h) + |h|(r(h)))) \\
&= g(f(x)) + (g'(f(x)))((f'(x))(h) + |h|(r(h))) \\
&\quad + |(f'(x))(h) + |h|(r(h))|(s((f'(x))(h) + |h|(r(h)))) \\
&= g(f(x)) + (g'(f(x)))((f'(x))(h)) \\
&\quad + (g'(f(x)))(|h|(r(h))) + |(f'(x))(h) + |h|(r(h))|(s((f'(x))(h) + |h|(r(h)))) \\
&= (g \circ f)(x) + ((g'(f(x))) \circ (f'(x)))(h) + |h|((g'(f(x)))(r(h))) \\
&\quad + |(f'(x))(h) + |h|(r(h))|(s((f'(x))(h) + |h|(r(h)))) \\
&= (g \circ f)(x) + ((g'(f(x)))(f'(x)))(h) \\
&\quad + |h|\left( (g'(f(x)))(r(h)) + \frac{1}{|h|}|(f'(x))(h) + |h|(r(h))|(s((f'(x))(h) + |h|(r(h)))) \right).
\end{aligned}
$$

So, for every $h$ in $\{h : x + h \in E\}$,

$$
\begin{aligned}
(g \circ f)(x + h) &= (g \circ f)(x) + ((g'(f(x)))(f'(x)))(h) \\
&\quad + |h|((g'(f(x)))(r(h))) \\
&\quad + \frac{1}{|h|}|(f'(x))(h) + |h|(r(h))|(s((f'(x))(h) + |h|(r(h))))).
\end{aligned}
$$

Now, let us define the function $t : \{h : x + h \in E\} \to \mathbb{R}^p$ as follows: For every $h$ in $\{h : x + h \in E\}$,

$$
t(h) \equiv \begin{cases} (g'(f(x)))(r(h)) + \frac{1}{|h|}|(f'(x))(h) + |h|(r(h))|(s((f'(x))(h) + |h|(r(h)))), & \text{if } h \neq 0 \\ 0, & \text{if } h = 0 \end{cases}.
$$

It suffices to prove that $t$ is continuous at 0, that is,

$$
\lim_{h \to 0} t(h) = 0.
$$

For this purpose, let us take any $\varepsilon > 0$.

Since $(g'(f(x)))(r(h)) = ((g'(f(x))) \circ r)(h)$, the linear transformation $g'(f(x))$ is continuous, and $r$ is continuous at 0, the mapping $h \mapsto (g'(f(x)))(r(h))$ is continuous at 0. Hence, $\lim_{h \to 0}(g'(f(x)))(r(h)) = (g'(f(x)))(r(0)) = (g'(f(x)))(0) = 0$. It follows that there exists $\delta_1 > 0$ such that for every $h$ in $\{h : x + h \in E\}$ satisfying $0 < |h| < \delta_1$, we have

$$
|(g'(f(x)))(r(h))| < \frac{\varepsilon}{2}.
$$

Since, for every nonzero $h$ in $\{h : x + h \in E\}$, we have

$$\frac{1}{|h|}|(f'(x))(h) + |h|(r(h))| \le \frac{1}{|h|}(|(f'(x))(h)| + ||h|(r(h))|)$$

$$= \frac{1}{|h|}(|(f'(x))(h)| + |h||r(h)|) \le \frac{1}{|h|}(||f'(x)|||h| + |h||r(h)|) = ||f'(x)|| + |r(h)|.$$

Since $\lim_{h \to 0} r(h) = 0$, there exists $\delta_2 > 0$ such that for every $h$ in $\{h : x + h \in E\}$ satisfying $0 < |h| < \delta_2$, we have

$$|r(h)| < 1.$$

So, for every nonzero $h$ in $\{h : x + h \in E\}$ satisfying $0 < |h| < \delta_2$, we have

$$\frac{1}{|h|}|(f'(x))(h) + |h|(r(h))| \le ||f'(x)|| + 1.$$

Since $\lim_{h \to 0} r(h) = 0$, and $\lim_{k \to 0} s(k) = 0$, $\lim_{h \to 0} s((f'(x))(h) + |h|(r(h))) = s((f'(x))(0) + |0|(r(0))) = s(0 + 0) = s(0) = 0$, and hence, there exists $\delta_3 > 0$ such that for every $h$ in $\{h : x + h \in E\}$ satisfying $0 < |h| < \delta_3$, we have $|s((f'(x))(h) + |h|(r(h)))| < \frac{\varepsilon}{2(||f'(x)||+1)}$.

Hence, for every $h$ in $\{h : x + h \in E\}$ satisfying $0 < |h| < \min\{\delta_1, \delta_2, \delta_3\}$, we have

$$\left|\frac{1}{|h|}|(f'(x))(h) + |h|(r(h))|(s((f'(x))(h) + |h|(r(h))))\right|$$

$$= \frac{1}{|h|}|(f'(x))(h) + |h|(r(h))||s((f'(x))(h) + |h|(r(h)))|$$

$$\le (||f'(x)|| + 1)|s((f'(x))(h) + |h|(r(h)))| < (||f'(x)|| + 1) \cdot \frac{\varepsilon}{2(||f'(x)|| + 1)} = \frac{\varepsilon}{2}.$$

Further,

$$|t(h) - 0| = |t(h)| = \left|(g'(f(x)))(r(h)) + \frac{1}{|h|}|(f'(x))(h) + |h|(r(h))|\right.$$

$$(s((f'(x))(h) + |h|(r(h))))| \le |(g'(f(x)))(r(h))|$$

$$\left. + \left|\frac{1}{|h|}|(f'(x))(h) + |h|(r(h))|(s((f'(x))(h) + |h|(r(h))))\right|\right| < \frac{\varepsilon}{2} + \frac{\varepsilon}{2} = \varepsilon.$$

This proves that $\lim_{h \to 0} t(h) = 0$. $\qquad\square$

**Note 3.19** Theorem 3.18 is known as the *chain rule of derivative*.

**Theorem 3.20** *Let $E$ be an open subset of $\mathbb{R}^3$. Let $f : E \to \mathbb{R}^2$. Let $f_1 : E \to \mathbb{R}$, $f_2 : E \to \mathbb{R}$ be the component functions of $f$, that is, $f(h) = (f_1(h), f_2(h))$ for every $h$ in $E$. Let $x \equiv (x_1, x_2, x_3)$ be in $E$. Let $f$ be differentiable at $x$.*

Then, all the partial derivatives $(D_1f_1)(x), (D_2f_1)(x), (D_3f_1)(x), (D_1f_2)(x), (D_2f_2)$
$(x), (D_3f_2)(x)$ exist.

Also,

$$(f'(x))((1,0,0)) = ((D_1f_1)(x), (D_1f_2)(x)),$$
$$(f'(x))((0,1,0)) = ((D_2f_1)(x), (D_2f_2)(x)),$$
$$(f'(x))((0,0,1)) = ((D_3f_1)(x), (D_3f_2)(x)),$$

and for every $h$ in $\mathbb{R}^3$,

$$(f'(x))(h) = (((\nabla f_1)(x)) \cdot h, ((\nabla f_2)(x)) \cdot h),$$

where

$$(\nabla f_1)(x) \equiv ((D_1f_1)(x), (D_2f_1)(x), (D_3f_1)(x)),$$
$$(\nabla f_2)(x) \equiv ((D_1f_2)(x), (D_2f_2)(x), (D_3f_2)(x)).$$

(Here, $(\nabla f_1)(x)$ is called the *gradient of* $f_1$, etc.)

*Proof* Since $f$ is differentiable at $x$, $f'(x)$ is in $L(\mathbb{R}^3, \mathbb{R}^2)$. Let

$$[f'(x)] = \begin{bmatrix} a_{11} & a_{12} & a_{13} \\ a_{21} & a_{22} & a_{23} \end{bmatrix}_{2 \times 3}$$

relative to the basis $\{(1,0,0),(0,1,0),(0,0,1)\}$ of $\mathbb{R}^3$ and the basis $\{(1,0),(0,1)\}$
of $\mathbb{R}^2$. By Theorem 3.14,

$$\lim_{(h_1,h_2,h_3)=h \to 0} \left( \frac{1}{|h|} (f_1(x+h) - f_1(x) - (h_1,h_2,h_3) \cdot (a_{11}, a_{12}, a_{13})) \right) = 0.$$

Hence,

$$\lim_{t \to 0} \left( \frac{1}{|(0,t,0)|} (f_1((x_1,x_2,x_3) + (0,t,0)) - f_1((x_1,x_2,x_3)) - (0,t,0) \cdot (a_{11}, a_{12}, a_{13})) \right) = 0$$

or,

$$\lim_{t \to 0} \left( \frac{1}{|t|} (f_1((x_1,x_2+t,x_3)) - f_1((x_1,x_2,x_3)) - ta_{12}) \right) = 0$$

or,

$$\lim_{t \to 0} \left( \frac{1}{t} (f_1((x_1,x_2+t,x_3)) - f_1((x_1,x_2,x_3)) - ta_{12}) \right) = 0$$

or,

$$\lim_{t \to 0} \left( \frac{1}{t} (f_1((x_1, x_2 + t, x_3)) - f_1((x_1, x_2, x_3))) - a_{12} \right) = 0$$

or,

$$(D_2 f_1)(x) = \lim_{t \to 0} \frac{f_1((x_1, x_2 + t, x_3)) - f_1((x_1, x_2, x_3))}{t} = a_{12}.$$

It follows that $(D_2 f_1)(x)$ exists, and its value is $a_{12}$.
Similarly, all $(D_j f_i)(x)$ exist, and $(D_j f_i)(x) = a_{ij}$. Hence,

$$[f'(x)] = \begin{bmatrix} (D_1 f_1)(x) & (D_2 f_1)(x) & (D_3 f_1)(x) \\ (D_1 f_2)(x) & (D_2 f_2)(x) & (D_3 f_2)(x) \end{bmatrix}_{2 \times 3}$$

relative to the basis $\{(1,0,0),(0,1,0),(0,0,1)\}$ of $\mathbb{R}^3$ and the basis $\{(1,0),(0,1)\}$ of $\mathbb{R}^2$. Also,

$$(f'(x))((1,0,0)) = ((D_1 f_1)(x), (D_1 f_2)(x)),$$
$$(f'(x))((0,1,0)) = ((D_2 f_1)(x), (D_2 f_2)(x)),$$
$$(f'(x))((0,0,1)) = ((D_3 f_1)(x), (D_3 f_2)(x)).$$

Also, for every $h$ in $\mathbb{R}^3$,

$$\begin{aligned}
\text{LHS} &= (f'(x))(h) = (f'(x))((h_1, h_2, h_3)) \\
&= (f'(x))(h_1(1,0,0) + h_2(0,1,0) + h_3(0,0,1)) \\
&= h_1((D_1 f_1)(x), (D_1 f_2)(x)) + h_2((D_2 f_1)(x), (D_2 f_2)(x)) \\
&\quad + h_3((D_3 f_1)(x), (D_3 f_2)(x)) \\
&= (h_1((D_1 f_1)(x)) + h_2((D_2 f_1)(x)) + h_3((D_3 f_1)(x)), h_1((D_1 f_2)(x)) \\
&\quad + h_2((D_2 f_2)(x)) + h_3((D_3 f_2)(x))) \\
&= (((D_1 f_1)(x), (D_2 f_1)(x), (D_3 f_1)(x)) \cdot (h_1, h_2, h_3), \\
&\quad ((D_1 f_2)(x), (D_2 f_2)(x), (D_3 f_2)(x)) \cdot (h_1, h_2, h_3)) \\
&= (((\nabla f_1)(x)) \cdot (h_1, h_2, h_3), ((\nabla f_2)(x)) \cdot (h_1, h_2, h_3)) \\
&= (((\nabla f_1)(x)) \cdot h, ((\nabla f_2)(x)) \cdot h) = \text{RHS}.
\end{aligned}$$

$\square$

**Note 3.21** The result similar to Theorem 3.20 can be proved as above for $\mathbb{R}^n$ in place of $\mathbb{R}^3$ and $\mathbb{R}^m$ in place of $\mathbb{R}^2$.

**Theorem 3.22** *Let $E$ be an open subset of $\mathbb{R}^3$. Let $f : E \to \mathbb{R}^2$. Let $f_1 : E \to \mathbb{R}$, $f_2 : E \to \mathbb{R}$ be the component functions of $f$, that is, $f(h) = (f_1(h), f_2(h))$ for every $h$ in $E$. Let $x \equiv (x_1, x_2, x_3)$ be in $E$. Let $f$ be differentiable at $x$. Let $u \equiv (u_1, u_2, u_3)$ be a unit vector in $\mathbb{R}^3$. Then,*

$$(D_u f_1)(x) = u \cdot ((\nabla f_1)(x)),$$
$$(D_u f_2)(x) = u \cdot ((\nabla f_2)(x)),$$

where

$$(D_u f_1)(x) \equiv \lim_{t \to 0} \frac{f_1((x_1 + tu_1, x_2 + tu_2, x_3 + tu_3)) - f_1((x_1, x_2, x_3))}{t},$$
$$(D_u f_2)(x) \equiv \lim_{t \to 0} \frac{f_2((x_1 + tu_1, x_2 + tu_2, x_3 + tu_3)) - f_2((x_1, x_2, x_3))}{t}.$$

(Here, $(D_u f_1)(x)$ is called the directional derivative of $f_1$ in the direction of u, etc.)

*Proof* Since $f$ is differentiable at $x$, $f'(x)$ is in $L(\mathbb{R}^3, \mathbb{R}^2)$. Let

$$[f'(x)] = \begin{bmatrix} a_{11} & a_{12} & a_{13} \\ a_{21} & a_{22} & a_{23} \end{bmatrix}_{2 \times 3}$$

relative to the basis $\{(1,0,0),(0,1,0),(0,0,1)\}$ of $\mathbb{R}^3$ and the basis $\{(1,0),(0,1)\}$ of $\mathbb{R}^2$. By Theorem 3.14,

$$\lim_{(h_1,h_2,h_3) \equiv h \to 0} \left( \frac{1}{|h|} (f_1(x+h) - f_1(x) - (h_1, h_2, h_3) \cdot (a_{11}, a_{12}, a_{13})) \right) = 0.$$

Hence,

$$\lim_{t \to 0} \left( \frac{1}{|t(u_1, u_2, u_3)|} (f_1((x_1, x_2, x_3) + t(u_1, u_2, u_3)) - f_1((x_1, x_2, x_3)) \right.$$
$$\left. - (t(u_1, u_2, u_3)) \cdot (a_{11}, a_{12}, a_{13})) \right) = 0$$

or,

$$\lim_{t \to 0} \left( \frac{1}{|t||(u_1, u_2, u_3)|} (f_1((x_1 + tu_1, x_2 + tu_2, x_3 + tu_3)) - f_1((x_1, x_2, x_3)) \right.$$
$$\left. - t((u_1, u_2, u_3) \cdot (a_{11}, a_{12}, a_{13}))) \right) = 0$$

or,

$$\lim_{t \to 0} \left( \frac{1}{|t| \cdot 1} (f_1((x_1 + tu_1, x_2 + tu_2, x_3 + tu_3)) - f_1((x_1, x_2, x_3)) \right.$$
$$\left. - t((u_1, u_2, u_3) \cdot (a_{11}, a_{12}, a_{13}))) \right) = 0$$

or,

$$\lim_{t \to 0} \left( \frac{1}{t} (f_1((x_1 + tu_1, x_2 + tu_2, x_3 + tu_3)) - f_1((x_1, x_2, x_3)) \right.$$
$$\left. -t((u_1, u_2, u_3) \cdot (a_{11}, a_{12}, a_{13}))) \right) = 0$$

or,

$$\lim_{t \to 0} \left( \frac{f_1((x_1 + tu_1, x_2 + tu_2, x_3 + tu_3)) - f_1((x_1, x_2, x_3))}{t} \right.$$
$$\left. ((u_1, u_2, u_3) \cdot (a_{11}, a_{12}, a_{13})) \right) = 0$$

or,

$$(D_u f_1)(x) = \lim_{t \to 0} \frac{f_1((x_1 + tu_1, x_2 + tu_2, x_3 + tu_3)) - f_1((x_1, x_2, x_3))}{t}$$
$$= (u_1, u_2, u_3) \cdot ((D_1 f_1)(x), (D_2 f_1)(x), (D_3 f_1)(x))$$
$$= (u_1, u_2, u_3) \cdot ((\nabla f_1)(x)) = u \cdot ((\nabla f_1)(x))$$

Thus,

$$(D_u f_1)(x) = u \cdot ((\nabla f_1)(x)).$$

Similarly,

$$(D_u f_2)(x) = u \cdot ((\nabla f_2)(x)). \qquad \square$$

**Note 3.23** The result similar to Theorem 3.22 can be proved as above for $\mathbb{R}^n$ in place of $\mathbb{R}^3$ and $\mathbb{R}^m$ in place of $\mathbb{R}^2$.

**Theorem 3.24** *Let $E$ be an open subset of $\mathbb{R}^n$. Let $f : E \to \mathbb{R}$. Let $x \equiv (x_1, \ldots, x_n)$ be in $E$ Let $f$ be differentiable at $x$. Then,*

$$\max\{(D_u f)(x) : u \in \mathbb{R}^n \text{ and } |u| = 1\} = \left( D_{\left( \frac{1}{|(\nabla f)(x)|}((\nabla f)(x)) \right)} f \right)(x).$$

*(In short, if $f$ is differentiable at $x$, then the maximum value of the directional derivative of $f$ at $x$ is obtained in the direction of the gradient of $f$ at $x$.)*

*Proof* Let $u \equiv (u_1, \ldots, u_n)$ be any unit vector in $\mathbb{R}^n$. We have to prove that

$$(D_u f)(x) \le \left( D_{\left( \frac{1}{|(\nabla f)(x)|}((\nabla f)(x)) \right)} f \right)(x).$$

By Theorem 3.22, $(D_u f)(x) = u \cdot ((\nabla f)(x))$. Hence,

$$(D_u f)(x) \le |(D_u f)(x)| = |u \cdot ((\nabla f)(x))| \le |u||(\nabla f)(x)|$$

$$= 1 \cdot |(\nabla f)(x)| = |(\nabla f)(x)| = \frac{1}{|(\nabla f)(x)|}|(\nabla f)(x)|^2$$

$$= \frac{1}{|(\nabla f)(x)|}(((\nabla f)(x)) \cdot ((\nabla f)(x)))$$

$$= \left(\frac{1}{|(\nabla f)(x)|}((\nabla f)(x))\right) \cdot ((\nabla f)(x))$$

$$= \left(D_{\left(\frac{1}{|(\nabla f)(x)|}((\nabla f)(x))\right)} f\right)(x).$$

Hence,

$$(D_u f)(x) \le \left(D_{\left(\frac{1}{|(\nabla f)(x)|}((\nabla f)(x))\right)} f\right)(x).$$

$\square$

**Theorem 3.25** *Let $a < b$. Let $\gamma : (a,b) \to \mathbb{R}^2$ be any function differentiable on the open interval $(a,b)$. Let $G$ be an open subset of $\mathbb{R}^2$, and $G$ contains the range of $\gamma$. Let $f : G \to \mathbb{R}$ be differentiable on $G$ Then, for every $t$ in the open interval $(a,b)$,*

$$(f \circ \gamma)'(t) = ((\nabla f)(\gamma(t))) \cdot \left(\frac{d}{dt}\gamma(t)\right).$$

*Proof* Let $\gamma_1 : (a,b) \to \mathbb{R}$, $\gamma_2 : (a,b) \to \mathbb{R}$ be the component functions of $\gamma$, that is, $\gamma(t) = (\gamma_1(t), \gamma_2(t))$ for every $t$ in $(a,b)$. Let us take any $t$ in the open interval $(a,b)$. By the chain rule of derivative, we have

$$(f \circ \gamma)'(t) = (f'(\gamma(t)))(\gamma'(t)).$$

Hence,

$$[(f \circ \gamma)'(t)] = [(f'(\gamma(t)))(\gamma'(t))] = [(f'(\gamma(t)))][(\gamma'(t))]$$

$$= [(D_1 f)(\gamma(t)) \quad (D_2 f)(\gamma(t))]_{1\times 2} \begin{bmatrix} \dfrac{d}{dt}\gamma_1(t) \\[2mm] \dfrac{d}{dt}\gamma_2(t) \end{bmatrix}_{2\times 1}$$

$$= \left[((D_1 f)(\gamma(t)))\left(\frac{d}{dt}\gamma_1(t)\right) + ((D_2 f)(\gamma(t)))\left(\frac{d}{dt}\gamma_2(t)\right)\right].$$

So,

$$
\begin{aligned}
\text{LHS} = (f \circ \gamma)'(t) &= ((D_1 f)(\gamma(t))) \left( \frac{d}{dt} \gamma_1(t) \right) + ((D_2 f)(\gamma(t))) \left( \frac{d}{dt} \gamma_2(t) \right) \\
&= ((D_1 f)(\gamma(t)), (D_2 f)(\gamma(t))) \cdot \left( \frac{d}{dt} \gamma_1(t), \frac{d}{dt} \gamma_2(t) \right) \\
&= ((D_1 f)(\gamma(t)), (D_2 f)(\gamma(t))) \cdot \frac{d}{dt} (\gamma_1(t), \gamma_2(t)) \\
&= ((D_1 f)(\gamma(t)), (D_2 f)(\gamma(t))) \cdot \left( \frac{d}{dt} \gamma(t) \right) \\
&= ((\nabla f)(\gamma(t))) \cdot \left( \frac{d}{dt} \gamma(t) \right) = \text{RHS}.
\end{aligned}
$$

$\square$

**Note 3.26** The result similar to Theorem 3.25 can be proved as above for $\mathbb{R}^m$ in place of $\mathbb{R}^2$.

**Note 3.27** Let $a$ be in $\mathbb{R}^n$. The function $t \mapsto ta$ from $\mathbb{R}$ to $\mathbb{R}^n$ is denoted by $a$, provided that there is no confusion. Here,

$$
\|a\| = \sup\{|ta| : t \in \mathbb{R} \text{ and } |t| \le 1\} = \sup\{|t||a| : t \in \mathbb{R} \text{ and } |t| \le 1\} \le |a| = |1a|
$$
$$
\le \sup\{|ta| : t \in \mathbb{R} \text{ and } |t| \le 1\} = \|a\|.
$$

This shows that

$$
\|a\| = |a|.
$$

**Theorem 3.28** *Let $E$ be an open convex subset of $\mathbb{R}^3$. Let $f : E \to \mathbb{R}^2$ be differentiable on $E$. Let the set $\{\|f'(x)\| : x \in E\}$ be bounded above. Then, for every $a$, $b$ in $E$,*

$$
|f(b) - f(a)| \le |b - a|(\sup\{\|f'(x)\| : x \in E\})
$$

*Proof* First of all, we shall prove the following lemma.  $\square$

**Lemma 3.29** *Let $a < b$. Let $f : [a, b] \to \mathbb{R}^n$ be any continuous function. If $f$ is differentiable on $(a, b)$, then there exists a real number $t$ in $(a, b)$ such that*

$$
|f(b) - f(a)| \le (b - a)|f'(t)|.
$$

*Proof of Lemma*

Case I: when $f(a) = f(b)$. The lemma is trivial in this case.

Case II: when $f(a) \ne f(b)$. Since $f(a) \ne f(b)$, $0 < |f(b) - f(a)|$, and hence, $\frac{1}{|f(b)-f(a)|}(f(b) - f(a))$ is in $\mathbb{R}^n$. Put $c \equiv \frac{1}{|f(b)-f(a)|}(f(b) - f(a))$. Clearly, $|c| = 1$. Let us define a function

$$g : [a, b] \rightarrow \mathbb{R}$$

as follows: For every $x$ in $[a, b]$,

$$g(x) \equiv c \cdot (f(x)).$$

Now, we want to apply Lagrange's mean value theorem on $g$. Since $f : [a, b] \rightarrow \mathbb{R}^n$ is continuous, $g : [a, b] \rightarrow \mathbb{R}$ is continuous. Since $f : [a, b] \rightarrow \mathbb{R}^n$ is differentiable on $(a, b)$, $g : [a, b] \rightarrow \mathbb{R}$ is differentiable on $(a, b)$, and for every $x$ in $(a, b)$,

$$g'(x) = c \cdot (f'(x)).$$

Hence, by the Lagrange's mean value theorem, there exists a real number $t$ in $(a, b)$ such that

$$
\begin{aligned}
|f(b) - f(a)| &= \left( \frac{1}{|f(b) - f(a)|} (f(b) - f(a)) \right) \cdot (f(b) - f(a)) = c \cdot (f(b) - f(a)) \\
&= c \cdot (f(b)) - c \cdot (f(a)) = g(b) - g(a) = (b - a)g'(t) = (b - a)(c \cdot (f'(t))) \\
&= c \cdot ((b - a)(f'(t))) \le |c \cdot ((b - a)(f'(t)))| \le |c||(b - a)(f'(t))| \\
&= 1|(b - a)(f'(t))| = (b - a)|f'(t)|.
\end{aligned}
$$

Thus, there exists a real number $t$ in $(a, b)$ such that

$$|f(b) - f(a)| \le (b - a)|f'(t)|.$$

$\square$

*Proof of the main theorem* Let us take any $a \equiv (a_1, a_2, a_3), b \equiv (b_1, b_2, b_3)$ in $E$. Let us take a nonzero element $c \equiv (c_1, c_2)$ in $\mathbb{R}^2$. Let $f_1, f_2$ be the component functions of $f$. Let us define a function

$$\gamma : [0, 1] \rightarrow \mathbb{R}^3$$

as follows: For every $t$ in the closed interval $[0, 1]$,

$$\gamma(t) \equiv (1 - t)a + tb.$$

Clearly, $\gamma$ is continuous on $[0,1]$. Also, it is clear that $\gamma$ is differentiable on the open interval $(0,1)$, and for every $t$ in $(0,1)$,

$$\gamma'(t) = b - a.$$

Since $E$ is convex, $a$ and $b$ are in $E$, the range of $\gamma$ is contained in $E$. Since $\gamma : [0, 1] \rightarrow \mathbb{R}^3$, the range of $\gamma$ is contained in $E$, and $f : E \rightarrow \mathbb{R}^2$, the composite function $f \circ \gamma$ is defined on $[0, 1]$.

We want to apply Lemma 3.29 on $f \circ \gamma$. Since $f : E \to \mathbb{R}^2$ is differentiable on $E$, $f$ is continuous on $E$. Since $\gamma$ is continuous on $[0, 1]$, the range of $\gamma$ is contained in $E$, and $f$ is continuous on $E$, $f \circ \gamma$ is continuous on $[0, 1]$. Since $\gamma$ is differentiable on the open interval $(0, 1)$, the range of $\gamma$ is contained in $E$, and $f : E \to \mathbb{R}^2$ is differentiable on $E$, by Theorem 3.18, the composite function $(f \circ \gamma) : [0, 1] \to \mathbb{R}^2$ is differentiable on the open interval $(0, 1)$, and for every $x$ in $(0, 1)$,

$$(f \circ \gamma)'(x) = (f'(\gamma(x)))(\gamma'(x)).$$

Hence, by Lemma 3.29 and Theorem 3.2, there exists a real number $t$ in $(0, 1)$ such that

$$
\begin{aligned}
|f(b) - f(a)| = |f(\gamma(1)) - f(\gamma(0))| &= |(f \circ \gamma)(1) - (f \circ \gamma)(0)| \\
&\leq (1 - 0)|(f \circ \gamma)'(t)| \\
&= |(f \circ \gamma)'(t)| = \|(f \circ \gamma)'(t)\| \\
&= \|(f'(\gamma(t)))(\gamma'(t))\| \leq \|f'(\gamma(t))\| \|\gamma'(t)\| \\
&= \|f'(\gamma(t))\| \|b - a\| = \|f'(\gamma(t))\| |b - a| \\
&= |b - a| \|f'(\gamma(t))\| \leq |b - a|(\sup\{\|f'(x)\| : x \in E\}).
\end{aligned}
$$

Thus,

$$|f(b) - f(a)| \leq |b - a|(\sup\{\|f'(x)\| : x \in E\}). \qquad \square$$

**Note 3.30** The result similar to Theorem 3.28 can be proved as above for $\mathbb{R}^n$ in place of $\mathbb{R}^3$ and $\mathbb{R}^m$ in place of $\mathbb{R}^2$.

**Theorem 3.31** *Let $E$ be an open subset of $\mathbb{R}^3$. Let $f : E \to \mathbb{R}^2$. Let $f_1 : E \to \mathbb{R}$, $f_2 : E \to \mathbb{R}$ be the component functions of $f$, that is, $f(x) = (f_1(x), f_2(x))$ for every $x$ in $E$. If $f : E \to \mathbb{R}^2$ is differentiable on $E$, and the mapping $x \mapsto f'(x)$ from $E(\subset \mathbb{R}^3)$ to the metric space $L(\mathbb{R}^3, \mathbb{R}^2)$ is continuous, then $f$ is $C^1$ function (i.e., for $i = 1, 2,$ and $j = 1, 2, 3,$ each function $D_j f_i : E \to \mathbb{R}$ exists and is continuous).*

*Proof* Let us take any $x$ in $E$. Since $x$ is in $E$, and $f : E \to \mathbb{R}^2$ is differentiable on $E$, $f$ is differentiable at $x$. Since $f$ is differentiable at $x$, by Theorem 3.20, all the partial derivatives $(D_j f_i)(x)$ exist. This shows that each $D_j f_i : E \to \mathbb{R}$ is a function for $i = 1$, 2, and $j = 1, 2, 3$. Now, we shall try to prove that $D_1 f_2 : E \to \mathbb{R}$ is continuous. For this purpose, let us fix any $x$ in $E$.

Let us take any $\varepsilon > 0$. Since the mapping $x \mapsto f'(x)$ from $E(\subset \mathbb{R}^3)$ to the metric space $L(\mathbb{R}^3, \mathbb{R}^2)$ is continuous, there exists $\delta > 0$ such that for every $y$ in $E$ satisfying $|y - x| < \delta$, we have $\|f'(y) - f'(x)\| < \varepsilon$.

Let us take any $y$ in $E$ satisfying $|y - x| < \delta$. It is enough to prove that $|(D_1 f_2)(y) - (D_1 f_2)(x)| < \varepsilon$. Since

$$[f'(y) - f'(x)] = [f'(y)] - [f'(x)]$$

$$= \begin{bmatrix} (D_1f_1)(y) & (D_2f_1)(y) & (D_3f_1)(y) \\ (D_1f_2)(y) & (D_2f_2)(y) & (D_3f_1)(y) \end{bmatrix}_{2\times3}$$

$$- \begin{bmatrix} (D_1f_1)(x) & (D_2f_1)(x) & (D_3f_1)(x) \\ (D_1f_2)(x) & (D_2f_2)(x) & (D_3f_1)(x) \end{bmatrix}_{2\times3}$$

$$= \begin{bmatrix} (D_1f_1)(y) - (D_1f_1)(x) & (D_2f_1)(y) - (D_2f_1)(x) & (D_3f_1)(y) - (D_3f_1)(x) \\ (D_1f_2)(y) - (D_1f_2)(x) & (D_2f_2)(y) - (D_2f_2)(x) & (D_3f_1)(y) - (D_3f_1)(x) \end{bmatrix}_{2\times3},$$

so

$$|(D_1f_2)(y) - (D_1f_2)(x)| \leq |((D_1f_1)(y) - (D_1f_1)(x), (D_1f_2)(y) - (D_1f_2)(x))|$$
$$= |(f'(y) - f'(x))((1,0,0))|$$
$$\leq \sup\{|(f'(y) - f'(x))(t)| : t \in \mathbb{R}^3, |t| \leq 1\}$$
$$= \|f'(y) - f'(x)\| < \varepsilon.$$

Hence,

$$|(D_1f_2)(y) - (D_1f_2)(x)| < \varepsilon.$$

This proves that $D_1f_2 : E \to \mathbb{R}$ is continuous. Similarly, all other partial derivatives are continuous. Hence, $f$ is a $C^1$ function.                     □

**Theorem 3.32** *Let $E$ be an open subset of $\mathbb{R}^3$. Let $f : E \to \mathbb{R}^2$. Let $f_1 : E \to \mathbb{R}$, $f_2 : E \to \mathbb{R}$ be the component functions of $f$, that is, $f(x) = (f_1(x), f_2(x))$ for every $x$ in $E$. If $f$ is a $C^1$ function (i.e., for $i = 1, 2$, and $j = 1, 2, 3$, each function $D_jf_i : E \to \mathbb{R}$ exists and is continuous), then $f : E \to \mathbb{R}^2$ is differentiable on $E$.*

*Proof* Let us take any $a \equiv (a_1, a_2, a_3)$ in $E$. We have to prove that $f : E \to \mathbb{R}^2$ is differentiable at $a$. By Theorem 3.14, it is enough to prove that

(i) $\lim\limits_{(h_1,h_2,h_3) \equiv h \to 0} \left( \dfrac{1}{|(h_1, h_2, h_3)|} (f_1((a_1, a_2, a_3)) \right.$

$$+ (h_1, h_2, h_3)) - f_1((a_1, a_2, a_3)) - (h_1, h_2, h_3)$$

$$\left. \cdot ((D_1f_1)((a_1, a_2, a_3)), (D_2f_1)((a_1, a_2, a_3)), (D_3f_1)((a_1, a_2, a_3))) \right) = 0,$$

and

(ii) $\lim\limits_{(h_1,h_2,h_3) \equiv h \to 0} \left( \dfrac{1}{|(h_1, h_2, h_3)|} (f_2((a_1, a_2, a_3)) \right.$

$$+ (h_1, h_2, h_3)) - f_2((a_1, a_2, a_3)) - (h_1, h_2, h_3)$$

$$\left. \cdot ((D_1f_2)((a_1, a_2, a_3)), (D_2f_2)((a_1, a_2, a_3)), (D_3f_2)((a_1, a_2, a_3))) \right) = 0.$$

For (i): We have to prove that

$$0 = \lim_{(h_1,h_2,h_3) \equiv h \to 0} \left( \frac{1}{|(h_1,h_2,h_3)|} (f_1((a_1,a_2,a_3) + (h_1,h_2,h_3)) - f_1((a_1,a_2,a_3)) \right.$$
$$\left. - (h_1,h_2,h_3) \cdot ((D_1 f_1)((a_1,a_2,a_3)), (D_2 f_1)((a_1,a_2,a_3)), (D_3 f_1)((a_1,a_2,a_3)))) \right),$$

that is,

$$0 = \lim_{(h_1,h_2,h_3) \equiv h \to 0} \left( \frac{1}{|(h_1,h_2,h_3)|} (f_1((a_1 + h_1, a_2 + h_2, a_3 + h_3)) - f_1((a_1,a_2,a_3)) \right.$$
$$\left. - (h_1((D_1 f_1)((a_1,a_2,a_3))) + h_2((D_2 f_1)((a_1,a_2,a_3))) + h_3((D_3 f_1)((a_1,a_2,a_3)))) ) \right)$$

For this purpose, let us take any $\varepsilon > 0$.

Since $D_1 f_1 : E \to \mathbb{R}$, $D_2 f_1 : E \to \mathbb{R}$, $D_3 f_1 : E \to \mathbb{R}$ are continuous at $(a_1, a_2, a_3)$, there exists $\delta^* > 0$ such that for every $(x_1, x_2, x_3)$ in $E$ satisfying $|(x_1, x_2, x_3) - (a_1, a_2, a_3)| < \delta^*$, we have

$$\left| (D_j f_1)((x_1, x_2, x_3)) - (D_j f_1)((a_1, a_2, a_3)) \right| < \frac{\varepsilon}{3} (j = 1, 2, 3) \cdots (*)$$

Since $(a_1, a_2, a_3)$ is in $E$, and $E$ is an open subset of $\mathbb{R}^3$, we can find $\delta > 0$ such that $0 < \delta < \delta^*$ and the Cartesian product $K \equiv [a_1 - \delta, a_1 + \delta] \times [a_2 - \delta, a_2 + \delta] \times [a_3 - \delta, a_3 + \delta] \subset E$.

Let us fix any $h \equiv (h_1, h_2, h_3)$ such that $|(h_1, h_2, h_3)| < \delta$. Since $|h_1| \le |(h_1, h_2, h_3)| < \delta$, $-\delta < h_1 < \delta$, or $a_1 - \delta < a_1 + h_1 < a_1 + \delta$, or $a_1 + h_1 \in [a_1 - \delta, a_1 + \delta]$. Similarly, $a_2 + h_2 \in [a_2 - \delta, a_2 + \delta]$, and $a_3 + h_3 \in [a_3 - \delta, a_3 + \delta]$. Since $a_1, a_1 + h_1 \in [a_1 - \delta, a_1 + \delta]$, and for every $t$ in $[a_1 - \delta, a_1 + \delta]$, $(t, a_2 + h_2, a_3 + h_3) \in [a_1 - \delta, a_1 + \delta] \times [a_2 - \delta, a_2 + \delta] \times [a_3 - \delta, a_3 + \delta] \subset E$, we can define a function

$$g_1 : [a_1 - \delta, a_1 + \delta] \to \mathbb{R}$$

as follows: For every $t$ in $[a_1 - \delta, a_1 + \delta]$,

$$g_1(t) = f_1((t, a_2 + h_2, a_3 + h_3)).$$

Now, we want to apply Lagrange's mean value theorem on $g_1$. For fixed $t_0$ in $[a_1 - \delta, a_1 + \delta]$, we have

$$\lim_{t\to 0}\frac{g_1(t_0+t)-g_1(t_0)}{t}=\lim_{t\to 0}\frac{f_1(t_0+t,a_2+h_2,a_3+h_3)-f_1((t_0,a_2+h_2,a_3+h_3))}{t}$$

$$=(D_1f_1)((t_0,a_2+h_2,a_3+h_3)).$$

Since

$\lim_{t\to 0}\frac{g_1(t_0+t)-g_1(t_0)}{t}=(D_1f_1)((t_0,a_2+h_2,a_3+h_3))$, $(t_0,a_2+h_2,a_3+h_3)$ is in $E$, and $D_1f_1:E\to\mathbb{R}$ exists, $g_1$ is differentiable at $t_0$. Further,

$$g'(t_0)=(D_1f_1)((t_0,a_2+h_2,a_3+h_3)).$$

Since $g_1$ is differentiable at $t_0$, $g_1$ is continuous at $t_0$.

Thus, we see that $g_1:[a_1-\delta,a_1+\delta]\to\mathbb{R}$ is continuous and differentiable on $[a_1-\delta,a_1+\delta]$. Hence, by the Lagrange's mean value theorem, there exists a real number $c_1$ lying between $a_1+h_1$ and $a_1$ such that

$$f_1((a_1+h_1,a_2+h_2,a_3+h_3))-f_1((a_1,a_2+h_2,a_3+h_3))$$
$$=g_1(a_1+h_1)-g_1(a_1)$$
$$=((a_1+h_1)-a_1)g'(c_1)=h_1(g'(c_1))$$
$$=h_1((D_1f_1)((c_1,a_2+h_2,a_3+h_3))).$$

Thus, we see that there exists a real number $c_1$ lying between $a_1+h_1$ and $a_1$ such that

$$f_1((a_1+h_1,a_2+h_2,a_3+h_3))=f_1((a_1,a_2+h_2,a_3+h_3))$$
$$+h_1((D_1f_1)((c_1,a_2+h_2,a_3+h_3))).$$

Similarly, there exists a real number $c_2$ lying between $a_2+h_2$ and $a_2$ such that

$$f_1((a_1,a_2+h_2,a_3+h_3))=f_1((a_1,a_2,a_3+h_3))+h_2((D_2f_1)((a_1,c_2,a_3+h_3))).$$

Again, there exists a real number $c_3$ lying between $a_3+h_3$ and $a_3$ such that

$$f_1((a_1,a_2,a_3+h_3))=f_1((a_1,a_2,a_3))+h_3((D_3f_1)((a_1,a_2,c_3))).$$

On adding the last three equations, we get

$$f_1((a_1+h_1,a_2+h_2,a_3+h_3))-f_1((a_1,a_2,a_3))$$
$$=h_1((D_1f_1)((c_1,a_2+h_2,a_3+h_3)))$$
$$+h_2((D_2f_1)((a_1,c_2,a_3+h_3)))+h_3((D_3f_1)((a_1,a_2,c_3)))$$

or,

$$f_1 q((a_1 + h_1, a_2 + h_2, a_3 + h_3)) - f_1((a_1, a_2, a_3))(h_1((D_1 f_1)((a_1, a_2, a_3)))$$
$$+ h_2((D_2 f_1)((a_1, a_2, a_3))) + h_3((D_3 f_1)((a_1, a_2, a_3))))$$
$$= (h_1((D_1 f_1)((c_1, a_2 + h_2, a_3 + h_3))) + h_2((D_2 f_1)((a_1, c_2, a_3 + h_3)))$$
$$+ h_3((D_3 f_1)((a_1, a_2, c_3)))) - (h_1((D_1 f_1)((a_1, a_2, a_3)))$$
$$+ h_2((D_2 f_1)((a_1, a_2, a_3))) + h_3((D_3 f_1)((a_1, a_2, a_3))))$$
$$= h_1((D_1 f_1)((c_1, a_2 + h_2, a_3 + h_3)) - (D_1 f_1)((a_1, a_2, a_3)))$$
$$+ h_2((D_2 f_1)((a_1, c_2, a_3 + h_3)) - (D_2 f_1)((a_1, a_2, a_3)))$$
$$+ h_3((D_3 f_1)((a_1, a_2, c_3)) - (D_3 f_1)((a_1, a_2, a_3))).$$

So,

$$= \left| \frac{1}{|(h_1, h_2, h_3)|} (f_1((a_1 + h_1, a_2 + h_2, a_3 + h_3)) - f_1((a_1, a_2, a_3)) \right.$$

$$\left. -(h_1((D_1 f_1)((a_1, a_2, a_3))) + h_2((D_2 f_1)((a_1, a_2, a_3))) + h_3((D_3 f_1)((a_1, a_2, a_3))))) \right|$$

$$= \left| \frac{1}{|(h_1, h_2, h_3)|} (h_1((D_1 f_1)((c_1, a_2 + h_2, a_3 + h_3)) - (D_1 f_1)((a_1, a_2, a_3))) \right.$$

$$+ h_2((D_2 f_1)((a_1, c_2, a_3 + h_3)) - (D_2 f_1)((a_1, a_2, a_3)))$$

$$\left. + h_3((D_3 f_1)((a_1, a_2, c_3)) - (D_3 f_1)((a_1, a_2, a_3)))) \right|$$

$$\leq \frac{|h_1|}{|(h_1, h_2, h_3)|} |(D_1 f_1)((c_1, a_2 + h_2, a_3 + h_3)) - (D_1 f_1)((a_1, a_2, a_3))|$$

$$+ \frac{|h_2|}{|(h_1, h_2, h_3)|} |(D_2 f_1)((a_1, c_2, a_3 + h_3)) - (D_2 f_1)((a_1, a_2, a_3))|$$

$$+ \frac{|h_3|}{|(h_1, h_2, h_3)|} |(D_3 f_1)((a_1, a_2, c_3)) - (D_3 f_1)((a_1, a_2, a_3))|$$

$$= |(D_1 f_1)((c_1, a_2 + h_2, a_3 + h_3)) - (D_1 f_1)((a_1, a_2, a_3))|$$
$$+ |(D_2 f_1)((a_1, c_2, a_3 + h_3)) - (D_2 f_1)((a_1, a_2, a_3))|$$
$$+ |(D_3 f_1)((a_1, a_2, c_3)) - (D_3 f_1)((a_1, a_2, a_3))|.$$

Since $c_1$ lies between $a_1 + h_1$ and $a_1$, $(c_1 - a_1)^2 \leq (h_1)^2$, and hence,

$$|(c_1, a_2 + h_2, a_3 + h_3) - (a_1, a_2, a_3)| = |(c_1 - a_1, h_2, h_3)|$$
$$= \sqrt{(c_1 - a_1)^2 + (h_2)^2 + (h_3)^2} \leq \sqrt{(h_1)^2 + (h_2)^2 + (h_3)^2}$$
$$= |(h_1, h_2, h_3) - (0, 0, 0)| < \delta < \delta^*.$$

Since
$|(c_1, a_2 + h_2, a_3 + h_3) - (a_1, a_2, a_3)| < \delta^*$, by $(*)$,

$$|(D_1f_1)((c_1, a_2 + h_2, a_3 + h_3)) - (D_1f_1)((a_1, a_2, a_3))| < \frac{\varepsilon}{3}.$$

Similarly,

$$|(D_2f_1)((a_1, c_2, a_3 + h_3)) - (D_2f_1)((a_1, a_2, a_3))| < \frac{\varepsilon}{3},$$

and

$$|(D_3f_1)((a_1, a_2, c_3)) - (D_3f_1)((a_1, a_2, a_3))| < \frac{\varepsilon}{3}.$$

Hence,

$$\left| \frac{1}{|(h_1, h_2, h_3)|} (f_1((a_1 + h_1, a_2 + h_2, a_3 + h_3)) - f_1((a_1, a_2, a_3)) \right.$$

$$\left. - (h_1((D_1f_1)((a_1, a_2, a_3))) + h_2((D_2f_1)((a_1, a_2, a_3))) + h_3((D_3f_1)((a_1, a_2, a_3)))) \right|$$

$$\leq |(D_1f_1)((c_1, a_2 + h_2, a_3 + h_3)) - (D_1f_1)((a_1, a_2, a_3))|$$
$$+ |(D_2f_1)((a_1, c_2, a_3 + h_3)) - (D_2f_1)((a_1, a_2, a_3))|$$
$$+ |(D_3f_1)((a_1, a_2, c_3)) - (D_3f_1)((a_1, a_2, a_3))|$$
$$< \frac{\varepsilon}{3} + \frac{\varepsilon}{3} + \frac{\varepsilon}{3} = \varepsilon.$$

This proves (i). Similarly, (ii) is proved.                                       □

**Theorem 3.33** *Let $E$ be an open subset of $\mathbb{R}^3$. Let $f : E \to \mathbb{R}^2$. Let $f_1 : E \to \mathbb{R}$, $f_2 : E \to \mathbb{R}$ be the component functions of $f$, that is, $f(x) = (f_1(x), f_2(x))$ for every $x$ in $E$. $f$ is a $C^1$ function (i.e., for $i = 1, 2$, and $j = 1, 2, 3$, each function $D_jf_i : E \to \mathbb{R}$ exists and is continuous), if and only if $f : E \to \mathbb{R}^2$ is differentiable on $E$ and $f' : E \to L(\mathbb{R}^3, \mathbb{R}^2)$ is continuous.*

*Proof* By using Theorems 3.31 and 3.32, we find that the only part that remains to be proved is the following:

if $f$ is a $C^1$ function, then the mapping $f' : E \to L(\mathbb{R}^3, \mathbb{R}^2)$ is continuous.

For this purpose, let us take any $x$ in $E$. We have to prove that $f'$ is continuous at $x$.

Let us take any $\varepsilon > 0$. Since, for $i = 1, 2$, and $j = 1, 2, 3$, each function $D_jf_i : E \to \mathbb{R}$ is continuous at $x$, there exists a real number $\delta > 0$ such that for every $y$ in $E$ satisfying $|y - x| < \delta$, we have

$$|(D_jf_i)(y) - (D_jf_i)(x)| < \frac{\varepsilon}{\sqrt{2 \times 3}} \quad \text{(for every } i = 1, 2, \text{ and } j = 1, 2, 3).$$

Now, let us take any $y$ in $E$ satisfying $|y - x| < \delta$. It is enough to prove that

$$\|f'(y) - f'(x)\| < \varepsilon.$$

Here,

$$[f'(x)] = \begin{bmatrix} (D_1f_1)(x) & (D_2f_1)(x) & (D_3f_1)(x) \\ (D_1f_2)(x) & (D_2f_2)(x) & (D_3f_3)(x) \end{bmatrix}_{2\times 3},$$

and

$$[f'(y)] = \begin{bmatrix} (D_1f_1)(y) & (D_2f_1)(y) & (D_3f_1)(y) \\ (D_1f_2)(y) & (D_2f_2)(y) & (D_3f_3)(y) \end{bmatrix}_{2\times 3},$$

so

$$[f'(y) - f'(x)] = \begin{bmatrix} (D_1f_1)(y) - (D_1f_1)(x) & (D_2f_1)(y) - (D_2f_1)(x) & (D_3f_1)(y) - (D_3f_1)(x) \\ (D_1f_2)(y) - (D_1f_2)(x) & (D_2f_2)(y) - (D_2f_2)(x) & (D_3f_3)(y) - (D_3f_3)(x) \end{bmatrix}_{2\times 3},$$

and hence,

$$\|f'(y) - f'(x)\| \le \sqrt{\sum_{j=1}^{3}\left(\sum_{i=1}^{2}|(D_jf_i)(y) - (D_jf_i)(x)|^2\right)} < \sqrt{\sum_{j=1}^{3}\left(\sum_{i=1}^{2}\left(\frac{\varepsilon}{\sqrt{2\times 3}}\right)^2\right)} = \varepsilon.$$

$\square$

**Note 3.34** The result similar to Theorem 3.33 can be proved as above for $\mathbb{R}^n$ in place of $\mathbb{R}^3$ and $\mathbb{R}^m$ in place of $\mathbb{R}^2$.

## 3.3 Inverse Function Theorem

**Theorem 3.35** *Let X be a complete metric space, with metric d. Let $f : X \to X$ be any function. Let c be a real number such that $0 < c < 1$, and for every x, y in X,*

$$d(f(x), f(y)) \le c(d(x, y)).$$

*Then, there exists a unique x in X such that $f(x) = x$.*

*Proof* **Existence**: Let us take any $a$ in $X$. Since $a$ is in $X$, and $f : X \to X, f(a)$ is in $X$. Similarly, $f(f(a))$ is in $X, f(f(f(a)))$ is in $X$, etc. Thus, we get a sequence

$$\{a, f(a), f(f(a)), f(f(f(a))), \ldots\}$$

in $X$. Put $a_0 \equiv a, a_1 \equiv f(a), a_2 \equiv f(f(a))$, $a_3 \equiv f(f(f(a)))$, etc. Here, $\{a_0, a_1, a_2, a_3, \ldots\}$ is a sequence in $X$. Also, $f(a_n) = a_{n+1}$. Let us observe that by using the given condition,

$$d(f(a), f(f(a))) \le c(d(a, f(a))).$$

Also,

$$d(f(f(a)), f(f(f(a)))) \le c(d(f(a), f(f(a)))) \le c(c(d(a, f(a)))) = c^2(d(a, f(a))),$$

so

$$d(f(f(a)), f(f(f(a)))) \le c^2(d(a, f(a))).$$

Similarly,

$$d(f(f(f(a))), f(f(f(f(a))))) \le c^3(d(a, f(a))),$$

etc. Now,

$$
\begin{aligned}
d(f(f(a)), f(f(f(f(f(f(a))))))) &\le c(d(f(a), f(f(f(f(f(a))))))) \\
&\le c(c(d(a, f(f(f(f(a))))))) = c^2(d(a, f(f(f(f(a)))))) \\
&\le c^2(d(a, f(a)) + d(f(a), f(f(a))) + d(f(f(a)), f(f(f(a)))) \\
&\quad + d(f(f(f(a))), f(f(f(f(a)))))) \\
&\le c^2(d(a, f(a)) + c(d(a, f(a))) + c^2(d(a, f(a))) + c^3(d(a, f(a)))) \\
&= c^2(1 + c + c^2 + c^3)(d(a, f(a))) \le c^2(1 + c + c^2 + c^3 + c^4 + \cdots)(d(a, f(a))) \\
&= c^2\left(\frac{1}{1 - c}\right)(d(a, f(a))).
\end{aligned}
$$

Thus, we have seen that

$$d(f(f(a)), f(f(f(f(f(f(a))))))) \le c^2\left(\frac{1}{1 - c}\right)(d(a, f(a))).$$

Similarly,

$$d(f(f(f(a))), f(f(f(f(f(f(f(f(a)))))))))) \le c^3\left(\frac{1}{1 - c}\right)(d(a, f(a))),$$

etc. Thus, $d(a_2, a_6) \le (\frac{1}{1-c})(d(a, f(a)))c^2$, $d(a_3, a_7) \le (\frac{1}{1-c})(d(a, f(a)))c^3$, etc.

We shall try to prove that the sequence $\{a_0, a_1, a_2, a_3, \ldots\}$ is Cauchy. For this purpose, let us take $\varepsilon > 0$. Since $0 < c < 1$, $\lim_{n \to \infty} c^n = 0$, and hence, $\lim_{n \to \infty} (\frac{1}{1-c})(d(a, f(a)))c^n = 0$. Therefore, there exists a positive integer $N$ such that for every $n \geq N$, we have $|(\frac{1}{1-c})(d(a,f(a)))c^n - 0| < \varepsilon$. Hence, $m \geq n \geq N$ implies

$$d(a_n, a_m) \leq \left(\frac{1}{1-c}\right)(d(a,f(a)))c^n = \left|\left(\frac{1}{1-c}\right)(d(a,f(a)))c^n - 0\right| < \varepsilon.$$

This proves that $\{a_0, a_1, a_2, a_3, \ldots\}$ is Cauchy in X. Since $\{a_0, a_1, a_2, a_3, \ldots\}$ is Cauchy in X, and X is a complete metric space, there exists an element b in X such that

$$\lim_{n \to \infty} a_n = b.$$

Now, we shall try to show that the function $f : X \to X$ is continuous at b. For this purpose, let us take any $\varepsilon > 0$. Now, let us take any y in X satisfying $d(y, b) < \varepsilon$. We have to show that $d(f(y), f(b)) < \varepsilon$. Here, by using the given conditions, we have

$$d(f(y), f(b)) \leq c(d(y,b)) < c(\varepsilon) < 1\varepsilon = \varepsilon.$$

This proves that the function $f : X \to X$ is continuous at b. Since $f : X \to X$ is continuous at b, and $\lim_{n \to \infty} a_n = b$, $\lim_{n \to \infty} f(a_n) = f(b)$. Hence, $f(b) = \lim_{n \to \infty} f(a_n) = \lim_{n \to \infty} a_{n+1} = \lim_{n \to \infty} a_n = b$. Thus, $f(b) = b$. This proves the existence part of the theorem.

*Uniqueness*: If not, otherwise, suppose that there exist x, y in X such that $x \neq y, f(x) = x$, and $f(y) = y$. We have to arrive at a contradiction. From the given condition, we have

$$0 \neq d(x, y) = d(f(x), f(y)) \leq c(d(x,y)).$$

This implies that $1 \leq c$, which contradicts the supposition. $\qquad \square$

**Theorem 3.36** *Let E be an open subset of $\mathbb{R}^3$. Let $f : E \to \mathbb{R}^3$. Let a be in E. If*

(i) *f is a $C^1$ function,*
(ii) *$f'(a)$ is invertible,*

*then there exists a connected open neighborhood U of a such that*

1. *U is contained in E,*
2. *f is 1–1 on U (i.e., if x and y are in U, and $f(x) = f(y)$, then $x = y$),*
3. *f(U) is connected and open in $\mathbb{R}^3$,*
4. *the 1–1 function $f^{-1}$ from f(U) onto U is a $C^1$ function,*
5. *if f is a $C^2$ function, then $f^{-1}$ from f(U) onto U is a $C^2$ function, etc.*

*Proof* Let $f_1 : E \to \mathbb{R}$, $f_2 : E \to \mathbb{R}$, $f_3 : E \to \mathbb{R}$ be the component functions of f, that is, $f(x) = (f_1(x), f_2(x), f_3(x))$ for every x in E. Let $a \equiv (a_1, a_2, a_3)$. Let $\Omega$ be the set of all invertible (i.e., 1–1 and onto) members of $L(\mathbb{R}^3)$. Since $f'(a)$ is invertible, $f'(a)$ is in $\Omega$ Since $f'(a)$ is in $\Omega$, by Theorem 3.3, the open sphere

$$S_{\frac{1}{\left\|(f'(a))^{-1}\right\|}}(f'(a)),$$

with center $f'(a)$ and radius

$$\frac{1}{\left\|(f'(a))^{-1}\right\|},$$

is contained in $\Omega$

Since $f$ is a $C^1$ function, by Theorem 3.33, $f' : E \rightarrow L(\mathbb{R}^3)$ is continuous. Since $f' : E \rightarrow L(\mathbb{R}^3)$ is continuous at $a$, and

$$S_{\frac{1}{2\left\|(f'(a))^{-1}\right\|}}(f'(a))$$

is an open neighborhood of $f'(a)$ in $\Omega$, there exists a real number $r > 0$ such that the open sphere $S_r(a)$ is contained in $E$, and for every $x$ in $S_r(a)$, we have $f'(x)$ in

$$S_{\frac{1}{2\left\|(f'(a))^{-1}\right\|}}(f'(a)).$$

For every $x$ in $S_r(a)$, we have $f'(x)$ in

$$S_{\frac{1}{2\left\|(f'(a))^{-1}\right\|}}(f'(a)),$$

so

$$\|f'(x) - f'(a)\| < \frac{1}{2\left\|(f'(a))^{-1}\right\|}$$

or,

$$\left\|I - \left((f'(a))^{-1}(f'(x))\right)\right\| = \left\|\left((f'(a))^{-1}(f'(x))\right) - I\right\| = \left\|\left((f'(a))^{-1} \circ (f'(x))\right) - I\right\|$$
$$= \left\|\left((f'(a))^{-1} \circ (f'(x))\right) - \left((f'(a))^{-1} \circ (f'(a))\right)\right\|$$
$$= \left\|(f'(a))^{-1} \circ (f'(x) - f'(a))\right\| = \left\|(f'(a))^{-1}(f'(x) - f'(a))\right\|$$
$$\leq \left\|(f'(a))^{-1}\right\|\|f'(x) - f'(a)\| < \frac{1}{2}.$$

Thus, we see that for every $x$ in $S_r(a)$, we have

$$\left\| I - \left( (f'(a))^{-1}(f'(x)) \right) \right\| \le \frac{1}{2} \quad \cdots (*)$$

For every $k \equiv (k_1, k_2, k_3)$ in $\mathbb{R}^3$, let us define a function

$$\varphi_k : S_r(a) \to \mathbb{R}^3$$

as follows: For every $x \equiv (x_1, x_2, x_3)$ in $S_r(a)$,

$$\varphi_k(x) = x - \left( (f'(a))^{-1} \right)(f(x)) + k.$$

Now, we shall try to show that for every $x$ in $S_r(a)$,

$$\varphi_k'(x) = I - (f'(a))^{-1}(f'(x)).$$

Let

$$\left[ (f'(a))^{-1} \right] \equiv \begin{bmatrix} a_{11} & a_{12} & a_{13} \\ a_{21} & a_{22} & a_{23} \\ a_{31} & a_{32} & a_{33} \end{bmatrix}_{3 \times 3}$$

relative to the basis $\{(1,0,0), (0,1,0), (0,0,1)\}$ of $\mathbb{R}^3$.
Since

$$\begin{aligned}
\varphi_k(x) &= x - \left( (f'(a))^{-1} \right)(f(x)) + k \\
&= (x_1, x_2, x_3) - \left( (f'(a))^{-1} \right)((f_1(x), f_2(x), f_3(x))) + (k_1, k_2, k_3) \\
&= (x_1, x_2, x_3) - ((f_1(x))(a_{11}, a_{21}, a_{31}) + (f_2(x))(a_{12}, a_{22}, a_{32}) \\
&\quad + (f_3(x))(a_{13}, a_{23}, a_{33})) + (k_1, k_2, k_3) \\
&= (x_1 - (a_{11}(f_1(x)) + a_{12}(f_2(x)) + a_{13}(f_3(x))) + k_1, x_2 \\
&\quad - (a_{21}(f_1(x)) + a_{22}(f_2(x)) + a_{23}(f_3(x))) + k_2, x_3 \\
&\quad - (a_{31}(f_1(x)) + a_{32}(f_2(x)) + a_{33}(f_3(x))) + k_3)
\end{aligned}$$

or,

$$\begin{aligned}
\varphi_k(x) = (x_1 &- (a_{11}(f_1(x)) + a_{12}(f_2(x)) + a_{13}(f_3(x))) + k_1, x_2 \\
&- (a_{21}(f_1(x)) + a_{22}(f_2(x)) + a_{23}(f_3(x))) + k_2, x_3 \\
&- (a_{31}(f_1(x)) + a_{32}(f_2(x)) + a_{33}(f_3(x))) + k_3)
\end{aligned}$$

so,

$$
\begin{aligned}
\left(\varphi_k'(x)\right)((1,0,0)) = &(D_1(x_1 - (a_{11}(f_1(x)) + a_{12}(f_2(x)) + a_{13}(f_3(x))) + k_1),\\
&D_1(x_2 - (a_{21}(f_1(x)) + a_{22}(f_2(x)) + a_{23}(f_3(x))) + k_2),\\
&D_1(x_3 - (a_{31}(f_1(x)) + a_{32}(f_2(x)) + a_{33}(f_3(x))) + k_3))\\
= &(1 - (a_{11}((D_1f_1)(x)) + a_{12}((D_1f_2)(x)) + a_{13}((D_1f_3)(x))),\\
&- (a_{21}((D_1f_1)(x)) + a_{22}((D_1f_2)(x)) + a_{23}((D_1f_3)(x))),\\
&-(a_{31}((D_1f_1)(x)) + a_{32}((D_1f_2)(x)) + a_{33}((D_1f_3)(x)))).
\end{aligned}
$$

Also,

$$
\begin{aligned}
&\left(I - (f'(a))^{-1}(f'(x))\right)((1,0,0)) = I((1,0,0)) - \left((f'(a))^{-1}\right)((f'(x))((1,0,0)))\\
&= (1,0,0) - \left((f'(a))^{-1}\right)((D_1f_1)(x), (D_1f_2)(x), (D_1f_3)(x))\\
&= (1,0,0) - \left(((D_1f_1)(x))\left(\left((f'(a))^{-1}\right)((1,0,0))\right)\right.\\
&\quad\left. + ((D_1f_2)(x))\left(\left((f'(a))^{-1}\right)((0,1,0))\right) + ((D_1f_3)(x))\left(\left((f'(a))^{-1}\right)((0,0,1))\right)\right)\\
&= (1,0,0) - (((D_1f_1)(x))(a_{11}, a_{21}, a_{31}) + ((D_1f_2)(x))(a_{12}, a_{22}, a_{32})\\
&\quad + ((D_1f_3)(x))(a_{13}, a_{23}, a_{33}))\\
&= (1 - (a_{11}((D_1f_1)(x)) + a_{12}((D_1f_2)(x)) + a_{13}((D_1f_3)(x))),\\
&\quad - (a_{21}((D_1f_1)(x)) + a_{22}((D_1f_2)(x)) + a_{23}((D_1f_3)(x))),\\
&\quad - (a_{31}((D_1f_1)(x)) + a_{32}((D_1f_2)(x)) + a_{33}((D_1f_3)(x)))).
\end{aligned}
$$

This shows that

$$
(\varphi_k'(x))((1,0,0)) = \left(I - (f'(a))^{-1}(f'(x))\right)((1,0,0)).
$$

Similarly,

$$
\begin{aligned}
(\varphi_k'(x))((0,1,0)) &= \left(I - (f'(a))^{-1}(f'(x))\right)((0,1,0)),\\
(\varphi_k'(x))((0,0,1)) &= \left(I - (f'(a))^{-1}(f'(x))\right)((0,0,1)).
\end{aligned}
$$

Hence, for every $y$ in $\mathbb{R}^3$, we have

$$
(\varphi_k'(x))(y) = \left(I - (f'(a))^{-1}(f'(x))\right)(y).
$$

This shows that for every $x$ in $S_r(a)$,

$$
\varphi_k'(x) = I - (f'(a))^{-1}(f'(x)),
$$

and hence, by $(*)$,

$$\|\varphi_k'(x)\| \le \frac{1}{2}.$$

Therefore, $\frac{1}{2}$ is an upper bound of the set $\{\|\varphi_k'(x)\| : x \in S_r(a)\}$, and hence, $\{\|\varphi_k'(x)\| : x \in S_r(a)\}$ is bounded above.

Now, we want to apply Theorem 3.28. Since $S_r(a)$ is an open convex subset of $\mathbb{R}^3$, $\varphi_k : S_r(a) \to \mathbb{R}^3$ is differentiable on $S_r(a)$, and the set $\{\|\varphi_k'(x)\| : x \in S_r(a)\}$ is bounded above, for every $x$, $y$ in $S_r(a)$,

$$|\varphi_k(y) - \varphi_k(x)| \le |y - x|(\sup\{\|\varphi_k'(x)\| : x \in S_r(a)\}) \le |y - x|\frac{1}{2}.$$

Hence, for every $k$ in $\mathbb{R}^3$, and for every $x$, $y$ in $S_r(a)$, we have

$$|\varphi_k(y) - \varphi_k(x)| \le \frac{1}{2}|y - x| \quad \cdots (**).$$

Next, for every $x$, $y$ in $S_r(a)$, we have

$$
\begin{aligned}
&|y - x| - \left|\left((f'(a))^{-1}\right)(f(y) - f(x))\right| \\
&= |y - x| - \left|\left((f'(a))^{-1}\right)(f(y)) - \left((f'(a))^{-1}\right)(f(x))\right| \\
&\le \left|(y - x) - \left(\left((f'(a))^{-1}\right)(f(y)) - \left((f'(a))^{-1}\right)(f(x))\right)\right| \\
&= \left|\left(y - \left((f'(a))^{-1}\right)(f(y)) + k\right) - \left(x - \left((f'(a))^{-1}\right)(f(x)) + k\right)\right| \\
&= |\varphi_k(y) - \varphi_k(x)| \le \frac{1}{2}|y - x|.
\end{aligned}
$$

Hence, for every $x$, $y$ in $S_r(a)$, we have

$$\frac{1}{2}|y - x| \le \left|\left((f'(a))^{-1}\right)(f(y) - f(x))\right| \quad \cdots (***)$$

For 1: We have seen that the open neighborhood $S_r(a)$ of $a$ is contained in $E$. We will take $S_r(a)$ for $U$. Since $S_r(a)$ is connected, $U$ is connected.

For 2: We have to show that $f$ is 1–1 on $S_r(a)$. If not, otherwise, let $f$ be not 1–1 on $S_r(a)$. We have to arrive at a contradiction. Since $f$ is not 1–1 on $S_r(a)$, there exist $b$ and $c$ in $S_r(a)$ such that $b \ne c$, and $f(b) = f(c)$. So, from $(***)$,

$$
\begin{aligned}
0 < \frac{1}{2}|c - b| &\le \left|\left((f'(a))^{-1}\right)(f(c) - f(b))\right| = \left|\left((f'(a))^{-1}\right)(f(b) - f(b))\right| \\
&= \left|\left((f'(a))^{-1}\right)(0)\right| = |0| = 0,
\end{aligned}
$$

which is a contradiction. Hence, $f$ is 1–1 on $S_r(a)$.

For 3: We have to show that $f(S_r(a))$ is an open subset of $\mathbb{R}^3$ and is connected. Since $S_r(a)$ is connected, and $f$ is continuous, $f(S_r(a))$ is connected. Now, we want to show that $f(S_r(a))$ is an open subset of $\mathbb{R}^3$. For this purpose, let us take any $f(b)$ in $f(S_r(a))$, where $b$ is in $S_r(a)$. We have to find a real number $\varepsilon > 0$ such that the sphere

$$S_{\frac{\varepsilon}{2\|(f'(a))^{-1}\|}}(f(b))$$

is contained in $f(S_r(a))$, that is, we have to find a real number $\varepsilon > 0$ such that if

$$|y - f(b)| < \frac{\varepsilon}{2\|(f'(a))^{-1}\|}$$

then there exists $x^*$ in $\mathbb{R}^3$ such that $|x^* - a| < r$, and $f(x^*) = y$.

Since $b$ is in $S_r(a)$, we can find a real number $\varepsilon > 0$ such that $S_\varepsilon[b] \equiv \{x : x \in \mathbb{R}^3 \text{ and } |x - b| \le \varepsilon\} \subset S_r(a)$. Next, let us take any $y$ in $\mathbb{R}^3$ such that

$$|y - f(b)| < \frac{\varepsilon}{2\|(f'(a))^{-1}\|}.$$

Now, we will try to show that

$$\varphi_{((f'(a))^{-1})(y)} : S_\varepsilon[b] \to S_\varepsilon[b].$$

Observe that

$$
\begin{aligned}
\left|\varphi_{((f'(a))^{-1})(y)}(b) - b\right| &= \left|\left(b - \left((f'(a))^{-1}\right)(f(b)) + \left((f'(a))^{-1}\right)(y)\right) - b\right| \\
&= \left|\left((f'(a))^{-1}\right)(y) - \left((f'(a))^{-1}\right)(f(b))\right| = \left|\left((f'(a))^{-1}\right)(y - f(b))\right| \\
&\le \left\|(f'(a))^{-1}\right\| |y - f(b)| \le \left\|(f'(a))^{-1}\right\| \frac{\varepsilon}{2\|(f'(a))^{-1}\|} = \frac{\varepsilon}{2}.
\end{aligned}
$$

Thus,

$$\left|\varphi_{((f'(a))^{-1})(y)}(b) - b\right| \le \frac{\varepsilon}{2}.$$

Next, let us take any $z$ in $S_\varepsilon[b]$.
We have to prove that $\varphi_{((f'(a))^{-1})(y)}(z)$ is in $S_\varepsilon[b]$, that is,

$$\left|\varphi_{((f'(a))^{-1})(y)}(z) - b\right| \le \varepsilon.$$

Since $z$ is in $S_\varepsilon[b]$, $|z - b| \leq \varepsilon$, and hence, by $(**)$,

$$
\begin{aligned}
\left| \varphi_{((f'(a))^{-1})(y)}(z) - b \right| &\leq \left| \varphi_{((f'(a))^{-1})(y)}(z) - \varphi_{((f'(a))^{-1})(y)}(b) \right| + \left| \varphi_{((f'(a))^{-1})(y)}(b) - b \right| \\
&\leq \frac{1}{2}|z - b| + \left| \varphi_{((f'(a))^{-1})(y)}(b) - b \right| \\
&\leq \frac{1}{2}\varepsilon + \left| \varphi_{((f'(a))^{-1})(y)}(b) - b \right| \leq \frac{1}{2}\varepsilon + \frac{\varepsilon}{2} = \varepsilon.
\end{aligned}
$$

This proves that

$$
\varphi_{((f'(a))^{-1})(y)} : S_\varepsilon[b] \to S_\varepsilon[b].
$$

Since $\{x : x \in \mathbb{R}^3$ and $|x - b| \leq \varepsilon\}$ is a closed subset of the complete metric space $\mathbb{R}^3$, $S_\varepsilon[b]$ is a complete metric space. Also, from $(**)$, for every $s$, $t$ in $S_\varepsilon[b]$, we have

$$
\left| \varphi_{((f'(a))^{-1})(y)}(s) - \varphi_{((f'(a))^{-1})(y)}(t) \right| \leq \frac{1}{2}|s - t|.
$$

Hence, by Theorem 3.35, there exists a unique $x^*$ in $S_\varepsilon[b](\subset S_r(a))$ such that $\varphi_{((f'(a))^{-1})(y)}(x^*) = x^*$. Hence, $|x^* - a| < r$, and

$$
x^* = \varphi_{((f'(a))^{-1})(y)}(x^*) = x^* - \left( (f'(a))^{-1} \right)(f(x^*)) + \left( (f'(a))^{-1} \right)(y).
$$

Therefore,

$$
\begin{aligned}
\left( (f'(a))^{-1} \right)(0) = 0 &= -\left( (f'(a))^{-1} \right)(f(x^*)) + \left( (f'(a))^{-1} \right)(y) \\
&= \left( (f'(a))^{-1} \right)(y - f(x^*)).
\end{aligned}
$$

Now, since $(f'(a))^{-1}$ is invertible, $0 = y - f(x^*)$, or $f(x^*) = y$. Thus, we have shown that $f(S_r(a))$ is an open subset of $\mathbb{R}^3$.

For 4: We first prove that the 1–1 function $f^{-1}$ from $f(S_r(a))$ onto $S_r(a)$ is differentiable at every point of $f(S_r(a))$. For this purpose, let us take any $f(x)$ in $f(S_r(a))$, where $x$ is in $S_r(a)$. Since $f : E \to \mathbb{R}^3$ is a $C^1$ function, by Theorem 3.33, $f$ is differentiable at every point of $E$. Since $f$ is differentiable at every point of $E$, and $x$ is in $S_r(a) (\subset E)$, $f'(x)$ exists. Observe that if $t$ is in $S_r(a)$, then

$$
f'(t) \in S_{\frac{1}{2\|(f'(a))^{-1}\|}}(f'(a)) \subset S_{\frac{1}{\|(f'(a))^{-1}\|}}(f'(a)) \subset \Omega,
$$

and hence, $(f'(t))^{-1}$ exists. Thus, if $t$ is in $S_r(a)$, then

$$f'(t), (f'(t))^{-1} \in \Omega. \quad \cdots (*')$$

Since $x$ is in $S_r(a)$, from $(*')$, $f'(x)$, $(f'(x))^{-1}$ exist. Now, we will try to prove that $f^{-1}$ is differentiable at $f(x)$ and $(f^{-1})'(f(x)) = (f'(x))^{-1}$, that is,

$$\lim_{k \to 0} \left( \frac{1}{|k|} \left( f^{-1}(f(x) + k) - f^{-1}(f(x)) - \left( (f'(x))^{-1} \right)(k) \right) \right) = 0,$$

that is,

$$\lim_{\substack{y \to f(x) \\ y \in f(S_r(a))}} \left( \frac{1}{|y - f(x)|} \left( f^{-1}(y) - f^{-1}(f(x)) - \left( (f'(x))^{-1} \right)(y - f(x)) \right) \right) = 0,$$

that is,

$$\lim_{\substack{y \to f(x) \\ y \in f(S_r(a))}} \frac{1}{|y - f(x)|} \left| f^{-1}(y) - f^{-1}(f(x)) - \left( (f'(x))^{-1} \right)(y - f(x)) \right| = 0,$$

that is,

$$\lim_{\substack{y \to f(x) \\ y \in f(S_r(a))}} \frac{1}{|y - f(x)|} \left| f^{-1}(y) - x - \left( (f'(x))^{-1} \right)(y - f(x)) \right| = 0.$$

For this purpose, let us take a sequence $\{f(a_n)\}$ in $f(S_r(a))$ such that each $a_n$ is in $S_r(a)$, each $f(a_n)$ is different from $f(x)$, and $\lim_{n \to \infty} f(a_n) = f(x)$. We have to prove that

$$\lim_{n \to \infty} \frac{1}{|f(a_n) - f(x)|} \left| f^{-1}(f(a_n)) - x - \left( (f'(x))^{-1} \right)(f(a_n) - f(x)) \right| = 0,$$

that is,

$$\lim_{n \to \infty} \frac{1}{|f(a_n) - f(x)|} \left| a_n - x - \left( (f'(x))^{-1} \right)(f(a_n) - f(x)) \right| = 0.$$

Since $x$ and $a_n$ are in $S_r(a)$, from $(* * *)$,

$$\frac{1}{2} |a_n - x| \le \left| \left( (f'(a))^{-1} \right)(f(a_n) - f(x)) \right| \le \left\| (f'(a))^{-1} \right\| |f(a_n) - f(x)|,$$

and hence,

$$0 \le |a_n - x| \le 2\left\|(f'(a))^{-1}\right\|\left|f(a_n) - f(x)\right| \quad (n = 1, 2, 3, \cdots).$$

This, together with $\lim_{n\to\infty} f(a_n) = f(x)$, implies that $\lim_{n\to\infty} |a_n - x| = 0$. Thus,

$$\lim_{n\to\infty} a_n = x.$$

Since each $f(a_n)$ is different from $f(x)$, $x \ne a_n$. Since $f'(x)$ exists, each $a_n$ is in $S_r(a)$, $x$ is in $S_r(a)$, each $a_n$ is different from $x$, and $\lim_{n\to\infty} a_n = x$,

$$\lim_{n\to\infty} \frac{1}{|a_n - x|} |f(a_n) - f(x) - ((f'(x))(a_n - x))| = 0.$$

Since each $a_n \ne x$, $|a_n - x| \ne 0$. Also, each $|f(a_n) - f(x)| \ne 0$. It follows that for every positive integer $n$,

$$\frac{1}{|f(a_n) - f(x)|} \le 2\left\|(f'(a))^{-1}\right\| \frac{1}{|a_n - x|}.$$

Now, for every positive integer $n$, we have

$$\frac{1}{|f(a_n) - f(x)|} \left|a_n - x - \left((f'(x))^{-1}\right)(f(a_n) - f(x))\right|$$

$$\le \left(2\left\|(f'(a))^{-1}\right\| \frac{1}{|a_n - x|}\right) \left|a_n - x - \left((f'(x))^{-1}\right)(f(a_n) - f(x))\right|$$

$$\le \left(2\left\|(f'(a))^{-1}\right\| \frac{1}{|a_n - x|}\right) \left|\left(\left((f'(x))^{-1}\right)(f'(x))\right)(a_n - x) - \left((f'(x))^{-1}\right)(f(a_n) - f(x))\right|$$

$$\le \left(2\left\|(f'(a))^{-1}\right\| \frac{1}{|a_n - x|}\right) \left|\left((f'(x))^{-1}\right)((f'(x))(a_n - x)) - \left((f'(x))^{-1}\right)(f(a_n) - f(x))\right|$$

$$\le \left(2\left\|(f'(a))^{-1}\right\| \frac{1}{|a_n - x|}\right) \left|\left((f'(x))^{-1}\right)(((f'(x))(a_n - x)) - (f(a_n) - f(x)))\right|$$

$$\le \left(2\left\|(f'(a))^{-1}\right\| \frac{1}{|a_n - x|}\right) \left\|(f'(x))^{-1}\right\| |((f'(x))(a_n - x)) - (f(a_n) - f(x))|$$

$$= \left(2\left\|(f'(a))^{-1}\right\|\left\|(f'(x))^{-1}\right\|\right) \left(\frac{1}{|a_n - x|} |f(a_n) - f(x) - ((f'(x))(a_n - x))|\right).$$

Since

$$0 \le \frac{1}{|f(a_n) - f(x)|} \left|a_n - x - \left((f'(x))^{-1}\right)(f(a_n) - f(x))\right|$$

$$\le \left(2\left\|(f'(a))^{-1}\right\|\left\|(f'(x))^{-1}\right\|\right) \left(\frac{1}{|a_n - x|} |f(a_n) - f(x) - ((f'(x))(a_n - x))|\right),$$

and

$$\lim_{n\to\infty} \frac{1}{|a_n - x|} |f(a_n) - f(x) - ((f'(x))(a_n - x))| = 0,$$

so,

$$\lim_{n\to\infty} \frac{1}{|f(a_n) - f(x)|} \left| a_n - x - \left((f'(x))^{-1}\right)(f(a_n) - f(x)) \right| = 0.$$

This proves that $f^{-1}$ is differentiable at $f(x)$ and

$$\left(f^{-1}\right)'(f(x)) = (f'(x))^{-1} \quad \cdots (*'')$$

for every $x$ in $S_r(a)$. Hence, $f^{-1}$ is differentiable on the open subset $f(S_r(a))$ of $\mathbb{R}^3$.

Since $f^{-1}$ is differentiable on the open subset $f(S_r(a))$ of $\mathbb{R}^3$, $f^{-1}$ is continuous on the open subset $f(S_r(a))$ of $\mathbb{R}^3$. Now, we want to prove that the mapping $(f^{-1})'$ : $f(S_r(a)) \to L(\mathbb{R}^3)$ is continuous. Let us take any $y$ in $f(S_r(a))$. Since $y$ is in $f(S_r(a))$, there exists $x$ in $S_r(a)$, such that $f(x) = y$. Since $f(x) = y \in f(S_r(a))$, $f^{-1}(y) = x$. Since $x$ is in $S_r(a)$, by $(*'')$,

$$\left(f^{-1}\right)'(y) = \left(f^{-1}\right)'(f(x)) = (f'(x))^{-1} = \left(f'(f^{-1}(y))\right)^{-1}.$$

Hence, for every $y$ in $f(S_r(a))$,

$$\left(f^{-1}\right)'(y) = \left(f'(f^{-1}(y))\right)^{-1}.$$

We have seen that $y \mapsto f^{-1}(y)$ is differentiable on $f(S_r(a))$, so $y \mapsto f^{-1}(y)$ is a continuous function from $f(S_r(a))$ to $S_r(a)$. Since $f$ is a $C^1$ function on $E$, by Theorem 3.33, $x \mapsto f'(x)$ is continuous on $E$. Since $y \mapsto f^{-1}(y)$ is a continuous function from $f(S_r(a))$ to $S_r(a)$, and $x \mapsto f'(x)$ is continuous on $E(\supset S_r(a))$, by $(*')$, their composite function $y \mapsto f'(f^{-1}(y))$ is continuous from $f(S_r(a))$ to $\Omega$. Since $y \mapsto f'(f^{-1}(y))$ is continuous from $f(S_r(a))$ to $\Omega$, and by Theorem 3.3, the mapping $A \mapsto A^{-1}$ from $\Omega$ to $\Omega$ is continuous, their composite function $y \mapsto (f'(f^{-1}(y)))^{-1}$ is continuous from $f(S_r(a))$ to $\Omega$. Since $y \mapsto (f'(f^{-1}(y)))^{-1}$ is continuous from $f(S_r(a))$ to $\Omega$, and $(f'(f^{-1}(y)))^{-1} = (f^{-1})'(y)$, the function $y \mapsto (f^{-1})'(y)$ is continuous from $f(S_r(a))$ to $\Omega$, and hence, by Theorem 3.33, $f^{-1}$ is a $C^1$ function on the open subset $f(S_r(a))$ of $\mathbb{R}^3$.

For 5: Let $f$ be a $C^2$ function. Let $g_1 : f(S_r(a)) \to \mathbb{R}$, $g_2 : f(S_r(a)) \to \mathbb{R}$, $g_3 : f(S_r(a)) \to \mathbb{R}$ be the component functions of $f^{-1}$ that is, $f^{-1}(x) = (g_1(x), g_2(x), g_3(x))$ for every $x$ in $f(S_r(a))$. We have to prove that $f^{-1}$ from $f(S_r(a))$ onto $S_r(a)$ is a $C^2$ function, that is, each of the nine second-order partial derivatives $D_{jk}(g_i) : f(S_r(a)) \to \mathbb{R}$ exists and is continuous. Let us take any $y$ in $f(S_r(a))$. We will try to show that $(D_{12}(g_3))(y)$ exists. Since $y$ is in $f(S_r(a))$,

$$\left(f^{-1}\right)'(y) = \left(f'\left(f^{-1}(y)\right)\right)^{-1},$$

or,

$$\left(\left(f^{-1}\right)'(y)\right)\left(f'\left(f^{-1}(y)\right)\right) = I,$$

and hence, by Theorem 3.6,

$$\begin{bmatrix} 1 & 0 & 0 \\ 0 & 1 & 0 \\ 0 & 0 & 1 \end{bmatrix}_{3\times3} = [I] = \left[\left(\left(f^{-1}\right)'(y)\right)\left(f'\left(f^{-1}(y)\right)\right)\right] = \left[\left(f^{-1}\right)'(y)\right]\left[f'\left(f^{-1}(y)\right)\right]$$

$$= \begin{bmatrix} (D_1(g_1))(y) & (D_2(g_1))(y) & (D_3(g_1))(y) \\ (D_1(g_2))(y) & (D_2(g_2))(y) & (D_3(g_2))(y) \\ (D_1(g_3))(y) & (D_2(g_3))(y) & (D_3(g_3))(y) \end{bmatrix}_{3\times3}$$

$$\times \begin{bmatrix} (D_1(f_1))(f^{-1}(y)) & (D_2(f_1))(f^{-1}(y)) & (D_3(f_1))(f^{-1}(y)) \\ (D_1(f_2))(f^{-1}(y)) & (D_2(f_2))(f^{-1}(y)) & (D_3(f_2))(f^{-1}(y)) \\ (D_1(f_3))(f^{-1}(y)) & (D_2(f_3))(f^{-1}(y)) & (D_3(f_3))(f^{-1}(y)) \end{bmatrix}_{3\times3}.$$

This shows that

$$\begin{bmatrix} (D_1(g_1))(y) & (D_2(g_1))(y) & (D_3(g_1))(y) \\ (D_1(g_2))(y) & (D_2(g_2))(y) & (D_3(g_2))(y) \\ (D_1(g_3))(y) & (D_2(g_3))(y) & (D_3(g_3))(y) \end{bmatrix}_{3\times3}$$

$$= \begin{bmatrix} (D_1(f_1))(f^{-1}(y)) & (D_2(f_1))(f^{-1}(y)) & (D_3(f_1))(f^{-1}(y)) \\ (D_1(f_2))(f^{-1}(y)) & (D_2(f_2))(f^{-1}(y)) & (D_3(f_2))(f^{-1}(y)) \\ (D_1(f_3))(f^{-1}(y)) & (D_2(f_3))(f^{-1}(y)) & (D_3(f_3))(f^{-1}(y)) \end{bmatrix}^{-1}.$$

Hence,

$$(D_2(g_3))(y) = \cfrac{1}{\begin{vmatrix} (D_1(f_1))(f^{-1}(y)) & (D_2(f_1))(f^{-1}(y)) & (D_3(f_1))(f^{-1}(y)) \\ (D_1(f_2))(f^{-1}(y)) & (D_2(f_2))(f^{-1}(y)) & (D_3(f_2))(f^{-1}(y)) \\ (D_1(f_3))(f^{-1}(y)) & (D_2(f_3))(f^{-1}(y)) & (D_3(f_3))(f^{-1}(y)) \end{vmatrix}}$$

$$\times \left(-\begin{vmatrix} (D_1(f_1))(f^{-1}(y)) & (D_2(f_1))(f^{-1}(y)) \\ (D_1(f_3))(f^{-1}(y)) & (D_2(f_3))(f^{-1}(y)) \end{vmatrix}\right)$$

$$= -\cfrac{\begin{vmatrix} (D_1(f_1))(f^{-1}(y)) & (D_2(f_1))(f^{-1}(y)) \\ (D_1(f_3))(f^{-1}(y)) & (D_2(f_3))(f^{-1}(y)) \end{vmatrix}}{\begin{vmatrix} (D_1(f_1))(f^{-1}(y)) & (D_2(f_1))(f^{-1}(y)) & (D_3(f_1))(f^{-1}(y)) \\ (D_1(f_2))(f^{-1}(y)) & (D_2(f_2))(f^{-1}(y)) & (D_3(f_2))(f^{-1}(y)) \\ (D_1(f_3))(f^{-1}(y)) & (D_2(f_3))(f^{-1}(y)) & (D_3(f_3))(f^{-1}(y)) \end{vmatrix}}.$$

Now, since $f$ is a $C^2$ function, each of the nine second-order partial derivatives $D_{jk}(f_i) : S_r(a) \to \mathbb{R}$ of $f_1$ exists, and for every $y$ in $f(S_r(a))$, we have

$$(D_{12}(g_3))(y) = D_1 \left( -\frac{\begin{vmatrix} (D_1(f_1))(f^{-1}(y)) & (D_2(f_1))(f^{-1}(y)) \\ (D_1(f_3))(f^{-1}(y)) & (D_2(f_3))(f^{-1}(y)) \end{vmatrix}}{\begin{vmatrix} (D_1(f_1))(f^{-1}(y)) & (D_2(f_1))(f^{-1}(y)) & (D_3(f_1))(f^{-1}(y)) \\ (D_1(f_2))(f^{-1}(y)) & (D_2(f_2))(f^{-1}(y)) & (D_3(f_2))(f^{-1}(y)) \\ (D_1(f_3))(f^{-1}(y)) & (D_2(f_3))(f^{-1}(y)) & (D_3(f_3))(f^{-1}(y)) \end{vmatrix}} \right) = -\frac{\text{num}}{\text{denom}},$$

where

$$\text{num} \equiv \left( D_1 \begin{vmatrix} (D_1(f_1))(f^{-1}(y)) & (D_2(f_1))(f^{-1}(y)) \\ (D_1(f_3))(f^{-1}(y)) & (D_2(f_3))(f^{-1}(y)) \end{vmatrix} \right)$$

$$\times \begin{vmatrix} (D_1(f_1))(f^{-1}(y)) & (D_2(f_1))(f^{-1}(y)) & (D_3(f_1))(f^{-1}(y)) \\ (D_1(f_2))(f^{-1}(y)) & (D_2(f_2))(f^{-1}(y)) & (D_3(f_2))(f^{-1}(y)) \\ (D_1(f_3))(f^{-1}(y)) & (D_2(f_3))(f^{-1}(y)) & (D_3(f_3))(f^{-1}(y)) \end{vmatrix}$$

$$- \begin{vmatrix} (D_1(f_1))(f^{-1}(y)) & (D_2(f_1))(f^{-1}(y)) \\ (D_1(f_3))(f^{-1}(y)) & (D_2(f_3))(f^{-1}(y)) \end{vmatrix}$$

$$\times \left( D_1 \begin{vmatrix} (D_1(f_1))(f^{-1}(y)) & (D_2(f_1))(f^{-1}(y)) & (D_3(f_1))(f^{-1}(y)) \\ (D_1(f_2))(f^{-1}(y)) & (D_2(f_2))(f^{-1}(y)) & (D_3(f_2))(f^{-1}(y)) \\ (D_1(f_3))(f^{-1}(y)) & (D_2(f_3))(f^{-1}(y)) & (D_3(f_3))(f^{-1}(y)) \end{vmatrix} \right)$$

$$= \left( \begin{vmatrix} (D_{11}(f_1))(f^{-1}(y)) & (D_2(f_1))(f^{-1}(y)) \\ (D_{11}(f_3))(f^{-1}(y)) & (D_2(f_3))(f^{-1}(y)) \end{vmatrix} + \begin{vmatrix} (D_1(f_1))(f^{-1}(y)) & (D_{12}(f_1))(f^{-1}(y)) \\ (D_1(f_3))(f^{-1}(y)) & (D_{12}(f_3))(f^{-1}(y)) \end{vmatrix} \right)$$

$$\times \begin{vmatrix} (D_1(f_1))(f^{-1}(y)) & (D_2(f_1))(f^{-1}(y)) & (D_3(f_1))(f^{-1}(y)) \\ (D_1(f_2))(f^{-1}(y)) & (D_2(f_2))(f^{-1}(y)) & (D_3(f_2))(f^{-1}(y)) \\ (D_1(f_3))(f^{-1}(y)) & (D_2(f_3))(f^{-1}(y)) & (D_3(f_3))(f^{-1}(y)) \end{vmatrix}$$

$$- \begin{vmatrix} (D_1(f_1))(f^{-1}(y)) & (D_2(f_1))(f^{-1}(y)) \\ (D_1(f_3))(f^{-1}(y)) & (D_2(f_3))(f^{-1}(y)) \end{vmatrix}$$

$$\times \left( \begin{vmatrix} (D_{11}(f_1))(f^{-1}(y)) & (D_2(f_1))(f^{-1}(y)) & (D_3(f_1))(f^{-1}(y)) \\ (D_{11}(f_2))(f^{-1}(y)) & (D_2(f_2))(f^{-1}(y)) & (D_3(f_2))(f^{-1}(y)) \\ (D_{11}(f_3))(f^{-1}(y)) & (D_2(f_3))(f^{-1}(y)) & (D_3(f_3))(f^{-1}(y)) \end{vmatrix} \right.$$

$$+ \begin{vmatrix} (D_1(f_1))(f^{-1}(y)) & (D_{12}(f_1))(f^{-1}(y)) & (D_3(f_1))(f^{-1}(y)) \\ (D_1(f_2))(f^{-1}(y)) & (D_{12}(f_2))(f^{-1}(y)) & (D_3(f_2))(f^{-1}(y)) \\ (D_1(f_3))(f^{-1}(y)) & (D_{12}(f_3))(f^{-1}(y)) & (D_3(f_3))(f^{-1}(y)) \end{vmatrix}$$

$$\times \left. \begin{vmatrix} (D_1(f_1))(f^{-1}(y)) & (D_2(f_1))(f^{-1}(y)) & (D_{13}(f_1))(f^{-1}(y)) \\ (D_1(f_2))(f^{-1}(y)) & (D_2(f_2))(f^{-1}(y)) & (D_{13}(f_2))(f^{-1}(y)) \\ (D_1(f_3))(f^{-1}(y)) & (D_2(f_3))(f^{-1}(y)) & (D_{13}(f_3))(f^{-1}(y)) \end{vmatrix} \right),$$

and

$$\text{denom} \equiv \begin{vmatrix} (D_1(f_1))(f^{-1}(y)) & (D_2(f_1))(f^{-1}(y)) & (D_3(f_1))(f^{-1}(y)) \\ (D_1(f_2))(f^{-1}(y)) & (D_2(f_2))(f^{-1}(y)) & (D_3(f_2))(f^{-1}(y)) \\ (D_1(f_3))(f^{-1}(y)) & (D_2(f_3))(f^{-1}(y)) & (D_3(f_3))(f^{-1}(y)) \end{vmatrix}^2.$$

It follows that $(D_{12}(g_3))(y)$ exists. So $D_{12}(g_3) : f(S_r(a)) \to \mathbb{R}$.

Next, since each of the nine second-order partial derivatives $D_{jk}(f_i) : S_r(a) \to \mathbb{R}$ of $f_i$ is continuous, and $f^{-1}$ is continuous on $f(S_r(a))$, by the above formula for *num* and *denom*, we find that *num* and *denom* are continuous functions on $f(S_r(a))$, and hence, $D_{12}(g_3)(= -\frac{\text{num}}{\text{denom}})$ is continuous on $f(S_r(a))$.

Similarly, all the other $D_{jk}(g_i) : f(S_r(a)) \to \mathbb{R}$ are continuous. This proves that $f^{-1}$ from $f(U)$ onto $U$ is a $C^2$ function, etc. $\qquad\qquad\square$

**Note 3.37** The result similar to Theorem 3.36 can be proved as above for $\mathbb{R}^n$ in place of $\mathbb{R}^3$. This theorem is known as the *inverse function theorem*.

**Theorem 3.38** *Let E be an open subset of $\mathbb{R}^n$. Let $f : E \to \mathbb{R}^n$ be a function. If*

1. *f is a $C^1$ function,*
2. *$f'(a)$ is invertible for every a in E,*

*then f is an open mapping (i.e., f-image of every open subset of E in an open set).*

*Proof* Let us take any open subset $G$ of $E$. We have to prove that $f(G)$ is an open set. For this purpose, let us take any $f(a)$ in $f(G)$ where $a$ is in $G$. Since $a \in G \subset E$, by the given assumption, $f'(a)$ is invertible. Now, by the inverse function theorem, there exists an open neighborhood $U$ of $a$ such that $U$ is contained in $G$ and $f(U)$ is open in $\mathbb{R}^3$. Since $a \in U \subset G, f(a) \in f(U) \subset f(G)$. Since $f(a) \in f(U)$, and $f(U)$ is open in $\mathbb{R}^3$, $f(U)$ is an open neighborhood of $f(a)$. Since $f(U)$ is an open neighborhood of $f(a)$ and $f(U) \subset f(G)$, $f(a)$ is an interior point of $f(G)$, and hence, $f(G)$ is an open set. $\qquad\qquad\square$

## 3.4   Implicit Function Theorem

**Definition** Let $a \equiv (a_1, a_2, a_3) \in \mathbb{R}^3, b \equiv (b_1, b_2) \in \mathbb{R}^2$. By the ordered pair $(a, b)$, we mean $(a_1, a_2, a_3, b_1, b_2)$. Clearly, if $a \in \mathbb{R}^3, b \in \mathbb{R}^2$, then $(a, b) \in \mathbb{R}^{3+2}$. Let $A \in L(\mathbb{R}^{3+2}, \mathbb{R}^3)$. By $A_x$, we mean the function

$$A_x : \mathbb{R}^3 \to \mathbb{R}^3$$

defined as follows: For every $(a_1, a_2, a_3)$ in $\mathbb{R}^3$,

$$A_x((a_1, a_2, a_3)) \equiv A(a_1, a_2, a_3, 0, 0).$$

We shall show that $A_x \in L(\mathbb{R}^3, \mathbb{R}^3)$. We must prove

(i)  $A_x(a + b) = A_x(a) + A_x(b)$ for every $a$, $b$ in $\mathbb{R}^3$,

(ii)  $A_x(ta) = t(A_x(a))$ for every $a$ in $\mathbb{R}^3$ and for every real $t$.

For (i): Let us take any $a \equiv (a_1, a_2, a_3), b \equiv (b_1, b_2, b_3)$ in $\mathbb{R}^3$.

$$\begin{aligned}
\text{LHS} = A_x(a + b) &= A_x((a_1, a_2, a_3) + (b_1, b_2, b_3)) \\
&= A_x((a_1 + b_1, a_2 + b_2, a_3 + b_3)) \\
&= A((a_1 + b_1, a_2 + b_2, a_3 + b_3, 0, 0)) \\
&= A((a_1, a_2, a_3, 0, 0) + (b_1, b_2, b_3, 0, 0)) \\
&= A((a_1, a_2, a_3, 0, 0)) + A((b_1, b_2, b_3, 0, 0)) \\
&= A_x((a_1, a_2, a_3)) + A_x((b_1, b_2, b_3)) \\
&= A_x(a) + A_x(b) = \text{RHS}.
\end{aligned}$$

For (ii): Let us take any $a \equiv (a_1, a_2, a_3)$ in $\mathbb{R}^3$ and any real $t$.

$$\begin{aligned}
\text{LHS} = A_x(ta) = A_x(t(a_1, a_2, a_3)) &= A_x((ta_1, ta_2, ta_3)) \\
&= A((ta_1, ta_2, ta_3, 0, 0)) = A(t(a_1, a_2, a_3, 0, 0)) \\
&= t(A((a_1, a_2, a_3, 0, 0))) = t(A_x((a_1, a_2, a_3))) \\
&= t(A_x(a)) = \text{RHS}.
\end{aligned}$$

Thus, we have shown that if $A \in L(\mathbb{R}^{3+2}, \mathbb{R}^3)$, then $A_x \in L(\mathbb{R}^3, \mathbb{R}^3)$. Again, by $A_y$, we mean the function

$$A_y : \mathbb{R}^2 \to \mathbb{R}^3$$

defined as follows: For every $(a_1, a_2)$ in $\mathbb{R}^2$,

$$A_y(a) \equiv A(0, 0, 0, a_1, a_2).$$

We shall show that $A_y \in L(\mathbb{R}^2, \mathbb{R}^3)$. We must prove

(i)  $A_y(a + b) = A_y(a) + A_y(b)$ for every $a$, $b$ in $\mathbb{R}^2$,

(ii)  $A_y(ta) = t(A_y(a))$ for every $a$ in $\mathbb{R}^2$ and for every real $t$.

For (i): Let us take any $a \equiv (a_1, a_2), b \equiv (b_1, b_2)$ in $\mathbb{R}^2$.

$$
\begin{aligned}
\text{LHS} &= A_y(a + b) = A_y((a_1, a_2) + (b_1, b_2)) = A_y((a_1 + b_1, a_2 + b_2)) \\
&= A((0, 0, 0, a_1 + b_1, a_2 + b_2)) = A((0, 0, 0, a_1, a_2) + (0, 0, 0, b_1, b_2)) \\
&= A((0, 0, 0, a_1, a_2)) + A((0, 0, 0, b_1, b_2)) = A_y((a_1, a_2)) + A_y((b_1, b_2)) \\
&= A_y(a) + A_y(b) = \text{RHS}.
\end{aligned}
$$

For (ii): Let us take any $a \equiv (a_1, a_2)$ in $\mathbb{R}^2$ and any real $t$.

$$
\begin{aligned}
\text{LHS} &= A_y(ta) = A_y(t(a_1, a_2)) = A_y((ta_1, ta_2)) \\
&= A((0, 0, 0, ta_1, ta_2)) = A(t(0, 0, 0, a_1, a_2)) \\
&= t(A((0, 0, 0, a_1, a_2))) = t(A_y((a_1, a_2))) \\
&= t(A_y(a)) = \text{RHS}.
\end{aligned}
$$

Thus, we have shown that if $A \in L(\mathbb{R}^{3+2}, \mathbb{R}^3)$, then $A_y \in L(\mathbb{R}^2, \mathbb{R}^3)$.
Now, we will try to prove that for every $a$ in $\mathbb{R}^3$, and for every $b$ in $\mathbb{R}^2$,

$$
A((a, b)) = A_x(a) + A_y(b) \cdots (*).
$$

For this purpose, let $a \equiv (a_1, a_2, a_3) \in \mathbb{R}^3$ and $b \equiv (b_1, b_2) \in \mathbb{R}^2$.

$$
\begin{aligned}
\text{LHS} &= A((a, b)) = A((a_1, a_2, a_3, b_1, b_2)) = A((a_1, a_2, a_3, 0, 0) + (0, 0, 0, b_1, b_2)) \\
&= A((a_1, a_2, a_3, 0, 0)) + A((0, 0, 0, b_1, b_2)) \\
&= A_x((a_1, a_2, a_3)) + A_y((b_1, b_2)) = A_x(a) + A_y(b) = \text{RHS}.
\end{aligned}
$$

**Note 3.39** Similar notations and results as above can be supplied for $\mathbb{R}^n$ in place of $\mathbb{R}^3$ and $\mathbb{R}^m$ in place of $\mathbb{R}^2$.

**Note 3.40** Let $A \in L(\mathbb{R}^{3+2}, \mathbb{R}^3)$, and let

$$
[A] = \begin{bmatrix} a_{11} & a_{12} & a_{13} & a_{14} & a_{15} \\ a_{21} & a_{22} & a_{23} & a_{24} & a_{25} \\ a_{31} & a_{32} & a_{33} & a_{34} & a_{35} \end{bmatrix}_{3 \times 5}.
$$

So,

$$
\begin{aligned}
A_x((1, 0, 0)) &= A((1, 0, 0, 0, 0)) = (a_{11}, a_{21}, a_{31}), \\
A_x((0, 1, 0)) &= A((0, 1, 0, 0, 0)) = (a_{12}, a_{22}, a_{32}), \\
A_x((0, 0, 1)) &= A((0, 0, 1, 0, 0)) = (a_{13}, a_{23}, a_{33}).
\end{aligned}
$$

Hence,

$$[A_x] = \begin{bmatrix} a_{11} & a_{12} & a_{13} \\ a_{21} & a_{22} & a_{23} \\ a_{31} & a_{32} & a_{33} \end{bmatrix}_{3\times3}.$$

Next,

$$A_y((1,0)) = A((0,0,0,1,0)) = (a_{14}, a_{24}, a_{34}),$$
$$A_y((0,1)) = A((0,0,0,0,1)) = (a_{15}, a_{25}, a_{35}).$$

Hence,

$$[A_y] = \begin{bmatrix} a_{14} & a_{15} \\ a_{24} & a_{25} \\ a_{34} & a_{35} \end{bmatrix}_{3\times2}.$$

Conclusion: If $A \in L(\mathbb{R}^{3+2}, \mathbb{R}^3)$, and

$$[A] = \begin{bmatrix} a_{11} & a_{12} & a_{13} & a_{14} & a_{15} \\ a_{21} & a_{22} & a_{23} & a_{24} & a_{25} \\ a_{31} & a_{32} & a_{33} & a_{34} & a_{35} \end{bmatrix}_{3\times5},$$

then

$$[A_x] = \begin{bmatrix} a_{11} & a_{12} & a_{13} \\ a_{21} & a_{22} & a_{23} \\ a_{31} & a_{32} & a_{33} \end{bmatrix}_{3\times3} \quad \text{and} \quad [A_y] = \begin{bmatrix} a_{14} & a_{15} \\ a_{24} & a_{25} \\ a_{34} & a_{35} \end{bmatrix}_{3\times2}.$$

Also, if $A_x$ is invertible, then with the usual meaning of symbols,

$$\left[ (A_x)^{-1} \right] = \begin{bmatrix} a_{11} & a_{12} & a_{13} \\ a_{21} & a_{22} & a_{23} \\ a_{31} & a_{32} & a_{33} \end{bmatrix}^{-1} = \frac{1}{\Delta} \begin{bmatrix} A_{11} & A_{12} & A_{13} \\ A_{21} & A_{22} & A_{23} \\ A_{31} & A_{32} & A_{33} \end{bmatrix}^T = \frac{1}{\Delta} \begin{bmatrix} A_{11} & A_{21} & A_{31} \\ A_{12} & A_{22} & A_{32} \\ A_{13} & A_{23} & A_{33} \end{bmatrix},$$

and hence,

$$\left( (A_x)^{-1} \right)((1,0,0)) = \frac{1}{\Delta}(A_{11}, A_{12}, A_{13}),$$
$$\left( (A_x)^{-1} \right)((0,1,0)) = \frac{1}{\Delta}(A_{21}, A_{22}, A_{23}),$$
$$\left( (A_x)^{-1} \right)((0,0,1)) = \frac{1}{\Delta}(A_{31}, A_{32}, A_{33}).$$

Conclusion: If $A \in L(\mathbb{R}^{3+2}, \mathbb{R}^3)$, $A_x$ is invertible, and

$$[A] = \begin{bmatrix} a_{11} & a_{12} & a_{13} & a_{14} & a_{15} \\ a_{21} & a_{22} & a_{23} & a_{24} & a_{25} \\ a_{31} & a_{32} & a_{33} & a_{34} & a_{35} \end{bmatrix}_{3 \times 5},$$

then

$$\left((A_x)^{-1}\right)((1,0,0)) = \frac{1}{\Delta}(A_{11}, A_{12}, A_{13}),$$

$$\left((A_x)^{-1}\right)((0,1,0)) = \frac{1}{\Delta}(A_{21}, A_{22}, A_{23}),$$

$$\left((A_x)^{-1}\right)((0,0,1)) = \frac{1}{\Delta}(A_{31}, A_{32}, A_{33}).$$

**Theorem 3.41** *Let $A \in L(\mathbb{R}^{3+2}, \mathbb{R}^3)$. If $A_x$ is invertible, then for every $k$ in $\mathbb{R}^2$, there exists a unique $h$ in $\mathbb{R}^3$, such that $A(h,k) = 0$. The value of $h$ is $-(((A_x)^{-1})(A_y(k)))$.*

*Proof* **Existence**: Let us take any $k$ in $\mathbb{R}^2$. Since $k$ is in $\mathbb{R}^2$, and $A_y : \mathbb{R}^2 \to \mathbb{R}^3$, $A_y(k)$ is in $\mathbb{R}^3$. Since $A_y(k)$ is in $\mathbb{R}^3$, $A_x : \mathbb{R}^3 \to \mathbb{R}^3$, and $A_x$ is invertible, $-(((A_x)^{-1})(A_y(k)))$ is in $\mathbb{R}^3$. Also, from the formula $(*)$,

$$\begin{aligned} A\left(\left(-\left(\left((A_x)^{-1}\right)(A_y(k))\right), k\right)\right) &= A_x\left(-\left(\left((A_x)^{-1}\right)(A_y(k))\right)\right) + A_y(k) \\ &= -\left(A_x\left(\left((A_x)^{-1}\right)(A_y(k))\right)\right) + A_y(k) \\ &= -\left(\left(A_x \circ (A_x)^{-1}\right)(A_y(k))\right) + A_y(k) \\ &= -(A_y(k)) + A_y(k) = 0. \end{aligned}$$

**Uniqueness**: If not, otherwise, let there exist $k$ in $\mathbb{R}^2$, $h$ in $\mathbb{R}^3$, and $h_1$ in $\mathbb{R}^3$ such that $h \neq h_1$, $A((h,k)) = 0$, and $A((h_1,k)) = 0$. We have to arrive at a contradiction. Since

$$0 = A((h,k)) = A_x(h) + A_y(k),$$

and $A_x$ is invertible,

$$\begin{aligned} 0 = \left((A_x)^{-1}\right)(0) &= \left((A_x)^{-1}\right)(A_x(h) + A_y(k)) \\ &= \left((A_x)^{-1}\right)(A_x(h)) + \left((A_x)^{-1}\right)(A_y(k)) \\ &= \left(\left((A_x)^{-1}\right) \circ A_x\right)(h) + \left((A_x)^{-1}\right)(A_y(k)) = I(h) + \left((A_x)^{-1}\right)(A_y(k)) \\ &= h + \left((A_x)^{-1}\right)(A_y(k)). \end{aligned}$$

It follows that

$$h = -\left(\left((A_x)^{-1}\right)(A_y(k))\right).$$

Similarly,

$$h_1 = -\left(\left((A_x)^{-1}\right)(A_y(k))\right),$$

which contradicts the assumption $h \neq h_1$. Further, we have shown that for a given $k$, the value of $h$ is $-(((A_x)^{-1})(A_y(k)))$.                                    $\square$

**Note 3.42** The result similar to Theorem 3.41 can be proved as above for $\mathbb{R}^n$ in place of $\mathbb{R}^3$ and $\mathbb{R}^m$ in place of $\mathbb{R}^2$.

**Theorem 3.43** *Let $E$ be an open subset of $\mathbb{R}^{3+2}$. Let $f : E \to \mathbb{R}^3$. Let $a \in \mathbb{R}^3, b \in \mathbb{R}^2$, and $(a, b) \in E$. If*

(i) *$f$ is a $C^1$ function,*
(ii) *$(f'((a, b)))_x$ is invertible,*
(iii) *$f((a, b)) = 0$,*

*then there exist an open neighborhood $U(\subset E)$ of $(a, b)$ in $\mathbb{R}^{3+2}$, an open neighborhood $W$ of $b$ in $\mathbb{R}^2$, and a function*

$$g : W \to \mathbb{R}^3$$

*such that*

1. *for every $y$ in $W$, $(g(y), y) \in U$,*
2. *for every $y$ in $W$, $f((g(y), y)) = 0$,*
3. *$g(b) = a$,*
4. *$g$ is a $C^1$ function,*
5. *for every $y^*$ in $W$, $g'(y^*) = -((f'((g(y^*), y^*)))_x)^{-1}(f'((g(y^*), y^*)))_y$*
6. *$g'(b) = -((f'((a, b)))_x)^{-1}(f'((a, b)))_y$*
7. *if $f$ is a $C^2$ function, then $g$ is a $C^2$ function, etc.*

*Proof* Let $f_1 : E \to \mathbb{R}$, $f_2 : E \to \mathbb{R}$, $f_3 : E \to \mathbb{R}$, be the component functions of $f$, that is, $f((x, y)) = (f_1((x, y)), f_2((x, y)), f_3((x, y)))$ for every $(x, y)$ in $E(\subset \mathbb{R}^{3+2})$. Let us define a function

$$F : E \to \mathbb{R}^{3+2}$$

as follows: For every $(x, y)$ in $E$, where $x \equiv (x_1, x_2, x_3)$ is in $\mathbb{R}^3$ and $y \equiv (y_1, y_2)$ is in $\mathbb{R}^2$,

$$F((x,y)) \equiv (f((x,y)),y) = ((f_1((x,y)),f_2((x,y)),f_3((x,y))),(y_1,y_2))$$
$$= (f_1((x,y)),f_2((x,y)),f_3((x,y)),y_1,y_2).$$

Clearly, the component functions of $F$ are

$$(x,y) \mapsto f_1((x,y)), \quad (x,y) \mapsto f_2((x,y)), \quad (x,y) \mapsto f_3((x,y)), \quad (x,y) \mapsto y_1,$$
$$(x,y) \mapsto y_2$$

from $E$ to $\mathbb{R}$.

We want to apply the inverse function theorem on $F$. For this purpose, we first prove

(a) $F$ is a $C^1$ function,
(b) $F'((a,b))$ is invertible.

For (a): Since $f$ is a $C^1$ function, for $i = 1,2,3$ and $j = 1,2,3,4,5(= 3+2)$, each function $D_j f_i : E \to \mathbb{R}$ exists and is continuous. Also, $(x,y) \mapsto y_1$ and $(x,y) \mapsto y_2$ are smooth functions. This shows that $F$ is a $C^1$ function.

For (b): Here,

$\det[F'((a,b))]$

$$= \det \begin{bmatrix} (D_1f_1)((a,b)) & (D_2f_1)((a,b)) & (D_3f_1)((a,b)) & (D_4f_1)((a,b)) & (D_5f_1)((a,b)) \\ (D_1f_2)((a,b)) & (D_2f_2)((a,b)) & (D_3f_2)((a,b)) & (D_4f_2)((a,b)) & (D_5f_2)((a,b)) \\ (D_1f_3)((a,b)) & (D_2f_3)((a,b)) & (D_3f_3)((a,b)) & (D_4f_3)((a,b)) & (D_5f_3)((a,b)) \\ 0 & 0 & 0 & 1 & 0 \\ 0 & 0 & 0 & 0 & 1 \end{bmatrix}_{5\times5}$$

$$= \det \begin{bmatrix} (D_1f_1)((a,b)) & (D_2f_1)((a,b)) & (D_3f_1)((a,b)) \\ (D_1f_2)((a,b)) & (D_2f_2)((a,b)) & (D_3f_2)((a,b)) \\ (D_1f_3)((a,b)) & (D_2f_3)((a,b)) & (D_3f_3)((a,b)) \end{bmatrix}.$$

Since $(f'((a,b))) \in L(\mathbb{R}^{3+2},\mathbb{R}^3)$, $(f'((a,b)))_x : \mathbb{R}^3 \to \mathbb{R}^3$.
Also,

$$(f'((a,b)))_x((1,0,0)) = (f'((a,b)))((1,0,0,0,0))$$
$$= ((D_1f_1)((a,b)),(D_1f_2)((a,b)),(D_1f_3)((a,b))),$$
$$(f'((a,b)))_x((0,1,0)) = (f'((a,b)))((0,1,0,0,0))$$
$$= ((D_2f_1)((a,b)),(D_2f_2)((a,b)),(D_2f_3)((a,b))),$$

and

$$(f'((a,b)))_x((0,0,1)) = (f'((a,b)))((0,0,1,0,0))$$
$$= ((D_3f_1)((a,b)),(D_3f_2)((a,b)),(D_3f_3)((a,b))),$$

so,

$$[(f'((a,b)))_x] = \begin{bmatrix} (D_1f_1)((a,b)) & (D_2f_1)((a,b)) & (D_3f_1)((a,b)) \\ (D_1f_2)((a,b)) & (D_2f_2)((a,b)) & (D_3f_2)((a,b)) \\ (D_1f_3)((a,b)) & (D_2f_3)((a,b)) & (D_3f_3)((a,b)) \end{bmatrix}.$$

Next, since $(f'((a,b)))_x$ is invertible,

$$0 \neq \det[(f'((a,b)))_x] = \det \begin{bmatrix} (D_1f_1)((a,b)) & (D_2f_1)((a,b)) & (D_3f_1)((a,b)) \\ (D_1f_2)((a,b)) & (D_2f_2)((a,b)) & (D_3f_2)((a,b)) \\ (D_1f_3)((a,b)) & (D_2f_3)((a,b)) & (D_3f_3)((a,b)) \end{bmatrix}$$

$$= \det[F'((a,b))].$$

Since $\det[F'((a,b))] \neq 0$, so $F'((a,b))$ is invertible. Since $f$ is a $C^1$ function,

$$(x,y) \mapsto \det \begin{bmatrix} (D_1f_1)((x,y)) & (D_2f_1)((x,y)) & (D_3f_1)((x,y)) \\ (D_1f_2)((x,y)) & (D_2f_2)((x,y)) & (D_3f_2)((x,y)) \\ (D_1f_3)((x,y)) & (D_2f_3)((x,y)) & (D_3f_3)((x,y)) \end{bmatrix}$$

is continuous. Further, since $\det[(f'((a,b)))_x]$ is nonzero, there exists a neighborhood $U^*$ of $(a,b)$ such that $\det[(f'((x,y)))_x]$ is nonzero for every $(x,y)$ in $U^*$. It follows that $(f'((x,y)))_x$ is invertible for every $(x,y)$ in $U^*$. Now, by the inverse function theorem and Theorem 3.38, there exists an open neighborhood $U$ of $(a,b)$ such that

(1)   $U$ is contained in $E$,
(2)   $F$ is 1–1 on $U$,
(3)   $F(U)$ is open in $\mathbb{R}^{3+2}$,
(4)   the 1–1 function $F^{-1}$ from $F(U)$ onto $U$ is a $C^1$ function,
(5)   $U$ is contained in $U^*$.

Put

$$W \equiv \{ y : y \in \mathbb{R}^2 \text{ and for some } x^* \text{ in } \mathbb{R}^3(x^*,y) \in U, f((x^*,y)) = 0 \}.$$

Since $(a,b) \in U$, and $f((a,b)) = 0$, $b \in W$. Let us observe that

$$\begin{aligned} W &= \{ y : y \in \mathbb{R}^2 \text{ and for some } x^* \text{ in } \mathbb{R}^3, (x^*,y) \in U, f((x^*,y)) = 0 \} \\ &= \{ y : y \in \mathbb{R}^2 \text{ and } (0,y) = (f(x^*,y^*),y^*) \text{ for some } (x^*,y^*) \text{ in } U \} \\ &= \{ y : y \in \mathbb{R}^2 \text{ and } (0,y) = F((x^*,y^*)) \text{ for some } (x^*,y^*) \text{ in } U \} \\ &= \{ y : y \in \mathbb{R}^2 \text{ and } (0,y) \in F(U) \}. \end{aligned}$$

Thus,

$$W = \{ y : y \in \mathbb{R}^2 \text{ and } (0,y) \in F(U) \}.$$

Now, we will show that $W$ is an open subset of $\mathbb{R}^2$. For this purpose, let us take any $y_0 \in W$. Since $y_0 \in W$, $(0,y_0) \in F(U)$. Since $(0,y_0) \in F(U)$, and $F(U)$ is open in $\mathbb{R}^{3+2}$, there exist open neighborhood $V_1$ of 0 in $\mathbb{R}^3$ and open neighborhood $V_2$ of $y_0$ in $\mathbb{R}^2$ such that $V_1 \times V_2 \subset F(U)$. Since $0 \in V_1$, and $V_1 \times V_2 \subset F(U)$, $\{0\} \times V_2 \subset V_1 \times V_2 \subset F(U)$, and hence, $(0, y) \in F(U)$ for every $y$ in $V_2$. It follows that $V_2 \subset W$. Since $V_2 \subset W$, and $V_2$ is an open neighborhood of $y_0$, $y_0$ is an interior point of $W$. Hence, $W$ is an open subset of $\mathbb{R}^2$. Since $W$ is an open subset of $\mathbb{R}^2$, and $b \in W$, $W$ is an open neighborhood of $b$ in $\mathbb{R}^2$. Now, we will try to prove that if $y \in W$, then there exists a unique $x^*$ in $\mathbb{R}$ such that

$$(x^*, y) \in U, f((x^*, y)) = 0.$$

For this purpose, let us take any $y \in W$. The existence of $x^*$ is clear from the definition of $W$.

*Uniqueness*: If not, otherwise, let there exist $x^*, x^{**}$ in $\mathbb{R}^3$ such that $(x^*, y) \in U, f((x^*, y)) = 0, (x^{**}, y) \in U, f((x^{**}, y)) = 0, x^* \neq x^{**}$. We have to arrive at a contradiction. Here, $x^* \neq x^{**}$, so $(x^*, y) \neq (x^{**}, y)$. Since $(x^*, y) \neq (x^{**}, y), (x^*, y) \in U$, $(x^{**}, y) \in U$, and $F$ is 1–1 on $U$, $(0, y) = (f(x^*, y), y) = F((x^*, y)) \neq F((x^{**}, y)) = (f((x^{**}, y)), y) = (0, y)$. Thus, $(0, y) \neq (0, y)$, which is a contradiction. Thus, we have shown that if $y \in W$, then there exists a unique $x^*$ in $\mathbb{R}^3$ such that

$$(x^*, y) \in U, f((x^*, y)) = 0.$$

Let us denote $x^*$ by $g(y)$. Thus, $g : W \to \mathbb{R}^3$ such that for every $y$ in $W$,

$$(g(y), y) \in U, f((g(y), y)) = 0.$$

This proves 1, 2.

Further, since $b \in W$, $(g(b), b) \in U$, and $f((a, b)) = 0 = f((g(b), b))$. Since $f((a, b)) = f((g(b), b))$, $F((a, b)) = (f((a, b)), b) = (f((g(b), b)), b) = F((g(b), b))$. Since $F((a, b)) = F((g(b), b))$, $(a, b) \in U$, $(g(b), b) \in U$, and $F$ is 1–1 on $U$, $(a, b) = (g(b), b)$, and hence, $g(b) = a$.

This proves 3.

4: Let $g_1 : W \to \mathbb{R}$, $g_2 : W \to \mathbb{R}$, $g_3 : W \to \mathbb{R}$ be the component functions of $g$, that is, $g((y_1, y_2)) = (g_1((y_1, y_2)), g_2((y_1, y_2)), g_3((y_1, y_2)))$ for every $(y_1, y_2)$ in $W(\subset \mathbb{R}^2)$. Let

$$h_1 : F(U) \to \mathbb{R}, \ h_2 : F(U) \to \mathbb{R}, \ h_3 : F(U) \to \mathbb{R}, h_4 : F(U) \to \mathbb{R}, \ h_5 : F(U) \\ \to \mathbb{R}$$

be the component functions of $F^{-1}$, that is,

$$\left(F^{-1}\right)((x, y)) = (h_1((x, y)), h_2((x, y)), h_3((x, y)), h_4((x, y)), h_5((x, y)))$$

for every $(x, y)$ in $F(U)(\subset \mathbb{R}^{3+2})$.

Since the function $F^{-1}$ from $F(U)$ onto $U$ is $C^1$, each of the 25 first-order partial derivatives $D_j h_i : F(U) \to \mathbb{R}$ exists and is continuous. We have to prove that $g$ is a

$C^1$ function, that is, each of the six first-order partial derivatives $D_j g_i : W \to \mathbb{R}$ exists and is continuous. Here, we will try to prove that $D_2 g_1 : W \to \mathbb{R}$ exists. For this purpose, let us fix any $y^* \equiv (y_1^*, y_2^*)$ in $W$. We will try to prove that $(D_2 g_1)((y_1^*, y_2^*))$ exists, that is, $\lim_{t \to 0} \frac{g_1((y_1^*, y_2^* + t)) - g_1((y_1^*, y_2^*))}{t}$ exists.

Since $y^*$ is in $W$, $(g(y^*), y^*) \in U$, $0 = f((g(y^*), y^*))$, and hence,

$$(0, y^*) = (f((g(y^*), y^*)), y^*) = F((g(y^*), y^*)) \in F(U).$$

Since $(0, y^*) \in F(U)$, and $D_5 h_1 : F(U) \to \mathbb{R}$ exists, $(D_5 h_1)((0, y^*))$ exists. Since $F^{-1}$ is a 1–1 function from $F(U)$ onto $U$, and for every $(x, y)$ in $U$, where $x \equiv (x_1, x_2, x_3)$ is in $\mathbb{R}^3$ and $y \equiv (y_1, y_2)$ is in $\mathbb{R}^2$,

$$F((x, y)) = (f((x, y)), y),$$
$$(x, y) = (F^{-1})((f((x, y)), y)).$$

Further, since for every $y \equiv (y_1, y_2)$ in $W$, we have $(g(y), y) \in U$, $0 = f((g(y), y))$,

$$(g_1((y_1, y_2)), g_2((y_1, y_2)), g_3((y_1, y_2)), y_1, y_2)$$
$$= ((g_1((y_1, y_2)), g_2((y_1, y_2)), g_3((y_1, y_2))), (y_1, y_2))$$
$$= (g(y), y) = (F^{-1})((f((g(y), y)), y)) = (F^{-1})((0, y))$$
$$= (h_1((0, y)), h_2((0, y)), h_3((0, y)), h_4((0, y)), h_5((0, y))).$$

Hence, for every $y \equiv (y_1, y_2)$ in $W$,

$$g_1((y_1, y_2)) = h_1((0, y)) = h_1((0, 0, 0, y_1, y_2)).$$

Now,

$$\lim_{t \to 0} \frac{g_1((y_1^*, y_2^* + t)) - g_1((y_1^*, y_2^*))}{t} = \lim_{t \to 0} \frac{h_1((0, 0, 0, y_1^*, y_2^* + t)) - h_1((0, 0, 0, y_1^*, y_2^*))}{t}$$
$$= (D_5 h_1)(((0, 0, 0), (y_1^*, y_2^*))) = (D_5 h_1)((0, y^*)).$$

This proves that $D_2 g_1 : W \to \mathbb{R}$ exists. We have seen that for every $y^*$ in $W$,

$$(D_2 g_1)((y^*)) = \lim_{t \to 0} \frac{g_1((y_1^*, y_2^* + t)) - g_1((y_1^*, y_2^*))}{t} = (D_5 h_1)((0, y^*)).$$

Further, since $y^* \mapsto (0, y^*)$ and $(x, y) \mapsto (D_5 h_1)((x, y))$ are continuous functions, their composite function $y^* \mapsto (D_5 h_1)((0, y^*)) (= (D_2 g_1)((y^*)))$ is continuous. Thus, we have shown that $D_2 g_1 : W \to \mathbb{R}$ exists and is continuous. Similarly, all other $D_j g_i : W \to \mathbb{R}$ exist and are continuous.

This proves 4.

5: Let us fix any $y^* \equiv (y_1^*, y_2^*)$ in $W$. Since $y^* = (y_1^*, y_2^*)$ is in $W$, $(g(y^*), y^*) \in U, f((g(y^*), y^*)) = 0$. Let us define a function

$$h_{y^*} : W \to \mathbb{R}^{3+2}$$

as follows: For every $y \equiv (y_1, y_2)$ in $W$,

$$h_{y^*}(y) \equiv (g(y), y) = (g_1((y_1, y_2)), g_2((y_1, y_2)), g_3((y_1, y_2)), y_1, y_2).$$

If $y$ is in $W$, then $(g(y), y) \in U, f((g(y), y)) = 0$. Hence, $h_{y^*} : W \to U(\subset E)$. Here, component functions of $h_{y^*}$ are $y \mapsto g_1(y), y \mapsto g_2(y), y \mapsto g_3(y), y \mapsto y_1,$ $y \mapsto y_2$. Since $g$ is a $C^1$ function, $y \mapsto g_1(y), y \mapsto g_2(y), y \mapsto g_3(y)$ are continuously differentiable functions. Further, $y \mapsto y_1, y \mapsto y_2$ are smooth functions. Thus, we see that all component functions of $h_{y^*}$ are continuously differentiable functions, and hence, $h_{y^*}$ is a $C^1$ function on $W$. Since $h_{y^*}$ is a $C^1$ function on $W$, and $y^*$ is in $W$, $h_{y^*}$ is differentiable at $y^*$. Since $f$ is a $C^1$ function on $E$, and $h_{y^*}(y^*) = (g(y^*), y^*) \in U \subset E, f$ is differentiable at $h_{y^*}(y^*)$. For every $y$ in $W$, we have $(g(y), y) \in U$, and

$$f(h_{y^*}(y)) = f((g(y), y)) = 0.$$

Let us take any $y$ in $W$. Since $h_{y^*}$ is a $C^1$ function on $W$, and $y$ is in $W$, $h_{y^*}$ is differentiable at $y$. Since $y$ is in $W$, $h_{y^*} : W \to U(\subset E)$, $h_{y^*}(y)$ is in $E$. Since $h_{y^*}(y)$ is in $E$, and $f$ is a $C^1$ function, $f$ is differentiable at $h_{y^*}(y)$. Now, since $y$ is in $W$, $f$ is differentiable at $h_{y^*}(y)$, and $h_{y^*}$ is differentiable at $y$, by the chain rule of derivative,

$$(f'((g(y), y))) \left( (h_{y^*})'(y) \right) = (f'(h_{y^*}(y))) \left( (h_{y^*})'(y) \right) = 0.$$

Hence,

$$
\begin{aligned}
(0,0,0) = 0((1,0)) &= \left( (f'((g(y), y))) \left( (h_{y^*})'(y) \right) \right)((1,0)) \\
&= (f'((g(y), y))) \left( \left( (h_{y^*})'(y) \right)((1,0)) \right) \\
&= (f'((g(y), y)))(((D_1 g_1)(y), (D_1 g_2)(y), (D_1 g_3)(y), 1, 0)) \\
&= ((D_1 g_1)(y))((f'((g(y), y)))((1, 0, 0, 0, 0))) \\
&\quad + ((D_1 g_2)(y))((f'((g(y), y)))((0, 1, 0, 0, 0))) \\
&\quad + ((D_1 g_3)(y))((f'((g(y), y)))((0, 0, 1, 0, 0))) \\
&\quad + 1(f'((g(y), y)))((0, 0, 0, 1, 0)) + 0 \\
&= ((D_1 g_1)(y))((D_1 f_1)((g(y), y)), (D_1 f_2)((g(y), y)), (D_1 f_3)((g(y), y))) \\
&\quad + ((D_1 g_2)(y))((D_2 f_1)((g(y), y)), (D_2 f_2)((g(y), y)), (D_2 f_3)((g(y), y))) \\
&\quad + ((D_1 g_3)(y))((D_3 f_1)((g(y), y)), (D_3 f_2)((g(y), y)), (D_3 f_3)((g(y), y))) \\
&\quad + ((D_4 f_1)((g(y), y)), (D_4 f_2)((g(y), y)), (D_4 f_3)((g(y), y)))
\end{aligned}
$$

or,

$$(0,0,0) = ((D_1g_1)(y))((D_1f_1)((g(y),y)), (D_1f_2)((g(y),y)), (D_1f_3)((g(y),y)))$$
$$+ ((D_1g_2)(y))((D_2f_1)((g(y),y)), (D_2f_2)((g(y),y)), (D_2f_3)((g(y),y)))$$
$$+ ((D_1g_3)(y))((D_3f_1)((g(y),y)), (D_3f_2)((g(y),y)), (D_3f_3)((g(y),y)))$$
$$+ ((D_4f_1)((g(y),y)), (D_4f_2)((g(y),y)), (D_4f_3)((g(y),y))).$$

So, for every $i = 1,2,3$, and for every $y$ in $W$,

$$0 = ((D_1g_1)(y))((D_1f_i)((g(y),y))) + ((D_1g_2)(y))((D_2f_i)((g(y),y)))$$
$$+ ((D_1g_3)(y))((D_3f_i)((g(y),y))) + (D_4f_i)((g(y),y)).$$

Now, since $y^*$ is in $W$, for every $i = 1,2,3$,

$$0 = ((D_1g_1)(y^*))((D_1f_i)((g(y^*),y^*))) + ((D_1g_2)(y^*))((D_2f_i)((g(y^*),y^*)))$$
$$+ ((D_1g_3)(y^*))((D_3f_i)((g(y^*),y^*))) + (D_4f_i)((g(y^*),y^*)) \cdots (*).$$

Next, we will prove that

$$g'(y^*) = -\left((f'((g(y^*),y^*)))_x\right)^{-1}(f'((g(y^*),y^*)))_y$$

that is,

$$((f'((g(y^*),y^*)))_x)(g'(y^*)) = -(f'((g(y^*),y^*)))_y$$

that is,

(i)  $(((f'((g(y^*),y^*)))_x)(g'(y^*)))((1,0)) = (-(f'((g(y^*),y^*)))_y)((1,0)),$

and

(ii)  $(((f'((g(y^*),y^*)))_x)(g'(y^*)))((0,1)) = (-(f'((g(y^*),y^*)))_y)((0,1)).$
For (i):

LHS $= (((f'((g(y^*),y^*)))_x)(g'(y^*)))((1,0)) = ((f'((g(y^*),y^*)))_x)((g'(y^*))((1,0)))$
$= ((f'((g(y^*),y^*)))_x)(((D_1g_1)(y^*),(D_1g_2)(y^*),(D_1g_3)(y^*)))$
$= (f'((g(y^*),y^*)))(((D_1g_1)(y^*),(D_1g_2)(y^*),(D_1g_3)(y^*),0,0))$
$= ((D_1g_1)(y^*))((D_1f_1)((g(y^*),y^*)),(D_1f_2)((g(y^*),y^*)),(D_1f_3)((g(y^*),y^*)))$
$\quad + ((D_1g_2)(y^*))((D_2f_1)((g(y^*),y^*)),(D_2f_2)((g(y^*),y^*)),(D_2f_3)((g(y^*),y^*)))$
$\quad + ((D_1g_3)(y^*))((D_3f_1)((g(y^*),y^*)),(D_3f_2)((g(y^*),y^*)),(D_3f_3)((g(y^*),y^*)))$
$= (-(D_4f_1)((g(y^*),y^*)), -(D_4f_2)((g(y^*),y^*)), -(D_4f_3)((g(y^*),y^*))),$

by $(*)$. Now,

$$
\begin{aligned}
\text{RHS} &= \Big(-(f'((g(y^*),y^*)))_y\Big)((1,0)) = -\Big((f'((g(y^*),y^*)))_y((1,0))\Big) \\
&= -((f'((g(y^*),y^*)))((0,0,01,0))) \\
&= -((D_4f_1)((g(y^*),y^*)),(D_4f_2)((g(y^*),y^*)),(D_4f_3)((g(y^*),y^*))) \\
&= (-(D_4f_1)((g(y^*),y^*)),-(D_4f_2)((g(y^*),y^*)),-(D_4f_3)((g(y^*),y^*))).
\end{aligned}
$$

This proves (i). Similarly, (ii) holds.

This proves 5.

6: Since $b \in W$, and $g(b) = a$, by 5,

$$
g'(b) = -\big((f'((g(b),b)))_x\big)^{-1}(f'((g(b),b)))_y = -\big((f'((a,b)))_x\big)^{-1}(f'((a,b)))_y.
$$

This proves 6.

7: Let $f$ be a $C^2$ function. We have to prove that $g : W \to \mathbb{R}^3$ is a $C^2$ function, that is, each of the 12 second-order partial derivatives $D_{jk}g_i : W \to \mathbb{R}$ exists and is continuous. We will try to prove that $D_{12}g_3 : W \to \mathbb{R}$ exists and is continuous. Let us take any $y$ in $W$.

By 5,

$$
g'(y) = -\big((f'((g(y),y)))_x\big)^{-1}(f'((g(y),y)))_y
$$

so,

$$
\begin{aligned}
((D_1g_1)(y),(D_1g_2)(y),(D_1g_3)(y)) &= (g'(y))((1,0)) \\
&= \Big(-((f'((g(y),y)))_x)^{-1}(f'((g(y),y)))_y\Big)((1,0)) \\
&= -\Big(((f'((g(y),y)))_x)^{-1}\Big)\Big((f'((g(y),y)))_y((1,0))\Big) \\
&= -\Big(((f'((g(y),y)))_x)^{-1}\Big)((f'((g(y),y)))((0,0,0,1,0))) \\
&= -\Big(((f'((g(y),y)))_x)^{-1}\Big)((D_4f_1)((g(y),y)),(D_4f_2)((g(y),y)),(D_4f_3)((g(y),y))) \\
&= -((D_4f_1)((g(y),y)))\Big(\big(((f'((g(y),y)))_x)^{-1}\big)(1,0,0)\Big) \\
&\quad - ((D_4f_2)((g(y),y)))\Big(\big(((f'((g(y),y)))_x)^{-1}\big)(0,1,0)\Big) \\
&\quad - ((D_4f_3)((g(y),y)))\Big(\big(((f'((g(y),y)))_x)^{-1}\big)(0,0,1)\Big) \\
&= -((D_4f_1)((g(y),y)))\frac{1}{\Delta}(A_{11},A_{12},A_{13}) - ((D_4f_2)((g(y),y)))\frac{1}{\Delta}(A_{21},A_{22},A_{23}) \\
&\quad - ((D_4f_3)((g(y),y)))\frac{1}{\Delta}(A_{31},A_{32},A_{33}),
\end{aligned}
$$

where

$$\Delta \equiv \begin{vmatrix} (D_1f_1)((g(y),y)) & (D_2f_1)((g(y),y)) & (D_3f_1)((g(y),y)) \\ (D_1f_2)((g(y),y)) & (D_2f_2)((g(y),y)) & (D_3f_2)((g(y),y)) \\ (D_1f_3)((g(y),y)) & (D_2f_3)((g(y),y)) & (D_3f_3)((g(y),y)) \end{vmatrix},$$

and $A_{ij}$ denotes the cofactor of $(D_jf_i)((g(y),y))$ in the above determinant. Hence,

$$(D_1g_3)(y) = -\frac{1}{\Delta}(((D_4f_1)((g(y),y)))A_{13} + ((D_4f_2)((g(y),y)))A_{23} + ((D_4f_3)((g(y),y)))A_{33}).$$

Now, since $f$ is a $C^2$ function, and $g$ is a $C^1$ function, the second-order partial derivative $(D_{12}g_3)(y)$ exists and is continuous. Similarly, all other $(D_{jk}g_i)(y)$ exist and are continuous. This proves that $g$ is a $C^2$ function.  □

**Note 3.44** The result similar to Theorem 3.43 can be proved as above for $\mathbb{R}^n$ in place of $\mathbb{R}^3$ and $\mathbb{R}^m$ in place of $\mathbb{R}^2$. This theorem is known as the *implicit function theorem*. As above, we can prove the following theorem:

**Theorem 3.45** Let $E$ be an open subset of $\mathbb{R}^{3+2}$. Let $f : E \to \mathbb{R}^2$. Let $a \in \mathbb{R}^3, b \in \mathbb{R}^2$, and $(a, b) \in E$. If

(i)  $f$ is a $C^1$ function,
(ii)  $(f'((a,b)))_y$ is invertible,
(iii)  $f((a,b)) = 0$,
   then there exist an open neighborhood $U(\subset E)$ of $(a, b)$ in $\mathbb{R}^{3+2}$, an open neighborhood $W$ of $a$ in $\mathbb{R}^3$, and a function

$$g : W \to \mathbb{R}^2$$

such that

1. for every $x$ in $W$, $(x, g(x)) \in U$,
2. for every $x$ in $W$, $f(x, g(x)) = 0$,
3. $g(a) = b$,
4. $g$ is a $C^1$ function,
5. for every $x^*$ in $W$, $g'(x^*) = -((f'((x^*, g(x^*))))_y)^{-1}(f'((x^*, g(x^*))))_x$
6. $g'(a) = -((f'((a,b)))_y)^{-1}(f'((a,b)))_x$
7. if $f$ is a $C^2$ function, then $g$ is a $C^2$ function, etc.

## 3.5  Constant Rank Theorem (Easy Version)

**Definition** Let $\iota_1 : \mathbb{R}^m \to \mathbb{R}^m \times \mathbb{R}^n$ be the function defined as follows: For every $x$ in $\mathbb{R}^m$

$$\iota_1(x) \equiv (x, 0).$$

Here, $\iota_1$ is called the *first injection of factor into product*. Similarly, by the *second injection* $\iota_2$ *of factor into product*, we mean the function $\iota_2 : \mathbb{R}^n \to \mathbb{R}^m \times \mathbb{R}^n$ defined as follows: For every $y$ in $\mathbb{R}^n$

$$\iota_2(y) \equiv (0, y).$$

Let $\pi_1 : \mathbb{R}^m \times \mathbb{R}^n \to \mathbb{R}^m$ be the function defined as follows: For every $x$ in $\mathbb{R}^m$ and for every $y$ in $\mathbb{R}^n$

$$\pi_1(x, y) \equiv x.$$

Here, $\pi_1$ is called the *first projection from product onto factor*. Similarly, by the *second projection* $\pi_2$ *from product onto factor*, we mean the function $\pi_2 : \mathbb{R}^m \times \mathbb{R}^n \to \mathbb{R}^n$ defined as follows: For every $x$ in $\mathbb{R}^m$ and for every $y$ in $\mathbb{R}^n$

$$\pi_2(x, y) \equiv y.$$

**Theorem 3.46** *Let $U$ be an open neighborhood of $(a; b)(\equiv (a_1, a_2, a_3; b_1, b_2))$ in $\mathbb{R}^3 \times \mathbb{R}^2$. Let $f : U \to \mathbb{R}^2$ be a smooth function. Let $f(a_1, a_2, a_3; b_1, b_2) = (0, 0)$. If*

$$(f'(a_1, a_2, a_3; b_1, b_2)) \circ \iota_2 : \mathbb{R}^2 \to \mathbb{R}^2$$

*is a linear isomorphism, then*

1. *there exists a local diffeomorphism $\varphi$ at $(a_1, a_2, a_3; b_1, b_2)$ in $\mathbb{R}^3 \times \mathbb{R}^2$ such that $f \circ \varphi = \pi_2$.*
   *(i.e., there exist an open neighborhood $V$ of $(a_1, a_2, a_3)$ in $\mathbb{R}^3$, an open neighborhood $W$ of $(b_1, b_2)$ in $\mathbb{R}^2$, an open neighborhood $U_1$ of $(a_1, a_2, a_3; b_1, b_2)$ in $\mathbb{R}^3 \times \mathbb{R}^2$, and a function $\varphi : U_1 \to V \times W$ such that $V \times W \subset U, \varphi$ is a diffeomorphism, and for every $(x_1, x_2, x_3; y_1, y_2)$ in $U_1$, $f(\varphi(x_1, x_2, x_3; y_1, y_2)) = (y_1, y_2).$)*
2. *there exist an open neighborhood $W_1$ of $(a_1, a_2, a_3)$ in $\mathbb{R}^3$, an open neighborhood $W_2$ of $(b_1, b_2)$ in $\mathbb{R}^2$, and a smooth function $(g_1, g_2) \equiv g : W_1 \to \mathbb{R}^2$ such that*

   (a) *for every $(x_1, x_2, x_3)$ in $W_1$, $(x_1, x_2, x_3; g_1(x_1, x_2, x_3), g_2(x_1, x_2, x_3)) \in U$,*
   (b) *for every $(x_1, x_2, x_3)$ in $W_1, f(x_1, x_2, x_3; g_1(x_1, x_2, x_3), g_2(x_1, x_2, x_3)) = (0, 0)$,*

(c) *g is unique, (i.e., if $(h_1, h_2) \equiv h : W_1 \to \mathbb{R}^2$ is a smooth function such that for every $(x_1, x_2, x_3)$ in $W_1$, $(x_1, x_2, x_3; h_1(x_1, x_2, x_3), h_2(x_1, x_2, x_3)) \in U$, and $f(x_1, x_2, x_3; h_1(x_1, x_2, x_3), h_2(x_1, x_2, x_3)) = (0, 0)$, then $g = h$.)*

(d) *if $\alpha : W_1 \to f^{-1}(0, 0)$ is the function defined such that for every $(x_1, x_2, x_3)$ in $W_1$*

$$\alpha(x_1, x_2, x_3) \equiv (x_1, x_2, x_3; g_1(x_1, x_2, x_3), g_2(x_1, x_2, x_3)),$$

*then*

(i)    *$\alpha(W_1)$ is an open neighborhood of $(a; b)$ in $f^{-1}(0, 0)$,*

(ii)   *$\alpha$ is a homeomorphism from $W_1$ onto $\alpha(W_1)$,*

(iii) *$\alpha = \varphi \circ \tau_1$ and $\alpha^{-1} = \pi_1$ on their common domain,*

(iv) *$\alpha, \alpha^{-1}$ are smooth functions.*

*Proof* Let $f \equiv (f_1, f_2)$. Since $(f'(a_1, a_2, a_3; b_1, b_2)) \circ \iota_2 : \mathbb{R}^2 \to \mathbb{R}^2$ is an isomorphism, and for every $(y_1, y_2)$ in $\mathbb{R}^2$,

$$((f'(a_1, a_2, a_3; b_1, b_2)) \circ \iota_2)(y_1, y_2) = (f'(a_1, a_2, a_3; b_1, b_2))(\iota_2(y_1, y_2))$$
$$= (f'(a_1, a_2, a_3; b_1, b_2))(0, 0, 0; y_1, y_2)$$

$$= \begin{bmatrix} (D_1 f_1)(a, b)(D_2 f_1)(a, b)(D_3 f_1)(a, b)(D_4 f_1)(a, b)(D_5 f_1)(a, b) \\ (D_1 f_2)(a, b)(D_2 f_2)(a, b)(D_3 f_2)(a, b)(D_4 f_2)(a, b)(D_5 f_2)(a, b) \end{bmatrix} \begin{bmatrix} 0 \\ 0 \\ 0 \\ y_1 \\ y_2 \end{bmatrix}$$

$$= \begin{bmatrix} ((D_4 f_1)(a, b))y_1 + ((D_5 f_1)(a, b))y_2 \\ ((D_4 f_2)(a, b))y_1 + ((D_5 f_2)(a, b))y_2 \end{bmatrix} = \begin{bmatrix} (D_4 f_1)(a, b)(D_5 f_1)(a, b) \\ (D_4 f_2)(a, b)(D_5 f_2)(a, b) \end{bmatrix} \begin{bmatrix} y_1 \\ y_2 \end{bmatrix},$$

it follows that

$$\det \begin{bmatrix} (D_4 f_1)(a, b) \ (D_5 f_1)(a, b) \\ (D_4 f_2)(a, b) \ (D_5 f_2)(a, b) \end{bmatrix} \neq 0.$$

Let us define a function $\psi : U \to \mathbb{R}^3 \times \mathbb{R}^2$ as follows: For every $(x, y)$ in $U$ where $x \equiv (x_1, x_2, x_3)$ is in $\mathbb{R}^3$ and $y \equiv (y_1, y_2)$ is in $\mathbb{R}^2$,

$$\psi(x_1, x_2, x_3; y_1, y_2) = \psi(x, y) \equiv (x, f(x, y))$$
$$= (x_1, x_2, x_3, f_1(x_1, x_2, x_3; y_1, y_2), f_2(x_1, x_2, x_3; y_1, y_2)).$$

Now, since $f : U \to \mathbb{R}^2$ is a smooth function, $\psi : U \to \mathbb{R}^3 \times \mathbb{R}^2$ is also a smooth function. Further, since

$$\det(\psi'(a_1, a_2, a_3; b_1, b_2)) = \det \begin{bmatrix} 1 & 0 & 0 & 0 & & 0 \\ 0 & 1 & 0 & 0 & & 0 \\ 0 & 0 & 1 & 0 & & 0 \\ 0 & 0 & 0 & (D_4f_1)(a,b) & (D_5f_1)(a,b) \\ 0 & 0 & 0 & (D_4f_2)(a,b) & (D_5f_2)(a,b) \end{bmatrix}$$

$$= \det \begin{bmatrix} (D_4f_1)(a,b) & (D_5f_1)(a,b) \\ (D_4f_2)(a,b) & (D_5f_2)(a,b) \end{bmatrix}$$

is nonzero, $\psi'(a_1, a_2, a_3; b_1, b_2)$ is invertible. Hence, by the inverse function theorem, there exist an open neighborhood $V$ of $(a_1, a_2, a_3)$ in $\mathbb{R}^3$, an open neighborhood $W$ of $(b_1, b_2)$ in $\mathbb{R}^2$, and an open subset $U_1$ of $\mathbb{R}^3 \times \mathbb{R}^2$ such that $V \times W \subset U$ and $\psi$ acts as a diffeomorphism from $V \times W$ onto $U_1$. Hence, $\psi^{-1} : U_1 \to V \times W$ is a diffeomorphism. Put $\varphi \equiv \psi^{-1}$. Clearly, $\varphi : U_1 \to V \times W$ is a diffeomorphism. Now, for every $x$ in $V$, and for every $y$ in $W$, we have

$$(\pi_2 \circ \psi)(x, y) = \pi_2(\psi(x, y)) = \pi_2((x, f(x, y))) = f(x, y).$$

So, $\pi_2 \circ \psi = f$, and hence, $f \circ \varphi = (\pi_2 \circ \psi) \circ \varphi = \pi_2 \circ (\psi \circ \varphi) = \pi_2 \circ (\psi \circ \psi^{-1}) = \pi_2$.

This proves 1.

Since $(f'(a_1, a_2, a_3; b_1, b_2)) \circ \iota_2 : \mathbb{R}^2 \to \mathbb{R}^2$ is an isomorphism,

$$\det \left[ (f'(a_1, a_2, a_3; b_1, b_2))_y \right] = \det \begin{bmatrix} (D_4f_1)(a,b) & (D_5f_1)(a,b) \\ (D_4f_2)(a,b) & (D_5f_2)(a,b) \end{bmatrix} \neq 0,$$

and hence, $(f'(a_1, a_2, a_3; b_1, b_2))_y$ is invertible. Now, by the implicit function theorem, there exist an open neighborhood $W_1$ of $(a_1, a_2, a_3)$ in $\mathbb{R}^3$, an open neighborhood $W_2$ of $(b_1, b_2)$ in $\mathbb{R}^2$, and a smooth function $(g_1, g_2) \equiv g : W_1 \to \mathbb{R}^2$ such that $W_1 \times W_2 \subset U$ and for every $(x_1, x_2, x_3)$ in $W_1$,

$$(x_1, x_2, x_3; g_1(x_1, x_2, x_3), g_2(x_1, x_2, x_3)) \in W_1 \times W_2,$$

and

$$f(x_1, x_2, x_3; g_1(x_1, x_2, x_3), g_2(x_1, x_2, x_3)) = (0, 0).$$

This proves 2(a) and 2(b).

Next, let $h : W_1 \to \mathbb{R}^2$ be a smooth function such that for every $x \equiv (x_1, x_2, x_3)$ in $W_1$, $(x_1, x_2, x_3; h_1(x_1, x_2, x_3), h_2(x_1, x_2, x_3)) \in U$, and $f(x_1, x_2, x_3; h_1(x_1, x_2, x_3),$

$h_2(x_1, x_2, x_3)) = (0, 0)$, where $h \equiv (h_1, h_2)$. We have to prove that $g = h$, that is, for every $(x_1, x_2, x_3)$ in $W_1$,

$$g(x_1, x_2, x_3) = h(x_1, x_2, x_3).$$

Since

$$(f_1(x_1, x_2, x_3; g_1(x_1, x_2, x_3), g_2(x_1, x_2, x_3)), f_2(x_1, x_2, x_3; g_1(x_1, x_2, x_3), g_2(x_1, x_2, x_3)))$$
$$= f(x_1, x_2, x_3; g_1(x_1, x_2, x_3), g_2(x_1, x_2, x_3)) = (0, 0)$$

so,

$$\psi(x_1, x_2, x_3; g_1(x_1, x_2, x_3), g_2(x_1, x_2, x_3))$$
$$= (x_1, x_2, x_3, f_1(x_1, x_2, x_3; g_1(x_1, x_2, x_3), g_2(x_1, x_2, x_3)),$$
$$f_2(x_1, x_2, x_3; g_1(x_1, x_2, x_3), g_2(x_1, x_2, x_3)))$$
$$= (x_1, x_2, x_3, 0, 0),$$

and hence,

$$(x_1, x_2, x_3; g_1(x_1, x_2, x_3), g_2(x_1, x_2, x_3)) = \psi^{-1}(x_1, x_2, x_3, 0, 0) = \varphi(x_1, x_2, x_3, 0, 0)$$

for every $(x_1, x_2, x_3)$ in $W_1$. Similarly,

$$(x_1, x_2, x_3; h_1(x_1, x_2, x_3), h_2(x_1, x_2, x_3)) = \varphi(x_1, x_2, x_3, 0, 0)$$

for every $(x_1, x_2, x_3)$ in $W_1$. It follows that for every $x \equiv (x_1, x_2, x_3)$ in $W_1$,

$$(x, g(x)) = (x_1, x_2, x_3; g_1(x_1, x_2, x_3), g_2(x_1, x_2, x_3))$$
$$= (x_1, x_2, x_3; h_1(x_1, x_2, x_3), h_2(x_1, x_2, x_3)) = (x, h(x)).$$

Hence, $g = h$.

This proves 2(c).

(d) (i): For every $(x_1, x_2, x_3)$ in $W_1$, $f(x_1, x_2, x_3; g_1(x_1, x_2, x_3), g_2(x_1, x_2, x_3)) = (0, 0)$, so

$$\alpha(x_1, x_2, x_3) = (x_1, x_2, x_3; g_1(x_1, x_2, x_3), g_2(x_1, x_2, x_3)) \in f^{-1}(0, 0).$$

Hence, $\alpha : W_1 \to f^{-1}(0, 0)$ is a function. Since $(a_1, a_2, a_3)$ is in $W_1$,

$$\alpha(a_1, a_2, a_3) = (a_1, a_2, a_3; g_1(a_1, a_2, a_3), g_2(a_1, a_2, a_3)) = \varphi(a_1, a_2, a_3, 0, 0).$$

Further, since

$$\psi(a_1, a_2, a_3; b_1, b_2) = (a_1, a_2, a_3, f_1(a_1, a_2, a_3; b_1, b_2), f_2(a_1, a_2, a_3; b_1, b_2))$$
$$= (a_1, a_2, a_3; 0, 0)$$

so

$$(a, b) = (a_1, a_2, a_3; b_1, b_2) = \varphi(a_1, a_2, a_3, 0, 0) = \alpha(a_1, a_2, a_3) \in \alpha(W_1).$$

Now, we want to show that $\alpha(W_1)$ is an open set.
Let us observe that

$$\alpha(W_1) = \varphi(W_1 \times \mathbb{R}^2) \cap f^{-1}(0, 0).$$

(*Reason*: For this purpose, let us take any $\alpha(x_1, x_2, x_3) \in$ LHS, where $(x_1, x_2, x_3)$ is in $W_1$. Here,

$$\alpha(x_1, x_2, x_3) = (x_1, x_2, x_3; g_1(x_1, x_2, x_3), g_2(x_1, x_2, x_3)) = \varphi(x_1, x_2, x_3, 0, 0)$$
$$\in \varphi(W_1 \times \mathbb{R}^2).$$

Also, since $f(\alpha(x_1, x_2, x_3)) = f(x_1, x_2, x_3; g_1(x_1, x_2, x_3), g_2(x_1, x_2, x_3)) = (0, 0)$, $\alpha(x_1, x_2, x_3) \in f^{-1}(0, 0)$. It follows that $\alpha(x_1, x_2, x_3) \in \varphi(W_1 \times \mathbb{R}^2) \cap f^{-1}(0, 0) =$ RHS. Hence, LHS $\subset$ RHS.

Next, let us take any $(x_1, x_2, x_3; y_1, y_2) \in$ RHS $= \varphi(W_1 \times \mathbb{R}^2) \cap f^{-1}(0, 0)$. We have to prove that $(x_1, x_2, x_3; y_1, y_2) \in \alpha(W_1)$. Since $(x_1, x_2, x_3; y_1, y_2) \in \varphi(W_1 \times \mathbb{R}^2) \cap f^{-1}(0, 0)$, $(x_1, x_2, x_3; y_1, y_2) \in U$, $(x_1, x_2, x_3; y_1, y_2) \in V \times W$,

$$(f_1(x_1, x_2, x_3; y_1, y_2), f_2(x_1, x_2, x_3; y_1, y_2)) = f(x_1, x_2, x_3; y_1, y_2) = (0, 0),$$

and there exists $(w_1, w_2, w_3; t_1, t_2) \in W_1 \times \mathbb{R}^2$ such that $\varphi(w_1, w_2, w_3; t_1, t_2) = (x_1, x_2, x_3; y_1, y_2)$. Thus,

$$(w_1, w_2, w_3; t_1, t_2) = \psi(x_1, x_2, x_3; y_1, y_2)$$
$$= (x_1, x_2, x_3, f_1(x_1, x_2, x_3; y_1, y_2), f_2(x_1, x_2, x_3; y_1, y_2))$$
$$= (x_1, x_2, x_3; 0, 0).$$

Hence, $(x_1, x_2, x_3) = (w_1, w_2, w_3) \in W_1$ .Also

$$(x_1, x_2, x_3; y_1, y_2) = \varphi(w_1, w_2, w_3; t_1, t_2) = \varphi(x_1, x_2, x_3; 0, 0)$$
$$= (x_1, x_2, x_3; g_1(x_1, x_2, x_3), g_2(x_1, x_2, x_3)) = \alpha(x_1, x_2, x_3) \in \alpha(W_1).$$

This proves that $\alpha(W_1) = \varphi(W_1 \times \mathbb{R}^2) \cap f^{-1}(0, 0)$.)

Since $W_1$ is an open subset of $\mathbb{R}^3$, $W_1 \times \mathbb{R}^2$ is an open subset of $\mathbb{R}^3 \times \mathbb{R}^2$. Further, since $\varphi : U_1 \to V \times W$ is a diffeomorphism, and $W_1 \times \mathbb{R}^2$ is an open subset of $\mathbb{R}^3 \times \mathbb{R}^2$, $\varphi(W_1 \times \mathbb{R}^2)$ is an open subset of $\mathbb{R}^3 \times \mathbb{R}^2$, and hence, $\alpha(W_1)(= \varphi(W_1 \times \mathbb{R}^2) \cap f^{-1}(0,0))$ is open in $f^{-1}(0,0)$. This proves (d)(i).

(d)(ii): Here, we must prove that

(A) $\alpha : W_1 \to \alpha(W_1)$ is 1–1,

(B) $\alpha$ is continuous,

(C) $\alpha^{-1} : \alpha(W_1) \to W_1$ is continuous. For (A): Let $\alpha(x_1, x_2, x_3) = \alpha(y_1, y_2, y_3)$. We have to prove that $(x_1, x_2, x_3) = (y_1, y_2, y_3)$. Since

$$(x_1, x_2, x_3; g_1(x_1, x_2, x_3), g_2(x_1, x_2, x_3)) = \alpha(x_1, x_2, x_3) = \alpha(y_1, y_2, y_3)$$
$$= (y_1, y_2, y_3; g_1(y_1, y_2, y_3), g_2(y_1, y_2, y_3)),$$

$(x_1, x_2, x_3) = (y_1, y_2, y_3)$.

For (B): Since $g : W_1 \to \mathbb{R}^2$ is a smooth function, and

$$\alpha : (x_1, x_2, x_3) \mapsto (x_1, x_2, x_3; g_1(x_1, x_2, x_3), g_2(x_1, x_2, x_3)),$$

each of the component functions of $\alpha$ is continuous, and hence, $\alpha$ is continuous.

For (C): Since for every $x \equiv (x_1, x_2, x_3) \in W_1$,

$$\alpha^{-1}(x_1, x_2, x_3; g_1(x_1, x_2, x_3), g_2(x_1, x_2, x_3)) = (x_1, x_2, x_3)$$
$$= \pi_1(x_1, x_2, x_3; g_1(x_1, x_2, x_3), g_2(x_1, x_2, x_3)),$$

and $\pi_1$ is continuous, $\alpha^{-1}$ is continuous. This proves (d)(ii).

(d)(iii) Since for every $x \equiv (x_1, x_2, x_3) \in W_1$,

$$\alpha(x_1, x_2, x_3) = \varphi(x_1, x_2, x_3; 0, 0) = \varphi(\tau_1(x_1, x_2, x_3)) = (\varphi \circ \tau_1)(x_1, x_2, x_3),$$

$\alpha = \varphi \circ \tau_1$. Next, since

$$\alpha^{-1}(x_1, x_2, x_3; g_1(x_1, x_2, x_3), g_2(x_1, x_2, x_3)) = (x_1, x_2, x_3)$$
$$= \pi_1(x_1, x_2, x_3; g_1(x_1, x_2, x_3), g_2(x_1, x_2, x_3)),$$

$\alpha^{-1} = \pi_1$ on their common domain. This proves (d)(iii).

(iv) Since $\varphi$ and $\tau_1$ are smooth functions, their composite $\varphi \circ \tau_1 (= \alpha)$ is smooth, and hence, $\alpha$ is smooth. Since $\pi_1(= \alpha^{-1})$ is smooth, $\alpha^{-1}$ is smooth. □

**Theorem 3.47** *Let $U$ be an open neighborhood of $(a; b)$ in $\mathbb{R}^m \times \mathbb{R}^p$. Let $f : U \to \mathbb{R}^p$ be a smooth function. Let $f(a; b) = 0$. If*

$$(f'(a; b)) \circ \iota_2 : \mathbb{R}^p \to \mathbb{R}^p$$

*is a linear isomorphism, then*

1. *there exists a local diffeomorphism $\varphi$ at $(a;b)$ in $\mathbb{R}^m \times \mathbb{R}^p$ such that $f \circ \varphi = \pi_2$. (i.e., there exist an open neighborhood $V$ of $a$ in $\mathbb{R}^m$, an open neighborhood $W$ of $b$ in $\mathbb{R}^p$, an open neighborhood $U_1$ of $(a;b)$ in $\mathbb{R}^m \times \mathbb{R}^p$, and a function $\varphi : U_1 \to V \times W$ such that $V \times W \subset U, \varphi$ is a diffeomorphism, and for every $(x;y)$ in $U_1, f(\varphi(x;y)) = y$.)*
2. *there exist an open neighborhood $W_1$ of $a$ in $\mathbb{R}^m$, an open neighborhood $W_2$ of $b$ in $\mathbb{R}^p$, and a smooth function $g : W_1 \to \mathbb{R}^p$ such that*

   (a) *for every $x$ in $W_1$, $(x;g(x)) \in U$,*
   (b) *for every $x$ in $W_1$, $f(x;g(x)) = 0$,*
   (c) *$g$ is unique, (i.e., if $h : W_1 \to \mathbb{R}^p$ is a smooth function such that for every $x$ in $W_1$, $(x;h(x)) \in U$, and $f(x;h(x)) = 0$, then $g = h$.)*
   (d) *if $\alpha : W_1 \to f^{-1}(0)$ is the function defined such that for every $x$ in $W_1$*

$$\alpha(x) \equiv (x;g(x)),$$

*then*

   (i) *$\alpha(W_1)$ is an open neighborhood of $(a;b)$ in $f^{-1}(0)$,*
   (ii) *$\alpha$ is a homeomorphism from $W_1$ onto $\alpha(W_1)$,*
   (iii) *$\alpha = \varphi \circ \tau_1$ and $\alpha^{-1} = \pi_1$ on their common domain,*
   (iv) *$\alpha, \alpha^{-1}$ are smooth functions.*

*Proof* Its proof is quite similar to the proof of Theorem 3.46. $\square$

**Theorem 3.48** *Let $f : \mathbb{R}^m \times \mathbb{R}^p \to \mathbb{R}^p$ be a smooth function. Let $a$ be in $\mathbb{R}^m$ and $b$ be in $\mathbb{R}^p$. Let $f(a;b) = 0$. If for every $(x,y)$ in $f^{-1}(0)$, the rank of linear transformation $f'(x;y)$ is $p$, then $f^{-1}(0)$ is a smooth manifold of dimension $m$.*

*Proof* Since $f : \mathbb{R}^m \times \mathbb{R}^p \to \mathbb{R}^p$ is a smooth function, $f$ is a continuous function. Since $f : \mathbb{R}^m \times \mathbb{R}^p \to \mathbb{R}^p$ is a continuous function, and $\{0\}$ is a closed subset of $\mathbb{R}^p$, $f^{-1}(0)$ is closed subset of $\mathbb{R}^m \times \mathbb{R}^p$. Since $\mathbb{R}^m \times \mathbb{R}^p$ is homeomorphic to $\mathbb{R}^{m+p}$, and $\mathbb{R}^{m+p}$ is Hausdorff and second countable space, $\mathbb{R}^m \times \mathbb{R}^p$ is Hausdorff and second countable. Since $\mathbb{R}^m \times \mathbb{R}^p$ is Hausdorff and second countable, and $f^{-1}(0)$ is a subspace of $\mathbb{R}^m \times \mathbb{R}^p$, $f^{-1}(0)$ is Hausdorff and second countable.

Let us take any $(a,b)$ in $f^{-1}(0)$. Since $(a,b)$ is in $f^{-1}(0)$, $f(a,b) = 0$. By assumption, the rank of linear transformation $f'(a;b) : \mathbb{R}^m \times \mathbb{R}^p \to \mathbb{R}^p$ is $p$, and hence, $(f'(a;b)) \circ \iota_2 : \mathbb{R}^p \to \mathbb{R}^p$ is a linear isomorphism. Now, by Theorem 3.47, there exist an open neighborhood $W_1$ of $a$ in $\mathbb{R}^m$ and a function $\alpha : W_1 \to f^{-1}(0)$ such that

   (i) $\alpha(W_1)$ is an open neighborhood of $(a;b)$ in $f^{-1}(0)$,
   (ii) $\alpha$ is a homeomorphism from $W_1$ onto $\alpha(W_1)$,
   (iii) $\alpha, \alpha^{-1}$ are smooth functions.

Hence, $\alpha(W_1)$ is an open neighborhood of $(a; b)$ in $f^{-1}(0)$, and $\alpha^{-1} : \alpha(W_1) \to W_1$ is a diffeomorphism. This shows that $f^{-1}(0)$ is a smooth manifold of dimension $m$. $\square$

**Definition** Let $N$ be a smooth manifold of dimension $m + p$. Let $M \subset N$. If for every $x$ in $M$, there exists an admissible coordinate chart $(U, \varphi_U)$ in $N$ satisfying $x \in U$ such that

$$\varphi_U(U \cap M) = (\varphi_U(U)) \cap (\mathbb{R}^m \times \{0\})$$

then we say that $M$ is a smooth $m$-submanifold of $N$.

**Theorem 3.49** *Let $N$ be a smooth manifold of dimension $m + p$. Let $M \subset N$. If $M$ is a smooth $m$-submanifold of $N$, then $M$ is a smooth manifold of dimension $m$.*

*Proof* Let us take any $a$ in $M$. Since $a$ is in $M$, and $M$ is a smooth $m$-submanifold of $N$, there exists an admissible coordinate chart $(U, \varphi_U)$ in $N$ satisfying $a \in U$, and

$$\varphi_U(U \cap M) = (\varphi_U(U)) \cap (\mathbb{R}^m \times \{0\}).$$

Since $a$ is in $M$, and $a \in U$, $a$ is in $U \cap M$. Since $U$ is open in $N$, and $M \subset N$, $U \cap M$ is open in $M$. Since $a$ is in $U \cap M$, and $U \cap M$ is open in $M$, $U \cap M$ is an open neighborhood of $a$ in $M$. Since $(U, \varphi_U)$ is an admissible coordinate chart in the $m + p$-dimensional smooth manifold $N$, $\varphi_U(U)$ is open in $\mathbb{R}^{m+p}(\cong (\mathbb{R}^m \times \mathbb{R}^p) \supset (\mathbb{R}^m \times \{0\}))$, and hence, $\varphi_U(U \cap M)(= (\varphi_U(U)) \cap (\mathbb{R}^m \times \{0\}))$ is open in $\mathbb{R}^m(\cong (\mathbb{R}^m \times \{0\}))$. Since $(U, \varphi_U)$ is an admissible coordinate chart in $N$, $\varphi_U$ is 1–1, and hence, the restriction $\varphi_U|_{U \cap M}$ of $\varphi_U$ to $U \cap M$, is 1–1.

Thus, the restriction $\varphi_U|_{U \cap M}$ of $\varphi_U$ to $U \cap M$ is a 1–1 mapping from the open neighborhood $U \cap M$ of $a$ in $M$ onto the open set $\varphi_U(U \cap M)$ of $\mathbb{R}^m$.

Since $(U, \varphi_U)$ is an admissible coordinate chart in $N$, $\varphi_U$ is continuous, and hence, the restriction $\varphi_U|_{U \cap M}$ of $\varphi_U$ to $U \cap M$ is continuous. Since $(U, \varphi_U)$ is an admissible coordinate chart in $N$, $(\varphi_U)^{-1} : \varphi_U(U) \to U$ is continuous, and hence, the restriction $((\varphi_U)^{-1})|_{\varphi_U(U \cap M)}$ of $(\varphi_U)^{-1}$ to $\varphi_U(U \cap M)(= (\varphi_U(U)) \cap (\mathbb{R}^m \times \{0\}))$ is continuous. Further, since $(\varphi_U|_{U \cap M})^{-1} = ((\varphi_U)^{-1})|_{\varphi_U(U \cap M)}$, it follows that the inverse function $(\varphi_U|_{U \cap M})^{-1}$ is continuous.

Thus, the ordered pair $(U \cap M, \varphi_U|_{U \cap M})$ is a coordinate chart of $M$ satisfying $a \in M$. Next, let $(U \cap M, \varphi_U|_{U \cap M})$ and $(V \cap M, \psi_V|_{V \cap M})$ be two coordinate charts of $M$, where $(U, \varphi_U)$ and $(V, \psi_V)$ are admissible coordinate charts in $N$, and $(U \cap M) \cap (V \cap M)$ is nonempty. We have to prove that $(\psi_V|_{V \cap M}) \circ (\varphi_U|_{U \cap M})^{-1}$ is smooth. Since $(U, \varphi_U)$ and $(V, \psi_V)$ are admissible coordinate charts in $N$, $\psi_V \circ (\varphi_U)^{-1}$ is smooth, and hence, $(\psi_V|_{V \cap M}) \circ (\varphi_U|_{U \cap M})^{-1}(= (\psi_V \circ (\varphi_U)^{-1})|_{(U \cap M) \cap (V \cap M)})$ is smooth. Hence, $M$ is a smooth manifold of dimension $m$. $\square$

## 3.6   Smooth Bump Functions

**Theorem 3.50** *Let* $\varepsilon > 0$. *Let* $a$ *be in* $\mathbb{R}^3$. *Let* $B_\varepsilon(a) \equiv \{x : x \in \mathbb{R}^3, |x - a| < \varepsilon\}$. *Then, there exists a function* $h : \mathbb{R}^3 \to [0, 1]$ *such that*

1. $h$ *is smooth,*
2. $h$ *is onto,*
3. *for every* $x$ *in* $B_\varepsilon(a)$, $h(x) > 0$,
4. *for every* $x$ *in* $\{x : x \in \mathbb{R}^3, |x - a| < \frac{\varepsilon}{2}\}$, $h(x) = 1$,
5. *for every* $x \notin B_\varepsilon(a)$, $h(x) = 0$.

*Proof* Let us define a function $f : \mathbb{R} \to \mathbb{R}$ as follows: For every $x$ in $\mathbb{R}$,

$$f(x) \equiv \begin{cases} e^{-\frac{1}{x}} & \text{if} \quad 0 < x, \\ 0 & \text{if} \quad x \le 0. \end{cases}$$

Thus, $f(x)$ is positive if and only if $x$ is positive, and $f(x)$ is never negative. We shall first try to prove by induction (a): For every $x > 0$, and for every nonnegative integer $k$, the $k$th derivative $f^{(k)}(x)$ is of the form $(p_{2k}(\frac{1}{x}))e^{-\frac{1}{x}}$ for some polynomial $p_{2k}(y)$ of degree $2k$ in $y$. Let us denote the statement, $f^{(k)}(x)$ is of the form $p_{2k}(\frac{1}{x})e^{-\frac{1}{x}}$ for some polynomial $p_{2k}(y)$ of degree $2k$ in $y$, by $P(k)$. We must prove

1. $P(1)$ holds,
2. if $P(k)$ holds, then $P(k + 1)$ holds.

For 1: $P(1)$ means $f'(x)$ is of the form $p_2(\frac{1}{x})e^{-\frac{1}{x}}$ for some polynomial $p_2(y)$ of degree 2 in $y$. From the given definition of $f$,

$$f'(x) = \frac{1}{x^2} e^{-\frac{1}{x}} = p_2\left(\frac{1}{x}\right) e^{-\frac{1}{x}},$$

where $p_2(y) \equiv y^2$. Here, $p_2(y)$ is a polynomial of degree 2 in $y$. This proves 1. For 2: Let $P(k)$ be true, that is, $f^{(k)}(x)$ is of the form $p_{2k}(\frac{1}{x})e^{-\frac{1}{x}}$ for some polynomial $p_{2k}(y)$ of degree $2k$ in $y$. We have to prove that $P(k + 1)$ is true, that is, $f^{(k+1)}(x)$ is of the form $p_{2k+2}(\frac{1}{x})e^{-\frac{1}{x}}$ for some polynomial $p_{2k+2}(y)$ of degree $2k + 2$ in $y$. Here,

$$f^{(k+1)}(x) = \left(f^{(k)}(x)\right)' = \frac{d}{dx} f^{(k)}(x) = \frac{d}{dx}\left(p_{2k}\left(\frac{1}{x}\right)e^{-\frac{1}{x}}\right)$$

$$= \left(p'_{2k}\left(\frac{1}{x}\right)\right)\frac{-1}{x^2} e^{-\frac{1}{x}} + p_{2k}\left(\frac{1}{x}\right)\left(e^{-\frac{1}{x}}\frac{1}{x^2}\right)$$

$$= \left(p_{2k}\left(\frac{1}{x}\right) - p'_{2k}\left(\frac{1}{x}\right)\right)\left(e^{-\frac{1}{x}}\frac{1}{x^2}\right)$$

$$= q_{2k+2}\left(\frac{1}{x}\right)e^{-\frac{1}{x}},$$

where $q_{2k+2}(y) \equiv (p_{2k}(y) - p'_{2k}(y))y^2$, which is a polynomial of degree $2k + 2$ in $y$. This proves 2.

Hence, by the principle of mathematical induction, $P(k)$ is true for every positive integer $k$. This proves (a). Now, we shall try to prove (b): $f$ is $C^\infty$ on $\mathbb{R}$ and that $f^{(k)}(0) = 0$ for all $k \geq 0$. Here, for $x < 0$, $f'(x) = 0$, and for $0 < x$, $f'(x) = \frac{\mathrm{d}}{\mathrm{d}x} e^{-\frac{1}{x}} = \frac{1}{x^2} e^{-\frac{1}{x}}$. Next,

$$\lim_{\substack{\varepsilon \to 0 \\ \varepsilon > 0}} \frac{f(0 - \varepsilon) - f(0)}{-\varepsilon} = \lim_{\substack{\varepsilon \to 0 \\ \varepsilon > 0}} \frac{f(-\varepsilon) - 0}{-\varepsilon} = -\lim_{\substack{\varepsilon \to 0 \\ \varepsilon > 0}} \frac{f(-\varepsilon)}{\varepsilon} = -\lim_{\substack{\varepsilon \to 0 \\ \varepsilon > 0}} \frac{0}{\varepsilon} = 0,$$

and

$$\lim_{\substack{\varepsilon \to 0 \\ \varepsilon > 0}} \frac{f(0 + \varepsilon) - f(0)}{\varepsilon} = \lim_{\substack{\varepsilon \to 0 \\ \varepsilon > 0}} \frac{f(\varepsilon) - 0}{\varepsilon}$$

$$= \lim_{\substack{\varepsilon \to 0 \\ \varepsilon > 0}} \frac{f(\varepsilon)}{\varepsilon} = \lim_{\substack{\varepsilon \to 0 \\ \varepsilon > 0}} \frac{e^{-\frac{1}{\varepsilon}}}{\varepsilon}$$

$$= \lim_{t \to \infty} \frac{t}{e^t} = \lim_{t \to \infty} \frac{1}{e^t} = 0.$$

Hence,

$$\lim_{h \to 0} \frac{f(0 + h) - f(0)}{h} = 0,$$

that is, $f'(0) = 0$. Thus,

$$f'(x) = \begin{cases} 0 & \text{if } x \leq 0, \\ \frac{1}{x^2} e^{-\frac{1}{x}} & \text{if } 0 < x. \end{cases}$$

Here, for $x < 0$, $f''(x) = 0$, and for $0 < x$,

$$f''(x) = \frac{\mathrm{d}}{\mathrm{d}x} f'(x) = \frac{\mathrm{d}}{\mathrm{d}x} \left( \frac{1}{x^2} e^{-\frac{1}{x}} \right) = \frac{-2}{x^3} e^{-\frac{1}{x}} + \frac{1}{x^2} \left( \frac{1}{x^2} e^{-\frac{1}{x}} \right) = \left( \frac{-2}{x^3} + \frac{1}{x^4} \right) e^{-\frac{1}{x}}.$$

Now,

$$\lim_{\substack{\varepsilon \to 0 \\ \varepsilon > 0}} \frac{f'(0-\varepsilon) - f'(0)}{-\varepsilon} = \lim_{\substack{\varepsilon \to 0 \\ \varepsilon > 0}} \frac{f'(-\varepsilon) - 0}{-\varepsilon} = -\lim_{\substack{\varepsilon \to 0 \\ \varepsilon > 0}} \frac{f'(-\varepsilon)}{\varepsilon} = -\lim_{\substack{\varepsilon \to 0 \\ \varepsilon > 0}} \frac{0}{\varepsilon}$$

$$= 0,$$

and

$$\lim_{\substack{\varepsilon \to 0 \\ \varepsilon > 0}} \frac{f'(0+\varepsilon) - f'(0)}{\varepsilon} = \lim_{\substack{\varepsilon \to 0 \\ \varepsilon > 0}} \frac{f'(\varepsilon) - 0}{\varepsilon} = \lim_{\substack{\varepsilon \to 0 \\ \varepsilon > 0}} \frac{f'(\varepsilon)}{\varepsilon} = \lim_{\substack{\varepsilon \to 0 \\ \varepsilon > 0}} \frac{\frac{1}{\varepsilon^2}e^{-\frac{1}{\varepsilon}}}{\varepsilon} = \lim_{\substack{\varepsilon \to 0 \\ \varepsilon > 0}} \frac{1}{\varepsilon^3 e^{\frac{1}{\varepsilon}}} = \lim_{t \to \infty} \frac{t^3}{e^t} = \lim_{t \to \infty} \frac{3t^2}{e^t}$$

$$= \lim_{t \to \infty} \frac{3 \cdot 2t}{e^t} = \lim_{t \to \infty} \frac{3 \cdot 2 \cdot 1}{e^t} = 0.$$

Hence,

$$\lim_{h \to 0} \frac{f'(0+h) - f'(0)}{h} = 0,$$

that is, $f''(0) = 0$. Thus,

$$f''(x) = \begin{cases} 0 & \text{if } x \le 0, \\ \left(\frac{-2}{x^3} + \frac{1}{x^4}\right)e^{-\frac{1}{x}} & \text{if } 0 < x. \end{cases}$$

Here, for $x < 0$ $f'''(x) = 0$ and for $0 < x$,

$$f'''(x) = \frac{d}{dx} f''(x) = \frac{d}{dx}\left(\left(\frac{-2}{x^3} + \frac{1}{x^4}\right)e^{-\frac{1}{x}}\right) = \left(\frac{6}{x^4} + \frac{-4}{x^5}\right)e^{-\frac{1}{x}} + \frac{1}{x^2}e^{-\frac{1}{x}}\left(\frac{-2}{x^3} + \frac{1}{x^4}\right)$$

$$= \left(\frac{6}{x^4} + \frac{-6}{x^5} + \frac{1}{x^6}\right)e^{-\frac{1}{x}}.$$

Now,

$$\lim_{\substack{\varepsilon \to 0 \\ \varepsilon > 0}} \frac{f''(0-\varepsilon) - f''(0)}{-\varepsilon} = \lim_{\substack{\varepsilon \to 0 \\ \varepsilon > 0}} \frac{f''(-\varepsilon) - 0}{-\varepsilon} = -\lim_{\substack{\varepsilon \to 0 \\ \varepsilon > 0}} \frac{f''(-\varepsilon)}{\varepsilon} = -\lim_{\substack{\varepsilon \to 0 \\ \varepsilon > 0}} \frac{0}{\varepsilon}$$

$$= 0,$$

and

$$\lim_{\substack{\varepsilon \to 0 \\ \varepsilon > 0}} \frac{f''(0+\varepsilon)-f''(0)}{\varepsilon} = \lim_{\substack{\varepsilon \to 0 \\ \varepsilon > 0}} \frac{f''(\varepsilon)-0}{\varepsilon} = \lim_{\substack{\varepsilon \to 0 \\ \varepsilon > 0}} \frac{f''(\varepsilon)}{\varepsilon} = \lim_{\substack{\varepsilon \to 0 \\ \varepsilon > 0}} \frac{\left(\frac{-2}{\varepsilon^3}+\frac{1}{\varepsilon^4}\right)e^{-\frac{1}{\varepsilon}}}{\varepsilon}$$

$$= \lim_{\substack{\varepsilon \to 0 \\ \varepsilon > 0}} \left(\frac{-2}{\varepsilon^4}+\frac{1}{\varepsilon^5}\right)\frac{1}{e^{\frac{1}{\varepsilon}}} = \lim_{t \to \infty} \frac{-2t^4+t^5}{e^t} = \lim_{t \to \infty} \frac{5!}{e^t} = 0.$$

Hence,

$$\lim_{h \to 0} \frac{f''(0+h)-f''(0)}{h} = 0,$$

that is, $f'''(0) = 0$. Thus,

$$f'''(x) = \begin{cases} 0 & \text{if } x \leq 0, \\ \left(\frac{6}{x^4}+\frac{-6}{x^5}+\frac{1}{x^6}\right)e^{-\frac{1}{x}} & \text{if } 0 < x. \end{cases}$$

Similarly, $f^{(4)}(0) = 0, f^5(0) = 0$, etc. This completes the proof of (b). Thus, $f : \mathbb{R} \to \mathbb{R}$ is a smooth function.

Now, let us define a function $g : \mathbb{R}^3 \to \mathbb{R}$ as follows: For every $x \equiv (x_1, x_2, x_3)$ in $\mathbb{R}^3$,

$$g(x) \equiv \frac{f\left(\varepsilon^2 - \left((x_1)^2+(x_2)^2+(x_3)^2\right)\right)}{f\left(\varepsilon^2 - \left((x_1)^2+(x_2)^2+(x_3)^2\right)\right) + f\left(\left((x_1)^2+(x_2)^2+(x_3)^2\right) - \frac{1}{4}\varepsilon^2\right)}.$$

Clearly, in the expression for $g(x)$, the denominator

$$f\left(\varepsilon^2 - \left((x_1)^2+(x_2)^2+(x_3)^2\right)\right) + f\left(\left((x_1)^2+(x_2)^2+(x_3)^2\right) - \frac{1}{4}\varepsilon^2\right)$$

is nonzero, and hence, $g : \mathbb{R}^3 \to \mathbb{R}$ is a well-defined function. Since $0 \leq f$, it is clear that $0 \leq g \leq 1$. Further, since $f$ is $C^\infty$ on $\mathbb{R}$, $g$ is $C^\infty$ on $\mathbb{R}$. Next, let us observe that if $x \equiv (x_1, x_2, x_3)$ satisfies

$$\varepsilon^2 - \left((x_1)^2+(x_2)^2+(x_3)^2\right) \leq 0, \text{ then } f\left(\varepsilon^2 - \left((x_1)^2+(x_2)^2+(x_3)^2\right)\right) = 0,$$

and hence,

$$g(x)\left(=\frac{f\left(\varepsilon^2-\left((x_1)^2+(x_2)^2+(x_3)^2\right)\right)}{f\left(\varepsilon^2-\left((x_1)^2+(x_2)^2+(x_3)^2\right)\right)+f\left(\left((x_1)^2+(x_2)^2+(x_3)^2\right)-\frac{1}{4}\varepsilon^2\right)}\right)\text{ is }0.$$

For every $x \equiv (x_1, x_2, x_3)$ satisfying

$$\left((x_1)^2+(x_2)^2+(x_3)^2\right)-\frac{1}{4}\varepsilon^2\le 0,\quad f\left(\left((x_1)^2+(x_2)^2+(x_3)^2\right)-\frac{1}{4}\varepsilon^2\right)=0,$$

and hence,

$$g(x)\left(=\frac{f\left(\varepsilon^2-\left((x_1)^2+(x_2)^2+(x_3)^2\right)\right)}{f\left(\varepsilon^2-\left((x_1)^2+(x_2)^2+(x_3)^2\right)\right)+f\left(\left((x_1)^2+(x_2)^2+(x_3)^2\right)-\frac{1}{4}\varepsilon^2\right)}\right.$$

$$\left.=\frac{f\left(\varepsilon^2-\left((x_1)^2+(x_2)^2+(x_3)^2\right)\right)}{f\left(\varepsilon^2-\left((x_1)^2+(x_2)^2+(x_3)^2\right)\right)+0}\right)\text{ is }1.$$

Finally, for every $x \equiv (x_1, x_2, x_3)$ satisfying $(x_1)^2 + (x_2)^2 + (x_3)^2 < \varepsilon^2$, we have $\varepsilon^2 - ((x_1)^2 + (x_2)^2 + (x_3)^2)$ is positive, and hence, $f(\varepsilon^2 - ((x_1)^2 + (x_2)^2 + (x_3)^2))$ is positive. Thus, for every $x \equiv (x_1, x_2, x_3)$ satisfying $(x_1)^2 + (x_2)^2 + (x_3)^2 < \varepsilon^2$,

$$g(x)\left(=\frac{f\left(\varepsilon^2-\left((x_1)^2+(x_2)^2+(x_3)^2\right)\right)}{f\left(\varepsilon^2-\left((x_1)^2+(x_2)^2+(x_3)^2\right)\right)+f\left(\left((x_1)^2+(x_2)^2+(x_3)^2\right)-\frac{1}{4}\varepsilon^2\right)}\right)$$

is positive. Since $g : \mathbb{R}^3 \to [0, 1]$, and $g$ is $C^\infty$ on $\mathbb{R}$, $g$ is continuous. Since $g : \mathbb{R}^3 \to [0, 1]$ is continuous, $g$ assumes 0, and $g$ assumes 1, $g$ assumes all values of $[0, 1]$. Hence, $g : \mathbb{R}^3 \to [0, 1]$ is onto. Thus, we have seen that $g$ is a smooth function from $\mathbb{R}^3$ onto $[0, 1]$ such that for every $x$ in $\mathbb{R}^3$,

$$g(x) = \begin{cases} 1 & \text{if } |x| \le \frac{\varepsilon}{2}, \\ 0 & \text{if } \varepsilon \le |x|, \\ \text{positive} & \text{if } |x| < \varepsilon. \end{cases}$$

Next, let us define a function $h : \mathbb{R}^3 \to [0, 1]$ as follows: For every $x$ in $\mathbb{R}^3$,

$$h(x) \equiv g(x - a).$$

1. Since $g$ is a smooth function, and $x \mapsto (x - a)$ is a smooth function, their composite $h : x \mapsto g(x - a)$ is also a smooth function.
2. Since $x \mapsto (x - a)$ is a function from $\mathbb{R}^3$ onto $\mathbb{R}^3$, and $g$ from $\mathbb{R}^3$ onto $[0, 1]$, their composite $h : x \mapsto g(x - a)$ is onto.
3. If $x$ is in $B_\varepsilon(a)$, then $|x - a| < \varepsilon$, and hence, $h(x)(= g(x - a))$ is positive.
4. For every $x$ in $\{x : x \in \mathbb{R}^3, |x - a| < \frac{\varepsilon}{2}\}$, we have $|x - a| < \frac{\varepsilon}{2}$, and therefore, $h(x) = g(x - a) = 1$.
5. For every $x \notin B_\varepsilon(a)$, we have $\varepsilon \leq |x - a|$, and hence, $h(x) = g(x - a) = 0$. $\square$

**Note 3.51** We shall try to prove that there exists a smooth function $h : \mathbb{R} \to \mathbb{R}$ such that

1. $h(t) = 1$ if $t \leq 1$,
2. $0 < h(t) < 1$ if $1 < t < 2$,
3. $h(t) = 0$ if $2 \leq t$.

In the proof of Theorem 3.50, we have seen that the function $f : \mathbb{R} \to \mathbb{R}$ is defined as follows: For every $t$ in $\mathbb{R}$,

$$f(t) \equiv \begin{cases} e^{-\frac{1}{t}} & \text{if } 0 < t \\ 0 & \text{if } t \leq 0, \end{cases}$$

is smooth. Now, let us define the function $h : \mathbb{R} \to \mathbb{R}$ as follows: For every $t$ in $\mathbb{R}$,

$$h(t) \equiv \frac{f(2 - t)}{f(2 - t) + f(t - 1)}.$$

If $t \leq 1$, then $-1 \leq -t$, and hence, $1 \leq 2 - t$. If $t \leq 1$, then $t - 1 \leq 0$, and hence, $f(t - 1) = 0$. Thus, for $t \leq 1$,

$$h(t) = \frac{e^{-\frac{1}{2-t}}}{e^{-\frac{1}{2-t}} + 0} = 1.$$

If $2 \leq t$, then $2 - t \leq 0$, and hence, $f(2 - t) = 0$. If $2 \leq t$, then $1 \leq t - 1$. Thus, for $2 \leq t$,

$$h(t) = \frac{0}{0 + e^{-\frac{1}{t-1}}} = 0.$$

If $1 < t < 2$, then $0 < t - 1$, and hence,

$$0 < e^{-\frac{1}{t-1}} = f(t - 1).$$

If $1 < t < 2$, then $0 < 2 - t$, and hence, $0 < e^{-\frac{1}{2-t}} = f(2 - t)$. It follows that for $1 < t < 2$,

$$h(t) = \frac{e^{-\frac{1}{2-t}}}{e^{-\frac{1}{2-t}} + e^{-\frac{1}{t-1}}} \in (0, 1).$$

It remains to show that $h$ is smooth. Since $f$ is smooth, and $t \mapsto 2 - t$ is smooth, their composite $t \mapsto f(2 - t)$ is smooth. Similarly, $t \mapsto f(t - 1)$ is smooth. This shows that

$$t \mapsto \frac{f(2 - t)}{f(2 - t) + f(t - 1)} (= h(t))$$

is smooth, and hence, $h$ is smooth. $\qquad\square$

Here, we observe that

1. $h(t) = 1$ if and only if $t \le 1$,
2. $h(t) = 0$ if and only if $2 \le t$,
3. $0 < h(t) < 1$ if and only if $1 < t < 2$.

**Note 3.52** We shall try to prove that there exists a function $H : \mathbb{R}^3 \to [0, 1]$ such that

1. $H$ is smooth,
2. $H(x) = 1$, if $x$ is in the closed ball $B_1[0] (= \{x : x \in \mathbb{R}^3, |x| \le 1\})$,
3. the closure $(H^{-1}((0, 1]))^-$ of $H^{-1}((0, 1])$ is equal to the closed ball $B_2[0] (= \{x : x \in \mathbb{R}^3, |x| \le 2\})$.

Let us define the function $H : \mathbb{R}^3 \to [0, 1]$ as follows: For every $x \equiv (x_1, x_2, x_3)$ in $\mathbb{R}^3$,

$$H(x) \equiv h\left( \sqrt{(x_1)^2 + (x_2)^2 + (x_3)^2} \right),$$

where $h$ denotes the function as defined in Note 3.51.

For 1: Since $x \equiv (x_1, x_2, x_3) \mapsto (\sqrt{(x_1)^2 + (x_2)^2 + (x_3)^2})$ is a smooth function over $\mathbb{R}^3 - \{(0, 0, 0)\}$, and $h : \mathbb{R} \to [0, 1]$ is a smooth function, their composite $x \mapsto h(\sqrt{(x_1)^2 + (x_2)^2 + (x_3)^2})(= H(x))$ is a smooth function from

$\mathbb{R}^3 - \{(0,0,0)\}$  to  $[0,1]$.  It remains to show that  $(x_1, x_2, x_3) \mapsto h$ $(\sqrt{(x_1)^2 + (x_2)^2 + (x_3)^2})$ is smooth at $(0,0,0)$. Here,

$$\lim_{\substack{t \to 0 \\ t > 0}} \frac{h\left(\sqrt{(0+t)^2 + (0)^2 + (0)^2}\right) - h\left(\sqrt{(0)^2 + (0)^2 + (0)^2}\right)}{t}$$

$$= \lim_{\substack{t \to 0 \\ t > 0}} \frac{h(t) - h(0)}{t} = \lim_{\substack{t \to 0 \\ t > 0}} \frac{1-1}{t} = \lim_{\substack{t \to 0 \\ t > 0}} 0 = 0,$$

and

$$\lim_{\substack{t \to 0 \\ t > 0}} \frac{h\left(\sqrt{(0-t)^2 + (0)^2 + (0)^2}\right) - h\left(\sqrt{(0)^2 + (0)^2 + (0)^2}\right)}{-t}$$

$$= \lim_{\substack{t \to 0 \\ t > 0}} \frac{h(t) - h(0)}{-t} = \lim_{\substack{t \to 0 \\ t > 0}} \frac{1-1}{-t} = \lim_{\substack{t \to 0 \\ t > 0}} 0 = 0.$$

Since

$$\lim_{\substack{t \to 0 \\ t > 0}} \frac{h\left(\sqrt{(0+t)^2 + (0)^2 + (0)^2}\right) - h\left(\sqrt{(0)^2 + (0)^2 + (0)^2}\right)}{t} = 0$$

$$= \lim_{\substack{t \to 0 \\ t > 0}} \frac{h\left(\sqrt{(0-t)^2 + (0)^2 + (0)^2}\right) - h\left(\sqrt{(0)^2 + (0)^2 + (0)^2}\right)}{-t},$$

$(D_1 H)(0,0,0)$ exists, and its value is 0. Also, for $(x_1, x_2, x_3) \neq (0,0,0)$,

$$(D_1 H)(x_1, x_2, x_3) = \left(h'\left(\sqrt{(x_1)^2 + (x_2)^2 + (x_3)^2}\right)\right) \frac{1}{2\sqrt{(x_1)^2 + (x_2)^2 + (x_3)^2}} (2x_1 + 0 + 0)$$

$$= \frac{x_1}{\sqrt{(x_1)^2 + (x_2)^2 + (x_3)^2}} h'\left(\sqrt{(x_1)^2 + (x_2)^2 + (x_3)^2}\right).$$

Thus,

$$(D_1 H)(x_1, x_2, x_3) = \begin{cases} \dfrac{x_1}{\sqrt{(x_1)^2+(x_2)^2+(x_3)^2}} h'\left(\sqrt{(x_1)^2+(x_2)^2+(x_3)^2}\right) & \text{if} \quad (x_1, x_2, x_3) \neq (0,0,0) \\ 0 & \text{if} \quad (x_1, x_2, x_3) = (0,0,0). \end{cases}$$

Now, we want to show that $D_1 H$ is continuous at $(0,0,0)$.
We recall that the smooth function $h : \mathbb{R} \to \mathbb{R}$ is defined as follows: For every $t$ in $\mathbb{R}$,

$$h(t) = \frac{f(2-t)}{f(2-t)+f(t-1)}. \text{ Also } f(x) \equiv \begin{cases} e^{-\frac{1}{x}} & \text{if} \quad 0<x, \\ 0 & \text{if} \quad x \leq 0, \end{cases} \text{ and } f'(x)$$
$$= \begin{cases} 0 & \text{if} \quad x \leq 0, \\ \frac{1}{x^2} e^{-\frac{1}{x}} & \text{if} \quad 0<x. \end{cases}$$

So

$$h'(t) = \frac{(f'(2-t))(-1)(f(2-t)+f(t-1)) - (f(2-t))(-f'(2-t)+f'(t-1))}{(f(2-t)+f(t-1))^2},$$

and hence,

$$\begin{aligned} h'(0) &= \frac{(f'(2-0))(-1)(f(2-0)+f(0-1)) - (f(2-0))(-f'(2-0)+f'(0-1))}{(f(2-0)+f(0-1))^2} \\ &= \frac{(f'(2))(-1)(f(2)+f(-1)) - (f(2))(-f'(2)+f'(-1))}{(f(2)+f(-1))^2} \\ &= \frac{(f'(2))(-1)(f(2)+0) - (f(2))(-f'(2)+0)}{(f(2)+0)^2} = \frac{0}{\left(e^{-\frac{1}{2}}\right)^2} = 0. \end{aligned}$$

Thus, $h'(0) = 0$. Since for $(x_1, x_2, x_3) \neq (0,0,0)$, we have

$$0 \leq \left| \frac{x_1}{\sqrt{(x_1)^2+(x_2)^2+(x_3)^2}} \right| \leq 1,$$

and $h'(0) = 0$,

$$\lim_{(x_1,x_2,x_3) \to (0,0,0)} (D_1 H)(x_1, x_2, x_3)$$

$$= \lim_{(x_1,x_2,x_3) \to (0,0,0)} \frac{x_1}{\sqrt{(x_1)^2+(x_2)^2+(x_3)^2}} h'\left(\sqrt{(x_1)^2+(x_2)^2+(x_3)^2}\right) = 0 = (D_1 H)(0,0,0).$$

This shows that $D_1 H$ is continuous at $(0, 0, 0)$. Similarly, $D_2 H$ is continuous at $(0, 0, 0)$, and $D_3 H$ is continuous at $(0, 0, 0)$. Similarly, higher partial derivatives of $H$ exist at $(0, 0, 0)$. This shows that $H$ is smooth.

For 2: Let us take any $x \equiv (x_1, x_2, x_3)$ satisfying $|x| \leq 1$. So $H(x) = h(\sqrt{(x_1)^2 + (x_2)^2 + (x_3)^2}) = h(|x|) = 1$.

For 3: Let us take any $a \equiv (a_1, a_2, a_3)$ in $(H^{-1}((0, 1]))^-$. We have to show that $\sqrt{(a_1)^2 + (a_2)^2 + (a_3)^2} \leq 2$. If not, otherwise, let $2 < \sqrt{(a_1)^2 + (a_2)^2 + (a_3)^2}$.

We have to arrive at a contradiction. Since $2 < \sqrt{(a_1)^2 + (a_2)^2 + (a_3)^2} = |a|$, there exists a real number $\varepsilon > 0$ such that the open ball $B_\varepsilon(a)$ contains no point of $B_2[0]$. Since $B_\varepsilon(a)$ is an open neighborhood of $a$, and $a$ is in $(H^{-1}((0, 1]))^-$, there exists $b$ in $H^{-1}((0, 1])$ such that $b$ is in $B_\varepsilon(a)$. Since $b$ is in $H^{-1}((0, 1])$, $h(|b|) = H(b) \in (0, 1]$. Since $h(|b|) \in (0, 1]$, and $h(t) = 0$ if and only if $2 \leq t$, $|b| < 2$, and hence, $b$ is in $B_2[0]$. Since $b$ is in $B_2[0]$, and $B_\varepsilon(a)$ contains no point of $B_2[0]$, $b$ is not in $B_\varepsilon(a)$, a contradiction. Thus, we have shown that $(H^{-1}((0, 1]))^- \subset B_2[0]$. It remains to prove that $B_2[0] \subset (H^{-1}((0, 1]))^-$. Let us take any $a$ in $B_2[0]$. We have to prove that $a \in (H^{-1}((0, 1]))^-$. Since $a$ is in $B_2[0]$, $|a| \leq 2$.

Case I: when $|a| < 2$. Since $|a| < 2$, and $h(t) = 0$ if and only if $2 \leq t$, $0 < h(|a|) = H(a) \leq 1$, and hence, $a \in H^{-1}((0, 1]) \subset (H^{-1}((0, 1]))^-$. Hence, $a \in (H^{-1}((0, 1]))^-$.

Case II: when $|a| = 2$. For every positive integer $n > 1$,

$$\lim_{n \to \infty} \left(1 - \frac{1}{2n}\right) a = a,$$

and

$$\left|\left(1 - \frac{1}{2n}\right) a\right| = \left(1 - \frac{1}{2n}\right) |a| = \left(1 - \frac{1}{2n}\right) 2 = 2 - \frac{1}{n} \in (1, 2).$$

Since

$$\left|\left(1 - \frac{1}{2n}\right) a\right| \in (1, 2),$$

$$H\left(\left(1 - \frac{1}{2n}\right) a\right) = h\left(\left|\left(1 - \frac{1}{2n}\right) a\right|\right) \in (0, 1) \subset (0, 1],$$

and hence,

$$\left(1 - \frac{1}{2n}\right) a \in H^{-1}((0, 1]).$$

Since, for every positive integer, $n > 1$,

$$\left(1 - \frac{1}{2n}\right)a \in H^{-1}((0, 1]),$$

and

$$\lim_{n \to \infty} \left(1 - \frac{1}{2n}\right)a = a, \text{ so } a \in \left(H^{-1}((0, 1])\right)^-.$$

Thus, we see that in all cases, $a \in (H^{-1}((0, 1]))^-$. $\qquad\qquad\qquad\square$

As above, we can show that there exists a function $H : \mathbb{R}^n \to [0, 1]$ such that

1. $H$ is smooth,
2. $H(x) = 1$, if $x$ is in the closed ball $B_1[0] (= \{x : x \in \mathbb{R}^n, |x| \le 1\})$,
3. the closure $(H^{-1}((0, 1]))^-$ of $H^{-1}((0, 1])$ is equal to the closed ball $B_2[0] (= \{x : x \in \mathbb{R}^n, |x| \le 2\})$.

**Definition** Let $h : \mathbb{R} \to \mathbb{R}$ be a smooth function. If

1. $h(t) = 1$ if $t \le 1$,
2. $0 < h(t) < 1$ if $1 < t < 2$,
3. $h(t) = 0$ if $2 \le t$,
then we say that $h$ is a *cutoff function*. By Note 3.51, there exists a cutoff function.

**Definition** Let $X$ be a topological space. Let $f : X \mapsto \mathbb{R}$ be any function. The closure $(f^{-1}(\mathbb{R} - \{0\}))^-$ of $f^{-1}(\mathbb{R} - \{0\})$ is called the *support of* $f$ and is denoted by $\text{supp} f$. By *$f$ is compactly supported*, we mean that $\text{supp} f$ is a compact set. By the note 3.52, there exists a function $H : \mathbb{R}^n \to [0, 1]$ such that

1. $H$ is smooth,
2. $H(x) = 1$, if $x$ is in the closed ball $B_1[0]$,
3. $\text{supp } H = B_2[0]$.

**Definition** Let $X$ be a topological space. Let $f : X \mapsto \mathbb{R}$ be any function. Let $S$ be any nonempty subset of $X$. If $\text{supp} f$ is contained in $S$, then we say that $f$ is *supported in S*.

**Definition** Let $H : \mathbb{R}^n \to [0, 1]$ be a function. Let $F$ be a closed subset of $\mathbb{R}^n$, and let $G$ be an open subset of $\mathbb{R}^n$. If

1. $H$ is smooth,
2. $H(x) = 1$, if $x$ is in $F$,
3. $H$ is supported in $G$,

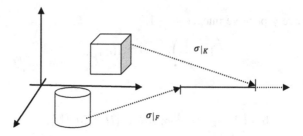

**Fig. 3.1** $F$ and $K$ are disjoint

then we say that $H$ *is a smooth bump function corresponding to closed set* $F$ *and open set* $G$. By Note 3.52, there exists a smooth bump function corresponding to closed set $B_1[0]$ and open set $B_3(0)$.

**Theorem 3.53** *Let $F$ be a nonempty closed subset of $\mathbb{R}^3$, and let $K$ be a nonempty compact subset of $\mathbb{R}^3$. If $F$ and $K$ are disjoint, then there exists a function $\sigma$ : $\mathbb{R}^3 \to [0,1]$ such that*

1. *$\sigma$ is smooth,*
2. *for every $x$ in $F$, $\sigma(x) = 0$,*
3. *for every $x$ in $K$, $\sigma(x) = 1$,*
4. *$\sigma$ is onto.*

*Proof* Since $F$ and $K$ are disjoint, $K \subset F^c$ where $F^c$ denotes the complement of $F$. Since $F$ is a closed subset of $\mathbb{R}^3$, $F^c$ is an open set. Let us take any $x$ in $K$. Since $x \in K \subset F^c, x \in F^c$. Since $x \in F^c$, and $F^c$ is an open set, there exists a real number $\delta_x > 0$ such that $x \in B_{\frac{1}{2}\delta_x}(x) \subset F^c$. It follows that the collection $\{B_{\frac{1}{2}\delta_x}(x) : x \in K\}$ is an open cover of $K$. Further, since $K$ is compact, there exist finite many $x_1, x_2, \ldots, x_n$ in $K$ such that $K$ is contained in $B_{\frac{1}{2}\delta_{x_1}}(x_1) \cup B_{\frac{1}{2}\delta_{x_2}}(x_2) \cup \cdots \cup B_{\frac{1}{2}\delta_{x_n}}(x_n)$. Since (see Fig. 3.1)

$$B_{\frac{1}{2}\delta_{x_1}}(x_1) \subset F^c, \ B_{\frac{1}{2}\delta_{x_2}}(x_2) \subset F^c, \ldots, B_{\frac{1}{2}\delta_{x_n}}(x_n) \subset F^c,$$

$(B_{\frac{1}{2}\delta_{x_1}}(x_1) \cup B_{\frac{1}{2}\delta_{x_2}}(x_2) \cup \cdots \cup B_{\frac{1}{2}\delta_{x_n}}(x_n)) \subset F^c$. Now, for every $i = 1, 2, \ldots, n$, by Theorem 3.50, there exist functions $h_i : \mathbb{R}^3 \to [0,1]$ such that

1. *$h_i$ is smooth,*
2. *$h_i$ is onto,*
3. *for every $x$ in $B_{\frac{1}{2}\delta_{x_i}}(x_i)$, $h_i(x) > 0$,*
4. *for every $x$ in $\{x : x \in \mathbb{R}^3, |x - x_i| < \frac{1}{2}\delta_{x_i}\}$, $h_i(x) = 1$,*
5. *for every $x \notin B_{\frac{1}{2}\delta_{x_i}}(x_i)$, $h_i(x) = 0$.*

Now, let us define a function $\sigma : \mathbb{R}^3 \to [0, 1]$ as follows: For every $x$ in $\mathbb{R}^3$,

$$\sigma(x) \equiv 1 - ((1 - h_1(x))(1 - h_2(x)) \cdots (1 - h_n(x))).$$

Since $0 \leq h_1 \leq 1$, for every $x$ in $\mathbb{R}^3$, $0 \leq 1 - h_1(x) \leq 1$. Similarly, for every $x$ in $\mathbb{R}^3$, $0 \leq 1 - h_2(x) \leq 1, \ldots, 0 \leq 1 - h_n(x) \leq 1$. Hence, for every $x$ in $\mathbb{R}^3$, $0 \leq (1 - h_1(x))(1 - h_2(x)) \cdots (1 - h_n(x)) \leq 1$. It follows that for every $x$ in $\mathbb{R}^3$, $0 \leq 1 - ((1 - h_1(x))(1 - h_2(x)) \cdots (1 - h_n(x))) \leq 1$. Thus, we have shown that $\sigma : \mathbb{R}^3 \to [0, 1]$ is a well-defined function.

1. Since each $h_i$ is smooth, $\sigma : x \mapsto 1 - ((1 - h_1(x))(1 - h_2(x)) \cdots (1 - h_n(x)))$ is smooth.
2. Let us take any $x$ in $F$. Since $x$ is in $F$, and
   $(B_{\frac{1}{2}\delta_{x_1}}(x_1) \cup B_{\frac{1}{2}\delta_{x_2}}(x_2) \cup \cdots \cup B_{\frac{1}{2}\delta_{x_n}}(x_n)) \subset F^c$, $x \notin (B_{\frac{1}{2}\delta_{x_1}}(x_1) \cup B_{\frac{1}{2}\delta_{x_2}}(x_2) \cup \cdots \cup B_{\frac{1}{2}\delta_{x_n}}(x_n))$, and hence, for every $i = 1, 2, \ldots, n$, we have $x \notin B_{\frac{1}{2}\delta_{x_i}}(x_i)$. Thus, $\sigma(x) = 1 - ((1 - 0)(1 - 0) \cdots (1 - 0)) = 0$.
3. Let us take any $x$ in $K$. Since $x$ is in $K$, and $K$ is contained in $B_{\frac{1}{2}\delta_{x_1}}(x_1) \cup B_{\frac{1}{2}\delta_{x_2}}(x_2) \cup \cdots \cup B_{\frac{1}{2}\delta_{x_n}}(x_n)$, there exists $j \in \{1, 2, \ldots, n\}$ such that $x \in B_{\frac{1}{2}\delta_{x_j}}(x_j)$. For simplicity, let us take 1 for $j$. Now, since $x \in B_{\frac{1}{2}\delta_{x_1}}(x_1)$, $h_1(x) = 1$, and hence, $\sigma(x) = 1 - ((1 - 1)(1 - h_2(x)) \cdots (1 - h_n(x))) = 1$.
4. Since $F$ is nonempty, there exists $a$ in $F$. Hence, $\sigma(a) = 0$. Thus, the function $\sigma$ assumes 0. Similarly, $\sigma$ assumes 1. Since $\sigma : \mathbb{R}^3 \to [0, 1]$, and $\sigma$ is smooth, $\sigma$ is continuous. Since $\sigma : \mathbb{R}^3 \to [0, 1]$ is continuous, $\sigma$ assumes 0, and $\sigma$ assumes 1, $\sigma$ assumes all values of $[0, 1]$. Hence, $\sigma : \mathbb{R}^3 \to [0, 1]$ is onto. $\qquad\square$

**Note 3.54** The result similar to Theorem 3.53 can be proved as above for $\mathbb{R}^n$ in place of $\mathbb{R}^3$.

**Theorem 3.55** *Let $a$ be in $\mathbb{R}^3$. Let $U$ be an open neighborhood of $a$ in $\mathbb{R}^3$. Let $f : U \to \mathbb{R}$ be any function. If $f$ is smooth, then there exist an open neighborhood $V$ of $a$ and a function $F : \mathbb{R}^3 \to \mathbb{R}$ such that*

1. *$V$ is contained in $U$,*
2. *for every $x$ in $V$, $F(x) = f(x)$,*
3. *for every $x \notin U$, $F(x) = 0$,*
4. *$F$ is smooth.*

*Proof* Since $U$ is an open neighborhood of $a$, there exists a real number $\varepsilon > 0$ such that $B_\varepsilon(a) \subset U$. Since $B_{\varepsilon/3}[a] (\equiv \{x : x \in \mathbb{R}^3, |x - a| \leq \frac{\varepsilon}{3}\})$ is a closed and bounded subset of $\mathbb{R}^3$, $B_{\varepsilon/3}[a]$ is compact. Since $B_{\varepsilon/3}[a] \subset B_{\varepsilon/2}(a)$, $B_{\varepsilon/3}[a]$ and $\mathbb{R}^3 - B_{\varepsilon/2}(a)$

are disjoint sets. Since $B_{\varepsilon/2}(a)$ is an open set, $\mathbb{R}^3 - B_{\varepsilon/2}(a)$ is a closed set. Now, by Theorem 3.53, there exists a function $\sigma : \mathbb{R}^3 \to [0,1]$ such that

1. $\sigma$ is smooth,
2. for every $x$ in $\mathbb{R}^3 - B_{\varepsilon/2}(a)$, $\sigma(x) = 0$,
3. for every $x$ in $B_{\varepsilon/3}[a]$, $\sigma(x) = 1$,
4. $\sigma$ is onto.

Let us take $B_{\varepsilon/3}(a)$ for $V$. Let us define a function $F : \mathbb{R}^3 \to \mathbb{R}$ as follows: For every $x$ in $\mathbb{R}^3$,

$$F(x) \equiv \begin{cases} f(x)\sigma(x) & \text{if } x \in U, \\ 0 & \text{if } x \notin U. \end{cases}$$

1. Since $V = B_{\varepsilon/3}(a) \subset B_{\varepsilon}(a) \subset U$, $V$ is contained in $U$.
2. For every $x$ in $V$, we have $x \in V = B_{\varepsilon/3}(a) \subset B_{\varepsilon/3}[a] \subset B_{\varepsilon}(a) \subset U$, and hence, $F(x) = f(x)\sigma(x) = f(x)1 = f(x)$.
3. Let us take any $x \notin U$. Hence, by the definition of $F$, we have $F(x) = 0$.
4. Since $f$ and $\sigma$ both are smooth on the open set $U$, their product $x \mapsto f(x)\sigma(x)$ is smooth on $U$. Further, for $x \in U$, we have $F(x) = f(x)\sigma(x)$, so $F$ is smooth over $U$. Since for every $x$ in $U - B_{\varepsilon/2}(a)$, we have $F(x) = f(x)\sigma(x) = f(x)0 = 0$. Hence, by 3, for every $x$ in $\mathbb{R}^3 - B_{\varepsilon/2}(a)$, $F(x) = 0$. This shows that $F$ is smooth over the open set $\mathbb{R}^3 - B_{\varepsilon/2}[a]$. Since $F$ is smooth over the open set $U$, and $F$ is smooth over the open set $\mathbb{R}^3 - B_{\varepsilon/2}[a]$, $F$ is smooth over the open set $U \cup (\mathbb{R}^3 - B_{\varepsilon/2}[a])(= \mathbb{R}^3)$. Thus, $F : \mathbb{R}^3 \to \mathbb{R}$ is a smooth function.  $\square$

**Note 3.56** The result similar to Theorem 3.55 can be proved as above for $\mathbb{R}^n$ in place of $\mathbb{R}^3$.

## 3.7 The Constant Rank Theorem in $\mathbb{R}^n$

**Theorem 3.57** *Let $A_0$ be an open neighborhood of $(0,0,0)$ in $\mathbb{R}^3$. Let $B_0$ be an open subset of $\mathbb{R}^4$. Let $F : A_0 \to B_0$ be a function. If*

(i)  *$F$ is a $C^2$ function,*
(ii)  *for every $x$ in $A_0$, the linear transformation $F'(x)$ has rank 2,*
(iii)  *$F(0,0,0) = (0,0,0,0)$,*

*then there exist an open neighborhood $A$ of $(0,0,0)$ in $\mathbb{R}^3$, an open neighborhood $B$ of $(0,0,0,0)$ in $\mathbb{R}^4$, an open subset $U$ of $\mathbb{R}^3$, an open subset $V$ of $\mathbb{R}^4$, a function $G : A \to U$, and a function $H : B \to V$ such that*

1. *A is contained in $A_0$, and B is contained in $B_0$,*
2. *G is a $C^2$ diffeomorphism (i.e., G is 1–1 onto, G is a $C^2$ function, and $G^{-1}$ is a $C^2$ function),*
3. *H is a $C^2$ diffeomorphism,*
4. *$F(A) \subset B$,*
5. *For every $x \equiv (x_1, x_2, x_3)$ in $U$, $(H \circ F \circ G^{-1})(x_1, x_2, x_3) = (x_1, x_2, 0, 0)$.*

*Proof* Let $F \equiv (F_1, F_2, F_3, F_4)$. Since $(0, 0, 0)$ is in $A_0$, the linear transformation $F'(0, 0, 0)$ has rank 2. Hence,

$$\text{rank} \begin{bmatrix} (D_1F_1)(0,0,0) & (D_2F_1)(0,0,0) & (D_3F_1)(0,0,0) \\ (D_1F_2)(0,0,0) & (D_2F_2)(0,0,0) & (D_3F_2)(0,0,0) \\ (D_1F_3)(0,0,0) & (D_2F_3)(0,0,0) & (D_3F_3)(0,0,0) \\ (D_1F_4)(0,0,0) & (D_2F_4)(0,0,0) & (D_3F_4)(0,0,0) \end{bmatrix} = 2.$$

This shows that there exists a nonzero $2 \times 2$ minor determinant of

$$\begin{bmatrix} (D_1F_1)(0,0,0) & (D_2F_1)(0,0,0) & (D_3F_1)(0,0,0) \\ (D_1F_2)(0,0,0) & (D_2F_2)(0,0,0) & (D_3F_2)(0,0,0) \\ (D_1F_3)(0,0,0) & (D_2F_3)(0,0,0) & (D_3F_3)(0,0,0) \\ (D_1F_4)(0,0,0) & (D_2F_4)(0,0,0) & (D_3F_4)(0,0,0) \end{bmatrix}.$$

For simplicity, let

$$\det \begin{bmatrix} (D_1F_1)(0,0,0) & (D_2F_1)(0,0,0) \\ (D_1F_2)(0,0,0) & (D_2F_2)(0,0,0) \end{bmatrix} \neq 0.$$

Now, let us define a function $G : A_0 \to \mathbb{R}^3$ as follows: For every $(x_1, x_2, x_3)$ in $A_0$,

$$G(x_1, x_2, x_3) = (F_1(x_1, x_2, x_3), F_2(x_1, x_2, x_3), x_3).$$

Since $(F_1, F_2, F_3, F_4) \equiv F : A_0 \to B_0 \subset \mathbb{R}^4$ is a $C^2$ function, each $F_i : A_0 \to \mathbb{R}$ is a $C^2$ function, and hence, each component function of $G$ is $C^2$. This shows that $G$ is a $C^2$ function. Also, for every $x \equiv (x_1, x_2, x_3)$ in $A_0$,

$$G'(x) = \begin{bmatrix} (D_1F_1)(x_1,x_2,x_3) & (D_2F_1)(x_1,x_2,x_3) & (D_3F_1)(x_1,x_2,x_3) \\ (D_1F_2)(x_1,x_2,x_3) & (D_2F_2)(x_1,x_2,x_3) & (D_3F_2)(x_1,x_2,x_3) \\ 0 & 0 & 1 \end{bmatrix}.$$

Hence,

$$\det(G'(0,0,0)) = \det \begin{bmatrix} (D_1F_1)(0,0,0) & (D_2F_1)(0,0,0) & (D_3F_1)(0,0,0) \\ (D_1F_2)(0,0,0) & (D_2F_2)(0,0,0) & (D_3F_2)(0,0,0) \\ 0 & 0 & 1 \end{bmatrix}$$

$$= \det \begin{bmatrix} (D_1F_1)(0,0,0) & (D_2F_1)(0,0,0) \\ (D_1F_2)(0,0,0) & (D_2F_2)(0,0,0) \end{bmatrix} \neq 0.$$

Since $\det(G'(0,0,0))$ is nonzero, $G'(0,0,0)$ is invertible. Now, we can apply the inverse function theorem on $G : A_0 \to \mathbb{R}^3$. There exists an open neighborhood $A_1$ of $(0,0,0)$ such that

1. $A_1$ is contained in $A_0$,
2. $G$ is 1–1 on $A_1$ (i.e., if $x$ and $y$ are in $A_1$, and $G(x) = G(y)$, then $x = y$),
3. $G(A_1)$ is open in $\mathbb{R}^3$,
4. the 1–1 function $G^{-1}$ from $G(A_1)$ onto $A_1$ is a $C^2$ function.

Now,  since  $(F_1(0,0,0), F_2(0,0,0), F_3(0,0,0), F_4(0,0,0)) = F(0,0,0) = (0,0,0,0)$, each $F_i(0,0,0) = 0$. Next, since $(0,0,0)$ is in $A_0$, and $G : A_0 \to \mathbb{R}^3$, $G(0,0,0) = (F_1(0,0,0), F_2(0,0,0), 0) = (0,0,0)$.  Thus,  $G(0,0,0) = (0,0,0)$. Since $G$ is 1–1 on $A_1$, $G(0,0,0) = (0,0,0)$, and $(0,0,0)$ is in $A_1$, $G^{-1}(0,0,0) = (0,0,0)$.  Now,  $(F \circ G^{-1})(0,0,0) = F(G^{-1}(0,0,0)) = F(0,0,0) = (0,0,0,0)$. Thus, $(F \circ G^{-1})(0,0,0) = (0,0,0,0)$. Since $A_1$ is contained in $A_0$, and $F : A_0 \to B_0$, $(F \circ G^{-1})(G(A_1)) = F(A_1) \subset B_0$.  Thus,  $F \circ G^{-1} : G(A_1) \to B_0$.  For  every $(x_1, x_2, x_3)$ in $G(A_1)$, there  exists $(y_1, y_2, y_3)$ in $A_1$ such that $(x_1, x_2, x_3) = G(y_1, y_2, y_3) = (F_1(y_1, y_2, y_3), F_2(y_1, y_2, y_3), y_3)$. Further, since $G$ is 1–1 on $A_1$,

$$\begin{aligned} (F \circ G^{-1})(x_1, x_2, x_3) &= F(G^{-1}(x_1, x_2, x_3)) = F(y_1, y_2, y_3) \\ &= (F_1(y_1, y_2, y_3), F_2(y_1, y_2, y_3), F_3(y_1, y_2, y_3), F_4(y_1, y_2, y_3)) \\ &= (x_1, x_2, F_3(y_1, y_2, y_3), F_4(y_1, y_2, y_3)) \\ &= (x_1, x_2, F_3(G^{-1}(x_1, x_2, x_3)), F_4(G^{-1}(x_1, x_2, x_3))) \\ &= (x_1, x_2, (F_3 \circ G^{-1})(x_1, x_2, x_3), (F_4 \circ G^{-1})(x_1, x_2, x_3)). \end{aligned}$$

Thus, for every $(x_1, x_2, x_3)$ in $G(A_1)$,

$$(F \circ G^{-1})(x_1, x_2, x_3) = (x_1, x_2, (F_3 \circ G^{-1})(x_1, x_2, x_3), (F_4 \circ G^{-1})(x_1, x_2, x_3)).$$

Hence, for every $x \equiv (x_1, x_2, x_3)$ in $G(A_1)$,

$$(F \circ G^{-1})'(x) = \begin{bmatrix} 1 & 0 & 0 \\ 0 & 1 & 0 \\ (D_1(F_3 \circ G^{-1}))(x_1, x_2, x_3) & (D_2(F_3 \circ G^{-1}))(x_1, x_2, x_3) & (D_3(F_3 \circ G^{-1}))(x_1, x_2, x_3) \\ (D_1(F_4 \circ G^{-1}))(x_1, x_2, x_3) & (D_2(F_4 \circ G^{-1}))(x_1, x_2, x_3) & (D_3(F_4 \circ G^{-1}))(x_1, x_2, x_3) \end{bmatrix}.$$

Also, for every $x \equiv (x_1, x_2, x_3)$ in $G(A_1)$,

$$\begin{bmatrix} 1 & 0 & 0 \\ 0 & 1 & 0 \\ 0 & 0 & 1 \end{bmatrix} = I'(x) = (G \circ G^{-1})'(x) = (G'(G^{-1}(x))) \circ ((G^{-1})'(x)),$$

so

$$0 \neq 1 = \det \begin{bmatrix} 1 & 0 & 0 \\ 0 & 1 & 0 \\ 0 & 0 & 1 \end{bmatrix} = \det \left( (G'(G^{-1}(x))) \circ ((G^{-1})'(x)) \right)$$
$$= \left( \det(G'(G^{-1}(x))) \right) \left( \det \left( (G^{-1})'(x) \right) \right).$$

It follows that $\det((G^{-1})'(x)) \neq 0$, and hence, $(G^{-1})'(x)$ is invertible for every $x$ in $G(A_1)$. Since for every $x \equiv (x_1, x_2, x_3)$ in $G(A_1)$, $(F \circ G^{-1})'(x) = (F'(G^{-1}(x))) \circ ((G^{-1})'(x))$, and $(G^{-1})'(x)$ is invertible,

$$\text{rank} \begin{bmatrix} 1 & 0 & 0 \\ 0 & 1 & 0 \\ (D_1(F_3 \circ G^{-1}))(x_1, x_2, x_3) & (D_2(F_3 \circ G^{-1}))(x_1, x_2, x_3) & (D_3(F_3 \circ G^{-1}))(x_1, x_2, x_3) \\ (D_1(F_4 \circ G^{-1}))(x_1, x_2, x_3) & (D_2(F_4 \circ G^{-1}))(x_1, x_2, x_3) & (D_3(F_4 \circ G^{-1}))(x_1, x_2, x_3) \end{bmatrix}$$
$$= \text{rank} \left( (F \circ G^{-1})'(x) \right) = \text{rank} \left( (F'(G^{-1}(x))) \circ ((G^{-1})'(x)) \right) = \text{rank}(F'(G^{-1}(x))) = 2.$$

Thus, for every $x \equiv (x_1, x_2, x_3)$ in $G(A_1)$,

$$(D_3(F_3 \circ G^{-1}))(x_1, x_2, x_3)$$
$$= \det \begin{bmatrix} 1 & 0 & 0 \\ 0 & 1 & 0 \\ (D_1(F_3 \circ G^{-1}))(x_1, x_2, x_3) & (D_2(F_3 \circ G^{-1}))(x_1, x_2, x_3) & (D_3(F_3 \circ G^{-1}))(x_1, x_2, x_3) \end{bmatrix} = 0,$$

and

$$(D_3(F_4 \circ G^{-1}))(x_1, x_2, x_3)$$
$$= \det \begin{bmatrix} 1 & 0 & 0 \\ 0 & 1 & 0 \\ (D_1(F_4 \circ G^{-1}))(x_1, x_2, x_3) & (D_2(F_4 \circ G^{-1}))(x_1, x_2, x_3) & (D_3(F_4 \circ G^{-1}))(x_1, x_2, x_3) \end{bmatrix} = 0.$$

It follows that for every $x \equiv (x_1, x_2, x_3)$ in $G(A_1)$, $F_3 \circ G^{-1}$ and $F_4 \circ G^{-1}$ are functions of $x_1, x_2$ only.

Since $G(0,0,0) = (0,0,0)$, $A_1$ is an open neighborhood of $(0,0,0)$, and $G(A_1)$ is open in $\mathbb{R}^3$, $G(A_1)$ is an open neighborhood of $(0,0,0)$, and hence, $G(A_1) \times \mathbb{R}$ is an open neighborhood of $(0,0,0,0)$ in $\mathbb{R}^4$. Let us define a function $T : G(A_1) \times \mathbb{R} \to \mathbb{R}^4$ as follows: For every $(y_1, y_2, y_3, y_4)$ in $G(A_1) \times \mathbb{R}$,

$$T(y_1, y_2, y_3, y_4) \equiv \left(y_1, y_2, y_3 + (F_3 \circ G^{-1})(y_1, y_2, y_3), y_4 + (F_4 \circ G^{-1})(y_1, y_2, y_3)\right).$$

Here, each component function of $T$ is $C^2$, so $T$ is a $C^2$ function, and hence, $T$ is continuous. Here,

$$\left((F_1 \circ G^{-1})(0,0,0), (F_2 \circ G^{-1})(0,0,0), (F_3 \circ G^{-1})(0,0,0), (F_4 \circ G^{-1})(0,0,0)\right)$$
$$= \left(F_1(G^{-1}(0,0,0)), F_2(G^{-1}(0,0,0)), F_3(G^{-1}(0,0,0)), F_4(G^{-1}(0,0,0))\right)$$
$$= F(G^{-1}(0,0,0)) = (F \circ G^{-1})(0,0,0) = (0,0,0,0),$$

so

$$T(0,0,0,0) = \left(0,0,0 + (F_3 \circ G^{-1})(0,0,0), 0 + (F_4 \circ G^{-1})(0,0,0)\right)$$
$$= \left(0,0, (F_3 \circ G^{-1})(0,0,0), (F_4 \circ G^{-1})(0,0,0)\right) = (0,0,0,0).$$

Since $T : G(A_1) \times \mathbb{R} \to \mathbb{R}^4$ is continuous, $T(0,0,0,0) = (0,0,0,0)$, and $B_0$ be an open neighborhood of $(0,0,0,0)$ in $\mathbb{R}^4$, there exists an open neighborhood $V_1$ of $(0,0,0,0)$ such that $V_1$ is contained in $G(A_1) \times \mathbb{R}$, and $T(V_1)$ is contained in $B_0$.

Now, for every $y \equiv (y_1, y_2, y_3, y_4)$ in $V_1$,

$$T'(y_1, y_2, y_3, y_4)$$
$$= \begin{bmatrix} 1 & 0 \\ 0 & 1 \\ D_1(y_3 + (F_3 \circ G^{-1})(y_1, y_2, y_3)) & D_2(y_3 + (F_3 \circ G^{-1})(y_1, y_2, y_3)) \\ D_1(y_4 + (F_4 \circ G^{-1})(y_1, y_2, y_3)) & D_2(y_4 + (F_4 \circ G^{-1})(y_1, y_2, y_3)) \\ 0 & 0 \\ 0 & 0 \\ D_3(y_3 + (F_3 \circ G^{-1})(y_1, y_2, y_3)) & D_4(y_3 + (F_3 \circ G^{-1})(y_1, y_2, y_3)) \\ D_3(y_4 + (F_4 \circ G^{-1})(y_1, y_2, y_3)) & D_4(y_4 + (F_4 \circ G^{-1})(y_1, y_2, y_3)) \end{bmatrix}$$

$$= \begin{bmatrix} 1 & 0 \\ 0 & 1 \\ 0+D_1(F_3 \circ G^{-1})(y_1,y_2,y_3) & 0+D_2(F_3 \circ G^{-1})(y_1,y_2,y_3) \\ 0+D_1(F_3 \circ G^{-1})(y_1,y_2,y_3) & 0+D_2(F_4 \circ G^{-1})(y_1,y_2,y_3) \end{bmatrix}$$

$$\begin{matrix} 0 & 0 \\ 0 & 0 \\ 1+D_3(F_3 \circ G^{-1})(y_1,y_2,y_3) & 0 \\ 0+D_3(F_4 \circ G^{-1})(y_1,y_2,y_3) & 1+D_4(F_4 \circ G^{-1})(y_1,y_2,y_3) \end{matrix}$$

$$= \begin{bmatrix} 1 & 0 \\ 0 & 1 \\ D_1(F_3 \circ G^{-1})(y_1,y_2,y_3) & D_2(F_3 \circ G^{-1})(y_1,y_2,y_3) \\ D_1(F_3 \circ G^{-1})(y_1,y_2,y_3) & D_2(F_4 \circ G^{-1})(y_1,y_2,y_3) \end{bmatrix}$$

$$\begin{matrix} 0 & 0 \\ 0 & 0 \\ 1+D_3(F_3 \circ G^{-1})(y_1,y_2,y_3) & 0 \\ D_3(F_4 \circ G^{-1})(y_1,y_2,y_3) & 1+D_4(F_4 \circ G^{-1})(y_1,y_2,y_3) \end{matrix}$$

$$= \begin{bmatrix} 1 & 0 & 0 & 0 \\ 0 & 1 & 0 & 0 \\ D_1(F_3 \circ G^{-1})(y_1,y_2,y_3) & D_2(F_3 \circ G^{-1})(y_1,y_2,y_3) & 1+0 & 0 \\ D_1(F_3 \circ G^{-1})(y_1,y_2,y_3) & D_2(F_4 \circ G^{-1})(y_1,y_2,y_3) & 0 & 1+0 \end{bmatrix}$$

$$= \begin{bmatrix} 1 & 0 & 0 & 0 \\ 0 & 1 & 0 & 0 \\ D_1(F_3 \circ G^{-1})(y_1,y_2,y_3) & D_2(F_3 \circ G^{-1})(y_1,y_2,y_3) & 1 & 0 \\ D_1(F_3 \circ G^{-1})(y_1,y_2,y_3) & D_2(F_4 \circ G^{-1})(y_1,y_2,y_3) & 0 & 1 \end{bmatrix}$$

Hence, for every $y \equiv (y_1,y_2,y_3,y_4)$ in $V_1$,

$$\det(T'(y_1,y_2,y_3,y_4)) = \det \begin{bmatrix} 1 & 0 & 0 & 0 \\ 0 & 1 & 0 & 0 \\ D_1(F_3 \circ G^{-1})(y_1,y_2,y_3) & D_2(F_3 \circ G^{-1})(y_1,y_2,y_3) & 1 & 0 \\ D_1(F_3 \circ G^{-1})(y_1,y_2,y_3) & D_2(F_4 \circ G^{-1})(y_1,y_2,y_3) & 0 & 1 \end{bmatrix} = 1 \neq 0.$$

Since for every $y \equiv (y_1,y_2,y_3,y_4)$ in $V_1$, $\det(T'(y_1,y_2,y_3,y_4))$ is nonzero, $T'(y_1,y_2,y_3,y_4)$ is invertible, and hence, $T'(0,0,0,0)$ is invertible. Now, we can apply the inverse function theorem on $T : V_1 \to \mathbb{R}^4$. There exists an open neighborhood $V$ of $(0,0,0,0)$ such that

1. $V$ is contained in $V_1$,
2. $T$ is 1–1 on $V$ (i.e., if $x$ and $y$ are in $V$, and $T(x) = T(y)$, then $x = y$),
3. $T(V)$ is open in $\mathbb{R}^4$,
4. the 1–1 function $T^{-1}$ from $T(V)$ onto $V$ is a $C^2$ function.

Here, $(0,0,0,0)$ is in $V$, and $T(0,0,0,0) = (0,0,0,0)$, so $(0,0,0,0)$ is in $T(V)$. Further, since $T(V)$ is open in $\mathbb{R}^4$, $T(V)$ is an open neighborhood of $(0,0,0,0)$ in $\mathbb{R}^4$. Since $G^{-1}$ is a $C^2$ function from $G(A_1)$ onto $A_1$, $A_1$ is contained in $A_0$, and $F : A_0 \rightarrow B_0$ is a $C^2$ function, their composite function $F \circ G^{-1}\, G(A_1) \rightarrow B_0$ is $C^2$. Since $G(A_1)$ is an open neighborhood of $(0,0,0)$, $(0,0,0)$ is in $G(A_1)$. Since $(0,0,0)$ is in $G(A_1)$, and $F \circ G^{-1} : G(A_1) \rightarrow B_0$ is $C^2$, $F \circ G^{-1} : G(A_1) \rightarrow B_0$ is continuous at $(0,0,0)$. Since $F \circ G^{-1} : G(A_1) \rightarrow B_0$ is continuous at $(0,0,0)$, $(F \circ G^{-1})(0,0,0) = (0,0,0,0)$, and $T(V)$ is an open neighborhood of $T(0,0,0,0)(= (0,0,0,0))$, there exists an open neighborhood $U$ of $(0,0,0)$ such that $U$ is contained in $G(A_1)$ and $(F \circ G^{-1})(U)$ is contained in $T(V)$.

Now, let us take the 1–1 function $T^{-1}$ from $T(V)$ onto $V$ for $H$, the open set $T(V)$ for $B$, and $G^{-1}(U)$ for $A$. It remains to prove the following:

(i)    $A$ is an open neighborhood of $(0,0,0)$,

(ii)   $B$ is an open neighborhood of $(0,0,0,0)$,

(iii)  $U$ is an open subset of $\mathbb{R}^3$,

(iv)   $V$ is an open subset of $\mathbb{R}^4$,

(v)    $G : A \rightarrow U$ is 1–1,

(vi)   $G : A \rightarrow U$ is onto,

(vii)  $G : A \rightarrow U$ is $C^2$,

(viii) $G^{-1} : U \rightarrow A$ is $C^2$,

(ix)   $H : B \rightarrow V$ is 1–1,

(x)    $H : B \rightarrow V$ is onto,

(xi)   $H : B \rightarrow V$ is $C^2$,

(xii)  $H^{-1} : V \rightarrow B$ is $C^2$,

(xiii) $A$ is contained in $A_0$,

(xiv)  $B$ is contained in $B_0$,

(xv)   $F(A) \subset B$

(xvi)  For every $x \equiv (x_1, x_2, x_3)$ in $U$, $(H \circ F \circ G^{-1})(x_1, x_2, x_3) = (x_1, x_2, 0, 0)$.

For (i): Since $G : A_1 \rightarrow G(A_1)$ is a $C^2$ function, and $(0,0,0)$ is in $A_1$, $G : A_1 \rightarrow G(A_1)$ is continuous at $(0,0,0)$. Since $G$ is 1–1 on $A_1$, $G : A_1 \rightarrow G(A_1)$ is continuous at $(0,0,0)$, and $U$ is an open neighborhood of $G(0,0,0)(= (0,0,0))$ such that $U$ is contained in the open set $G(A_1)$, $G^{-1}(U)$ is an open neighborhood of $(0,0,0)$ and is contained in $A_1$. Thus, $A$ is an open neighborhood of $(0,0,0)$.
For (ii): Since $T(V)$ is an open neighborhood of $(0,0,0,0)$ in $\mathbb{R}^4$, and $B$ stands for $T(V)$, $B$ is an open neighborhood of $(0,0,0,0)$ in $\mathbb{R}^4$.
For (iii): This has been shown above.
For (iv): This has been shown above.
For (v): Since $U$ is contained in $G(A_1)$, and $G$ is 1–1 on $A_1$, $G^{-1}(U)$ is contained in $A_1$. Since $G$ is 1–1 on $A_1$, $G^{-1}(U)$ is contained in $A_1$, and $A$ stands for $G^{-1}(U)$, $G : A \rightarrow U$ is 1–1.
For (vi): Since $G : A \rightarrow U$ is 1–1, $G(A) = G(G^{-1}(U)) = U$. So $G : A \rightarrow U$ is onto.

For (vii): Since $G$ is $C^2$ on $A_0$, and $A = G^{-1}(U) \subset A_1 \subset A_0$, $G$ is $C^2$ on $A$.

For (viii): Since $G^{-1}(U) \subset A_1$, and $G$ is 1–1 on $A_1$, $U \subset G(A_1)$. Since $U \subset G(A_1)$, and the function $G^{-1}$ from $G(A_1)$ onto $A_1$ is a $C^2$ function, $G^{-1} : U \to G^{-1}(U)$ is $C^2$, that is, $G^{-1} : U \to A$ is $C^2$.

For (ix): Since $T$ is 1–1 on $V$, $T^{-1} : T(V) \to V$ is a 1–1 function. Since $H$ stands for the function $T^{-1}$ and $B$ stands for $T(V)$, $H : B \to V$ is 1–1.

For (x): Since $T$ is 1–1 on $V$, $T^{-1} : T(V) \to V$ is onto, and hence, $H : B \to V$ is onto.

For (xi): Since the function $T^{-1}$ from $T(V)$ onto $V$ is a $C^2$ function, $H : B \to V$ is $C^2$.

For (xii): Since $T : G(A_1) \times \mathbb{R} \to \mathbb{R}^4$ is a $C^2$ function, and $V$ is an open subset of the open set $G(A_1) \times \mathbb{R}$, $T : V \to T(V)$ is a $C^2$ function, and hence, $H^{-1} : V \to B$ is $C^2$.

For (xiii): Since $U$ is contained in $G(A_1)$, and $G$ is 1–1 on $A_1$, $G^{-1}(U)$ is contained in $A_1 (\subset A_0)$ and hence, $A$ is contained in $A_0$.

For (xiv): Since $V$ is contained in $V_1$, and $T(V_1)$ is contained in $B_0$, $T(V)$ is contained in $B_0$, and hence, $B$ is contained in $B_0$.

For (xv): Here, $F(A) = F(G^{-1}(U)) = (F \circ G^{-1})(U) \subset T(V) = B$. Thus, $F(A) \subset B$.

For (xvi): Let us take any $x \equiv (x_1, x_2, x_3)$ in $U$. Since $(x_1, x_2, x_3)$ is in $U$, $(x_1, x_2, (F_3 \circ G^{-1})(x_1, x_2, x_3), (F_4 \circ G^{-1})(x_1, x_2, x_3)) = (F \circ G^{-1})(x_1, x_2, x_3) \in (F \circ G^{-1})(U) \subset T(V)$. Next, since $T : V \to T(V)$ is 1–1 onto,

$$
\begin{aligned}
(H \circ F \circ G^{-1})(x_1, x_2, x_3) &= H\big((F \circ G^{-1})(x_1, x_2, x_3)\big) \\
&= H\big(x_1, x_2, (F_3 \circ G^{-1})(x_1, x_2, x_3), (F_4 \circ G^{-1})(x_1, x_2, x_3)\big) \\
&= T^{-1}\big(x_1, x_2, (F_3 \circ G^{-1})(x_1, x_2, x_3), (F_4 \circ G^{-1})(x_1, x_2, x_3)\big).
\end{aligned}
$$

It remains to prove that

$$
T^{-1}\big(x_1, x_2, (F_3 \circ G^{-1})(x_1, x_2, x_3), (F_4 \circ G^{-1})(x_1, x_2, x_3)\big) = (x_1, x_2, 0, 0).
$$

Let $T^{-1}(x_1, x_2, (F_3 \circ G^{-1})(x_1, x_2, x_3), (F_4 \circ G^{-1})(x_1, x_2, x_3)) = (y_1, y_2, y_3, y_4) \in V \subset V_1 \subset G(A_1) \times \mathbb{R}$.

It follows that $(y_1, y_2, y_3) \in G(A_1)$. Since $(y_1, y_2, y_3) \in G(A_1)$, and $(x_1, x_2, x_3) \in U \subset G(A_1)$,

$$
\begin{aligned}
\big(x_1, x_2, (F_3 \circ G^{-1})(x_1, x_2, x_3), (F_4 \circ G^{-1})(x_1, x_2, x_3)\big) &= T(y_1, y_2, y_3, y_4) \\
&= \big(y_1, y_2, y_3 + (F_3 \circ G^{-1})(y_1, y_2, y_3), y_4 + (F_4 \circ G^{-1})(y_1, y_2, y_3)\big).
\end{aligned}
$$

Hence,

$$y_1 = x_1, \ y_2 = x_2, \ \left(F_3 \circ G^{-1}\right)(x_1, x_2, x_3) = y_3 + \left(F_3 \circ G^{-1}\right)(y_1, y_2, y_3),$$

and

$$\left(F_4 \circ G^{-1}\right)(x_1, x_2, x_3) = y_4 + \left(F_4 \circ G^{-1}\right)(y_1, y_2, y_3).$$

It follows that

$$\left(F_3 \circ G^{-1}\right)(x_1, x_2, x_3) = y_3 + \left(F_3 \circ G^{-1}\right)(x_1, x_2, y_3),$$

and

$$\left(F_4 \circ G^{-1}\right)(x_1, x_2, x_3) = y_4 + \left(F_4 \circ G^{-1}\right)(x_1, x_2, y_3).$$

Since $(y_1, y_2, y_3) \in G(A_1)$, $y_1 = x_1$, $y_2 = x_2$, $(x_1, x_2, y_3) \in G(A_1)$. Since for every $(\xi, \eta, \varsigma)$ in $G(A_1)$, $F_3 \circ G^{-1}$ is a function of $\xi, \eta$ only, and $(x_1, x_2, y_3), (x_1, x_2, x_3) \in G(A_1)$, $(F_3 \circ G^{-1})(x_1, x_2, y_3) = (F_3 \circ G^{-1})(x_1, x_2, x_3)$. Since $(F_3 \circ G^{-1})(x_1, x_2, y_3) - (F_3 \circ G^{-1})(x_1, x_2, x_3)$, and $(F_3 \circ G^{-1})(x_1, x_2, x_3) - y_3 + (F_3 \circ G^{-1})(x_1, x_2, y_3)$, $y_3 = 0$. Since for every $(\xi, \eta, \varsigma)$ in $G(A_1)$, $F_4 \circ G^{-1}$ is a function of $\xi, \eta$ only, and $(x_1, x_2, y_3), (x_1, x_2, x_3) \in G(A_1)$, $(F_4 \circ G^{-1})(x_1, x_2, y_3) = (F_4 \circ G^{-1})(x_1, x_2, x_3)$.

Now, since $(F_4 \circ G^{-1})(x_1, x_2, y_3) = (F_4 \circ G^{-1})(x_1, x_2, x_3)$, and $(F_4 \circ G^{-1})(x_1, x_2, x_3) = y_4 + (F_4 \circ G^{-1})(x_1, x_2, y_3)$, $y_4 = 0$. Since $y_3 = 0, y_4 = 0, y_1 = x_1$, $y_2 = x_2$, and $T^{-1}(x_1, x_2, (F_3 \circ G^{-1})(x_1, x_2, x_3), (F_4 \circ G^{-1})(x_1, x_2, x_3)) = (y_1, y_2, y_3, y_4)$,

$$T^{-1}\left(x_1, x_2, \left(F_3 \circ G^{-1}\right)(x_1, x_2, x_3), \left(F_4 \circ G^{-1}\right)(x_1, x_2, x_3)\right) = (x_1, x_2, 0, 0). \quad \square$$

**Theorem 3.58** *Let $A_0$ be an open neighborhood of $(0,0,0)$ in $\mathbb{R}^3$. Let $B_0$ be an open neighborhood of $b \equiv (b_1, b_2, b_3, b_4)$ in $\mathbb{R}^4$. Let $F : A_0 \to B_0$ be a function. If*

(i) *$F$ is a $C^2$ function,*
(ii) *for every $x$ in $A_0$, the linear transformation $F'(x)$ has rank 2,*
(iii) *$F(0,0,0) = b$,*

*then there exist an open neighborhood $A$ of $(0,0,0)$ in $\mathbb{R}^3$, an open neighborhood $B$ of $b$ in $\mathbb{R}^4$, an open subset $U$ of $\mathbb{R}^3$, an open subset $V$ of $\mathbb{R}^4$, a function $G : A \to U$, and a function $H : B \to V$ such that*

1. *$A$ is contained in $A_0$, and $B$ is contained in $B_0$,*
2. *$G$ is a $C^2$ diffeomorphism (i.e., $G$ is 1–1 onto, $G$ is a $C^2$ function, and $G^{-1}$ is a $C^2$ function),*
3. *$H$ is a $C^2$ diffeomorphism,*

4. $F(A) \subset B$,
5. For every $x \equiv (x_1, x_2, x_3)$ in $U$, $(H \circ F \circ G^{-1})(x_1, x_2, x_3) = (x_1, x_2, 0, 0)$.

*Proof* Put $\hat{A}_0 \equiv A_0$, and $\hat{B}_0 \equiv B_0 - b$. Let us define a function $\hat{F} : \hat{A}_0 \to \hat{B}_0$ as follows: For every $x$ in $\hat{A}_0$ (since $x$ is in $\hat{A}_0(= A_0)$, $x$ is in $A_0$, and hence, $F(x)$ is defined.)

$$\hat{F}(x) \equiv F(x) - b = (T_{-b} \circ F)(x),$$

where $T_{-b} : y \mapsto (y - b)$ is a translation. Thus, $\hat{F} = T_{-b} \circ F$. Hence, $T_b \circ \hat{F} = F$. Also, $\hat{F}(0,0,0) = F(0,0,0) - b = b - b = (0,0,0,0)$. Thus, $\hat{F}(0,0,0) = (0,0,0,0)$. Since $A_0$ is an open neighborhood of $(0,0,0)$, $\hat{A}_0(= A_0)$ is an open neighborhood of 0. Since $B_0$ is an open neighborhood of $b$, $\hat{B}_0(= B_0 - b)$ is an open neighborhood of 0. Since $F$ is a $C^2$ function, and $T_{-b}$ is a $C^2$ function, their composite $T_{-b} \circ F(= \hat{F})$ is a $C^2$ function. Thus, $\hat{F}$ is a $C^2$ function. For every $x$ in $\hat{A}_0$, $\hat{F}'(x) = (T_{-b} \circ F)'(x) = ((T_{-b})'(F(x))) \circ (F'(x)) = I_4 \circ (F'(x)) = F'(x)$. Thus, for every $x$ in $\hat{A}_0$, $\hat{F}'(x) = F'(x)$. If $x$ is in $\hat{A}_0$, then $x$ is in $A_0$, and hence, $\hat{F}'(x)(= F'(x))$ has rank 2.

Thus, we see that all the conditions of Theorem 3.57 are satisfied for $\hat{F} : \hat{A}_0 \to \hat{B}_0$. Hence, by Theorem 3.57, there exist an open neighborhood $\hat{A}$ of $(0,0,0)$ in $\mathbb{R}^3$, an open neighborhood $\hat{B}$ of $(0,0,0,0)$ in $\mathbb{R}^4$, an open subset $\hat{U}$ of $\mathbb{R}^3$, an open subset $\hat{V}$ of $\mathbb{R}^4$, a function $\hat{G} : \hat{A} \to \hat{U}$, and a function $\hat{H} : \hat{B} \to \hat{V}$ such that

1. $\hat{A}$ is contained in $\hat{A}_0$, and $\hat{B}$ is contained in $\hat{B}_0$,
2. $\hat{G}$ is a $C^2$ diffeomorphism (i.e., $\hat{G}$ is 1–1 onto, $\hat{G}$ is a $C^2$ function, and $\hat{G}^{-1}$ is a $C^2$ function),
3. $\hat{H}$ is a $C^2$ diffeomorphism,
4. $\hat{F}(\hat{A}) \subset \hat{B}$,
5. For every $x \equiv (x_1, x_2, x_3)$ in $\hat{U}$, $(\hat{H} \circ \hat{F} \circ \hat{G}^{-1})(x_1, x_2, x_3) = (x_1, x_2, 0, 0)$.

Put $A \equiv \hat{A}$, $B \equiv \hat{B} + b$, $U \equiv \hat{U}$, and $V \equiv \hat{V}$. Let us define a function $G : A \to U$ as follows: For every $x$ in $A$, $G(x) \equiv \hat{G}(x)$. Let us define a function $H : B \to V$ as follows: For every $y$ in $B$, $H(y) \equiv \hat{H}(y - b) = (\hat{H} \circ T_{-b})(y)$. Hence, $G = \hat{G}$, and $H = \hat{H} \circ T_{-b}$. Further, $\hat{H} = H \circ (T_{-b})^{-1} = H \circ T_b$.

It remains to prove the following:

(i) $A$ is an open neighborhood of $(0,0,0)$,
(ii) $B$ is an open neighborhood of $b$,
(iii) $U$ is an open subset of $\mathbb{R}^3$,
(iv) $V$ is an open subset of $\mathbb{R}^4$,
(v) $G : A \to U$ is 1–1,
(vi) $G : A \to U$ is onto,

(vii)   $G : A \to U$ is $C^2$,

(viii)  $G^{-1} : U \to A$ is $C^2$,

 (ix)   $H : B \to V$ is 1–1,

  (x)   $H : B \to V$ is onto,

 (xi)   $H : B \to V$ is $C^2$,

(xii)   $H^{-1} : V \to B$ is $C^2$,

(xiii)  $A$ is contained in $A_0$,

(xiv)   $B$ is contained in $B_0$,

 (xv)   $F(A) \subset B$,

(xvi)   For every $x \equiv (x_1, x_2, x_3)$ in $U$, $(H \circ F \circ G^{-1})(x_1, x_2, x_3) = (x_1, x_2, 0, 0)$.

For (i): Since $\hat{A}$ is an open neighborhood of $(0,0,0)$, $A(= \hat{A})$ is an open neighborhood of $(0,0,0)$.

For (ii): Since $\hat{B}$ is an open neighborhood of $0$, $B(= \hat{B} + b)$ is an open neighborhood of $b$.

For (iii): Since $\hat{U}$ is an open subset of $\mathbb{R}^3$, $U(= \hat{U})$ is an open subset of $\mathbb{R}^3$.

For (iv): Since $\hat{V}$ is an open subset of $\mathbb{R}^4$, $V(= \hat{V})$ is an open subset of $\mathbb{R}^4$.

For (v): Since $\hat{G} : \hat{A} \to \hat{U}$ is 1–1, $\hat{G} = G$, $A = \hat{A}$, and $U = \hat{U}$, $G : A \to U$ is 1–1.

For (vi): Since $\hat{G} : \hat{A} \to \hat{U}$ is onto, $G = \hat{G}$, $A = \hat{A}$, and $U = \hat{U}$, $G : A \to U$ is onto.

For (vii): Since $\hat{G}$ is $C^2$, and $G = \hat{G}$, $G$ is $C^2$.

For (viii): Since $\hat{G}^{-1}$ is $C^2$, and $G = \hat{G}$, $G^{-1}$ is $C^2$.

For (ix): Since $\hat{H} : \hat{B} \to \hat{V}$ is 1–1, and $T_{-b} : \hat{B} + b \to \hat{B}$ is 1–1, $(H =)\hat{H} \circ T_{-b} :$ $\hat{B} + b(= B) \to \hat{V}(= V)$ is 1–1. Hence, $H : B \to V$ is 1–1.

For (x): Since $\hat{H} : \hat{B} \to \hat{V}$ is onto, and $T_{-b} : \hat{B} + b \to \hat{B}$ is onto, $(H =)\hat{H} \circ T_{-b} :$ $\hat{B} + b(= B) \to \hat{V}(= V)$ is onto. Hence, $H : B \to V$ is onto.

For (xi): Since $\hat{H} : \hat{B} \to \hat{V}$ is $C^2$, and $T_{-b} : \hat{B} + b \to \hat{B}$ is $C^2$, $(H =)\hat{H} \circ T_{-b} :$ $\hat{B} + b(= B) \to \hat{V}(= V)$ is $C^2$. Hence, $H : B \to V$ is $C^2$.

For (xii): Here $H^{-1} = (\hat{H} \circ T_{-b})^{-1} = (T_{-b})^{-1} \circ \hat{H}^{-1} = T_b \circ \hat{H}^{-1}$. Since $\hat{H}^{-1} :$ $\hat{V}(= V) \to \hat{B}$ is $C^2$, and the translation $T_b : \hat{B} \to \hat{B} + b(= B)$ is $C^2$, their composite $(H^{-1} =)T_b \circ \hat{H}^{-1} : V \to B$ is $C^2$, and hence, $H^{-1} : V \to B$ is $C^2$.

For (xiii): Since $\hat{A}$ is contained in $\hat{A}_0$, $\hat{A} = A$, and $\hat{A}_0 = A_0$, $A$ is contained in $A_0$.

For (xiv): Since $\hat{B}$ is contained in $\hat{B}_0$, $\hat{B} = B - b$, and $\hat{B}_0 = B_0 - b$, $B - b$ is contained in $B_0 - b$, and hence, $B$ is contained in $B_0$.

For (xv): Let us take any $x$ in $A$. We have to prove that $F(x)$ is in $B$. Here, $x$ is in $A$, and $A = \hat{A}$, so $x$ is in $\hat{A}$. Since $x$ is in $\hat{A}$, and $\hat{F}(\hat{A}) \subset \hat{B} = B - b$, $F(x) - b = \hat{F}(x) \in B - b$, and hence, $F(x)$ is in $B$.

For (xvi): Let us take any $x \equiv (x_1, x_2, x_3)$ in $U$. We have to show that $(H \circ F \circ G^{-1})(x_1, x_2, x_3) = (x_1, x_2, 0, 0)$.

$$\begin{aligned} \text{LHS} &= \left(H \circ F \circ G^{-1}\right)(x_1, x_2, x_3) = \left(H \circ \left(T_b \circ \hat{F}\right) \circ G^{-1}\right)(x_1, x_2, x_3) \\ &= \left((H \circ T_b) \circ \hat{F} \circ G^{-1}\right)(x_1, x_2, x_3) = \left(\hat{H} \circ \hat{F} \circ \hat{G}^{-1}\right)(x_1, x_2, x_3) \\ &= (x_1, x_2, 0, 0) = \text{RHS}. \end{aligned}$$

$\square$

**Theorem 3.59** *Let $A_0$ be an open neighborhood of $a \equiv (a_1, a_2, a_3)$ in $\mathbb{R}^3$. Let $B_0$ be an open neighborhood of $b \equiv (b_1, b_2, b_3, b_4)$ in $\mathbb{R}^4$. Let $F : A_0 \to B_0$ be a function. If*

(i) *$F$ is a $C^2$ function,*
(ii) *for every $x$ in $A_0$, the linear transformation $F'(x)$ has rank 2,*
(iii) *$F(a) = b$,*

*then there exist an open neighborhood $A$ of $a$ in $\mathbb{R}^3$, an open neighborhood $B$ of $b$ in $\mathbb{R}^4$, an open subset $U$ of $\mathbb{R}^3$, an open subset $V$ of $\mathbb{R}^4$, a function $G : A \to U$, and a function $H : B \to V$ such that*

1. *$A$ is contained in $A_0$, and $B$ is contained in $B_0$.*
2. *$G$ is a $C^2$ diffeomorphism (i.e., $G$ is 1–1 onto, $G$ is a $C^2$ function, and $G^{-1}$ is a $C^2$ function),*
3. *$H$ is a $C^2$ diffeomorphism,*
4. *$F(A) \subset B$,*
5. *For every $x \equiv (x_1, x_2, x_3)$ in $U$, $(H \circ F \circ G^{-1})(x_1, x_2, x_3) = (x_1, x_2, 0, 0)$.*

*Proof* Put $\hat{A}_0 \equiv A_0 - a$, and $\hat{B}_0 \equiv B_0$. Let us define a function $\hat{F} : \hat{A}_0 \to \hat{B}_0$ as follows: For every $x$ in $\hat{A}_0$ (since $x$ is in $\hat{A}_0 = A_0 - a$, $x + a$ is in $A_0$, and hence, $F(x + a)$ is defined.)

$$\hat{F}(x) \equiv F(x + a) = (F \circ T_a)(x),$$

where $T_a : x \mapsto (x + a)$ is a translation. Thus, $\hat{F} = F \circ T_a$. Hence, $\hat{F} \circ T_{-a} = F$. Also, $\hat{F}(0, 0, 0) = F(a) = b$. Thus, $\hat{F}(0, 0, 0) = b$. Since $A_0$ is an open neighborhood of $a$, $\hat{A}_0 (= A_0 - a)$ is an open neighborhood of 0. Since $B_0$ is an open neighborhood of $b$, $\hat{B}_0 (= B_0)$ is an open neighborhood of $b$. Since $F$ is a $C^2$ function, and $T_a$ is a $C^2$ function, their composite $F \circ T_a (= \hat{F})$ is a $C^2$ function. Thus, $\hat{F}$ is a $C^2$ function. For every $x$ in $\hat{A}_0$,

$$\hat{F}'(x) = (F \circ T_a)'(x) = (F'(T_a(x))) \circ ((T_a)'(x)) = (F'(x + a)) \circ I_3 = F'(x + a).$$

Thus, for every $x$ in $\hat{A}_0$, $\hat{F}'(x) = F'(x)$. If $x$ is in $\hat{A}_0$, then $x + a$ is in $A_0$, and hence, $\hat{F}'(x)(= F'(x + a))$ has rank 2. Thus, we see that all the conditions of

Theorem 3.58 are satisfied for $\hat{F} : \hat{A}_0 \to \hat{B}_0$. Hence, by Theorem 3.58, there exist an open neighborhood $\hat{A}$ of $(0, 0, 0)$ in $\mathbb{R}^3$, an open neighborhood $\hat{B}$ of $b$ in $\mathbb{R}^4$, an open subset $\hat{U}$ of $\mathbb{R}^3$, an open subset $\hat{V}$ of $\mathbb{R}^4$, a function $\hat{G} : \hat{A} \to \hat{U}$, and a function $\hat{H} : \hat{B} \to \hat{V}$ such that

1. $\hat{A}$ is contained in $\hat{A}_0$, and $\hat{B}$ is contained in $\hat{B}_0$,
2. $\hat{G}$ is a $C^2$ diffeomorphism (i.e., $\hat{G}$ is 1–1 onto, $\hat{G}$ is a $C^2$ function, and $\hat{G}^{-1}$ is a $C^2$ function),
3. $\hat{H}$ is a $C^2$ diffeomorphism,
4. $\hat{F}(\hat{A}) \subset \hat{B}$,
5. For every $x \equiv (x_1, x_2, x_3)$ in $\hat{U}$, $(\hat{H} \circ \hat{F} \circ \hat{G}^{-1})(x_1, x_2, x_3) = (x_1, x_2, 0, 0)$.

Put $A \equiv \hat{A} + a$, $B \equiv \hat{B}$, $U \equiv \hat{U}$, and $V \equiv \hat{V}$. Let us define a function $G : A \to U$ as follows: For every $x$ in $A$, $G(x) \equiv \hat{G}(x - a) = (\hat{G} \circ T_{-a})(x)$. Let us define a function $H : B \to V$ as follows: For every $y$ in $B$, $H(y) \equiv \hat{H}(y)$. Hence, $G = \hat{G} \circ T_{-a}$, and $H = \hat{H}$. Further, $\hat{G} = G \circ T_a$.

It remains to prove the following:

(i)   $A$ is an open neighborhood of $a$,
(ii)   $B$ is an open neighborhood of $b$,
(iii)   $U$ is an open subset of $\mathbb{R}^3$,
(iv)   $V$ is an open subset of $\mathbb{R}^4$,
(v)   $G : A \to U$ is 1–1,
(vi)   $G : A \to U$ is onto,
(vii)   $G : A \to U$ is $C^2$,
(viii)   $G^{-1} : U \to A$ is $C^2$,
(ix)   $H : B \to V$ is 1–1,
(x)   $H : B \to V$ is onto,
(xi)   $H : B \to V$ is $C^2$,
(xii)   $H^{-1} : V \to B$ is $C^2$,
(xiii)   $A$ is contained in $A_0$,
(xiv)   $B$ is contained in $B_0$,
(xv)   $F(A) \subset B$,
(xvi)   For every $x \equiv (x_1, x_2, x_3)$ in $U$, $(H \circ F \circ G^{-1})(x_1, x_2, x_3) = (x_1, x_2, 0, 0)$.

For (i): Since $\hat{A}$ is an open neighborhood of $(0, 0, 0)$, $A(= \hat{A} + a)$ is an open neighborhood of $a$.

For (ii): Since $\hat{B}$ is an open neighborhood of $b$, $B(= \hat{B})$ is an open neighborhood of $b$.

For (iii): Since $\hat{U}$ is an open subset of $\mathbb{R}^3$, $U(= \hat{U})$ is an open subset of $\mathbb{R}^3$.

For (iv): Since $\hat{V}$ is an open subset of $\mathbb{R}^4$, $V(= \hat{V})$ is an open subset of $\mathbb{R}^4$.

For (v): Since $\hat{G} : \hat{A} \to \hat{U}$ is 1–1, and $T_{-a} : \hat{A} + a \to \hat{A}$ is 1–1, $\hat{G} \circ T_{-a} : \hat{A} + a \to \hat{U}$ is 1–1, and hence, $\hat{G} \circ T_{-a} : A \to U$ is 1–1.

For (vi): Since $\hat{G} : \hat{A} \to \hat{U}$ is onto, and $T_{-a} : \hat{A} + a \to \hat{A}$ is onto, $\hat{G} \circ T_{-a} :$ $\hat{A} + a \to \hat{U}$ is onto, and hence, $\hat{G} \circ T_{-a} : A \to U$ is onto.

For (vii): Since $\hat{G} : \hat{A} \to \hat{U}$ is $C^2$, and $T_{-a} : \hat{A} + a \to \hat{A}$ is $C^2$, $\hat{G} \circ T_{-a} :$ $\hat{A} + a \to \hat{U}$ is$C^2$, and hence, $\hat{G} \circ T_{-a} : A \to U$ is $C^2$.

For (viii): Since $\hat{G}^{-1} = (G \circ T_a)^{-1} = T_{-a} \circ G^{-1}$, $G^{-1} = T_a \circ \hat{G}^{-1}$. Since $\hat{G}^{-1} : \hat{U} \to \hat{A}$is $C^2$, and $T_a : \hat{A} \to \hat{A} + a$ is $C^2$, their composite $G^{-1} = T_a \circ$ $\hat{G}^{-1} : \hat{U} \to \hat{A} + a$ is $C^2$, and hence, $G^{-1} : U \to A$ is $C^2$.

For (ix): Since $\hat{H} : \hat{B} \to \hat{V}$ is 1–1, $H = \hat{H}$, $B = \hat{B}$, and $V = \hat{V}$, $H : B \to V$ is 1–1.

For (x): Since $\hat{H} : \hat{B} \to \hat{V}$ is onto, $H = \hat{H}$, $B = \hat{B}$, and $V = \hat{V}$, $H : B \to V$ is onto.

For (xi): Since $\hat{H} : \hat{B} \to \hat{V}$ is $C^2$, $H = \hat{H}$, $B = \hat{B}$, and $V = \hat{V}$, $H : B \to V$ is $C^2$.

For (xii): Since $\hat{H}^{-1} : \hat{V} \to \hat{B}$ is $C^2$, $H = \hat{H}$, $B = \hat{B}$, and $V = \hat{V}$, $H^{-1} : V \to B$ is $C^2$.

For (xiii): Since $\hat{A}(= A - a)$ is contained in $\hat{A}_0(= A_0 - a)$, $A$ is contained in $A_0$.

For (xiv): Since $\hat{B}(= B)$ is contained in $\hat{B}_0(= B_0)$, $B$ is contained in $B_0$.

For (xv): Let us take any $x$ in $A$. We have to prove that $F(x) \in B$. Since $x$ is in $A = \hat{A} + a$, $x - a \in \hat{A}$. Now, since $\hat{F}(\hat{A}) \subset \hat{B}, F(x) = (\hat{F} \circ T_{-a})(x) = \hat{F}(x - a)$ $\in \hat{B}$.

For (xvi): Let us take any $x \equiv (x_1, x_2, x_3)$ in $U$. Now, since $U = \hat{U}$, $(x_1, x_2, x_3)$ is in $\hat{U}$, and hence,

$$\text{LHS} = (H \circ F \circ G^{-1})(x_1, x_2, x_3) = \left( \hat{H} \circ (\hat{F} \circ T_{-a}) \circ (\hat{G} \circ T_{-a})^{-1} \right)(x_1, x_2, x_3)$$
$$= \left( \hat{H} \circ (\hat{F} \circ T_{-a}) \circ (T_a \circ \hat{G}^{-1}) \right)(x_1, x_2, x_3) = \left( \hat{H} \circ \hat{F} \circ \hat{G}^{-1} \right)(x_1, x_2, x_3)$$
$$= (x_1, x_2, 0, 0) = \text{RHS}.$$

$\square$

**Theorem 3.60** *Let $A_0$ be an open neighborhood of $a \equiv (a_1, a_2, \ldots, a_n)$ in $\mathbb{R}^n$. Let $B_0$ be an open neighborhood of $b \equiv (b_1, b_2, \ldots, b_m)$ in $\mathbb{R}^m$. Let $F : A_0 \to B_0$ be a function. If*

(i)  *$F$ is a $C^r$ function,*
(ii)  *For every $x$ in $A_0$, the linear transformation $F'(x)$ has rank $k$,*
(iii)  *$F(a) = b$,*

*then there exist an open neighborhood $A$ of $a$ in $\mathbb{R}^n$, an open neighborhood $B$ of $b$ in $\mathbb{R}^m$, an open subset $U$ of $\mathbb{R}^n$, an open subset $V$ of $\mathbb{R}^m$, a function $G : A \to U$, and a function $H : B \to V$ such that*

1. *$A$ is contained in $A_0$, and $B$ is contained in $B_0$,*
2. *$G$ is a $C^r$ diffeomorphism (i.e., $G$ is 1–1 onto, $G$ is a $C^k$ function, and $G^{-1}$ is a $C^k$ function),*

3. *H is a $C^k$ diffeomorphism,*
4. *$F(A) \subset B$,*
5. *For every $x \equiv (x_1, x_2, \ldots, x_n)$ in $U$, $(H \circ F \circ G^{-1})(x_1, x_2, \ldots, x_n) = (x_1, x_2, \ldots$*
   *$x_k, \underbrace{0, \ldots, 0}_{m-k})$.*

*Proof* Its proof is quite similar to the proof of Theorem 3.59.    □

**Note 3.61** Theorem 3.60 is known as the **constant rank theorem**.


## 3.8 Taylor's Theorem

**Definition** Let $p \in S \subset \mathbb{R}^n$. If for every $x$ in $S$, the line segment $[p, x] \equiv \{(1 - \lambda)p + \lambda x : 0 \le \lambda \le 1\} \subset S$, then we say that $S$ *is star-shaped with respect to point p.*

**Theorem 3.62** *Let $p \in U \subset \mathbb{R}^2$. Let $U$ be star-shaped with respect to point $p \equiv (p^1, p^2)$, and let $U$ be an open set. Let $f : U \to \mathbb{R}$ be $C^\infty$ on $U$. Then, there exist $C^\infty$ functions $g_1(x), g_2(x)$ on $U$ such that*

1. *For every $x \equiv (x^1, x^2)$ in $U, f(x) = f(p) + (g_1(x))(x^1 - p^1) + (g_2(x))(x^2 - p^2)$,*
2. *$g_1(p) = \frac{\partial f}{\partial x^1}(p)$ and $g_2(p) = \frac{\partial f}{\partial x^2}(p)$.*

*Proof* Let us take any $x$ in $U$ and any $t$ satisfying $0 \le t \le 1$. Now, since $U$ is star-shaped with respect to point $p$, $(1 - t)p + tx$ is in $S$, and hence, $f((1 - t)p + tx)$ is a real number. Here,

$$\frac{d}{dt}f((1-t)p+tx) = \frac{d}{dt}f(p^1 + t(x^1 - p^1), p^2 + t(x^2 - p^2))$$

$$= \left(\frac{\partial f}{\partial x^1}((1-t)p+tx)\right)\left(\frac{d}{dt}(p^1 + t(x^1 - p^1))\right)$$

$$+ \left(\frac{\partial f}{\partial x^2}((1-t)p+tx)\right)\left(\frac{d}{dt}(p^2 + t(x^2 - p^2))\right)$$

$$= \left(\frac{\partial f}{\partial x^1}((1-t)p+tx)\right)(x^1 - p^1)$$

$$+ \left(\frac{\partial f}{\partial x^2}((1-t)p+tx)\right)(x^2 - p^2),$$

hence,

$$(x^1 - p^1) \int_0^1 \left( \frac{\partial f}{\partial x^1}((1-t)p + tx) \right) dt + (x^2 - p^2) \int_0^1 \left( \frac{\partial f}{\partial x^2}((1-t)p + tx) \right) dt$$

$$= \int_0^1 \left( \left( \frac{\partial f}{\partial x^1}((1-t)p + tx) \right)(x^1 - p^1) + \left( \frac{\partial f}{\partial x^2}((1-t)p + tx) \right)(x^2 - p^2) \right) dt$$

$$= f((1-t)p + tx)|_{t=0}^{t=1} = f(x) - f(p).$$

Now,

$$f(x) = f(p) + (x^1 - p^1)(g_1(x)) + (x^2 - p^2)(g_2(x)),$$

where

$$g_1(x) \equiv \int_0^1 \left( \frac{\partial f}{\partial x^1}((1-t)p + tx) \right) dt, \text{ and } g_2(x) \equiv \int_0^1 \left( \frac{\partial f}{\partial x^2}((1-t)p + tx) \right) dt.$$

It remains to prove that $g_1, g_2$ are $C^\infty$ on $U$, and $g_i(p) = \frac{\partial f}{\partial x^i}(p)(i = 1, 2)$. Here,

$$\frac{\partial g_1}{\partial x^1} = \frac{\partial}{\partial x^1} \left( \int_0^1 \left( \frac{\partial f}{\partial x^1}((1-t)p + tx) \right) dt \right)$$

$$= \int_0^1 \left( \frac{\partial}{\partial x^1} \left( \frac{\partial f}{\partial x^1}((1-t)p + tx) \right) \right) dt$$

$$= \int_0^1 \left( \frac{\partial}{\partial x^1} \left( \frac{\partial f}{\partial x^1}(p^1 + t(x^1 - p^1), p^2 + t(x^2 - p^2)) \right) \right) dt$$

$$= \int_0^1 \left( \left( \frac{\partial}{\partial x^1} \left( \frac{\partial f}{\partial x^1}(p^1 + t(x^1 - p^1), p^2 + t(x^2 - p^2)) \right) \right) \frac{\partial(p^1 + t(x^1 - p^1))}{\partial x^1} \right.$$

$$\left. + \left( \frac{\partial}{\partial x^2} \left( \frac{\partial f}{\partial x^1}(p^1 + t(x^1 - p^1), p^2 + t(x^2 - p^2)) \right) \right) \frac{\partial(p^1 + t(x^1 - p^1))}{\partial x^2} \right) dt$$

$$= \int_0^1 \left( \left( \frac{\partial}{\partial x^1} \left( \frac{\partial f}{\partial x^1}(p^1 + t(x^1 - p^1), p^2 + t(x^2 - p^2)) \right) \right) t \right.$$

$$\left. + \left( \frac{\partial}{\partial x^2} \left( \frac{\partial f}{\partial x^1}(p^1 + t(x^1 - p^1), p^2 + t(x^2 - p^2)) \right) \right) 0 \right) dt$$

$$= \int_0^1 \left( \left( \frac{\partial^2 f}{\partial(x^1)^2}(p^1 + t(x^1 - p^1), p^2 + t(x^2 - p^2)) \right) t \right) dt$$

$$= \int_0^1 t \left( \frac{\partial^2 f}{\partial(x^1)^2}((1-t)p + tx) \right) dt.$$

Thus,

$$\frac{\partial g_1}{\partial x^1} = \int_0^1 t\left(\frac{\partial^2 f}{\partial (x^1)^2}((1-t)p + tx)\right)dt.$$

Similarly, $\dfrac{\partial g_1}{\partial x^2}, \dfrac{\partial^2 g}{\partial (x^1)^2}$, etc. can be obtained. This proves that $g_1$ is $C^\infty$ on $U$.

Similarly, $g_2$ is $C^\infty$ on $U$. Now,

$$g_1(p) = \int_0^1 \left(\frac{\partial f}{\partial x^1}((1-t)p + tp)\right)dt = \int_0^1 \left(\frac{\partial f}{\partial x^1}(p)\right)dt = \left(\frac{\partial f}{\partial x^1}(p)\right)\int_0^1 dt$$
$$= \left(\frac{\partial f}{\partial x^1}(p)\right)1 = \frac{\partial f}{\partial x^1}(p).$$

Thus, $g_1(p) = \frac{\partial f}{\partial x^1}(p)$. Similarly, $g_2(p) = \frac{\partial f}{\partial x^2}(p)$.                                    □

**Note 3.63** The result similar to above is also valid for $\mathbb{R}^3, \mathbb{R}^4, \ldots$.

**Theorem 3.64** *Prove that if $f : \mathbb{R}^2 \to \mathbb{R}$ is $C^\infty$, then there exist $C^\infty$ functions $g_{11}, g_{12}, g_{22}$ on $\mathbb{R}^2$ such that*

$$f(x,y) = f(0,0) + \left(\frac{\partial f}{\partial x}(0,0)\right)x + \left(\frac{\partial f}{\partial y}(0,0)\right)y + x^2 g_{11}(x,y) + xy g_{12}(x,y)$$
$$+ y^2 g_{22}(x,y).$$

*Proof* Here, $f : \mathbb{R}^2 \to \mathbb{R}$ is $C^\infty$, by Theorem 3.62, there exist $C^\infty$ functions $g_1(x,y)$, $g_2(x,y)$ on $\mathbb{R}^2$ such that for every $(x,y)$ in $\mathbb{R}^2$, $f(x,y) = f(0,0) + (g_1(x,y))(x-0) + (g_2(x,y))(y-0) = f(0,0) + x g_1(x,y) + y g_2(x,y)$ and $g_1(0,0) = \frac{\partial f}{\partial x}(0,0)$, $g_2(0,0) = \frac{\partial f}{\partial y}(0,0)$. Since $g_1(x,y)$ is a $C^\infty$ function, there exist $C^\infty$ functions $g_{11}(x,y), g_{12}(x,y)$ on $\mathbb{R}^2$ such that for every $(x,y)$ in $\mathbb{R}^2$, $g_1(x,y) = g_1(0,0) + (g_{11}(x,y))(x-0) + (g_{12}(x,y))(y-0) = \frac{\partial f}{\partial x}(0,0) + x g_{11}(x,y) + y g_{12}(x,y)$ and $g_{11}(0,0) = (\frac{\partial g_1}{\partial x})(0,0), g_{12}(0,0) = \frac{\partial g_1}{\partial y}(0,0)$. Similarly, there exist $C^\infty$ functions $g_{21}(x,y), g_{22}(x,y)$ on $\mathbb{R}^2$ such that for every $(x,y)$ in $\mathbb{R}^2$, $g_2(x,y) = g_2(0,0) + (g_{21}(x,y))(x-0) + (g_{22}(x,y))(y-0) = \frac{\partial f}{\partial y}(0,0) + x g_{21}(x,y) + y g_{22}(x,y)$ and $g_{21}(0,0) = (\frac{\partial g_2}{\partial x})(0,0), g_{22}(0,0) = \frac{\partial g_2}{\partial y}(0,0)$. Hence,

$$f(x,y) = f(0,0) + x\left(\frac{\partial f}{\partial x}(0,0) + xg_{11}(x,y) + yg_{12}(x,y)\right)$$

$$+ y\left(\frac{\partial f}{\partial y}(0,0) + xg_{21}(x,y) + yg_{22}(x,y)\right)$$

$$= f(0,0) + \left(\frac{\partial f}{\partial x}(0,0)\right)x + \left(\frac{\partial f}{\partial y}(0,0)\right)y$$

$$+ (g_{11}(x,y))x^2 + (g_{12}(x,y) + g_{21}(x,y))xy + (g_{22}(x,y))y^2.$$

Since $g_{12}(x,y), g_{21}(x,y)$ are $C^\infty$ functions, $g_{12}(x,y) + g_{21}(x,y)$ is also a $C^\infty$ function. $\qquad\square$

# Chapter 4
# Topological Properties of Smooth Manifolds

In real analysis, its deep theorems (like Lagrange's mean value theorem, fundamental theorem of integral calculus, etc.) are quite difficult to prove without the help of topological theorems (like Weierstrass theorem, Heine–Borel theorem, etc.) for real line. Similar is the situation in complex analysis. Here, in the study of smooth manifolds, in order to prove its deep rooted theorems, we will need its topological properties vigorously. So, for smoothening later work, it is better to collect all the results, to be needed in future, relating to topological properties of smooth manifolds at one place. This chapter aims at this end and is largely self-contained. The pace of this chapter is moderate so that the topic is well assimilated.

## 4.1 Constant Rank Theorem

**Theorem 4.1** *Let $E$ be an open subset of $\mathbb{R}^3$. Let $f : E \to \mathbb{R}^3$. Let $a$ be in $E$. If*

(i) *$f$ is a smooth function,*
(ii) *the determinant of the Jacobian matrix of $f$ at $a$ is nonzero, that is,*

$$\det \begin{bmatrix} (D_1 f^1)(a) & (D_2 f^1)(a) & (D_3 f^1)(a) \\ (D_1 f^2)(a) & (D_2 f^2)(a) & (D_3 f^2)(a) \\ (D_1 f^3)(a) & (D_2 f^3)(a) & (D_3 f^3)(a) \end{bmatrix}_{3 \times 3} \neq 0,$$

*where $f^1, f^2, f^3$ are the component functions of $f$, then there exists a connected open neighborhood $U$ of $a$ such that*

1. $U$ is contained in $E$,
2. $f(U)$ is connected and open in $\mathbb{R}^3$,
3. $f$ has a smooth inverse on $f(U)$.

© Springer India 2014
R. Sinha, *Smooth Manifolds*, DOI 10.1007/978-81-322-2104-3_4

*Proof* We want to apply Theorem 3.36. Since $f$ is a smooth function, $f$ is a $C^1$ function. Since

$$\det \begin{bmatrix} (D_1f^1)(a) & (D_2f^1)(a) & (D_3f^1)(a) \\ (D_1f^2)(a) & (D_2f^2)(a) & (D_3f^2)(a) \\ (D_1f^3)(a) & (D_2f^3)(a) & (D_3f^3)(a) \end{bmatrix}_{3\times3} \neq 0,$$

$$\begin{bmatrix} (D_1f^1)(a) & (D_2f^1)(a) & (D_3f^1)(a) \\ (D_1f^2)(a) & (D_2f^2)(a) & (D_3f^2)(a) \\ (D_1f^3)(a) & (D_2f^3)(a) & (D_3f^3)(a) \end{bmatrix}^{-1}$$

exists, and hence, $f'(a)$ is invertible. Hence, by the Theorem 3.36, there exists a connected open neighborhood $U$ of $a$ such that

1. $U$ is contained in $E$,
2. $f$ is 1–1 on $U$ (that is, if $x$, $y$ are in $U$, and $f(x) = f(y)$, then $x = y$),
3. $f(U)$ is connected and open in $\mathbb{R}^3$,
4. the 1–1 function $f^{-1}$ from $f(U)$ onto $U$ is a $C^1$ function,
5. if $f$ is a $C^2$ function, then $f^{-1}$ from $f(U)$ onto $U$ is a $C^2$ function, etc.

By the conclusions 1, 2, 3, and 4, we find the $f^{-1}$ is a $C^1$ function on $f(U)$. Now, it remains to prove that $f^{-1}$ is smooth on $f(U)$, that is, $f^{-1}$ is a $C^n$ function for every $n = 2, 3, 4, \ldots$. Since $f$ is a smooth function, $f$ is a $C^2$ function, and hence by conclusion 5, $f^{-1}$ is a $C^2$ function. Again, since $f$ is a smooth function, $f$ is a $C^3$ function, and hence by conclusion 5, $f^{-1}$ is a $C^3$ function, etc. Hence, $f^{-1}$ is smooth on $f(U)$. □

**Note 4.2** The result similar to Theorem 4.1 can be proved as above for $\mathbb{R}^n$ in place of $\mathbb{R}^3$.

**Theorem 4.3** *Let $E$ be an open subset of $\mathbb{R}^3$. Let $f : E \to \mathbb{R}^3$. Let $a$ be in $E$. If*

(i)  *$f$ is a smooth function,*
(ii)  *the tangent map $f_*$ of $f$ at $a$ is an isomorphism,*

*then there exists an open neighborhood $U$ of $a$, and an open neighborhood $V$ of $f(a)$ such that the restriction of $f$ to $U$ is a diffeomorphism from $U$ onto $V$.*

*Proof* By the Note 2.77, the tangent map $f_*$ of $f$ at $a$ is an isomorphism if and only if

$$\det \begin{bmatrix} (D_1f^1)(a) & (D_2f^1)(a) & (D_3f^1)(a) \\ (D_1f^2)(a) & (D_2f^2)(a) & (D_3f^2)(a) \\ (D_1f^3)(a) & (D_2f^3)(a) & (D_3f^3)(a) \end{bmatrix} \neq 0.$$

This shows that we can apply Theorem 4.1. By Theorem 4.1, there exists an open neighborhood $U$ of $a$ such that

1. $U$ is contained in $E$,
2. $f(U)$ is open in $\mathbb{R}^3$,
3. $f$ has a smooth inverse on $f(U)$.

This shows that $f(U)$ is an open neighborhood of $f(a)$ such that the restriction of $f$ to $U$ is a diffeomorphism from $U$ onto $f(U)$. □

**Note 4.4** The result similar to Theorem 4.3 can be proved as above for $\mathbb{R}^n$ in place of $\mathbb{R}^3$.

**Theorem 4.5** *Let $M$ and $N$ be 3-dimensional smooth manifolds. Let $f : M \to N$. Let $p$ be in $M$. If*

(i) *$f$ is a smooth map,*
(ii) *the tangent map $f_* : T_p(M) \to T_{f(p)}(N)$ is an isomorphism,*

*then there exists an open neighborhood $U$ of $p$ in $M$ such that*

1. *$V \equiv f(U)$ is an open neighborhood of $f(p)$ in $N$,*
2. *the restriction of $f$ to $U$ is a diffeomorphism from $U$ onto $V$.*

*Proof* Here, $p$ is in $M$, and $f : M \to N$, so $f(p)$ is in $N$. Since $f(p)$ is in $N$, and $N$ is a 3-dimensional smooth manifold, there exists an admissible coordinate chart $(V, \psi_V)$ in $N$ satisfying $f(p) \in V$. Since $f : M \to N$ is a smooth map, $f : M \to N$ is a continuous map. Since $f : M \to N$ is a continuous map, and $V$ is an open neighborhood of $F(p)$, there exists an open neighborhood $U_1$ of $p$ in $M$ such that $f(U_1)$ is contained in $V$. Here, $p$ is in $M$, and $M$ is a 3-dimensional smooth manifold, so there exists an admissible coordinate chart $(U, \varphi_U)$ in $M$ satisfying $p \in U$. Since $U$ and $U_1$ are open neighborhoods of $p$ in $M$, their intersection $U \cap U_1$ is an open neighborhood of $p$ in $M$. Since $\varphi_U$ is a homeomorphism from open set $U$ onto open set $\varphi_U(U)$, and $U \cap U_1$ is an open set, $\varphi_U(U \cap U_1)$ is an open set.

Here $f : M \to N$ is a smooth map, $(\varphi_U)^{-1}$ is a smooth map, and $\psi_V$ is a smooth map, so their composite $\psi_V \circ (f \circ (\varphi_U)^{-1})$ is a smooth function on the open neighborhood $\varphi_U(U \cap U_1)$ of $\varphi_U(p)$ in $\mathbb{R}^3$. We want to apply Theorem 4.1 on the function $\psi_V \circ (f \circ (\varphi_U)^{-1})$. Since the tangent map $f_* : T_p(M) \to T_{f(p)}(N)$ is an isomorphism so, by the Note 2.77,

$$
\det
\begin{bmatrix}
\left(D_1\left(\left(\psi_V\circ\left(f\circ(\varphi_U)^{-1}\right)\right)^1\right)\right)(\varphi_U(p)) & \left(D_2\left(\left(\psi_V\circ\left(f\circ(\varphi_U)^{-1}\right)\right)^1\right)\right)(\varphi_U(p)) \\
\left(D_1\left(\left(\psi_V\circ\left(f\circ(\varphi_U)^{-1}\right)\right)^2\right)\right)(\varphi_U(p)) & \left(D_2\left(\left(\psi_V\circ\left(f\circ(\varphi_U)^{-1}\right)\right)^2\right)\right)(\varphi_U(p)) \\
\left(D_1\left(\left(\psi_V\circ\left(f\circ(\varphi_U)^{-1}\right)\right)^3\right)\right)(\varphi_U(p)) & \left(D_2\left(\left(\psi_V\circ\left(f\circ(\varphi_U)^{-1}\right)\right)^3\right)\right)(\varphi_U(p)) \\
\left(D_3\left(\left(\psi_V\circ\left(f\circ(\varphi_U)^{-1}\right)\right)^1\right)\right)(\varphi_U(p)) \\
\left(D_3\left(\left(\psi_V\circ\left(f\circ(\varphi_U)^{-1}\right)\right)^2\right)\right)(\varphi_U(p)) \\
\left(D_3\left(\left(\psi_V\circ\left(f\circ(\varphi_U)^{-1}\right)\right)^3\right)\right)(\varphi_U(p))
\end{bmatrix}
\neq 0,
$$

and hence, the determinant of the Jacobian matrix of $\psi_V\circ(f\circ(\varphi_U)^{-1})$ at $\varphi_U(p)$ is nonzero. Hence, by Theorem 4.1, there exists an open neighborhood $W$ of $\varphi_U(p)$ such that

1. $W$ is contained in $\varphi_U(U\cap U_1)$,
2. $(\psi_V\circ(f\circ(\varphi_U)^{-1}))(W)$ is open in $\mathbb{R}^3$,
3. $\psi_V\circ(f\circ(\varphi_U)^{-1})$ has a smooth inverse on $(\psi_V\circ(f\circ(\varphi_U)^{-1}))(W)$.

Since $W$ is contained in $\varphi_U(U\cap U_1)$, and $\varphi_U$ is 1–1, $(\varphi_U)^{-1}(W)$ is contained in $U\cap U_1$. Since $\varphi_U(p)$ is in $W$, $p$ is in $(\varphi_U)^{-1}(W)$. Since $\varphi_U$ is a homeomorphism, and $W$ is an open subset of open set $\varphi_U(U)$, $(\varphi_U)^{-1}(W)$ is open. Thus, $(\varphi_U)^{-1}(W)$ is an open neighborhood of $p$ in $M$.

For 1: We have to prove that $f((\varphi_U)^{-1}(W))$ is open in $N$. Since $(\psi_V\circ(f\circ(\varphi_U)^{-1}))(W)$ is open, and $\psi_V$ is a homeomorphism, $(\psi_V)^{-1}((\psi_V\circ(f\circ(\varphi_U)^{-1}))(W))(=f((\varphi_U)^{-1}(W)))$ is open.

For 2: Since $\psi_V\circ(f\circ(\varphi_U)^{-1})$ is 1–1, $(\psi_V)^{-1}$ is 1–1 and $\varphi_U$ is 1–1, their composite $(\psi_V)^{-1}\circ(\psi_V\circ(f\circ(\varphi_U)^{-1}))\circ\varphi_U(=f)$ is 1–1. Since $f$ is 1–1, the function $f^{-1}$ exists. It remains to prove that $f^{-1}$ is smooth. Since $(\psi_V\circ(f\circ(\varphi_U)^{-1}))^{-1}(=\varphi_U\circ f^{-1}\circ(\psi_V)^{-1})$ is smooth, $(\varphi_U)^{-1}$ is smooth, and $\psi_V$ is smooth, their composite function $(\varphi_U)^{-1}\circ(\varphi_U\circ f^{-1}\circ(\psi_V)^{-1})\circ\psi_V(=f^{-1})$ is smooth.  □

**Note 4.6** The result similar to Theorem 4.5 can be proved as above for $\mathbb{R}^n$ in place of $\mathbb{R}^3$.

**Theorem 4.7** *Let $M$ be an $m$-dimensional differentiable manifold, $N$ be an $n$-dimensional smooth manifold, and $F : N \to M$ be a smooth function. Let $p$ be in $N$. Let $(U,\varphi_U)$ be an admissible coordinate chart in $N$ satisfying $p \in U$, $(V,\psi_V)$ be an admissible coordinate chart in $M$ satisfying $F(p) \in V$, and $F(U) \subset V$. Let $(\widehat{U},\varphi_{\widehat{U}})$*

*be an admissible coordinate chart in N satisfying $p \in \widehat{U}$, $(\widehat{V}, \psi_{\widehat{V}})$ be an admissible coordinate chart in M satisfying $F(p) \in \widehat{V}$, and $F(\widehat{U}) \subset \widehat{V}$. Then, the rank of $(\psi_V \circ F \circ (\varphi_U)^{-1})'(\varphi_U(p))$ and the rank of $(\psi_{\widehat{V}} \circ F \circ (\varphi_{\widehat{U}})^{-1})'(\varphi_{\widehat{U}}(p))$ are equal.*

*Proof* Here,

$$
\left( \psi_{\widehat{V}} \circ F \circ \left( \varphi_{\widehat{U}} \right)^{-1} \right)' \left( \varphi_{\widehat{U}}(p) \right) = \left( \psi_{\widehat{V}} \circ F \circ \left( (\varphi_U)^{-1} \circ \varphi_U \right) \circ \left( \varphi_{\widehat{U}} \right)^{-1} \right)' \left( \varphi_{\widehat{U}}(p) \right)
$$

$$
= \left( \psi_{\widehat{V}} \circ \left( (\psi_V)^{-1} \circ \psi_V \right) \circ F \circ \left( (\varphi_U)^{-1} \circ \varphi_U \right) \circ \left( \varphi_{\widehat{U}} \right)^{-1} \right)' \left( \varphi_{\widehat{U}}(p) \right)
$$

$$
= \left( \left( \psi_{\widehat{V}} \circ (\psi_V)^{-1} \right) \circ \left( \psi_V \circ F \circ (\varphi_U)^{-1} \right) \circ \left( \varphi_U \circ \left( \varphi_{\widehat{U}} \right)^{-1} \right) \right)' \left( \varphi_{\widehat{U}}(p) \right)
$$

$$
= \left( \left( \psi_{\widehat{V}} \circ (\psi_V)^{-1} \right)' \left( \left( \left( \psi_V \circ F \circ (\varphi_U)^{-1} \right) \circ \left( \varphi_U \circ \left( \varphi_{\widehat{U}} \right)^{-1} \right) \right) (\varphi_{\widehat{U}}(p)) \right) \right)
$$

$$
\circ \left( \left( \psi_V \circ F \circ (\varphi_U)^{-1} \right)' \left( \left( \varphi_U \circ \left( \varphi_{\widehat{U}} \right)^{-1} \right) (\varphi_{\widehat{U}}(p)) \right) \right) \circ \left( \left( \varphi_U \circ \left( \varphi_{\widehat{U}} \right)^{-1} \right)' (\varphi_{\widehat{U}}(p)) \right)
$$

$$
= \left( \left( \psi_{\widehat{V}} \circ (\psi_V)^{-1} \right)' \left( \left( \left( \psi_V \circ F \circ (\varphi_U)^{-1} \right) \circ \left( \varphi_U \circ \left( \varphi_{\widehat{U}} \right)^{-1} \right) \right) (\varphi_{\widehat{U}}(p)) \right) \right)
$$

$$
\circ \left( \left( \psi_V \circ F \circ (\varphi_U)^{-1} \right)' (\varphi_U(p)) \right) \circ \left( \left( \varphi_U \circ \left( \varphi_{\widehat{U}} \right)^{-1} \right)' (\varphi_{\widehat{U}}(p)) \right)
$$

$$
= \left( \left( \psi_{\widehat{V}} \circ (\psi_V)^{-1} \right)' ((\psi_V \circ F)(p)) \right) \circ \left( \left( \psi_V \circ F \circ (\varphi_U)^{-1} \right)' (\varphi_U(p)) \right)
$$

$$
\circ \left( \left( \varphi_U \circ \left( \varphi_{\widehat{U}} \right)^{-1} \right)' \left( \varphi_{\widehat{U}}(p) \right) \right).
$$

Since $\psi_{\widehat{V}} \circ (\psi_V)^{-1} : \psi_V(V \cap \widehat{V}) \to \psi_{\widehat{V}}(V \cap \widehat{V})$ is a smooth function, and $(\psi_{\widehat{V}} \circ (\psi_V)^{-1})^{-1} = \psi_V \circ (\psi_{\widehat{V}})^{-1} : \psi_{\widehat{V}}(V \cap \widehat{V}) \to \psi_V(V \cap \widehat{V})$ is a smooth function, $\psi_{\widehat{V}} \circ (\psi_V)^{-1} : \psi_V(V \cap \widehat{V}) \to \psi_{\widehat{V}}(V \cap \widehat{V})$ is a diffeomorphism, and hence $(\psi_{\widehat{V}} \circ (\psi_V)^{-1})'((\psi_V \circ F)(p))$ is invertible. It follows that the rank of composite

$$
\left( \left( \psi_{\widehat{V}} \circ (\psi_V)^{-1} \right)' ((\psi_V \circ F)(p)) \right) \circ \left( \left( \psi_V \circ F \circ (\varphi_U)^{-1} \right)' (\varphi_U(p)) \right)
$$

and the rank of $((\psi_V \circ F \circ (\varphi_U)^{-1})'(\varphi_U(p)))$ are equal. Similarly the rank of

$$
\left( \left( \psi_{\widehat{V}} \circ (\psi_V)^{-1} \right)' ((\psi_V \circ F)(p)) \right) \circ \left( \left( \psi_V \circ F \circ (\varphi_U)^{-1} \right)' (\varphi_U(p)) \right)
$$

$$
\circ \left( \left( \varphi_U \circ \left( \varphi_{\widehat{U}} \right)^{-1} \right)' \left( \varphi_{\widehat{U}}(p) \right) \right)
$$

is equal to the rank of

$$\left(\left(\psi_{\widehat{V}} \circ (\psi_V)^{-1}\right)'((\psi_V \circ F)(p))\right) \circ \left(\left(\psi_V \circ F \circ (\varphi_U)^{-1}\right)'(\varphi_U(p))\right).$$

Hence, rank of

$$\left(\left(\psi_{\widehat{V}} \circ (\psi_V)^{-1}\right)'((\psi_V \circ F)(p))\right) \circ \left(\left(\psi_V \circ F \circ (\varphi_U)^{-1}\right)'(\varphi_U(p))\right)$$

is equal to rank of $((\psi_V \circ F \circ (\varphi_U)^{-1})'(\varphi_U(p)))$. □

In light of the above theorem, the following definition is well defined:

**Definition** Let $M$ be an $m$-dimensional smooth manifold, $N$ be an $n$-dimensional smooth manifold, and $F : N \to M$ be a smooth function. Let $p$ be in $N$. Since $F : N \to M$ is a smooth function, there exist an admissible coordinate chart $(U, \varphi_U)$ in $N$ satisfying $p \in U$, and an admissible coordinate chart $(V, \psi_V)$ in $M$ satisfying $F(p) \in V$, such that $F(U) \subset V$, and $\psi_V \circ (F \circ (\varphi_U)^{-1}) : \varphi_U(U) \to \psi_V(V)$ is $C^\infty$ at the point $\varphi_U(p)$ in $\mathbb{R}^3$. Hence, the linear transformation $(\psi_V \circ F \circ (\varphi_U)^{-1})'(\varphi_U(p))$ from $\mathbb{R}^n$ to $\mathbb{R}^m$ exists. The rank of $(\psi_V \circ F \circ (\varphi_U)^{-1})'(\varphi_U(p))$ is called the *rank of $F$ at $p$*. Since $\mathrm{rank}((\psi_V \circ F \circ (\varphi_U)^{-1})'(\varphi_U(p))) \leq \min\{m, n\}$, the rank of $F$ at $p$ is less than or equal to $\min\{m, n\}$.

**Definition** Let $M$ be an $m$-dimensional differentiable manifold, $N$ be an $n$-dimensional smooth manifold, and $F : N \to M$ be a smooth function. Let $k$ be any non-negative integer. By *F has rank k*, we mean that $k$ is the rank of $F$ at every point of $N$.

**Theorem 4.8** *Let $N$ be a 4-dimensional smooth manifold, $M$ be a 3-dimensional smooth manifold, $F : M \to N$ be a smooth map, and 2 be the rank of $F$. Let $p$ be in $M$. Then, there exist admissible coordinate chart $(U^*, \varphi_{U^*})$ in $M$ satisfying $p \in U^*$, and admissible coordinate chart $(V^*, \psi_{V^*})$ in $N$ satisfying $F(p) \in V^*$, such that $F(U^*) \subset V^*$, and for every $(x_1, x_2, x_3)$ in $\varphi_{U^*}(U^*)$,*

$$\left(\psi_{V^*} \circ F \circ (\varphi_{U^*})^{-1}\right)(x_1, x_2, x_3) = (x_1, x_2, 0, 0).$$

*Proof* Since $p$ is in $M$, and $F : M \to N$ is a smooth map, there exist an admissible coordinate chart $(U, \varphi_U)$ in $M$ satisfying $p \in U$, and an admissible coordinate chart $(V, \psi_V)$ in $N$ satisfying $F(p) \in V$ such that $F(U) \subset V$, and each of the 4 component functions of the mapping

$$\psi_V \circ \left(F \circ (\varphi_U)^{-1}\right) : \varphi_U(U) \to \psi_V(V)$$

is $C^\infty$ at the point $\varphi_U(p)$ in $\mathbb{R}^3$.

Now, we want to apply Theorem 3.59. Here, $\varphi_U(U)$ acts for $A_0$, $\varphi_U(p)$ acts for $a$, $\psi_V(V)$ acts for $B_0$, $\psi_V(F(p))$ acts for $b$, $\psi_V \circ (F \circ (\varphi_U)^{-1})$ acts for $F$. Now, we must verify the following conditions of the theorem:

1. $\varphi_U(U)$ is an open neighborhood of $\varphi_U(p)$ in $\mathbb{R}^3$,
2. $\psi_V(V)$ is an open neighborhood of $\psi_V(F(p))$ in $\mathbb{R}^4$,
3. $\psi_V \circ (F \circ (\varphi_U)^{-1}) : \varphi_U(U) \to \psi_V(V)$ is a $C^\infty$ function,
4. for every $x$ in $\varphi_U(U)$, the linear transformation $(\psi_V \circ (F \circ (\varphi_U)^{-1}))'(x)$ has rank 2,
5. $(\psi_V \circ (F \circ (\varphi_U)^{-1}))(\varphi_U(p)) = \psi_V(F(p))$.

For 1: Since $(U, \varphi_U)$ is an admissible coordinate chart in $M$ satisfying $p \in U$, $\varphi_U(U)$ is an open neighborhood of $\varphi_U(p)$ in $\mathbb{R}^3$.
For 2: Since $(V, \psi_V)$ is an admissible coordinate chart in $N$ satisfying $F(p) \in V$, $\psi_V(V)$ is an open neighborhood of $\psi_V(F(p))$ in $\mathbb{R}^4$.
For 3: Since $F : M \to N$ is a smooth function, $(U, \varphi_U)$ is an admissible coordinate chart in $M$ satisfying $p \in U$, and $(V, \psi_V)$ an admissible coordinate chart in $N$ satisfying $F(p) \in V$, $\psi_V \circ (F \circ (\varphi_U)^{-1}) : \varphi_U(U) \to \psi_V(V)$ is a $C^\infty$ function.
For 4: Let us take any $\varphi_U(q)$ in $\varphi_U(U)$ where $q$ is in $U$. We have to prove that the rank of $(\psi_V \circ F \circ (\varphi_U)^{-1})'(\varphi_U(q))$ is 2. Since 2 is the rank of $F : M \to N$, and $p$ is in $N$, the rank of $F$ at $p$ is 2, and hence, the rank of $(\psi_V \circ F \circ (\varphi_U)^{-1})'(\varphi_U(p))$ is 2.
For 5: Here, LHS $= (\psi_V \circ (F \circ (\varphi_U)^{-1}))(\varphi_U(p)) = (\psi_V \circ F)(p) = \psi_V(F(p)) =$ RHS. Hence, by Theorem 3.59, there exists an open neighborhood $A$ of $\varphi_U(p)$ in $\mathbb{R}^3$, an open neighborhood $B$ of $(\psi_V \circ F)(p)$ in $\mathbb{R}^4$, an open subset $\widehat{U}$ of $\mathbb{R}^3$, an open subset $\widehat{V}$ of $\mathbb{R}^4$, a function $\widehat{G} : A \to \widehat{U}$, and a function $\widehat{H} : B \to \widehat{V}$ such that

1. $A$ is contained in $\varphi_U(U)$, and $B$ is contained in $\psi_V(V)$,
2. $\widehat{G}$ is a $C^\infty$ diffeomorphism (that is, $\widehat{G}$ is 1–1 onto, $\widehat{G}$ is a $C^\infty$ function, and $\widehat{G}^{-1}$ is a $C^\infty$ function),
3. $\widehat{H}$ is a $C^\infty$ diffeomorphism,
4. $(\psi_V \circ (F \circ (\varphi_U)^{-1}))(A) \subset B$,
5. For every $x \equiv (x_1, x_2, x_3)$ in $\widehat{U}$, $(\widehat{H} \circ (\psi_V \circ (F \circ (\varphi_U)^{-1})) \circ \widehat{G}^{-1})(x_1, x_2, x_3) = (x_1, x_2, 0, 0)$.

Now, let us take $\widehat{G} \circ \varphi_U$ for $\varphi_{U^*}$, $\widehat{H} \circ \psi_V$ for $\psi_{V^*}$, $(\varphi_U)^{-1}(A)$ for $U^*$, $(\psi_V)^{-1}(B)$ for $V^*$. It remains to prove:

(i) $(\varphi_U)^{-1}(A)$ is open in $M$,
(ii) $(\widehat{G} \circ \varphi_U)((\varphi_U)^{-1}(A))$ is open in $\mathbb{R}^3$,
(iii) $\widehat{G} \circ \varphi_U : (\varphi_U)^{-1}(A) \to (\widehat{G} \circ \varphi_U)((\varphi_U)^{-1}(A))$ is a homeomorphism,

(iv)  $((\varphi_U)^{-1}(A), \widehat{G} \circ \varphi_U)$ is an admissible coordinate chart in $M$ satisfying $p \in (\varphi_U)^{-1}(A)$,

(v)  $((\psi_V)^{-1}(B), \widehat{H} \circ \psi_V)$ is an admissible coordinate chart in $N$ satisfying $F(p) \in (\psi_V)^{-1}(B)$,

(vi)  for every $(x_1, x_2, x_3)$ in $(\widehat{G} \circ \varphi_U)((\varphi_U)^{-1}(A))$, $((\widehat{H} \circ \psi_V) \circ F \circ (\widehat{G} \circ \varphi_U)^{-1})(x_1, x_2, x_3) = (x_1, x_2, 0, 0)$,

(vii)  $F((\varphi_U)^{-1}(A)) \subset (\psi_V)^{-1}(B)$.

For (i): Since $A$ is an open neighborhood of $\varphi_U(p)$ in $\mathbb{R}^3$, $A$ is contained in $\varphi_U(U)$, and $(U, \varphi_U)$ is an admissible coordinate chart in $M$ satisfying $p \in U$, $(\varphi_U)^{-1}(A)$ is open in $M$.

For (ii): Since $\widehat{G} : A \to \widehat{U}$ is a diffeomorphism, $\widehat{G} : A \to \widehat{U}$ is a homeomorphism, and hence, $(\widehat{G} \circ \varphi_U)((\varphi_U)^{-1}(A))(= \widehat{G}(A))$ is open in $\mathbb{R}^3$.

For (iii): Since $\widehat{G} : A \to \widehat{U}$ is a $C^\infty$ diffeomorphism, $\widehat{G} : A \to \widehat{U}$ is a homeomorphism. Since $\varphi_U : (\varphi_U)^{-1}(A) \to A$ is a homeomorphism, and $\widehat{G} : A \to \widehat{U}$ is a homeomorphism, $\widehat{G} \circ \varphi_U : (\varphi_U)^{-1}(A) \to (\widehat{G} \circ \varphi_U)((\varphi_U)^{-1}(A))(= \widehat{G}(A) = \widehat{U})$ is a homeomorphism.

For (iv): Let $(W, \chi_W)$ be any admissible coordinate chart in $M$ satisfying $W \cap ((\varphi_U)^{-1}(A))$ is nonempty.

Since $A$ is an open neighborhood of $\varphi_U(p)$, $\varphi_U(p) \in A$, and hence $p \in (\varphi_U)^{-1}(A)$. Now, we have to prove that $\chi_W \circ (\widehat{G} \circ \varphi_U)^{-1}$, and $(\widehat{G} \circ \varphi_U) \circ (\chi_W)^{-1}$ are $C^\infty$. Since $(W, \chi_W), (U, \varphi_U)$ are admissible coordinate charts in $M$, $\chi_W \circ (\varphi_U)^{-1}$ is $C^\infty$. Now, since $\widehat{G}$ is a $C^\infty$ diffeomorphism, $\widehat{G}^{-1}$ is $C^\infty$, and hence, $\chi_W \circ (\widehat{G} \circ \varphi_U)^{-1}(= (\chi_W \circ (\varphi_U)^{-1}) \circ \widehat{G}^{-1})$ is $C^\infty$. Similarly, $(\widehat{G} \circ \varphi_U) \circ (\chi_W)^{-1}$ is $C^\infty$.

For (v): Its proof is similar to that of (iv).

For (vi): Let us take any $(x_1, x_2, x_3)$ in $(\widehat{G} \circ \varphi_U)((\varphi_U)^{-1}(A))$. We have to prove that

$$\left( \left( \widehat{H} \circ \psi_V \right) \circ F \circ \left( \widehat{G} \circ \varphi_U \right)^{-1} \right)(x_1, x_2, x_3) = (x_1, x_2, 0, 0).$$

Since $(x_1, x_2, x_3) \in (\widehat{G} \circ \varphi_U)((\varphi_U)^{-1}(A)) = \widehat{G}(A) = \widehat{U}$,

$$\text{LHS} = \left( \left( \widehat{H} \circ \psi_V \right) \circ F \circ \left( \widehat{G} \circ \varphi_U \right)^{-1} \right)(x_1, x_2, x_3)$$
$$= \left( \widehat{H} \circ \left( \psi_V \circ \left( F \circ (\varphi_U)^{-1} \right) \right) \circ \widehat{G}^{-1} \right)(x_1, x_2, x_3)$$
$$= (x_1, x_2, 0, 0) = \text{RHS}.$$

For (vii): Since $\psi_V((F \circ (\varphi_U)^{-1})(A)) = (\psi_V \circ (F \circ (\varphi_U)^{-1}))(A) \subset B$, and $\psi_V$ is 1–1, $F((\varphi_U)^{-1}(A)) \subset (\psi_V)^{-1}(B)$. $\qquad\square$

**Note 4.9** As in Theorem 4.8, we can prove the following result:

Let $N$ be an $n$-dimensional smooth manifold, $M$ be an $m$-dimensional smooth manifold, $F : M \to N$ be a smooth map, and $r$ be the rank of $F$. Then, for every $p$ in $M$, there exist admissible coordinate chart $(U, \varphi)$ in $M$ satisfying $p \in U$, and admissible coordinate chart $(V, \psi)$ in $N$ satisfying $F(p) \in V$ such that $F(U) \subset V$, and for every $(x_1, \ldots, x_r, x_{r+1}, \ldots, x_m)$ in $\varphi(U)$,

$$
\left(\psi \circ F \circ \varphi^{-1}\right)(x_1, \ldots, x_r, x_{r+1}, \ldots, x_m) = \left(x_1, \ldots, x_r, \underbrace{0, \ldots, 0}_{n-r}\right).
$$

This theorem is also known as the *constant rank theorem*.

**Lemma 4.10** *Let $M$ be an $m$-dimensional smooth manifold. Let $(U, \varphi_U)$ be an admissible coordinate chart of $M$. Let $V$ be a nonempty open subset of $U$. Then, $(V, \psi_V)$ is an admissible coordinate chart of $M$, where $\psi_V$ denotes the restriction of $\varphi_U$ on $V$.*

*Proof* Here, we must prove:

1. $V$ is open,
2. $\psi_V(V)$ is open,
3. $\psi_V : V \to \psi_V(V)$ is 1–1 onto,
4. $\psi_V$ is continuous,
5. $(\psi_V)^{-1}$ is continuous,
6. if $(W, \chi_W)$ is an admissible coordinate chart of $M$, then $\chi_W \circ (\psi_V)^{-1} : \psi_V(W \cap V) \to \chi_W(W \cap V)$ is a smooth function,
7. if $(W, \chi_W)$ is an admissible coordinate chart of $M$, then $\psi_V \circ (\chi_W)^{-1} : \chi_W(W \cap V) \to \psi_V(W \cap V)$ is a smooth function.

For (1): It is given.

For (2): Since $(U, \varphi_U)$ is a coordinate chart of $M$, $\varphi_U(U)$ is open in $\mathbb{R}^4$, and $\varphi_U : U \to \varphi_U(U)$ is an open mapping. Since $\varphi_U : U \to \varphi_U(U)$ is an open mapping, and $V$ is a nonempty open subset of $U$, $\varphi_U(V)$ is an open subset of the open set $\varphi_U(U)$, and hence, $\varphi_U(V)$ is an open set. Since $\varphi_U(V)$ is an open set, and $\psi_V$ is the restriction of $\varphi_U$ on $V$, $\psi_V(V)$ is an open set.

For (3): Since $(U, \varphi_U)$ is a coordinate chart of $M$, $\varphi_U : U \to \varphi_U(U)$ is 1–1. Since $\varphi_U : U \to \varphi_U(U)$ is 1–1, and $\psi_V$ is the restriction of $\varphi_U$ on $V$, $\psi_V : V \to \psi_V(V)$ is 1–1. Further, clearly $\psi_V : V \to \psi_V(V)$ is onto.

For (4): Since $(U, \varphi_U)$ is a coordinate chart of $M$, $\varphi_U : U \to \varphi_U(U)$ is continuous, and hence, its restriction $\psi_V : V \to \psi_V(V)$ is continuous.

For (5): Since $V$ is contained in $U$, $\varphi_U(V)(= \psi_V(V))$ is contained in $\varphi_U(U)$. Since $(U, \varphi_U)$ is a coordinate chart of $M$, $(\varphi_U)^{-1} : \varphi_U(U) \to U$ is continuous, and hence, its restriction $(\varphi_U)^{-1}|_{\psi_V(V)}(= (\psi_V)^{-1})$ is continuous.

For (6): Let us take any admissible coordinate chart $(W, \chi_W)$ of $M$.

We shall try to prove that $\chi_W \circ (\psi_V)^{-1} : \psi_V(W \cap V) \to \chi_W(W \cap V)$ is a smooth function. Since $(W, \chi_W)$, and $(U, \varphi_U)$ are admissible coordinate charts of $M$, $\chi_W \circ (\varphi_U)^{-1} : \varphi_U(W \cap U) \to \chi_W(W \cap U)$ is a smooth function. Since $V$ is contained in $U$, $W \cap U$, is contained in $W \cap U$, and hence, $\varphi_U(W \cap V)$ is contained in $\varphi_U(W \cap U)$. Since $\psi_V$ is the restriction of $\varphi_U : U \to \varphi_U(U)$ on $V$, and $W \cap V$ is contained in $V$, $\psi_V(W \cap V) = \varphi_U(W \cap V)$. Since $\psi_V(W \cap V) = \varphi_U(W \cap V)$, and $\varphi_U(W \cap V)$ is contained in $\varphi_U(W \cap U)$, $\psi_V(W \cap V)$ is contained in $\varphi_U(W \cap U)$. Since $\psi_V(W \cap V)$ is contained in $\varphi_U(W \cap U)$, and $\chi_W \circ (\varphi_U)^{-1} : \varphi_U(W \cap U) \to \chi_W(W \cap U)$ is a smooth function, the restriction of $\chi_W \circ (\varphi_U)^{-1}$ to $\psi_V(W \cap V)$ is a smooth function. If $\psi_V(x) \in \psi_V(W \cap V)$, where $x \in W \cap V(\subset V)$, then, $(\chi_W \circ (\varphi_U)^{-1})(\psi_V(x)) = (\chi_W \circ (\varphi_U)^{-1})(\varphi_U(x)) = \chi_W(x) = (\chi_W \circ (\psi_V)^{-1})(\psi_V(x))$.

It follows that the restriction of $\chi_W \circ (\varphi_U)^{-1}$ to $\psi_V(W \cap V)$, and $\chi_W \circ (\psi_V)^{-1} : \psi_V(W \cap V) \to \chi_W(W \cap V)$ are the same functions. Since the restriction of $\chi_W \circ (\varphi_U)^{-1}$ to $\psi_V(W \cap V)$, and $\chi_W \circ (\psi_V)^{-1} : \psi_V(W \cap V) \to \chi_W(W \cap V)$ are the same functions, and the restriction of $\chi_W \circ (\varphi_U)^{-1}$ to $\psi_V(W \cap V)$ is a smooth function, $\chi_W \circ (\psi_V)^{-1} : \psi_V(W \cap V) \to \chi_W(W \cap V)$ is a smooth function.

For (7): Let us take any admissible coordinate chart $(W, \chi_W)$ of $M$. We shall try to prove that $\psi_V \circ (\chi_W)^{-1} : \chi_W(W \cap V) \to \psi_V(W \cap V)$ is a smooth function.

Since $(W, \chi_W)$, and $(U, \varphi_U)$ are admissible coordinate charts of $M$, $\varphi_U \circ (\chi_W)^{-1} : \chi_W(W \cap U) \to \varphi_U(W \cap U)$ is a smooth function. Since $V$ is contained in $U$, $W \cap V$ is contained in $W \cap U$, and hence, $\chi_W(W \cap V)$ is contained in $\chi_W(W \cap U)$. Since $\chi_W(W \cap V)$ is contained in $\chi_W(W \cap U)$, and $\varphi_U \circ (\chi_W)^{-1} : \chi_W(W \cap U) \to \varphi_U(W \cap U)$ is a smooth function, the restriction of $\varphi_U \circ (\chi_W)^{-1}$ to $\chi_W(W \cap V)$ is a smooth function. If $\chi_W(x) \in \chi_W(W \cap V)$, where $x \in W \cap V(\subset V)$, then $\varphi_U(x) = \psi_V(x)$, and hence, $(\varphi_U \circ (\chi_W)^{-1})(\chi_W(x)) = \varphi_U(x) = \psi_V(x) = (\psi_V \circ (\chi_W)^{-1})(\chi_W(x))$.

It follows that restriction of $\varphi_U \circ (\chi_W)^{-1}$ to $\chi_W(W \cap V)$, and $\psi_V \circ (\chi_W)^{-1} : \chi_W(W \cap V) \to \psi_V(W \cap V)$ are the same the functions. Since the restriction of $\varphi_U \circ (\chi_W)^{-1}$ to $\chi_W(W \cap V)$, and the function $\psi_V \circ (\chi_W)^{-1} : \chi_W(W \cap V) \to \psi_V(W \cap V)$ are the same, and the restriction of $\varphi_U \circ (\chi_W)^{-1}$ to $\chi_W(W \cap V)$ is a smooth function, $\psi_V \circ (\chi_W)^{-1} : \chi_W(W \cap V) \to \psi_V(W \cap V)$ is a smooth function. $\quad\square$

**Lemma 4.11** *Let $M$ be an $m$-dimensional smooth manifold. Let $(U, \varphi_U)$ be an admissible coordinate chart of $M$. Let $G$ be a nonempty open subset of the open set*

$\varphi_U(U)(\subset \mathbb{R}^m)$. *Then* $((\varphi_U)^{-1}(G), \psi_{(\varphi_U)^{-1}(G)})$ *is an admissible coordinate chart of* M, *where* $\psi_{(\varphi_U)^{-1}(G)}$ *denotes the restriction of* $\varphi_U$ *on* $(\varphi_U)^{-1}(G)$.

*Proof* Since $(U, \varphi_U)$ is an admissible coordinate chart of M, and G is a nonempty open subset of the open set $\varphi_U(U)$, $(\varphi_U)^{-1}(G)$ is a nonempty open subset of U. So, by Lemma 4.10, $((\varphi_U)^{-1}(G), \psi_{(\varphi_U)^{-1}(G)})$ is an admissible coordinate chart of M, where $\psi_{(\varphi_U)^{-1}(G)}$ denotes the restriction of $\varphi_U$ on $(\varphi_U)^{-1}(G)$. $\qquad\square$

**Theorem 4.12** *Let* N *be a 4-dimensional smooth manifold,* M *be a 3-dimensional smooth manifold,* $F : M \to N$ *be a smooth function, and 2 be the rank of* F. *Let* p *be in* M. *Then, there exist admissible coordinate chart* $(U, \varphi_U)$ *in* M *satisfying* $p \in U$, *and admissible coordinate chart* $(V, \psi_V)$ *in* N *satisfying* $F(p) \in V$ *such that* $\varphi_U(p) = (0,0,0), \psi_V(F(p)) = (0,0,0,0)$ *and for every* $(x_1, x_2, x_3)$ *in* $\varphi_U(U)$,

$$\left(\psi_V \circ F \circ (\varphi_U)^{-1}\right)(x_1, x_2, x_3) = (x_1, x_2, 0, 0).$$

*Further, we may assume* $\varphi_U(U) = (-\varepsilon, \varepsilon) \times (-\varepsilon, \varepsilon) \times (-\varepsilon, \varepsilon)$, *and* $\psi_V(V) = (-\varepsilon, \varepsilon) \times (-\varepsilon, \varepsilon) \times (-\varepsilon, \varepsilon) \times (-\varepsilon, \varepsilon)$ *for some* $\varepsilon > 0$. *In short,* $\varphi_U(U) = C_\varepsilon^3(0)$, *and* $\psi_V(V) = C_\varepsilon^4(0)$ *for some* $\varepsilon > 0$.

*Proof* By Theorem 4.8, there exist admissible coordinate chart $(U^*, \varphi_{U^*})$ in M satisfying $p \in U^*$, and admissible coordinate chart $(V^*, \psi_{V^*})$ in N satisfying $F(p) \in V^*$ such that for every $(x_1, x_2, x_3)$ in $\varphi_{U^*}(U^*)$,

$$\left(\psi_{V^*} \circ F \circ (\varphi_{U^*})^{-1}\right)(x_1, x_2, x_3) = (x_1, x_2, 0, 0).$$

Put $U \equiv U^*, \varphi_U \equiv \varphi_{U^*} - \varphi_{U^*}(p), V \equiv V^*, \psi_V \equiv \psi_{V^*} - ((\varphi_{U^*}(p))^1, (\varphi_{U^*}(p))^2, 0, 0)$. It remains to prove:

(i) $(U, \varphi_U)$ is an admissible coordinate chart in N satisfying $p \in U$,
(ii) $(V, \psi_V)$ is an admissible coordinate chart in M satisfying $F(p) \in V$,
(iii) $\varphi_U(p) = (0,0,0)$,
(iv) for every $(x_1, x_2, x_3)$ in $\varphi_U(U)$, $(\psi_V \circ F \circ (\varphi_U)^{-1})(x_1, x_2, x_3) = (x_1, x_2, 0, 0)$,
(v) $\psi_V(F(p)) = (0,0,0,0)$.

For (i): Since $(U^*, \varphi_{U^*})$ is an admissible coordinate chart in N satisfying $p \in U^*$, $U^*(= U)$ is an open neighborhood of p, and hence, U is an open neighborhood of p.

Since $(U^*, \varphi_{U^*})$ is an admissible coordinate chart in N, $\varphi_{U^*}(U^*)$ is open in $\mathbb{R}^3$. Now, since $U = U^*$, $\varphi_{U^*}(U)$ is open in $\mathbb{R}^3$, and hence, $(\varphi_U + \varphi_{U^*}(p))(U)(= \varphi_U(U) + \varphi_{U^*}(p))$ is open in $\mathbb{R}^3$. Since $\varphi_U(U) + \varphi_{U^*}(p)$ is open in $\mathbb{R}^3$, $\varphi_U(U)$ is open in $\mathbb{R}^3$. Since $(U^*, \varphi_{U^*})$ is an admissible coordinate chart in M,

$\varphi_{U^*} : U^* \to \varphi_{U^*}(U^*)$ is a homeomorphism. Since $\varphi_{U^*} : U^* \to \varphi_{U^*}(U^*)$ is a homeomorphism, and the translation

$$T_{-\varphi_{U^*}(p)} : \varphi_{U^*}(U^*) \to (\varphi_{U^*}(U^*) - \varphi_{U^*}(p))$$

is a homeomorphism, their composite

$$(\varphi_U =)T_{-\varphi_{U^*}(p)} \circ \varphi_{U^*} : U^*(= U) \to (\varphi_{U^*}(U^*) - \varphi_{U^*}(p))(= (\varphi_{U^*} - \varphi_{U^*}(p))(U^*) = \varphi_U(U))$$

is a homeomorphism. Thus, we have shown that $(U, \varphi_U)$ is a coordinate chart in $M$ satisfying $p \in U$.

Next, let $(W, \chi_W)$ be an admissible coordinate chart in $M$ such that $U \cap W$ is nonempty. Now, we have to prove that $\chi_W \circ (\varphi_U)^{-1}$, and $\varphi_U \circ (\chi_W)^{-1}$ are $C^\infty$. Here,

$$\chi_W \circ (\varphi_U)^{-1} = \chi_W \circ (\varphi_{U^*} - \varphi_{U^*}(p))^{-1} = \chi_W \circ (T_{-\varphi_{U^*}(p)} \circ \varphi_{U^*})^{-1}$$
$$= \chi_W \circ ((\varphi_{U^*})^{-1} \circ T_{\varphi_{U^*}(p)}) = (\chi_W \circ (\varphi_{U^*})^{-1}) \circ T_{\varphi_{U^*}(p)},$$

so $\chi_W \circ (\varphi_U)^{-1} = (\chi_W \circ (\varphi_{U^*})^{-1}) \circ T_{\varphi_{U^*}(p)}$. Since $(W, \chi_W)$, and $(U^*, \varphi_{U^*})$ are admissible coordinate charts in $M$, $\chi_W \circ (\varphi_{U^*})^{-1}$ is $C^\infty$. Since $\chi_W \circ (\varphi_{U^*})^{-1}$ is $C^\infty$, and translation $T_{\varphi_{U^*}(p)}$ is $C^\infty$, their composite $(\chi_W \circ (\varphi_{U^*})^{-1}) \circ T_{\varphi_{U^*}(p)}(= \chi_W \circ (\varphi_U)^{-1})$ is $C^\infty$, and hence, $\chi_W \circ (\varphi_U)^{-1}$ is $C^\infty$. Similarly, $\varphi_U \circ (\chi_W)^{-1}$ is $C^\infty$.

For (ii): Its proof is similar to that of (i).

For (iii): $\varphi_U(p) = (\varphi_{U^*} - \varphi_{U^*}(p))(p) = \varphi_{U^*}(p) - \varphi_{U^*}(p) = (0,0,0)$. Thus $\varphi_U(p) = (0,0,0)$.

For (iv): Let us take any $(x_1, x_2, x_3)$ in $\varphi_U(U)$. We have to prove that $(\psi_V \circ F \circ (\varphi_U)^{-1})(x_1, x_2, x_3) = (x_1, x_2, 0, 0)$. Here, $(\varphi_U)^{-1} = (\varphi_{U^*} - \varphi_{U^*}(p))^{-1} = (T_{-\varphi_{U^*}(p)} \circ \varphi_{U^*})^{-1} = (\varphi_{U^*})^{-1} \circ T_{\varphi_{U^*}(p)}$. Since $\varphi_U = \varphi_{U^*} - \varphi_{U^*}(p)$, $\varphi_U(U) = \varphi_{U^*}(U) - \varphi_{U^*}(p) = \varphi_{U^*}(U^*) - \varphi_{U^*}(p)$. Now, since $(x_1, x_2, x_3)$ is in $\varphi_U(U)$, $(x_1 + (\varphi_{U^*}(p))^1, x_2 + (\varphi_{U^*}(p))^2, x_3 + (\varphi_{U^*}(p))^3)$ is in $\varphi_{U^*}(U^*)$, and hence,

$$\left(\psi_{V^*} \circ F \circ (\varphi_{U^*})^{-1}\right)\left(x_1 + (\varphi_{U^*}(p))^1, x_2 + (\varphi_{U^*}(p))^2, x_3 + (\varphi_{U^*}(p))^3\right)$$
$$= \left(x_1 + (\varphi_{U^*}(p))^1, x_2 + (\varphi_{U^*}(p))^2, 0, 0\right).$$

Now,

$$
\begin{aligned}
\text{LHS} &= \left( \psi_V \circ F \circ (\varphi_U)^{-1} \right)(x_1, x_2, x_3) = \left( \psi_V \circ F \circ \left( (\varphi_{U^*})^{-1}_{\varphi_{U^*}(p)} \right) \right)(x_1, x_2, x_3) \\
&= \left( \left( \psi_{V^*} - \left( (\varphi_{U^*}(p))^1, (\varphi_{U^*}(p))^2, 0, 0 \right) \right) \circ F \circ \left( (\varphi_{U^*})^{-1} \circ T_{\varphi_{U^*}(p)} \right) \right)(x_1, x_2, x_3) \\
&= \left( \left( T_{-\left( (\varphi_{U^*}(p))^1, (\varphi_{U^*}(p))^2, 0, 0 \right)} \circ \psi_{V^*} \right) \circ F \circ \left( (\varphi_{U^*})^{-1} \circ T_{\varphi_{U^*}(p)} \right) \right)(x_1, x_2, x_3) \\
&= \left( T_{-\left( (\varphi_{U^*}(p))^1, (\varphi_{U^*}(p))^2, 0, 0 \right)} \circ \left( \psi_{V^*} \circ F \circ (\varphi_{U^*})^{-1} \right) \circ T_{\varphi_{U^*}(p)} \right)(x_1, x_2, x_3) \\
&= \left( T_{-\left( (\varphi_{U^*}(p))^1, (\varphi_{U^*}(p))^2, 0, 0 \right)} \circ \left( \psi_{V^*} \circ F \circ (\varphi_{U^*})^{-1} \right) \right) \\
&\quad \left( x_1 + (\varphi_{U^*}(p))^1, x_2 + (\varphi_{U^*}(p))^2, x_3 + (\varphi_{U^*}(p))^3 \right) \\
&= \left( T_{-\left( (\varphi_{U^*}(p))^1, (\varphi_{U^*}(p))^2, 0, 0 \right)} \right) \left( x_1 + (\varphi_{U^*}(p))^1, x_2 + (\varphi_{U^*}(p))^2, 0, 0 \right) \\
&= (x_1, x_2, 0, 0) = \text{RHS}.
\end{aligned}
$$

For (v): Since $p \in U$, $\varphi_U(p) \in \varphi_U(U)$, and $((\varphi_U(p))^1, (\varphi_U(p))^2, (\varphi_U(p))^2) = \varphi_U(p) = (0, 0, 0)$ so

$$
\begin{aligned}
(0, 0, 0, 0) &= \left( (\varphi_U(p))^1, (\varphi_U(p))^2, 0, 0 \right) = \left( \psi_V \circ F \circ (\varphi_U)^{-1} \right)(\varphi_U(p)) \\
&= (\psi_V \circ F)(p) = \psi_V(F(p)).
\end{aligned}
$$

This completes the proof of the main part.

Since $\psi_V(V)$ is an open neighborhood of $\psi_V(F(p))(= (0, 0, 0, 0))$ in $\mathbb{R}^4$, and $\varphi_U(U)$ is an open neighborhood of $\varphi_U(p)(= (0, 0, 0))$, there exists $\varepsilon > 0$ such that $(0, 0, 0, 0) \in (-\varepsilon, \varepsilon) \times (-\varepsilon, \varepsilon) \times (-\varepsilon, \varepsilon) \times (-\varepsilon, \varepsilon)(= C_\varepsilon^4(0)) \subset \psi_V(V)$, and $(0, 0, 0) \in (-\varepsilon, \varepsilon) \times (-\varepsilon, \varepsilon) \times (-\varepsilon, \varepsilon)(= C_\varepsilon^3(0)) \subset \varphi_U(U)$. Now, by Lemma 4.11, $((\psi_V)^{-1}(C_\varepsilon^4(0)), \psi_V)$ is an admissible coordinate chart in $N$ satisfying $F(p) \in (\psi_V)^{-1}(C_\varepsilon^4(0))$, and $((\varphi_U)^{-1}(C_\varepsilon^3(0)), \varphi_U)$ is an admissible coordinate chart in $M$ satisfying $p \in (\varphi_U)^{-1}(C_\varepsilon^3(0))$. Since $\varphi_U(p) = (0, 0, 0) \in C_\varepsilon^3(0)$, $p \in (\varphi_U)^{-1}(C_\varepsilon^3(0))$. Next, since $\psi_V(F(p)) = (0, 0, 0, 0) \in C_\varepsilon^4(0)$, so $F(p) \in (\psi_V)^{-1}(C_\varepsilon^4(0))$. Also since $\varphi_U$ is 1-1, $\varphi_U((\varphi_U)^{-1}(C_\varepsilon^3(0))) = C_\varepsilon^3(0)$. Similarly, $\psi_V((\psi_V)^{-1}(C_\varepsilon^4(0))) = C_\varepsilon^4(0)$.

Now, it remains to prove: for every $(x_1, x_2, x_3)$ in $C_\varepsilon^3(0)$, $(\psi_V \circ F \circ (\varphi_U)^{-1})$ $(x_1, x_2, x_3) = (x_1, x_2, 0, 0)$. For this purpose, let us take any $(x_1, x_2, x_3)$ in $C_\varepsilon^3(0)$. Since $(x_1, x_2, x_3)$ is in $C_\varepsilon^3(0)$, and $C_\varepsilon^3(0) \subset \varphi_U(U)$, $(x_1, x_2, x_3)$ is in $\varphi_U(U)$, and hence, $(\psi_V \circ F \circ (\varphi_U)^{-1})(x_1, x_2, x_3) = (x_1, x_2, 0, 0)$. $\qquad \square$

**Theorem 4.13** *Let $M$ be an $m$-dimensional smooth manifold, $N$ be an $n$-dimensional smooth manifold, $F : M \to N$ be a smooth function, and $k$ be the rank of $F$. Let $p$ be in $M$. Then, there exist admissible coordinate chart $(U, \varphi_U)$ in $M$ satisfying*

$p \in U$, and admissible coordinate chart $(V, \psi_V)$ in $N$ satisfying $F(p) \in V$ such that $\varphi_U(p) = 0, \psi_V(F(p)) = 0$, and for every, $(x_1, x_2, \ldots, x_n)$ in $\varphi_U(U)$,

$$\left(\psi_V \circ F \circ (\varphi_U)^{-1}\right)(x_1, x_2, \ldots, x_m) = \left(x_1, x_2, \ldots, x_k, \underbrace{0, \ldots, 0}_{n-k}\right).$$

Further, we may assume $\varphi_U(U) = C_\varepsilon^m(0)$, and $\psi_V(V) = C_\varepsilon^n(0)$ for some $\varepsilon > 0$.

*Proof* Its proof is quite similar to the proof of Theorem 4.12. □

**Lemma 4.14** *Let $M$ be an $m$-dimensional smooth manifold, $N$ be an $n$-dimensional smooth manifold, and $F : N \rightarrow M$ be a 1–1 smooth function. Let $\mathcal{O}_1$ be the topology of $N$. Let $\mathcal{A}_1$ be the differential structure on $N$. Put $\mathcal{O} \equiv \{F(G) : G \in \mathcal{O}_1\}$ and $\mathcal{A} \equiv \{(F(U), \varphi_U \circ F^{-1}) : (U, \varphi_U) \in \mathcal{A}_1\}$. Then,*

  I. *$\mathcal{O}$ is a topology over $F(N)$,*
 II. *$\mathcal{A}$ determines a unique differential structure on $F(N)$, and*
III. *$F$ is a diffeomorphism from $N$ onto $F(N)$.*

*Proof of I*

 (i) Since $\phi \in \mathcal{O}_1$, $F(\phi)(= \phi)$ is in $\mathcal{O}$, and hence, $\phi \in \mathcal{O}$. Next, since $N \in \mathcal{O}_1$, so $F(N)$ is in $\mathcal{O}$.

 (ii) Let $F(G) \in \mathcal{O}$, where $G \in \mathcal{O}_1$. Let $F(H) \in \mathcal{O}$, where $H \in \mathcal{O}_1$. We have to prove that $F(G) \cap F(H) \in \mathcal{O}$. Since $G \in \mathcal{O}_1, H \in \mathcal{O}_1$, and $\mathcal{O}_1$ is a topology, $G \cap H \in \mathcal{O}_1$, and hence, $F(G \cap H) \in \mathcal{O}$. Since $F : N \rightarrow M$ is 1–1, $F(G \cap H) = F(G) \cap F(H)$. Since $F(G \cap H) = F(G) \cap F(H)$, and $F(G \cap H) \in \mathcal{O}$, $F(G) \cap F(H) \in \mathcal{O}$.

(iii) Let $F(G_i) \in \mathcal{O}$, where $G_i \in \mathcal{O}_1$ for every index $i$ in index set $I$. We have to prove that $\cup_{i \in I} F(G_i) \in \mathcal{O}$. Since for every index $i$, $G_i \in \mathcal{O}_1$, and $\mathcal{O}_1$ is a topology, $\cup_{i \in I} G_i \in \mathcal{O}_1$, and hence, $F(\cup_{i \in I} G_i)(= \cup_{i \in I} F(G_i)) \in \mathcal{O}$. Thus, $\cup_{i \in I} F(G_i) \in \mathcal{O}$. Hence, $\mathcal{O}$ is a topology over $F(N)$. □

*Proof of II* It is clear that $F$ is a homeomorphism from topological space $(N, \mathcal{O}_1)$ onto topological space $(F(N), \mathcal{O})$. Since $N$ is an $n$-dimensional smooth manifold, and $\mathcal{O}_1$ is the topology of $N$, $(N, \mathcal{O}_1)$ is a Hausdorff topological space. Since $(N, \mathcal{O}_1)$ is a Hausdorff topological space, and $(N, \mathcal{O}_1)$ is homeomorphic onto topological space $(F(N), \mathcal{O})$, $(F(N), \mathcal{O})$ is a Hausdorff topological space. Since $N$ is an $n$-dimensional smooth manifold, and $\mathcal{O}_1$ is the topology of $N$, $(N, \mathcal{O}_1)$ is a second countable space. Since $(N, \mathcal{O}_1)$ is a second countable space, and $(N, \mathcal{O}_1)$ is homeomorphic onto topological space $(F(N), \mathcal{O})$, $(F(N), \mathcal{O})$ is a second countable space.

We want to prove that $(F(N), \mathcal{O})$ is an $n$-dimensional topological manifold. For this purpose, let us take any $F(x)$ in $F(N)$, where $x$ is in $N$. Since $x$ is in $N$, and $\mathcal{A}_1$ is a differential structure on $N$, there exists $(U, \varphi_U) \in \mathcal{A}_1$ such that $x \in U$. It follows that $F(x) \in F(U)$. Since $(U, \varphi_U) \in \mathcal{A}_1$, $(F(U), \varphi_U \circ F^{-1}) \in \mathcal{A}$. It remains to prove:

(a) $F(U) \in \mathcal{O}$,
(b) $(\varphi_U \circ F^{-1})(F(U))$ is an open subset of $\mathbb{R}^n$,
(c) $\varphi_U \circ F^{-1}$ is a homeomorphism from $F(U)$ (relative to topology $\mathcal{O}$) onto $(\varphi_U \circ F^{-1})(F(U))$.

For (a): Here, $(U, \varphi_U) \in \mathcal{A}_1$, and $\mathcal{A}_1$ is a differential structure on $N$, $U \in \mathcal{O}_1$, and hence, $F(U) \in \mathcal{O}$.

For (b): Here, $U \in \mathcal{O}_1$, and $\mathcal{O}_1$ is the topology of $N$, $U$ is contained in $N$. Since $U$ is contained in $N$, $F$ is 1–1, $(\varphi_U \circ F^{-1})(F(U)) = \varphi_U(U)$. Since $(U, \varphi_U) \in \mathcal{A}_1$, and $\mathcal{A}_1$ is a differential structure on $N$, $\varphi_U(U)(= (\varphi_U \circ F^{-1})(F(U)))$ is open in $\mathbb{R}^n$. Hence, $(\varphi_U \circ F^{-1})(F(U))$ is an open subset of $\mathbb{R}^n$.

For (c): Since $(U, \varphi_U) \in \mathcal{A}_1$, $\varphi_U$ is 1–1. Since $\varphi_U$ is 1–1, and $F$ is 1–1, the composite $\varphi_U \circ F^{-1}$ is 1–1. Thus, $\varphi_U \circ F^{-1} : F(U) \to (\varphi_U \circ F^{-1})(F(U))$ is a 1–1 onto function.

Since $(U, \varphi_U) \in \mathcal{A}_1$, $\varphi_U$ is a homeomorphism from $U$ onto $\varphi_U(U)$. Since $F$ is a homeomorphism from topological space $(N, \mathcal{O}_1)$ onto topological space $(F(N), \mathcal{O})$, $F^{-1}$ is a homeomorphism from topological space $(F(N), \mathcal{O})$ onto topological space $(N, \mathcal{O}_1)$, and hence, the restriction of $F^{-1}$ on $F(U)$ (relative to topology $\mathcal{O}$) is a homeomorphism from $F(U)$ onto $U$. It follows that $\varphi_U \circ F^{-1}$ is a homeomorphism from $F(U)$ (relative to topology $\mathcal{O}$) onto $\varphi_U(U)(= (\varphi_U \circ F^{-1})(F(U)))$. Thus, we have shown that $(F(N), \mathcal{O})$ is an $n$-dimensional topological manifold.

Now we want to prove: All pairs of members in $\mathcal{A}$ are $C^\infty$-compatible. For this purpose, let us take any $(F(U), \varphi_U \circ F^{-1}) \in \mathcal{A}$, where $(U, \varphi_U)$ is in $\mathcal{A}_1$, and $(F(V), \psi_V \circ F^{-1}) \in \mathcal{A}$ where $(V, \psi_V)$ is in $\mathcal{A}_1$ satisfying $F(U) \cap F(V)$ is nonempty. Since $F$ is 1–1, $F(U) \cap F(V) = F(U \cap V)$. Since $F(U) \cap F(V) = F(U \cap V)$, and $F(U) \cap F(V)$ is nonempty, $F(U \cap V)$is nonempty, and hence, $U \cap V$ is nonempty. Since $U \cap V$ is nonempty, $\mathcal{A}_1$ is a differential structure on $N$, $(U, \varphi_U)$ is in $\mathcal{A}_1$, and $(V, \psi_V)$ is in $\mathcal{A}_1$, $(U, \varphi_U)$ and $(V, \psi_V)$ are $C^\infty$-compatible, that is, $(\psi_V \circ (\varphi_U)^{-1}) : \varphi_U(U \cap V) \to \varphi_V(U \cap V)$ is a smooth function.

We have to show that $(F(U), \varphi_U \circ F^{-1})$ and $(F(V), \psi_V \circ F^{-1})$ are $C^\infty$-compatible, that is, $((\psi_V \circ F^{-1}) \circ (\varphi_U \circ F^{-1})^{-1}) : (\varphi_U \circ F^{-1})(F(U) \cap F(V)) \to (\psi_V \circ F^{-1})(F(U) \cap F(V))$ is a smooth function, that is, $(\psi_V \circ F^{-1}) \circ (F \circ (\varphi_U)^{-1}) : (\varphi_U \circ F^{-1})(F(U \cap V)) \to (\psi_V \circ F^{-1})(F(U \cap V))$ is a smooth function, that is, $(\psi_V \circ (\varphi_U)^{-1}) : \varphi_U(U \cap V) \to \psi_V(U \cap V)$ is a smooth function. This has already been shown. Thus, we have shown that $\mathcal{A}$ determines a unique differentiable structure on $F(N)$ that contains $\mathcal{A}$. In short, $F(N)$ is an $n$-dimensional smooth manifold.   $\square$

*Proof of III*   Now, we want to prove that $F$ is a diffeomorphism from $N$ onto $F(N)$. For this purpose, let us take any $p$ in $N$. We have to prove that $F$ is $C^\infty$ at $p$. Now let us take any admissible coordinate chart $(U, \varphi_U)$ of $N$ satisfying $p \in U$, an admissible coordinate chart $(F(V), \psi_V \circ F^{-1})$ of $F(N)$ satisfying $F(p) \in F(V)$ where $(V, \psi_V)$ is in $\mathcal{A}_1$. Since $F$ is 1–1, and $F(p) \in F(V), p \in V$. Since $p \in U$, and $p \in V$, $p \in U \cap V$, and hence, $U \cap V$ is nonempty. We shall try to prove that $(\psi_V \circ F^{-1}) \circ F \circ (\varphi_U)^{-1}$ is $C^\infty$ at $\varphi_U(p)$. Since $(U, \varphi_U)$ is in $\mathcal{A}_1$, $(V, \psi_V)$ is in $\mathcal{A}_1$, and $U \cap V$ is nonempty, $\psi_V \circ (\varphi_U)^{-1}(= (\psi_V \circ F^{-1}) \circ F \circ (\varphi_U)^{-1})$ is $C^\infty$ at $\varphi_U(p)$. Thus $(\psi_V \circ F^{-1}) \circ F \circ (\varphi_U)^{-1}$ is $C^\infty$ at $\varphi_U(p)$. Thus $F$ is $C^\infty$. Similarly, $F^{-1}$ is $C^\infty$. This proves that $F$ is a diffeomorphism from $N$ onto $F(N)$.                    $\square$

## 4.2 Lie Groups

**Definition** Let $G$ be an $n$-dimensional smooth manifold that is also a group. Let $m : G \times G \to G$ be the mapping defined as follows: for every $g, h$ in $G$,

$$m(g, h) \equiv gh.$$

Let $i : G \to G$ be the mapping defined as follows: for every $g$ in $G$,

$$i(g) \equiv g^{-1}.$$

If $m$ is a smooth map from $2n$-dimensional product manifold $G \times G$ to $n$-dimensional smooth manifold $G$, and $i$ is a smooth map from $n$-dimensional smooth manifold $G$ to itself, then we say that $G$ is a *Lie group*.

Here, $m$ is the binary operation of the group $G$ and is smooth. Further, $i$ is the (algebraic) inverse mapping of the group $G$, which is smooth. Since smooth mappings are continuous mappings, and $m, i$ are smooth mappings, $m, i$ are continuous mappings. Since the binary operation $m$ of group $G$ is continuous, and the inverse mapping $i$ of the group $G$ is continuous, $G$ is a topological group. Thus, every Lie group is also a topological group.

**Note 4.15** Let $G$ be an $n$-dimensional smooth manifold, and $M$ be an $m$-dimensional smooth manifold. Let $a$ be in $G$. Let us define a mapping $F : M \to G \times M$ as follows: for every $h$ in $M$,

$$F(h) \equiv (a, h).$$

We shall try to show that $F$ is a smooth mapping. For this purpose, let us fix any $h_0$ in $M$. We have to prove that $F$ is smooth at $h_0$ in $M$.

For this purpose, let us take any admissible coordinate chart $(V, \psi_V)$ of $M$ satisfying $h_0 \in V$, and any admissible coordinate chart $(U, \varphi_U)$ of $G$ satisfying $a \in U$. Here, $(U, \varphi_U)$ is an admissible coordinate chart of $G$ satisfying $a \in U$, $(V, \psi_V)$ is an admissible coordinate chart of $M$ satisfying $h_0 \in V$, and $F(h_0) = (a, h_0) \in U \times V$, $(U \times V, (\varphi_U \times \psi_V))$ is an admissible coordinate chart of $G \times M$ satisfying $F(h_0) \in U \times V$. We have to prove that each of the 2 component functions of the mapping

$$(\varphi_U \times \psi_V) \circ \left(F \circ (\psi_V)^{-1}\right) : \psi_V(V \cap F^{-1}(U \times V)) \to (\varphi_U \times \psi_V)(U \times V)$$

is $C^\infty$ at the point $\psi_V(h_0)$ in $\mathbb{R}^m$, that is, $(r,s) \mapsto \pi_1(((\varphi_U \times \psi_V) \circ (F \circ (\psi_V)^{-1}))$ $(r,s))$, and $(r,s) \mapsto \pi_2(((\varphi_U \times \psi_V) \circ (F \circ (\psi_V)^{-1}))(r,s))$ are $C^\infty$ at the point $\psi_V(h_0)$ in $\mathbb{R}^m$. For every $(r,s)$ in $\psi_v(V \cap F^{-1}(U \times V))$, there exists $y$ in $V$ such that $\psi_V(y) = (r,s)$. So $y = (\psi_V)^{-1}(r,s)$. Now, since

$$\begin{aligned}
\left((\varphi_U \times \psi_V) \circ \left(F \circ (\psi_V)^{-1}\right)\right)(r,s) &= (\varphi_U \times \psi_V)\left(F\left((\psi_V)^{-1}(r,s)\right)\right) \\
&= (\varphi_U \times \psi_V)(F(y)) \\
&= (\varphi_U \times \psi_V)(a,y) = (\varphi_U(a), \psi_V(y)),
\end{aligned}$$

$(r,s) \mapsto \pi_1(((\varphi_U \times \psi_V) \circ (F \circ (\psi_V)^{-1}))(r,s)) = \pi_1(\varphi_U(a), \psi_V(y)) = \varphi_U(a)$, which is a constant function. Hence, $(r,s) \mapsto \pi_1(((\varphi_U \times \psi_V) \circ (F \circ (\psi_V)^{-1}))(r,s))$ is a smooth function. Next, since $(r,s) \mapsto \pi_2(((\varphi_U \times \psi_V) \circ (F \circ (\psi_V)^{-1}))(r,s)) = \pi_2(\varphi_U(a), \psi_V(y)) = \psi_V(y) = (r,s)$ is the identity function, $(r,s) \mapsto \pi_2(((\varphi_U \times \psi_V) \circ (F \circ (\psi_V)^{-1}))(r,s))$ is a smooth function. Thus, we have shown that $F$ is a smooth mapping, that is, $h \mapsto (a,h)$ is a smooth mapping.

Similarly, $h \mapsto (h,a)$ is a smooth mapping.

**Note 4.16** Let $G, H, K$ be smooth manifolds. Let $f : H \to K$ be a smooth mapping. Let $F : G \times H \to G \times K$ be the mapping defined as follows: for every $(x,y)$ in $G \times H$,

$$F(x,y) \equiv (x, f(y)).$$

We shall try to show that $F$ is a smooth mapping. For simplicity, let each of $G, H, K$ have dimension 2. Let us fix any $(x_0, y_0)$ in $G \times H$. We have to prove that $F$ is smooth at $(x_0, y_0)$.

For this purpose, let us take any admissible coordinate chart $(U \times V, (\varphi_U \times \psi_V))$ of the product manifold $G \times H$, where $(U, \varphi_U)$ is an admissible coordinate chart of $G$ satisfying $x_0 \in U$, and $(V, \psi_V)$ is an admissible coordinate chart of $H$ satisfying $y_0 \in V$. Next, let us take any admissible coordinate chart $(U \times \tilde{V}, (\varphi_U \times \psi_{\tilde{V}}))$ of $G \times K$, where $(\tilde{V}, \psi_{\tilde{V}})$ is an admissible coordinate chart of $K$ satisfying $f(y_0) \in \tilde{V}$. Here, since $x_0 \in U$, and $f(y_0) \in \tilde{V}$, $F(x_0, y_0) = (x_0, f(y_0)) \in U \times \tilde{V}$. Thus, $(U \times \tilde{V}, (\varphi_U \times \psi_{\tilde{V}}))$ is an admissible coordinate chart of $G \times K$ satisfying $F(x_0, y_0) \in U \times \tilde{V}$. Now, we have to prove that each of the 4 component functions of the mapping

$$\begin{aligned}
(\varphi_U \times \psi_{\tilde{V}}) \circ \left(F \circ (\varphi_U \times \psi_V)^{-1}\right) : (\varphi_U \times \psi_V)((U \times V) \cap F^{-1}(U \times \tilde{V})) \\
\to (\varphi_U \times \psi_{\tilde{V}})(U \times \tilde{V})
\end{aligned}$$

is $C^\infty$ at the point $(\varphi_U \times \psi_V)(x_0, y_0)$ in $\mathbb{R}^4$, that is, $(r, s, t, u) \mapsto \pi_1(((\varphi_U \times \psi_{\tilde{V}}) \circ (F \circ (\varphi_U \times \psi_V)^{-1}))(r, s, t, u))$, $(r, s, t, u) \mapsto \pi_2(((\varphi_U \times \psi_{\tilde{V}}) \circ (F \circ (\varphi_U \times \psi_V)^{-1}))(r, s, t, u))$, $(r, s, t, u) \mapsto \pi_3(((\varphi_U \times \psi_{\tilde{V}}) \circ (F \circ (\varphi_U \times \psi_V)^{-1}))(r, s, t, u))$, and $(r, s, t, u) \mapsto \pi_4(((\varphi_U \times \psi_{\tilde{V}}) \circ (F \circ (\varphi_U \times \psi_V)^{-1}))(r, s, t, u))$ are $C^\infty$ at the point $(\varphi_U \times \psi_V)(x_0, y_0)$ in $\mathbb{R}^4$.

For every $(r, s, t, u)$ in $(\varphi_U \times \psi_V)((U \times V) \cap F^{-1}(U \times \tilde{V}))$, there exist $x$ in $U$, and $y$ in $V$ such that $(\varphi_U \times \psi_V)(x, y) = (r, s, t, u)$. So $(r, s, t, u) = (\varphi_U \times \psi_V)(x, y) = (\varphi_U(x), \psi_V(y))$, and hence, $\varphi_U(x) = (r, s)$, and $\psi_V(y) = (t, u)$. Now, since

$$
\begin{aligned}
\left((\varphi_U \times \psi_{\tilde{V}}) \circ \left(F \circ (\varphi_U \times \psi_V)^{-1}\right)\right) & (r, s, t, u) \\
&= \left((\varphi_U \times \psi_{\tilde{V}}) \circ \left(F \circ (\varphi_U \times \psi_V)^{-1}\right)\right)((\varphi_U \times \psi_V)(x, y)) \\
&= (\varphi_U \times \psi_{\tilde{V}})(F(x, y)) = (\varphi_U \times \psi_{\tilde{V}})(x, f(y)) \\
&= (\varphi_U(x), \psi_{\tilde{V}}(f(y))) = (\varphi_U(x), (\psi_{\tilde{V}} \circ f)(y)) \\
&= (r, s, (\psi_{\tilde{V}} \circ f)(y)) = \left(r, s, (\psi_{\tilde{V}} \circ f)\left((\psi_V)^{-1}(t, u)\right)\right) \\
&= \left(r, s, \left(\psi_{\tilde{V}} \circ f \circ (\psi_V)^{-1}\right)(t, u)\right),
\end{aligned}
$$

so

$$
\begin{aligned}
(r, s, t, u) \mapsto \pi_1 &\left(\left((\varphi_U \times \psi_{\tilde{V}}) \circ \left(F \circ (\varphi_U \times \psi_V)^{-1}\right)\right)(r, s, t, u)\right) \\
&= \pi_1\left(r, s, \left(\psi_{\tilde{V}} \circ f \circ (\psi_V)^{-1}\right)(t, u)\right) = r,
\end{aligned}
$$

which is a smooth function. Hence, $(r, s) \mapsto \pi_1(((\varphi_U \times \psi_{\tilde{V}}) \circ (F \circ (\varphi_U \times \psi_V)^{-1}))(r, s, t, u))$ is a smooth function. Next, since

$$
\begin{aligned}
(r, s, t, u) \mapsto \pi_2 &\left(\left((\varphi_U \times \psi_{\tilde{V}}) \circ \left(F \circ (\varphi_U \times \psi_V)^{-1}\right)\right)(r, s, t, u)\right) \\
\pi_2 &\left(r, s, \left(\psi_{\tilde{V}} \circ f \circ (\psi_V)^{-1}\right)(t, u)\right) = s,
\end{aligned}
$$

which is a smooth function, $(r, s, t, u) \mapsto \pi_2(((\varphi_U \times \psi_{\tilde{V}}) \circ (F \circ (\varphi_U \times \psi_V)^{-1}))(r, s, t, u))$ is a smooth function. Next,

$$
\begin{aligned}
(r, s, t, u) \mapsto \pi_3 &\left(\left((\varphi_U \times \psi_{\tilde{V}}) \circ \left(F \circ (\varphi_U \times \psi_V)^{-1}\right)\right)(r, s, t, u)\right) \\
&= \pi_3\left(r, s, \left(\psi_{\tilde{V}} \circ f \circ (\psi_V)^{-1}\right)(t, u)\right) \\
&= \pi_1\left(\left(\psi_{\tilde{V}} \circ f \circ (\psi_V)^{-1}\right)(t, u)\right).
\end{aligned}
$$

Now, since $f : H \to K$ is a smooth mapping, $(V, \psi_V)$ is an admissible coordinate chart of $H$ satisfying $y_0 \in V$, and $(\tilde{V}, \psi_{\tilde{V}})$ is an admissible coordinate chart of $K$ satisfying $f(y_0) \in \tilde{V}$,

$$\psi_{\tilde{V}} \circ \left( f \circ (\psi_V)^{-1} \right) : \psi_V \left( V \cap f^{-1}(\tilde{V}) \right) \to \psi_{\tilde{V}}(\tilde{V})$$

is $C^\infty$ at the point $\psi_V(y_0)$ in $\mathbb{R}^2$, that is, $(t, u) \mapsto \pi_1(\psi_{\tilde{V}} \circ (f \circ (\psi_V)^{-1})(t, u))$, and $(t, u) \mapsto \pi_2(\psi_{\tilde{V}} \circ (f \circ (\psi_V)^{-1})(r, s))$ are $C^\infty$ at the point $\psi_V(y_0)$ in $\mathbb{R}^2$. Since $(r, s, t, u) \mapsto (t, u)$ is smooth, and $(t, u) \mapsto \pi_1(\psi_{\tilde{V}} \circ (f \circ (\psi_V)^{-1})(t, u))$ is smooth, their composite $(r, s, t, u) \mapsto \pi_1(\psi_{\tilde{V}} \circ (f \circ (\psi_V)^{-1})(t, u))$ is smooth, and hence,

$$(r, s, t, u) \mapsto \pi_3\left( \left( (\varphi_U \times \psi_{\tilde{V}}) \circ \left( F \circ (\varphi_U \times \psi_V)^{-1} \right) \right)(r, s, t, u) \right)$$

is a smooth map. Similarly, $(r, s, t, u) \mapsto \pi_4(((\varphi_U \times \psi_{\tilde{V}}) \circ (F \circ (\varphi_U \times \psi_V)^{-1}))$ $(r, s, t, u))$ is a smooth function.

Thus, we have shown that $F$ is a smooth mapping. In short, if $y \mapsto f(y)$ is smooth map, then $(x, y) \mapsto (x, f(y))$ is smooth.

Similarly, if $x \mapsto f(x)$ is a smooth map, then $(x, y) \mapsto (f(x), y)$ is smooth.

**Lemma 4.17** *Let $G$ be a Lie group. Let $a$ be in $G$. Then, the mapping $F : y \mapsto ay$ is a smooth mapping from smooth manifold $G$ to itself. The mapping $F$ is denoted by $L_a$ and is called the left translation by $a$. Thus, for every $y \in G$, $L_a(y) = ay$. Observe that $L_a : G \to G$ is a 1–1, onto mapping, and $(L_a)^{-1} = L_{a^{-1}}$. Now, since $L_a, L_{a^{-1}}(= (L_a)^{-1})$ are smooth, $L_a$ is a diffeomorphism.*

*Proof* Since $G$ is a Lie group, $G$ is a smooth manifold. Hence $y \mapsto (a, y)$ is a smooth map. Since $G$ is a Lie group, $(x, y) \mapsto xy$ is a smooth map. Since $y \mapsto (a, y)$ is a smooth map, and $(x, y) \mapsto xy$ is a smooth map, their composite map $y \mapsto ay$ is a smooth map. $\qquad \square$

**Note 4.18** Let $G$ be a Lie group. Let $a$ be in $G$. Then, as above, it can be shown that the mapping $F : x \mapsto xa$ is a smooth mapping from smooth manifold $G$ to itself. The mapping $F$ is denoted by $R_a$ and is called the *right translation by $a$*. Thus, for every $x \in G, R_a(y) = ya$. Observe that $R_a : G \to G$ is a 1–1, onto mapping, and $(R_a)^{-1} = R_{a^{-1}}$. Now, since $R_a, R_{a^{-1}}(= (R_a)^{-1})$ are smooth, $R_a$ is a diffeomorphism.

**Lemma 4.19** *Let $G$ be an $n$-dimensional smooth manifold that is also a group. If $(g, h) \mapsto gh^{-1}$ is a smooth mapping from $2n$-dimensional product manifold $G \times G$ to $n$-dimensional smooth manifold $G$, then $G$ is a Lie group.*

*Proof* For simplicity, let $n = 2$. Next, let $e$ be the identity element of the group. Let us define a mapping $m : G \times G \to G$ as follows: for every $g, h$ in $G$,

$$m(g, h) \equiv gh.$$

Let us define a mapping $i : G \to G$ as follows: for every $g$ in $G$,

$$i(g) \equiv g^{-1}.$$

We have to show that

(i)   $i$ is smooth,
(ii)  $m$ is smooth.

For (i): Since $h \mapsto (e, h)$ is a smooth mapping, and $(g, h) \mapsto gh^{-1}$ is a smooth mapping, their composite $h \mapsto eh^{-1} = h^{-1} = i(h)$ is a smooth mapping, and hence, $i$ is a smooth mapping.

For (ii): Since $h \mapsto h^{-1}$ is a smooth mapping, $(g, h) \mapsto (g, h^{-1})$ is a smooth mapping. Since $(g, h) \mapsto (g, h^{-1})$ is a smooth mapping, and $(g, h) \mapsto gh^{-1}$ is a smooth mapping, their composite $(g, h) \mapsto g(h^{-1})^{-1} = gh = m(gh)$ is a smooth mapping, and hence, $m$ is a smooth mapping.                                               □

**Note 4.20** Let $M$ be an $m$-dimensional smooth manifold, whose topology is $\mathcal{O}$, and differential structure is $\mathcal{A}$. Let $G$ be a nonempty open subset of $M$. Let $\mathcal{O}_1$ be the induced topology over $G$, that is, $\mathcal{O}_1 = \{G_1 : G_1 \subset G, \text{ and } G_1 \in \mathcal{O}\}$. Put

$$\mathcal{A}_G \equiv \{(U, \varphi_U) : (U, \varphi_U) \in \mathcal{A}, \text{ and } U \subset G\}.$$

Since $M$ is an $m$-dimensional smooth manifold, $M$ is a Hausdorff space. Since $M$ is a Hausdorff space, and $G$ is a subspace of $M$, $G$ with the induced topology is a Hausdorff space. Since $M$ is an $m$-dimensional smooth manifold, $M$ is a second countable space. Since $M$ is a second countable space, and $G$ is a subspace of $M$, $G$ with the induced topology is a second countable space.

Now, we shall try to show that $\mathcal{A}_G$ is an atlas on $G$ that is,

1. $\cup\{U : (U, \varphi_U) \text{ is in } \mathcal{A}_G\} = G$,
2. all pairs of members in $\mathcal{A}_G$ are $C^\infty$-compatible.

For (1): By the definition of $\mathcal{A}_G$, it is clear that $\cup\{U : (U, \varphi_U) \text{ is in } \mathcal{A}_G\} \subset G$. So, it remains to prove that $G \subset \cup\{U : (U, \varphi_U) \text{ is in } \mathcal{A}_G\}$.

For this purpose, let us take any $p$ in $G$. Since $p$ is in $G$, and $G$ is a subset of $M$, $p$ is in $M$. Since $p$ is in $M$, and $M$ is an $m$-dimensional smooth manifold, there exists $(U, \varphi_U)$ in $\mathcal{A}$ such that $p$ is in $U$. Since $(U, \varphi_U)$ is in $\mathcal{A}$, $U$ is an open subset of $M$. Since $U$ is open, and $G$ is open, $U \cap G$ is open. Thus, $U \cap G$ is an open neighborhood of $p$.

We shall try to show that $(U \cap G, \varphi_U|_{(U \cap G)})$ is in $\mathcal{A}_G$, that is, $(U \cap G, \varphi_U|_{(U \cap G)})$ $\in \mathcal{A}$, and $U \cap G \subset G$. Since $U \cap G \subset G$, it remains to show that $(U \cap G, \varphi_U|_{(U \cap G)})$ $\in \mathcal{A}$. Since $(U, \varphi_U)$ is in $\mathcal{A}$, $(U, \varphi_U)$ is an admissible coordinate chart of $M$. Now, since $U \cap G$ is a nonempty open subset of $U$, by the Lemma 4.10, $(U \cap G, \varphi_U|_{(U \cap G)})$ is an admissible coordinate chart of $M$, and hence, $(U \cap G, \varphi_U|_{(U \cap G)}) \in \mathcal{A}$.

For (2): Let us take any $(U, \varphi_U), (V, \psi_V) \in \mathcal{A}_G$. We have to show that $(U, \varphi_U)$ and $(V, \psi_V)$ are $C^\infty$-compatible. Since $(U, \varphi_U) \in \mathcal{A}_G$, by the definition of $\mathcal{A}_G$, $(U, \varphi_U) \in \mathcal{A}$. Similarly $(V, \psi_V) \in \mathcal{A}$. Since $(U, \varphi_U) \in \mathcal{A}, (V, \psi_V) \in \mathcal{A}$, and $\mathcal{A}$ is a differential structure, $(U, \varphi_U)$ and $(V, \psi_V)$ are $C^\infty$-compatible. Thus, we have shown that $\mathcal{A}_G$ is an atlas on $G$. Hence, $\mathcal{A}_G$ determines a unique smooth structure on $G$. Thus, the open subset $G$ of smooth manifold $M$ becomes an $m$-dimensional smooth manifold. Here, $G$ is called an *open submanifold of M*.

*Example 4.21* Let $M(2 \times 3, \mathbb{R})$ be the collection of all $2 \times 3$ matrices with real entries, that is,

$$M(2 \times 3, \mathbb{R}) \equiv \left\{ \begin{bmatrix} a\,b\,c \\ d\,e\,f \end{bmatrix} : a, b, c, d, e, f \in \mathbb{R} \right\}.$$

We define addition of matrices as follows:

$$\begin{bmatrix} a & b & c \\ d & e & f \end{bmatrix} + \begin{bmatrix} a_1 & b_1 & c_1 \\ d_1 & e_1 & f_1 \end{bmatrix} \equiv \begin{bmatrix} a + a_1 & b + b_1 & c + c_1 \\ d + d_1 & e + e_1 & f + f_1 \end{bmatrix}.$$

Clearly, $M(2 \times 3, \mathbb{R})$ is a commutative group under addition as binary operation. We define scalar multiplication of matrix as follows:

$$t \begin{bmatrix} a & b & c \\ d & e & f \end{bmatrix} \equiv \begin{bmatrix} ta & tb & tc \\ td & te & tf \end{bmatrix}.$$

Clearly, $M(2 \times 3, \mathbb{R})$ becomes a real linear space of dimension $(2 \times 3)$. Here, one of the basis of $M(2 \times 3, \mathbb{R})$ is

$$\left\{ \begin{bmatrix} 1 & 0 & 0 \\ 0 & 0 & 0 \end{bmatrix}, \begin{bmatrix} 0 & 1 & 0 \\ 0 & 0 & 0 \end{bmatrix}, \begin{bmatrix} 0 & 0 & 1 \\ 0 & 0 & 0 \end{bmatrix}, \begin{bmatrix} 0 & 0 & 0 \\ 1 & 0 & 0 \end{bmatrix}, \begin{bmatrix} 0 & 0 & 0 \\ 0 & 1 & 0 \end{bmatrix}, \begin{bmatrix} 0 & 0 & 0 \\ 0 & 0 & 1 \end{bmatrix} \right\}.$$

Clearly, the real linear space $\mathbb{R}^{2 \times 3}$ is isomorphic to the real linear space $M(2 \times 3, \mathbb{R})$. Since $\mathbb{R}^{2 \times 3}$ is isomorphic to $M(2 \times 3, \mathbb{R})$, we will not distinguish between

$$\begin{bmatrix} a & b & c \\ d & e & f \end{bmatrix} \text{ and } (a,b,c,d,e,f).$$

Now since $\mathbb{R}^{2\times3}$ is a $(2 \times 3)$-dimensional smooth manifold, $M(2 \times 3, \mathbb{R})$ is also a $(2 \times 3)$-dimensional smooth manifold. Similarly, $M(m \times n, \mathbb{R})$ is a $mn$-dimensional smooth manifold. Clearly,

$$M(2 \times 3, \mathbb{C})\left( \equiv \left\{ \begin{bmatrix} a & b & c \\ d & e & f \end{bmatrix} : a,b,c,d,e,f \text{ are complex numbers} \right\} \right)$$

is a $2 \times (2 \times 3)$-dimensional smooth manifold. Similarly, $M(m \times n, \mathbb{C})$ is a $2mn$-dimensional smooth manifold. The $n^2$-dimensional smooth manifold $M(n \times n, \mathbb{R})$ is denoted by $M(n, \mathbb{R})$. The $2n^2$-dimensional smooth manifold $M(n \times n, \mathbb{C})$ is denoted by $M(n, \mathbb{C})$. Let $GL(2, \mathbb{R})$ be the collection of all invertible $2 \times 2$ matrices with real entries, that is,

$$GL(2, \mathbb{R}) \equiv \left\{ \begin{bmatrix} x_1 & x_2 \\ x_3 & x_4 \end{bmatrix} : \begin{bmatrix} x_1 & x_2 \\ x_3 & x_4 \end{bmatrix} \in M(2, \mathbb{R}) \text{ and } x_1x_4 - x_2x_3 \neq 0 \right\}.$$

It is clear that $\begin{bmatrix} x_1 & x_2 \\ x_3 & x_4 \end{bmatrix} \mapsto (x_1, x_2, x_3, x_4)$ is in 1–1 correspondence from $GL(2, \mathbb{R})$ onto $\{(x_1, x_2, x_3, x_4) : (x_1, x_2, x_3, x_4) \in \mathbb{R}^4 \text{ and } x_1x_4 - x_2x_3 \neq 0\}$. Let us define a function $f : \mathbb{R}^4 \to \mathbb{R}$ as follows: for every $(x_1, x_2, x_3, x_4)$ in $\mathbb{R}^4$

$$f(x_1, x_2, x_3, x_4) \equiv x_1x_4 - x_2x_3.$$

We shall try to show that $f$ is a continuous function.

Since $(x_1, x_2, x_3, x_4) \mapsto x_1$, and $(x_1, x_2, x_3, x_4) \mapsto x_4$ are continuous functions, $(x_1, x_2, x_3, x_4) \mapsto (x_1, x_4)$ is continuous. Since $(x_1, x_2, x_3, x_4) \mapsto (x_1, x_4)$ is continuous, and the multiplication operation $(x_1, x_4) \mapsto x_1x_4$ is continuous, their composite function $(x_1, x_2, x_3, x_4) \mapsto x_1x_4$ is continuous. Similarly, $(x_1, x_2, x_3, x_4) \mapsto x_2x_3$ is continuous. Since $(x_1, x_2, x_3, x_4) \mapsto x_1x_4$ is continuous, and $(x_1, x_2, x_3, x_4) \mapsto x_2x_3$ is continuous, $(x_1, x_2, x_3, x_4) \mapsto (x_1x_4, x_2x_3)$ is continuous. Since $(x_1, x_2, x_3, x_4) \mapsto (x_1x_4, x_2x_3)$ is continuous, and the difference operation $(y, z) \mapsto y - z$ is continuous, their composite function $(x_1, x_2, x_3, x_4) \mapsto x_1x_4 - x_2x_3 = f(x_1, x_2, x_3, x_4)$ is continuous, and hence, $f$ is a continuous function.

Since $\{0\}$ is a closed subset of $\mathbb{R}$, $\mathbb{R} - \{0\}$ is an open subset of $\mathbb{R}$. Since $\mathbb{R} - \{0\}$ is an open subset of $\mathbb{R}$, and $f : \mathbb{R}^4 \to \mathbb{R}$ is a continuous function, $f^{-1}(\mathbb{R} - \{0\})$ $(= \{(x_1, x_2, x_3, x_4) : (x_1, x_2, x_3, x_4) \in \mathbb{R}^4 \text{ and } x_1x_4 - x_2x_3 \neq 0\})$ is an open subset of $\mathbb{R}^4$, and hence, $\{(x_1, x_2, x_3, x_4) : (x_1, x_2, x_3, x_4) \in \mathbb{R}^4 \text{ and } x_1x_4 - x_2x_3 \neq 0\}$ is an open subset of $\mathbb{R}^4$. Since $\{(x_1, x_2, x_3, x_4) : (x_1, x_2, x_3, x_4) \in \mathbb{R}^4 \text{ and } x_1x_4 - x_2x_3 \neq 0\}$ is an open subset of $\mathbb{R}^4$, and $\begin{bmatrix} x_1 & x_2 \\ x_3 & x_4 \end{bmatrix} \mapsto (x_1, x_2, x_3, x_4)$ is in 1–1 correspondence from $GL(2, \mathbb{R})$ onto $\{(x_1, x_2, x_3, x_4) : (x_1, x_2, x_3, x_4) \in \mathbb{R}^4 \text{ and } x_1x_4 - x_2x_3 \neq 0\}$, $GL(2, \mathbb{R})$

is an open subset of $\mathrm{M}(2,\mathbb{R})(=\mathbb{R}^4)$. Since $\mathrm{GL}(2,\mathbb{R})$ is an open subset of $\mathbb{R}^4$, and $\mathbb{R}^4$ is a 4-dimensional smooth manifold, $\mathrm{GL}(2,\mathbb{R})$ is a 4-dimensional smooth manifold.

Clearly, $\{(\mathrm{GL}(2,\mathbb{R}),\mathrm{Id}_{\mathrm{GL}(2,\mathbb{R})})\}$ is an atlas of $\mathrm{GL}(2,\mathbb{R})$. This shows that $(\mathrm{GL}(2,\mathbb{R})\times\mathrm{GL}(2,\mathbb{R}),(\mathrm{Id}_{\mathrm{GL}(2,\mathbb{R})}\times\mathrm{Id}_{\mathrm{GL}(2,\mathbb{R})}))$ is an admissible coordinate chart of the product manifold $\mathrm{GL}(2,\mathbb{R})\times\mathrm{GL}(2,\mathbb{R})$. Since, for every

$$\left(\begin{bmatrix} a & b \\ c & d \end{bmatrix},\begin{bmatrix} e & f \\ g & h \end{bmatrix}\right)\text{ in }\mathrm{GL}(2,\mathbb{R})\times\mathrm{GL}(2,\mathbb{R}),$$

we have

$$(\mathrm{Id}_{\mathrm{GL}(2,\mathbb{R})}\times\mathrm{Id}_{\mathrm{GL}(2,\mathbb{R})})\left(\begin{bmatrix} a & b \\ c & d \end{bmatrix},\begin{bmatrix} e & f \\ g & h \end{bmatrix}\right)$$

$$=\left(\mathrm{Id}_{\mathrm{GL}(2,\mathbb{R})}\left(\begin{bmatrix} a & b \\ c & d \end{bmatrix}\right),\mathrm{Id}_{\mathrm{GL}(2,\mathbb{R})}\left(\begin{bmatrix} e & f \\ g & h \end{bmatrix}\right)\right)$$

$$=\left(\begin{bmatrix} a & b \\ c & d \end{bmatrix},\begin{bmatrix} e & f \\ g & h \end{bmatrix}\right)=\mathrm{Id}_{\mathrm{GL}(2,\mathbb{R})\times\mathrm{GL}(2,\mathbb{R})}\left(\begin{bmatrix} a & b \\ c & d \end{bmatrix},\begin{bmatrix} e & f \\ g & h \end{bmatrix}\right),$$

$\mathrm{Id}_{\mathrm{GL}(2,\mathbb{R})}\times\mathrm{Id}_{\mathrm{GL}(2,\mathbb{R})}=\mathrm{Id}_{\mathrm{GL}(2,\mathbb{R})\times\mathrm{GL}(2,\mathbb{R})}$, and hence, $(\mathrm{GL}(2,\mathbb{R})\times\mathrm{GL}(2,\mathbb{R}),$ $\mathrm{Id}_{\mathrm{GL}(2,\mathbb{R})\times\mathrm{GL}(2,\mathbb{R})})$ is an admissible coordinate chart of the product manifold $\mathrm{GL}(2,\mathbb{R})\times\mathrm{GL}(2,\mathbb{R})$.

We define multiplication of matrices as follows:

$$\begin{bmatrix} a & b \\ c & d \end{bmatrix}\begin{bmatrix} e & f \\ g & h \end{bmatrix}\equiv\begin{bmatrix} ae+bg & af+bh \\ ce+dg & cf+dh \end{bmatrix}.$$

We know that $\mathrm{GL}(2,\mathbb{R})$ becomes a group under multiplication as binary operation. Thus, $\mathrm{GL}(2,\mathbb{R})$ is a 4-dimensional smooth manifold that is also a group. We want to show that $\mathrm{GL}(2,\mathbb{R})$ is a Lie group. Let us define $m:\mathrm{GL}(2,\mathbb{R})\times\mathrm{GL}(2,\mathbb{R})\to\mathrm{GL}(2,\mathbb{R})$ as follows: for every

$$\left(\begin{bmatrix} a & b \\ c & d \end{bmatrix},\begin{bmatrix} e & f \\ g & h \end{bmatrix}\right)\text{ in }\mathrm{GL}(2,\mathbb{R})\times\mathrm{GL}(2,\mathbb{R}),$$

we have

$$m\left(\begin{bmatrix} a & b \\ c & d \end{bmatrix},\begin{bmatrix} e & f \\ g & h \end{bmatrix}\right)\equiv\begin{bmatrix} a & b \\ c & d \end{bmatrix}\begin{bmatrix} e & f \\ g & h \end{bmatrix}=\begin{bmatrix} ae+bg & af+bh \\ ce+dg & cf+dh \end{bmatrix}.$$

Let us define $i:\mathrm{GL}(2,\mathbb{R})\to\mathrm{GL}(2,\mathbb{R})$ as follows: for every $\begin{bmatrix} a & b \\ c & d \end{bmatrix}$ in $\mathrm{GL}(2,\mathbb{R})$,

$$i\left(\begin{bmatrix} a & b \\ c & d \end{bmatrix}\right) \equiv \frac{1}{ad-bc}\begin{bmatrix} d & -b \\ -c & a \end{bmatrix} = \begin{bmatrix} \dfrac{d}{ad-bc} & \dfrac{-b}{ad-bc} \\ \dfrac{-c}{ad-bc} & \dfrac{a}{ad-bc} \end{bmatrix}.$$

We have to show:

(i)  $m$ is a smooth map from $(4+4)$-dimensional product manifold $\mathrm{GL}(2,\mathbb{R}) \times \mathrm{GL}(2,\mathbb{R})$ to 4-dimensional smooth manifold $\mathrm{GL}(2,\mathbb{R})$,

(ii)  $i$ is a smooth map from 4-dimensional smooth manifold $\mathrm{GL}(2,\mathbb{R})$ to 4-dimensional smooth manifold $\mathrm{GL}(2,\mathbb{R})$.

For (i): For this purpose, let us fix any

$$\left(\begin{bmatrix} a & b \\ c & d \end{bmatrix}, \begin{bmatrix} e & f \\ g & h \end{bmatrix}\right) \text{ in } \mathrm{GL}(2,\mathbb{R}) \times \mathrm{GL}(2,\mathbb{R}).$$

We have to prove that $m$ is smooth at

$$\left(\begin{bmatrix} a & b \\ c & d \end{bmatrix}, \begin{bmatrix} e & f \\ g & h \end{bmatrix}\right).$$

For this purpose, let us take the admissible coordinate chart $(\mathrm{GL}(2,\mathbb{R}) \times \mathrm{GL}(2,\mathbb{R}), \mathrm{Id}_{\mathrm{GL}(2,\mathbb{R})\times\mathrm{GL}(2,\mathbb{R})})$ of the product manifold $\mathrm{GL}(2,\mathbb{R}) \times \mathrm{GL}(2,\mathbb{R})$. Here,

$$\left(\begin{bmatrix} a & b \\ c & d \end{bmatrix}, \begin{bmatrix} e & f \\ g & h \end{bmatrix}\right) \in \mathrm{GL}(2,\mathbb{R}) \times \mathrm{GL}(2,\mathbb{R}).$$

Next, let us take the admissible coordinate chart $(\mathrm{GL}(2,\mathbb{R}), \mathrm{Id}_{\mathrm{GL}(2,\mathbb{R})})$ of $\mathrm{GL}(2,\mathbb{R})$. Clearly,

$$m\left(\begin{bmatrix} a & b \\ c & d \end{bmatrix}, \begin{bmatrix} e & f \\ g & h \end{bmatrix}\right) = \begin{bmatrix} ae+bg & af+bh \\ ce+dg & cf+dh \end{bmatrix} \in \mathrm{GL}(2,\mathbb{R}).$$

Now, we have to prove that each of the 4 component functions of the mapping

$$\mathrm{Id}_{\mathrm{GL}(2,\mathbb{R})} \circ \left(m \circ \left(\mathrm{Id}_{\mathrm{GL}(2,\mathbb{R})\times\mathrm{GL}(2,\mathbb{R})}\right)^{-1}\right) : \left(\mathrm{Id}_{\mathrm{GL}(2,\mathbb{R})\times\mathrm{GL}(2,\mathbb{R})}\right)$$
$$\left((\mathrm{GL}(2,\mathbb{R}) \times \mathrm{GL}(2,\mathbb{R})) \cap m^{-1}(\mathrm{GL}(2,\mathbb{R}))\right) \to \mathrm{Id}_{\mathrm{GL}(2,\mathbb{R})}(\mathrm{GL}(2,\mathbb{R}))$$

is $C^\infty$ at the point

$$\mathrm{Id}_{\mathrm{GL}(2,\mathbb{R})\times\mathrm{GL}(2,\mathbb{R})}\left(\begin{bmatrix} a & b \\ c & d \end{bmatrix}, \begin{bmatrix} e & f \\ g & h \end{bmatrix}\right) \left(= \left(\begin{bmatrix} a & b \\ c & d \end{bmatrix}, \begin{bmatrix} e & f \\ g & h \end{bmatrix}\right)\right) \text{ in } \mathbb{R}^8.$$

Let us observe that $\text{Id}_{\text{GL}(2,\mathbb{R})} \circ (m \circ (\text{Id}_{\text{GL}(2,\mathbb{R}) \times \text{GL}(2,\mathbb{R})})^{-1}) = m$, $(\text{Id}_{\text{GL}(2,\mathbb{R}) \times \text{GL}(2,\mathbb{R})})$ $((\text{GL}(2,\mathbb{R}) \times \text{GL}(2,\mathbb{R})) \cap m^{-1}(\text{GL}(2,\mathbb{R}))) = (\text{GL}(2,\mathbb{R}) \times \text{GL}(2,\mathbb{R})) \cap m^{-1}(\text{GL}(2,\mathbb{R})) = (\text{GL}(2,\mathbb{R}) \times \text{GL}(2,\mathbb{R})) \cap (\text{GL}(2,\mathbb{R}) \times \text{GL}(2,\mathbb{R})) = (\text{GL}(2,\mathbb{R}) \times \text{GL}(2,\mathbb{R}))$, and $\text{Id}_{\text{GL}(2,\mathbb{R})}(\text{GL}(2,\mathbb{R})) = \text{GL}(2,\mathbb{R})$, so we have to prove that each of the 4 component functions of the mapping

$$m : \text{GL}(2,\mathbb{R}) \times \text{GL}(2,\mathbb{R}) \rightarrow \text{GL}(2,\mathbb{R})$$

is $C^\infty$ at the point

$$\left( \begin{bmatrix} a & b \\ c & d \end{bmatrix}, \begin{bmatrix} e & f \\ g & h \end{bmatrix} \right),$$

that is,

$$\left( \begin{bmatrix} x_1 & x_2 \\ x_3 & x_4 \end{bmatrix} \begin{bmatrix} y_1 & y_2 \\ y_3 & y_4 \end{bmatrix} \right) \mapsto x_1 y_1 + x_2 y_3, \quad \left( \begin{bmatrix} x_1 & x_2 \\ x_3 & x_4 \end{bmatrix} \begin{bmatrix} y_1 & y_2 \\ y_3 & y_4 \end{bmatrix} \right) \mapsto x_1 y_2 + x_2 y_4,$$

$$\left( \begin{bmatrix} x_1 & x_2 \\ x_3 & x_4 \end{bmatrix} \begin{bmatrix} y_1 & y_2 \\ y_3 & y_4 \end{bmatrix} \right) \mapsto x_3 y_1 + x_4 y_3, \quad \left( \begin{bmatrix} x_1 & x_2 \\ x_3 & x_4 \end{bmatrix} \begin{bmatrix} y_1 & y_2 \\ y_3 & y_4 \end{bmatrix} \right) \mapsto x_3 y_2 + x_4 y_4$$

are smooth, that is, $(x_1, x_2, x_3, x_4, y_1, y_2, y_3, y_4) \mapsto x_1 y_1 + x_2 y_3$, $(x_1, x_2, x_3, x_4, y_1, y_2, y_3, y_4) \mapsto x_1 y_2 + x_2 y_4$, $(x_1, x_2, x_3, x_4, y_1, y_2, y_3, y_4) \mapsto x_3 y_1 + x_4 y_3$, $(x_1, x_2, x_3, x_4, y_1, y_2, y_3, y_4) \mapsto x_3 y_2 + x_4 y_4$ are smooth. Clearly, all these four functions are smooth. Thus, we have shown that $m$ is a smooth map.

For (ii): For this purpose, let us fix any

$$\begin{bmatrix} a & b \\ c & d \end{bmatrix} \text{ in GL}(2,\mathbb{R}).$$

We have to prove that $i$ is smooth at

$$\begin{bmatrix} a & b \\ c & d \end{bmatrix}.$$

For this purpose, let us take the admissible coordinate chart $(\text{GL}(2,\mathbb{R}), \text{Id}_{\text{GL}(2,\mathbb{R})})$ of the smooth manifold $\text{GL}(2,\mathbb{R})$. Here,

$$\begin{bmatrix} a & b \\ c & d \end{bmatrix} \in \text{GL}(2,\mathbb{R}).$$

Next, let us take the admissible coordinate chart $(GL(2, \mathbb{R}), \mathrm{Id}_{GL(2,\mathbb{R})})$ of the smooth manifold $GL(2, \mathbb{R})$. Clearly,

$$i\left(\begin{bmatrix} a & b \\ c & d \end{bmatrix}\right) = \frac{1}{ad - bc}\begin{bmatrix} d & -b \\ -c & a \end{bmatrix} = \begin{bmatrix} \dfrac{d}{ad-bc} & \dfrac{-b}{ad-bc} \\ \dfrac{-c}{ad-bc} & \dfrac{a}{ad-bc} \end{bmatrix} \in GL(2, \mathbb{R}).$$

Now, we have to prove that each of the 4 component functions of the mapping

$$\mathrm{Id}_{GL(2,\mathbb{R})} \circ \left(i \circ \left(\mathrm{Id}_{GL(2,\mathbb{R})}\right)^{-1}\right) : \left(\mathrm{Id}_{GL(2,\mathbb{R})}\right)\left((GL(2, \mathbb{R})) \cap i^{-1}(GL(2, \mathbb{R}))\right)$$
$$\rightarrow \mathrm{Id}_{GL(2,\mathbb{R})}(GL(2, \mathbb{R}))$$

is $C^\infty$ at the point

$$\mathrm{Id}_{GL(2,\mathbb{R})}\left(\begin{bmatrix} a & b \\ c & d \end{bmatrix}\right)\left(= \begin{bmatrix} a & b \\ c & d \end{bmatrix}\right) \text{ in } \mathbb{R}^4.$$

Let us observe that $\mathrm{Id}_{GL(2,\mathbb{R})} \circ (i \circ (\mathrm{Id}_{GL(2,\mathbb{R})})^{-1}) = i$, $(\mathrm{Id}_{GL(2,\mathbb{R})})((GL(2, \mathbb{R})) \cap i^{-1}(GL(2, \mathbb{R}))) = (GL(2, \mathbb{R})) \cap i^{-1}(GL(2, \mathbb{R})) = (GL(2, \mathbb{R})) \cap (GL(2, \mathbb{R})) = GL(2, \mathbb{R})$, and $\mathrm{Id}_{GL(2,\mathbb{R})}(GL(2, \mathbb{R})) = GL(2, \mathbb{R})$, so we have to prove that each of the 4 component functions of the mapping

$$i : GL(2, \mathbb{R}) \rightarrow GL(2, \mathbb{R})$$

is $C^\infty$ at the point

$$\begin{bmatrix} a & b \\ c & d \end{bmatrix},$$

that is,

$$\begin{bmatrix} x_1 & x_2 \\ x_3 & x_4 \end{bmatrix} \mapsto \frac{x_4}{x_1 x_4 - x_2 x_3}, \quad \begin{bmatrix} x_1 & x_2 \\ x_3 & x_4 \end{bmatrix} \mapsto \frac{-x_2}{x_1 x_4 - x_2 x_3},$$
$$\begin{bmatrix} x_1 & x_2 \\ x_3 & x_4 \end{bmatrix} \mapsto \frac{-x_3}{x_1 x_4 - x_2 x_3}, \quad \begin{bmatrix} x_1 & x_2 \\ x_3 & x_4 \end{bmatrix} \mapsto \frac{x_1}{x_1 x_4 - x_2 x_3}$$

are smooth, that is,

$$(x_1, x_2, x_3, x_4) \mapsto \frac{x_4}{x_1 x_4 - x_2 x_3}, \quad (x_1, x_2, x_3, x_4) \mapsto \frac{-x_2}{x_1 x_4 - x_2 x_3},$$
$$(x_1, x_2, x_3, x_4) \mapsto \frac{-x_3}{x_1 x_4 - x_2 x_3}, \quad (x_1, x_2, x_3, x_4) \mapsto \frac{-x_2}{x_1 x_4 - x_2 x_3}$$

are smooth. Clearly, all these four functions are smooth. Thus, we have shown that $i$ is a smooth map. Hence, $GL(2, \mathbb{R})$ is a 4-dimensional Lie group. Similarly,

$GL(n, \mathbb{R})$ is a $n^2$-dimensional Lie group. $GL(n, \mathbb{R})$ is known as the *general linear group*.

As above, it can be shown that $GL(n, \mathbb{C})$ is a $2n^2$-dimensional Lie group. $GL(n, \mathbb{C})$ is known as the *complex general linear group*.

*Example 4.22* Let $V$ be a real linear space. Let $GL(V)$ be the collection of all invertible linear transformations from $V$ onto $V$. We shall try to show that $GL(V)$ is a group under composition $\circ$ of mappings as binary operation.

(1) Closure law: Let us take any $T_1, T_2$ in $GL(V)$. We have to show that $T_1 \circ T_2$ is in $GL(V)$, that is,

    (a) $T_1 \circ T_2$ is 1–1 from $V$ to $V$,
    (b) $T_1 \circ T_2$ maps $V$ onto $V$,
    (c) for every $x, y$ in $V$, and for every $s, t$ in $\mathbb{R}$, $(T_1 \circ T_2)(sx + ty) = s((T_1 \circ T_2)(x)) + t((T_1 \circ T_2)(y))$.

For (a): Since $T_1$ is in $GL(V)$, $T_1$ is invertible, and hence, $T_1$ is 1–1. Similarly, $T_2$ is 1–1. Since $T_1$ is 1–1, and $T_2$ is 1–1, their composite $T_1 \circ T_2$ is 1–1.
For (b): Since $T_1$ is in $GL(V)$, $T_1$ is invertible, and hence, $T_1$ maps $V$ onto $V$. Similarly, $T_2$ maps $V$ onto $V$. Since $T_1$ maps $V$ onto $V$, and $T_2$ maps $V$ onto $V$, their composite $T_1 \circ T_2$ maps $V$ onto $V$.
For (c): LHS $= (T_1 \circ T_2)(sx + ty) = T_1(T_2(sx + ty)) = T_1(s(T_2(x)) + t(T_2(y))) = s(T_1(T_2(x))) + t(T_1(T_2(y))) = s((T_1 \circ T_2)(x)) + t((T_1 \circ T_2)(y)) = $ RHS.
Thus, we have shown that $T_1 \circ T_2$ is in $GL(V)$.

(2) $\circ$ is associative: Let $T_1, T_2, T_3$ be in $GL(V)$. We have to show that $(T_1 \circ T_2) \circ T_3 = T_1 \circ (T_2 \circ T_3)$. Since the composition of mappings is associative, $(T_1 \circ T_2) \circ T_3 = T_1 \circ (T_2 \circ T_3)$.

(3) Existence of identity element: Let us define the mapping $\mathrm{Id}_V : V \to V$ as follows: for every $x$ in $V$, $\mathrm{Id}_V(x) = x$. It is clear that $\mathrm{Id}_V$ is an invertible linear transformation from $V$ onto $V$, and hence, $\mathrm{Id}_V$ is in $GL(V)$. Also, for any $T$ in $GL(V)$, it is clear that $\mathrm{Id}_V \circ T = T \circ \mathrm{Id}_V = T$. Hence, $\mathrm{Id}_V$ serves the purpose of identity element in $GL(V)$.

(4) Existence of inverse element: Let us take any $T$ in $GL(V)$. Since $T$ is in $GL(V)$, $T$ is invertible, and hence, $T^{-1}$ exists. Now, we shall try to show that $T^{-1}$ is in $GL(V)$, that is, $T^{-1} : V \mapsto V$ is linear. For this purpose, let us take any $x, y$ in $V$, and any real $s, t$. We have to show that $T^{-1}(sx + ty) = s(T^{-1}(x)) + t(T^{-1}(y))$. Since $x$ is in $V$, and $T$ maps $V$ onto $V$, there exists $x_1$ in $V$ such that $T(x_1) = x$. Hence, $T^{-1}(x) = x_1$. Similarly $T(y_1) = y$, and $T^{-1}(y) = y_1$. Now,

$$LHS = T^{-1}(sx + ty) = T^{-1}(s(T(x_1)) + t(T(y_1))) = T^{-1}(T(sx_1 + ty_1)) = sx_1 + ty_1$$
$$= s(T^{-1}(x)) + t(T^{-1}(y)) = \text{RHS}.$$

Thus, $T^{-1}$ is in GL($V$). It is clear that $T^{-1} \circ T = T \circ T^{-1} = \mathrm{Id}_V$. Hence, $T^{-1}$ serves the purpose of the inverse element of $T$ in GL($V$). Thus, we have shown that GL($V$) is a group.

Next, let $V$ be finite-dimensional real linear space. Let $n$ be the dimension of $V$. For simplicity, we shall take 2 for $n$. Let $\{e_1, e_2\}$ be a basis of the real linear space $V$. Let us take any $T$ in GL($V$). So, $T : V \mapsto V$. Let $T(e_1) \equiv a_{11}e_1 + a_{21}e_2$, and $T(e_2) \equiv a_{12}e_1 + a_{22}e_2$. Since $T$ is in GL($V$), $T$ is invertible, and hence, there exists unique $(t_1, t_2)$ such that

$$
\begin{aligned}
(a_{11}t_1 + a_{12}t_2)e_1 + (a_{21}t_1 + a_{22}t_2)e_2 &= t_1(a_{11}e_1 + a_{21}e_2) + t_2(a_{12}e_1 + a_{22}e_2) \\
&= t_1(T(e_1)) + t_2(T(e_2)) = T(t_1e_1 + t_2e_2) \\
&= e_1 = 1e_1 + 0e_2.
\end{aligned}
$$

Thus, the system of equations

$$
\begin{cases}
a_{11}t_1 + a_{12}t_2 = 1 \\
a_{21}t_1 + a_{22}t_2 = 0
\end{cases}
$$

has a unique solution for $(t_1, t_2)$, and hence, $a_{11}a_{22} - a_{12}a_{21} \neq 0$. Since $a_{11}a_{22} - a_{12}u_{21} \neq 0$,

$$
\begin{bmatrix} a_{11} & a_{12} \\ a_{21} & a_{22} \end{bmatrix} \text{ is in GL}(2, \mathbb{R}).
$$

Now, we can define a mapping $\eta : \text{GL}(V) \to \text{GL}(2, \mathbb{R})$ as follows:

$$
\eta(T) \equiv \begin{bmatrix} a_{11} & a_{12} \\ a_{21} & a_{22} \end{bmatrix}.
$$

We shall try to show that $\eta$ is an isomorphism from group GL($V$) onto group GL($2, \mathbb{R}$). For this purpose, we must show:

(i)   $\eta$ is 1–1,
(ii)   $\eta$ maps GL($V$) onto GL($2, \mathbb{R}$),
(iii)   for every $S, T$ in GL($V$), $\eta(S \circ T) = (\eta(S))(\eta(T))$.

For (i): Let $\eta(S) = \eta(T)$, where $S(e_1) = a_{11}e_1 + a_{21}e_2$, $S(e_2) = a_{12}e_1 + a_{22}e_2$, $T(e_1) = b_{11}e_1 + b_{21}e_2$, and $T(e_2) = b_{12}e_1 + b_{22}e_2$. We have to show that $S = T$, that is, $S(e_1) = T(e_1)$, and $S(e_2) = T(e_2)$. Since

$$
\begin{bmatrix} a_{11} & a_{12} \\ a_{21} & a_{22} \end{bmatrix} = \eta(S) = \eta(T) = \begin{bmatrix} b_{11} & b_{12} \\ b_{21} & b_{22} \end{bmatrix},
$$

$a_{11} = b_{11}$, and $a_{21} = b_{21}$, and hence, $S(e_1) = a_{11}e_1 + a_{21}e_2 = b_{11}e_1 + b_{21}e_2 = T(e_1)$. Thus, $S(e_1) = T(e_1)$. Similarly, $S(e_2) = T(e_2)$.

For (ii): We have to show that $\eta : \mathrm{GL}(V) \to \mathrm{GL}(2, \mathbb{R})$ is onto. For this purpose, let us take any

$$\begin{bmatrix} a_{11} & a_{12} \\ a_{21} & a_{22} \end{bmatrix} \text{ in } \mathrm{GL}(2, \mathbb{R}).$$

We have to construct invertible linear transformation $T$ from $V$ to $V$ such that

$$\eta(T) = \begin{bmatrix} a_{11} & a_{12} \\ a_{21} & a_{22} \end{bmatrix}.$$

Let us define $T : V \to V$ as follows: for every $t_1, t_2 \in \mathbb{R}$,

$$T(t_1 e_1 + t_2 e_2) = (a_{11} t_1 + a_{12} t_2) e_1 + (a_{21} t_1 + a_{22} t_2) e_2.$$

It is clear that $T$ is linear. Also $T(e_1) = T(1e_1 + 0e_2) = (a_{11} 1 + a_{12} 0) e_1 + (a_{21} 1 + a_{22} 0) e_2 = a_{11} e_1 + a_{21} e_2$. Thus, $T(e_1) = a_{11} e_1 + a_{21} e_2$. Similarly, $T(e_2) = a_{12} e_1 + a_{22} e_2$. Now, since $\det \begin{bmatrix} a_{11} & a_{12} \\ a_{21} & a_{22} \end{bmatrix} = a_{11} a_{22} - a_{12} a_{21} \neq 0$, $T$ is invertible. Since $T$ is an invertible linear transformation from $V$ onto $V$, $T$ is in $\mathrm{GL}(V)$. Since $T(e_1) = a_{11} e_1 + a_{21} e_2$, and $T(e_2) = a_{12} e_1 + a_{22} e_2$,

$$\eta(T) = \begin{bmatrix} a_{11} & a_{12} \\ a_{21} & a_{22} \end{bmatrix}.$$

For (iii): Let

$$\eta(S) = \begin{bmatrix} a_{11} & a_{12} \\ a_{21} & a_{22} \end{bmatrix},$$

where $S(e_1) = a_{11} e_1 + a_{21} e_2$, and $S(e_2) = a_{12} e_1 + a_{22} e_2$. Let

$$\eta(T) = \begin{bmatrix} b_{11} & b_{12} \\ b_{21} & b_{22} \end{bmatrix},$$

where $T(e_1) = b_{11} e_1 + b_{21} e_2$, and $T(e_2) = b_{12} e_1 + b_{22} e_2$. So,

$$
\begin{aligned}
(S \circ T)(e_1) &= S(T(e_1)) = S(b_{11} e_1 + b_{21} e_2) = b_{11}(S(e_1)) + b_{21}(S(e_2)) \\
&= b_{11}(a_{11} e_1 + a_{21} e_2) + b_{21}(a_{12} e_1 + a_{22} e_2) \\
&= (a_{11} b_{11} + a_{12} b_{21}) e_1 + (a_{21} b_{11} + a_{22} b_{21}) e_2.
\end{aligned}
$$

Thus,

$$(S \circ T)(e_1) = (a_{11}b_{11} + a_{12}b_{21})e_1 + (a_{21}b_{11} + a_{22}b_{21})e_2.$$

Next,

$$
\begin{aligned}
(S \circ T)(e_2) &= S(T(e_2)) = S(b_{12}e_1 + b_{22}e_2) = b_{12}(S(e_1)) + b_{22}(S(e_2)) \\
&= b_{12}(a_{11}e_1 + a_{21}e_2) + b_{22}(a_{12}e_1 + a_{22}e_2) \\
&= (a_{11}b_{12} + a_{12}b_{22})e_1 + (a_{21}b_{12} + a_{22}b_{22})e_2.
\end{aligned}
$$

Thus,

$$(S \circ T)(e_2) = (a_{11}b_{12} + a_{12}b_{22})e_1 + (a_{21}b_{12} + a_{22}b_{22})e_2.$$

It follows that

$$
\begin{aligned}
\text{LHS} = \eta(S \circ T) &= \begin{bmatrix} a_{11}b_{11} + a_{12}b_{21} & a_{11}b_{12} + a_{12}b_{22} \\ a_{21}b_{11} + a_{22}b_{21} & a_{21}b_{12} + a_{22}b_{22} \end{bmatrix} = \begin{bmatrix} a_{11} & a_{12} \\ a_{21} & a_{22} \end{bmatrix} \begin{bmatrix} b_{11} & b_{12} \\ b_{21} & b_{22} \end{bmatrix} \\
&= (\eta(S))(\eta(T)) = \text{RHS}.
\end{aligned}
$$

Thus, we have shown that $\eta$ is an isomorphism between group $GL(V)$ onto group $GL(2, \mathbb{R})$. Hence, group $GL(V)$ is isomorphic onto group $GL(2, \mathbb{R})$. Since $GL(V)$ and $GL(2, \mathbb{R})$ are isomorphic groups, and $GL(2, \mathbb{R})$ is a Lie group, we can bestow a topology on $GL(V)$ so that $GL(V)$ is also a Lie group. It also shows that the smooth structure on $GL(V)$ is independent of the choice of basis of $V$. Thus, if $V$ is a real linear space of dimension $n$, then $GL(V)$ is isomorphic to the Lie group $GL(n, \mathbb{R})$. Similarly, if $V$ is a complex linear space of dimension $n$, then $GL(V)$ is isomorphic to the Lie group $GL(n, \mathbb{C})$.

*Example 4.23* Observe that $(\mathbb{R}, +)$ is a group, and $\mathbb{R}$ is a 1-dimensional smooth manifold. We want to show that $\mathbb{R}$ is a Lie group.

By Lemma 4.19, it is enough to show that the mapping $f : (x, y) \mapsto x - y$ is a smooth mapping from 2-dimensional product manifold $\mathbb{R} \times \mathbb{R}$ to 1-dimensional smooth manifold $\mathbb{R}$. For this purpose, let us fix any $(a, b)$ in $\mathbb{R} \times \mathbb{R}$. We have to prove that $f$ is smooth at $(a, b)$. For this purpose, let us take the admissible coordinate chart $(\mathbb{R} \times \mathbb{R}, \text{Id}_{\mathbb{R} \times \mathbb{R}})$ of the product manifold $\mathbb{R} \times \mathbb{R}$. Here, $(a, b) \in \mathbb{R} \times \mathbb{R}$. Next, let us take the admissible coordinate chart $(\mathbb{R}, \text{Id}_{\mathbb{R}})$ of $\mathbb{R}$. Clearly $f(a, b) = a - b \in \mathbb{R}$. Now, we have to prove that the function

$$\text{Id}_{\mathbb{R}} \circ \left( f \circ (\text{Id}_{\mathbb{R} \times \mathbb{R}})^{-1} \right) : (\text{Id}_{\mathbb{R} \times \mathbb{R}}) \left( (\mathbb{R} \times \mathbb{R}) \cap f^{-1}(\mathbb{R}) \right) \to \text{Id}_{\mathbb{R}}(\mathbb{R})$$

is $C^\infty$ at the point $\text{Id}_{\mathbb{R} \times \mathbb{R}}(a, b) (= (a, b))$ in $\mathbb{R}^2$. Let us observe that $\text{Id}_{\mathbb{R}} \circ (f \circ (\text{Id}_{\mathbb{R} \times \mathbb{R}})^{-1}) = f$, $(\text{Id}_{\mathbb{R} \times \mathbb{R}})((\mathbb{R} \times \mathbb{R}) \cap f^{-1}(\mathbb{R})) = (\mathbb{R} \times \mathbb{R}) \cap f^{-1}(\mathbb{R}) = (\mathbb{R} \times \mathbb{R}) \cap (\mathbb{R} \times \mathbb{R}) = \mathbb{R} \times \mathbb{R}$, and $\text{Id}_{\mathbb{R}}(\mathbb{R}) = \mathbb{R}$, so we have to prove that the function

$$f : \mathbb{R} \times \mathbb{R} \to \mathbb{R}$$

is $C^\infty$ at the point $(a, b)$. But this is known to be true. Thus, we have shown that $(x, y) \mapsto x - y$ is a smooth map. This proves that $\mathbb{R}$ is a Lie group under addition.

Similarly, $\mathbb{R}^2$ is a Lie group under addition, $\mathbb{R}^3$ is a Lie group under addition, etc. Since each ordered pair $(x, y)$ of real numbers $x, y$ can be identified with the complex number $x + \sqrt{-1}y$, and $\mathbb{R}^2$ is a Lie group under addition, clearly $\mathbb{C}$ is a Lie group under addition.

*Example 4.24* By $\mathbb{R}^*$ we shall mean the set of all nonzero real numbers. We know that $\mathbb{R}^*$ is a group under multiplication as binary operation. Since the elements of $\mathbb{R}^*$ can be identified, in a natural way, with the elements of $GL(1, \mathbb{R})$, and $GL(1, \mathbb{R})$, is a 1-dimensional Lie group, $\mathbb{R}^*$ is also a 1-dimensional Lie group under multiplication.

Next, since $\mathbb{R}^*$ is a 1-dimensional Lie group, $\mathbb{R}^*$ is a smooth manifold. Since $\mathbb{R}^*$ is a smooth manifold, and $\{x : x \in \mathbb{R} \text{ and } 0 < x\}$ is an open subset of $\mathbb{R}^*$, $\{x : x \in \mathbb{R} \text{ and } 0 < x\}$ is a smooth manifold. We know that $\{x : x \in \mathbb{R} \text{ and } 0 < x\}$ is a subgroup of the multiplicative group $\mathbb{R}^*$. Thus, $\{x : x \in \mathbb{R} \text{ and } 0 < x\}$ is a 1-dimensional smooth manifold that is also a group. Since the function $(x, y) \mapsto \frac{x}{y}$ is smooth so, by Lemma 4.19, $\{x : x \in \mathbb{R} \text{ and } 0 < x\}$ is a 1-dimensional Lie group under multiplication.

*Example 4.25* By $\mathbb{C}^*$ we shall mean the set of all nonzero complex numbers. We know that $\mathbb{C}^*$ is a group under multiplication as binary operation. Since the elements of $\mathbb{R}^*$ can be identified, in a natural way, with the elements of $GL(1, \mathbb{C})$, and $GL(1, \mathbb{C})$ is a 2-dimensional Lie group, $\mathbb{C}^*$ is also a 2-dimensional Lie group under multiplication.

*Example 4.26* We know that the unit circle $S^1 (= \{(x, y) : (x, y) \in \mathbb{R}^2 \text{ and } \sqrt{x^2 + y^2} = 1\})$ is a 1-dimensional smooth manifold. Observe that we can identify $(x, y)$ with the complex number $x + \sqrt{-1}y$. Thus, we can identify the unit circle $S^1$ with $\{z : z \in \mathbb{C}^* \text{ and } |z| = 1\}$. Further, we know that $\{z : z \in \mathbb{C}^* \text{ and } |z| = 1\}$ is a multiplicative group. Thus, unit circle $S^1$ is a 1-dimensional smooth manifold that is also a group. Since the function $(z, w) \mapsto \frac{z}{w}$ is smooth, by Lemma 4.19, the circle $S^1$ is a 1-dimensional Lie group. The Lie group $S^1$ is called the *circle group*.

*Example 4.27* Let $G_1$ and $G_2$ be Lie groups. Since $G_1$ and $G_2$ are Lie groups, $G_1$ and $G_2$ are groups. We know that the Cartesian product $G_1 \times G_2$ is also a group under the binary operation given by $(g_1, h_1)(g_2, h_2) \equiv (g_1 g_2, h_1 h_2)$. Since $G_1$ and $G_2$ are Lie groups, $G_1$ and $G_2$ are smooth manifolds, and hence, their Cartesian product $G_1 \times G_2$ is a smooth manifold (called the *product manifold of $G_1$ and $G_2$*). Thus, $G_1 \times G_2$ is a 1-dimensional smooth manifold that is also a group.

We want to show that $G_1 \times G_2$ is a Lie group. By Lemma 4.19, it is enough to show that the mapping $((x,y),(z,w)) \mapsto (xz^{-1}, yw^{-1})$ is smooth. Since the projection map $((x,y),(z,w)) \mapsto (x,y)$ is smooth, and $(x,y) \mapsto x$ is smooth, their composite $((x,y),(z,w)) \mapsto x$ is smooth. Similarly, $((x,y),(z,w)) \mapsto y$ is smooth. Since $((x,y),(z,w)) \mapsto x$ and $((x,y),(z,w)) \mapsto z$ are smooth, $((x,y),(z,w)) \mapsto (x,z)$ is smooth. Since $G_1$ is a Lie group, and $x, z$ are in $G_1$, $(x,z) \mapsto xz^{-1}$ is smooth. Since $((x,y),(z,w)) \mapsto (x,z)$ is smooth, and $(x,z) \mapsto xz^{-1}$ is smooth, their composite $((x,y),(z,w)) \mapsto xz^{-1}$ is smooth. Similarly, $((x,y),(z,w)) \mapsto yw^{-1}$ is smooth. Since $((x,y),(z,w)) \mapsto xz^{-1}$ and $((x,y),(z,w)) \mapsto yw^{-1}$ are smooth, $((x,y),(z,w)) \mapsto (xz^{-1}, yw^{-1})$ is smooth. Thus, we have shown that $G_1 \times G_2$ is a Lie group.

Similarly, if $G_1, G_2, G_3$ are Lie groups, then $G_1 \times G_2 \times G_3$ is a Lie group, etc. Here, $G_1 \times G_2 \times G_3$ is called the *direct product* of Lie groups $G_1, G_2, G_3$; etc. The direct product of circle group $S_1$ with itself, that is, $S_1 \times S_1$ is an abelian Lie group under pointwise multiplication. $S_1 \times S_1$ is called the 2-*torus* and is denoted by $\mathbb{T}^2$. Similarly, by 3-*torus* $\mathbb{T}^3$, we mean the abelian Lie group $S_1 \times S_1 \times S_1$, etc.

*Example 4.28* Let $G$ be a finite group. Let us bestow discrete topology over $G$. Observe that discrete topology over $G$ is Hausdorff and second countable. Here, $G$ is called a 0-*dimensional Lie group* or *discrete group*.

**Definition** Let $G$ and $H$ be Lie groups. Let $F : G \to H$ be a mapping. If $F$ is a smooth mapping, and $F$ is a group homomorphism, then we say that $F$ *is a Lie group homomorphism from $G$ to $H$*.

**Definition** Let $G$ and $H$ be Lie groups. Let $F : G \to H$ be a mapping. If $F$ is a group isomorphism, and $F$ is a diffeomorphism, then we say that $F$ *is a Lie group isomorphism from $G$ onto $H$*.

**Definition** Let $G$ and $H$ be Lie groups. If there exists a mapping $F : G \to H$ such that $F$ is a Lie group isomorphism from $G$ onto $H$, then we say that $G$ *and $H$ are isomorphic Lie groups*.

*Example 4.29* We have seen that the circle $S^1$, and the set $\mathbb{C}^*$ of all nonzero complex numbers are Lie groups. Let us define a function $I : S^1 \to \mathbb{C}^*$ as follows: for every $z$ in $S^1$, $I(z) = z$. Clearly, the *inclusion map $I$* is a Lie group homomorphism from $S^1$ to $\mathbb{C}^*$.

*Example 4.30* We have seen that $\mathbb{R}$ is a Lie group under addition, and $\mathbb{R}^*$ is a Lie group under multiplication. Consider the function $\exp : \mathbb{R} \to \mathbb{R}^*$ defined as follows: for every $x$ in $\mathbb{R}$, $\exp(x) \equiv e^x$. Since $x \mapsto e^x$ is a smooth function and is a group homomorphism from $\mathbb{R}$ to $\mathbb{R}^*$, $\exp$ is a Lie group homomorphism from $\mathbb{R}$ to $\mathbb{R}^*$.

Observe that the range of $\exp$ is $\{x : x \in \mathbb{R} \text{ and } 0 < x\}$. We have seen that $\{x : x \in \mathbb{R} \text{ and } 0 < x\}$ is a Lie group under multiplication. Here, the inverse function of $\exp$ is $\ln : \{x : x \in \mathbb{R} \text{ and } 0 < x\} \to \mathbb{R}$. Since $\ln$ is a smooth function and is 1–1 onto, $\exp$ is a Lie group isomorphism from $\mathbb{R}$ to $\{x : x \in \mathbb{R} \text{ and } 0 < x\}$.

*Example 4.31* We have seen that $\mathbb{C}$ is a Lie group under addition, and $\mathbb{C}^*$ is a Lie group under multiplication. Consider the function $\exp : \mathbb{C} \to \mathbb{C}^*$ defined as follows: for every $z$ in $\mathbb{C}$,

$$\exp(z) \equiv e^z = e^{\mathrm{Re}(z)}\Big(\cos(\mathrm{Im}(z)) + \sqrt{-1}\,\sin(\mathrm{Im}(z))\Big).$$

Since $(x, y) = z \mapsto e^z = (e^x \cos y, e^x \sin y)$ is a smooth function and is a group homomorphism from $\mathbb{C}$ to $\mathbb{C}^*$, $\exp$ is a Lie group homomorphism from $\mathbb{C}$ to $\mathbb{C}^*$. Observe that the range of $\exp$ is $\mathbb{C}^*$, $\exp$ is a function from $\mathbb{C}$ onto $\mathbb{C}^*$. Since $\exp$ is not 1–1, $\exp$ is not a Lie group isomorphism from $\mathbb{C}$ to $\mathbb{C}^*$.

*Example 4.32* We have seen that $\mathbb{R}$ is a Lie group under addition, and the circle $S^1$ is a Lie group under multiplication. Consider the function $\varepsilon : \mathbb{R} \to S^1$ defined as follows: for every $t$ in $\mathbb{R}$,

$$\varepsilon(t) \equiv \cos(2\pi t) + \sqrt{-1}\,\sin(2\pi t).$$

Since $t \mapsto (\cos(2\pi t) + \sqrt{-1}\,\sin(2\pi t)) = (\cos(2\pi t), \sin(2\pi t))$ is a smooth function and is a group homomorphism from $\mathbb{R}$ onto $S^1$, $\varepsilon$ is a Lie group homomorphism from $\mathbb{R}$ onto $S^1$.

Here, the kernel $\ker(\varepsilon)$ of homomorphism $\varepsilon$ is given by $\ker(\varepsilon) = \varepsilon^{-1}(1) = \{t : t \in \mathbb{R} \text{ and } \varepsilon(t) = 1\} = \{t : t \in \mathbb{R} \text{ and } \cos(2\pi t) + \sqrt{-1}\,\sin(2\pi t) = 1\} = \{0, \pm 1, \pm 2, \pm 3, \ldots\} = \mathbb{Z}$. We have seen that $\mathbb{R}^2$ is a Lie group under addition and the 2-torus $\mathbb{T}^2(= S_1 \times S_1)$ is a Lie group under pointwise multiplication.

Consider the function $\varepsilon^2 : \mathbb{R}^2 \to S_1 \times S_1$ defined as follows: for every $s, t$ in $\mathbb{R}$,

$$\varepsilon^2(s, t) \equiv \Big(\cos(2\pi s) + \sqrt{-1}\,\sin(2\pi s), \cos(2\pi t) + \sqrt{-1}\,\sin(2\pi t)\Big).$$

Since

$$\begin{aligned}
(s, t) \mapsto &\Big(\cos(2\pi s) + \sqrt{-1}\,\sin(2\pi s), \cos(2\pi t) + \sqrt{-1}\,\sin(2\pi t)\Big) \\
&= ((\cos(2\pi s), \sin(2\pi s)), (\cos(2\pi t), \sin(2\pi t))) \\
&= (\cos(2\pi s), \sin(2\pi s), \cos(2\pi t), \sin(2\pi t))
\end{aligned}$$

is a smooth function and is a group homomorphism from $\mathbb{R}^2$ onto $S_1 \times S_1$, $\varepsilon^2$ is a Lie group homomorphism from $\mathbb{R}^2$ onto 2-torus $\mathbb{T}^2(= S_1 \times S_1)$.

Here, the kernel $\ker(\varepsilon^2)$ of homomorphism $\varepsilon^2$ is given by $\ker(\varepsilon^2) = (\varepsilon^2)^{-1}(1, 1) = \{(s, t) : (s, t) \in \mathbb{R}^2 \text{ and } \varepsilon^2(s, t) = (1, 1)\} = \{(s, t) : (s, t) \in \mathbb{R}^2, \cos(2\pi s) + \sqrt{-1}\,\sin(2\pi s) = 1, \text{ and } \cos(2\pi t) + \sqrt{-1}\,\sin(2\pi t) = 1\} = \{0, \pm 1, \pm 2, \pm 3, \ldots\} \times \{0, \pm 1, \pm 2, \pm 3, \ldots\} = \mathbb{Z}^2$.

Similar discussion can be given for $\varepsilon^3, \varepsilon^4$, etc.

*Example 4.33* We have seen that

$$\mathrm{GL}(2,\mathbb{R}) \left( = \left\{ \begin{bmatrix} x_1 x_2 \\ x_3 x_4 \end{bmatrix} : \begin{bmatrix} x_1 x_2 \\ x_3 x_4 \end{bmatrix} \in \mathrm{M}(2,\mathbb{R}) \text{ and } x_1 x_4 - x_2 x_3 \neq 0 \right\} \right)$$

is a Lie group under matrix multiplication, and $\mathbb{R}^*$ is a Lie group under ordinary multiplication. Consider the function $\det : \mathrm{GL}(2,\mathbb{R}) \mapsto \mathbb{R}^*$ defined as follows: For every

$$\begin{bmatrix} x_1 & x_2 \\ x_3 & x_4 \end{bmatrix} \in \mathrm{M}(2,\mathbb{R}),$$

we have

$$\det \left( \begin{bmatrix} x_1 & x_2 \\ x_3 & x_4 \end{bmatrix} \right) \equiv x_1 x_4 - x_2 x_3.$$

Since

$$(x_1, x_2, x_3, x_4) = \begin{bmatrix} x_1 & x_2 \\ x_3 & x_4 \end{bmatrix} \mapsto (x_1 x_4 - x_2 x_3)$$

is a smooth function, and by a property of determinant det is a group homomorphism from $\mathrm{GL}(2,\mathbb{R})$ onto $\mathbb{R}^*$, det is a Lie group homomorphism from $\mathrm{GL}(2,\mathbb{R})$ onto $\mathbb{R}^*$.

*Example 4.34* Let us take any *n*-dimensional Lie group $G$. Let us fix any $a$ in $G$. For simplicity, let us take 2 for $n$. Consider the mapping $C_a : G \mapsto G$ defined as follows: for every $x$ in $G$, $C_a(x) \equiv axa^{-1}$. We want to show that $C_a$ is a Lie group homomorphism from $G$ onto $G$, that is,

1. $C_a$ is a group homomorphism,
2. $C_a$ is onto,
3. $C_a$ is smooth.

For 1: Since for every $x, y$ in $G$, $C_a(xy) \equiv a(xy)a^{-1} = (axa^{-1})(aya^{-1}) = (C_a(x))(C_a(y))$, $C_a$ is a group homomorphism.

For 2: For this purpose, let us take any $y$ in $G$. Since $a^{-1}ya$ is in $G$, and $C_a(a^{-1}ya) = a(a^{-1}ya)a^{-1} = y$, it follows that $C_a$ is onto.

For 3: Since $G$ is a Lie group, and $a$ is in group $G$, the map $x \mapsto ax$ is smooth. Similarly $x \mapsto xa^{-1}$, is smooth. Since $x \mapsto ax$ is smooth, and $x \mapsto xa^{-1}$, their composite $x \mapsto axa^{-1} (= C_a(x))$ is smooth, and hence $C_a$ is smooth.

Thus we have shown that $C_a$ is a Lie group homomorphism from $G$ onto $G$.

## 4.3  Locally Path Connected Spaces

**Note 4.35** Let $X$ be a topological space. Let $S$ be a connected subset of $X$. We shall try to show that the closure $S^-$ of $S$ is also connected. We claim that $S^-$ is connected. If not, otherwise, let $S^-$ be not connected. We have to arrive at a contradiction. Since $S^-$ is not connected, there exist open subsets $G_1, G_2$ of $X$ such that $G_1 \cap (S^-)$ is nonempty, $G_2 \cap (S^-)$ is nonempty, $G_1 \cap G_2 \cap (S^-)$ is empty, and $S^- \subset G_1 \cup G_2$. Since $G_1 \cap (S^-)$ is nonempty, there exists $a \in G_1 \cap (S^-)$, and hence $a \in G_1$. Since $a \in G_1$ and $G_1$ is open, $G_1$ is an open neighborhood of $a$. Since $a \in G_1 \cap (S^-)$, $a \in S^-$. Since $a \in S^-$ and $G_1$ is an open neighborhood of $a$, $G_1 \cap S$ is nonempty. Similarly, $G_2 \cap S$ is nonempty. Since $S \subset S^-$, and $G_1 \cap G_2 \cap (S^-)$ is empty, $G_1 \cap G_2 \cap S$ is empty. Since $S \subset S^- \subset G_1 \cup G_2$, $S \subset G_1 \cup G_2$. Since $G_1, G_2$ are open subsets of $X$, $G_1 \cap S$ is nonempty, $G_2 \cap S$ is nonempty, $G_1 \cap G_2 \cap S$ is empty, and $S \subset G_1 \cup G_2$, $S$ is not connected, a contradiction.

**Note 4.36** Let $X$ be a topological space. Let $S$ be a nonempty subset of $X$. For every $x, y$ in $S$, by $x \sim y$ we shall mean that there exists a connected subset $C$ of $X$ such that $C \subset S, x \in C$, and $y \in C$. We shall show that $\sim$ is an equivalence relation over $S$, that is,

1. for every $x$ in $S$, $x \sim x$,
2. if $x \sim y$ then $y \sim x$,
3. if $x \sim y$, and $y \sim z$ then $x \sim z$.

For 1: Let us take any $x$ in $S$. Since the singleton set $\{x\}$ is connected subset of $X$, and $x \in \{x\} \subset S$, by the definition of $\sim$, $x \sim x$.

For 2: Let $x \sim y$. Now, by the definition of $\sim$, there exists a connected subset $C$ of $X$ such that $C \subset S, x \in C$, and $y \in C$. Thus, $C$ is a connected subset of $X$ such that $C \subset S, y \in C$, and $x \in C$, and hence, by the definition of $\sim$, $y \sim x$.

For 3: Let $x \sim y$, and $y \sim z$. Since $x \sim y$, so by the definition of $\sim$, there exists a connected subset $C$ of $X$ such that $C \subset S, x \in C$, and $y \in C$. Similarly, there exists a connected subset $C_1$ of $X$ such that $C_1 \subset S, y \in C_1$, and $z \in C_1$. Since $C$ is a connected subset of $X$, $C_1$ is a connected subset of $X$, and $y \in C \cap C_1$, $C \cup C_1$ is a connected subset of $X$. Also $x \in C \subset C \cup C_1$, and $z \in C_1 \subset C \cup C_1$. Since $C \subset S$, and $C_1 \subset S$, so $C \cup C_1 \subset S$. It follows, by the definition of $\sim$, that $x \sim z$.

Thus, we have shown that $\sim$ is an equivalence relation over $S$. Hence, $S$ is partitioned into equivalence classes. Here, each equivalence class is called a *component of S*.

Now, we shall try to show that each component of $S$ is a connected subset of $X$. Let us take any component $[a]$ of $S$, where $a$ is in $S$, and $[a] \equiv \{x : x \in S, \text{ and } x \sim a\}$. We claim that $[a]$ is connected. If not, otherwise, let $[a]$ be not connected, that is, there exist open subsets $G_1, G_2$ of $X$ such that $G_1 \cap [a]$ is nonempty, $G_2 \cap [a]$ is nonempty, $G_1 \cap G_2 \cap [a]$ is empty, and $[a] \subset G_1 \cup G_2$.

We have to arrive at a contradiction. Since $G_1 \cap [a]$ is nonempty, there exists $b$ in $G_1 \cap [a]$. Similarly, there exists $c$ in $G_2 \cap [a]$. Here, $b$ and $c$ are in $[a]$, and $[a]$ is an equivalence class, so $b \sim c$, and hence, by the definition of $\sim$, there exists a connected subset $C$ of $X$ such that $C \subset S, b \in C$, and $c \in C$. Thus, $b$ is in $G_1 \cap C$, and hence, $G_1 \cap C$ is nonempty. Similarly, $G_2 \cap C$ is nonempty. Clearly, $C$ is contained in $[a]$.

(Reason: If not, otherwise, there exists $t \in C$ such that $t \notin [a]$. Since $C$ is a connected subset of $X$, $C \subset S, b \in C$, and $t \in C$, $b \sim t$. Since $b$ is in $G_1 \cap [a]$, $b \sim a$. Since $b \sim t$, and $b \sim a$, $t \sim a$, and hence $t \in [a]$, a contradiction.)

Since $C$ is contained in $[a]$, and $G_1 \cap G_2 \cap [a]$ is empty, $G_1 \cap G_2 \cap C$ is empty. Next since $C$ is contained in $[a]$ and $[a] \subset G_1 \cup G_2$, $C \subset G_1 \cup G_2$. Since $G_1, G_2$ are open subsets of $X$, $G_1 \cap C$ is nonempty, $G_2 \cap C$ is nonempty, $G_1 \cap G_2 \cap C$ is empty, and $C \subset G_1 \cup G_2$, $C$ is not connected, a contradiction.

Thus, we have shown that $[a]$ is a connected subset of $S$. Clearly, $[a]$ is a maximal connected subset of $S$.

(Reason: If not, otherwise, let $[a]$ be not a maximal connected subset of $S$. So, there exists a connected subset $C$ of $S$ such that $[a]$ is a proper subset of $C$. So, there exists $t$ in $C$ such that $t$ is not in $[a]$. Since $a \in [a] \subset C \subset S$, $t$ is in $C$, and $C$ is a connected subset of $X$, $t \sim a$, and hence, $t$ is in $[a]$, a contradiction.)

Finally, we want to show that every component of $X$ is a closed set. Let us take any component $C$ of $X$. Since $C$ is a component of $X$, $C$ is connected, and hence, $C^-$ is connected. Since $C$ is a component of $X$, $C^-$ is connected, and $C \subset C^- \subset X$, $C = C^-$, and hence, $C$ is closed.

Thus, we have seen that

(i)   every nonempty subset $S$ topological space $X$ is partitioned into components, and
(ii)  every component of $S$ is connected.
(iii) if $C$ is a component of $S$, $C_1$ is connected, and $C \subset C_1 \subset S$ then $C = C_1$.
(iv)  every component of $X$ is a closed set.

**Note 4.37** Let $X$ be a topological space. Let $S$ be a nonempty subset of $X$. Let $a$, $b$ be elements of $S$. Let $f$ be a function from closed interval $[0,1]$ to $S$. If

(i)   $f$ is continuous,
(ii)  $f(0) = a$ and $f(1) = b$,

then, we say that $f$ is a path in $S$ from $a$ to $b$. Let $X$ be a topological space. Let $S$ be a nonempty subset of $X$. If for every $a$, $b$ in $S$, there exists a path $f$ in $S$ from $a$ to $b$, then we say that $S$ is a path connected subset of $X$. By $X$ is path connected we mean: $X$ is a path connected subset of $X$. In other words, a topological space $X$ is path connected means for every $x, y$ in $X$, there exists a continuous function $f : [0, 1] \mapsto X$ such that $f(0) = x$ and $f(1) = y$.

Let $X$ be a topological space. Let $S$ be a nonempty subset of $X$. For every $x, y$ in $S$, by $x \sim y$, we shall mean that the following statement is true:

there exists a path $f$ in $S$ from $x$ to $y$.

We shall try to show that $\sim$ is an equivalence relation over $S$. By the definition of equivalence relation, we must prove:

(1) $x \sim x$ for every $x$ in $S$,
(2) if $x \sim y$, then $y \sim x$,
(3) if $x \sim y$ and $y \sim z$, then $x \sim z$.

For (1): Let us take any $x$ in $S$. By the definition of $\sim$, we must find a path $f$ in $S$ from $x$ to $x$. For this purpose, consider the constant function $f : [0, 1] \to S$ defined by

$$f(t) \equiv x$$

for every $t$ in $[0,1]$. Since every constant function is a continuous function, and $f$ is a constant function, $f$ is a continuous function. Further, by the definition of $f$, $f(0) = x$ and $f(1) = x$. Hence, by the definition of path, $f$ is a path in $S$ from $x$ to $x$. This proves (1).

For (2): Let $x \sim y$. So, by the definition of $\sim$, there exists a path $f$ in $S$ from $x$ to $y$. Since $f$ is a path in $S$ from $x$ to $y$ so, by the definition of path, $f$ is a function from the closed interval $[0,1]$ to $S$ such that

(i) $f$ is continuous,
(ii) $f(0) = x$ and $f(1) = y$.

We have to prove: $y \sim x$. So, by the definition of $\sim$, we must find a path in $S$ from $y$ to $x$, and hence, by the definition of path, we must find a function $g : [0, 1] \mapsto S$ such that

(1') $g$ is continuous,
(2') $g(0) = y$ and $g(1) = x$.

Let us define $g : [0, 1] \to X$ as follows: $g(t) \equiv f(1 - t)$ for every $t$ in $[0, 1]$. Clearly, the range of $g$ is equal to the range of $f$. Since the range of $g$ is equal to the range of $f$, and $f : [0, 1] \to S$, $g : [0, 1] \to S$.

For (1'): Since $t \mapsto 1 - t$ is a polynomial function, this is continuous. Since $t \mapsto 1 - t$ is a continuous function, and $f$ is a continuous function, their composite function $t \mapsto f(1 - t) = g(t)$ is continuous, and hence, $g$ is continuous. This proves (1').

For (2') : Here, by the definition of $g$, $g(0) = f(1 - 0) = f(1) = y$, and $g(1) = f(1 - 1) = f(0) = x$. This proves 2'.

Thus we have shown: If $x \sim y$, then $y \sim x$. This proves (2).

For (3): Given that $x \sim y$ and $y \sim z$. We have to prove: $x \sim z$, that is, by the definition of $\sim$, we must find a path $h$ in $S$ from $x$ to $z$, that is, we must find a function $h : [0, 1] \to S$ such that

(1') $h$ is continuous,

(2') $h(0) = x$ and $h(1) = z$.

Since $x \sim y$, by the definition of $\sim$, there exists a path in $S$ from $x$ to $y$, and hence, there exists a function $f : [0, 1] \to S$ such that

(1'') $f$ is continuous,

(2'') $f(0) = x$ and $f(1) = y$.

Since $y \sim z$, by the definition of $\sim$, there exists a path in $S$ from $y$ to $z$, and hence, there exists a function $g : [0, 1] \mapsto S$ such that

(1''') $g$ is continuous,

(2''') $g(0) = y$ and $g(1) = z$. Let us define a function $h : [0, 1] \mapsto X$ as follows:

$$h(t) \equiv \begin{cases} f(2t) & \text{if } 0 \le t < \frac{1}{2} \\ g(2t - 1) & \text{if } \frac{1}{2} \le t \le 1. \end{cases}$$

Clearly, the range of $h$ is the union of the range of $f$, and the range of $g$. Further, since $f : [0, 1] \to S$, and $g : [0, 1] \to S$, the range of $h$ is contained $S$, and hence $h : [0, 1] \mapsto S$.

For (1'): we have to prove that $h$ is continuous over $[0, 1]$, that is, $h$ is continuous at all points of $[0, 1]$. By the definition of $h$ over $0 \le t < \frac{1}{2}$, $h$ is the composite function of two continuous functions, namely $t \mapsto 2t$ and $f$ so, $h$ is continuous at all points of $\{t : 0 \le t < \frac{1}{2}\}$. Similarly, $h$ is continuous at all points of $\{t : \frac{1}{2} < t \le 1\}$. So, it remains to prove that $h$ is continuous at $\frac{1}{2}$. For this purpose, we should calculate the left-hand limit of $h$ at $\frac{1}{2}$, the right-hand limit of $h$ at $\frac{1}{2}$ and $h(\frac{1}{2})$. If all the three values are equal, then we can say that $h$ is continuous at $\frac{1}{2}$. In short, we have to prove:

$$\lim_{t \to \left(\frac{1}{2}\right)^-} h(t) = \lim_{t \to \left(\frac{1}{2}\right)^+} h(t) = h\left(\frac{1}{2}\right).$$

Here, $\lim_{t \to \left(\frac{1}{2}\right)^-} h(t) = \lim_{t \to \frac{1}{2}} f(2t)$, by the definition of $h$. Since $t \mapsto f(2t)$ is a continuous function over $[0, \frac{1}{2}]$, $\lim_{t \to \frac{1}{2}} f(2t) = f(2 \times \frac{1}{2}) = f(1) = y$. Hence, $\lim_{t \to \left(\frac{1}{2}\right)^-} h(t) = y$. Similarly, $\lim_{t \to \left(\frac{1}{2}\right)^+} h(t) = \lim_{t \to \frac{1}{2}} g(2t - 1) = g(2(\frac{1}{2}) - 1) = g(0) = y$. Hence $\lim_{t \to \left(\frac{1}{2}\right)^-} h(t) = \lim_{t \to \left(\frac{1}{2}\right)^+} h(t) = y$. Further, by the definition of $h$, $h(\frac{1}{2}) = g(2(\frac{1}{2}) - 1) = g(0) = y$. Hence, $\lim_{t \to \left(\frac{1}{2}\right)^-} h(t) = \lim_{t \to \left(\frac{1}{2}\right)^+} h(t) = h(\frac{1}{2})$.

Thus, we have shown that $h$ is continuous over $[0,1]$. This proves (1')

For (2'): Here $h(0) = f(2(0)) = f(0) = x$, and $h(1) = g(2(1) - 1) = g(1) = z$. This proves (2'). Thus, we have shown: If $x \sim y$ and $y \sim z$, then $x \sim z$. This proves (3).

Hence, $\sim$ is an equivalence relation over $S$. Hence, $S$ is partitioned into equivalence classes. Here, each equivalence class is called a *path component of S*.

**Note 4.38** Let $X$ be a topological space. Let $S$ be a nonempty subset of $X$. Now, we shall try to prove that each path component of $S$ is a path connected subset of $X$. For this purpose, let us take any path component $[a]$ of $S$, where $a \in S, [a] \equiv \{x : x \in S, \text{ and } x \sim a\}$, and $x \sim a$ stands for: There exists a path in $S$ from $x$ to $a$. We have to show that $[a]$ is a path connected subset of $S$. For this purpose, let us take any $x, y$ in $[a]$. We have to find a path in $[a]$ from $x$ to $y$. Since $[a]$ is a path component of $S$, $[a]$ is an equivalence class. Since $[a]$ is an equivalence class, and $x$ is in $[a]$, $x \sim a$, and hence, there exists a function $f : [0, 1] \to S$ such that

(1$'$) $f : [0, 1] \to S$ is continuous,
(2$'$) $f(0) = x$ and $f(1) = a$. Clearly, the range of $f$ is contained in $[a]$.

(Reason: Let us take any $b \equiv f(t_0) \in \text{ran } f$, where $t_0 \in (0, 1)$. We have to prove that $b \in [a]$, that is, $a \sim b$. We must find a function $F : [0, 1] \to S$ such that (i) $F$ is continuous, (ii) $F(0) = a$ and $F(1) = b$.

Let us define $F : [0, 1] \to S$ as follows: for every $t$ in $[0,1]$, $F(t) \equiv f(1 + (t_0 - 1)t)$. Clearly, the range of $[0,1]$ under $t \mapsto (1 + (t_0 - 1)t)$ is $[t_0, 1]$, and hence, the ran $F \subset f([t_0, 1]) \subset S$. Since ran $F \subset f([t_0, 1]) \subset S$, $F : [0, 1] \to S$. Since $F$ is the composite of two continuous functions $t \mapsto (1 + (t_0 - 1)t)$, and $f$, $F$ is continuous. Clearly, $F(0) = f(1 + (t_0 - 1)0) = f(1) = a$, and $F(1) = f(1 + (t_0 - 1)1) = f(t_0) = b$.)

Since the range of $f$ is contained in $[a]$, and $f : [0, 1] \to S, f : [0, 1] \to [a]$. Now, since $f : [0, 1] \to S$, is continuous, $f : [0, 1] \to [a]$. is continuous. Further, $f(0) = x$ and $f(1) = a$. Thus we get a path in $[a]$ from $x$ to $a$. Similarly, we get a path in $[a]$ from $a$ to $y$. Since there exist a path in $[a]$ from $x$ to $a$, and a path in $[a]$ from $a$ to $y$, there exists a path in $[a]$ from $x$ to $y$.

Let $X$ be a topological space. Let $S$ be a nonempty subset of $X$. Now, we shall try to prove that if $C$ is a path component of $S$, $C_1$ is path connected, and $C \subset C_1 \subset S$, then $C = C_1$. If $C$ is a component of $S$, $C_1$ is connected, and $C \subset C_1 \subset S$, then $C = C_1$. Let us take a path component $[a]$ of $S$, and a path connected set $C$ satisfying $[a] \subset C \subset S$. We claim that $[a] = C_1$. If not, otherwise, let $[a]$ be a proper subset of $C$. So, there exists $t$ in $C$ such that $t$ is not in $[a]$. Since $a \in [a] \subset C$, $a$ is in $C$. Since $a$ is in $C$, $t$ is in $C$, and $C$ is a path connected subset of $X$, there exists a path $f$ in $C$ from $t$ to $a$. Since $f$ is a path in $C$, and $C \subset S, f$ is a path in $S$, and hence, $t$ is in $[a]$, a contradiction.

Thus, we have seen that

(i)   every nonempty subset $S$ of a topological space $X$ is partitioned into path components,
(ii)  every path component of $S$ is path connected,
(iii) if $C$ is a path component of $S$, $C_1$ is path connected, and $C \subset C_1 \subset S$ then $C = C_1$.

**Note 4.39** Let $X$ be a topological space. Let $S$ be a nonempty subset of $X$. We shall try to prove: Each component of $S$ is a union of path components of $S$. Let us take any component $\{y : y \in S \text{ and } y \sim a\}$ of $S$, where $a$ is in $S$, and $y \sim a$ stands for:

There exists a connected subset $A$ of $X$ such that $A \subset S$, and $y$, $a$ are in $A$. It is enough to show that

$$\{y : y \in S \text{ and } y \sim' a\} \subset \{y : y \in S \text{ and } y \sim a\}$$

where $y \sim' a$ stands for: There exists a path in $S$ from $y$ to $a$. For this purpose, let us take any $y$ in *LHS*. We have to prove that $y$ is in *RHS*, that is, $y \sim a$. By the definition of $\sim$, we should find a connected subset $A$ of $X$ such that $A \subset S$, and $y$, $a$ are in $A$. Since $y$ is in *LHS*, $y \sim' a$. Since $y \sim' a$, by the definition of $\sim'$, there exists a path $f$ in $S$ from $y$ to $a$. Since $f$ is a path in $S$ from $y$ to $a$, $f$ is a function from closed interval $[0,1]$ to $S$ satisfying

(I)   $f$ is continuous,

(II)  $f(0) = y$ and $f(1) = a$. Since $f : [0, 1] \to S$ is continuous, and $[0,1]$ is connected, so the $f$-image set $f([0, 1])$ of $[0, 1]$ is a connected subset of $X$. Since $0$ is in $[0, 1]$, $f(0)$ is in $f([0, 1])$. Similarly, $f(1)$ is in $f([0, 1])$. Further, $f(0) = y$ and $f(1) = a$. Thus, $f([0, 1])$ is a connected subset of $X$ such that $f([0, 1]) \subset S$, and $y$, $a$ are in $f([0, 1])$.

**Note 4.40** Let $X$ be a topological space. We shall try to show: If $X$ is path connected, then $X$ is connected. If not, otherwise, let $X$ be not connected. We have to arrive at a contradiction. Since $X$ is not connected, there exist open subsets $G_1, G_2$ of $X$ such that $G_1$ is nonempty, $G_2$ is nonempty, $G_1 \cap G_2$ is empty, and $X = G_1 \cup G_2$. Since $G_1$ is nonempty, there exists $a$ in $G_1$. Similarly, there exists $b$ in $G_2$. Now, since $X$ is path connected, there exists a continuous function $f : [0, 1] \to X$ such that $f(0) = a$, and $f(1) = b$. Since $f : [0, 1] \to X$ is continuous, and $[0, 1]$ is connected, the image set $f([0, 1])$ is connected. Here, $a$ is in $G_1$, and $a = f(0) \in f([0, 1])$, so $a \in G_1 \cap f([0, 1])$, and hence, $G_1 \cap f([0, 1])$ is nonempty. Similarly, $G_2 \cap f([0, 1])$ is nonempty. Next, since $G_1 \cap G_2$ is empty, $G_1 \cap G_2 \cap f([0, 1])$ is empty. Since $G_1$, $G_2$ are open subsets of $X$ such that $G_1 \cap f([0, 1])$ is nonempty, $G_2 \cap f([0, 1])$ is nonempty, $G_1 \cap G_2 \cap f([0, 1])$ is empty, and $f([0, 1]) \subset X = G_1 \cup G_2$, $f([0, 1])$ is not connected, a contradiction.

**Definition** Let $X$ be a topological space. If for every $x$ in $X$, and for every open neighborhood $U$ of $x$, there exists an open neighborhood $V$ of $x$ such that

(i)   $x \in V \subset U$,

(ii)  $V$ is a connected subset of $X$,

then we say that $X$ is a *locally connected space*.

**Definition** Let $X$ be a topological space. If for every $x$ in $X$, and for every open neighborhood $U$ of $x$, there exists an open neighborhood $V$ of $x$ such that

(i)   $x \in V \subset U$,

(ii)  $V$ is a path connected subset of $X$,

then we say that $X$ is a *locally path connected space*.

**Note 4.41** Let $X$ be a topological space. We shall try to prove: If $X$ is locally connected space, and $G$ is a nonempty open subset of $X$, then each component of $G$ is open. Let $X$ be a locally connected space, and let $G$ be an open subset of $X$. We have to show that each component of $G$ is open in $X$. By the definition of component, every component of $G$ is of the form

$$[a] \equiv \{x : x \in G \text{ and } x \sim a\}$$

for some $a \in G$, where $x \sim a$ stands for: There exists a connected subset $A$ of $X$ such that $A \subset G$, $x \in A$, and $a \in A$. We have to show that $[a]$ is an open subset of $X$, that is, every point of $[a]$ is an interior point of $[a]$. For this purpose, let us take any $x \in [a]$. We must find an open neighborhood $V$ of $x$ such that $V \subset [a]$. Since $x \in [a]$ and $[a] \subset G$, $x \in G$. Since $x \in G$ and $G$ is open in $X$, $G$ is an open neighborhood of $x$. Since $X$ is a locally connected space, and $G$ is an open neighborhood of $x$, there exists an open neighborhood $V$ of $x$ such that

(I)   $x \in V \subset G$,
(II)  $V$ is a connected subset of $X$.

We claim that $V \subset [a]$.

If not, otherwise, let $V \not\subset [a]$. So, there exists $b$ in $V$ such that $b \notin [a]$. Since $[a]$ is an equivalence class of $a$, and $b \notin [a]$, $b \nsim a$. Here, $b \in V \subset G$ so $b \in G$. Similarly, $x \in G$. Further, since $V$ is a connected subset of $X$, $V \subset G$, and $x$, $b$ are in $V$, by the definition of $\sim$, $x \sim b$. Since $x \sim b$, and $\sim$ is an equivalence relation, $[x] = [b]$. Since $x \in [a]$, and $\sim$ is an equivalence relation, $[x] = [a]$. Since $[x] = [b]$, and $[x] = [a]$, $[a] = [b]$. Since $[a] = [b]$, $a \sim b$. This is a contradiction. So, our claim is true, that is, $V \subset [a]$.

Clearly, if $X$ is locally connected space, then each component of $X$ is open.

**Note 4.42** Let $X$ be a topological space. We shall try to prove: If $X$ is locally path connected space, and $G$ is a nonempty open subset of $X$, then each path component of $G$ is open. Let $X$ be a locally path connected space, and let $G$ be an open subset of $X$. We have to show that each path component of $G$ is open in $X$. By the definition of path component, every path component of $G$ is of the form

$$[a] \equiv \{x : x \in G \text{ and } x \sim a\}$$

for some $a \in G$, where $x \sim a$ stands for: There exists a path in $G$ from $x$ to $a$. We have to show that $[a]$ is an open subset of $X$, that is, every point of $[a]$ is an interior point of $[a]$. For this purpose, let us take any $x \in [a]$. We must find an open neighborhood $V$ of $x$ such that $V \subset [a]$.

Since $x \in [a]$, and $[a] \subset G$, $x \in G$. Since $x \in G$ and $G$ is open in $X$, $G$ is an open neighborhood of $x$. Since $X$ is a locally path connected space, and $G$ is an open neighborhood of $x$, there exists an open neighborhood $V$ of $x$ such that

(I)   $x \in V \subset G$,
(II)  $V$ is a path connected subset of $X$.

It remains to prove: $V \subset [a]$. If not, otherwise, let $V \not\subset [a]$. We have to arrive at a contradiction. Since $V \not\subset [a]$, there exists $b$ in $V$ such that $b \notin [a]$. Since $[a]$ is an equivalence class of $a$, and $b \notin [a]$, $b \not\sim a$. Here, $b \in V \subset G$ so $b \in G$. Similarly, $x \in G$. Further, since $V$ is a path connected subset of $X$ and $x$, $b$ are in $V$, there exists a path in $V$ from $x$ to $b$. Since there exists a path in $V$ from $x$ to $b$, and $V \subset G$, by the definition of $\sim$, $x \sim b$. Since $x \sim b$ and $\sim$ is an equivalence relation, $[x] = [b]$. Since $x \in [a]$, and $\sim$ is an equivalence relation, $[x] = [a]$. Since $[x] = [b]$, and $[x] = [a]$, $[a] = [b]$. Since $[a] = [b]$, $a \sim b$. Thus, we get a contradiction.

Clearly, if $X$ is a locally path connected space, then each path component of $X$ is open.

**Note 4.43** Let $X$ be a locally path connected space. We shall try to prove that the collection of all path components of $X$, and the collection of all components of $X$ are equal. For any $a$ in $X$, let $\{y : y \in X \text{ and } y \sim' a\}$ be a path component of $X$, where $y \sim' a$ stands for: There exists a path in $X$ from $y$ to $a$. It is enough to prove that

$$\{y : y \in X \text{ and } y \sim' a\} = \{y : y \in X \text{ and } y \sim a\}$$

where $y \sim a$ stands for: There exists a connected subset $A$ of $X$ such that $A \subset G, y \in A$, and $a \in A$. We know that $\{y : y \in X \text{ and } y \sim' a\} \subset \{y : y \in X \text{ and } y \sim a\}$. So, it remains to prove that

$$\{y : y \in X \text{ and } y \sim a\} \subset \{y : y \in X \text{ and } y \sim' a\} \quad \ldots (*)$$

If not, otherwise, let there exists $y^* \in \{y : y \in X \text{ and } y \sim a\}$ such that $y^* \notin \{y : y \in X \text{ and } y \sim' a\}$. We have to arrive at a contradiction. Here, $y^* \sim a$ is true and $y^* \sim' a$ is false. Since $y^* \sim a$ is true, by the definition of $\sim$, there exists a connected subset $A$ of $X$ such that $y^*$, $a$ are in $A$. Since $\{y : y \in X \text{ and } y \sim' y^*\}$ is the path component of $X$ containing $y^*$, and $X$ is a locally path connected space,

$$P \equiv \{y : y \in X \text{ and } y \sim' y^*\}$$

is an open set containing $y^*$. Since $P$ contains $y^*$, and $y^*$ is in $A$, $y^* \in A \cap P$, and therefore, $A \cap P$ is nonempty.

Since $X$ is partitioned into path components, complement of each path component in $X$ is a union of path components of $X$. Since $X$ is a locally path connected space, each path component is open. Since union of open sets is open, complement of each path component in $X$ is union of path components, and each path component is open, complement of each path component in $X$ is open. Since $\{y : y \in X \text{ and } y \sim' y^*\}$ is a path component, and complement of each path component in $X$ is open, the complement $P'$ of $P$ is open. Thus, we have obtained two open sets $P$ and $P'$ such that $A \cap P$ is nonempty.

Now, we shall try to prove that $A \cap (P')$ in nonempty. Since $y^* \notin \{y : y \in X \text{ and } y \sim' a\}$, and $\sim'$ is an equivalence relation over $X$, $\{y : y \in X \text{ and } y \sim' y^*\}(= P)$ and $\{y : y \in X \text{ and } y \sim' a\}$ are disjoint sets, and therefore, $\{y : y \in X \text{ and } y \sim' a\}$ is contained in $P'$. Since $\sim'$ is reflexive over $X$,

$a$ is in $\{y : y \in X \text{ and } y \sim 'a\}$. Since $a \in \{y : y \in X \text{ and } y \sim 'a\} \subset P'$, $a$ is in $P'$. Since $a$ is in $P'$ and $a$ is in $A$, $a$ is in $A \cap (P')$, and therefore, $A \cap (P')$ is nonempty.

Since $P$ and $P'$ are two open sets such that

(I) $P \cap A$ and $P' \cap A$ are nonempty sets,
(II) $P \cap P' \cap A$ is empty,
(III) $A \subset P \cup (P')$,

by the definition of connected set, $A$ is not a connected set. This is a contradiction. This proves (*). $\qquad \square$

## 4.4 Smooth Manifold as Paracompact Space

**Definition** Let $n$ be a fixed positive integer. Let $M$ be a topological space. If the following conditions are satisfied:

1. $M$ is a Hausdorff space,
2. $M$ is a second countable space,
3. for every $x$ in $M$, there exists an open neighborhood $U$ of $x$ in $M$, and an open subset $\widetilde{U}$ of Euclidean space $\mathbb{R}^n$ such that $U$ and $\widetilde{U}$ are homeomorphic, then we say that *M is a topological manifold of dimension n*. In other words, *M is a topological manifold of dimension n* means $M$ is a topological space that satisfies the following conditions:

$1'$. If $x$ and $y$ are distinct points of $M$, then there exist an open neighborhood $U$ of $x$ and an open neighborhood $V$ of $y$ such that $U$ and $V$ have no common element,
$2'$. There exists a countable collection $\mathcal{C}$ of open subsets of $M$ such that for every neighborhood $U$ of $x$, there exists a member $G$ of $\mathcal{C}$ such that $x \in G \subset U$,
$3'$. For every $x$ in $M$, there exist an open neighborhood $U$ of $x$ in $M$, an open subset $\widetilde{U}$ of Euclidean space $\mathbb{R}^n$, and a 1–1 mapping $\varphi$ from $U$ onto $\widetilde{U}$ such that $\varphi$ and its inverse $\varphi^{-1}$, both are continuous functions.

Here, $(U, \varphi)$ is referred as a *coordinate chart of M.*

**Note 4.44** Let $M$ be a topological manifold of dimension $n$. We shall try to show that $M$ is locally path connected. For this purpose, let us take an element $x$ in $M$ and an open neighborhood $U$ of $x$ in $M$. We have to find an open set $V$ of $x$ such that

(I) $x \in V \subset U$,
(II) $V$ is a path connected subset of $M$.

Since $M$ is a topological manifold, and $x$ is in $M$ so, there exist an open neighborhood $U_1$ of $x$ in $M$ and an open subset $\widetilde{U}_1$ of Euclidean space $\mathbb{R}^n$ such that $U_1$ and $\widetilde{U}_1$ are homeomorphic.

Since $U$ and $U_1$ are open neighborhoods of $x$, their intersection $U \cap U_1$ is also an open neighborhood of $x$. Here that $U_1$ and $\widetilde{U}_1$ are homeomorphic so, there exists a 1–1 mapping $\varphi$ from $U_1$ onto $\widetilde{U}_1$ such that $\varphi$ and its inverse $\varphi^{-1}$, both are continuous functions. Since $U \cap U_1$ is an open subset of $U_1$, and $\varphi^{-1} : \widetilde{U}_1 \overset{1-1,\,\text{onto}}{\longrightarrow} U_1$ is continuous, $\varphi(U \cap U_1)$ is an open subset of $\widetilde{U}_1$ Since $\varphi(U \cap U_1)$ is an open subset of $\widetilde{U}_1$, and $\widetilde{U}_1$ is an open subset of $\mathbb{R}^n$, $\varphi(U \cap U_1)$ is an open subset of $\mathbb{R}^n$. Since $U \cap U_1$ is an open neighborhood of $x$, $x$ is in $U \cap U_1$, and hence, $\varphi(x)$ is in $\varphi(U \cap U_1)$. Since $\varphi(x)$ is in $\varphi(U \cap U_1)$ and $\varphi(U \cap U_1)$ is an open subset of $\mathbb{R}^n$, $\varphi(U \cap U_1)$ is an open neighborhood of $\varphi(x)$ in $\mathbb{R}^n$. So, there exists $\varepsilon > 0$ such that the open ball

$$B \equiv \{y \in \mathbb{R}^n : |y - \varphi(x)| < \varepsilon\}$$

is contained in $\varphi(U \cap U_1)$. Since $B \subset \varphi(U \cap U_1) \subset \widetilde{U}_1$, $B$ is contained in $\widetilde{U}_1$. We know that $B$ is open in $\mathbb{R}^n$. Since $\varphi$ is a continuous function from $U_1$ onto $\widetilde{U}_1$, and $B$ is an open subset of $\widetilde{U}_1$, $\varphi^{-1}(B)$ is an open subset of $U_1$. Let us take $\varphi^{-1}(B)$ for $V$ in (I) and (II). So, we must prove:

(I)   $x \in \varphi^{-1}(B) \subset U$,
(II)  $\varphi^{-1}(B)$ is a path connected subset of $M$.

For (I): Since $|\varphi(x) - \varphi(x)| = 0 < \varepsilon$, $\varphi(x) \in \{y \in \mathbb{R}^n : |y - \varphi(x)| < \varepsilon\} = B$, and hence, $x \in \varphi^{-1}(B)$. Since $U \cap U_1$ is contained in $U_1$, $\varphi(U \cap U_1) \subset \varphi(U)$. Since $B \subset \varphi(U \cap U_1) \subset \varphi(U)$, $B \subset \varphi(U)$ and hence $\varphi^{-1}(B) \subset U$. This proves (I).

For (II): We have to prove that $\varphi^{-1}(B)$ is a path connected subset of $M$. For this purpose, let us take any $a$ and $b$ in $\varphi^{-1}(B)$. We have to find a path $f$ in $M$ from $a$ to $b$, that is, we have to find a continuous function $f$ from closed interval $[0, 1]$ to $M$ such that $f(0) = a$ and $f(1) = b$. Consider the function $f : [0, 1] \mapsto M$ defined by $f(t) \equiv \varphi^{-1}((1 - t)\varphi(a) + t\varphi(b))$ for every $t$ in $[0, 1]$.

Here, $t \mapsto \varphi^{-1}((1 - t)\varphi(a) + t\varphi(b))$ is a composite function of two functions $t \mapsto (1 - t)\varphi(a) + t\varphi(b)$ and $\varphi^{-1}$. Since $a$ is in $\varphi^{-1}(B)$, $\varphi(a) \in B$, and hence, by the definition of $B$, $|\varphi(a) - \varphi(x)| < \varepsilon$. Similarly, $|\varphi(b) - \varphi(x)| < \varepsilon$. Since $|\varphi(a) - \varphi(x)| < \varepsilon$, and $|\varphi(b) - \varphi(x)| < \varepsilon$, for every $t$ satisfying $0 \leq t \leq 1$, we have

$$|((1 - t)\varphi(a) + t\varphi(b)) - \varphi(x)| = |(1 - t)(\varphi(a) - \varphi(x)) + t(\varphi(b) - \varphi(x))|$$
$$\leq (1 - t)|\varphi(a) - \varphi(x)| + t|\varphi(b) - \varphi(x)|$$
$$< (1 - t)\varepsilon + t\varepsilon = \varepsilon.$$

It follows, by the definition of $B$, that $t \mapsto (1 - t)\varphi(a) + t\varphi(b)$ is a function from $[0,1]$ to $B$. Since $t \mapsto (1 - t)\varphi(a) + t\varphi(b)$ is a function from $[0,1]$ to $B(\subset \widetilde{U}_1)$, and $\varphi^{-1} : \widetilde{U}_1 \mapsto U_1 \subset M$, their composite $t \mapsto \varphi^{-1}((1 - t)\varphi(a) + t\varphi(b))$ maps $[0, 1]$ to $M$. Since $t \mapsto (1 - t)\varphi(a) + t\varphi(b)$ and $\varphi^{-1}$ both are continuous, their composite function $t \mapsto \varphi^{-1}((1-t)\varphi(a) + t\varphi(b))$ is continuous. Further, since $f(t) = \varphi^{-1}((1-t)\varphi(a) + t\varphi(b))$ and , $t \mapsto \varphi^{-1}((1-t)\varphi(a) + t\varphi(b))$ is continuous, $f :$ $[0,1] \mapsto M$ is continuous. Also $f(0) = \varphi^{-1}((1-0)\varphi(a) + 0\varphi(b)) = \varphi^{-1}(\varphi(a)) = a$, and $f(1) = \varphi^{-1}((1-1)\varphi(a) + 1\varphi(b)) = \varphi^{-1}(\varphi(b)) = b$. Thus $f$ is a path in $M$ from $a$ to $b$. This proves (II).

**Note 4.45** Let $M$ be a topological manifold of dimension $n$. We shall try to prove:

   (i)  $M$ is connected if and only if it is path connected.
  (ii)  The collection of all path components of $M$ is equal to the collection of all components of $M$.
 (iii)  $M$ has countable many components, each of which is an open subset of $M$, and a connected topological manifold.

For (i): Let $M$ be path connected. We have to prove that $M$ is connected. Since $M$ is path connected, $M$ is a path component. Since $M$ is a topological manifold, $M$ is locally path connected. Since $M$ is locally path connected, and $M$ is a path component, $M$ is a component. Since $M$ is a component, $M$ is connected.
*Converse*: Let $M$ be connected. We have to prove that $M$ is path connected. Since $M$ is connected, $M$ is a component. Since $M$ is a topological manifold, $M$ is locally path connected. Since $M$ is locally path connected, and $M$ is a component, $M$ is a path component. Since $M$ is a path component, $M$ is path connected.
This completes the proof of (i).
For (ii): Since $M$ is a topological manifold, $M$ is locally path connected. Since $M$ is locally path connected, the collection of all path components of $M$ is equal to the collection of all components of $M$. This proves (ii).
For (iii): We have to prove that $M$ has countable many components. If not, otherwise, let $M$ has uncountable many components. We have to arrive at a contradiction.

Let $C$ be a component of $M$. Since $M$ is a topological manifold, $M$ is locally path connected. Since $M$ is locally path connected, and $C$ is a component of $M$, $C$ is a path component. Since $M$ is locally path connected, and $C$ is a path component, $C$ is open in $M$. Since $C$ is open in $M$, and $a$ is in $C$, $C$ is an open neighborhood of $a$ in $M$. Since $M$ is a topological manifold, $M$ is a second countable space. Since $M$ is a second countable space, there exists a countable collection $\mathcal{C}$ of open subsets of $M$ such that for every neighborhood $U$ of $x$, there exists a member $G$ of $\mathcal{C}$ such that $x \in G \subset U$. Now, since $C$ is an open neighborhood of $a$, there exists a member $G$ of $\mathcal{C}$ such that $x \in G \subset C$.

Thus, we see that corresponding to each component $C$, there exists a member $G$ of $\mathcal{C}$ such that $G$ is contained in $C$. Since $M$ is partitioned into components, each component contains a member of $\mathcal{C}$, and the collection of all components is uncountable, the collection $\mathcal{C}$ is uncountable. This is a contradiction. This proves that $M$ has countable many components.

Now, let us take a component $C$. We have to prove that $C$ is open and connected. Since $M$ is a topological manifold, $M$ is locally path connected. Since $M$ is locally path connected and $C$ is a component of $M$, $C$ is a path component. Since $M$ is locally path connected, and $C$ is a path component, $C$ is open in $M$. Since $C$ is a component, $C$ is connected. Hence, $C$ is open and connected.

Lastly, we have to show that $C$ is a topological manifold of dimension $n$, where topology on $C$ is the induced topology of $M$. Since $C$ is an open subset of topological space $M$, under induced topology of $M$, open subsets of $C$ are exactly those subsets of $C$ which are open in $M$. By the definition of topological manifold, we must prove:

1. $C$ is a Hausdorff space,
2. $C$ is a second countable space,
3. for every $x$ in $C$, there exist a neighborhood $U_1$ of $x$ in $C$, and an open subset $\widetilde{U}_1$ of Euclidean space $\mathbb{R}^n$ such that $U_1$ and $\widetilde{U}_1$ are homeomorphic:

   For 1: Since $M$ is a topological manifold, $M$ is a Hausdorff space. Since $M$ is a Hausdorff space, and $C$ is a subspace of $M$, $C$ is a Hausdorff space. This proves 1.
   For 2: Since $M$ is a topological manifold, $M$ is a second countable space. Since $M$ is a second countable space, and $C$ is a subspace of $M$, $C$ is a second countable. This proves 2.
   For 3: Let us fix any $a$ in $C$. We have to find an open neighborhood $U_1$ of $a$ in $C$, and an open subset $\widetilde{U}_1$ of Euclidean space $\mathbb{R}^n$ such that $U_1$ and $\widetilde{U}_1$ are homeomorphic.

Since $a$ is in $C$, and $C$ is contained in $M$ so, $a$ is in $M$. Since $a$ is in $M$, and $M$ is a topological manifold, there exists an open neighborhood $U$ of $a$ in $M$, an open subset $\widetilde{U}$ of Euclidean space $\mathbb{R}^n$, and a 1–1 mapping $\varphi$ from $U$ onto $\widetilde{U}$ such that $\varphi$ and its inverse $\varphi^{-1}$, both are continuous functions. Since $a$ is in $C$, and $C$ is open, $C$ is an open neighborhood of $a$. Since $C$ is an open neighborhood of $a$, and $U$ is an open neighborhood of $a$, $C \cap U$ is an open neighborhood of $a$, and $C \cap U$ is contained in $U$. Here, function $\varphi$ from $U$ to $\widetilde{U}$ is 1–1, so its restriction $\varphi|_{C \cap U}$ to $C \cap U$ is also 1–1. Since function $\varphi$ from $U$ to $\widetilde{U}$ is continuous, its restriction $\varphi|_{C \cap U}$ to $C \cap U$ is also continuous. Since the mapping $\varphi^{-1} : \widetilde{U} \overset{1-1,\,\text{onto}}{\longrightarrow} U$ is continuous, and $C \cap U$ is an open subset of $U$, $\varphi(C \cap U)$ is open in $\widetilde{U}$. Since $\varphi(C \cap U)$ is open in $\widetilde{U}$, and $\widetilde{U}$ is open in $\mathbb{R}^n$, $\varphi(C \cap U)$ is open in $\mathbb{R}^n$ Since $\varphi : U \to \widetilde{U}$ is a homeomorphism, $C \cap U$ is an open subset of $U$, and $\varphi(C \cap U)$ is an open subset of $\widetilde{U}$, $\varphi|_{C \cap U} : C \cap U \to \varphi(C \cap U)$ is a homeomorphism. This proves 3.

**Note 4.46** Let $X$ be a second countable space. We shall try to show that every open cover of $X$ has a countable subcover. Let $\mathcal{G}$ be an open cover of $X$. We have to find a countable subcover of $\mathcal{G}$. Since $X$ is a second countable space, there exists a basis $\mathcal{B}$ of $X$ such that $\mathcal{B}$ is countable. Put

$$\mathcal{B}_1 \equiv \{U : U \in \mathcal{B}, \text{and there exists } G \in \mathcal{G} \text{ such that } U \subset G\}.$$

Since $\mathcal{B}_1 \subset \mathcal{B}$, and $\mathcal{B}$ is countable, $\mathcal{B}_1$ is countable. Now, we shall try to show that $\mathcal{B}_1$ is a basis of $X$. For this purpose, let us take any open neighborhood $G$ of $x$ in $X$. We have to find a member $U$ of $\mathcal{B}_1$ such that $x \in U \subset G$.

Since $x$ is in $X$, and $\mathcal{G}$ is a cover of $X$, there exists $G_1$ in $\mathcal{G}$ such that $x$ is in $G_1$. Since $G_1$ is in $\mathcal{G}$, and $\mathcal{G}$ is an open cover of $X$, $G_1$ is open. Since $G_1$ is open, and $x$ is in $G_1$, $G_1$ is an open neighborhood of $x$. Since $G_1$ is an open neighborhood of $x$, and $G$ is an open neighborhood of $x$, their intersection $G_1 \cap G$ is an open neighborhood of $x$. Since $G_1 \cap G$ is an open neighborhood of $x$, and $\mathcal{B}$ is a basis of $X$, there exists $U$ in $\mathcal{B}$ such that $x \in U \subset G_1 \cap G \subset G$. Since $U \subset G_1 \cap G \subset G_1$, $U \subset G_1$ Since $U$ is in $\mathcal{B}$, $G_1 \in \mathcal{G}$, and $U \subset G_1$, by the definition of $\mathcal{B}_1$, $U$ is in $\mathcal{B}_1$. Thus, we have shown that $\mathcal{B}_1$ is a basis of $X$.

By the definition of $\mathcal{B}_1$, for each $U$ in $\mathcal{B}_1$, the set $\{G : G \in \mathcal{G}, \text{and } U \subset G\}$ is nonempty, and hence, we can choose a unique $G_U$ in $\mathcal{G}$ such that $U \subset G_U$. Put

$$\mathcal{G}_1 \equiv \{G_U : U \in \mathcal{B}_1\}.$$

We shall try to show that $\mathcal{G}_1$ is a countable subcover of $\mathcal{G}$. Since $\mathcal{B}_1$ is countable, by the definition of $\mathcal{G}_1$, $\mathcal{G}_1$ is countable. Clearly, $\mathcal{G}_1 \subset \mathcal{G}$. It remains to prove that $\mathcal{G}_1$ is a cover of $X$. For this purpose, let us take any $x$ in $X$. We have to find a member of $\mathcal{G}_1$ that contains $x$.

Since $X$ is an open neighborhood of $x$, and $\mathcal{B}_1$ is a basis of $X$, there exists $U$ in $\mathcal{B}_1$ such that $x \in U \subset X$. Since $U$ is in $\mathcal{B}_1$, by the definition of $\mathcal{G}_1, G_U \in \mathcal{G}_1$, and hence $U \subset G_U$. Since $x \in U$, and $U \subset G_U, x \in G_U$, where $G_U \in \mathcal{G}_1$. Thus, $\mathcal{G}_1$ is a cover of $X$. Thus, we have shown that $\mathcal{G}_1$ is a countable subcover of $\mathcal{G}$.

**Lemma 4.47** *Let $M$ be a topological manifold of dimension $n$. Then, there exists a countable collection $\{(U_1, \varphi_1), (U_2, \varphi_2), (U_3, \varphi_3), \ldots\}$ of coordinate charts of $M$ such that*

1. *$\{U_1, U_2, U_3, \ldots\}$ is a basis of $M$,*
2. *each $\varphi_i(U_i)$ is an open ball in $\mathbb{R}^n$,*
3. *each closure $U_i^-$ of $U_i$ is a compact subset of $M$.*

*Proof* Let $\mathcal{C}$ be the collection of all coordinate charts $(U, \varphi)$ of $M$. Since $M$ is a topological manifold, $\{U : (U, \varphi) \in \mathcal{C}\}$ is an open cover of $M$. Since $M$ is a topological manifold, $M$ is second countable. Since $M$ is a second countable space, and $\{U : (U, \varphi) \in \mathcal{C}\}$ is an open cover of $M$, there exists a countable subcover $\{U_1, U_2, U_3, \ldots\}$ of $\{U : (U, \varphi) \in \mathcal{C}\}$. Here, $U_1 \in \{U_1, U_2, U_3, \ldots\} \subset \{U : (U, \varphi) \in \mathcal{C}\}$, $U_1 \in \{U : (U, \varphi) \in \mathcal{C}\}$, and hence, there exists a function $\varphi_1$ such that $(U_1, \varphi_1) \in \mathcal{C}$. It follows that there exists an open subset $\tilde{U}_1$ of $\mathbb{R}^n$ such that $\varphi_1 : U_1 \mapsto \tilde{U}_1$ is a homeomorphism.

Let $\mathcal{B}$ be the collection of all open balls $B_r(x)$ with center $x$ and radius $r$ such that $r$ is a positive rational, and each coordinate of $x$ is a rational number. We know that

$\mathcal{B}$ is a basis of $\mathbb{R}^n$. Since the set of all rational numbers is countable, the collection $\mathcal{B}$ is countable. Let $\mathcal{B}_1$ be the collection of all $B_r(x) \in \mathcal{B}$ such that the closure $(B_r(x))^-$ of $B_r(x)$ is contained in $\widetilde{U}_1$. We shall try to show that $\mathcal{B}_1$ is a basis of $\widetilde{U}_1$.

For this purpose, let us take any open neighborhood $\widetilde{V}$ of $y$ such that $\widetilde{V}$ is contained in $\widetilde{U}_1$. Since $\widetilde{V}$ is an open neighborhood of $y$, and $\mathcal{B}$ is a basis of $\mathbb{R}^n$, there exists an open ball $B_{r_y}(x_y)$ in $\mathcal{B}$ such that $y \in B_{r_y}(x_y) \subset \widetilde{V}$, where $r_y$ is a positive rational, and each coordinate of $x_y$ is a rational number. Since $y \in B_{r_y}(x_y)$, $|y - x_y| < r_y$. There exists a positive rational number $s_y$ such that $|y - x_y| < s_y < r_y$. Hence, $y \in B_{s_y}(x_y) \subset (B_{s_y}(x_y))^- \subset \{z : |z - x_y| \le s_y\} \subset B_{r_y}(x_y) \subset \widetilde{V} \subset \widetilde{U}_1$. Since $s_y$ is a positive rational, and each coordinate of $x_y$ is a rational number, $B_{s_y}(x_y)$ is in $\mathcal{B}$. Since $B_{s_y}(x_y)$ is in $\mathcal{B}$, and $(B_{s_y}(x_y))^- \subset \widetilde{U}_1$, $B_{s_y}(x_y)$ is in $\mathcal{B}_1$. Also $y \in B_{s_y}(x_y) \subset \widetilde{V}$, and $B_{s_y}(x_y)$ is an open neighborhood of $y$, $\mathcal{B}_1$ is a basis of $\widetilde{U}_1$. Thus, we have shown that $\mathcal{B}_1$ is a basis of $\widetilde{U}_1$.

Since $\mathcal{B}$ is countable, and $\mathcal{B}_1$ is contained in $\mathcal{B}$, $\mathcal{B}_1$ is countable. Thus, $\mathcal{B}_1$ is a countable basis of $\widetilde{U}_1$. Let us take any $B_r(x)$ in $\mathcal{B}_1$. Since $B_r(x)$ is in $\mathcal{B}_1$, and $\mathcal{B}_1$ is a basis of $\widetilde{U}_1$, $B_r(x)$ is an open subset of the open set $\widetilde{U}_1$. Since $B_r(x)$ is an open subset of $\widetilde{U}_1$, and $\varphi_1 : U_1 \to \widetilde{U}_1$ is a homeomorphism, $\varphi_1^{-1}(B_r(x))$ is open in $\widetilde{U}_1$. Since $\varphi_1^{-1}(B_r(x))$ is open in $\widetilde{U}_1$, and $\varphi_1 : U_1 \to \widetilde{U}_1$ is a homeomorphism, the restriction $\varphi_1|_{\varphi_1^{-1}(B_r(x))} : \varphi_1^{-1}(B_r(x)) \to B_r(x)$ is a homomorphism. Thus, the ordered pair $(\varphi_1^{-1}(B_r(x)), \varphi_1|_{\varphi_1^{-1}(B_r(x))})$ is a coordinate chart of $U_1$.

Since $\mathcal{B}_1$ is countable, the collection $\{(\varphi_1^{-1}(B_r(x)), \varphi_1|_{\varphi_1^{-1}(B_r(x))}) : B_r(x) \text{ is in } \mathcal{B}_1\}$ of coordinate charts is countable. Thus, $\{(\varphi_1^{-1}(B_r(x)), \varphi_1|_{\varphi_1^{-1}(B_r(x))}) : B_r(x) \text{ is in } \mathcal{B}_1\}$ is a countable collection of coordinate charts of $M$.

Now, we shall try to prove that

1. $\{\varphi_1^{-1}(B_r(x)) : B_r(x) \text{ is in } \mathcal{B}_1\}$ is a basis of $U_1$,
2. for each $B_r(x)$ in $\mathcal{B}_1$, $(\varphi_1|_{\varphi_1^{-1}(B_r(x))})(\varphi_1^{-1}(B_r(x)))$ is an open ball in $\mathbb{R}^n$,
3. for each $B_r(x)$ in $\mathcal{B}_1$, the closure $(\varphi_1^{-1}(B_r(x)))^-$ of $\varphi_1^{-1}(B_r(x))$ is a compact subset of $U_1$.

For 1: Since $\varphi_1 : U_1 \mapsto \widetilde{U}_1$ is a homeomorphism, $\mathcal{B}_1$ is a basis of $\widetilde{U}_1$, $\{\varphi_1^{-1}(B_r(x)) : B_r(x) \text{ is in } \mathcal{B}_1\}$ is a basis of $U_1$.

For 2: Since $(\varphi_1|_{\varphi_1^{-1}(B_r(x))})(\varphi_1^{-1}(B_r(x))) = B_r(x)$, and $B_r(x)$ is an open ball in $\mathbb{R}^n$, $(\varphi_1|_{\varphi_1^{-1}(B_r(x))})(\varphi_1^{-1}(B_r(x)))$ is an open ball in $\mathbb{R}^n$.

For 3: Since, for each $B_r(x)$ in $\mathcal{B}_1$, $(B_r(x))^- \subset \widetilde{U}_1$, $\varphi_1^{-1} : \widetilde{U}_1 \to U_1$ is continuous, and, by Heine–Borel theorem, $(B_r(x))^-$ is compact, $\varphi_1^{-1}((B_r(x))^-)$ $(= (\varphi_1^{-1}(B_r(x)))^-)$ is compact, and hence, $(\varphi_1^{-1}(B_r(x)))^-$ is a compact subset of $U_1$. Here, $U_2 \in \{U_1, U_2, U_3, \ldots\} \subset \{U : (U, \varphi) \in \mathcal{C}\}$, so $U_2 \in \{U : (U, \varphi) \in \mathcal{C}\}$, and hence, there exists a function $\varphi_2$ such that $(U_2, \varphi_2) \in \mathcal{C}$, and hence, there exists an open subset $\widetilde{U}_2$ of $\mathbb{R}^n$, such that $\varphi_2 : U_2 \mapsto \widetilde{U}_2$ is a homeomorphism. Let $\mathcal{B}_2$ be the collection of all $B_r(x) \in \mathcal{B}$ such that the closure $(B_r(x))^-$ of $B_r(x)$ is contained in $\widetilde{U}_2$

As above, $\{(\varphi_2^{-1}(B_r(x)), \varphi_2|_{\varphi_2^{-1}(B_r(x))}) : B_r(x)\text{is in } \mathcal{B}_2\}$ is a countable collection of coordinate charts of $M$ such that

(a) $\{\varphi_2^{-1}(B_r(x)) : B_r(x)\text{is in } \mathcal{B}_2\}$ is a basis of $U_2$,
(b) for each $B_r(x)$ in $\mathcal{B}_2$, $(\varphi_2|_{\varphi_2^{-1}(B_r(x))})(\varphi_2^{-1}(B_r(x)))$ is an open ball in $\mathbb{R}^n$,
(c) for each $B_r(x)$ in $\mathcal{B}_2$, the closure $(\varphi_2^{-1}(B_r(x)))^-$ of $\varphi_2^{-1}(B_r(x))$ is a compact subset of $U_2$, etc.

Since the countable union of countable sets is countable, the collection $\cup_{i=1}^{\infty}\{(\varphi_i^{-1}(B_r(x)), \varphi_i|_{\varphi_i^{-1}(B_r(x))}) : B_r(x)\text{is in } \mathcal{B}_i\}$ of coordinates charts is countable. It remains to prove that

I. $\cup_{i=1}^{\infty}\{\varphi_i^{-1}(B_r(x)) : B_r(x)\text{is in } \mathcal{B}_i\}$ is a basis of $M$,
II. for each $i = 1, 2, 3, \ldots$, and for each $B_r(x)$ in $\mathcal{B}_i$, $(\varphi_i|_{\varphi_i^{-1}(B_r(x))})(\varphi_i^{-1}(B_r(x)))$ is an open ball in $\mathbb{R}^n$,
III. for each $i = 1, 2, 3, \ldots$, and for each $B_r(x)$ in $\mathcal{B}_i$, the closure $(\varphi_i^{-1}(B_r(x)))^-$ of $\varphi_i^{-1}(B_r(x))$ is a compact subset of $M$, etc.

For I: Let $G$ be any open neighborhood of $a$ in $M$. Since $\{U_1, U_2, U_3, \ldots\}$ is a cover of $M$, there exists a positive integer $i$ such that $a \in U_i$. Since $U_i \cap G$ is an open neighborhood of $a$, $U_i \cap G$ is contained in $U_i$, and $\{\varphi_i^{-1}(B_r(x)) : B_r(x)$ is in $\mathcal{B}_i\}$ is a basis of $U_i$, there exists $B_r(x)$ in $\mathcal{B}_i$ such that $a \in \varphi_i^{-1}(B_r(x)) \subset U_i \cap G \subset G$. This shows that $\cup_{i=1}^{\infty}\{\varphi_i^{-1}(B_r(x)) : B_r(x)$ is in $\mathcal{B}_i\}$ is a basis of $M$.
For II: This is clear.
For III: This is clear. $\qquad\qquad\square$

**Note 4.48** Let $M$ be an $m$-dimensional topological manifold. Let $p \in M$. Let $U$ be an open neighborhood of $p$. We shall try to prove: There exists an open neighborhood $V$ of $p$ such that $V^- \subset U$. Since $M$ is an $m$-dimensional topological manifold so, by Lemma 4.47, there exists a countable collection $\{(U_1, \varphi_1), (U_2, \varphi_2), (U_3, \varphi_3), \ldots\}$ of coordinate charts of $M$ such that

1. $\{U_1, U_2, U_3, \ldots\}$ is a basis of $M$,
2. each $\varphi_i(U_i)$ is an open ball in $\mathbb{R}^m$,
3. each closure $U_i^-$ of $U_i$ is a compact subset of $M$.

Since $U$ is an open neighborhood of $p$, and $\{U_1, U_2, U_3, \ldots\}$ is a basis of $M$, there exists a positive integer $k$ such that $p \in U_k \subset U$. Here $\varphi_k(U_k)$ is an open ball in $\mathbb{R}^m$, and $\varphi_k(p) \in \varphi_k(U_k)$, there exists a real number $r > 0$ such that the closed ball $B_r[\varphi_k(p)] \subset \varphi_k(U_k)$, and hence, $\varphi_k^{-1}(B_r[\varphi_k(p)]) \subset U_k$. Since $(U_k, \varphi_k)$ is a coordinate chart of $M$, $\varphi_k : U_k \to \varphi_k(U_k)$ is a homeomorphism. Since $\varphi_k : U_k \to \varphi_k(U_k)$ is a homeomorphism, $\varphi_k^{-1} : \varphi_k(U_k) \to U_k$ is continuous. Since $\varphi_k^{-1} : \varphi_k(U_k) \to U_k$ is continuous, and $B_r[\varphi_k(p)]$ is compact, $\varphi_k^{-1}(B_r[\varphi_k(p)])$ is compact. Since $M$ is topological manifold, the topology of $M$ is Hausdorff. Since the topology of $M$ is Hausdorff, and $\varphi_k^{-1}(B_r[\varphi_k(p)])$ is compact, $\varphi_k^{-1}(B_r[\varphi_k(p)])$ is closed in $M$. Clearly, $\varphi_k^{-1}(B_r(\varphi_k(p)))$ is an open neighborhood of $p$, and $\varphi_k^{-1}(B_r(\varphi_k(p))) \subset \varphi_k^{-1}(B_r[\varphi_k(p)])$. Let us take $\varphi_k^{-1}(B_r(\varphi_k(p)))$ for $V$. Thus, $V$ is an open neighborhood of $p$, and $V \subset \varphi_k^{-1}(B_r[\varphi_k(p)])$. Since $V \subset \varphi_k^{-1}(B_r[\varphi_k(p)])$, and $\varphi_k^{-1}(B_r[\varphi_k(p)])$ is closed in $M$, $V^- \subset \varphi_k^{-1}(B_r[\varphi_k(p)]) \subset U_k \subset U$. Thus, $V^- \subset U$.

**Lemma 4.49** *Let $M$ be an $n$-dimensional smooth manifold. Then, there exists a countable collection $\{(U_1, \varphi_1), (U_2, \varphi_2), (U_3, \varphi_3), \ldots\}$ of admissible coordinate charts of $M$ such that*

1. *$\{U_1, U_2, U_3, \ldots\}$ is a basis of $M$.*
2. *each $\varphi_i(U_i)$ is an open ball in $\mathbb{R}^n$,*
3. *each closure $U_i^-$ of $U_i$ is a compact subset of $M$.*

*Proof* Let $\mathcal{C}$ be the collection of all admissible coordinate charts $(U, \varphi)$ of $M$. Since $M$ is a smooth manifold, $\{U : (U, \varphi) \in \mathcal{C}\}$ is an open cover of $M$. Since $M$ is a smooth manifold, $M$ is second countable. Since $M$ is a second countable space, and $\{U : (U, \varphi) \in \mathcal{C}\}$ is an open cover of $M$, there exists a countable subcover $\{U_1, U_2, U_3, \ldots\}$ of $\{U : (U, \varphi) \in \mathcal{C}\}$.

Here, $U_1 \in \{U_1, U_2, U_3, \ldots\} \subset \{U : (U, \varphi) \in \mathcal{C}\}$, $U_1 \in \{U : (U, \varphi) \in \mathcal{C}\}$, and hence, there exists a function $\varphi_1$, and an open subset $\widetilde{U}_1$ of $\mathbb{R}^n$ such that $\varphi_1 : U_1 \to \widetilde{U}_1$, and $(U_1, \varphi_1) \in \mathcal{C}$. Let $\mathcal{B}$ be the collection of all open balls $B_r(x)$ with center $x$ and radius $r$ such that $r$ is a positive rational, and each coordinate of $x$ is a rational number. We know that $\mathcal{B}$ is a basis of $\mathbb{R}^n$. Since the set of all rational numbers is countable, the collection $\mathcal{B}$ is countable. Let $\mathcal{B}_1$ be the collection of all $B_r(x) \in \mathcal{B}$ such that the closure $(B_r(x))^-$ of $B_r(x)$ is contained in $\widetilde{U}_1$.

We shall try to show that $\mathcal{B}_1$ is a basis of $\widetilde{U}_1$. For this purpose, let us take any open neighborhood $\widetilde{V}$ of $y$ such that $\widetilde{V}$ is contained in $\widetilde{U}_1$. Since $\widetilde{V}$ is an open neighborhood of $y$, and $\mathcal{B}$ is a basis of $\mathbb{R}^n$, there exists an open ball $B_{r_y}(x_y)$ in $\mathcal{B}$ such that $y \in B_{r_y}(x_y) \subset \widetilde{V}$, where $r_y$ is a positive rational, and each coordinate of $x_y$ is a rational number. Since $y \in B_{r_y}(x_y)$, $|y - x_y| < r_y$. There exists a positive rational number $s_y$ such that $|y - x_y| < s_y < r_y$. Hence, $y \in B_{s_y}(x_y) \subset (B_{s_y}(x_y))^- \subset \{y : |y - x_y| \leq s_y\} \subset B_{r_y}(x_y) \subset \widetilde{V} \subset \widetilde{U}_1$. Since $s_y$ is a positive rational, and each coordinate of $x_y$ is a rational number, $B_{s_y}(x_y)$ is in $\mathcal{B}$. Since $B_{s_y}(x_y)$ is in $\mathcal{B}$, and $(B_{s_y}(x_y))^- \subset \widetilde{U}_1$, $B_{s_y}(x_y)$ is in $\mathcal{B}_1$. Also $y \in B_{s_y}(x_y) \subset \widetilde{V}$, and $B_{s_y}(x_y)$ is an open neighborhood of $y$, $\mathcal{B}_1$ is a basis of $\widetilde{U}_1$.

Thus, we have shown that $\mathcal{B}_1$ is a basis of $\widetilde{U}_1$. Since $\mathcal{B}$ is countable, and $\mathcal{B}_1$ is contained in $\mathcal{B}$, $\mathcal{B}_1$ is countable. Thus, $\mathcal{B}_1$ is a countable basis of $\widetilde{U}_1$. Let us take any

$B_r(x)$ in $\mathcal{B}_1$. Since $B_r(x)$ is in $\mathcal{B}_1$, and $\mathcal{B}_1$ is a basis of $\widetilde{U}_1$, $B_r(x)$ is an open subset of the open set $\widetilde{U}_1$. Since $B_r(x)$ is an open subset of $\widetilde{U}_1$, and $\varphi_1 : U_1 \to \widetilde{U}_1$ is a homeomorphism, $\varphi_1^{-1}(B_r(x))$ is open in $U_1$. Since $\varphi_1^{-1}(B_r(x))$ is open in $U_1$, and $\varphi_1 : U_1 \to \widetilde{U}_1$ is a homeomorphism, the restriction $\varphi_1|_{\varphi_1^{-1}(B_r(x))} : \varphi_1^{-1}(B_r(x)) \to B_r(x)$ is a homomorphism. Thus, the ordered pair $(\varphi_1^{-1}(B_r(x)), \varphi_1|_{\varphi_1^{-1}(B_r(x))})$ is a coordinate chart of $U_1$.

Now, we want to show that $(\varphi_1^{-1}(B_r(x)), \varphi_1|_{\varphi_1^{-1}(B_r(x))}) \in \mathcal{C}$. For this purpose, let us take any $(W, \psi) \in \mathcal{C}$. Now, we must prove that

$$\left(\left(\varphi_1|_{\varphi_1^{-1}(B_r(x))}\right) \circ \psi^{-1}\right) : \psi\left((\varphi_1^{-1}(B_r(x))) \cap W\right)$$
$$\to \left(\varphi_1|_{\varphi_1^{-1}(B_r(x))}\right)\left((\varphi_1^{-1}(B_r(x))) \cap W\right),$$

and

$$\left(\psi \circ \left(\varphi_1|_{\varphi_1^{-1}(B_r(x))}\right)^{-1}\right) : \left(\varphi_1|_{\varphi_1^{-1}(B_r(x))}\right)\left((\varphi_1^{-1}(B_r(x))) \cap W\right)$$
$$\to \psi\left((\varphi_1^{-1}(B_r(x))) \cap W\right)$$

are $C^\infty$. Let us take any $(y_1, \ldots, y_n) \in \psi((\varphi_1^{-1}(B_r(x))) \cap W)$. So, there exists $p$ in $(\varphi_1^{-1}(B_r(x))) \cap W$ such that $\psi(p) = (y_1, \ldots, y_n)$, and hence $\psi^{-1}(y_1, \ldots, y_n) = p$, and $p \in \varphi_1^{-1}(B_r(x))$. Now,

$$\left(\left(\varphi_1|_{\varphi_1^{-1}(B_r(x))}\right) \circ \psi^{-1}\right)(y_1, \ldots, y_n)$$
$$= \left(\varphi_1|_{\varphi_1^{-1}(B_r(x))}\right)(\psi^{-1}(y_1, \ldots, y_n)) = \left(\varphi_1|_{\varphi_1^{-1}(B_r(x))}\right)(p)$$
$$= \varphi_1(p) = \varphi_1(\psi^{-1}(y_1, \ldots, y_n)) = (\varphi_1 \circ \psi^{-1})(y_1, \ldots, y_n)$$
$$= \left((\varphi_1 \circ \psi^{-1})|_{\psi((\varphi_1^{-1}(B_r(x))) \cap W)}\right)(y_1, \cdots, y_n),$$

so $(\varphi_1|_{\varphi_1^{-1}(B_r(x))}) \circ \psi^{-1} = (\varphi_1 \circ \psi^{-1})|_{\psi((\varphi_1^{-1}(B_r(x))) \cap W)}$. Since $(U_1, \varphi_1) \in \mathcal{C}$, $(W, \psi) \in \mathcal{C}$, and $\mathcal{C}$ is the collection of all admissible coordinate charts of $M$, $(\varphi_1 \circ \psi^{-1}) : \psi(U_1 \cap W) \to \varphi_1(U_1 \cap W)$ is $C^\infty$. Since $\varphi_1^{-1}(B_r(x))$ is an open subset of $U_1$, $\psi((\varphi_1^{-1}(B_r(x))) \cap W)$ is an open subset of $\psi(U_1 \cap W)$. Since $\psi((\varphi_1^{-1}(B_r(x))) \cap W)$ is an open subset of $\psi(U_1 \cap W)$, and $(\varphi_1 \circ \psi^{-1}) : \psi(U_1 \cap W) \to \varphi_1(U_1 \cap W)$ is $C^\infty$, $(\varphi_1 \circ \psi^{-1})|_{\psi((\varphi_1^{-1}(B_r(x))) \cap W)}(= (\varphi_1|_{\varphi_1^{-1}(B_r(x))}) \circ \psi^{-1})$ is $C^\infty$, and hence, $(\varphi_1|_{\varphi_1^{-1}(B_r(x))}) \circ \psi^{-1}$ is $C^\infty$. Thus, we have shown that $((\varphi_1|_{\varphi_1^{-1}(B_r(x))}) \circ \psi^{-1}) : \psi((\varphi_1^{-1}(B_r(x))) \cap W) \to (\varphi_1|_{\varphi_1^{-1}(B_r(x))})((\varphi_1^{-1}(B_r(x))) \cap W)$ is $C^\infty$. Similarly, $(\psi \circ (\varphi_1|_{\varphi_1^{-1}(B_r(x))})^{-1}) : (\varphi_1|_{\varphi_1^{-1}(B_r(x))})((\varphi_1^{-1}(B_r(x))) \cap W) \to \psi((\varphi_1^{-1}(B_r(x))) \cap W)$ is $C^\infty$. Thus, $(\varphi_1^{-1}(B_r(x)), \varphi_1|_{\varphi_1^{-1}(B_r(x))}) \in \mathcal{C}$.

Since $\mathcal{B}_1$ is countable, the collection $\{(\varphi_1^{-1}(B_r(x)), \varphi_1|_{\varphi_1^{-1}(B_r(x))}) : B_r(x) \text{ is in } \mathcal{B}_1\}$ of coordinate charts is countable. Thus, $\{(\varphi_1^{-1}(B_r(x)), \varphi_1|_{\varphi_1^{-1}(B_r(x))}) : B_r(x) \text{ is in } \mathcal{B}_1\}$ is a countable collection of admissible coordinate charts of $M$. Now, we shall try to prove that

1. $\{\varphi_1^{-1}(B_r(x)) : B_r(x) \text{ is in } \mathcal{B}_1\}$ is a basis of $U_1$,
2. for each $B_r(x)$ in $\mathcal{B}_1$, $(\varphi_1|_{\varphi_1^{-1}(B_r(x))})(\varphi_1^{-1}(B_r(x)))$ is an open ball in $\mathbb{R}^n$,
3. for each $B_r(x)$ in $\mathcal{B}_1$, the closure $(\varphi_1^{-1}(B_r(x)))^-$ of $\varphi_1^{-1}(B_r(x))$ is a compact subset of $U_1$.

For 1: Since $\varphi_1 : U_1 \to \tilde{U}_1$ is a homeomorphism, $\mathcal{B}_1$ is a basis of $\tilde{U}_1$, $\{\varphi_1^{-1}(B_r(x)) : B_r(x) \text{ is in } \mathcal{B}_1\}$ is a basis of $U_1$.

For 2: Since $(\varphi_1|_{\varphi_1^{-1}(B_r(x))})(\varphi_1^{-1}(B_r(x))) = B_r(x)$, and $B_r(x)$ is an open ball in $\mathbb{R}^n$, $(\varphi_1|_{\varphi_1^{-1}(B_r(x))})(\varphi_1^{-1}(B_r(x)))$ is an open ball in $\mathbb{R}^n$.

For 3: Since, for each $B_r(x)$ in $\mathcal{B}_1$, $(B_r(x))^- \subset \tilde{U}_1$, $\varphi_1^{-1} : \tilde{U}_1 \to U_1$ is continuous, and, by Heine–Borel theorem, $(B_r(x))^-$ is compact, $\varphi_1^{-1}((B_r(x))^-)(= (\varphi_1^{-1}(B_r(x)))^-)$ is compact, and hence, $(\varphi_1^{-1}(B_r(x)))^-$ is a compact subset of $U_1$.

Here, $U_2 \in \{U_1, U_2, U_3, \ldots\} \subset \{U : (U, \varphi) \in \mathcal{C}\}$, $U_2 \in \{U : (U, \varphi) \in \mathcal{C}\}$, and hence, there exists a function $\varphi_2$ such that $(U_2, \varphi_2) \in \mathcal{C}$, and hence, there exists an open subset $\tilde{U}_2$ of $\mathbb{R}^n$ such that $\varphi_2 : U_2 \to \tilde{U}_2$ is a homeomorphism. Let $\mathcal{B}_2$ be the collection of all $B_r(x) \in \mathcal{B}$ such that the closure $(B_r(x))^-$ of $B_r(x)$ is contained in $\tilde{U}_2$.

As above, $\{(\varphi_2^{-1}(B_r(x)), \varphi_2|_{\varphi_2^{-1}(B_r(x))}) : B_r(x) \text{ is in } \mathcal{B}_2\}$ is a countable collection of admissible coordinate charts of $M$ satisfying

(a) $\{\varphi_2^{-1}(B_r(x)) : B_r(x) \text{ is in } \mathcal{B}_2\}$ is a basis of $U_2$,
(b) for each $B_r(x)$ in $\mathcal{B}_2$, $(\varphi_2|_{\varphi_2^{-1}(B_r(x))})(\varphi_2^{-1}(B_r(x)))$ is an open ball in $\mathbb{R}^n$,
(c) for each $B_r(x)$ in $\mathcal{B}_2$, the closure $(\varphi_2^{-1}(B_r(x)))^-$ of $\varphi_2^{-1}(B_r(x))$ is a compact subset of $U_2$, etc.

Since the countable union of countable sets is countable, the collection $\cup_{i=1}^{\infty}\{(\varphi_i^{-1}(B_r(x)), \varphi_i|_{\varphi_i^{-1}(B_r(x))}) : B_r(x) \text{ is in } \mathcal{B}_i\}$ of admissible coordinates charts is countable. It remains to prove that

I. $\cup_{i=1}^{\infty}\{\varphi_i^{-1}(B_r(x)) : B_r(x) \text{ is in } \mathcal{B}_i\}$ is a basis of $M$,
II. for each $i = 1, 2, 3, \ldots$, and for each $B_r(x)$ in $\mathcal{B}_i$, $(\varphi_i|_{\varphi_i^{-1}(B_r(x))})(\varphi_i^{-1}(B_r(x)))$ is an open ball in $\mathbb{R}^n$,
III. for each $i = 1, 2, 3, \ldots$, and for each $B_r(x)$ in $\mathcal{B}_i$, the closure $(\varphi_i^{-1}(B_r(x)))^-$ of $\varphi_i^{-1}(B_r(x))$ is a compact subset of $M$, etc.

For I: Let $G$ be any open neighborhood of $a$ in $M$. Since $\{U_1, U_2, U_3, \ldots\}$ is a cover of $M$, there exists a positive integer $i$ such that $a \in U_i$. Since $U_i \cap G$ is an open neighborhood of $a$, $U_i \cap G$ is contained in $U_i$, and $\{\varphi_i^{-1}(B_r(x)) : B_r(x) \text{ is in } \mathcal{B}_i\}$ is a basis of $U_i$, there exists $B_r(x)$ is $\mathcal{B}_i$ such that $a \in \varphi_i^{-1}(B_r(x)) \subset U_i \cap G \subset G$. This shows that $\cup_{i=1}^{\infty} \{\varphi_i^{-1}(B_r(x)) : B_r(x) \text{ is in } \mathcal{B}_i\}$ is a basis of $M$.

For II: This is clear.

For III: This is clear. $\qquad\qquad\qquad\qquad\qquad\qquad\qquad\qquad\qquad\qquad\qquad\square$

**Definition** Let $X$ be a topological space. Let $\mathcal{U}$ be a collection of subsets of $X$. If for every $x$ in $X$, there exists an open neighborhood $G$ of $x$ such that

$$\{U : U \in \mathcal{U} \text{ and } U \cap G \text{ is nonempty}\}$$

is a finite set, then we say that $\mathcal{U}$ *is locally finite*. In other words, $\mathcal{U}$ *is locally finite* means for every $x$ in $X$, there exists an open neighborhood $G$ of $x$ such that all members of $\mathcal{U}$ except finite many are "outside" $G$. In short, if for every $x$ in $X$, there exists an open neighborhood $G$ of $x$ such that $G$ intersects only finite many members of $\mathcal{U}$, then we say that $\mathcal{U}$ *is locally finite*. Clearly, every subcollection of a locally finite collection is locally finite.

**Note 4.50** Let $X$ be a topological space. Let $\mathcal{U}$ be an open cover of $X$. Let each member of $\mathcal{U}$ intersects only finite many members of $\mathcal{U}$. We shall try to show that $\mathcal{U}$ is locally finite. For this purpose, let us take any $x$ in $X$. Since $x$ is in $X$, and $\mathcal{U}$ is a cover of $X$, there exists $G$ in $\mathcal{U}$ such that $x$ is in $G$. Since $G$ is in $\mathcal{U}$, and $\mathcal{U}$ is an open cover of $X$, $G$ is open. Since $G$ is open, and $x$ is in $G$, $G$ is an open neighborhood of $x$. Since $G$ is in $\mathcal{U}$, and each member of $\mathcal{U}$ intersects only finite many members of $\mathcal{U}$, $G$ intersects only finite many members of $\mathcal{U}$. This proves that $\mathcal{U}$ is locally finite.

**Lemma 4.51** *Let $M$ be a topological manifold. There exists a countable collection $\{G_1, G_2, G_3, \ldots\}$ of open sets such that*

1. *$\{G_1, G_2, G_3, \ldots\}$ is a cover of $M$,*
2. *$\{G_1, G_2, G_3, \ldots\}$ is locally finite,*
3. *for each $n = 1, 2, 3, \ldots$, the closure $G_n^-$ of $G_n$ is compact.*

*Proof* By Lemma 4.47, there exists a countable collection $\{(U_1, \varphi_1), (U_2, \varphi_2), (U_3, \varphi_3), \ldots\}$ of coordinate charts of $M$ such that

(i) $\{U_1, U_2, U_3, \ldots\}$ is a basis of $M$,
(ii) each closure $U_n^-$ of $U_n$ is a compact subset of $M$.

Put $B_1 \equiv U_1$.

By (i), clearly $\{U_1, U_2, U_3, \ldots\}$ is an open cover of $M$. Since $B_1 = U_1$, $B_1^- = U_1^-$. Since $B_1^- = U_1^-$, and $U_1^-$ is compact, $B_1^-$ is compact. Since $\{U_1, U_2, U_3, \ldots\}$ is an open cover of $M$, and $B_1^-$ is compact, there exists a positive integer $m_1 > 1$ such that $B_1^- \subset U_1 \cup U_2 \cup \cdots \cup U_{m_1}$.

Put $B_2 \equiv U_1 \cup U_2 \cup \cdots \cup U_{m_1}$. Since $B_2 = U_1 \cup U_2 \cup \cdots \cup U_{m_1}$, $B_2^- = (U_1 \cup U_2 \cup \cdots \cup U_{m_1})^- = (U_1^-) \cup (U_2^-) \cup \cdots \cup (U_{m_1}^-)$. Since each of $U_1^-, U_2^-, \ldots, U_{m_1}^-$ is compact, the finite union $(U_1^-) \cup (U_2^-) \cup \cdots \cup (U_{m_1}^-)(= B_2^-)$ is compact, and hence, $B_2^-$ is compact. Since $\{U_1, U_2, U_3, \ldots\}$ is an open cover of $M$, and $B_2^-$ is compact, there exists a positive integer $m_2 > m_1 > 1$ such that $B_2^- \subset U_1 \cup U_2 \cup \cdots \cup U_{m_1} \cup \cdots \cup U_{m_2}$.

Put $B_3 \equiv U_1 \cup U_2 \cup \cdots \cup U_{m_2}$. Since $B_3 = U_1 \cup U_2 \cup \cdots \cup U_{m_2}$, $B_3^- = (U_1 \cup U_2 \cup \cdots \cup U_{m_2})^- = (U_1^-) \cup (U_2^-) \cup \cdots \cup (U_{m_2}^-)$. Since each of $U_1^-, U_2^-, \ldots, U_{m_2}^-$ is compact, the finite union $(U_1^-) \cup (U_2^-) \cup \cdots \cup (U_{m_2}^-)(= B_3^-)$ is compact, and hence, $B_3^-$ is compact. Since $\{U_1, U_2, U_3, \ldots\}$ is an open cover of $M$, and $B_3^-$ is compact, there exists a positive integer $m_3 > m_2 > m_1 > 1$ such that $B_3^- \subset U_1 \cup U_2 \cup \cdots \cup U_{m_1} \cup \cdots \cup U_{m_2} \cup \cdots \cup U_{m_3}$.

Put $B_4 \equiv U_1 \cup U_2 \cup \cdots \cup U_{m_3}$, etc. Clearly

(a)  for every positive integer $n$, $U_n \subset B_n$,
(b)  $\{B_1, B_2, B_3, \ldots\}$ is an open cover of $M$,
(c)  $B_1 \subset B_2 \subset B_3 \subset \cdots$,
(d)  each closure $B_n^-$ of $B_n$ is a compact subset of $M$,
(e)  for every positive integer $n$, $B_n \subset B_n^- \subset B_{n+1}$.

Put $G_1 \equiv B_1, G_2 \equiv B_2, G_3 \equiv B_3 - (B_1^-), G_4 \equiv B_4 - (B_2^-), G_5 = B_5 - (B_3^-)$, etc. Clearly $\{G_1, G_2, G_3, \ldots\}$ is a countable collection of open sets (see Fig. 4.1).

It remains to prove that

1.  $\{G_1, G_2, G_3, \ldots\}$ is a cover of $M$,
2.  $\{G_1, G_2, G_3, \ldots\}$ is locally finite,
3.  for each $n = 1, 2, 3, \ldots$, the closure $G_n^-$ of $G_n$ is compact.

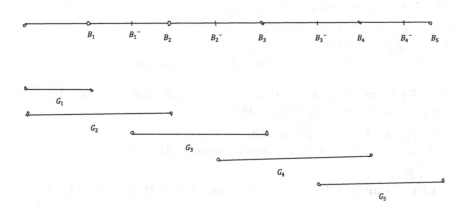

**Fig. 4.1**  A countable collection of open sets

For 1: This is clear.

For 2: Here, $\{G_1, G_2, G_3, \ldots\}$ is an open cover of $M$, and each member of $\{G_1, G_2, G_3, \ldots\}$ intersects only finite many members of $\{G_1, G_2, G_3, \ldots\}$, $\{G_1, G_2, G_3, \ldots\}$ is locally finite.

For 3: Since $G_1 = B_1$, $G_1^- = B_1^-$. Since $B_1^-$ is compact, $G_1^-$ is compact. Since $G_2 = B_2$, $G_2^- = B_2^-$. Since $B_2^-$ is compact, $G_2^-$ is compact. Since $G_3 \subset B_3^-$, $G_3^- \subset B_3^-$. Since the closed set $G_3^-$ is contained in the compact set $B_3^-$, $G_3^-$ is compact. Since $G_4 \subset B_4^-$, $G_4^- \subset B_4^-$. Since the closed set $G_4^-$ is contained in the compact set $B_4^-$, $G_4^-$ is compact, etc. This proves 3. $\qquad\square$

**Definition** Let $X$ be a topological space. Let $\mathcal{A}$, $\mathcal{B}$ be open covers of $X$. By $\mathcal{A}$ *is a refinement of* $\mathcal{B}$ we mean: For every $A$ in $\mathcal{A}$, there exists $B$ in $\mathcal{B}$ such that $A \subset B$.

**Lemma 4.52** *Let $M$ be an $m$-dimensional smooth manifold. Let $\mathcal{X}$ be any open cover of $M$. There exists a countable collection $\{(U_1, \varphi_1), (U_2, \varphi_2), (U_3, \varphi_3), \ldots\}$ of admissible coordinate charts of $M$ such that*

1. $\{U_1, U_2, U_3, \ldots\}$ *covers $M$,*
2. $\{U_1, U_2, U_3, \ldots\}$ *is locally finite,*
3. $\{U_1, U_2, U_3, \ldots\}$ *is a refinement of $\mathcal{X}$,*
4. *each $\varphi_n(U_n)$ is equal to the open ball $B_3(0)$,*
5. $\{\varphi_1^{-1}(B_1(0)), \varphi_2^{-1}(B_1(0)), \varphi_3^{-1}(B_1(0)), \ldots\}$ *covers $M$.*

*Proof* Since $M$ is a smooth manifold, $M$ is a topological manifold, and hence, by Lemma 4.51, there exists a countable collection $\{G_1, G_2, G_3, \ldots\}$ of open sets such that

a. $\{G_1, G_2, G_3, \ldots\}$ is a cover of $M$,
b. $\{G_1, G_2, G_3, \ldots\}$ is locally finite,
c. for each $n = 1, 2, 3, \ldots$, the closure $G_n^-$ of $G_n$ is compact.

Let us fix any $p$ in $M$. Since $\{G_1, G_2, G_3, \ldots\}$ is locally finite, and $p$ is in $M$, there exists an open neighborhood $W_p$ of $p$ such that

$$\{G_k : G_k \cap W_p \text{ is nonempty}\}$$

is a finite set.

Since $p \in W_p$, $\{G_k : p \in G_k\} = \{G_k : p \in G_k \cap W_p\} \subset \{G_k : G_k \cap W_p \text{ is nonempty}\}$. Since $\{G_k : p \in G_k\} \subset \{G_k : G_k \cap W_p \text{ is nonempty}\}$, and $\{G_k : G_k \cap W_p \text{ is nonempty}\}$ is finite, $\{G_k : p \in G_k\}$ is finite. Since $\{G_1, G_2, G_3, \ldots\}$ is a cover of $M$, and $p \in M$, $\{G_k : p \in G_k\}$ is nonempty. Since $\{G_k : p \in G_k\}$ is a nonempty finite collection of open neighborhoods of $p$, $\cap_{p \in G_k} G_k$ is an open neighborhood of $p$.

Since $\mathcal{X}$ is an open cover of $M$, and $p$ is in $M$, there exists $X_p$ in $\mathcal{X}$ such that $p \in X_p$. Since $\mathcal{X}$ is an open cover of $M$, and $X_p$ is in $\mathcal{X}$, $X_p$ is open. Since $X_p$ is open, and $p \in X_p$, $X_p$ is an open neighborhood of $p$.

Since $p$ is in $M$, and $M$ is a smooth manifold, there exists an admissible coordinate chart $(V_p, \varphi_p)$ of $M$ such that $V_p$ is an open neighborhood of $p$.

Since $\cap_{p \in G_k} G_k$ is an open neighborhood of $p$, $W_p$ is an open neighborhood of $p$, $X_p$ is an open neighborhood of $p$, and $V_p$ is an open neighborhood of $p$, their intersection $(\cap_{p \in G_k} G_k) \cap W_p \cap X_p \cap V_p$ is an open neighborhood of $p$.

Since $(V_p, \varphi_p)$ is an admissible coordinate chart of $M$, $(\cap_{p \in G_k} G_k) \cap W_p \cap X_p \cap V_p \subset V_p$, and $(\cap_{p \in G_k} G_k) \cap W_p \cap X_p \cap V_p$ is an open neighborhood of $p$, the pair $((\cap_{p \in G_k} G_k) \cap W_p \cap X_p \cap V_p, \varphi_p|_{(\cap_{p \in G_k} G_k) \cap W_p \cap X_p \cap V_p})$ is an admissible coordinate chart of $M$. Since $((\cap_{p \in G_k} G_k) \cap W_p \cap X_p \cap V_p, \varphi_p|_{(\cap_{p \in G_k} G_k) \cap W_p \cap X_p \cap V_p})$ is an admissible coordinate chart of $M$, and $p \in (\cap_{p \in G_k} G_k) \cap W_p \cap X_p \cap V_p$, there exists an admissible coordinate chart $(U_p, \psi_p)$ such that $U_p \subset (\cap_{p \in G_k} G_k) \cap W_p \cap X_p \cap V_p$, $U_p$ is an open neighborhood of $p$, $\psi_p(U_p) = B_3(0)$, and $\psi_p(p) = 0$. Thus, $\psi_p : U_p \to B_3(0)$ is a homeomorphism.

Since $U_p \subset (\cap_{p \in G_k} G_k) \cap W_p \cap X_p \cap V_p \subset W_p$, and $\{G_k : G_k \cap W_p \text{ is nonempty}\}$ is a finite set, $\{G_k : G_k \cap U_p \text{is nonempty}\}$ is a finite set.

It is clear that

$$\text{if } q \in G_{k_0} \text{ then } U_q \subset G_{k_0}.$$

(Reason: Let $q \in G_{k_0}$. Since $q \in G_{k_0}$, $U_q \subset (\cap_{q \in G_k} G_k) \cap W_q \cap X_q \cap V_q \subset (\cap_{q \in G_k} G_k) \subset G_{k_0}$.)

Since $(U_p, \psi_p)$ is a coordinate chart, and $\psi_p(U_p) = B_3(0)$, $\psi_p : U_p \to B_3(0)$ is continuous. Since $\psi_p : U_p \to B_3(0)$ is continuous, and $B_1(0)$ is an open subset of $B_3(0)$, $(\psi_p)^{-1}(B_1(0))$ is open in $U_p$. Since $(\psi_p)^{-1}(B_1(0))$ is open in $U_p$, and $U_p$ is open in $M$, $(\psi_p)^{-1}(B_1(0))$ is open in $M$. Since $\psi_p(p) = 0 \in B_1(0)$, $p \in (\psi_p)^{-1}(B_1(0))$. Thus, $(\psi_p)^{-1}(B_1(0))$ is an open neighborhood of $p$. Since $\psi_p : U_p \to B_3(0)$, and $B_1(0) \subset B_3(0)$, $(\psi_p)^{-1}(B_1(0)) \subset U_p$.

Since for every $p$ in $M$, $(\psi_p)^{-1}(B_1(0))$ is an open neighborhood of $p$, the collection $\{(\psi_p)^{-1}(B_1(0)) : p \in G_1^-\}$ is an open cover of $G_1^-$. Since $\{(\psi_p)^{-1}(B_1(0)) : p \in G_1^-\}$ is an open cover of $G_1^-$, and $G_1^-$ is compact, there exists finite many $p_{11}, p_{12}, \ldots, p_{1k_1} \in G_1^-$ such that $G_1^- \subset (\psi_{p_{11}})^{-1}(B_1(0)) \cup (\psi_{p_{12}})^{-1}(B_1(0)) \cup \cdots \cup (\psi_{p_{1k_1}})^{-1}(B_1(0)) \subset U_{p_{11}} \cup U_{p_{12}} \cup \cdots \cup U_{p_{1k_1}}$, where $(U_{p_{11}}, \psi_{p_{11}})$, $(U_{p_{12}}, \psi_{p_{12}}), \ldots, (U_{p_{1k_1}}, \psi_{p_{1k_1}})$ are admissible coordinate charts of $M$.

Similarly, there exists finite many $p_{21}, p_{22}, \ldots, p_{2k_2} \in G_2^-$ such that $G_2^- \subset U_{p_{21}} \cup U_{p_{22}} \cup \cdots \cup U_{p_{2k_2}}$, etc. Clearly, $(U_{p_{11}}, \psi_{p_{11}})$, $(U_{p_{12}}, \psi_{p_{12}}), \ldots, (U_{p_{1k_1}}, \psi_{p_{1k_1}})$, $(U_{p_{21}}, \psi_{p_{21}})$, $(U_{p_{22}}, \psi_{p_{22}}), \ldots, (U_{p_{2k_2}}, \psi_{p_{2k_2}})$, $(U_{p_{31}}, \psi_{p_{31}})$, $(U_{p_{32}}, \psi_{p_{32}}), \ldots, (U_{p_{3k_3}}, \psi_{p_{3k_3}}), \ldots$ constitute a countable collection of admissible coordinate charts of $M$.

For 1: Since $G_1 \subset G_1^- \subset (\psi_{p_{11}})^{-1}(B_1(0)) \cup (\psi_{p_{12}})^{-1}(B_1(0)) \cup \cdots \cup (\psi_{p_{1k_1}})^{-1}(B_1(0)) \subset U_{p_{11}} \cup U_{p_{12}} \cup \cdots \cup U_{p_{1k_1}}$, $G_2 \subset G_2^- \subset (\psi_{p_{21}})^{-1}(B_1(0)) \cup (\psi_{p_{22}})^{-1}(B_1(0)) \cup \cdots \cup (\psi_{p_{2k_2}})^{-1}(B_1(0)) \subset U_{p_{21}} \cup U_{p_{22}} \cup \cdots \cup U_{p_{2k_2}}, \ldots$, and $\{G_1, G_2, G_3, \cdots\}$ is a cover of $M$, $\{U_{p_{11}}, U_{p_{12}}, \ldots, U_{p_{1k_1}}, U_{p_{21}}, U_{p_{22}}, \ldots, U_{p_{2k_2}}, U_{p_{31}}, U_{p_{32}}, \ldots, U_{p_{3k_3}}, \ldots\}$ is an open cover of $M$.

For 2: We have to prove that $\{U_{p_{11}}, U_{p_{12}}, \ldots, U_{p_{1k_1}}, U_{p_{21}}, U_{p_{22}}, \ldots, U_{p_{2k_2}}, U_{p_{31}}, U_{p_{32}}, \ldots, U_{p_{3k_3}}, \ldots\}$ is locally finite.

Since $\{G_1, G_2, G_3, \ldots\}$ is a cover of $M$, and $p_{11} \in M$, there exists a positive integer $n_{11}$ such that $p_{11} \in G_{n_{11}}$. Since $p_{11} \in G_{n_{11}}$, $U_{p_{11}} \subset G_{n_{11}}$. Similarly, there exists a positive integer $n_{12}$ such that $U_{p_{12}} \subset G_{n_{12}}$, etc.

Since $\{U_{p_{11}}, U_{p_{12}}, \ldots, U_{p_{1k_1}}, U_{p_{21}}, U_{p_{22}}, \ldots, U_{p_{2k_2}}, U_{p_{31}}, U_{p_{32}}, \ldots, U_{p_{3k_3}}, \ldots\}$ is a cover of $M$, and each $U_{p_{ij}}$ is contained in $G_{n_{ij}}$, $\{G_{n_{11}}, G_{n_{12}}, \ldots, G_{n_{1k_1}}, G_{n_{21}}, G_{n_{22}}, \ldots, G_{n_{2k_2}}, G_{n_{31}}, G_{n_{32}}, \ldots, G_{n_{3k_3}}, \ldots\}$ is a cover of $M$. Since $\{G_1, G_2, G_3, \ldots\}$ is locally finite, and $\{G_{n_{11}}, G_{n_{12}}, \ldots, G_{n_{1k_1}}, G_{n_{21}}, G_{n_{22}}, \ldots, G_{n_{2k_2}}, G_{n_{31}}, G_{n_{32}}, \ldots, G_{n_{3k_3}}, \ldots\} \subset \{G_1, G_2, G_3, \ldots\}$, $\{G_{n_{11}}, G_{n_{12}}, \ldots, G_{n_{1k_1}}, G_{n_{21}}, G_{n_{22}}, \ldots, G_{n_{2k_2}}, G_{n_{31}}, G_{n_{32}}, \ldots, G_{n_{3k_3}}, \ldots\}$ is locally finite. Since $\{G_{n_{11}}, G_{n_{12}}, \ldots, G_{n_{1k_1}}, G_{n_{21}}, G_{n_{22}}, \ldots, G_{n_{2k_2}}, G_{n_{31}}, G_{n_{32}}, \ldots, G_{n_{3k_3}}, \ldots\}$ is locally finite, and each $U_{p_{ij}}$ is contained in $G_{n_{ij}}$, $\{U_{p_{11}}, U_{p_{12}}, \ldots, U_{p_{1k_1}}, U_{p_{21}}, U_{p_{22}}, \ldots, U_{p_{2k_2}}, U_{p_{31}}, U_{p_{32}}, \ldots, U_{p_{3k_3}}, \ldots\}$ is locally finite.

For 3: We have to show that the open cover $\{U_{p_{11}}, U_{p_{12}}, \ldots, U_{p_{1k_1}}, U_{p_{21}}, U_{p_{22}}, \ldots, U_{p_{2k_2}}, U_{p_{31}}, U_{p_{32}}, \ldots, U_{p_{3k_3}}, \ldots\}$ is a refinement of the open cover $\mathcal{X}$.

Here $U_{p_{11}} \subset (\cap_{p_{11} \in G_k} G_k) \cap W_{p_{11}} \cap X_{p_{11}} \cap V_{p_{11}} \subset X_{p_{11}} \in \mathcal{X}$. Thus, $U_{p_{11}} \subset X_{p_{11}} \in \mathcal{X}$. Similarly, $U_{p_{12}} \subset X_{p_{12}} \in \mathcal{X}$, etc.

For 4: This is clear.

For 5: We have to show that

$$\left\{\psi_{p_{11}}^{-1}(B_1(0)), \psi_{p_{12}}^{-1}(B_1(0)), \ldots, \psi_{p_{1k_1}}^{-1}(B_1(0)), \psi_{p_{21}}^{-1}(B_1(0)), \psi_{p_{22}}^{-1}(B_1(0)), \ldots, \psi_{p_{2k_2}}^{-1}(B_1(0)), \ldots\right\}$$

covers $M$.

Since $G_1 \subset G_1^- \subset (\psi_{p_{11}})^{-1}(B_1(0)) \cup (\psi_{p_{12}})^{-1}(B_1(0)) \cup \cdots \cup (\psi_{p_{1k_1}})^{-1}(B_1(0))$, $G_2 \subset G_2^- \subset (\psi_{p_{21}})^{-1}(B_1(0)) \cup (\psi_{p_{22}})^{-1}(B_1(0)) \cup \cdots \cup (\psi_{p_{2k_2}})^{-1}(B_1(0)), \ldots$, and

$\{G_1, G_2, G_3, \cdots\}$ is a cover of $M$, $\{\psi_{p_{11}}^{-1}(B_1(0)), \psi_{p_{12}}^{-1}(B_1(0)), \cdots, \psi_{p_{1k_1}}^{-1}(B_1(0)), \psi_{p_{21}}^{-1}(B_1(0)), \psi_{p_{22}}^{-1}(B_1(0)), \cdots, \psi_{p_{2k_2}}^{-1}(B_1(0)), \cdots\}$ is a cover of $M$. $\qquad\square$

**Definition** Let $X$ be a topological space. If for every open cover $\mathcal{A}$ of $X$, there exists an open cover $\mathcal{C}$ of $M$ such that $\mathcal{C}$ is locally finite, and $\mathcal{C}$ is a refinement of $\mathcal{A}$, then we say that $X$ *is a paracompact space.*

**Note 4.53** By the Lemma 4.52, we find that every smooth manifold is a para-compact space.

## 4.5 Partitions of Unity Theorem

**Note 4.54 (Pasting Lemma)** Let $X$ and $Y$ be topological spaces. Let $A$ and $B$ be subsets of $X$ such that the union of $A$ and $B$ is $X$. Let $f : A \to Y$ be a continuous function, where the topology of $A$ is the induced topology of $X$. Let $g : B \to Y$ be a continuous function, where the topology of $B$ is the induced topology of $X$. Let $f(x) = g(x)$ for every $x$ in $A \cap B$. Let $F : X \to Y$ be a function defined as follows:

$$F(x) \equiv \begin{cases} f(x) & \text{if } x \text{ is in } A \\ g(x) & \text{if } x \text{ is in } B. \end{cases}$$

Then, if $A$ and $B$ are open subsets of $X$, then $F$ is continuous.

*Proof* Since $f : A \to Y$ is a function, $g : B \to Y$ is a function, for every $x$ in $A \cap B$, $f(x) = g(x)$, and

$$F(x) \equiv \begin{cases} f(x) & \text{if } x \text{ is in } A \\ g(x) & \text{if } x \text{ is in } B, \end{cases}$$

$F$ is well defined on $A \cup B (= X)$. Let us take any $x$ in $X$. We have to prove that $F$ is continuous at $x$. Since $x$ is in $X$, and $X = A \cup B$, $x \in A$ or $x \in B$.

Case I: When $x \in A$. Since $f : A \to Y$ is a continuous function, and $x \in A$, $f$ is continuous at $x$. Since $f$ is continuous at $x$, and $F|_A = f$, $F|_A$ is continuous at $x$. Since $x \in A$, and $A$ is open, $x$ is an interior point of $A$. Since $x$ is an interior point of $A$, and $F|_A$ is continuous at $x$, $F$ is continuous at $x$.

Case II: When $x \in B$. As in case I, $F$ is continuous at $x$. $\qquad\square$

**Lemma 4.55** *Let $X$ be a topological space. Let $\mathcal{U}$ be a collection of subsets of $X$. If $\mathcal{U}$ is locally finite, then $(\cup\{U : U \in \mathcal{U}\})^- = \cup\{U^- : U \in \mathcal{U}\}$.*

*Proof* Let $\mathcal{U}$ be locally finite. For any $U \in \mathcal{U}$, $U \subset (\cup\{U : U \in \mathcal{U}\})$, so $U^- \subset (\cup\{U : U \in \mathcal{U}\})^-$. This shows that $\cup\{U^- : U \in \mathcal{U}\} \subset (\cup\{U : U \in \mathcal{U}\})^-$. It remains to show that $(\cup\{U : U \in \mathcal{U}\})^- \subset \cup\{U^- : U \in \mathcal{U}\}$. For this purpose, let us take any $a$ in $(\cup\{U : U \in \mathcal{U}\})^-$. We claim that $a$ is in $\cup\{U^- : U \in \mathcal{U}\}$. If not, otherwise, let for every $U \in \mathcal{U}$, $a \notin U^-$. We have to arrive at a contradiction. $\square$

Now, since $\mathcal{U}$ is locally finite, there exists an open neighborhood $G_a$ of $a$ such that all but finite many $U \in \mathcal{U}$ are outside $G_a$. For every $U \in \mathcal{U}$, we have $a \notin U^-$, there exists an open neighborhood $H_U$ of $a$ such that $H_U$ contains no point of $U^-$, and hence, $H_U$ contains no point of $U$. Since all but finite many $U \in \mathcal{U}$ are outside $G_a$, the finite intersection $(\cap_{G_a \cap U \neq \emptyset} H_U) \cap G_a$ is an open neighborhood of $a$. Since for every $U \in \mathcal{U}$, $H_U$ contains no point of $U$, and all but finite many $U \in \mathcal{U}$ are outside $G_a$, $(\cap_{G_a \cap U \neq \emptyset} H_U) \cap G_a$ contains no point of $\cup\{U : U \in \mathcal{U}\}$. It follows that $a$ is not in $(\cup\{U : U \in \mathcal{U}\})^-$, a contradiction. $\square$

**Definition** Let $X$ be a topological space. Let $f : X \to \mathbb{R}^k$ be any function. The closure $\{x : f(x) \neq 0\}^-$ of $\{x : f(x) \neq 0\}$ is denoted by supp $f$, and is called the *support of $f$*.

**Lemma 4.56** Let $M$ be a 3-dimensional smooth manifold. Let $\{X_i\}_{i \in I}$ be an open cover of $M$. Then, there exists a family $\{\psi_i\}_{i \in I}$ of functions $\psi_i : M \to [0, 1]$ such that

1. for each $i \in I$, $\psi_i$ is smooth,
2. for each $i \in I$, supp $\psi_i$ is contained in $X_i$,
3. $\{\text{supp } \psi_i\}_{i \in I}$ is locally finite,
4. $\sum_{i \in I} \psi_i = 1$.

*Proof* By the Lemma 4.52, there exists a countable collection $\{(U_1, \varphi_1), (U_2, \varphi_2), (U_3, \varphi_3), \ldots\}$ of admissible coordinate charts of $M$ such that

  i. $\{U_1, U_2, U_3, \ldots\}$ covers $M$,
 ii. $\{U_1, U_2, U_3, \ldots\}$ is locally finite,
iii. $\{U_1, U_2, U_3, \ldots\}$ is a refinement of $\{X_i\}_{i \in I}$,
 iv. each $\varphi_n(U_n)$ is equal to the open ball $B_3(0)$,
  v. $\{\varphi_1^{-1}(B_1(0)), \varphi_2^{-1}(B_1(0)), \varphi_3^{-1}(B_1(0)), \ldots\}$ covers $M$.

By the Note 3.52, there exists a function $H : \mathbb{R}^3 \to [0, 1]$ such that

  I. $H$ is smooth,
 II. $H(x) = 1$, if $x$ is in the closed ball $B_1[0]$,
III. supp $H = B_2[0]$.

Let us fix any positive integer $n$. Observe that

$$U_n = \varphi_n^{-1}(B_3(0)) \supset \varphi_n^{-1}(B_2[0]).$$

If $x \in U_n - \varphi_n^{-1}(B_2[0])$, then $x \notin \varphi_n^{-1}(B_2[0])$, and hence, $\varphi_n(x) \notin B_2[0] =$ supp $H = (H^{-1}(\mathbb{R} - \{0\}))^- \supset H^{-1}(\mathbb{R} - \{0\})$. It follows that if $x \in U_n -$

$\varphi_n^{-1}(B_2[0])$, then $\varphi_n(x) \notin H^{-1}(\mathbb{R} - \{0\})$, and hence, $H(\varphi_n(x)) = 0$. This shows that the function $f_n : M \to [0, 1]$ defined as follows: For every $x$ in $M$,

$$f_n(x) \equiv \begin{cases} H(\varphi_n(x)) & \text{if } x \in U_n, \\ 0 & \text{if } x \notin \varphi_n^{-1}(B_2[0]), \end{cases}$$

is well defined. Clearly, $f_n^{-1}(\mathbb{R} - \{0\}) \subset \varphi_n^{-1}(B_2[0])$, and hence, supp $f_n = (f_n^{-1}(\mathbb{R} - \{0\}))^- \subset (\varphi_n^{-1}(B_2[0]))^- = \varphi_n^{-1}(B_2[0]) \subset \varphi_n^{-1}(B_3(0)) = U_n$. Thus, supp $f_n \subset U_n$.

Since $(U_n, \varphi_n)$ is an admissible coordinate chart of 3-dimensional smooth manifold $M$, $\varphi_n(U_n)$ is open in $\mathbb{R}^3$, and $\varphi_n : U_n \to \varphi_n(U_n)(= B_3(0))$ is a homeomorphism. Since $\varphi_n : U_n \to \varphi_n(U_n)$, and $H : \mathbb{R}^3 \to [0, 1]$, $H \circ \varphi_n : U_n \to [0, 1]$. Clearly, $H \circ \varphi_n : U_n \to [0, 1]$ is smooth.

(Reason: Take any $p$ in $U_n$. We have to prove that $H \circ \varphi_n$ is smooth at $p$. For this purpose, let us take any admissible coordinate chart $(V, \psi)$ of $M$ such that $p \in V$. We have to prove that $(H \circ \varphi_n) \circ \psi^{-1}$ is smooth. Since $(U_n, \varphi_n)$, $(V, \psi)$ are admissible coordinate charts of $M$, and $U_n \cap V$ is nonempty, $\varphi_n \circ \psi^{-1}$ is smooth. Since $\varphi_n \circ \psi^{-1}$ is smooth, and $H$ is smooth, their composite $H \circ (\varphi_n \circ \psi^{-1})(= (H \circ \varphi_n) \circ \psi^{-1})$ is smooth, and hence, $(H \circ \varphi_n) \circ \psi^{-1}$ is smooth.)

Since

$$f_n(x) = \begin{cases} (H \circ \varphi_n)(x) & \text{if } x \in U_n, \\ 0 & \text{if } x \in M - (\varphi_n^{-1}(B_2[0])) \end{cases}$$

is a well-defined function, $U_n$ and $M - (\varphi_n^{-1}(B_2[0]))$ are open sets, the restriction $f_n|_{U_n}(= H \circ \varphi_n)$ is smooth, and the restriction $f_n|_{M-(\varphi_n^{-1}(B_2[0]))}(= 0)$ is smooth, $f_n$ is smooth.

Let us fix any $a$ in $M$. Since $\{U_1, U_2, U_3, \ldots\}$ is a locally finite cover of $M$, and $a$ is in $M$, there exists an open neighborhood $G_a$ of $a$, and a positive integer $N$ such that $k \geq N$ implies $U_k$ is outside $G_a$. For $k \geq N$, $U_k = \varphi_k^{-1}(B_3(0)) \supset \varphi_k^{-1}(B_2[0])$, and $U_k$ is outside $G_a$, $\varphi_k^{-1}(B_2[0])$ is outside $G_a$ for every $k \geq N$. Since for every $k \geq N$, $G_a$ is outside $\varphi_k^{-1}(B_2[0])$, by the definition of $f_k$, for every $x$ in the open neighborhood $G_a$ of $a$, and $k \geq N$, we have $f_k(x) = 0$. Now, since $a \in G_a$, $f_N(a) = 0 = f_{N+1}(a) = f_{N+2}(a) = \cdots$. It follows that $f_1(a) + f_2(a) + f_3(a) + \cdots$ is a real number.

Since $\{\varphi_1^{-1}(B_1(0)), \varphi_2^{-1}(B_1(0)), \varphi_3^{-1}(B_1(0)), \ldots\}$ covers $M$, and $a$ is in $M$, there exists a positive integer $N_0$ such that $a \in \varphi_{N_0}^{-1}(B_1(0)) \subset \varphi_{N_0}^{-1}(B_3(0)) = U_{N_0}$, and hence $\varphi_{N_0}(a) \in B_1(0) \subset B_1[0]$. Since $\varphi_{N_0}(a) \in B_1[0]$, $a \in U_{N_0}$, and $H(x) = 1$ for every $x$ in the closed ball $B_1[0]$, $f_{N_0}(a) = H(\varphi_{N_0}(a)) = 1$.

Since $f_1 : M \to [0, 1]$, $f_2 : M \to [0, 1]$, $f_3 : M \to [0, 1], \ldots$, and $f_{N_0}(a) = 1$, $f_1(a) + f_2(a) + f_3(a) + \cdots \geq f_{N_0}(a) = 1$.

Further, since $1 \leq f_1(a) + f_2(a) + f_3(a) + \cdots$, and $f_1(a) + f_2(a) + f_3(a) + \cdots$ is a real number,

$$\frac{f_n(a)}{f_1(a)+f_2(a)+f_3(a)+\cdots} \in [0,1].$$

Thus, the function $g_n : M \rightarrow [0,1]$ defined as follows: For every $x$ in $M$,

$$g_n(x) \equiv \frac{f_n(x)}{f_1(x)+f_2(x)+f_3(x)+\cdots}$$

is well defined. Further, we have shown that for every $a$ in $M$, there exists an open neighborhood $G_a$ of $a$, and a positive integer $N$ such that for every $x$ in the open neighborhood $G_a$,

$$g_n(x) = \frac{f_n(x)}{f_1(x)+f_2(x)+\cdots+f_N(x)}.$$

Further, since $f_1, f_2, f_3, \ldots$ are smooth,

$$x \mapsto \frac{f_n(x)}{f_1(x)+f_2(x)+\cdots+f_N(x)} (= g_n(x))$$

is smooth in $G_a$. This shows that $g_n$ is smooth. Clearly, supp $g_n =$ supp $f_n \subset U_n$.

Since for every $x$ in the open neighborhood $G_a$ of $a$, we have $f_N(x) = 0 = f_{N+1}(x) = f_{N+2}(x) = \cdots,$

$$g_N(x) = \frac{f_N(x)}{f_1(x)+f_2(x)+\cdots+f_N(x)} = \frac{0}{f_1(x)+f_2(x)+\cdots+f_N(x)} = 0,$$

$$g_{N+1}(x) = \frac{f_{N+1}(x)}{f_1(x)+f_2(x)+\cdots+f_N(x)} = \frac{0}{f_1(x)+f_2(x)+\cdots+f_N(x)} = 0,$$

$$g_{N+1}(x) = 0, \ldots.$$

Hence, $(g_1 + g_2 + g_3 + \cdots)|_{G_a} = (g_1 + g_2 + \cdots + g_N)|_{G_a}$. Clearly, for every $x$ in $M$,

$$g_1(x) + g_2(x) + g_3(x) + \cdots$$

$$= \frac{1}{f_1(x)+f_2(x)+f_3(x)+\cdots}f_1(x) + \frac{1}{f_1(x)+f_2(x)+f_3(x)+\cdots}f_2(x)$$

$$+ \frac{1}{f_1(x)+f_2(x)+f_3(x)+\cdots}f_3(x) + \cdots$$

$$= \frac{1}{f_1(x)+f_2(x)+f_3(x)+\cdots}(f_1(x)+f_2(x)+f_3(x)+\cdots) = 1.$$

In short, $g_1 + g_2 + g_3 + \cdots = 1$.

Since $\{U_1, U_2, U_3, \ldots\}$ is a refinement of $\{X_i\}_{i \in I}$, for every positive integer $m$, there exists an index $i(m) \in I$ such that $U_m \subset X_{i(m)}$. It follows that $\cup_{i^* \in I}\{m : i(m) = i^*\} = \{1, 2, 3, \ldots\}$.

Let us fix any $i^*$ in $I$. Since $\{x : 0 < f_1(x)\} \subset \{x : 0 < f_1(x)\}^- = \text{supp } f_1 \subset U_1$, $\{x : 0 < f_1(x)\} \subset U_1$. Similarly, $\{x : 0 < f_2(x)\} \subset U_2$, $\{x : 0 < f_3(x)\} \subset U_3$, .... Further, since $\{U_1, U_2, U_3, \ldots\}$ is locally finite, $\{\{x : 0 < f_1(x)\}, \{x : 0 < f_2(x)\}, \{x : 0 < f_3(x)\}, \ldots\}$ is locally finite, and hence, its subcollection $\{\{x : 0 < f_m(x)\} : i(m) = i^*\}$ is locally finite. It follows that $(\cup_{i(m)=i^*}\{x : 0 < f_m(x)\})^- = \cup_{i(m)=i^*}$ $(\{x : 0 < f_m(x)\}^-) = \cup_{i(m)=i^*}(\text{supp } f_m) \subset \cup_{i(m)=i^*} U_m \subset \cup_{i(m)=i^*} X_{i(m)} = \cup_{i(m)=i^*} X_{i^*} = X_{i^*}$. Thus $(\cup_{i(m)=i^*}\{x : 0 < f_m(x)\})^- \subset X_{i^*}$.

Since for every $x$ in $M$, $\sum_{i(m)=i^*} g_m(x) \leq g_1(x) + g_2(x) + g_3(x) + \cdots = 1$, it follows that the function $\psi_{i^*} : M \to [0, 1]$ defined as follows:

$$\psi_{i^*} = \begin{cases} \sum_{i(m)=i^*} g_m & \text{if } \{m : i(m) = i^*\} \text{ is nonempty,} \\ 0 & \text{if } \{m : i(m) = i^*\} \text{ is the empty set,} \end{cases}$$

is well defined.

For 1: Case I: When $\{m : i(m) = i^*\}$ is nonempty. Let us take any $a$ in $M$. We have to prove that $\sum_{i(m)=i^*} g_m (= \psi_{i^*})$ is smooth at $a$. We have seen that there exist an open neighborhood $G_a$ of $a$, and positive integers $m_1, m_2, \ldots, m_k$ such that $i^* = i(m_1) = i(m_2) = \cdots = i(m_k)$, and $(\sum_{i(m)=i^*} g_m)|_{G_a} = (g_{m_1} + g_{m_2} + \cdots + g_{m_k})|_{G_a}$. Since each $g_n$ is smooth, $(g_{m_1} + g_{m_2} + \cdots + g_{m_k})|_{G_a}$ is smooth at $a$. Since $(g_{m_1} + g_{m_2} + \cdots + g_{m_k})|_{G_a}$ is smooth at $a$, and $(\sum_{i(m)=i^*} g_m)|_{G_a} = (g_{m_1} + g_{m_2} + \cdots + g_{m_k})|_{G_a}$, $(\sum_{i(m)=i^*} g_m)|_{G_a}$ is smooth at $a$. This shows that $\psi_{i^*}(= \sum_{i(m)=i^*} g_m)$ is smooth.
Case 2: When $\{m : i(m) = i^*\}$ is the empty set. Since the constant function $0$ is smooth, and $\psi_{i^*} = 0$, $\psi_{i^*}$ is smooth. So, in all cases, $\psi_{i^*}$ is smooth.
For 2: Case I: When $\{m : i(m) = i^*\}$ is nonempty. We have to prove that $\text{supp } \psi_{i^*} \subset X_{i^*}$. Here

$$\text{supp } \psi_{i^*} = \text{supp}\left(\sum_{i(m)=i^*} g_m\right) = \left(\left(\sum_{i(m)=i^*} g_m\right)^{-1}(\mathbb{R} - \{0\})\right)^-$$

$$= \left\{x : 0 < \left(\sum_{i(m)=i^*} g_m\right)(x)\right\}^- = \left\{x : 0 < \sum_{i(m)=i^*} (g_m(x))\right\}^-$$

$$= \left\{x : 0 < \sum_{i(m)=i^*} \left(\frac{1}{f_1(x) + f_2(x) + f_3(x) + \cdots} f_m(x)\right)\right\}^-$$

$$= \left\{ x : 0 < \frac{1}{f_1(x) + f_2(x) + f_3(x) + \cdots} \sum_{i(m)=i^*} (f_m(x)) \right\}^-$$

$$= \left\{ x : 0 < \sum_{i(m)=i^*} f_m(x) \right\}^- = \left( \cup_{i(m)=i^*} \{x : 0 < f_m(x)\} \right)^- \subset X_{i^*}.$$

Hence, supp $\psi_{i^*} \subset X_{i^*}$.

Case II: When $\{m : i(m) = i^*\}$ is the empty set. Here, supp $\psi_{i^*} = $ supp $(0) = \emptyset \subset X_{i^*}$, and hence, supp $\psi_{i^*} \subset X_{i^*}$. Hence, in all cases, supp $\psi_{i^*} \subset X_{i^*}$.

For 3: We have seen that if $\{m : i(m) = i^*\}$ is nonempty, then supp $\psi_{i^*} = (\cup_{i(m)=i^*} \{x : 0 < f_m(x)\})^- = \cup_{i(m)=i^*} (\{x : 0 < f_m(x)\}^-) = \cup_{i(m)=i^*} (\text{supp } f_m) \subset \cup_{i(m)=i^*} U_m$. Thus, if $\{m : i(m) = i^*\}$ is nonempty, then supp $\psi_{i^*} \subset \cup_{i(m)=i^*} U_m$. Also, if $\{m : i(m) = i^*\}$ is the empty set, then supp $\psi_{i^*} = $ supp $0 = \emptyset$. Now, it suffices to show that $\{\cup_{i(m)=i^*} U_m : i^* \in I\}$ is locally finite. Since $\{\{m : i(m) = i^*\} : i^* \in I\}$ is a partition of $\{1, 2, 3, \ldots\}$, and $\{U_1, U_2, U_3, \ldots\}$ is locally finite, $\{\cup_{i(m)=i^*} U_m : i^* \in I\}$ is locally finite.

For 4: Here

$$\text{LHS} = \sum_{i \in I} \psi_i = \sum_{i^* \in I} \psi_{i^*} = \sum_{\substack{i^* \in I \\ \{m:i(m)=i^*\}\text{is non empty}}} \psi_{i^*} + \sum_{\substack{i^* \in I \\ \{m:i(m)=i^*\}\text{is empty set}}} \psi_{i^*}$$

$$= \sum_{\substack{i^* \in I \\ \{m:i(m)=i^*\}\text{is non empty}}} \psi_{i^*} + \sum_{\substack{i^* \in I \\ \{m:i(m)=i^*\}\text{is empty set}}} 0 = \sum_{\substack{i^* \in I \\ \{m:i(m)=i^*\}\text{is non empty}}} \psi_{i^*}$$

$$= \sum_{\substack{i^* \in I \\ \{m:i(m)=i^*\}\text{is non empty}}} \left( \sum_{i(m)=i^*} g_m \right) = g_1 + g_2 + g_3 + \cdots = 1 = \text{RHS.} \qquad \square$$

**Note 4.57** Clearly, the above lemma is also valid for $m$-dimensional smooth manifold $M$.

**Definition** Let $M$ be a topological space. Let $\{X_i\}_{i \in I}$ be an open cover of $M$. Let $\{\psi_i\}_{i \in I}$ be a family of functions $\psi_i : M \to [0, 1]$ satisfying the following conditions:

1. for each $i \in I$, $\psi_i$ is continuous,
2. for each $i \in I$, supp $\psi_i$ is contained in $X_i$,
3. $\{$supp $\psi_i\}_{i \in I}$ is locally finite,
4. $\sum_{i \in I} \psi_i = 1$.

Then, we say that $\{\psi_i\}_{i \in I}$ *is a partition of unity subordinate to* $\{X_i\}_{i \in I}$. Since $\{\text{supp } \psi_i\}_{i \in I}$ is locally finite, for every $x$ in $M$, there are only finite many nonzero $\psi_i(x)$'s in the sum $\sum_{i \in I} \psi_i(x)$, $\sum_{i \in I} \psi_i = 1$ is meaningful. Further, if $M$ is a smooth manifold, and each $\psi_i$ is smooth, then we say that $\{\psi_i\}_{i \in I}$ *is a smooth partition of unity subordinate to* $\{X_i\}_{i \in I}$. Now, the Lemma 4.56 can be abbreviated as: "If $M$ is an $n$-dimensional smooth manifold, and $\{X_i\}_{i \in I}$ is an open cover of $M$, then there exists a smooth partition of unity subordinate to $\{X_i\}_{i \in I}$." This result is known as the *partitions of unity theorem*.

**Definition** Let $M$ be a topological space. Let $A$ be a closed subset of $M$. Let $U$ be an open subset of $M$ such that $A \subset U$. Let $\psi : M \to [0, 1]$ be a continuous function. If

1. for every $x$ in $A$, $\psi(x) = 1$,
2. supp $\psi \subset U$,

then we say that $\psi$ is a *bump function for A supported in U*.

**Lemma 4.58** *Let $M$ be a smooth manifold. Let $A$ be a closed subset of $M$. Let $U$ be an open subset of $M$ such that $A \subset U$. Then, there exists a smooth function $\psi :$ $M \to [0, 1]$ such that $\psi$ is a bump function for $A$ supported in $U$.*

*Proof* Clearly, $\{M - A, U\}$ is an open cover of $M$. So, by the partitions of unity theorem, there exist functions $\psi_1 : M \to [0, 1], \psi : M \to [0, 1]$ such that

1. $\psi_1, \psi$ are smooth,
2. supp $\psi_1$ is contained in $M - A$, and supp $\psi$ is contained in $U$,
3. $\psi_1 + \psi = 1$.

It remains to prove: For every $x$ in $A$, $\psi(x) = 1$, that is, for every $x$ in $A$, $1 - \psi_1(x) = (1 - \psi_1)(x) = 1$, that is, for every $x$ in $A$, $\psi_1(x) = 0$. Since $\{x : 0 \neq \psi_1(x)\} \subset \text{supp } \psi_1 \subset M - A$,
$A \subset \{x : 0 = \psi_1(x)\} = \{x : 0 = 1 - \psi(x)\} = \{x : \psi(x) = 1\}$.                                          $\square$

**Lemma 4.59** *Let $M$ be an $m$-dimensional smooth manifold. Let $A$ be a nonempty closed subset of $M$. Let $U$ be an open subset of $M$ such that $A \subset U$. Let $f : A \to \mathbb{R}^k$ be a smooth function (that is, for every $p$ in $A$, there exists an open neighborhood $W_p$ of $p$, and a smooth function $F_p : W_p \to \mathbb{R}^k$ such that $f|_{W_p \cap A} = F_p|_{W_p \cap A}$). Then, there exists a smooth function $\tilde{f} : M \to \mathbb{R}^k$ such that*

1. $\tilde{f}_A = f$,
2. supp $\tilde{f} \subset U$.

*Proof* Let us fix any $p$ in $A$. Since $f : A \to \mathbb{R}^k$ is a smooth function, and $p$ is in $A$, there exists an open neighborhood $V_p$ of $p$, and a smooth function $F_p : V_p \to \mathbb{R}^k$ such that $f|_{V_p \cap A} = F_p|_{V_p \cap A}$. Since $p$ is in $A$, and $A \subset U$, $p$ is in $U$. Since $p$ is in $U$, and $U$ is open, $U$ is an open neighborhood of $p$. Since $U$ is an open neighborhood of $p$, and $V_p$ is an open neighborhood of $p$, $V_p \cap U$ is an open neighborhood of $p$.

Clearly, the collection $\{V_p \cap U : p \in A\} \cup \{M - A\}$ is an open cover of smooth manifold $M$. By the partitions of unity theorem, then there exist a collection $\{\psi_p : p \in A\} \cup \{\psi\}$ of smooth functions $\psi_p : M \to [0, 1]$, and $\psi : M \to [0, 1]$ such that

(a)  for each $p \in A$, supp $\psi_p$ is contained in $V_p \cap U$; supp $\psi$ is contained in $M - A$,
(b)  $\{$supp $\psi_p : p \in A\} \cup \{$supp $\psi\}$ is locally finite,
(c)  $\sum_{p \in A} \psi_p = 1 - \psi$.

Since for every $p \in A$, $\psi_p : M \to [0, 1]$ is smooth, and $V_p \cap U$ is an open neighborhood of $p$, $\psi_p|_{V_p \cap U}$ is smooth. Since $F_p : V_p \to \mathbb{R}^k$ is smooth, $V_p$ is an open neighborhood of $p$, $V_p \cap U$ is an open neighborhood of $p$, and $V_p \cap U \subset V_p$, $F_p|_{V_p \cap U}$ is smooth. Since $\psi_p|_{V_p \cap U}$ is smooth, $F_p|_{V_p \cap U}$ is smooth, and $V_p \cap U$ is an open neighborhood of $p$, $(\psi_p|_{V_p \cap U})(F_p|_{V_p \cap U})$ is smooth on $V_p \cap U$. For every $p$ in $A$, supp $\psi_p \subset V_p \cap U$, so $M - (V_p \cap U) \subset M - ($supp $\psi_p)$. If $t \in (V_p \cap U) \cap (M - ($supp $\psi_p))$ then $t \notin ($supp $\psi_p)$, and hence $(\psi_p|_{V_p \cap U})(t) = 0$. Thus, for every $t \in (V_p \cap U) \cap (M - ($supp $\psi_p))$, we have

$$\left( \left( \psi_p|_{V_p \cap U} \right) \left( F_p|_{V_p \cap U} \right) \right)(t) = \left( \left( \psi_p|_{V_p \cap U} \right)(t) \right) \left( \left( F_p|_{V_p \cap U} \right)(t) \right)$$
$$= 0 \left( \left( F_p|_{V_p \cap U} \right)(t) \right) = 0.$$

It follows that we can define a function $\chi_p : M \to \mathbb{R}^k$ as follows: For every $x$ in $M$,

$$\chi_p(x) \equiv \begin{cases} \left( \left( \psi_p|_{V_p \cap U} \right) \left( F_p|_{V_p \cap U} \right) \right)(x) & \text{if } x \in V_p \cap U \\ 0 & \text{if } x \in M - \left( \text{supp } \psi_p \right). \end{cases}$$

Since $M - ($supp $\psi_p) \subset \{x : \chi_p(x) = 0\}$, $\{x : \chi_p(x) \neq 0\} \subset$ supp $\psi_p$, and hence, supp $\chi_p = \{x : \chi_p(x) \neq 0\}^- \subset ($supp $\psi_p)^- =$ supp $\psi_p$. Thus, for every $p \in A$, supp $\chi_p \subset$ supp $\psi_p$. Since $\{$supp $\psi_p : p \in A\} \cup \{$supp $\psi\}$ is locally finite, its subcollection $\{$supp $\psi_p : p \in A\}$ is also locally finite. Since $\{$supp $\psi_p : p \in A\}$ is locally finite, and for every $p \in A$, supp $\chi_p \subset$ supp $\psi_p$, $\{$supp $\chi_p : p \in A\}$ is locally finite.

Further, since $V_p \cap U$ is open, $M - ($supp $\psi_p)$ is open, $(\psi_p|_{V_p \cap U})(F_p|_{V_p \cap U})$ is smooth on $V_p \cap U$, and $0$ is smooth on $M - ($supp $\psi_p)$, by the definition of $\chi_p$, $\chi_p : M \to \mathbb{R}^k$ is smooth.

Let us take any $x$ in $M$. Since $\{$supp $\psi_p : p \in A\}$ is also locally finite, and $x$ is in $M$, there exists an open neighborhood $G_x$ of $x$ such that all but finite many supp $\psi_p$'s are outside $G_x$. Since all but finite many supp $\psi_p$'s are outside $G_x$, there exists $p_1, p_2, \ldots, p_n \in A$ such that $(\sum_{p \in A} \chi_p)|_{G_x} = (\chi_{p_1} + \chi_{p_2} + \cdots + \chi_{p_n})|_{G_x}$. Since $\chi_{p_1}, \chi_{p_2}, \ldots, \chi_{p_n}$ are smooth, and $G_x$ is an open neighborhood of $x$, $(\chi_{p_1} + \chi_{p_2} + \cdots + \chi_{p_n})|_{G_x}$ is smooth on $G_x$. Since $(\chi_{p_1} + \chi_{p_2} + \cdots + \chi_{p_n})|_{G_x}$ is smooth on $G_x$, and $(\sum_{p \in A} \chi_p)|_{G_x} = (\chi_{p_1} + \chi_{p_2} + \cdots + \chi_{p_n})|_{G_x}$, $(\sum_{p \in A} \chi_p)|_{G_x}$ is smooth on $G_x$. It follows that $y \mapsto (\sum_{p \in A} \chi_p)(y)$ is smooth.

Now, we can define a function $\tilde{f} : M \to \mathbb{R}^k$ as follows:

$$\tilde{f} \equiv \sum_{p \in A} \chi_p.$$

Clearly, $\tilde{f}$ is smooth.

For 1: Let us fix any $x \in A$. We have to show that $\tilde{f}(x) = f(x)$.

Since $\{y : \psi(y) \neq 0\} \subset \text{supp } \psi \subset M - A$, $x \in A \subset \{y : \psi(y) = 0\}$, and hence $\psi(x) = 0$. Since $\text{supp } \psi_p \subset V_p \cap U$, $M - (V_p \cap U) \subset M - (\text{supp } \psi_p)$. If $x \notin V_p \cap U$, then $x \in M - (V_p \cap U) \subset M - (\text{supp } \psi_p)$, and hence $\chi_p(x) = 0$. Also if $x \notin V_p \cap U$ then $x \notin \text{supp } \psi_p = \{y : \psi_p(y) \neq 0\}^- \supset \{y : \psi_p(y) \neq 0\}$, and hence $\psi_p(x) = 0$. Thus, if $x \notin V_p \cap U$, then $\chi_p(x) = 0 = \psi_p(x)$. Now,

$$\tilde{f}(x) = \left(\sum_{p \in A} \chi_p\right)(x) = \sum_{\{p : p \in A\}} \chi_p(x) = \sum_{\{p : p \in A, x \in V_p \cap U\}} \chi_p(x) + \sum_{\{p : p \in A, x \notin V_p \cap U\}} \chi_p(x)$$

$$= \sum_{\{p : p \in A, x \in V_p \cap U\}} \chi_p(x) + \sum_{\{p : p \in A, x \notin V_p \cap U\}} 0 = \sum_{\{p : p \in A, x \in V_p \cap U\}} \chi_p(x)$$

$$= \sum_{\{p : p \in A, x \in V_p \cap U\}} \left(\left(\psi_p|_{V_p \cap U}\right)\left(F_p|_{V_p \cap U}\right)\right)(x) = \sum_{\{p : p \in A, x \in V_p \cap U\}} \left(\left(\psi_p|_{V_p \cap U}\right)(x)\right)\left(\left(F_p|_{V_p \cap U}\right)(x)\right)$$

$$= \sum_{\{p : p \in A, x \in V_p \cap U\}} \left(\left(\psi_p|_{V_p \cap U}\right)(x)\right)(F_p(x)) = \sum_{\{p : p \in A, x \in V_p \cap U\}} \left(\left(\psi_p|_{V_p \cap U}\right)(x)\right)(f(x))$$

$$= \left(\sum_{\{p : p \in A, x \in V_p \cap U\}} \left(\psi_p|_{V_p \cap U}\right)(x)\right)(f(x)) = \left(\sum_{\{p : p \in A, x \in V_p \cap U\}} \psi_p(x)\right)(f(x))$$

$$= \left(\sum_{\{p : p \in A, x \in V_p \cap U\}} \psi_p(x) + \sum_{\{p : p \in A, x \notin V_p \cap U\}} 0\right)(f(x))$$

$$= \left(\sum_{\{p : p \in A, x \in V_p \cap U\}} \psi_p(x) + \sum_{\{p : p \in A, x \notin V_p \cap U\}} \psi_p(x)\right)(f(x)) = \left(\sum_{\{p : p \in A\}} \psi_p(x)\right)(f(x))$$

$$= \left(\left(\sum_{\{p : p \in A\}} \psi_p\right)(x)\right)(f(x)) = ((1 - \psi)(x))(f(x)) = (1 - \psi(x))(f(x)) = (1 - 0)(f(x)) = f(x).$$

Thus, $\tilde{f}(x) = f(x)$. So, for any $x \in A$ we have $\tilde{f}(x) = f(x)$. Thus, $\tilde{f}|_A = f$. This proves 1.

For 2: Let us take any $y \in M$ such that $\tilde{f}(y) \neq 0$. Here,

$$
\begin{aligned}
0 \neq \tilde{f}(y) &= \left(\sum_{p \in A} \chi_p\right)(y) = \sum_{\{p : p \in A\}} \chi_p(y) = \sum_{\{p : p \in A, y \in V_p \cap U\}} \chi_p(y) + \sum_{\{p : p \in A, y \notin V_p \cap U\}} \chi_p(y) \\
&= \sum_{\{p : p \in A, y \in V_p \cap U\}} \chi_p(y) + \sum_{\{p : p \in A, y \notin V_p \cap U\}} 0 = \sum_{\{p : p \in A, y \in V_p \cap U\}} \chi_p(y) \\
&= \sum_{\{p : p \in A, y \in V_p \cap U\}} \left(\left(\psi_p|_{V_p \cap U}\right)\left(F_p|_{V_p \cap U}\right)\right)(y) = \sum_{\{p : p \in A, y \in V_p \cap U\}} \left(\left(\psi_p|_{V_p \cap U}\right)(y)\right)\left(\left(F_p|_{V_p \cap U}\right)(y)\right) \\
&= \sum_{\{p : p \in A, y \in V_p \cap U\}} \left(\psi_p(y)\right)\left(F_p(x)\right).
\end{aligned}
$$

Since $\sum_{\{p : p \in A, y \in V_p \cap U\}} (\psi_p(y))(F_p(x)) \neq 0$, there exists $p \in A$ such that $\psi_p(y) \neq 0$. It follows that $\{y : \tilde{f}(y) \neq 0\} \subset \cup_{p \in A}\{y : \psi_p(y) \neq 0\}$, and hence, $\operatorname{supp} \tilde{f} = \{y : \tilde{f}(y) \neq 0\}^{-} \subset (\cup_{p \in A}\{y : \psi_p(y) \neq 0\})^{-}$. Since $\{\operatorname{supp} \psi_p : p \in A\}$ is locally finite, $(\cup_{p \in A}\{y : \psi_p(y) \neq 0\})^{-} = \cup_{p \in A}(\{y : \psi_p(y) \neq 0\}^{-})$. It follows that $\operatorname{supp} \tilde{f} \subset \cup_{p \in A}(\{y : \psi_p(y) \neq 0\}^{-}) = \cup_{p \in A}(\operatorname{supp} \psi_p) \subset \cup_{p \in A}(V_p \cap U) = (\cup_{p \in A} V_p) \cap U \subset U$. Thus, $\operatorname{supp} \tilde{f} \subset U$. This proves 2. $\qquad \square$

## 4.6 Topological Manifolds With Boundary

**Definition** By $\mathbb{H}^3$ we mean the set $\{(x^1, x^2, x^3) : (x^1, x^2, x^3) \in \mathbb{R}^3 \text{ and } 0 \leq x^3\}$, and is called the *closed 3-dimensional upper half-space*. By $\operatorname{Int} \mathbb{H}^3$ we mean $\{(x^1, x^2, x^3) : (x^1, x^2, x^3) \in \mathbb{R}^3 \text{ and } 0 < x^3\}$. By $\partial \mathbb{H}^3$ we mean $\{(x^1, x^2, 0) : (x^1, x^2, 0) \in \mathbb{R}^3\}$. Observe that the collection of all open subsets of $\mathbb{H}^3$ which do not intersect $\partial \mathbb{H}^3$ is contained in the collection of all open subsets of $\mathbb{R}^3$.

Let $M$ be a topological space, whose topology is Hausdorff, and second countable. If for every $p$ in $M$, there exist an open neighborhood $U$ of $p$, "a subset $G$ of $\mathbb{H}^3$ which is open in $\mathbb{H}^3$ or a subset $G$ of $\mathbb{R}^3$ which is open in $\mathbb{R}^3$," and a homeomorphism $\varphi$ from $U$ onto $G$, then we say that *M is a 3-dimensional topological manifold with boundary*. Here, the ordered pair $(U, \varphi)$ is called a *chart of M*. If $\varphi(U)(= G)$ is open in $\mathbb{R}^3$, then we say that $(U, \varphi)$ is an *interior chart of M*. If $\varphi(U) (= G)$ is a subset of $\mathbb{H}^3$, $\varphi(U)$ is open in $\mathbb{H}^3$, and $\varphi(U) \cap \partial \mathbb{H}^3 \neq \emptyset$, then we say that $(U, \varphi)$ is a *boundary chart of M*.

Let $p \in M$. By $p$ *is an interior point of* $M$, we mean: There exists an interior chart $(U, \varphi)$ of $M$ such that $p \in U$. By $p$ *is a boundary point of* $M$, we mean: There exists a boundary chart $(U, \varphi)$ of $M$ such that $p \in U$, and $\varphi(p) \in \partial \mathbb{H}^3$. By Int $M$ we mean the set of all interior points of $M$, and is called the *interior of* $M$. By $\partial M$ we mean the set of all boundary points of $M$, and is called the *boundary of* $M$.

Similar definitions can be supplied for 4-dimensional topological manifolds with boundary, etc.

**Note 4.60** Let $M$ be a 3-dimensional topological manifold with boundary. We want to prove:

1. $M = (\text{Int } M) \cup (\partial M)$
2. $(\text{Int } M) \cap (\partial M) = \emptyset$.

For 1: Clearly, $(\text{Int } M) \subset M$, and $(\partial M) \subset M$. So, $(\text{Int } M) \cup (\partial M) \subset M$. It remains to prove that $M \subset (\text{Int } M) \cup (\partial M)$. If not, otherwise, let there exists $p \in M$ such that $p \notin \text{Int } M$, and $p \notin \partial M$. We have to arrive at a contradiction.

Since $p \in M$, and $M$ is a 3-dimensional topological manifold with boundary, there exist an open neighborhood $U$ of $p$, "a subset $G$ of $\mathbb{H}^3$ which is open in $\mathbb{H}^3$ or a subset $G$ of $\mathbb{R}^3$ which is open in $\mathbb{R}^3$," and a homeomorphism $\varphi$ from $U$ onto $G$. Thus, $(U, \varphi)$ is a chart for $M$. Since $p \notin \text{Int } M$, $p$ is not an interior point of $M$. Since $p$ is not an interior point of $M$, $(U, \varphi)$ is a chart for $M$, and $p \in U$, $(U, \varphi)$ is not an interior chart of $M$. Since $(U, \varphi)$ is not an interior chart of $M$, $\varphi(U) (= G)$ is not open in $\mathbb{R}^3$, and hence, $G$ is not open in $\mathbb{R}^3$. Since $G$ is not open in $\mathbb{R}^3$, and "$G$ is a subset of $\mathbb{H}^3$ which is open in $\mathbb{H}^3$ or $G$ is a subset of $\mathbb{R}^3$ which is open in $\mathbb{R}^3$," $G$ is a subset of $\mathbb{H}^3$ which is open in $\mathbb{H}^3$. Since $p \in U$, $\varphi(p) \in \varphi(U) (= G)$. Since $G$ is not open in $\mathbb{R}^3$, $G$ is a subset of $\mathbb{H}^3$, and $G$ is open in $\mathbb{H}^3$, $G \cap \partial \mathbb{H}^3 \neq \emptyset$.

(Reason: If not, otherwise, let $G \cap \partial \mathbb{H}^3 = \emptyset$. Since $G \subset \mathbb{H}^3 = \text{Int } \mathbb{H}^3 \cup \partial \mathbb{H}^3$, and $G \cap \partial \mathbb{H}^3 = \emptyset$, $G \subset \text{Int } \mathbb{H}^3$. Since $G$ is open in $\mathbb{H}^3$, there exists an open set $G_1$ in $\mathbb{R}^3$ such that $G_1 \cap \mathbb{H}^3 = G$. Since $G = G_1 \cap \mathbb{H}^3 \subset G_1$, and $G \subset \text{Int } \mathbb{H}^3$, $G \subset G_1 \cap \text{Int } \mathbb{H}^3$. Since $G_1 \cap \text{Int } \mathbb{H}^3 \subset \text{Int } \mathbb{H}^3 \subset \mathbb{H}^3$, and $G_1 \cap \text{Int } \mathbb{H}^3 \subset G_1$, $G_1 \cap \text{Int } \mathbb{H}^3 \subset G_1 \cap \mathbb{H}^3 = G$. Since $G_1 \cap \text{Int } \mathbb{H}^3 \subset G$, and $G \subset G_1 \cap \text{Int } \mathbb{H}^3$, $G = G_1 \cap \text{Int } \mathbb{H}^3$. Since $G_1$, $\text{Int } \mathbb{H}^3$ are open in $\mathbb{R}^3$, their intersection $G_1 \cap \text{Int } \mathbb{H}^3 (= G)$ is open in $\mathbb{R}^3$, and hence, $G$ is open in $\mathbb{R}^3$, a contradiction.)

Since $\varphi(U)(= G)$ is a subset of $\mathbb{H}^3$ which is open in $\mathbb{H}^3$, and $\varphi(U) \cap \partial \mathbb{H}^3 \neq \emptyset$, $(U, \varphi)$ is a boundary chart for $M$. Since $p \notin \partial M$, $p$ is not a boundary point of $M$. Since $p$ is not a boundary point of $M$, $(U, \varphi)$ is a chart for $M$ satisfying $p \in U$, and $(U, \varphi)$ is a boundary chart for $M$, $\varphi(p) \notin \partial \mathbb{H}^3$. Clearly, $\partial \mathbb{H}^3 (= \{(x^1, x^2, 0) : (x^1, x^2, 0) \in \mathbb{R}^3\})$ is closed in $\mathbb{H}^3$. It follows that $G - \partial \mathbb{H}^3$ is open in $\mathbb{H}^3$. Since

$\varphi(p) \notin \partial \mathbb{H}^3$, $\varphi(p) \in G$, and $G - \partial \mathbb{H}^3$ is open in $\mathbb{H}^3$, there exists an open neighborhood $H_{\varphi(p)}$ of $\varphi(p)$ in $\mathbb{H}^3$ such that $H_{\varphi(p)} \subset G - \partial \mathbb{H}^3 \subset G$, and hence $H_{\varphi(p)} \cap \partial \mathbb{H}^3 = \emptyset$. Since $H_{\varphi(p)} \cap \partial \mathbb{H}^3 = \emptyset$, and $H_{\varphi(p)}$ is open in $\mathbb{H}^3$, $H_{\varphi(p)}$ is open in $\mathbb{R}^3$. Since $\varphi$ is a homeomorphism from $U$ onto $G$, $G$ is open in $\mathbb{H}^3$, $H_{\varphi(p)} \subset G$, and $H_{\varphi(p)}$ is an open neighborhood of $\varphi(p)$ in $\mathbb{H}^3$, $\varphi^{-1}(H_{\varphi(p)})$ is an open neighborhood of $p$ in $M$, and $\varphi|_{\varphi^{-1}(H_{\varphi(p)})}$ is a homeomorphism from $\varphi^{-1}(H_{\varphi(p)})$ onto $H_{\varphi(p)}$. Thus, the ordered pair $(\varphi^{-1}(H_{\varphi(p)}), \varphi|_{\varphi^{-1}(H_{\varphi(p)})})$ is a chart for $M$. Now, since $(\varphi|_{\varphi^{-1}(H_{\varphi(p)})})$ $(\varphi^{-1}(H_{\varphi(p)})) = H_{\varphi(p)}$, and $H_{\varphi(p)}$ is open in $\mathbb{R}^3$, the chart $(\varphi^{-1}(H_{\varphi(p)}), \varphi|_{\varphi^{-1}(H_{\varphi(p)})})$ is an interior chart for $M$. Since $(\varphi^{-1}(H_{\varphi(p)}), \varphi|_{\varphi^{-1}(H_{\varphi(p)})})$ is an interior chart for $M$, and $p \in \varphi^{-1}(H_{\varphi(p)})$, $p$ is an interior point of $M$, and hence $p \in \text{Int } M$, a contradiction.

For 2: Its proof is postponed right now. This result is known as the *topological invariance of the boundary.*

Similar results can be supplied for 4-dimensional topological manifolds with boundary, etc.

**Note 4.61** Let $M$ be a topological $n$-manifold with boundary. We want to show that Int $M$ is an open subset of $M$. Here,

Int $M = \{p : p \in M$, there exists a chart$(U, \varphi)$ of $M$ such that $p \in U$, and $\varphi(U)$ is open in $\mathbb{R}^n\}$.

For this purpose, let us take any $p \in \text{Int } M$. We have to find an open neighborhood $U_p$ such that if $q \in U_p$ then $q \in \text{Int } M$. Since $p \in \text{Int } M$, $p \in M$, there exists a chart $(U, \varphi)$ of $M$ such that $p \in U$, and $\varphi(U)$ is open in $\mathbb{R}^n$. Since $(U, \varphi)$ is a chart, $U$ is open in $M$. Since $U$ is open in $M$, and $p \in U$, $U$ is an open neighborhood of $p$. Next, let us take any $q \in U$. Here, $(U, \varphi)$ is a chart of $M$ such that $q \in U$, and $\varphi(U)$ is open in $\mathbb{R}^n$. Hence, by definition of the Int $M$, $q \in \text{Int } M$.

This proves that Int $M$ is an open subset of $M$.

**Note 4.62** Clearly, a topological $n$-manifold $M$ "in the original sense" is a topological $n$-manifold $M$ with boundary.

If $M$ is a topological $n$-manifold $M$ "in the original sense," then Int $M = M$. Hence, by the topological invariance of the boundary, if $M$ is a topological $n$-manifold "in the original sense," then $\partial M = \emptyset$. Conversely, let $M$ be a topological $n$-manifold $M$ with boundary such that $\partial M = \emptyset$. Hence, $M = \text{Int } M \cup \partial M = \text{Int } M \cup \emptyset = \text{Int } M$. Since Int $M = M$, $M$ is a topological $n$-manifold "in the original sense."

Thus, we have shown that a topological $n$-manifold $M$ with boundary is a topological $n$-manifold if and only if $\partial M = \emptyset$. Sometimes, a manifold in the original sense is referred as *manifold without boundary*.

**Note 4.63** Let $M$ be a topological $n$-manifold with boundary. We want to show that Int $M$ is a topological $n$-manifold without boundary.

Here, Int $M$ is contained in $M$. The topology of Int $M$ is assumed to be the subspace topology of $M$. Since $M$ is a topological $n$-manifold with boundary, the topology of $M$ is Hausdorff, and second countable, and hence, the subspace topology over Int $M$ is Hausdorff, and second countable.

Next, let us take any $p \in$ Int $M$. Since $p \in$ Int $M$, $p$ is an interior point of $M$, and hence, there exists an interior chart $(U, \varphi)$ of $M$ such that $p \in U$. Since $(U, \varphi)$ is an internal chart of $M$, $(U, \varphi)$ is a chart of $M$, and $\varphi(U)$ is open in $\mathbb{R}^n$. Since $(U, \varphi)$ is a chart of $M$, $U$ is open in $M$, and $\varphi : U \to \varphi(U)$ is a homeomorphism. Since $U$ is open in $M$, and Int $M$ is open in $M$, $U \cap$ Int $M$ is open in the topological space Int $M$. Since $p \in U$, and $p \in$ Int $M$, $p \in U \cap$ Int $M$. Thus, $U \cap$ Int $M$ is an open neighborhood of $p$ in the topological space Int $M$. Since $U$ is open in $M$, and Int $M$ is open in $M$, $U \cap$ Int $M$ is open in $M$. Since $U \cap$ Int $M$, $U$ are open in $M$, and $U \cap$ Int $M \subset U$, $U \cap$ Int $M$ is open in $U$.

Since $\varphi : U \to \varphi(U)$ is a homeomorphism, and $U \cap$ Int $M$ is open in $U$, $\varphi(U \cap$ Int $M)$ is open in $\varphi(U)$. Since $\varphi(U \cap$ Int $M)$ is open in $\varphi(U)$, and $\varphi(U)$ is open in $\mathbb{R}^n$, $\varphi(U \cap$ Int $M)$ is open in $\mathbb{R}^n$. Since $U \cap$ Int $M$ is open in $U$, $\varphi(U \cap$ Int $M)$ is open in $\varphi(U)$, and $\varphi : U \to \varphi(U)$ is a homeomorphism, $\varphi|_{U \cap \text{Int}M}$ is a homeomorphism from $U \cap$ Int $M$ onto $\varphi(U \cap$ Int $M)$. Since $U \cap$ Int $M$ is an open neighborhood of $p$ in the topological space Int $M$, $\varphi(U \cap$ Int $M)$ is open in $\mathbb{R}^n$, and $\varphi|_{U \cap \text{Int}M}$ is a homeomorphism from $U \cap$ Int $M$ onto $\varphi(U \cap$ Int $M)$, the ordered pair $(U \cap$ Int $M, \varphi|_{U \cap \text{Int}M})$ is an interior chart of the topological space Int $M$. Also $p \in U \cap$ Int $M$. This shows that Int $M$ is a topological $n$-manifold without boundary.

**Note 4.64** Let $M$ be a topological 3-manifold with boundary. We want to show that

1. $\partial M$ is a closed subset of $M$,
2. $\partial M$ is a topological 2-manifold without boundary.

   For 1: Since Int $M$ is open in $M$, $M -$ Int $M$ is closed in $M$. By the topological invariance of the boundary, $\partial M = M -$ Int $M$. Since $\partial M = M -$ Int $M$, and $M -$ Int $M$ is closed in $M$, $\partial M$ is a closed subset of $M$.

   For 2: Here, $\partial M$ is contained in $M$. The topology of $\partial M$ is assumed to be the subspace topology of $M$. Since $M$ is a topological 3-manifold with boundary, the topology of $M$ is Hausdorff, and second countable, and hence, the subspace topology over $\partial M$ is Hausdorff, and second countable.

Next, let us fix any $p \in \partial M$. Since $p \in \partial M$, $p$ is a boundary point of $M$, and hence, there exists a boundary chart $(U, \varphi)$ of $M$ such that $p \in U$, and $\varphi(p) \in \partial \mathbb{H}^3 = \{(x^1, x^2, 0) : (x^1, x^2, 0) \in \mathbb{R}^3\}$. Since $(U, \varphi)$ is a boundary chart of $M$, $U$ is open in $M$, $\varphi(U)$ is open in $\mathbb{H}^3$, and $\varphi : U \to \varphi(U)$ is a homeomorphism. Since $p \in U$, $\varphi(p) \in \varphi(U)$. Since $\varphi(p) \in \varphi(U)$, and $\varphi(U)$ is open in $\mathbb{H}^3$, $\varphi(U)$ is an open neighborhood of $\varphi(p)$ in $\mathbb{H}^3$. Since $\varphi(U)$ is an open neighborhood of $\varphi(p)$ in $\mathbb{H}^3$, and $\varphi(p) \in \partial \mathbb{H}^3 \subset \mathbb{H}^3$, $\varphi(U) \cap \partial \mathbb{H}^3$ is an open neighborhood of $\varphi(p)$ in $\partial \mathbb{H}^3$. Since $\varphi : U \to \varphi(U)$ is a homeomorphism, and $\varphi(U) \cap \partial \mathbb{H}^3 \subset \varphi(U)$, the restriction $\varphi|_{\varphi^{-1}(\varphi(U) \cap \partial \mathbb{H}^3)}$ is a homeomorphism from $\varphi^{-1}(\varphi(U) \cap \partial \mathbb{H}^3)$ onto $\varphi(U) \cap \partial \mathbb{H}^3$. Since $\varphi(p) \in \varphi(U)$, and $\varphi(p) \in \partial \mathbb{H}^3$, $\varphi(p) \in \varphi(U) \cap \partial \mathbb{H}^3$, and hence $p \in \varphi^{-1}(\varphi(U) \cap \partial \mathbb{H}^3)$.

Now, we shall try to prove:

(a) $\varphi^{-1}(\varphi(U) \cap \partial \mathbb{H}^3) \subset \partial M$,
(b) $\varphi^{-1}(\varphi(U) \cap \partial \mathbb{H}^3)$ is open in $\partial M$,
(c) $\varphi(U) \cap \partial \mathbb{H}^3$ is open in $\partial \mathbb{H}^3$.

For a: Let us take any $a \in \varphi^{-1}(\varphi(U) \cap \partial \mathbb{H}^3)$. Since $a \in \varphi^{-1}(\varphi(U) \cap \partial \mathbb{H}^3)$, and $\varphi : U \to \varphi(U)$, $a \in U$. Since $a \in \varphi^{-1}(\varphi(U) \cap \partial \mathbb{H}^3)$, $\varphi(a) \in \varphi(U) \cap \partial \mathbb{H}^3$, and hence, $\varphi(a) \in \partial \mathbb{H}^3$. Since $(U, \varphi)$ is a boundary chart of $M$, $a \in U$, and $\varphi(a) \in \partial \mathbb{H}^3$, $a$ is a boundary point of $M$, and hence $a \in \partial M$.

For b: First of all, we want to prove that $\varphi^{-1}(\varphi(U) \cap \partial \mathbb{H}^3) = U \cap \partial M$. For this purpose, let us take any $a$ in $\varphi^{-1}(\varphi(U) \cap \partial \mathbb{H}^3)$. Since $a \in \varphi^{-1}(\varphi(U) \cap \partial \mathbb{H}^3)$, and $\varphi : U \to \varphi(U)$, $a \in U$. Hence, $\varphi^{-1}(\varphi(U) \cap \partial \mathbb{H}^3) \subset U$. Since $\varphi^{-1}(\varphi(U) \cap \partial \mathbb{H}^3) \subset U$, and from (a), $\varphi^{-1}(\varphi(U) \cap \partial \mathbb{H}^3) \subset \partial M$, $\varphi^{-1}(\varphi(U) \cap \partial \mathbb{H}^3) \subset U \cap \partial M$.

Next, let us take any $b$ in $U \cap \partial M$. Since $b$ is in $U \cap \partial M$, $b \in U$, and hence, $\varphi(b) \in \varphi(U)$. Clearly, $\varphi(b) \in \partial \mathbb{H}^3$.

(Reason: If not, otherwise, let $\varphi(b) \notin \partial \mathbb{H}^3$. We have to arrive at a contradiction. Since $b$ is in $U \cap \partial M$, $b$ is in $U$. Since $b$ is in $U$, and $\varphi : U \to \varphi(U)$, $\varphi(b) \in \varphi(U)$. Since $\varphi(b) \in \varphi(U)$, and $\varphi(U)$ is open in $\mathbb{H}^3$, $\varphi(b) \in \mathbb{H}^3 = \text{Int } \mathbb{H}^3 \cup \partial \mathbb{H}^3$. Since $\varphi(b) \in \text{Int } \mathbb{H}^3 \cup \partial \mathbb{H}^3$, and $\varphi(b) \notin \partial \mathbb{H}^3$, $\varphi(b) \in \text{Int } \mathbb{H}^3 = \{(x^1, x^2, x^3) : (x^1, x^2, x^3) \in \mathbb{R}^3 \text{ and } 0 < x^3\}$. Since $\varphi(b) \in \{(x^1, x^2, x^3) : (x^1, x^2, x^3) \in \mathbb{R}^3 \text{ and } 0 < x^3\}$, $\varphi(b) \in \varphi(U)$, and $\varphi(U)$ is open in $\mathbb{H}^3$, there exists a real number $r > 0$ such that the open ball $B_r(\varphi(b))$ is contained in $\varphi(U) \cap \text{Int } \mathbb{H}^3$. It follows that $B_r(\varphi(b))$ is open in $\varphi(U)$. Since $B_r(\varphi(b))$ is open in $\varphi(U)$, and $\varphi : U \to \varphi(U)$ is a homeomorphism, $\varphi^{-1}(B_r(\varphi(b)))$ is open in $U$, and the restriction $\varphi|_{\varphi^{-1}(B_r(\varphi(b)))}$ is a homeomorphism from $\varphi^{-1}(B_r(\varphi(b)))$ onto $B_r(\varphi(b))$. Since $\varphi(b) \in B_r(\varphi(b))$, $b \in \varphi^{-1}(B_r(\varphi(b)))$. Since $\varphi^{-1}(B_r(\varphi(b)))$ is open in $U$, and $U$ is open in $M$, $\varphi^{-1}(B_r(\varphi(b)))$ is open in $M$. Since $\varphi^{-1}(B_r(\varphi(b)))$ is open in $M$, and $b \in \varphi^{-1}(B_r(\varphi(b)))$, $\varphi^{-1}(B_r(\varphi(b)))$ is an open neighborhood of $b$ in $M$. It follows that the ordered pair

$(\varphi^{-1}(B_r(\varphi(b))), \varphi|_{\varphi^{-1}(B_r(\varphi(b)))})$ is a chart of $M$. Also since $b \in \varphi^{-1}(B_r(\varphi(b)))$, $(\varphi|_{\varphi^{-1}(B_r(\varphi(b)))})(\varphi^{-1}(B_r(\varphi(b)))) = B_r(\varphi(b))$, and $B_r(\varphi(b))$ is open in $\mathbb{R}^3$, $(\varphi^{-1}(B_r(\varphi(b))), \varphi|_{\varphi^{-1}(B_r(\varphi(b)))})$ is an interior chart of $M$, and hence $b \in \text{Int } M$. Since $b \in \text{Int } M$, by the topological invariance of the boundary, $b \notin \partial M$, and hence, $b \notin U \cap \partial M$, a contradiction.)

Since $\varphi(b) \in \partial \mathbb{H}^3$, and $\varphi(b) \in \varphi(U)$, $\varphi(b) \in \varphi(U) \cap \partial \mathbb{H}^3$, and hence, $b \in \varphi^{-1}(\varphi(U) \cap \partial \mathbb{H}^3)$. Thus, $U \cap \partial M \subset \varphi^{-1}(\varphi(U) \cap \partial \mathbb{H}^3)$.

Thus, we have shown that $\varphi^{-1}(\varphi(U) \cap \partial \mathbb{H}^3) = U \cap \partial M$. Since $U$ is open in $M$, and $\partial M \subset M$, $U \cap \partial M (= \varphi^{-1}(\varphi(U) \cap \partial \mathbb{H}^3))$ is open in $\partial M$, and hence, $\varphi^{-1}(\varphi(U) \cap \partial \mathbb{H}^3)$ is open in $\partial M$. This proves (b).

For c: Since $\varphi(U)$ is open in $\mathbb{H}^3$, and $\partial \mathbb{H}^3 \subset \mathbb{H}^3$, $\varphi(U) \cap \partial \mathbb{H}^3$ is open in $\partial \mathbb{H}^3$. This proves (c).

Since the mapping $\pi_3 : \partial \mathbb{H}^3 \to \mathbb{R}^2$ defined by $\pi_3(x^1, x^2, 0) \equiv (x^1, x^2)$ is a homeomorphism, and $\varphi|_{\varphi^{-1}(\varphi(U) \cap \partial \mathbb{H}^3)}$ is a homeomorphism from $\varphi^{-1}(\varphi(U) \cap \partial \mathbb{H}^3)$ onto $\varphi(U) \cap \partial \mathbb{H}^3$, their composite $\pi_3 \circ (\varphi|_{\varphi^{-1}(\varphi(U) \cap \partial \mathbb{H}^3)})$ is a homeomorphism from $\varphi^{-1}(\varphi(U) \cap \partial \mathbb{H}^3)$ onto $\pi_3(\varphi(U) \cap \partial \mathbb{H}^3)$. Since $\varphi(U) \cap \partial \mathbb{H}^3$ is open in $\partial \mathbb{H}^3$, and $\pi_3 : \partial \mathbb{H}^3 \to \mathbb{R}^2$ is a homeomorphism, $\pi_3(\varphi(U) \cap \partial \mathbb{H}^3)$ is open in $\mathbb{R}^2$. Here, $\pi_3 \circ (\varphi|_{\varphi^{-1}(\varphi(U) \cap \partial \mathbb{H}^3)})$ is a homeomorphism from $\varphi^{-1}(\varphi(U) \cap \partial \mathbb{H}^3)$ onto $\pi_3(\varphi(U) \cap \partial \mathbb{H}^3)$, $\varphi^{-1}(\varphi(U) \cap \partial \mathbb{H}^3)$ is open in $\partial M$, $p \in \varphi^{-1}(\varphi(U) \cap \partial \mathbb{H}^3)$, and $\pi_3(\varphi(U) \cap \partial \mathbb{H}^3)$ is open in $\mathbb{R}^2$, the ordered pair $(\varphi^{-1}(\varphi(U) \cap \partial \mathbb{H}^3), )\pi_3 \circ (\varphi|_{\varphi^{-1}(\varphi(U) \cap \partial \mathbb{H}^3)})$ is an internal chart of $\partial M$, where $p \in \varphi^{-1}(\varphi(U) \cap \partial \mathbb{H}^3)$. Hence, $\partial M$ is a topological 2-manifold without boundary.

This completes the proof of 2.

Similar results can be supplied for 4-dimensional topological manifolds with boundary, etc.

**Definition** Let $A$ be a nonempty subset of $\mathbb{R}^n$. Let $f : A \to \mathbb{R}^m$. By $f$ *is smooth* we mean: For every $x$ in $A$, there exist an open neighborhood $V_x$ of $x$ in $\mathbb{R}^n$, and a function $f_x : V_x \to \mathbb{R}^m$ such that $f_x|_{A \cap V_x} = f|_{A \cap V_x}$, and $f_x$ is smooth.

**Definition** Let $G$ be a nonempty subset of $\mathbb{H}^n$. Let $G$ be open in $\mathbb{H}^n$. Let $f : G \to \mathbb{R}^m$. By $f$ *is smooth* we mean: For every $x$ in $G$, there exist an open neighborhood $V_x$ of $x$ in $\mathbb{R}^n$, and a function $f_x : V_x \to \mathbb{R}^m$ such that $f_x|_{G \cap V_x} = f|_{G \cap V_x}$, and $f_x$ is smooth in the usual sense.

**Note 4.65** Let $G$ be a nonempty subset of $\mathbb{H}^n$. Let $G$ be open in $\mathbb{H}^n$. Let $f : G \to \mathbb{R}^m$. Let $f$ be smooth.

We first show that $G \cap (\text{Int } \mathbb{H}^n)$ is open in $\mathbb{R}^n$. Since $G$ is open in $\mathbb{H}^n$, there exists an open subset $U$ of $\mathbb{R}^n$ such that $G = U \cap \mathbb{H}^n$. Now, $G \cap (\text{Int } \mathbb{H}^n) = (U \cap \mathbb{H}^n) \cap (\text{Int } \mathbb{H}^n) = U \cap (\mathbb{H}^n \cap \text{Int } \mathbb{H}^n) = U \cap \text{Int } \mathbb{H}^n$. Since $U, \text{Int } \mathbb{H}^n$ are open in $\mathbb{R}^n$, $U \cap \text{Int } \mathbb{H}^n (= G \cap (\text{Int } \mathbb{H}^n))$ is open in $\mathbb{R}^n$, and hence, $G \cap (\text{Int } \mathbb{H}^n)$ is open in $\mathbb{R}^n$.

Since $G \cap (\text{Int } \mathbb{H}^n)(\subset G)$ is open in $\mathbb{R}^n$, and $f : G \to \mathbb{R}^m$, it is meaningful to say "$f|_{G \cap (\text{Int}\mathbb{H}^n)}$ is smooth in the usual sense." We want to show that $f|_{G \cap (\text{Int}\mathbb{H}^n)}$ is smooth in the usual sense. For this purpose, let us take any $a$ in $G \cap (\text{Int } \mathbb{H}^n)$. We have to show that $f|_{G \cap (\text{Int}\mathbb{H}^n)}$ is smooth at $a$ in the usual sense.

Since $a$ is in $G \cap (\text{Int } \mathbb{H}^n)$, $a$ is in $G$. Since $a$ is in $G$, and $f : G \to \mathbb{R}^m$ is smooth, there exist an open neighborhood $V_a$ of $a$ in $\mathbb{R}^n$, and a function $f_a : V_a \to \mathbb{R}^m$ such that $f_a|_{G \cap V_a} = f|_{G \cap V_a}$, and $f_a$ is smooth in the usual sense. Since $a$ is in $G \cap (\text{Int } \mathbb{H}^n)$, and $G \cap (\text{Int } \mathbb{H}^n)$ is open in $\mathbb{R}^n$, $G \cap (\text{Int } \mathbb{H}^n)$ is an open neighborhood of $a$ in $\mathbb{R}^n$. Since $G \cap (\text{Int } \mathbb{H}^n)$ is an open neighborhood of $a$ in $\mathbb{R}^n$, and $V_a$ is an open neighborhood of $a$ in $\mathbb{R}^n$, $G \cap (\text{Int } \mathbb{H}^n) \cap V_a$ is an open neighborhood of $a$ in $\mathbb{R}^n$, and is contained in the open set $V_a$. Now, since $f_a : V_a \to \mathbb{R}^m$ is smooth in the usual sense, the restriction $f_a|_{G \cap (\text{Int}\mathbb{H}^n) \cap V_a}$ is smooth at $a$ in the usual sense. Since $f_a|_{G \cap V_a} = f|_{G \cap V_a}$, $G \cap (\text{Int } \mathbb{H}^n) \cap V_a \subset G \cap V_a$, $f_a|_{G \cap (\text{Int}\mathbb{H}^n) \cap V_a} = f|_{G \cap (\text{Int}\mathbb{H}^n) \cap V_a}$. Since $f_a : V_a \to \mathbb{R}^m$ is smooth in the usual sense, $G \cap (\text{Int } \mathbb{H}^n) \cap V_a$ is an open neighborhood of $a$ in $\mathbb{R}^n$, and is contained in the open set $V_a$, $f_a|_{G \cap (\text{Int}\mathbb{H}^n) \cap V_a}$ is smooth at $a$ in the usual sense. Since $f_a|_{G \cap (\text{Int}\mathbb{H}^n) \cap V_a}$ is smooth at $a$ in the usual sense, and $f_a|_{G \cap (\text{Int}\mathbb{H}^n) \cap V_a} = f|_{G \cap (\text{Int}\mathbb{H}^n) \cap V_a}$, $f|_{G \cap (\text{Int}\mathbb{H}^n) \cap V_a}$ is smooth at $a$ in the usual sense. Since $f|_{G \cap (\text{Int}\mathbb{H}^n) \cap V_a}$ is smooth at $a$ in the usual sense, $G \cap (\text{Int } \mathbb{H}^n) \cap V_a$ is an open neighborhood of $a$ in $\mathbb{R}^n$, $G \cap (\text{Int } \mathbb{H}^n)$ is an open neighborhood of $a$ in $\mathbb{R}^n$, and $G \cap (\text{Int } \mathbb{H}^n) \cap V_a$ is contained in $G \cap (\text{Int } \mathbb{H}^n)$, $f|_{G \cap (\text{Int}\mathbb{H}^n)}$ is smooth at $a$ in the usual sense.

Thus, we have shown that if $G$ is open in $\mathbb{H}^n$, and $f : G \to \mathbb{R}^m$ is smooth, then $f|_{G \cap (\text{Int}\mathbb{H}^n)}$ is smooth in the usual sense.

**Note 4.66** Let $G$ be a nonempty subset of $\mathbb{H}^3$. Let $G$ be open in $\mathbb{H}^3$. Let $f : G \to \mathbb{R}^m$ be smooth.

We want to prove that $f : G \to \mathbb{R}^m$ is continuous. For this purpose, let us take any $a$ in $G$. Now, let us take any open neighborhood $V_{f(a)}$ of $f(a)$ in $\mathbb{R}^m$. We have to find an open neighborhood $U_a$ of $a$ in $\mathbb{R}^3$ such that

1. $U_a \cap \mathbb{H}^3 \subset G$,
2. $f(U_a \cap \mathbb{H}^3) \subset V_{f(a)}$.

Since $G$ is open in $\mathbb{H}^3$, there exists an open subset $W_1$ of $\mathbb{R}^3$ such that $G = W_1 \cap \mathbb{H}^3$. Since $a$ is in $G(= W_1 \cap \mathbb{H}^3)$, $a$ is in $W_1$. Since $a$ is in $W_1$, and $W_1$ is open in $\mathbb{R}^3$, $W_1$ is an open neighborhood of $a$ in $\mathbb{R}^3$.

Since $f : G \to \mathbb{R}^m$ is smooth, and $a$ is in $G$, there exist an open neighborhood $V_a$ of $a$ in $\mathbb{R}^n$, and a function $f_a : V_a \to \mathbb{R}^m$ such that $f_a|_{G \cap V_a} = f|_{G \cap V_a}$, and $f_a$ is smooth in the usual sense. Since $V_a$ is an open neighborhood of $a$ in $\mathbb{R}^n$, and $f_a : V_a \to \mathbb{R}^m$ is smooth in the usual sense, $f_a : V_a \to \mathbb{R}^m$ is continuous. Since $f_a : V_a \to \mathbb{R}^m$ is continuous, and $a$ is in $V_a$, $f_a : V_a \to \mathbb{R}^m$ is continuous at $a$.

Since $V_a$ is an open neighborhood of $a$ in $\mathbb{R}^3$, and $W_1$ is an open neighborhood of $a$ in $\mathbb{R}^3$, $V_a \cap W_1$ is an open neighborhood of $a$ in $\mathbb{R}^3$. Since $a \in G$, and $a \in V_a$, $a \in G \cap V_a$. Since $a \in G \cap V_a$, and $f_a|_{G \cap V_a} = f|_{G \cap V_a}$, $f_a(a) = f(a)$. Since $V_{f(a)}$ is an open neighborhood of $f(a)$, and $f_a(a) = f(a)$, $V_{f(a)}$ is an open neighborhood of $f_a(a)$.

Since $f_a : V_a \to \mathbb{R}^m$ is continuous at $a$, $V_a$ is an open neighborhood of $a$ in $\mathbb{R}^3$, $V_a \cap W_1$ is an open neighborhood of $a$ in $\mathbb{R}^3$, and $V_{f(a)}$ is an open neighborhood of $f_a(a)$ in $\mathbb{R}^m$, there exists an open neighborhood $U_a$ of $a$ in $\mathbb{R}^3$, such that $U_a \subset V_a \cap W_1$, and $f_a(U_a) \subset V_{f(a)}$.

For 1: Take any $x$ in $U_a \cap \mathbb{H}^3$. Since $x$ is in $U_a \cap \mathbb{H}^3$, $x$ is in $U_a \subset V_a \cap W_1$, and hence, $x$ is in $W_1$. Since $x$ is in $U_a \cap \mathbb{H}^3$, $x$ is in $\mathbb{H}^3$. Since $x$ is in $\mathbb{H}^3$, and $x$ is in $W_1$, $x$ is in $W_1 \cap \mathbb{H}^3 (= G)$, and hence, $x$ is in $G$. Thus, $U_a \cap \mathbb{H}^3 \subset G$.

For 2: Since $f_a|_{G \cap V_a} = f|_{G \cap V_a}$, and $G = W_1 \cap \mathbb{H}^3$, $f_a|_{W_1 \cap \mathbb{H}^3 \cap V_a} = f|_{W_1 \cap \mathbb{H}^3 \cap V_a}$. Since $U_a \subset V_a \cap W_1$, $U_a \cap \mathbb{H}^3 \subset V_a \cap W_1 \cap \mathbb{H}^3 = W_1 \cap \mathbb{H}^3 \cap V_a$. Since $U_a \cap \mathbb{H}^3 \subset W_1 \cap \mathbb{H}^3 \cap V_a$, and $f_a|_{W_1 \cap \mathbb{H}^3 \cap V_a} = f|_{W_1 \cap \mathbb{H}^3 \cap V_a}$, $f(U_a \cap \mathbb{H}^3) = f_a(U_a \cap \mathbb{H}^3) \subset f_a(U_a) \subset V_{f(a)}$. Thus $f(U_a \cap \mathbb{H}^3) \subset V_{f(a)}$.

## 4.7 Smooth Covering Maps

**Definition** Let $\widetilde{M}$ and $M$ be connected smooth manifolds. Let $\pi : \widetilde{M} \to M$ be a smooth map. If

1. $\pi$ maps $\widetilde{M}$ onto $M$,
2. for every $p$ in $M$, there exists a connected open neighborhood $U$ of $p$ such that for each component $C$ of the open set $\pi^{-1}(U)$, the restriction $\pi|_C : C \to U$ is a diffeomorphism, then we say that $\pi$ *is a smooth covering map, $M$ is a base of the covering,* and $\widetilde{M}$ *is a covering manifold of $M$.*

   Observe that, in the condition (2), since $\widetilde{M}$ is a locally path connected, and $\pi^{-1}(U)$ is an open subset of $\widetilde{M}$ each component of $\pi^{-1}(U)$ is open.

**Lemma 4.67** *Every smooth covering map is a topological covering map.*

*Proof* Let $\pi : \widetilde{M} \to M$ be a smooth covering map, where $\widetilde{M}$ and $M$ are connected smooth manifolds. Since $M$ is a smooth manifold, $M$ is a topological manifold, and hence, $M$ is locally path connected. Since $\pi : \widetilde{M} \to M$ is a smooth covering map, $\pi : \widetilde{M} \to M$ is a continuous map. Since $\widetilde{M}$ is a connected topological space, $M$ is a locally path connected topological space, and $\pi : \widetilde{M} \to M$ is a continuous map, it is meaningful to talk about "$\pi$ is a topological covering map," that is,

1. $\pi$ maps $\widetilde{M}$ onto $M$,
2. for every $p$ in $M$, there exists a connected open neighborhood $U$ of $p$ such that for each component $C$ of the open set $\pi^{-1}(U)$, the restriction $\pi|_C : C \to U$ is a homeomorphism,

For 1: Since $\pi : \widetilde{M} \to M$ is a smooth covering map, $\pi$ maps $\widetilde{M}$ onto $M$.

For 2: Let us take any $p$ in $M$. Since $\pi : \widetilde{M} \to M$ is a smooth covering map, there exists a connected open neighborhood $U$ of $p$ such that for each component $C$ of the open set $\pi^{-1}(U)$, the restriction $\pi|_C : C \to U$ is a diffeomorphism, and hence, $\pi|_C : C \to U$ is a homeomorphism. $\qquad\square$

**Lemma 4.68** *Let $\widetilde{M}$ and $M$ be connected smooth manifolds. Let $\pi : \widetilde{M} \to M$ be a smooth covering map. Then, $\pi$ is a local diffeomorphism, that is, for every $p$ in $\widetilde{M}$, there exists an open neighborhood $C$ of $p$ such that $\pi(C)$ is open in $M$, and the restriction $\pi|_C : C \to \pi(C)$ is a diffeomorphism.*

*Proof* Let us take an element $p$ in $\widetilde{M}$. We have to find an open neighborhood $C$ of $p$ such that $\pi(C)$ is open in $M$, and the restriction $\pi|_C : C \to \pi(C)$ is a diffeomorphism.

Since $\pi : \widetilde{M} \to M$, and $p$ is in $\widetilde{M}$, $\pi(p)$ is in $M$. Since $\pi(p)$ is in $M$, and $\pi : \widetilde{M} \to M$ is a smooth covering map, there exists a connected open neighborhood $U$ of $\pi(p)$ such that for each component $C$ of the open set $\pi^{-1}(U)$, the restriction $\pi|_C : C \to U$ is a diffeomorphism.

Since $\pi(p)$ is in $U$, $p \in \pi^{-1}(U)$. Since $p \in \pi^{-1}(U)$, and $\pi^{-1}(U)$ is partitioned into components, there exists a component $C$ of $\pi^{-1}(U)$ such that $p \in C$. Since $\pi : \widetilde{M} \to M$ is a smooth map, $\pi : \widetilde{M} \to M$ is a continuous map. Since $\pi : \widetilde{M} \to M$ is a continuous map, and $U$ is an open neighborhood of $\pi(p)$, $\pi^{-1}(U)$ is an open neighborhood of $p$. Since $\widetilde{M}$ is a smooth manifold, $\widetilde{M}$ is a topological manifold, and hence $\widetilde{M}$ is locally path connected. Since $\widetilde{M}$ is locally path connected, and $\pi^{-1}(U)$ is an open subset of $\widetilde{M}$ and $C$ is a component of $\pi^{-1}(U)$, $C$ is open. Since $C$ is open, and $p \in C$, $C$ is an open neighborhood of $p$ in $\widetilde{M}$.

Since $\pi|_C : C \to U$ is a diffeomorphism, $\pi(C) = (\pi|_C)(C) = U$. Since $\pi(C) = U$, and $U$ is open, $\pi(C)$ is open in $M$. Since $\pi|_C : C \to U$ is a diffeomorphism, $\pi$ is a local diffeomorphism. $\qquad\square$

# Chapter 5
# Immersions, Submersions, and Embeddings

The counterpart of "homeomorphism in topological spaces" and "linear isomorphism in real linear spaces" are the concepts of immersion, submersion, and embedding in smooth manifolds. Because a smooth manifold associates with a topological space together with a collection of linear spaces (i.e., tangent spaces at various points of manifold), there emerges different types of "homomorphisms" among smooth manifolds. Their definitions and various theorems on relationship between immersion, submersion, and embedding constitute a beautiful area of studies. In this chapter, we shall prove some of the important theorems on immersion, submersion, embedding, and its related notions. Definitely, the pace of this chapter is a bit slower for obvious reason.

## 5.1 Pointwise Pushforward

**Definition** Let $a \in \mathbb{R}^3$. By $\mathbb{R}^3_a$ we mean the Cartesian product $\{a\} \times \mathbb{R}^3$. Members of $\mathbb{R}^3_a$ are called *geometric tangent vectors in* $\mathbb{R}^3$ *at* $a$, and $\mathbb{R}^3_a$ is called the *geometric tangent space to* $\mathbb{R}^3$ *at* $a$. A geometric tangent vector $(a, v)$ in $\mathbb{R}^3$ at $a$ is denoted by $v_a$ or $v|_a$. We define vector addition and scalar multiplication over $\mathbb{R}^3_a$ as follows:

$$v_a + w_a \equiv (v + w)_a, \quad t(v_a) \equiv (tv)_a.$$

Clearly, $\mathbb{R}^3_a$ becomes a real linear space. Here, the *zero vector* is $(0, 0, 0)_a$, and the *negative vector* of $v_a$ is $(-v)_a$. It is easy to see that the mapping $v \mapsto v_a$ is an isomorphism from $\mathbb{R}^3$ onto $\mathbb{R}^3_a$, and hence, $\mathbb{R}^3$ and $\mathbb{R}^3_a$ are isomorphic.

**Note 5.1** Since members of a smooth manifold $M$ may not be added, the above definition of tangent space cannot be generalized. Observe that $C^\infty(\mathbb{R}^3)$ is a real linear space under pointwise vector addition and scalar multiplication. Also, corresponding to each geometric tangent vector $v_a$ in $\mathbb{R}^3$ at $a$, we can define a function $D_v|_a : C^\infty(\mathbb{R}^3) \to \mathbb{R}$ as follows: For every $f \in C^\infty(\mathbb{R}^3)$,

© Springer India 2014
R. Sinha, *Smooth Manifolds*, DOI 10.1007/978-81-322-2104-3_5

$$\left(D_v\big|_a\right)(f) \equiv (D_v)(f(a)) = \left(\frac{d}{dt}\bigg|_{t=0}\right)(f(a+tv)) = \lim_{t\to 0}\frac{f(a+tv)-f(a)}{t}.$$

We know that $D_v\big|_a$ is linear and satisfies the *Leibnitz rule*:

$$\left(D_v\big|_a\right)(f \cdot g) = \left(\left(D_v\big|_a\right)(f)\right)(g(a)) + (f(a))\left(\left(D_v\big|_a\right)(g)\right).$$

Also,

$$\left(D_{(v^1,v^2,v^3)}\big|_a\right)(f) = v^1\left(\left(D_{(1,0,0)}\big|_a\right)(f)\right) + v^2\left(\left(D_{(0,1,0)}\big|_a\right)(f)\right) + v^3\left(\left(D_{(0,0,1)}\big|_a\right)(f)\right)$$

$$= v^1 \lim_{t\to 0}\frac{f(a+t(1,0,0))-f(a)}{t} + v^2 \lim_{t\to 0}\frac{f(a+t(0,1,0))-f(a)}{t}$$

$$+ v^3 \lim_{t\to 0}\frac{f(a+t(0,0,1))-f(a)}{t}$$

$$= v^1((D_1 f)(a)) + v^2((D_2 f)(a))$$

$$+ v^3((D_3 f)(a)) = v^1\left(\frac{\partial f}{\partial x^1}(a)\right) + v^2\left(\frac{\partial f}{\partial x^2}(a)\right) + v^3\left(\frac{\partial f}{\partial x^3}(a)\right)$$

$$= \sum_{i=1}^{3} v^i\left(\frac{\partial f}{\partial x^i}(a)\right) = v^i\left(\frac{\partial f}{\partial x^i}(a)\right).$$

Thus,

$$\left(D_{(v^1,v^2,v^3)}\big|_a\right)(f) = v^i\left(\frac{\partial f}{\partial x^i}(a)\right).$$

It follows that

$$\left(D_{(1,0,0)}\big|_a\right)(f) = \frac{\partial f}{\partial x^1}(a), \quad \left(D_{(0,1,0)}\big|_a\right)(f) = \frac{\partial f}{\partial x^2}(a),$$

$$\text{and} \quad \left(D_{(0,0,1)}\big|_a\right)(f) = \frac{\partial f}{\partial x^3}(a).$$

**Definition** Let $a \in \mathbb{R}^3$. Let $w : C^\infty(\mathbb{R}^3) \to \mathbb{R}$ be a linear function. If $w$ satisfies the *Leibnitz rule*:

$$w(f \cdot g) = (w(f))(g(a)) + (f(a))(w(g))$$

for every $f$, $g$ in $C^\infty(\mathbb{R}^3)$, then we say that $w$ *is a derivation at a*. Let us denote the collection of all derivations at $a$ by $T_a\mathbb{R}^3$. Clearly, $T_a\mathbb{R}^3$ is a real linear space under pointwise vector addition and scalar multiplication. It is clear that if $v_a \in \mathbb{R}^3_a$, then the mapping $D_v\big|_a: C^\infty(\mathbb{R}^3) \to \mathbb{R}$, defined as above, is a derivation at $a$. Thus, for every $v_a \in \mathbb{R}^3_a$, $D_v\big|_a \in T_a\mathbb{R}^3$.

**Note 5.2** Let $a \in \mathbb{R}^3$. Let $w \in T_a\mathbb{R}^3$. Let $f, g \in C^\infty(\mathbb{R}^3)$. We shall try to show:

1. If $f$ is a constant function, then $w(f) = 0$,
2. If $f(a) = g(a) = 0$, then $w(f \cdot g) = 0$.

For 1: Here, $f$ is a constant function, so there exists a real number $c$ such that $f(x) = c$ for every $x$ in $\mathbb{R}^3$. Let us define a function $f_1 : \mathbb{R}^3 \to \mathbb{R}$ as follows: For every $x$ in $\mathbb{R}^3$, $f_1(x) = 1$. Clearly, $f_1 \in C^\infty(\mathbb{R}^3)$, and $f = cf_1$. Now, $w(f_1) = w$ $(f_1 \cdot f_1) = (w(f_1))\,(f_1(a)) + (f_1(a))\,(w(f_1)) = (w(f_1))\,1 + 1(w(f_1)) = 2(w(f_1))$, and hence, $w(f_1) = 0$. Next, $w(f) = w(cf_1) = c(w(f_1)) = c \cdot 0 = 0$. This proves 1. For 2: LHS $= w(f \cdot g) = (w(f))(g(a)) + (f(a))\,(w(g)) = (w(f)) \cdot 0 + 0 \cdot (w(g)) = 0 =$ RHS.

**Note 5.3** Let us fix any $a \equiv (a^1, a^2, a^3) \in \mathbb{R}^3$. Let $\eta : \mathbb{R}_a^3 \to T_a\mathbb{R}^3$ be the mapping defined as follows: For every $v_a \in \mathbb{R}_a^3$,

$$\eta(v_a) \equiv D_v|_a.$$

Then, $\eta$ is a linear isomorphism from real linear space $\mathbb{R}_a^3$ onto real linear space $T_a\mathbb{R}^3$.

Reason: Let $v_a, w_a \in \mathbb{R}_a^3$, where $v \equiv (v^1, v^2, v^3)$, and $w \equiv (w^1, w^2, w^3)$. Let $s, t \in \mathbb{R}$. We have to prove that $\eta(s(v_a) + t(w_a)) = s(\eta(v_a)) + t(\eta(w_a))$, that is, for every $f$ in $C^\infty(\mathbb{R}^3)$, $(\eta(s(v_a) + t(w_a)))\,(f) = (s(\eta(v_a)) + t(\eta(w_a)))(f)$.

$$\text{LHS} = (\eta(s(v_a) + t(w_a)))(f) = (\eta((sv)_a + (tw)_a))(f)$$
$$= (\eta((sv + tw)_a))(f) = \left(D_{(sv+tw)}|_a\right)(f)$$
$$= \left(D_{(sv^1+tw^1, sv^2+tw^2, sv^3+tw^3)}|_a\right)(f) = (sv^i + tw^i)\left(\frac{\partial f}{\partial x^i}(a)\right)$$
$$= s\left(v^i\left(\frac{\partial f}{\partial x^i}(a)\right)\right) + t\left(w^i\left(\frac{\partial f}{\partial x^i}(a)\right)\right) = s((D_v|_a)(f)) + t((D_w|_a)(f))$$
$$= s((\eta(v_a))(f)) + t((\eta(w_a))(f)) = (s(\eta(v_a)) + t(\eta(w_a)))(f) = \text{RHS}.$$

This proves that $\eta : \mathbb{R}_a^3 \to T_a\mathbb{R}^3$ is linear.

Now, we want to prove that $\eta$ is 1–1. For this purpose, let $\eta((v^1, v^2, v^3)_a) = \eta((w^1, w^2, w^3)_a)$. We have to prove that $(v^1, v^2, v^3) = (w^1, w^2, w^3)$. Let us define a function $\pi_1 : \mathbb{R}^3 \to \mathbb{R}$, as follows: For every $(x^1, x^2, x^3)$ in $\mathbb{R}^3$, $\pi_1(x^1, x^2, x^3) \equiv x^1$. Clearly, $\pi_1 \in C^\infty(\mathbb{R}^3)$. Since

$$D_{(v^1, v^2, v^3)}|_a = \eta\left((v^1, v^2, v^3)_a\right) = \eta\left((w^1, w^2, w^3)_a\right) = D_{(w^1, w^2, w^3)}|_a,$$

and $\pi_1 \in C^\infty(\mathbb{R}^3)$,

$$v^1 = v^1(1) + v^2(0) + v^3(0) = v^1\left(\frac{\partial f}{\partial x^1}(a)\right) + v^2\left(\frac{\partial f}{\partial x^2}(a)\right) + v^3\left(\frac{\partial f}{\partial x^3}(a)\right)$$

$$= \left(D_{(v^1,v^2,v^3)}\big|_a\right)(\pi_1) = \left(D_{(w^1,w^2,w^3)}\big|_a\right)(\pi_1) = w^1\left(\frac{\partial f}{\partial x^1}(a)\right)$$

$$+ w^2\left(\frac{\partial f}{\partial x^2}(a)\right) + w^3\left(\frac{\partial f}{\partial x^3}(a)\right)$$

$$= w^1(1) + w^2(0) + w^3(0) = w^1,$$

and hence, $v^1 = w^1$. Similarly, $v^2 = w^2, v^3 = w^3$. Hence, $(v^1, v^2, v^3) = (w^1, w^2, w^3)$.

Now, it remains to be proved that $\eta : \mathbb{R}^3_a \to T_a\mathbb{R}^3$ is onto. For this purpose, let us take any $w \in T_a\mathbb{R}^3$. We have to find an element $v_a \in \mathbb{R}^3_a$ such that $\eta(v_a) = w$, that is, $D_v|_a = w$, that is, for every $f \in C^\infty(\mathbb{R}^3)$, $(D_v|_a)(f) = w(f)$. For every $i = 1, 2, 3$, let us define a function $\pi_i : \mathbb{R}^3 \to \mathbb{R}$, as follows: For every $(x^1, x^2, x^3)$ in $\mathbb{R}^3$, $\pi_i(x^1, x^2, x^3) \equiv x^i$. Clearly, each $\pi_i \in C^\infty(\mathbb{R}^3)$. Since $w \in T_a\mathbb{R}^3$, $w : C^\infty(\mathbb{R}^3) \to \mathbb{R}$. Now, since $\pi_i \in C^\infty(\mathbb{R}^3)$, $w(\pi_i)$ is a real number. It follows that $(w(\pi_1), w(\pi_2), w(\pi_3)) \in \mathbb{R}^3$. Let us take $(w(\pi_1), w(\pi_2), w(\pi_3))$ for $v$. It remains to be proved that for every $f \in C^\infty(\mathbb{R}^3)$, $(D_{(w(\pi_1),w(\pi_2),w(\pi_3))}\big|_a)(f) = w(f)$, that is, for every $f \in C^\infty(\mathbb{R}^3)$,

$$(w(\pi_i))(\frac{\partial f}{\partial x^i}(a)) = w(f).$$

Since $f \in C^\infty(\mathbb{R}^3), f : \mathbb{R}^3 \to \mathbb{R}$, and is smooth. Now, by Theorem 3.62, there exist $C^\infty$ functions $g_1(x), g_2(x), g_3(x)$ on $\mathbb{R}^3$ such that

1. for every $x \equiv (x^1, x^2, x^3)$ in $\mathbb{R}^3$, $f(x) = f(a) + (g_1(x))(x^1 - a^1) + (g_2(x))$
   $(x^2 - a^2) + (g_3(x))(x^3 - a^3)$, that is, $f = f(a) + g_1 \cdot (\pi_1 - a^1) + g_2 \cdot (\pi_2 - a^1)$
   $+ g_3 \cdot (\pi_3 - a^1)$,
2. $g_1(a) = \frac{\partial f}{\partial x^1}(a), g_2(a) = \frac{\partial f}{\partial x^2}(a), g_3(a) = \frac{\partial f}{\partial x^3}(a)$.

$$\text{RHS} = w(f) = w\big(f(a) + g_1 \cdot (\pi_1 - a^1) + g_2 \cdot (\pi_2 - a^2) + g_3 \cdot (\pi_3 - a^3)\big)$$

$$= w(f(a)) + \big((w(g_1))((\pi_1 - a^1)(a)) + (w(\pi_1 - a^1))(g_1(a))\big)$$

$$+ \big((w(g_2))((\pi_2 - a^2)(a)) + (w(\pi_2 - a^2))(g_2(a))\big)$$

$$+ \big((w(g_3))((\pi_3 - a^3)(a)) + (w(\pi_3 - a^3))(g_1(a))\big)$$

$$= 0 + \big((w(g_1))((a^1 - a^1)) + (w(\pi_1) - 0)(g_1(a))\big)$$

$$+ \big((w(g_2))((a^2 - a^2)) + (w(\pi_2) - 0)(g_2(a))\big)$$

$$+ \big((w(g_3))((a^3 - a^3)) + (w(\pi_3) - 0)(g_3(a))\big)$$

$$= (w(\pi_1))(g_1(a)) + (w(\pi_2))(g_2(a)) + (w(\pi_3))(g_3(a))$$

$$= (w(\pi_i))(g_i(a)) = (w(\pi_i))\left(\frac{\partial f}{\partial x^i}(a)\right)$$

$$= \text{LHS}.$$

Thus, we have shown that $\eta$ is a linear isomorphism from real linear space $\mathbb{R}^3_a$ onto real linear space $T_a\mathbb{R}^3$. Hence, $\mathbb{R}^3_a$ is isomorphic to $T_a\mathbb{R}^3$.

Since $\mathbb{R}^3_a$ is a linear space of dimension 3, and $\mathbb{R}^3_a$ is isomorphic to $T_a\mathbb{R}^3$, $T_a\mathbb{R}^3$ is of dimension 3. Further, since $\{(1,0,0)_a, (0,1,0)_a, (0,0,1)_a\}$ is a basis of $\mathbb{R}^3_a$, and $\eta$ is a linear isomorphism from $\mathbb{R}^3_a$ onto $T_a\mathbb{R}^3$, $\{\eta((1,0,0)_a), \eta((0,1,0)_a), \eta((0,0,1)_a)\}$ is a basis of $T_a\mathbb{R}^3$, that is,

$$\left\{ D_{(1,0,0)}\big|_a, D_{(0,1,0)}\big|_a, D_{(0,0,1)}\big|_a \right\}$$

is a basis of $T_a\mathbb{R}^3$, that is,

$$\left\{ \frac{\partial}{\partial x^1}\bigg|_a, \frac{\partial}{\partial x^2}\bigg|_a, \frac{\partial}{\partial x^3}\bigg|_a \right\}$$

is a basis of $T_a\mathbb{R}^3$. Similarly,

$$\left\{ \frac{\partial}{\partial x^1}\bigg|_a, \frac{\partial}{\partial x^2}\bigg|_a, \ldots, \frac{\partial}{\partial x^n}\bigg|_a \right\}$$

is a basis of $T_a\mathbb{R}^n$.

**Note 5.4** Since $\mathbb{R}^3_a$ is the geometric tangent space to $\mathbb{R}^3$ at $a$, and $\mathbb{R}^3_a$ is isomorphic to $T_a\mathbb{R}^3$, $T_a\mathbb{R}^3$ can be thought of as the geometric tangent space to $\mathbb{R}^3$ at $a$. Similarly, for every $a$ in $\mathbb{R}^k$, $\mathbb{R}^k_a$ is isomorphic to $T_a\mathbb{R}^k$. The interesting thing is that $T_a\mathbb{R}^3$, but not $\mathbb{R}^3_a$, can be generalized on a smooth manifold.

**Definition** Let $M$ be an $m$-dimensional smooth manifold. Let $p \in M$. Let $v : C^\infty(M) \to \mathbb{R}$ be a linear function. If $v$ satisfies the *Leibnitz rule*:

$$v(f \cdot g) = (v(f))(g(p)) + (f(p))(v(g)),$$

for every $f, g$ in $C^\infty(M)$, then we say that $v$ *is a derivation at $p$*. The collection of all derivations at $p$ is denoted by $T_pM$. Clearly, $T_pM$ is a real linear space under pointwise vector addition and scalar multiplication. Here, the real linear space $T_pM$ is called the *tangent space to $M$ at $p$* and the members of $T_pM$ are called *tangent vectors at $p$*.

**Note 5.5** Let $M$ be an $m$-dimensional smooth manifold. Let $p \in M$. Let $v \in T_pM$. Let $f, g \in C^\infty(M)$. As above, it is easy to show:

1. If $f$ is a constant function, then $v(f) = 0$.
2. If $f(p) = g(p) = 0$, then $v(f \cdot g) = 0$.

**Note 5.6** Let $M$ be an $m$-dimensional smooth manifold and $N$ be an $n$-dimensional smooth manifold. Let $F : M \to N$ be a smooth map. Let $v \in T_pM$. Let $f \in C^\infty(N)$.

Since $f \in C^\infty(N), f : N \to \mathbb{R}$ is a smooth map. Since $F : M \to N$ is a smooth map, and $f : N \to \mathbb{R}$ is a smooth map, their composite $f \circ F : M \to \mathbb{R}$ is a smooth map, and hence, $f \circ F \in C^\infty(M)$. Since $v \in T_pM$, $v : C^\infty(M) \to \mathbb{R}$. Since $v : C^\infty(M) \to \mathbb{R}$, and $f \circ F \in C^\infty(M)$, $v(f \circ F)$ is a real number. Let us define the mapping $((dF_p)(v)) : C^\infty(N) \to \mathbb{R}$ as follows: For every $f \in C^\infty(N)$,

$$((dF_p)(v))(f) \equiv v(f \circ F).$$

We shall try to show that $((dF_p)(v)) \in T_{F(p)}N$, that is, $(dF_p)(v)$ is a derivation at $F(p)$. Observe that $(dF_p)(v) : C^\infty(N) \to \mathbb{R}$ is linear.

(Reason: Let $f, g \in C^\infty(N)$, and let $s$, $t$ be any real. We must prove:

$$((dF_p)(v))(sf + tg) = s(((dF_p)(v))(f)) + t(((dF_p)(v))(g)).$$

LHS $= ((dF_p)(v))(sf + tg) = v((sf + tg) \circ F) = v(s(f \circ F) + t(g \circ F)) = s(v(f \circ F)) + t(v(g \circ F)) = s(((dF_p)(v))(f)) + t(((dF_p)(v))(g)) = $ RHS.)

Now, we shall try to prove that $(dF_p)(v) : C^\infty(N) \to \mathbb{R}$ satisfies the *Leibnitz rule*:

$$((dF_p)(v))(f \cdot g) = (((dF_p)(v))(f))(g(F(p))) + (f(F(p)))(((dF_p)(v))(g))$$

for every $f, g$ in $C^\infty(N)$.

$$\begin{aligned}
\text{LHS} &= ((dF_p)(v))(f \cdot g) = v((f \cdot g) \circ F) = v((f \circ F) \cdot (g \circ F)) \\
&= (v(f \circ F))((g \circ F)(p)) + ((f \circ F)(p))(v(g \circ F)) \\
&= (((dF_p)(v))(f))((g \circ F)(p)) + ((f \circ F)(p))(((dF_p)(v))(g)) \\
&= (((dF_p)(v))(f))(g(F(p))) + (f(F(p)))(((dF_p)(v))(g)) = \text{RHS}.
\end{aligned}$$

Thus, we have shown that $(dF_p)(v) \in T_{F(p)}N$.

**Definition** Let $M$ be an $m$-dimensional smooth manifold, and $N$ be an $n$-dimensional smooth manifold. Let $F : M \to N$ be a smooth map. If $v \in T_pM$, then $(dF_p)(v) \in T_{F(p)}N$. Thus, $dF_p : T_pM \to T_{F(p)}N$. Here, $dF_p$ is called the *differential of F at p* (or *tangent map, total derivative*, or *derivative of F, or pointwise push-forward of F*). $dF_p$ is also denoted by $F'(p), DF, F_*$ (see Fig. 5.1).

**Note 5.7** Let $M$ be an $m$-dimensional smooth manifold, and $N$ be an $n$-dimensional smooth manifold. Let $F : M \to N$ be a smooth map. We shall try to show that $dF_p : T_pM \to T_{F(p)}N$ is a linear mapping from tangent space $T_pM$ to tangent space $T_{F(p)}N$. Let us take any $v_1, v_2$ in $T_pM$, and any real $s, t$. We have to show that

$$(dF_p)(sv_1 + tv_2) = s((dF_p)(v_1)) + t((dF_p)(v_2)),$$

that is, for every $f$ in $C^\infty(N)$,

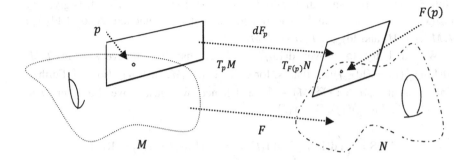

**Fig. 5.1** Tangent map

$$\big((dF_p)(sv_1 + tv_2)\big)(f) = \big(s((dF_p)(v_1)) + t((dF_p)(v_2))\big)(f).$$
$$\text{LHS} = \big((dF_p)(sv_1 + tv_2)\big)(f) = (sv_1 + tv_2)(f \circ F)$$
$$= s(v_1(f \circ F)) + t(v_2(f \circ F))$$
$$= s\big(((dF_p)(v_1))(f)\big) + t\big(((dF_p)(v_2))(f)\big)$$
$$= \big(s((dF_p)(v_1))\big)(f) + \big(t((dF_p)(v_2))\big)(f)$$
$$= \big(s((dF_p)(v_1)) + t((dF_p)(v_2))\big)(f) = \text{RHS}.$$

**Note 5.8** Let $M, N, P$ be smooth manifolds. Let $F : M \to N$, and $G : N \to P$ be smooth maps. Let $p \in M$. It follows that the composite map $G \circ F : M \to P$ is a smooth map. Here, $dF_p : T_pM \to T_{F(p)}N$, $dG_{F(p)} : T_{F(p)}N \to T_{G(F(p))}P$, and $d(G \circ F)_p : T_pM \to T_{(G \circ F)(p)}P$. Since $dF_p : T_pM \to T_{F(p)}N$, and $dG_{F(p)} : T_{F(p)}N \to T_{G(F(p))}P$, $(dG_{F(p)}) \circ (dF_p) : T_pM \to T_{G(F(p))}P$, that is, $(dG_{F(p)}) \circ (dF_p) : T_pM \to T_{(G \circ F)(p)}P$. Thus, $d(G \circ F)_p : T_pM \to T_{(G \circ F)(p)}P$, and $(dG_{F(p)}) \circ (dF_p) : T_pM \to T_{(G \circ F)(p)}P$.

We shall try to prove: $d(G \circ F)_p = (dG_{F(p)}) \circ (dF_p)$, that is, for every $v$ in $T_pM$, $(d(G \circ F)_p)(v) = ((dG_{F(p)}) \circ (dF_p))(v)$. Further, since $(d(G \circ F)_p)(v) \in T_{(G \circ F)(p)}P$, $(d(G \circ F)_p)(v) : C^{\infty}(P) \to \mathbb{R}$, and hence, we must prove: For every $f$ in $C^{\infty}(P)$,

$$\Big(\big(d(G \circ F)_p\big)(v)\Big)(f) = \big(((dG_{F(p)}) \circ (dF_p))(v)\big)(f).$$
$$\text{LHS} = \Big(\big(d(G \circ F)_p\big)(v)\Big)(f) = v(f \circ (G \circ F))$$
$$= v((f \circ G) \circ F) = ((dF_p)(v))(f \circ G)$$
$$= \big((dG_{F(p)})((dF_p)(v))\big)(f)$$
$$= \big(((dG_{F(p)}) \circ (dF_p))(v)\big)(f) = \text{RHS}.$$

Thus, we have shown that $d(G \circ F)_p = (dG_{F(p)}) \circ (dF_p)$.

**Note 5.9** Let $M$ be a smooth manifold. Let $p \in M$. Here, $\mathrm{Id}_M : M \to M$ is given by $\mathrm{Id}_M(x) = x$ for every $x$ in $M$. Clearly, $\mathrm{Id}_M$ is a smooth function. Also, $\mathrm{d}(\mathrm{Id}_M)_p :$ $T_p M \to T_p M$, and $\mathrm{Id}_{T_p M} : T_p M \to T_p M$.

We shall try to prove: $\mathrm{d}(\mathrm{Id}_M)_p = \mathrm{Id}_{T_p M}$, that is, for every $v$ in $T_p M$, $(\mathrm{d}(\mathrm{Id}_M)_p)(v) = (\mathrm{Id}_{T_p M})(v)$, that is, for every $v$ in $T_p M$, $(\mathrm{d}(\mathrm{Id}_M)_p)(v) = v$. Further, since $v$ is in $T_p M$, $v : C^\infty(M) \to \mathbb{R}$, and hence, we must prove: For every $f$ in $C^\infty(M)$, $((\mathrm{d}(\mathrm{Id}_M)_p)(v))(f) = v(f)$.

$$\mathrm{LHS} = \left(\left(\mathrm{d}(\mathrm{Id}_M)_p\right)(v)\right)(f) = v(f \circ (\mathrm{Id}_M)) = v(f) = \mathrm{RHS}.$$

Thus, we have shown that $\mathrm{d}(\mathrm{Id}_M)_p = \mathrm{Id}_{T_p M}$.

## 5.2 Open Submanifolds

**Note 5.10** Let $M, N$ be smooth manifolds. Let $F : M \to N$ be a diffeomorphism. Let $p \in M$. Since $F : M \to N$ is a diffeomorphism, $F : M \to N$ is a smooth map. We shall try to show: $\mathrm{d}F_p : T_p M \to T_{F(p)} N$ is an isomorphism.

Since $\mathrm{d}F_p$ is linear, it suffices to show that $\mathrm{d}F_p$ is 1–1 and onto.

$\mathrm{d}F_p$ is 1–1: Let $(\mathrm{d}F_p)(v) = (\mathrm{d}F_p)(w)$, where $v, w$ are in $T_p M$. We have to prove that $v = w$.

Since $\mathrm{d}F_p : T_p M \to T_{F(p)} N$, and $v$ is in $T_p M$, $(\mathrm{d}F_p)(v)$ is in $T_{F(p)} N$, and hence, $(\mathrm{d}F_p)(v) : C^\infty(N) \to \mathbb{R}$. Since $v$ is in $T_p M$, $v : C^\infty(M) \to \mathbb{R}$. Similarly, $w : C^\infty(M) \to \mathbb{R}$. Now, we have to prove that $v = w$, that is, for every $f$ in $C^\infty(M)$, $v(f) = w(f)$.

For this purpose, let us take any $f$ in $C^\infty(M)$. Since $f$ is in $C^\infty(M), f : M \to \mathbb{R}$ is smooth. Since $F : M \to N$ is a diffeomorphism, $F^{-1} : N \to M$ is smooth. Since $F^{-1} : N \to M$ is smooth, and $f : M \to \mathbb{R}$ is smooth, their composite $f \circ (F^{-1}) : N \to \mathbb{R}$ is smooth, and hence, $f \circ (F^{-1}) \in C^\infty(N)$. Now, since $(\mathrm{d}F_p)(v) = (\mathrm{d}F_p)(w)$, for every $g$ in $C^\infty(N)$, $v(g \circ F) = ((\mathrm{d}F_p)(v))(g) = ((\mathrm{d}F_p)(w))(g) = w(g \circ F)$. Since for every $g$ in $C^\infty(N)$, $v(g \circ F) = w(g \circ F)$, and $f \circ (F^{-1}) \in C^\infty(N)$, $\mathrm{LHS} = v(f) = v(f \circ \mathrm{Id}_M) = v(f \circ (F^{-1} \circ F)) = v((f \circ (F^{-1})) \circ F) = w((f \circ (F^{-1})) \circ F) = w(f \circ (F^{-1} \circ F)) = w(f \circ \mathrm{Id}_M) = w(f) = \mathrm{RHS}$.

Thus, we have shown that if $F : M \to N$ is a diffeomorphism, then $\mathrm{d}F_p :$ $T_p M \to T_{F(p)} N$ is an isomorphism.

**Note 5.11** Let $M, N$ be smooth manifolds. Let $F : M \to N$ be a diffeomorphism. Since $F : M \to N$ is a diffeomorphism, $F^{-1} : N \to M$ is a smooth map, and hence, $\mathrm{d}(F^{-1})_{F(p)} : T_{F(p)} N \to T_p M$ is an isomorphism. Since $\mathrm{d}F_p : T_p M \to T_{F(p)} N$ is an isomorphism, $(\mathrm{d}F_p)^{-1} : T_{F(p)} N \to T_p M$. Thus

$$\mathrm{d}\left(F^{-1}\right)_{F(p)} \colon T_{F(p)}N \to T_pM, \quad \text{and} \quad \left(\mathrm{d}F_p\right)^{-1} \colon T_{F(p)}N \to T_pM.$$

Now, we want to prove that $\mathrm{d}(F^{-1})_{F(p)} = (\mathrm{d}F_p)^{-1}$, that is, $(\mathrm{d}(F^{-1})_{F(p)}) \circ (\mathrm{d}F_p) = \mathrm{Id}_{T_pM}$.

$$\mathrm{LHS} = \left(\mathrm{d}\left(F^{-1}\right)_{F(p)}\right) \circ \left(\mathrm{d}F_p\right) = \mathrm{d}\left(\left(F^{-1}\right) \circ F\right)_p = \mathrm{d}(\mathrm{Id}_M)_p = \mathrm{Id}_{T_pM} = \mathrm{RHS}.$$

Thus, we have shown that if $F : M \to N$ is a diffeomorphism, then $\mathrm{d}(F^{-1})_{F(p)} = (\mathrm{d}F_p)^{-1}$.

**Theorem 5.12** *Let $M$ be a smooth manifold. Let $p \in M$, $v \in T_pM$. Let $f, g \in C^\infty(M)$. If there exists an open neighborhood $U$ of $p$ such that $f|_U = g|_U$, then $v(f) = v(g)$. In short, we say that tangent vectors act "locally."*

*Proof* Let $U$ be an open neighborhood of $p$ such that $f|_U = g|_U$. We want to show that $v(f) = v(g)$. Let us put $h \equiv f - g$. Since $f, g \in C^\infty(M)$, $h = f - g \in C^\infty(M)$, and hence, $h \in C^\infty(M)$. Since $f|_U = g|_U$, $h|_U = 0$. Clearly, $p \notin \mathrm{supp}\, h$. (Reason: If not, otherwise, let $p \in \mathrm{supp}\, h$. We have to arrive at a contradiction. Since $p \in \mathrm{supp}\, h = \{x : h(x) \neq 0\}^-$, and $U$ is an open neighborhood of $p$, there exists $t \in M$ such that $h(t) \neq 0$, and $t \in U$. Since $t \in U$, and $h|_U = 0$, $h(t) = 0$, a contradiction.) Since $p \notin \mathrm{supp}\, h$, $h(p) = 0$.

Here, $M$ is a smooth manifold, so $M$ is Hausdorff. Since $M$ is Hausdorff, and $p \in M$, $\{p\}$ is closed, and hence, $M - \{p\}$ is open. Since $p \notin \mathrm{supp}\, h$, $\mathrm{supp}\, h \subset M - \{p\}$. Here, $\mathrm{supp}\, h \subset M - \{p\}$, $\mathrm{supp}\, h$ is a closed subset of $M$, $M - \{p\}$ is an open subset of $M$, so, by Lemma 4.58, there exists a smooth function $\psi : M \to [0, 1]$ such that $\psi$ is a bump function for $\mathrm{supp}\, h$ supported in $M - \{p\}$, and hence,

1. for every $x$ in $\mathrm{supp}\, h$, $\psi(x) = 1$,
2. $\mathrm{supp}\, \psi \subset M - \{p\}$.

Since $\mathrm{supp}\, \psi \subset M - \{p\}$, $p \notin \mathrm{supp}\, \psi$, and hence, $\psi(p) = 0$. Since $\psi : M \to [0, 1]$ is smooth, $\psi \in C^\infty(M)$. Since $\psi \in C^\infty(M)$, and $h \in C^\infty(M)$, their product $(\psi \cdot h) \in C^\infty(M)$. Since $v \in T_pM$, $v : C^\infty(M) \to \mathbb{R}$. Now, $v(\psi \cdot h) = (v(\psi))(h(p)) + (\psi(p))(v(h)) = (v(\psi))(0) + (0)(v(h)) = 0$. Thus, $v(\psi \cdot h) = 0$.

Clearly, $\psi \cdot h = h$. (Reason: If $x \in \mathrm{supp}\, h$, then from condition 1, $(\psi \cdot h)(x) = (\psi(x))(h(x)) = 1(h(x)) = h(x)$. If $x \notin \mathrm{supp}\, h$, then $h(x) = 0$, and hence, $(\psi \cdot h)(x) = (\psi(x))(h(x)) = (\psi(x))0 = 0 = h(x)$. Thus, for all $x$ in $M$, $(\psi \cdot h)(x) = h(x)$.) Now, $0 = v(\psi \cdot h) = v(h) = v(f - g) = v(f) - v(g)$, so $v(f) = v(g)$. $\square$

**Theorem 5.13** *Let $M$ be an $m$-dimensional smooth manifold. Let $p \in M$. Let $U$ be an open neighborhood of $p$. We know that $U$ is also a smooth manifold (called open submanifold $U$ of $M$). Let $\imath : U \to M$ be the mapping defined as follows: For every $x$ in $U$, $\imath(x) \equiv x$. (Here, we say that $\imath$ is the inclusion map of $U$.) Then, the* *differential* $d\iota_p : T_pU \rightarrow T_{\iota(p)}M(= T_pM)$ *is an isomorphism, that is,* $d\iota_p : T_pU \rightarrow T_pM$ *is 1–1 and onto.*

*Proof* Since $M$ is a smooth manifold, $M$ is a topological manifold. Since $M$ is a topological manifold, and $U$ is an open neighborhood of $p$, there exists an open neighborhood $V$ of $p$ such that $V^- \subset U$. We know that $d\iota_p$ is linear.

$d\iota_p$ is 1–1: For this purpose, let $(d\iota_p)(v) = 0$ where $v \in T_pU$. We have to prove that $v = 0$. Since $v \in T_pU$, $v : C^\infty(U) \rightarrow \mathbb{R}$. We have to show that $v = 0$, that is, for every $f \in C^\infty(U)$, $v(f) = 0$. For this purpose, let us take any $f \in C^\infty(U)$. Since $f \in C^\infty(U), f : U \rightarrow \mathbb{R}$ is smooth. Since $f : U \rightarrow \mathbb{R}$ is smooth, and $V^- \subset U$, $f|_{V^-}$ is smooth. Here, $V^- \subset U$, $V^-$ is closed, $U$ is open, and $f|_{V^-}$ is smooth, so, by Lemma 4.59, there exists a smooth function $\tilde{f} : M \rightarrow \mathbb{R}^m$ such that $\tilde{f}|_{V^-} = f|_{V^-}$.

Since $\tilde{f}|_{V^-} = f|_{V^-}$, and $V \subset V^-, \tilde{f}|_V = f|_V$. Since $\tilde{f} : M \rightarrow \mathbb{R}^m$ is smooth, $\tilde{f}|_U$ is smooth on $U$, and hence, $\tilde{f}|_U \in C^\infty(U)$. Since $V \subset V^- \subset U$, $(\tilde{f}|_U)\big|_V = \tilde{f}|_V = f|_V$, and hence, $(\tilde{f}|_U)\big|_V = f|_V$. Since $U$ is a smooth manifold, $p \in U, v \in T_pU, \tilde{f}|_U, f \in C^\infty(U)$, $V$ is an open neighborhood of $p$ such that $V \subset U$, and $(\tilde{f}|_U)\big|_V = f|_V$, by Theorem 5.12, $v(\tilde{f}|_U) = v(f)$. Now,

$$\text{LHS} = v(f) = v\left(\tilde{f}|_U\right) = v(\tilde{f} \circ \iota) = ((d\iota_p)(v))(\tilde{f}) = 0(\tilde{f}) = 0 = \text{RHS}.$$

This proves that $d\iota_p$ is 1–1.

$d\iota_p : T_pU \rightarrow T_pM$ is onto: For this purpose, let us take any $w \in T_pM$. We have to find $v \in T_pU$ such that $(d\iota_p)(v) = w$. Since $w \in T_pM$, $w : C^\infty(M) \rightarrow \mathbb{R}$. It follows that we must find $v \in T_pU$ such that for every $f \in C^\infty(M)$, $((d\iota_p)(v))(f) = w(f)$, that is, for every $f \in C^\infty(M)$, $v(f \circ \iota) = w(f)$.

Now, let us take any $g \in C^\infty(U)$. Since $g \in C^\infty(U)$, $g : U \rightarrow \mathbb{R}$ is smooth. Since $M$ is a topological manifold, and $U$ is an open neighborhood of $p$, there exists an open neighborhood $V$ of $p$ such that $V^- \subset U$. Since $g : U \rightarrow \mathbb{R}$ is smooth, and $V^- \subset U$, $g|_{V^-}$ is smooth. Here, $V^- \subset U$, $V^-$ is closed, $U$ is open, and $g|_{V^-}$ is smooth, by Lemma 4.59, there exists a smooth function $g^\sim : M \rightarrow \mathbb{R}^m$ such that $g^\sim|_{V^-} = g|_{V^-}$. Next, let $g^\approx : M \rightarrow \mathbb{R}^m$ be another smooth function such that $g^\approx|_{V^-} = g|_{V^-}$. Since $g^\sim|_{V^-} = g|_{V^-} = g^\approx|_{V^-}$, and $V \subset V^-, g^\sim|_V = g^\approx|_V$. Since $M$ is a smooth manifold, $p \in M$, $w \in T_pM$, $g^\sim, g^\approx \in C^\infty(M)$, $V$ is an open neighborhood of $p$, and $g^\sim|_V = g^\approx|_V$, by Theorem 5.12, $w(g^\sim) = w(g^\approx)$.

Let us define a function $v : C^\infty(U) \rightarrow \mathbb{R}$ as follows: For every $g \in C^\infty(U)$, $v(g) \equiv w(g^\sim)$ where $g^\sim \in C^\infty(M)$, and $g^\sim|_{V^-} = g|_{V^-}$. From the above discussion, we find that $v : C^\infty(U) \rightarrow \mathbb{R}$ is well defined. Now, we want to prove that $v$ is linear. For this purpose, let us take any $g_1, g_2 \in C^\infty(U)$, and $s, t$ any real numbers. We have to prove that $v(sg_1 + tg_2) = s(v(g_1)) + t(v(g_2))$.

Let $g_1^\sim \in C^\infty(M)$ such that $g_1^\sim|_{V^-} = g_1|_{V^-}$, and $g_2^\sim \in C^\infty(M)$ such that $g_2^\sim|_{V^-} = g_2|_{V^-}$. It follows that $s(g_1^\sim) + t(g_2^\sim) \in C^\infty(M)$. Since $g_1^\sim|_{V^-} = g_1|_{V^-}$, and

$g_2^\sim|_{V^-} = g_2|_{V^-}$, $(s(g_1^\sim) + t(g_2^\sim))|_{V^-} = (sg_1 + tg_2)|_{V^-}$. Since $(sg_1 + tg_2)|_{V^-} = (s(g_1^\sim) + t(g_2^\sim))|_{V^-}$, and $s(g_1^\sim) + t(g_2^\sim) \in C^\infty(M)$,

$$\text{LHS} = v(sg_1 + tg_2) = w\big(s(g_1^\sim) + t(g_2^\sim)\big) = s\big(w(g_1^\sim)\big) + t\big(w(g_2^\sim)\big)$$
$$= s(v(g_1)) + t(v(g_2)) = \text{RHS}.$$

Now, we want to prove the *Leibnitz rule*: For every $f, g \in C^\infty(U)$,

$$v(f \cdot g) = (v(f))(g(p)) + (f(p))(v(g)).$$

Let $f^\sim \in C^\infty(M)$ such that $f^\sim|_{V^-} = f|_{V^-}$, and $g^\sim \in C^\infty(M)$ such that $g^\sim|_{V^-} = g|_{V^-}$. It follows that $(f^\sim) \cdot (g^\sim) \in C^\infty(M)$. Since $f^\sim|_{V^-} = f|_{V^-}$, and $g^\sim|_{V^-} = g|_{V^-}$, $((f^\sim) \cdot (g^\sim))|_{V^-} = (f \cdot g)|_{V^-}$. Since $f^\sim|_{V^-} = f|_{V^-}$, and $p \in V \subset V^-$, $f^\sim(p) = f(p)$. Similarly, $g^\sim(p) = g(p)$. Now,

$$\text{LHS} = v(f \cdot g) = w((f^\sim) \cdot (g^\sim)) = (w(f^\sim))(g^\sim(p)) + (f^\sim(p))(w(g^\sim))$$
$$= (v(f))(g^\sim(p)) + (f^\sim(p))(v(g)) = (v(f))(g(p)) + (f(p))(v(g)) = \text{RHS}.$$

Thus, we have shown that $v : C^\infty(U) \to \mathbb{R}$ is a derivation at $p$, and hence, $v \in T_pU$. Next, let us take any $f \in C^\infty(M)$. It remains to be proved that $v(f \circ \iota) = w(f)$. Since $\iota : U \to M$ is smooth, and $f : M \to \mathbb{R}$ is smooth, $f \circ \iota : U \to \mathbb{R}$ is smooth, and hence, $f \circ \iota \in C^\infty(U)$. Now, since $V^- \subset U$, $f|_{V^-} = (f \circ \iota)|_{V^-}$. Since $f \in C^\infty(M)$, $f \circ \iota \in C^\infty(U)$, and $f|_{V^-} = (f \circ \iota)|_{V^-}$, by the definition of $v$, $v(f \circ \iota) = w(f)$. $\qquad\square$

**Theorem 5.14** *Let $M$ be an $m$-dimensional smooth manifold. Let $p \in M$. Then, the real linear space $T_pM$ is of dimension $m$.*

Here, $M$ is an $m$-dimensional smooth manifold, and $p \in M$, so there exists an admissible coordinate chart $(U, \varphi)$ of $M$ such that $p \in U$. Since $(U, \varphi)$ is an admissible coordinate chart of $M$, there exists an open neighborhood $G$ of $\varphi(p)$ in $\mathbb{R}^m$ such that $\varphi : U \to G$ is a diffeomorphism. Since $\varphi : U \to G$ is a diffeomorphism, $F : M \to N$ is an isomorphism, and hence, $T_pU$ is isomorphic onto $T_{\varphi(p)}G$.

Since $M$ is an $m$-dimensional smooth manifold, $p \in M$, and $U$ is an open neighborhood of $p$, by Theorem 5.13, the differential $d\iota_p : T_pU \to T_{\iota(p)}M(= T_pM)$ is an isomorphism, where $\iota$ denotes the inclusion map of $U$. Since $d\iota_p : T_pU \to T_pM$ is an isomorphism, $T_pU$ is isomorphic onto $T_pM$.

Since $\mathbb{R}^m$ is an $m$-dimensional smooth manifold, $\varphi(p) \in \mathbb{R}^m$, and $G$ is an open neighborhood of $\varphi(p)$, by Theorem 5.13, the differential $d\hat{\iota}_{\varphi(p)} : T_{\varphi(p)}G \to T_{\hat{\iota}(\varphi(p))}\mathbb{R}^m(= T_{\varphi(p)}\mathbb{R}^m)$ is an isomorphism, where $\hat{\iota}$ denotes the inclusion map of $G$. Since $d\hat{\iota}_{\varphi(p)} : T_{\varphi(p)}G \to T_{\varphi(p)}\mathbb{R}^m$ is an isomorphism, $T_{\varphi(p)}G$ is isomorphic onto $T_{\varphi(p)}\mathbb{R}^m$.

We have seen that $\mathbb{R}^m_{\varphi(p)}$ is isomorphic to $T_{\varphi(p)}\mathbb{R}^m$. Since $T_pU$ is isomorphic onto $T_{\varphi(p)}G$, $T_pU$ is linear isomorphic onto $T_pM$, $T_{\varphi(p)}G$ is isomorphic onto $T_{\varphi(p)}\mathbb{R}^m$, and

$\mathbb{R}^m_{\varphi(p)}$ is isomorphic to $T_{\varphi(p)}\mathbb{R}^m$, $T_pM$ is isomorphic to $\mathbb{R}^m_{\varphi(p)}$, and hence, $\dim T_pM = \dim \mathbb{R}^m_{\varphi(p)}$. Since $\dim T_pM = \dim \mathbb{R}^m_{\varphi(p)}$, and $\dim \mathbb{R}^m_{\varphi(p)} = m$, $\dim T_pM = m$.

**Note 5.15** Let $M$ be a two-dimensional smooth manifold with differentiable structure $\mathcal{A}$, and let $N$ be a three-dimensional smooth manifold with differentiable structure $\mathcal{B}$. Hence, the product manifold $M \times N$ is a $(2+3)$-dimensional smooth manifold. Let $(p, q) \in M \times N$.

Since $(p, q) \in M \times N$, $p \in M$, and $q \in N$. Since $p \in M$, and $M$ is a two-dimensional smooth manifold, the tangent space $T_pM$ is a two-dimensional real linear space. Similarly, $T_qN$ is a three-dimensional real linear space, and the tangent space $T_{(p,q)}(M \times N)$ is a $(2+3)$-dimensional real linear space. Since $T_pM$ is a two-dimensional real linear space, and $T_qN$ is a three-dimensional real linear space, their direct product $T_pM \oplus T_pN$ is a $(2+3)$-dimensional real linear space. Let $\pi_1 : M \times N \to M$ be the mapping defined as follows: For every $(x, y) \in M \times N$, $\pi_1(x, y) \equiv x$. Let $\pi_2 : M \times N \to N$ be the mapping defined as follows: For every $(x, y) \in M \times N$, $\pi_2(x, y) \equiv y$. We know that $\pi_1 : M \times N \to M$ is a smooth mapping. Since $\pi_1 : M \times N \to M$ is a smooth mapping, and $(p, q) \in M \times N$, its differential $d(\pi_1)_{(p,q)} : T_{(p,q)}(M \times N) \to T_{\pi_1(p,q)}M(= T_pM)$ is a linear map, that is, $d(\pi_1)_{(p,q)} : T_{(p,q)}(M \times N) \to T_pM$ is a linear map. Similarly, $d(\pi_2)_{(p,q)} : T_{(p,q)}(M \times N) \to T_qN$ is a linear map.

Let us define a mapping $\alpha : T_{(p,q)}(M \times N) \to T_pM \oplus T_qN$ as follows: For every $v$ in $T_{(p,q)}(M \times N)$,

$$\alpha(v) = \left( \left( d(\pi_1)_{(p,q)} \right)(v), \left( d(\pi_2)_{(p,q)} \right)(v) \right).$$

Since $v \mapsto (d(\pi_1)_{(p,q)})(v)$ is linear, and $v \mapsto (d(\pi_2)_{(p,q)})(v)$ is linear, $v \mapsto ((d(\pi_1)_{(p,q)})(v), (d(\pi_2)_{(p,q)})(v))$ is linear, and hence, $\alpha : T_{(p,q)}(M \times N) \to T_pM \oplus T_qN$ is linear.

Similarly, we can show that let $M$ be an $m$-dimensional smooth manifold, $N$ be an $n$-dimensional smooth manifold, and $(p, q) \in M \times N$. Define $\alpha : T_{(p,q)}(M \times N) \to T_pM \oplus T_qN$ as follows: For every $v$ in $T_{(p,q)}(M \times N)$, $\alpha(v) \equiv ((d(\pi_1)_{(p,q)})(v), (d(\pi_2)_{(p,q)})(v))$. Then, $\alpha$ is linear, etc.

**Note 5.16** Let $M$ be an $m$-dimensional smooth manifold. Let $p \in M$. Let $U$ be an open neighborhood of $p$. We know that $U$ is also a smooth manifold (called open submanifold $U$ of $M$). Let $\iota : U \to M$ be the mapping defined as follows: For every $x$ in $U$, $\iota(x) \equiv x$. We know that the differential $d\iota_p : T_pU \to T_pM$ is an isomorphism.

So, from now on, we shall not distinguish between a derivation $v$ in $T_pU$, and the derivation $(d\iota_p)(v)$ in $T_pM$. Also, we shall not distinguish between $T_pU$ and $T_pM$.

Here, let $v$ be in $T_pU$. So $v : C^\infty(U) \to \mathbb{R}$. Since $(d\iota_p)(v)$ is in $T_pM$, $(d\iota_p)(v) : C^\infty(M) \to \mathbb{R}$. Observe that the action of "derivation at a point" on a function depends only on the values of the function in any open neighborhood of the point.

Also, we shall not distinguish between a derivation $v$ in $T_pM$, and the derivation $(\mathrm{d}\iota_p)^{-1}(v)$ in $T_pU$.

**Note 5.17** Let $M$ be an $m$-dimensional smooth manifold. Let $(U, \varphi)$ be an admissible coordinate chart of $M$. Let us fix any $p \in U$.

Since $(U, \varphi)$ is an admissible coordinate chart of $M$ satisfying $p \in U$, there exists an open neighborhood $G$ of $\varphi(p)$ in $\mathbb{R}^m$ such that $\varphi : U \to G$ is a diffeomorphism. Since $\varphi : U \to G$ is a diffeomorphism, $\mathrm{d}\varphi_p : T_pU \to T_{\varphi(p)}G$ is an isomorphism. Thus, for every $v \in T_pU$, and for every $f \in C^\infty(G)$,

$$\left((\mathrm{d}\varphi_p)(v)\right)(f) = v(f \circ \varphi).$$

Since $M$ is an $m$-dimensional smooth manifold, $p \in M$, and $U$ is an open neighborhood of $p$, by Theorem 5.13, the differential $\mathrm{d}\iota_p : T_pU \to T_{\iota(p)}M(= T_pM)$ is an isomorphism, where $\iota$ denotes the inclusion map of $U$. Since $\mathrm{d}\iota_p : T_pU \to T_pM$ is an isomorphism, $(\mathrm{d}\iota_p)^{-1} : T_pM \to T_pU$ is an isomorphism. Thus, for every $v \in T_pU$, and for every $f \in C^\infty(M)$,

$$\left((\mathrm{d}\iota_p)(v)\right)(f) = v(f \circ \iota) = v\left(f|_U\right).$$

Since $\mathbb{R}^m$ is an $m$-dimensional smooth manifold, $\varphi(p) \in \mathbb{R}^m$, and $G$ is an open neighborhood of $\varphi(p)$, by Theorem 5.13, the differential $\mathrm{d}\hat{\imath}_{\varphi(p)} : T_{\varphi(p)}G \to T_{\hat{\imath}(\varphi(p))}\mathbb{R}^m(= T_{\varphi(p)}\mathbb{R}^m)$ is an isomorphism, where $\hat{\imath}$ denotes the inclusion map of $G$. Thus, for every $v \in T_{\varphi(p)}G$, and for every $f \in C^\infty(\mathbb{R}^m)$,

$$\left((\mathrm{d}\hat{\imath}_{\varphi(p)})(v)\right)(f) = v(f \circ \hat{\imath}) = v\left(f|_G\right).$$

Since $(\mathrm{d}\iota_p)^{-1} : T_pM \to T_pU$ is an isomorphism, $\mathrm{d}\varphi_p : T_pU \to T_{\varphi(p)}G$ is an isomorphism, and $\mathrm{d}\hat{\imath}_{\varphi(p)} : T_{\varphi(p)}G \to T_{\varphi(p)}\mathbb{R}^m$ is an isomorphism, their composite $(\mathrm{d}\hat{\imath}_{\varphi(p)}) \circ (\mathrm{d}\varphi_p) \circ (\mathrm{d}\iota_p)^{-1} : T_pM \to T_{\varphi(p)}\mathbb{R}^m$ is an isomorphism. It follows that $(\mathrm{d}\iota_p) \circ (\mathrm{d}\varphi_p)^{-1} \circ (\mathrm{d}\hat{\imath}_{\varphi(p)})^{-1} : T_{\varphi(p)}\mathbb{R}^m \to T_pM$ is an isomorphism. Further, since $\varphi : U \to G$ is a diffeomorphism, $(\mathrm{d}\varphi_p)^{-1} = \mathrm{d}(\varphi^{-1})_{\varphi(p)}$, and hence, $(\mathrm{d}\iota_p) \circ \mathrm{d}(\varphi^{-1})_{\varphi(p)} \circ (\mathrm{d}\hat{\imath}_{\varphi(p)})^{-1} : T_{\varphi(p)}\mathbb{R}^m \to T_pM$ is an isomorphism. Since $(\mathrm{d}\iota_p) \circ \mathrm{d}(\varphi^{-1})_{\varphi(p)} \circ (\mathrm{d}\hat{\imath}_{\varphi(p)})^{-1} : T_{\varphi(p)}\mathbb{R}^m \to T_pM$ is an isomorphism, and

$$\left\{ \left.\frac{\partial}{\partial x^1}\right|_{\varphi(p)}, \left.\frac{\partial}{\partial x^2}\right|_{\varphi(p)}, \ldots, \left.\frac{\partial}{\partial x^m}\right|_{\varphi(p)} \right\}$$

is a basis of $T_{\varphi(p)}\mathbb{R}^m$,

$$\left\{ \left( (\mathrm{d}\iota_p) \circ \mathrm{d}(\varphi^{-1})_{\varphi(p)} \circ (\mathrm{d}\hat{\iota}_{\varphi(p)})^{-1} \right) \left( \frac{\partial}{\partial x^1} \bigg|_{\varphi(p)} \right), \left( (\mathrm{d}\iota_p) \circ \mathrm{d}(\varphi^{-1})_{\varphi(p)} \circ (\mathrm{d}\hat{\iota}_{\varphi(p)})^{-1} \right) \right.$$
$$\left. \left( \frac{\partial}{\partial x^2} \bigg|_{\varphi(p)} \right), \dots, \left( (\mathrm{d}\iota_p) \circ \mathrm{d}(\varphi^{-1})_{\varphi(p)} \circ (\mathrm{d}\hat{\iota}_{\varphi(p)})^{-1} \right) \left( \frac{\partial}{\partial x^m} \bigg|_{\varphi(p)} \right) \right\}$$

is a basis of $T_p M$. Here, we can write:

$$\left( (\mathrm{d}\iota_p) \circ \mathrm{d}(\varphi^{-1})_{\varphi(p)} \circ (\mathrm{d}\hat{\iota}_{\varphi(p)})^{-1} \right) \left( \frac{\partial}{\partial x^1} \bigg|_{\varphi(p)} \right)$$

$$= \left( (\mathrm{d}\iota_p) \circ \mathrm{d}(\varphi^{-1})_{\varphi(p)} \right) \left( \left( (\mathrm{d}\hat{\iota}_{\varphi(p)})^{-1} \right) \left( \frac{\partial}{\partial x^1} \bigg|_{\varphi(p)} \right) \right)$$

$$= \left( (\mathrm{d}\iota_p) \circ \mathrm{d}(\varphi^{-1})_{\varphi(p)} \right) \left( \frac{\partial}{\partial x^1} \bigg|_{\varphi(p)} \right) = (\mathrm{d}\iota_p) \left( \left( \mathrm{d}(\varphi^{-1})_{\varphi(p)} \right) \left( \frac{\partial}{\partial x^1} \bigg|_{\varphi(p)} \right) \right)$$

$$= \left( \mathrm{d}(\varphi^{-1})_{\varphi(p)} \right) \left( \frac{\partial}{\partial x^1} \bigg|_{\varphi(p)} \right).$$

Thus,

$$\left( (\mathrm{d}\iota_p) \circ \mathrm{d}(\varphi^{-1})_{\varphi(p)} \circ (\mathrm{d}\hat{\iota}_{\varphi(p)})^{-1} \right) \left( \frac{\partial}{\partial x^1} \bigg|_{\varphi(p)} \right) = \left( \mathrm{d}(\varphi^{-1})_{\varphi(p)} \right) \left( \frac{\partial}{\partial x^1} \bigg|_{\varphi(p)} \right).$$

Similarly,

$$\left( (\mathrm{d}\iota_p) \circ \mathrm{d}(\varphi^{-1})_{\varphi(p)} \circ (\mathrm{d}\hat{\iota}_{\varphi(p)})^{-1} \right) \left( \frac{\partial}{\partial x^2} \bigg|_{\varphi(p)} \right) = \left( \mathrm{d}(\varphi^{-1})_{\varphi(p)} \right) \left( \frac{\partial}{\partial x^2} \bigg|_{\varphi(p)} \right), \text{ etc,}$$

Thus,

$$\left\{ \left( \mathrm{d}(\varphi^{-1})_{\varphi(p)} \right) \left( \frac{\partial}{\partial x^1} \bigg|_{\varphi(p)} \right), \left( \mathrm{d}(\varphi^{-1})_{\varphi(p)} \right) \left( \frac{\partial}{\partial x^2} \bigg|_{\varphi(p)} \right), \dots, \left( \mathrm{d}(\varphi^{-1})_{\varphi(p)} \right) \left( \frac{\partial}{\partial x^m} \bigg|_{\varphi(p)} \right) \right\}$$

is a basis of $T_p M$. For $i = 1, 2, \dots, m$, the tangent vector

$$\left( \mathrm{d}(\varphi^{-1})_{\varphi(p)} \right) \left( \frac{\partial}{\partial x^i} \bigg|_{\varphi(p)} \right)$$

at $p$ is denoted by $\frac{\partial}{\partial x^i}\big|_p$. Thus,

$$\left\{ \frac{\partial}{\partial x^1}\bigg|_p, \frac{\partial}{\partial x^2}\bigg|_p, \ldots, \frac{\partial}{\partial x^m}\bigg|_p \right\}$$

is a basis of $T_pM$.

**Conclusion**: Let $M$ be an $m$-dimensional smooth manifold. Let $(U, \varphi)$ be an admissible coordinate chart of $M$. Then, for every $p \in U$,

$$\left\{ \frac{\partial}{\partial x^1}\bigg|_p, \frac{\partial}{\partial x^2}\bigg|_p, \ldots, \frac{\partial}{\partial x^m}\bigg|_p \right\}$$

is a basis of $T_pM$, where $\frac{\partial}{\partial x^i}\big|_p$ stands for

$$\left( d(\varphi^{-1})_{\varphi(p)} \right) \left( \frac{\partial}{\partial x^i}\bigg|_{\varphi(p)} \right).$$

**Definition** Let $M$ be an $m$-dimensional smooth manifold. Let $(U, \varphi)$ be an admissible coordinate chart of $M$. For every $p \in M$, the ordered basis

$$\left( \frac{\partial}{\partial x^1}\bigg|_p, \frac{\partial}{\partial x^2}\bigg|_p, \ldots, \frac{\partial}{\partial x^m}\bigg|_p \right),$$

as defined in the Note 5.17, is called the *coordinate basis of $T_pM$*. If $v \in T_pM$, then there exists a unique $m$-tuple $(v^1, v^2, \ldots, v^m)$ of real numbers such that

$$v = v^1 \left( \frac{\partial}{\partial x^1}\bigg|_p \right) + v^2 \left( \frac{\partial}{\partial x^2}\bigg|_p \right) + \cdots + v^m \left( \frac{\partial}{\partial x^m}\bigg|_p \right).$$

Here, we say that $(v^1, v^2, \ldots, v^m)$ *are the components of $v$ with respect to coordinate basis*.

**Note 5.18** Let $U$ be an open subset of $\mathbb{R}^n$, $V$ be an open subset of $\mathbb{R}^m$, and $p \in U$. Let $F : U \to V$ be smooth, where $F \equiv (F^1, F^2, \ldots, F^m)$.

Since $U$ is a nonempty open subset of the $n$-dimensional smooth manifold $\mathbb{R}^n$, $U$ is also an $n$-dimensional smooth manifold. Similarly, $V$ is an $m$-dimensional smooth manifold. It follows that $dF_p : T_pU \to T_{F(p)}V$ is a linear map from real linear space $T_pU$ to real linear space $T_{F(p)}V$. Here,

$$\left(\left.\frac{\partial}{\partial x^1}\right|_p, \left.\frac{\partial}{\partial x^2}\right|_p, \ldots, \left.\frac{\partial}{\partial x^n}\right|_p\right)$$

is the coordinate basis of $T_p\mathbb{R}^n$, and we do not distinguish between $T_pU$ and $T_p\mathbb{R}^n$, so

$$\left(\left.\frac{\partial}{\partial x^1}\right|_p, \left.\frac{\partial}{\partial x^2}\right|_p, \ldots, \left.\frac{\partial}{\partial x^n}\right|_p\right)$$

is the coordinate basis of $T_pU$. Similarly,

$$\left(\left.\frac{\partial}{\partial y^1}\right|_{F(p)}, \left.\frac{\partial}{\partial y^2}\right|_{F(p)}, \ldots, \left.\frac{\partial}{\partial y^m}\right|_{F(p)}\right)$$

is the coordinate basis of $T_{F(p)}V$. Now, since $dF_p : T_pU \to T_{F(p)}V$, for every $f \in C^\infty(V)$,

$$\left((dF_p)\left(\left.\frac{\partial}{\partial x^1}\right|_p\right)\right)(f) = \left(\left.\frac{\partial}{\partial x^1}\right|_p\right)(f \circ F)$$

$$= \text{1st column of } [(D_1f)(F(p))$$

$$\cdots (D_mf)(F(p))] \begin{bmatrix} (D_1F^1)(p) & & (D_nF^1)(p) \\ \vdots & \ddots & \vdots \\ (D_1F^m)(p) & & (D_nF^m)(p) \end{bmatrix}$$

$$= ((D_1f)(F(p)))((D_1F^1)(p))$$
$$+ \cdots + ((D_mf)(F(p)))((D_1F^m)(p))$$

$$= \sum_{i=1}^m \left(\frac{\partial f}{\partial y^i}(F(p))\right)\left(\frac{\partial F^i}{\partial x^1}(p)\right)$$

$$= \sum_{i=1}^m \left(\left(\frac{\partial F^i}{\partial x^1}(p)\right)\left(\frac{\partial f}{\partial y^i}(F(p))\right)\right)$$

$$= \sum_{i=1}^m \left(\left(\frac{\partial F^i}{\partial x^1}(p)\right)\left(\left(\left.\frac{\partial}{\partial y^i}\right|_{F(p)}\right)f\right)\right)$$

$$= \sum_{i=1}^m \left(\left(\left(\frac{\partial F^i}{\partial x^1}(p)\right)\left(\left.\frac{\partial}{\partial y^i}\right|_{F(p)}\right)\right)f\right)$$

$$= \left(\sum_{i=1}^m \left(\frac{\partial F^i}{\partial x^1}(p)\right)\left(\left.\frac{\partial}{\partial y^i}\right|_{F(p)}\right)\right)f$$

and hence,

$$(\mathrm{d}F_p)\left(\left.\frac{\partial}{\partial x^1}\right|_p\right) = \sum_{i=1}^{m}\left(\frac{\partial F^i}{\partial x^1}(p)\right)\left(\left.\frac{\partial}{\partial y^i}\right|_{F(p)}\right).$$

Similarly,

$$(\mathrm{d}F_p)\left(\left.\frac{\partial}{\partial x^2}\right|_p\right) = \sum_{i=1}^{m}\left(\frac{\partial F^i}{\partial x^2}(p)\right)\left(\left.\frac{\partial}{\partial y^i}\right|_{F(p)}\right), \text{etc.}$$

Since for every $j = 1, 2, \ldots, n$,

$$(\mathrm{d}F_p)\left(\left.\frac{\partial}{\partial x^j}\right|_p\right) = \sum_{i=1}^{m}\left(\frac{\partial F^i}{\partial x^j}(p)\right)\left(\left.\frac{\partial}{\partial y^i}\right|_{F(p)}\right),$$

and

$$\left(\left.\frac{\partial}{\partial y^1}\right|_{F(p)}, \left.\frac{\partial}{\partial y^2}\right|_{F(p)}, \ldots, \left.\frac{\partial}{\partial y^m}\right|_{F(p)}\right)$$

is the coordinate basis of $T_{F(p)}V$, the matrix representation of linear map $\mathrm{d}F_p$ is the following $m \times n$ matrix:

$$\begin{bmatrix} \frac{\partial F^1}{\partial x^1}(p) & \frac{\partial F^1}{\partial x^2}(p) & \cdots & \frac{\partial F^1}{\partial x^n}(p) \\ \frac{\partial F^2}{\partial x^1}(p) & \frac{\partial F^2}{\partial x^2}(p) & \cdots & \frac{\partial F^2}{\partial x^n}(p) \\ \cdots & \cdots & \cdots & \cdots \\ \frac{\partial F^m}{\partial x^1}(p) & \frac{\partial F^m}{\partial x^2}(p) & \cdots & \frac{\partial F^m}{\partial x^n}(p) \end{bmatrix}.$$

Let us recall that this matrix is the matrix representation of the total derivative $DF(p) : \mathbb{R}^n \to \mathbb{R}^m$. Thus, we find that in this case, differential $\mathrm{d}F_p : T_p\mathbb{R}^n \to T_{F(p)}\mathbb{R}^m$ corresponds to the total derivative $DF(p) : \mathbb{R}^n \to \mathbb{R}^m$, tangent space $T_p\mathbb{R}^n$ corresponds $\mathbb{R}^n$, and tangent space $T_{F(p)}\mathbb{R}^m$ corresponds $\mathbb{R}^m$.

**Note 5.19** Let $M$ be an $m$-dimensional smooth manifold, and $N$ be an $n$-dimensional smooth manifold. Let $p \in M$. Let $F : M \to N$ be a smooth mapping. Let $(U, \varphi)$ be an admissible coordinate chart of $M$ such that $p \in U$. Let $(V, \psi)$ be an admissible coordinate chart of $N$ such that $F(p) \in V$. Since $F : M \to N$ is a smooth mapping, $(\psi \circ F \circ \varphi^{-1}) : \varphi(U \cap F^{-1}(V)) \to \psi(V)$ is smooth. Put $(\hat{F}^1, \hat{F}^2, \ldots, \hat{F}^n) \equiv \hat{F} \equiv \psi \circ F \circ \varphi^{-1}$. Thus, $\hat{F} : \varphi(U \cap F^{-1}(V)) \to \psi(V)$ is smooth, that is, each $\pi_i \circ (\psi \circ F \circ \varphi^{-1})$ is smooth, where $\pi_i$ denotes the $i$th projection function from $\mathbb{R}^n$ to $\mathbb{R}$.

Since $F : M \to N$ is a smooth, $F : M \to N$ is continuous. Since $F : M \to N$ is continuous, and $V$ is open in $N$, $F^{-1}(V)$ is open in $M$. Since $F^{-1}(V)$ is open in $M$, and $U$ is open in $M$, $U \cap F^{-1}(V)$ is open in $M$. Clearly, $p \in U \cap F^{-1}(V)$. Thus, $U \cap F^{-1}(V)$ is an open neighborhood of $p$. Since $U \cap F^{-1}(V) \subset U$, $U \cap F^{-1}(V)$ is open in $M$, and $U$ is open in $M$, $U \cap F^{-1}(V)$ is open in $U$. Since $U \cap F^{-1}(V)$ is an open neighborhood of $p$ in $U$, and $\varphi$ is a homeomorphism from $U$ onto an open subset of $\mathbb{R}^m$, $\varphi(U \cap F^{-1}(V))$ is an open neighborhood of $\varphi(p)$ in $\mathbb{R}^m$. Clearly, $\psi(V)$ is open in $\mathbb{R}^n$. Now, since $\hat{F} : \varphi(U \cap F^{-1}(V)) \to \psi(V)$ is smooth, as in the Note 5.18, the matrix representation of $d\hat{F}_{\varphi(p)} : T_{\varphi(p)}\mathbb{R}^m \to T_{\hat{F}(\varphi(p))}\mathbb{R}^n (= T_{(\hat{F} \circ \varphi)(p)}\mathbb{R}^n = T_{(\psi \circ F)(p)}\mathbb{R}^n)$ is

$$
\begin{bmatrix}
\frac{\partial \hat{F}^1}{\partial x^1}(\varphi(p)) & \frac{\partial \hat{F}^1}{\partial x^2}(\varphi(p)) & \cdots & \frac{\partial \hat{F}^1}{\partial x^m}(\varphi(p)) \\
\frac{\partial \hat{F}^2}{\partial x^1}(\varphi(p)) & \frac{\partial \hat{F}^2}{\partial x^2}(\varphi(p)) & \cdots & \frac{\partial \hat{F}^2}{\partial x^m}(\varphi(p)) \\
\cdots & \cdots & \cdots & \cdots \\
\frac{\partial \hat{F}^n}{\partial x^1}(\varphi(p)) & \frac{\partial \hat{F}^n}{\partial x^2}(\varphi(p)) & \cdots & \frac{\partial \hat{F}^n}{\partial x^m}(\varphi(p))
\end{bmatrix}.
$$

Since $\hat{F} \equiv \psi \circ F \circ \varphi^{-1}$, and $\varphi$, $\psi$ are 1–1, $\psi^{-1} \circ \hat{F} = F \circ \varphi^{-1}$. Here, $F \circ \varphi^{-1} : \varphi(U) \to N$, so $d(F \circ \varphi^{-1})_{\varphi(p)} : T_{\varphi(p)}\mathbb{R}^m \to T_{F(p)}N$. Similarly $d(\psi^{-1} \circ \hat{F})_{\varphi(p)} : T_{\varphi(p)}\mathbb{R}^m \to T_{F(p)}N$. Here

$$
\left( \frac{\partial}{\partial x^1} \bigg|_p, \frac{\partial}{\partial x^2} \bigg|_p, \ldots, \frac{\partial}{\partial x^m} \bigg|_p \right)
$$

is the coordinate basis of $T_pM$, where

$$
\frac{\partial}{\partial x^i} \bigg|_p \equiv \left( d(\varphi^{-1})_{\varphi(p)} \right) \left( \frac{\partial}{\partial x^i} \bigg|_{\varphi(p)} \right).
$$

Also,

$$
\left( \frac{\partial}{\partial y^1} \bigg|_{F(p)}, \frac{\partial}{\partial y^2} \bigg|_{F(p)}, \ldots, \frac{\partial}{\partial y^n} \bigg|_{F(p)} \right)
$$

is the coordinate basis of $T_{F(p)}N$, where

$$
\frac{\partial}{\partial y^j} \bigg|_{F(p)} \equiv \left( d(\psi^{-1})_{\psi(F(p))} \right) \left( \frac{\partial}{\partial y^j} \bigg|_{\psi(F(p))} \right).
$$

Now, we want to obtain the matrix representation of $dF_p$.

Since $F : M \to N$ is a smooth mapping, $\mathrm{d}F_p : T_pM \to T_{F(p)}N$. Next, for $i = 1, 2, \ldots, m$,

$$
(\mathrm{d}F_p)\left(\left.\frac{\partial}{\partial x^i}\right|_p\right) = (\mathrm{d}F_p)\left(\left(\mathrm{d}(\varphi^{-1})_{\varphi(p)}\right)\left(\left.\frac{\partial}{\partial x^i}\right|_{\varphi(p)}\right)\right)
$$

$$
= \left((\mathrm{d}F_p) \circ \left(\mathrm{d}(\varphi^{-1})_{\varphi(p)}\right)\right)\left(\left.\frac{\partial}{\partial x^i}\right|_{\varphi(p)}\right)
$$

$$
= \left((\mathrm{d}F_{\varphi^{-1}(\varphi(p))}) \circ \left(\mathrm{d}(\varphi^{-1})_{\varphi(p)}\right)\right)\left(\left.\frac{\partial}{\partial x^i}\right|_{\varphi(p)}\right)
$$

$$
= \left(\mathrm{d}(F \circ (\varphi^{-1}))_{\varphi(p)}\right)\left(\left.\frac{\partial}{\partial x^i}\right|_{\varphi(p)}\right)
$$

$$
= \left(\mathrm{d}(\psi^{-1} \circ \hat{F})_{\varphi(p)}\right)\left(\left.\frac{\partial}{\partial x^i}\right|_{\varphi(p)}\right)
$$

$$
= \left(\left(\mathrm{d}(\psi^{-1})_{\hat{F}(\varphi(p))}\right) \circ (\mathrm{d}\hat{F}_{\varphi(p)})\right)\left(\left.\frac{\partial}{\partial x^i}\right|_{\varphi(p)}\right)
$$

$$
= \left(\mathrm{d}(\psi^{-1})_{\hat{F}(\varphi(p))}\right)\left((\mathrm{d}\hat{F}_{\varphi(p)})\left(\left.\frac{\partial}{\partial x^i}\right|_{\varphi(p)}\right)\right)
$$

$$
= \left(\mathrm{d}(\psi^{-1})_{F(\varphi(p))}\right)(i\text{th column of }
$$

$$
\begin{bmatrix}
\frac{\partial \hat{F}^1}{\partial x^1}(\varphi(p)) & \frac{\partial \hat{F}^1}{\partial x^2}(\varphi(p)) & \cdots & \frac{\partial \hat{F}^1}{\partial x^m}(\varphi(p)) \\
\frac{\partial \hat{F}^2}{\partial x^1}(\varphi(p)) & \frac{\partial \hat{F}^2}{\partial x^2}(\varphi(p)) & \cdots & \frac{\partial \hat{F}^2}{\partial x^m}(\varphi(p)) \\
\cdots & \cdots & \cdots\cdots & \cdots \\
\frac{\partial \hat{F}^n}{\partial x^1}(\varphi(p)) & \frac{\partial \hat{F}^n}{\partial x^2}(\varphi(p)) & \cdots & \frac{\partial \hat{F}^n}{\partial x^m}(\varphi(p))
\end{bmatrix})
$$

$$
= \left(\mathrm{d}(\psi^{-1})_{\hat{F}(\varphi(p))}\right)\left(\sum_{j=1}^{n}\left(\frac{\partial \hat{F}^j}{\partial x^i}(\varphi(p))\right)\left(\left.\frac{\partial}{\partial y^j}\right|_{\hat{F}(\varphi(p))}\right)\right)
$$

$$
= \left(\mathrm{d}(\psi^{-1})_{(\hat{F} \circ \varphi)(p)}\right)\left(\sum_{j=1}^{n}\left(\frac{\partial \hat{F}^j}{\partial x^i}(\varphi(p))\right)\left(\left.\frac{\partial}{\partial y^j}\right|_{(\hat{F} \circ \varphi)(p)}\right)\right)
$$

$$
= \left(\mathrm{d}(\psi^{-1})_{(\psi \circ F)(p)}\right)\left(\sum_{j=1}^{n}\left(\frac{\partial \hat{F}^j}{\partial x^i}(\varphi(p))\right)\left(\left.\frac{\partial}{\partial y^j}\right|_{(\psi \circ F)(p)}\right)\right)
$$

$$
= \sum_{j=1}^{n}\left(\frac{\partial \hat{F}^j}{\partial x^i}(\varphi(p))\right)\left(\left(\mathrm{d}(\psi^{-1})_{(\psi \circ F)(p)}\right)\left(\left.\frac{\partial}{\partial y^j}\right|_{(\psi \circ F)(p)}\right)\right)
$$

$$
= \sum_{j=1}^{n}\left(\frac{\partial \hat{F}^j}{\partial x^i}(\varphi(p))\right)\left(\left(\mathrm{d}(\psi^{-1})_{\psi(F(p))}\right)\left(\left.\frac{\partial}{\partial y^j}\right|_{\psi(F(p))}\right)\right)
$$

$$
= \sum_{j=1}^{n}\left(\frac{\partial \hat{F}^j}{\partial x^i}(\varphi(p))\right)\left(\left.\frac{\partial}{\partial y^j}\right|_{F(p)}\right).
$$

Thus,

$$(dF_p)\left(\frac{\partial}{\partial x^i}\Big|_p\right) = \sum_{j=1}^{n}\left(\frac{\partial \hat{F}^j}{\partial x^i}(\varphi(p))\right)\left(\frac{\partial}{\partial y^j}\Big|_{F(p)}\right).$$

Since for every $i = 1, 2, \ldots, m,$

$$(dF_p)\left(\frac{\partial}{\partial x^i}\Big|_p\right) = \sum_{j=1}^{n}\left(\frac{\partial \hat{F}^j}{\partial x^i}(\varphi(p))\right)\left(\frac{\partial}{\partial y^j}\Big|_{F(p)}\right),$$

and

$$\left(\frac{\partial}{\partial y^1}\Big|_{F(p)}, \frac{\partial}{\partial y^2}\Big|_{F(p)}, \ldots, \frac{\partial}{\partial y^n}\Big|_{F(p)}\right)$$

is the coordinate basis of $T_{F(p)}N$, the matrix representation of linear map $dF_p$ is the following $n \times m$ matrix:

$$\begin{bmatrix} \frac{\partial \hat{F}^1}{\partial x^1}(\varphi(p)) & \frac{\partial \hat{F}^1}{\partial x^2}(\varphi(p)) & \cdots & \frac{\partial \hat{F}^1}{\partial x^m}(\varphi(p)) \\ \frac{\partial \hat{F}^2}{\partial x^1}(\varphi(p)) & \frac{\partial \hat{F}^2}{\partial x^2}(\varphi(p)) & \cdots & \frac{\partial \hat{F}^2}{\partial x^m}(\varphi(p)) \\ \cdots & \cdots & \cdots & \cdots \\ \frac{\partial \hat{F}^n}{\partial x^1}(\varphi(p)) & \frac{\partial \hat{F}^n}{\partial x^2}(\varphi(p)) & \cdots & \frac{\partial \hat{F}^n}{\partial x^m}(\varphi(p)) \end{bmatrix}$$

$$= \begin{bmatrix} \frac{\partial\left(\pi_1 \circ \left(\psi \circ F \circ \varphi^{-1}\right)\right)}{\partial x^1}(\varphi(p)) & \frac{\partial\left(\pi_1 \circ \left(\psi \circ F \circ \varphi^{-1}\right)\right)}{\partial x^2}(\varphi(p)) & \cdots & \frac{\partial\left(\pi_1 \circ \left(\psi \circ F \circ \varphi^{-1}\right)\right)}{\partial x^m}(\varphi(p)) \\ \frac{\partial\left(\pi_2 \circ \left(\psi \circ F \circ \varphi^{-1}\right)\right)}{\partial x^1}(\varphi(p)) & \frac{\partial\left(\pi_2 \circ \left(\psi \circ F \circ \varphi^{-1}\right)\right)}{\partial x^2}(\varphi(p)) & \cdots & \frac{\partial\left(\pi_2 \circ \left(\psi \circ F \circ \varphi^{-1}\right)\right)}{\partial x^m}(\varphi(p)) \\ \cdots & \cdots & \cdots & \cdots \\ \frac{\partial\left(\pi_n \circ \left(\psi \circ F \circ \varphi^{-1}\right)\right)}{\partial x^1}(\varphi(p)) & \frac{\partial\left(\pi_n \circ \left(\psi \circ F \circ \varphi^{-1}\right)\right)}{\partial x^2}(\varphi(p)) & \cdots & \frac{\partial\left(\pi_n \circ \left(\psi \circ F \circ \varphi^{-1}\right)\right)}{\partial x^m}(\varphi(p)) \end{bmatrix}.$$

It follows that the matrix representation of $dF_p$ is the same as the matrix representation of $d(\psi \circ F \circ \varphi^{-1})_{\varphi(p)}$.

**Note 5.20** Let $M$ be an $m$-dimensional smooth manifold. Let $p \in M$. Let $(U, \varphi)$ be an admissible coordinate chart of $M$ such that $p \in U$. Let $(V, \psi)$ be an admissible coordinate chart of $M$ such that $p \in V$. Let $\varphi(p) \equiv (x^1, x^2, \ldots, x^m)$, and $\psi(p) \equiv \tilde{x}^1, \tilde{x}^2, \ldots, \tilde{x}^m$. Let

$$\left(\frac{\partial}{\partial x^1}\Big|_p, \frac{\partial}{\partial x^2}\Big|_p, \ldots, \frac{\partial}{\partial x^m}\Big|_p\right)$$

be the coordinate basis of $T_pM$ corresponding to $(U, \varphi)$, where

$$\frac{\partial}{\partial x^i}\bigg|_p \equiv \left(d(\varphi^{-1})_{\varphi(p)}\right)\left(\frac{\partial}{\partial x^i}\bigg|_{\varphi(p)}\right).$$

Let

$$\left(\frac{\partial}{\partial \tilde{x}^1}\bigg|_p, \frac{\partial}{\partial \tilde{x}^2}\bigg|_p, \dots, \frac{\partial}{\partial \tilde{x}^m}\bigg|_p\right)$$

be the coordinate basis of $T_pM$ corresponding to $(V, \psi)$, where

$$\frac{\partial}{\partial \tilde{x}^i}\bigg|_p \equiv \left(d(\psi^{-1})_{\psi(p)}\right)\left(\frac{\partial}{\partial \tilde{x}^i}\bigg|_{\psi(p)}\right).$$

Let $v \in T_pM$. Let $(v^1, v^2, \dots, v^m)$ be the components of $v$ with respect to coordinate basis

$$\left(\frac{\partial}{\partial x^1}\bigg|_p, \frac{\partial}{\partial x^2}\bigg|_p, \dots, \frac{\partial}{\partial x^m}\bigg|_p\right)$$

of $T_pM$. Let $(\tilde{v}^1, \tilde{v}^2, \dots, \tilde{v}^m)$ be the components of $v$ with respect to coordinate basis

$$\left(\frac{\partial}{\partial \tilde{x}^1}\bigg|_p, \frac{\partial}{\partial \tilde{x}^2}\bigg|_p, \dots, \frac{\partial}{\partial \tilde{x}^m}\bigg|_p\right)$$

of $T_pM$. Thus,

$$\tilde{v}^1\left(\frac{\partial}{\partial \tilde{x}^1}\bigg|_p\right) + \tilde{v}^2\left(\frac{\partial}{\partial \tilde{x}^2}\bigg|_p\right) + \cdots + \tilde{v}^m\left(\frac{\partial}{\partial \tilde{x}^m}\bigg|_p\right) = v$$

$$= v^1\left(\frac{\partial}{\partial x^1}\bigg|_p\right) + v^2\left(\frac{\partial}{\partial x^2}\bigg|_p\right) + \cdots + v^m\left(\frac{\partial}{\partial x^m}\bigg|_p\right).$$

Since $(U, \varphi)$ and $(V, \psi)$ are admissible coordinate charts of $M$ such that $p \in U \cap V$, $\psi \circ \varphi^{-1} : \varphi(U \cap V) \to \psi(U \cap V)$ is smooth.

Here, with the hope that there will be no confusion, for every $i = 1, 2, \dots, m$, the $i$th component of the "transition map" $\psi \circ \varphi^{-1}$ is generally denoted by $\tilde{x}^i$. Thus, $\tilde{x}^i$ has two meanings: the $i$th component of $\psi(p)$, and the $i$th component function of $\psi \circ \varphi^{-1}$. The situation indicates which meaning is to be attached with. Here, for every $x \in \varphi(U \cap V)$, $(\psi \circ \varphi^{-1})(x) = (\tilde{x}^1(x), \tilde{x}^2(x), \dots, \tilde{x}^m(x))$.

Since $\varphi(U \cap V)$ is an open neighborhood of $\varphi(p)$ in $\mathbb{R}^m$, $\psi(U \cap V)$ is an open neighborhood of $\psi(p)$ in $\mathbb{R}^m$, and $\psi \circ \varphi^{-1} : \varphi(U \cap V) \to \psi(U \cap V)$ is a diffeomorphism, the differential $\mathrm{d}(\psi \circ \varphi^{-1})_{\varphi(p)} : T_{\varphi(p)}\varphi(U \cap V) \to T_{\psi(p)}\psi(U \cap V)$ is an isomorphism, and hence, $\mathrm{d}(\psi \circ \varphi^{-1})_{\varphi(p)} : T_{\varphi(p)}\mathbb{R}^m \to T_{\psi(p)}\mathbb{R}^m$ is an isomorphism. Here,

$$
\left( \left.\frac{\partial}{\partial x^1}\right|_{\varphi(p)}, \left.\frac{\partial}{\partial x^2}\right|_{\varphi(p)}, \ldots, \left.\frac{\partial}{\partial x^m}\right|_{\varphi(p)} \right)
$$

is a basis of $T_{\varphi(p)}\mathbb{R}^m$. Hence, for $j = 1, 2, \ldots, m$,

$$
\left.\frac{\partial}{\partial x^j}\right|_p \equiv \left(\mathrm{d}(\varphi^{-1})_{\varphi(p)}\right)\left(\left.\frac{\partial}{\partial x^j}\right|_{\varphi(p)}\right) = \left(\mathrm{d}(\psi^{-1} \circ (\psi \circ \varphi^{-1}))_{\varphi(p)}\right)\left(\left.\frac{\partial}{\partial x^j}\right|_{\varphi(p)}\right)
$$

$$
= \left(\left(\mathrm{d}(\psi^{-1})_{(\psi \circ \varphi^{-1})(\varphi(p))}\right) \circ \left(\mathrm{d}(\psi \circ \varphi^{-1})_{\varphi(p)}\right)\right)\left(\left.\frac{\partial}{\partial x^j}\right|_{\varphi(p)}\right)
$$

$$
= \left(\left(\mathrm{d}(\psi^{-1})_{\psi(p)}\right) \circ \left(\mathrm{d}(\psi \circ \varphi^{-1})_{\varphi(p)}\right)\right)\left(\left.\frac{\partial}{\partial x^j}\right|_{\varphi(p)}\right)
$$

$$
= \left(\mathrm{d}(\psi^{-1})_{\psi(p)}\right)\left(\left(\mathrm{d}(\psi \circ \varphi^{-1})_{\varphi(p)}\right)\left(\left.\frac{\partial}{\partial x^j}\right|_{\varphi(p)}\right)\right)
$$

$$
= \left(\mathrm{d}(\psi^{-1})_{\psi(p)}\right)
$$
$$
\left( jth \text{ column of } \begin{bmatrix} \frac{\partial(\psi\circ\varphi^{-1})^1}{\partial x^1}(\varphi(p)) & \frac{\partial(\psi\circ\varphi^{-1})^1}{\partial x^2}(\varphi(p)) \cdots \frac{\partial(\psi\circ\varphi^{-1})^1}{\partial x^m}(\varphi(p)) \\ \frac{\partial(\psi\circ\varphi^{-1})^2}{\partial x^1}(\varphi(p)) & \frac{\partial(\psi\circ\varphi^{-1})^2}{\partial x^2}(\varphi(p)) \cdots \frac{\partial(\psi\circ\varphi^{-1})^2}{\partial x^m}(\varphi(p)) \\ \cdots\cdots\cdots\cdots \\ \frac{\partial(\psi\circ\varphi^{-1})^m}{\partial x^1}(\varphi(p)) & \frac{\partial(\psi\circ\varphi^{-1})^m}{\partial x^2}(\varphi(p)) \cdots \frac{\partial(\psi\circ\varphi^{-1})^m}{\partial x^m}(\varphi(p)) \end{bmatrix} \right)
$$

$$
= \left(\mathrm{d}(\psi^{-1})_{\psi(p)}\right)\left(\sum_{i=1}^{m}\left(\frac{\partial(\psi \circ \varphi^{-1})^i}{\partial x^j}(\varphi(p))\right)\left(\left.\frac{\partial}{\partial \tilde{x}^i}\right|_{(\psi\circ\varphi^{-1})(\varphi(p))}\right)\right)
$$

$$
= \left(\mathrm{d}(\psi^{-1})_{\psi(p)}\right)\left(\sum_{i=1}^{m}\left(\frac{\partial(\psi \circ \varphi^{-1})^i}{\partial x^j}(\varphi(p))\right)\left(\left.\frac{\partial}{\partial \tilde{x}^i}\right|_{\psi(p)}\right)\right)
$$

$$
= \left(\mathrm{d}(\psi^{-1})_{\psi(p)}\right)\left(\sum_{i=1}^{m}\left(\frac{\partial \overset{i}{x}}{\partial x^j}(\varphi(p))\right)\left(\left.\frac{\partial}{\partial \tilde{x}^i}\right|_{\psi(p)}\right)\right)
$$

$$
= \sum_{i=1}^{m}\left(\frac{\partial \tilde{x}^i}{\partial x^j}(\varphi(p))\right)\left(\left(\mathrm{d}(\psi^{-1})_{\psi(p)}\right)\left(\left.\frac{\partial}{\partial \tilde{x}^i}\right|_{\psi(p)}\right)\right)
$$

$$
= \sum_{i=1}^{m}\left(\frac{\partial \tilde{x}^i}{\partial x^j}(\varphi(p))\right)\left(\left.\frac{\partial}{\partial \tilde{x}^i}\right|_p\right).
$$

Thus, for $j = 1, 2, \ldots, m$,

$$\frac{\partial}{\partial x^j}\bigg|_p = \sum_{i=1}^{m} \left(\frac{\partial \tilde{x}^i}{\partial x^j}(\varphi(p))\right)\left(\frac{\partial}{\partial \tilde{x}^i}\bigg|_p\right).$$

Now,

$$\sum_{i=1}^{m} \tilde{v}^i \left(\frac{\partial}{\partial \tilde{x}^i}\bigg|_p\right) = \tilde{v}^1\left(\frac{\partial}{\partial \tilde{x}^1}\bigg|_p\right) + \tilde{v}^2\left(\frac{\partial}{\partial \tilde{x}^2}\bigg|_p\right) + \cdots + \tilde{v}^m\left(\frac{\partial}{\partial \tilde{x}^m}\bigg|_p\right) = v$$

$$= v^1\left(\frac{\partial}{\partial x^1}\bigg|_p\right) + v^2\left(\frac{\partial}{\partial x^2}\bigg|_p\right) + \cdots + v^m\left(\frac{\partial}{\partial x^m}\bigg|_p\right)$$

$$= \sum_{j=1}^{m} v^j\left(\frac{\partial}{\partial x^j}\bigg|_p\right) = \sum_{j=1}^{m} v^j\left(\sum_{i=1}^{m}\left(\frac{\partial \tilde{x}^i}{\partial x^j}(\varphi(p))\right)\left(\frac{\partial}{\partial \tilde{x}^i}\bigg|_p\right)\right)$$

$$= \sum_{j=1}^{m}\left(\sum_{i=1}^{m}\left(\frac{\partial \tilde{x}^i}{\partial x^j}(\varphi(p))\right)\left(v^j\left(\frac{\partial}{\partial \tilde{x}^i}\bigg|_p\right)\right)\right)$$

$$= \sum_{i=1}^{m}\left(\sum_{j=1}^{m}\left(\frac{\partial \tilde{x}^i}{\partial x^j}(\varphi(p))\right)\left(v^j\left(\frac{\partial}{\partial \tilde{x}^i}\bigg|_p\right)\right)\right)$$

$$= \sum_{i=1}^{m}\left(\left(\sum_{j=1}^{m}\left(\frac{\partial \tilde{x}^i}{\partial x^j}(\varphi(p))\right)v^j\right)\left(\frac{\partial}{\partial \tilde{x}^i}\bigg|_p\right)\right)$$

$$= \sum_{i=1}^{m}\left(\sum_{j=1}^{m}\left(\frac{\partial \tilde{x}^i}{\partial x^j}(\varphi(p))\right)v^j\right)\left(\frac{\partial}{\partial \tilde{x}^i}\bigg|_p\right),$$

so

$$\sum_{i=1}^{m} \tilde{v}^i\left(\frac{\partial}{\partial \tilde{x}^i}\bigg|_p\right) = \sum_{i=1}^{m}\left(\sum_{j=1}^{m}\left(\frac{\partial \tilde{x}^i}{\partial x^j}(\varphi(p))\right)v^j\right)\left(\frac{\partial}{\partial \tilde{x}^i}\bigg|_p\right),$$

and hence

$$\tilde{v}^i = \sum_{j=1}^{m}\left(\frac{\partial \tilde{x}^i}{\partial x^j}(\varphi(p))\right)v^j.$$

In short,

$$\tilde{v}^i = \left(\frac{\partial \tilde{x}^i}{\partial x^j}(\varphi(p))\right)v^j.$$

## 5.3 Tangent Bundles

**Lemma 5.21** *Let $M$ be a nonempty set. Let $\{U_i\}_{i \in I}$ be a nonempty family of subsets of $M$. Let $\{\varphi_i\}_{i \in I}$ be a family of* maps $\varphi_i$ *from $U_i$ to $\mathbb{R}^m$. If*

1. *for each $i \in I$, $\varphi_i$ is a 1–1 mapping from $U_i$ onto an open subset $\varphi_i(U_i)$ of $\mathbb{R}^m$,*
2. *for each $i, j \in I$, $\varphi_i(U_i \cap U_j)$, and $\varphi_j(U_i \cap U_j)$ are open in $\mathbb{R}^m$,*
3. *if $U_i \cap U_j \neq \emptyset$, then the map $\varphi_j \circ \varphi_i^{-1} : \varphi_i(U_i \cap U_j) \to \varphi_j(U_i \cap U_j)$ is smooth,*
4. *there exists a countable subcollection of $\{U_i\}_{i \in I}$ that covers $M$,*
5. *if $p, q$ are distinct points of $M$, then either (there exists $i \in I$ such that $p, q \in U_i$), or (there exist $i, j \in I$ such that $p \in U_i, q \in U_j$, and $U_i \cap U_j = \emptyset$), then there exists a unique smooth differential structure $\mathcal{A}$ over $M$ such that for each $i \in I$, $(U_i, \varphi_i) \in \mathcal{A}$.*

*Proof* First of all, we shall try to show that the collection $\{\varphi_k^{-1}(G) : k \in I,$ and $G$ is open in $\mathbb{R}^m\}$ is closed with respect to finite intersection.

For this purpose, let us take any $\varphi_i^{-1}(G_1), \varphi_j^{-1}(G_2) \in \{\varphi_k^{-1}(G) : k \in I,$ and $G$ is open in $\mathbb{R}^m\}$, where $i, j \in I$, $G_1, G_2$ are open in $\mathbb{R}^m$.

Case I: when $i = j$. Since $i = j$, $\varphi_i^{-1}(G_1) \cap \varphi_j^{-1}(G_2) = \varphi_i^{-1}(G_1) \cap \varphi_i^{-1}(G_2) = \varphi_i^{-1}(G_1 \cap G_2)$. Since $G_1, G_2$ are open in $\mathbb{R}^m$, $G_1 \cap G_2$ is open in $\mathbb{R}^m$. Now, since $i \in I$, and $G_1 \cap G_2$ is open in $\mathbb{R}^m$, $\varphi_i^{-1}(G_1 \cap G_2) \in \{\varphi_k^{-1}(G) : k \in I,$ and $G$ is open in $\mathbb{R}^m\}$, and hence, $\varphi_i^{-1}(G_1) \cap \varphi_j^{-1}(G_2) \in \{\varphi_k^{-1}(G) : k \in I,$ and $G$ is open in $\mathbb{R}^m\}$.

Case II: when $i \neq j$.

Subcase I: when $\varphi_i^{-1}(G_1) \cap \varphi_j^{-1}(G_2) = \emptyset$. Since $\{U_i\}_{i \in I}$ is a nonempty family of subsets of $M$, $I$ is nonempty. Since $I$ is nonempty, there exists $i_0 \in I$. Since $i_0 \in I$, and the empty set $\emptyset$ is open in $\mathbb{R}^m$, $\varphi_i^{-1}(G_1) \cap \varphi_j^{-1}(G_2) = \emptyset = \varphi_{i_0}^{-1}(\emptyset) \in \{\varphi_k^{-1}(G) : k \in I,$ and $G$ is open in $\mathbb{R}^m\}$, and hence $\varphi_i^{-1}(G_1) \cap \varphi_j^{-1}(G_2) \in \{\varphi_k^{-1}(G) : k \in I,$ and $G$ is open in $\mathbb{R}^m\}$.

Subcase II: when $\varphi_i^{-1}(G_1) \cap \varphi_j^{-1}(G_2) \neq \emptyset$.

Since $\varphi_i^{-1}(G_1) \cap \varphi_j^{-1}(G_2) \neq \emptyset$, there exists $x \in \varphi_i^{-1}(G_1) \cap \varphi_j^{-1}(G_2)$, and hence, $\varphi_i(x) \in G_1$, $\varphi_j(x) \in G_2$, $x \in U_i$, and $x \in U_j$. Since $x \in U_i$, and $x \in U_j$, $U_i \cap U_j \neq \emptyset$, and hence, by condition 3, $(\varphi_j \circ \varphi_i^{-1})^{-1}(G_2)$ is open in $\varphi_i(U_i \cap U_j)$. Since $(\varphi_j \circ \varphi_i^{-1})^{-1}(G_2)$ is open in $\varphi_i(U_i \cap U_j)$ and, by condition 2, $\varphi_i(U_i \cap U_j)$ is open in $\mathbb{R}^m$, $(\varphi_j \circ \varphi_i^{-1})^{-1}(G_2)(= (\varphi_i \circ \varphi_j^{-1})(G_2))$ is open in $\mathbb{R}^m$, and hence, $(\varphi_i \circ \varphi_j^{-1})(G_2)$ is open in $\mathbb{R}^m$. Since $(\varphi_i \circ \varphi_j^{-1})(G_2)$ is open in $\mathbb{R}^m$, and $G_1$ is open in $\mathbb{R}^m$, $G_1 \cap (\varphi_i \circ \varphi_j^{-1})(G_2)$ is open in $\mathbb{R}^m$. Since $G_1 \cap (\varphi_i \circ \varphi_j^{-1})(G_2)$ is open in $\mathbb{R}^m$, $\varphi_i^{-1}(G_1 \cap (\varphi_i \circ \varphi_j^{-1})(G_2)) \in \{\varphi_k^{-1}(G) : k \in I,$ and $G$ is open in $\mathbb{R}^m\}$. Since $\{\varphi_k^{-1}(G) : k \in I,$ and $G$ is open in $\mathbb{R}^m\} \ni \varphi_i^{-1}(G_1 \cap (\varphi_i \circ \varphi_j^{-1})(G_2)) = \varphi_i^{-1}(G_1) \cap \varphi_i^{-1}((\varphi_i \circ \varphi_j^{-1})(G_2)) = \varphi_i^{-1}(G_1) \cap \varphi_j^{-1}(G_2)$, $\varphi_i^{-1}(G_1) \cap \varphi_j^{-1}(G_2) \in \{\varphi_k^{-1}(G) : k \in I,$ and $G$ is open in $\mathbb{R}^m\}$.

Thus, we see that, in all cases, $\varphi_i^{-1}(G_1) \cap \varphi_j^{-1}(G_2) \in \{\varphi_k^{-1}(G) : k \in I,$ and $G$ is open in $\mathbb{R}^m\}$. Hence, $\{\varphi_k^{-1}(G) : k \in I,$ and $G$ is open in $\mathbb{R}^m\}$ is closed with respect to finite intersection. Let $\mathcal{O}$ be the collection of arbitrary unions of the members of $\{\varphi_i^{-1}(G) : i \in I,$ and $G$ is open in $\mathbb{R}^m\}$.

We shall show that $\mathcal{O}$ is a topology over $M$.

(a) Since $\{U_i\}_{i \in I}$ is a nonempty collection of subsets of $M$, $I$ is nonempty. Since $I$ is nonempty, there exists $i_0 \in I$. Since $i_0 \in I$, and the empty set $\emptyset$ is open in $\mathbb{R}^m$, $\emptyset = \varphi_{i_0}^{-1}(\emptyset) \in \{\varphi_i^{-1}(G) : i \in I,$ and $G$ is open in $\mathbb{R}^m\} \subset \mathcal{O}$, and hence, $\emptyset \in \mathcal{O}$.

(b) Since for each $i \in I$, $\varphi_i$ maps from $U_i$ to $\mathbb{R}^m$, $\varphi_i^{-1}(\mathbb{R}^m) = U_i$. Now, since $\mathbb{R}^m$ is open in $\mathbb{R}^m$, $\{U_i : i \in I\} = \{\varphi_i^{-1}(\mathbb{R}^m) : i \in I\} \subset \{\varphi_i^{-1}(G) : i \in I,$ and $G$ is open in $\mathbb{R}^m\}$, and hence, $\cup_{i \in I} U_i \in \mathcal{O}$. By condition 4, there exists a countable subcollection of $\{U_i\}_{i \in I}$ that covers $M$, so $\{U_i\}_{i \in I}$ also covers $M$, and hence, $\cup_{i \in I} U_i = M$. Since $M = \cup_{i \in I} U_i \in \mathcal{O}$, $M \in \mathcal{O}$.

(c) Clearly, $\mathcal{O}$ is closed with respect to arbitrary union.

(d) Let $\cup\{\varphi_i^{-1}(G_{i1}) : i \in I,$ and $G_{i1}$ is open in $\mathbb{R}^m\} \in \mathcal{O}$, and $\cup\{\varphi_i^{-1}(G_{i2}) : i \in I,$ and $G_{i2}$ is open in $\mathbb{R}^m\} \in \mathcal{O}$. We have to prove that $(\cup\{\varphi_i^{-1}(G_{i1}) : i \in I,$ and $G_{i1}$ is open in $\mathbb{R}^m\}) \cap (\cup\{\varphi_i^{-1}(G_{i2}) : i \in I,$ and $G_{i2}$ is open in $\mathbb{R}^m\}) \in \mathcal{O}$.

Here, $(\cup\{\varphi_i^{-1}(G_{i1}) : i \in I,$ and $G_{i1}$ is open in $\mathbb{R}^m\}) \cap (\cup\{\varphi_i^{-1}(G_{i2}) : i \in I,$ and $G_{i2}$ is open in $\mathbb{R}^\triangleright\}) = \cup\{\varphi_j^{-1}(G_{j1}) \cap \varphi_k^{-1}(G_{k2}) : (j,k) \in I \times I, G_{j1}, G_{k2}$ are open in $\mathbb{R}^m\}$.

Since $\{\varphi_i^{-1}(G) : i \in I,$ and $G$ is open in $\mathbb{R}^m\}$ is closed with respect to finite intersection, $\{\varphi_j^{-1}(G_{j1}) \cap \varphi_k^{-1}(G_{k2}) : (j,k) \in I \times I, G_{j1}, G_{k2}$ are open in $\mathbb{R}^m\} \subset \{\varphi_i^{-1}(G) : i \in I,$ and $G$ is open in $\mathbb{R}^m\}$, and hence, $\cup\{\varphi_j^{-1}(G_{j1}) \cap \varphi_k^{-1}(G_{k2}) : (j,k) \in I \times I, G_{j1}, G_{k2}$ are open in $\mathbb{R}^m\} \in \mathcal{O}$. It follows that $(\cup\{\varphi_i^{-1}(G_{i1}) : i \in I,$ and $G_{i1}$ is open in $\mathbb{R}^m\}) \cap (\cup\{\varphi_i^{-1}(G_{i2}) : i \in I,$ and $G_{i2}$ is open in $\mathbb{R}^m\}) \in \mathcal{O}$. Thus, $\mathcal{O}$ is a topology over $M$.

It is clear that $\{\varphi_i^{-1}(G) : i \in I,$ and $G$ is open in $\mathbb{R}^m\}$ is a basis of the topological space $(M, \mathcal{O})$. Now, we want to prove that for each $i \in I$, $U_i \in \mathcal{O}$. For this purpose, let us fix any $i_0 \in I$. We have to prove that $U_{i_0} \in \mathcal{O}$. Clearly, $\{\varphi_i^{-1}(G) : i \in I,$ and $G$ is open in $\mathbb{R}^m\} \subset \mathcal{O}$, and $U_{i_0} = \varphi_{i_0}^{-1}(\mathbb{R}^m) \in \{\varphi_i^{-1}(G) : i \in I,$ and $G$ is open in $\mathbb{R}^m\}$, so $U_{i_0} \in \mathcal{O}$.

Now, we want to prove that for each $i \in I$, $\varphi_i : U_i \to \varphi_i(U_i)$ is continuous. For this purpose, let us take any open set $G$ in $\varphi_i(U_i)$. We have to show that $\varphi_i^{-1}(G)$ is open in $U_i$. Since $G$ is open in $\varphi_i(U_i)$, and $\varphi_i(U_i)$ is open in $\mathbb{R}^m$, $G$ is open in $\mathbb{R}^m$, and hence, by the definition of $\mathcal{O}$, for every $i \in I$, $\varphi_i^{-1}(G) \in \mathcal{O}$. Since $\varphi_i^{-1}(G) \in \mathcal{O}$, $\varphi_i^{-1}(G) \cap U_i$ is open in $U_i$. Since $\varphi_i : U_i \to \varphi_i(U_i)$, $\varphi_i^{-1}(G) \subset U_i$, and hence, $\varphi_i^{-1}(G) \cap U_i = \varphi_i^{-1}(G)$. Since $\varphi_i^{-1}(G) \cap U_i = \varphi_i^{-1}(G)$, and $\varphi_i^{-1}(G) \cap U_i$ is open in $U_i$, $\varphi_i^{-1}(G)$ is open in $U_i$.

Thus, we have shown that for each $i \in I$, $\varphi_i : U_i \to \varphi_i(U_i)$ is continuous. Now, we want to show that $\varphi_i$ is an open mapping. For this purpose, let us take any open set $G$ in $U_i$. We have to show that $\varphi_i(G)$ is open in $\varphi_i(U_i)$. Since $\{\varphi_j^{-1}(G) : j \in I,$ and $G$ is open in $\mathbb{R}^m\}$ is a basis of the topological space $(M, \mathcal{O})$, $\{\varphi_j^{-1}(G) \cap U_i : j \in I,$ and $G$ is open in $\mathbb{R}^m\}$ is a basis of $U_i$. Now, it suffices to prove that $\varphi_i(\varphi_j^{-1}(G) \cap U_i)$ is open in $\varphi_i(U_i)$ for every $j \in I$, and for every $G$ which is open in $\mathbb{R}^m$.

Case I: when $\varphi_j^{-1}(G) \cap U_i = \emptyset$. Here, $\varphi_i(\varphi_j^{-1}(G) \cap U_i) = \varphi_i(\emptyset) = \emptyset$, which is open in $\varphi_i(U_i)$.

Case II: when $\varphi_j^{-1}(G) \cap U_i \neq \emptyset$. Since $\varphi_j^{-1}(G) \subset U_j$, and $\varphi_j^{-1}(G) \cap U_i \neq \emptyset$, $U_j \cap U_i \neq \emptyset$. Here, $\varphi_i$ is 1–1, so $\varphi_i(\varphi_j^{-1}(G) \cap U_i) = \varphi_i(\varphi_j^{-1}(G)) \cap \varphi_i(U_i) = (\varphi_i \circ \varphi_j^{-1})(G) \cap \varphi_i(U_i)$. Since $U_j \cap U_i \neq \emptyset$, by the condition 3, $\varphi_i \circ \varphi_j^{-1}$ is open. Since $G$ is open in $\mathbb{R}^m$, and $\varphi_i \circ \varphi_j^{-1}$ is open, $(\varphi_i \circ \varphi_j^{-1})(G)$ is open in $\mathbb{R}^m$, and hence, $(\varphi_i \circ \varphi_j^{-1})(G) \cap \varphi_i(U_i)$ is open in $\varphi_i(U_i)$. Since $(\varphi_i \circ \varphi_j^{-1})(G) \cap \varphi_i(U_i)$ $(= \varphi_i(\varphi_j^{-1}(G) \cap U_i))$ is open in $\varphi_i(U_i)$, $\varphi_i(\varphi_j^{-1}(G) \cap U_i)$ is open in $\varphi_i(U_i)$.

Thus, in all cases, $\varphi_i(\varphi_j^{-1}(G) \cap U_i)$ is open in $\varphi_i(U_i)$. This shows that $\varphi_i$ is an open mapping. Thus, we have shown that $\varphi_i : U_i \to \varphi_i(U_i)$ is a homeomorphism. It follows that for each $i \in I$, the ordered pair $(U_i, \varphi_i)$ is a coordinate chart of $M$.

Now, we want to show that the topology $\mathcal{O}$ of $M$ is Hausdorff. For this purpose, let us take any two distinct points $p, q$ in $M$. We have to find an open neighborhood $G$ of $p$, and an open neighborhood $G_1$ of $q$ such that $G, G_1$ are disjoint. Since $p, q$ are distinct points of $M$, by condition 5, either (there exists $i \in I$ such that $p, q \in U_i$) or (there exist $i, j \in I$ such that $p \in U_i, q \in U_j$, and $U_i \cap U_j = \emptyset$).

Case I: when there exist $i, j \in I$ such that $p \in U_i, q \in U_j$, and $U_i \cap U_j = \emptyset$. Since $i \in I, U_i \in \mathcal{O}$. Since $U_i \in \mathcal{O}$, and $p \in U_i$, $U_i$ is an open neighborhood of $p$ in $M$. Similarly, $U_j$ is an open neighborhood of $q$ in $M$. Also, $U_i \cap U_j = \emptyset$. Thus, we have found an open neighborhood $U_i$ of $p$, and an open neighborhood $U_j$ of $q$ such that $U_i, U_j$ are disjoint.

Case II: when there exists $i \in I$ such that $p, q \in U_i$. Since $\varphi_i : U_i \to \varphi_i(U_i)$ is continuous, $\varphi_i$ is 1–1, and $p, q$ are distinct elements of $U_i$, $\varphi_i(p), \varphi_i(q)$ are distinct points of $\varphi_i(U_i)$. Since, by condition 1, $\varphi_i(U_i)$ is open in $\mathbb{R}^m$, and $\varphi_i(p), \varphi_i(q)$ are distinct points of $\varphi_i(U_i)$, there exist an open neighborhood $G_1$ of $\varphi_i(p)$ in $\mathbb{R}^m$ and an open neighborhood $G_2$ of $\varphi_i(q)$ in $\mathbb{R}^m$ such that $G_1 \subset \varphi_i(U_i), G_2 \subset \varphi_i(U_i)$, and $G_1, G_2$ are disjoint. Since $\varphi_i : U_i \to \varphi_i(U_i)$ is continuous, and $G_1$ is an open neighborhood of $\varphi_i(p)$ in $\mathbb{R}^m$, $\varphi_i^{-1}(G_1)$ is an open neighborhood of $p$ in $U_i$. Since $\varphi_i^{-1}(G_1)$ is an open neighborhood of $p$ in $U_i$, and $U_i \in \mathcal{O}$, $\varphi_i^{-1}(G_1)$ is an open neighborhood of $p$ in $M$. Similarly, $\varphi_i^{-1}(G_2)$ is an open neighborhood of $q$ in $M$. Also, $\varphi_i^{-1}(G_1) \cap \varphi_i^{-1}(G_2) = \varphi_i^{-1}(G_1 \cap G_2) = \varphi_i^{-1}(\emptyset) = \emptyset$. Thus, we have found an open neighborhood $\varphi_i^{-1}(G_1)$ of $p$, and an open neighborhood $\varphi_i^{-1}(G_2)$ of $q$ such that $\varphi_i^{-1}(G_1), \varphi_i^{-1}(G_2)$ are disjoint.

Thus, we see that, in all cases, there exist disjoint open neighborhoods of $p$ and $q$. Hence, the topology $\mathcal{O}$ over $M$ is Hausdorff.

Now, we want to show that the topology $\mathcal{O}$ of $M$ is second countable. Observe that for every $j \in I$, the subspace topology of $U_j$ is second countable.

(Reason: For every $j \in I$, $\varphi_j(U_j)$ is a subset of $\mathbb{R}^m$. Since each $\varphi_j(U_j)$ is a subset of $\mathbb{R}^m$, and the topology of $\mathbb{R}^m$ is second countable, the subspace topology of $\varphi_j(U_j)$ is second countable. Since the topology of $\varphi_j(U_j)$ is second countable, and $\varphi_j : U_j \to \varphi_j(U_j)$ is a homeomorphism, the topology of $U_j$ is second countable.)

By condition 4, there exists a countable subcollection $\{U_{i_1}, U_{i_2}, U_{i_3}, \ldots\}$ of $\{U_i\}_{i \in I}$ such that $\{U_{i_1}, U_{i_2}, U_{i_3}, \ldots\}$ covers $M$. Since $\{U_{i_1}, U_{i_2}, U_{i_3}, \ldots\}$ is a sub-collection of $\{U_i\}_{i \in I}$, and for each $i \in I$, the topology of $U_i$ is second countable, each $U_{i_k} (k = 1, 2, 3, \ldots)$ is a second countable space, and hence, there exists a countable basis $\{V_{k1}, V_{k2}, V_{k3}, \ldots\}$ of $U_{i_k}$. Since each $V_{kl}$ is open in $U_{i_k}$, and $U_{i_k} \in \mathcal{O}$, each $V_{kl} \in \mathcal{O}$. Observe that $\{V_{kl} : k, l \text{ are positive integers}\}$ is a countable col-lection of open subsets of $M$. We shall try to show that $\{V_{kl} : k, l \text{ are positive integers}\}$ is a basis of $M$. For this purpose, let us take any open neighborhood $G$ of $p$ in $M$.

Since $p \in M$, and $\{U_{i_1}, U_{i_2}, U_{i_3}, \ldots\}$ covers $M$, there exists a positive integer $k$ such that $p \in U_{i_k}$. Since $p \in U_{i_k}$, $p \in G$, $p \in U_{i_k} \cap G$. Since $G$ are open in $M$, $U_{i_k} \cap G$ is open in $U_{i_k}$. Thus, $U_{i_k} \cap G$ is an open neighborhood of $p$ in $U_{i_k}$. Since $U_{i_k} \cap G$ is an open neighborhood of $p$ in $U_{i_k}$, and $\{V_{k1}, V_{k2}, V_{k3}, \ldots\}$ is a basis of $U_{i_k}$, there exists a positive integer $l$ such that $p \in V_{kl} \subset U_{i_k} \cap G \subset G$.

This shows that $\{V_{kl} : k, l \text{ are positive integers}\}$ is a countable basis of $M$. Hence, the topology of $M$ is second countable.

Now, we want to show that $M$ is an $m$-dimensional topological manifold. For this purpose, let us take any $p \in M$. Since $p \in M$, and $\{U_i\}_{i \in I}$ is a cover of $M$, there exists $i_0 \in I$ such that $p \in U_{i_0}$. Here, $(U_{i_0}, \varphi_{i_0})$ is a coordinate chart of $M$, and $p \in U_{i_0}$. It follows that $M$ is an $m$-dimensional topological manifold.

Next, let $\mathcal{B}$ be the collection of all coordinate charts $(U_i, \varphi_i)$, where $i \in I$. We shall try to prove that $\mathcal{B}$ is an atlas on $M$. For this purpose, let us take any coordinate charts $(U_i, \varphi_i), (U_j, \varphi_j)$, where $U_i \cap U_j \neq \emptyset$. We have to prove that $\varphi_j \circ \varphi_i^{-1} : \varphi_i(U_i \cap U_j) \to \varphi_j(U_i \cap U_j)$ is smooth. Since $U_i \cap U_j \neq \emptyset$, by the condition 3, $\varphi_j \circ \varphi_i^{-1} : \varphi_i(U_i \cap U_j) \to \varphi_j(U_i \cap U_j)$ is smooth. This proves that $\mathcal{B}$ is an atlas on $M$.

Since $\mathcal{B}$ is an atlas on an $m$-dimensional topological manifold $M$, by Note 1.5, there exists a unique smooth differential structure $\mathcal{A}$ on $M$ which contains $\mathcal{B}$. $\qquad \square$

**Definition** Let $\{X_i\}_{i \in I}$ be any family of sets. The set $\{(i, x) : i \in I \text{ and } x \in X_i\}$ is denoted by $\coprod_{i \in I} X_i$, and is called the *disjoint union* of $\{X_i\}_{i \in I}$. Thus, $\coprod_{i \in I} X_i = \cup_{i \in I} (\{i\} \times X_i)$.

**Note 5.22** Let $M$ be an $m$-dimensional smooth manifold. Let $\mathcal{A}$ be the smooth structure of $M$. Let $(U, \varphi) \in \mathcal{A}$. For every $p \in U$, let

$$\left( \frac{\partial}{\partial x^1}\bigg|_p, \frac{\partial}{\partial x^2}\bigg|_p, \ldots, \frac{\partial}{\partial x^m}\bigg|_p \right)$$

be the coordinate basis of $T_pM$ corresponding to $(U, \varphi)$, where

$$\frac{\partial}{\partial x^i}\bigg|_p \equiv \left(\mathrm{d}\left(\varphi^{-1}\right)_{\varphi(p)}\right)\left(\frac{\partial}{\partial x^i}\bigg|_{\varphi(p)}\right).$$

Here, $\coprod_{r \in U} T_rM \subset \coprod_{r \in M} T_rM$, and $\varphi(U) \times \mathbb{R}^m \subset \mathbb{R}^m \times \mathbb{R}^m = \mathbb{R}^{2m}$. Let us define a function $\tilde{\varphi} : \coprod_{r \in U} T_rM \rightarrow \varphi(U) \times \mathbb{R}^m$ as follows: For every

$$\left(p, v^1\left(\frac{\partial}{\partial x^1}\bigg|_p\right) + \cdots + v^m\left(\frac{\partial}{\partial x^m}\bigg|_p\right)\right) \in \coprod_{r \in U} T_rM,$$

$$\tilde{\varphi}\left(p, v^1\left(\frac{\partial}{\partial x^1}\bigg|_p\right) + \cdots + v^m\left(\frac{\partial}{\partial x^m}\bigg|_p\right)\right) \equiv \left((\pi_1 \circ \varphi)(p), \ldots, (\pi_m \circ \varphi)(p), v^1, \ldots, v^m\right).$$

First of all, observe that $\tilde{\varphi}$ is well defined.

(Reason: Let $(p, v^1(\frac{\partial}{\partial x^1}\big|_p) + \cdots + v^m(\frac{\partial}{\partial x^m}\big|_p)) = (q, w^1(\frac{\partial}{\partial x^1}\big|_q) + \cdots + w^m(\frac{\partial}{\partial x^m}\big|_q))$ where $(p, v^1(\frac{\partial}{\partial x^1}\big|_p) + \cdots + v^m(\frac{\partial}{\partial x^m}\big|_p)) \in \coprod_{r \in U} T_rM$, and

$$\left(q, w^1\left(\frac{\partial}{\partial x^1}\bigg|_q\right) + \cdots + w^m\left(\frac{\partial}{\partial x^m}\bigg|_q\right)\right) \in \coprod_{r \in U} T_rM.$$

We have to show that

$$\left((\pi_1 \circ \varphi)(p), \ldots, (\pi_m \circ \varphi)(p), v^1, \ldots v^m\right) = \left((\pi_1 \circ \varphi)(q), \ldots, (\pi_m \circ \varphi)(q), w^1, \ldots, w^m\right).$$

Since

$$\left(p, v^1\left(\frac{\partial}{\partial x^1}\bigg|_p\right) + \cdots + v^m\left(\frac{\partial}{\partial x^m}\bigg|_p\right)\right) = \left(q, w^1\left(\frac{\partial}{\partial x^1}\bigg|_q\right) + \cdots + w^m\left(\frac{\partial}{\partial x^m}\bigg|_q\right)\right),$$

$p = q$ and

$$v^1\left(\frac{\partial}{\partial x^1}\bigg|_p\right) + \cdots + v^m\left(\frac{\partial}{\partial x^m}\bigg|_p\right) = w^1\left(\frac{\partial}{\partial x^1}\bigg|_q\right) + \cdots + w^m\left(\frac{\partial}{\partial x^m}\bigg|_q\right).$$

Since $p = q$, for every $i = 1, \ldots, m$, $(\pi_i \circ \varphi)(p) = (\pi_i \circ \varphi)(q)$. Since $p = q$, and

$$v^1 \left( \frac{\partial}{\partial x^1} \bigg|_p \right) + \cdots + v^m \left( \frac{\partial}{\partial x^m} \bigg|_p \right) = w^1 \left( \frac{\partial}{\partial x^1} \bigg|_q \right) + \cdots + w^m \left( \frac{\partial}{\partial x^m} \bigg|_q \right),$$

$$\text{so } v^1 \left( \frac{\partial}{\partial x^1} \bigg|_p \right) + \cdots + v^m \left( \frac{\partial}{\partial x^m} \bigg|_p \right) = w^1 \left( \frac{\partial}{\partial x^1} \bigg|_p \right) + \cdots + w^m \left( \frac{\partial}{\partial x^m} \bigg|_p \right).$$

Since

$$v^1 \left( \frac{\partial}{\partial x^1} \bigg|_p \right) + \cdots + v^m \left( \frac{\partial}{\partial x^m} \bigg|_p \right) = w^1 \left( \frac{\partial}{\partial x^1} \bigg|_p \right) + \cdots + w^m \left( \frac{\partial}{\partial x^m} \bigg|_p \right),$$

and

$$\left( \frac{\partial}{\partial x^1} \bigg|_p , \frac{\partial}{\partial x^2} \bigg|_p , \ldots, \frac{\partial}{\partial x^m} \bigg|_p \right)$$

is a basis of $T_pM$, for every $i = 1, \ldots, m$, $v^i = w^i$. Since for every $i = 1, \ldots, m$, $v^i = w^i$, and $(\pi_i \circ \varphi)(p) = (\pi_i \circ \varphi)(q)$, $((\pi_1 \circ \varphi)(p), \ldots, (\pi_m \circ \varphi)(p), v^1, \ldots v^m) = ((\pi_1 \circ \varphi)(q), \ldots, (\pi_m \circ \varphi)(q), w^1, \ldots, w^m)$.) Thus, $\{ \coprod_{r \in U} T_rM \}_{(U, \varphi) \in \mathcal{A}}$ is a family of subsets of $\coprod_{r \in M} T_rM$, and $\{ \tilde{\varphi} \}_{(U, \varphi) \in \mathcal{A}}$ is a family of functions from $\coprod_{r \in U} T_rM$ onto $\varphi(U) \times \mathbb{R}^m (\subset \mathbb{R}^{2m})$.

Now, we shall verify all the conditions 1–5 of Lemma 5.21:

1. $\tilde{\varphi} : \coprod_{r \in U} T_rM \to \varphi(U) \times \mathbb{R}^m$ is 1–1: For this purpose, let

$$\tilde{\varphi} \left( p, v^1 \left( \frac{\partial}{\partial x^1} \bigg|_p \right) + \cdots + v^m \left( \frac{\partial}{\partial x^m} \bigg|_p \right) \right) = \tilde{\varphi} \left( q, w^1 \left( \frac{\partial}{\partial x^1} \bigg|_q \right) + \cdots + w^m \left( \frac{\partial}{\partial x^m} \bigg|_q \right) \right).$$

We have to show that

$$\left( p, v^1 \left( \frac{\partial}{\partial x^1} \bigg|_p \right) + \cdots + v^m \left( \frac{\partial}{\partial x^m} \bigg|_p \right) \right) = \left( q, w^1 \left( \frac{\partial}{\partial x^1} \bigg|_q \right) + \cdots + w^m \left( \frac{\partial}{\partial x^m} \bigg|_q \right) \right).$$

Here,

$$((\pi_1 \circ \varphi)(p), \ldots, (\pi_m \circ \varphi)(p), v^1, \ldots, v^m) = \tilde{\varphi} \left( p, v^1 \left( \frac{\partial}{\partial x^1} \bigg|_p \right) + \cdots + v^m \left( \frac{\partial}{\partial x^m} \bigg|_p \right) \right)$$

$$= \tilde{\varphi} \left( q, w^1 \left( \frac{\partial}{\partial x^1} \bigg|_q \right) + \cdots + w^m \left( \frac{\partial}{\partial x^m} \bigg|_q \right) \right)$$

$$= ((\pi_1 \circ \varphi)(q), \ldots, (\pi_m \circ \varphi)(q), w^1, \ldots, w^m),$$

and hence, for every $i = 1, \ldots, m$, $(\pi_i \circ \varphi)(p) = (\pi_i \circ \varphi)(q)$, and $v^i = w^i$. Since for every $i = 1, \ldots, m$, $(\pi_i \circ \varphi)(p) = (\pi_i \circ \varphi)(q)$, $\varphi(p) = \varphi(q)$. Since $\varphi(p) = \varphi(q)$, and $\varphi$ is 1–1, $p = q$. Since for every $i = 1, \ldots, m$, $v^i = w^i$,

$$v^1\left(\left.\frac{\partial}{\partial x^1}\right|_p\right) + \cdots + v^m\left(\left.\frac{\partial}{\partial x^m}\right|_p\right) = w^1\left(\left.\frac{\partial}{\partial x^1}\right|_p\right) + \cdots + w^m\left(\left.\frac{\partial}{\partial x^m}\right|_p\right).$$

Since

$$v^1\left(\left.\frac{\partial}{\partial x^1}\right|_p\right) + \cdots + v^m\left(\left.\frac{\partial}{\partial x^m}\right|_p\right) = w^1\left(\left.\frac{\partial}{\partial x^1}\right|_p\right) + \cdots + w^m\left(\left.\frac{\partial}{\partial x^m}\right|_p\right), \text{ and } p = q,$$

so $$v^1\left(\left.\frac{\partial}{\partial x^1}\right|_p\right) + \cdots + v^m\left(\left.\frac{\partial}{\partial x^m}\right|_p\right) = w^1\left(\left.\frac{\partial}{\partial x^1}\right|_q\right) + \cdots + w^m\left(\left.\frac{\partial}{\partial x^m}\right|_q\right).$$

It follows that

$$\left(p, v^1\left(\left.\frac{\partial}{\partial x^1}\right|_p\right) + \cdots + v^m\left(\left.\frac{\partial}{\partial x^m}\right|_p\right)\right) = \left(q, w^1\left(\left.\frac{\partial}{\partial x^1}\right|_q\right) + \cdots + w^m\left(\left.\frac{\partial}{\partial x^m}\right|_q\right)\right).$$

Thus, we have shown that $\tilde{\varphi} : \coprod_{r \in U} T_r M \to \varphi(U) \times \mathbb{R}^m$ is 1–1. Now, we want to show that $\tilde{\varphi} : \coprod_{r \in U} T_r M \to \varphi(U) \times \mathbb{R}^m$ is onto. For this purpose, let us take any $((\pi_1 \circ \varphi)(p), \ldots, (\pi_m \circ \varphi)(p), v^1, \ldots, v^m) \in \varphi(U) \times \mathbb{R}^m$, where $p \in U$. Here, $p \in U$, and

$$v^1\left(\left.\frac{\partial}{\partial x^1}\right|_p\right) + \cdots + v^m\left(\left.\frac{\partial}{\partial x^m}\right|_p\right) \in T_p M,$$

so

$$v^1\left(\left.\frac{\partial}{\partial x^1}\right|_p\right) + \cdots + v^m\left(\left.\frac{\partial}{\partial x^m}\right|_p\right) \in \coprod_{r \in U} T_r M.$$

Also,

$$\tilde{\varphi}\left(v^1\left(\left.\frac{\partial}{\partial x^1}\right|_p\right) + \cdots + v^m\left(\left.\frac{\partial}{\partial x^m}\right|_p\right)\right) = ((\pi_1 \circ \varphi)(p), \ldots, (\pi_m \circ \varphi)(p), v^1, \ldots, v^m).$$

This shows that $\tilde{\varphi} : \coprod_{r \in U} T_r M \to \varphi(U) \times \mathbb{R}^m$ is a 1–1 onto mapping.

Now, we want to show that $\varphi(U) \times \mathbb{R}^m$ is open in $\mathbb{R}^{2m}$. Since $(U, \varphi) \in \mathcal{A}$, and $\mathcal{A}$ is a smooth structure of $M$, $\varphi(U)$ is open in $\mathbb{R}^m$, and hence, $\varphi(U) \times \mathbb{R}^m$ is open in $\mathbb{R}^m \times \mathbb{R}^m (= \mathbb{R}^{2m})$.

2. Let us take any $(U, \varphi) \in \mathcal{A}$, and $(V, \psi) \in \mathcal{A}$. We have to prove that $\tilde{\varphi}((\coprod_{r \in U} T_r M) \cap (\coprod_{r \in V} T_r M))$ is open in $\mathbb{R}^{2m}$. Here, $\tilde{\varphi}((\coprod_{r \in U} T_r M) \cap (\coprod_{r \in V} T_r M)) = \tilde{\varphi}(\coprod_{r \in U \cap V} T_r M)$. Here, $\tilde{\varphi} : \coprod_{r \in U} T_r M \to \varphi(U) \times \mathbb{R}^m$ is defined as follows: For every

$$\left( p, v^1 \left( \left. \frac{\partial}{\partial x^1} \right|_p \right) + \cdots + v^m \left( \left. \frac{\partial}{\partial x^m} \right|_p \right) \right) \in \coprod_{r \in U} T_r M,$$

$$\tilde{\varphi} \left( p, v^1 \left( \left. \frac{\partial}{\partial x^1} \right|_p \right) + \cdots + v^m \left( \left. \frac{\partial}{\partial x^m} \right|_p \right) \right) \equiv \left( (\pi_1 \circ \varphi)(p), \ldots, (\pi_m \circ \varphi)(p), v^1, \ldots, v^m \right)$$

It follows that $\tilde{\varphi}(\coprod_{r \in U \cap V} T_r M) = \varphi(U \cap V) \times \mathbb{R}^m$, and hence, $\tilde{\varphi}((\coprod_{r \in U} T_r M) \cap (\coprod_{r \in V} T_r M)) = \varphi(U \cap V) \times \mathbb{R}^m$. Since $(U, \varphi) \in \mathcal{A}$, and $(V, \psi) \in \mathcal{A}$, $\varphi(U \cap V)$ is open in $\mathbb{R}^m$, and hence, $\varphi(U \cap V) \times \mathbb{R}^m$ is open in $\mathbb{R}^{2m}$. This shows that $\tilde{\varphi}((\coprod_{r \in U} T_r M) \cap (\coprod_{r \in V} T_r M))$ is open in $\mathbb{R}^{2m}$. Similarly, $\tilde{\psi}((\coprod_{r \in U} T_r M) \cap (\coprod_{r \in V} T_r M))$ is open in $\mathbb{R}^{2m}$.

3. Let us take any $(U, \varphi) \in \mathcal{A}$, and $(V, \psi) \in \mathcal{A}$. Let $(\coprod_{r \in U} T_r M) \cap (\coprod_{r \in V} T_r M) \neq \emptyset$. We have to prove that

$$\tilde{\psi} \circ \tilde{\varphi}^{-1} : \tilde{\varphi} \left( \left( \coprod_{r \in U} T_r M \right) \cap \left( \coprod_{r \in V} T_r M \right) \right) \to \tilde{\psi} \left( \left( \coprod_{r \in U} T_r M \right) \cap \left( \coprod_{r \in V} T_r M \right) \right)$$

is smooth, that is,

$$\tilde{\psi} \circ \tilde{\varphi}^{-1} : \tilde{\varphi} \left( \coprod_{r \in U \cap V} T_r M \right) \to \tilde{\psi} \left( \coprod_{r \in U \cap V} T_r M \right)$$

is smooth. For every $p \in U \cap V$, let

$$\left( \left. \frac{\partial}{\partial x^1} \right|_p, \left. \frac{\partial}{\partial x^2} \right|_p, \ldots, \left. \frac{\partial}{\partial x^m} \right|_p \right)$$

be the coordinate basis of $T_p M$ corresponding to $(U, \varphi)$, where

$$\left. \frac{\partial}{\partial x^i} \right|_p \equiv \left( d(\varphi^{-1})_{\varphi(p)} \right) \left( \left. \frac{\partial}{\partial x^i} \right|_{\varphi(p)} \right),$$

and let

$$\left(\left.\frac{\partial}{\partial \tilde{x}^1}\right|_p, \left.\frac{\partial}{\partial \tilde{x}^2}\right|_p, \dots, \left.\frac{\partial}{\partial \tilde{x}^m}\right|_p\right)$$

be the coordinate basis of $T_pM$ corresponding to $(V, \psi)$, where

$$\left.\frac{\partial}{\partial \tilde{x}^i}\right|_p \equiv \left(\mathrm{d}(\psi^{-1})_{\psi(p)}\right)\left(\left.\frac{\partial}{\partial \tilde{x}^i}\right|_{\psi(p)}\right).$$

Also,

$$\left.\frac{\partial}{\partial x^j}\right|_p = \left(\frac{\partial \tilde{x}^i}{\partial x^j}(\varphi(p))\right)\left(\left.\frac{\partial}{\partial \tilde{x}^i}\right|_p\right).$$

Here, for every $(x^1, \dots, x^m, v^1, \dots, v^m) \in \tilde{\varphi}(\coprod_{r \in U \cap V} T_r M)$, we have

$$\left(\tilde{\psi} \circ \tilde{\varphi}^{-1}\right)(x^1, \dots, x^m, v^1, \dots, v^m) = \tilde{\psi}(\tilde{\varphi}^{-1}(x^1, \dots, x^m, v^1, \dots, v^m))$$

$$= \tilde{\psi}\left(\varphi^{-1}(x^1, \dots, x^m), v^1\left(\left.\frac{\partial}{\partial x^1}\right|_{\varphi^{-1}(x^1, \dots, x^m)}\right.\right.$$

$$+ \dots + v^m\left.\left.\left(\left.\frac{\partial}{\partial x^m}\right|_{\varphi^{-1}(x^1, \dots, x^m)}\right)\right)\right)$$

$$= \tilde{\psi}\left(\varphi^{-1}(x^1, \dots, x^m), v^i\left(\left.\frac{\partial}{\partial x^i}\right|_{\varphi^{-1}(x^1, \dots, x^m)}\right)\right)$$

$$= \tilde{\psi}\left(\varphi^{-1}(x^1, \dots, x^m), v^i\left(\left(\frac{\partial \tilde{x}^j}{\partial x^i}(x^1, \dots, x^m)\right)\left(\left.\frac{\partial}{\partial \tilde{x}^j}\right|_{\varphi^{-1}(x^1, \dots, x^m)}\right)\right)\right)$$

$$= \tilde{\psi}\left(\varphi^{-1}(x^1, \dots, x^m), \left(v^i\left(\frac{\partial \tilde{x}^j}{\partial x^i}(x^1, \dots, x^m)\right)\right)\left(\left.\frac{\partial}{\partial \tilde{x}^j}\right|_{\varphi^{-1}(x^1, \dots, x^m)}\right)\right)$$

$$= ((\pi_1 \circ \psi)(\varphi^{-1}(x^1, \dots, x^m)), \dots, (\pi_m \circ \psi)(\varphi^{-1}(x^1, \dots, x^m)),$$

$$v^i\left(\frac{\partial \tilde{x}^1}{\partial x^i}(x^1, \dots, x^m)\right), \dots, v^i\left(\frac{\partial \tilde{x}^m}{\partial x^i}(x^1, \dots, x^m)\right))$$

$$= ((\pi_1 \circ (\psi \circ \varphi^{-1}))(x^1, \dots, x^m), \dots, (\pi_m \circ (\psi \circ \varphi^{-1}))(x^1, \dots, x^m),$$

$$v^i\left(\frac{\partial \tilde{x}^1}{\partial x^i}(x^1, \dots, x^m)\right), \dots, v^i\left(\frac{\partial \tilde{x}^m}{\partial x^i}(x^1, \dots, x^m)\right))$$

$$= ((\pi_1 \circ (\psi \circ \varphi^{-1}))(x^1, \dots, x^m), \dots, (\pi_m \circ (\psi \circ \varphi^{-1}))(x^1, \dots, x^m),$$

$$v^i\left(\frac{\partial(\psi \circ \varphi^{-1})^1}{\partial x^i}(x^1, \dots, x^m)\right), \dots, v^i\left(\frac{\partial(\psi \circ \varphi^{-1})^m}{\partial x^i}(x^1, \dots, x^m)\right))$$

$$= ((\psi \circ \varphi^{-1})^1(x^1, \dots, x^m), \dots, (\psi \circ \varphi^{-1})^m(x^1, \dots, x^m),$$

$$v^i\left(\frac{\partial(\psi \circ \varphi^{-1})^1}{\partial x^i}(x^1, \dots, x^m)\right), \dots, v^i\left(\frac{\partial(\psi \circ \varphi^{-1})^m}{\partial x^i}(x^1, \dots, x^m)\right)).$$

Thus,

$$\left(\tilde{\psi} \circ \tilde{\varphi}^{-1}\right)\left(x^1, \ldots, x^m, v^1, \ldots, v^m\right) = \left(\left(\psi \circ \varphi^{-1}\right)^1\left(x^1, \ldots, x^m\right), \ldots, \left(\psi \circ \varphi^{-1}\right)^m\left(x^1, \ldots, x^m\right),\right.$$

$$\left.v^i\left(\frac{\partial(\psi \circ \varphi^{-1})^1}{\partial x^i}\left(x^1, \ldots, x^m\right)\right), \ldots, v^i\left(\frac{\partial(\psi \circ \varphi^{-1})^m}{\partial x^i}\left(x^1, \ldots, x^m\right)\right)\right).$$

Here, $\psi \circ \varphi^{-1}$ is smooth, so each of $(x^1, \ldots, x^m, v^1, \ldots, v^m) \mapsto (\psi \circ \varphi^{-1})^i$ $(x^1, \ldots, x^m)$ is smooth, and each of

$$\left(x^1, \ldots, x^m, v^1, \ldots, v^m\right) \mapsto v^i\left(\frac{\partial(\psi \circ \varphi^{-1})^j}{\partial x^i}\left(x^1, \ldots, x^m\right)\right)$$

is smooth, and hence, $\tilde{\psi} \circ \tilde{\varphi}^{-1}$ is smooth.

4. Since $\{M\}$ is an open cover of $M$, by the Lemma 4.52, there exists a countable collection $\{(U_1, \varphi_1), (U_2, \varphi_2), (U_3, \varphi_3), \cdots\}$ of admissible coordinate charts of $M$ such that $\{U_1, U_2, U_3, \cdots\}$ covers $M$. Here, each $(U_k, \varphi_k)$ is an admissible coordinate charts of $M$, so each $(U_k, \varphi_k) \in \mathcal{A}$, and hence, $\{\coprod_{r \in U_k} T_r M\}_{k \in \mathbb{N}}$ is a countable subcollection of $\{\coprod_{r \in U} T_r M\}_{(U, \varphi) \in \mathcal{A}}$.

Now, we want to prove that $\{\coprod_{r \in U_k} T_r M\}_{k \in \mathbb{N}}$ covers $\coprod_{r \in M} T_r M$. Since $\{U_1, U_2, U_3, \ldots\}$ covers $M$, $M = \cup_{k=1}^{\infty} U_k$, and hence, $\cup_{k=1}^{\infty}\left(\coprod_{q \in U_k} T_q M\right) = \coprod_{q \in \cup_{k=1}^{\infty} U_k} T_q M = \coprod_{q \in M} T_q M = \coprod_{r \in M} T_r M$. It follows that $\{\coprod_{q \in U_1} T_q M, \coprod_{q \in U_2} T_q M, \coprod_{q \in U_3} T_q M, \ldots\}$ covers $\coprod_{r \in M} T_r M$.

5. Let $(p, v) \in TM$, $(q, w) \in \coprod_{r \in M} T_r M$, and $(p, v) \neq (q, w)$. Since $(p, v) \in \coprod_{r \in M} T_r M$, $v \in T_p M$. Similarly, $w \in T_q M$.

Case I: when $p = q$. Since $p = q$, and $(p, v) \neq (q, w)$, $v \neq w$. Since $p = q$, and $w \in T_q M$, $w \in T_p M$. Since $p \in M$, and $M$ is an $m$-dimensional smooth manifold, there exists $(V, \psi) \in \mathcal{A}$ such that $p \in V$. Since $(V, \psi) \in \mathcal{A}$, $\coprod_{r \in V} T_r M \in \{\coprod_{r \in U} T_r M\}_{(U, \varphi) \in \mathcal{A}}$. Since $p \in V$, and $v \in T_p M$, $(p, v) \in \coprod_{r \in V} T_r M$. Since $p \in V$, and $p = q$, $q \in V$. Since $q \in V$, and $w \in T_q M$, $(q, w) \in \coprod_{r \in V} T_r M$. Thus, there exists $\coprod_{r \in V} T_r M \in \{\coprod_{r \in U} T_r M\}_{(U, \varphi) \in \mathcal{A}}$ such that $(p, v), (q, w) \in \coprod_{r \in V} T_r M$.

Case II: when $p \neq q$. Since $M$ is an $m$-dimensional smooth manifold, its topology is Hausdorff. Since the topology of $M$ is Hausdorff, $p \neq q$, and $p, q \in M$, there exist an open neighborhood $G_1$ of $p$ and an open neighborhood $G_2$ of $q$ such that $G_1 \cap G_2 = \emptyset$. Since $p \in M$, and $M$ is a smooth manifold, there exists $(V, \psi) \in \mathcal{A}$ such that $p \in V$. Since $(V, \psi) \in \mathcal{A}$, and $p \in V$, $V$ is an open neighborhood of $p$. Since $V$ is an open neighborhood of $p$, and $G_1$ is an open neighborhood of $p$, $G_1 \cap V$ is an open neighborhood of $p$. Since $(V, \psi) \in \mathcal{A}$, $G_1 \cap V$ is an open subset of $V$, and $\mathcal{A}$ is a smooth structure on $M$, $(G_1 \cap V, \psi|_{G_1 \cap V}) \in \mathcal{A}$. Similarly, there exists $(W, \chi) \in \mathcal{A}$ such that $(G_2 \cap W, \chi|_{G_2 \cap W}) \in \mathcal{A}$. Since $(G_1 \cap V,$ $\psi|_{G_1 \cap V}) \in \mathcal{A}$, $\coprod_{r \in G_1 \cap V} T_r M \in \{\coprod_{r \in U} T_r M\}_{(U, \varphi) \in \mathcal{A}}$. Similarly, $\coprod_{r \in G_2 \cap W} T_r M \in$ $\{\coprod_{r \in U} T_r M\}_{(U, \varphi) \in \mathcal{A}}$. Here, $G_1 \cap V$ is an open neighborhood of $p$, so $p \in G_1 \cap V$. Since $p \in G_1 \cap V$, and $v \in T_p M$, $(p, v) \in \coprod_{r \in G_1 \cap V} T_r M$. Similarly, $(q, w) \in$ $\coprod_{r \in G_2 \cap W} T_r M$. Next, $(\coprod_{r \in G_1 \cap V} T_r M) \cap (\coprod_{r \in G_2 \cap W} T_r M) = \coprod_{r \in (G_1 \cap V) \cap (G_2 \cap W)}$ $T_r M = \coprod_{r \in (G_1 \cap G_2) \cap (V \cap W)} T_r M = \coprod_{r \in \emptyset \cap (V \cap W)} T_r M = \coprod_{r \in \emptyset} T_r M = \emptyset$.

Hence, the condition 5 is satisfied.

Now, by Lemma 5.21, there exists a unique smooth differential structure $\mathcal{B}$ over $\coprod_{r \in M} T_r M$ such that for each $(U, \varphi) \in \mathcal{A}$, $(\coprod_{r \in U} T_r M, \tilde{\varphi}) \in \mathcal{B}$. Thus, $\coprod_{r \in M} T_r M$ becomes a $2m$-dimensional smooth manifold. This $2m$-dimensional smooth manifold $\coprod_{r \in M} T_r M$ is called the *tangent bundle of M*. It is denoted by $TM$. Here, corresponding to each admissible coordinate chart $(U, \varphi)$ of $M$, $(\coprod_{r \in U} T_r M, \tilde{\varphi})$ is an admissible coordinate chart of $TM$.

**Theorem 5.23** *Let M be an m-dimensional smooth manifold. Let $(M, \varphi)$ be an admissible coordinate chart of M. Then, TM is diffeomorphic onto $M \times \mathbb{R}^m$.*

*Proof* Since $(M, \varphi)$ is an admissible coordinate chart of $M$, by the Note 5.22, $(\coprod_{r \in M} T_r M, \tilde{\varphi})$ is an admissible coordinate chart of $TM$, where $\tilde{\varphi} : \coprod_{r \in M} T_r M \to$ $\varphi(M) \times \mathbb{R}^m$. Now, since $\coprod_{r \in M} T_r M = TM$, $(TM, \tilde{\varphi})$ is an admissible coordinate chart of $TM$, where $\tilde{\varphi} : TM \to \varphi(M) \times \mathbb{R}^m$. Since $(M, \varphi)$ is an admissible coordinate chart of $M$, $\varphi$ is a diffeomorphism from $M$ onto $\varphi(M)$. Since $\varphi$ is a diffeomorphism from $M$ onto $\varphi(M)$, $M$ is diffeomorphic onto $\varphi(M)$. Since $(TM, \tilde{\varphi})$ is an admissible coordinate chart of $TM$, where $\tilde{\varphi} : TM \to \varphi(M) \times \mathbb{R}^m$, $\tilde{\varphi} : TM \to$ $\varphi(M) \times \mathbb{R}^m$ is a diffeomorphism, and hence, $TM$ is diffeomorphic onto $\varphi(M) \times$ $\mathbb{R}^m$. Since $TM$ is diffeomorphic onto $\varphi(M) \times \mathbb{R}^m$, and $M$ is diffeomorphic onto $\varphi(M)$, $TM$ is diffeomorphic onto $M \times \mathbb{R}^m$. $\square$

**Definition** Let $M$ be an $m$-dimensional smooth manifold, $N$ be an $n$-dimensional smooth manifold, and $F : M \to N$ be smooth. The mapping $dF : TM \to TN$ is defined as follows: For every $(p, v) \in TM(= \coprod_{r \in M} T_r M)$, where $p \in M$, and $v \in T_p M$,

$$(dF)(p, v) \equiv (F(p), (dF_p)(v)).$$

Here, $dF$ is called the *global differential of F* (or, *global tangent map of F*).

**Theorem 5.24** *Let M be an m-dimensional smooth manifold, N be an n-dimensional smooth manifold, and $F : M \to N$ be smooth. Then, $dF : TM \to TN$ is smooth.*

*Proof* Let us take any $(p, v) \in TM(= \coprod_{r \in M} T_r M)$ where $p \in M$, and $(p, v) \in T_p M$. Let us take any admissible coordinate chart $(U, \varphi)$ of $M$ such that $p \in U$. For every $q \in U$, let

$$\left(\left.\frac{\partial}{\partial x^1}\right|_q, \left.\frac{\partial}{\partial x^2}\right|_q, \ldots, \left.\frac{\partial}{\partial x^m}\right|_q\right)$$

be the coordinate basis of $T_q M$ corresponding to $(U, \varphi)$, where

$$\left.\frac{\partial}{\partial x^i}\right|_q \equiv \left(\mathrm{d}(\varphi^{-1})_{\varphi(q)}\right)\left(\left.\frac{\partial}{\partial x^i}\right|_{\varphi(q)}\right).$$

Here, $\coprod_{r \in U} T_r M (\subset \coprod_{r \in M} T_r M = TM)$ is an open neighborhood of $(p, v)$ in $TM$, and $\varphi(U) \times \mathbb{R}^m (\subset \mathbb{R}^m \times \mathbb{R}^m = \mathbb{R}^{2m})$ is open in $\mathbb{R}^{2m}$.

Let us define a function $\tilde{\varphi} : \coprod_{r \in U} T_r M \to \varphi(U) \times \mathbb{R}^m$ as follows: For every

$$\left(q, v^1\left(\left.\frac{\partial}{\partial x^1}\right|_q\right) + \cdots + v^m\left(\left.\frac{\partial}{\partial x^m}\right|_q\right)\right) \in \coprod_{r \in U} T_r M,$$

$$\tilde{\varphi}\left(q, v^1\left(\left.\frac{\partial}{\partial x^1}\right|_q\right) + \cdots + v^m\left(\left.\frac{\partial}{\partial x^m}\right|_q\right)\right) \equiv ((\pi_1 \circ \varphi)(q), \ldots, (\pi_m \circ \varphi)(q), v^1, \ldots, v^m),$$

where $\pi_i : \mathbb{R}^m \to \mathbb{R}$ denotes the $i$th projection function of $\mathbb{R}^m$.

We know that $(\coprod_{r \in U} T_r M, \tilde{\varphi})$ is an admissible coordinate chart of $TM$. Here, $F : M \to N$, and $p \in M$, so $F(p) \in N$. Let us take any admissible coordinate chart $(V, \psi)$ of $N$ such that $F(p) \in V$. For every $r \in V$, let

$$\left(\left.\frac{\partial}{\partial y^1}\right|_r, \left.\frac{\partial}{\partial y^2}\right|_r, \ldots, \left.\frac{\partial}{\partial y^n}\right|_r\right)$$

be the coordinate basis of $T_r N$ corresponding to $(V, \psi)$, where

$$\left.\frac{\partial}{\partial y^i}\right|_r \equiv \left(\mathrm{d}(\psi^{-1})_{\psi(r)}\right)\left(\left.\frac{\partial}{\partial y^i}\right|_{\psi(r)}\right).$$

Here, $\coprod_{r \in V} T_r N (\subset \coprod_{r \in N} T_r N = TN)$ is an open neighborhood of $(\mathrm{d}F)(p, v)$ $(= (F(p), (\mathrm{d}F_p)(v)))$, and $\psi(V) \times \mathbb{R}^n (\subset \mathbb{R}^n \times \mathbb{R}^n = \mathbb{R}^{2n})$ is open in $\mathbb{R}^{2n}$. Let us define a function $\tilde{\psi} : \coprod_{r \in V} T_r N \to \psi(V) \times \mathbb{R}^n$ as follows: For every

$$\left(r, w^1\left(\left.\frac{\partial}{\partial y^1}\right|_r\right) + \cdots + w^n\left(\left.\frac{\partial}{\partial y^n}\right|_r\right)\right) \in \coprod_{r \in V} T_r N,$$

$$\tilde{\psi}\left(r, w^1\left(\left.\frac{\partial}{\partial y^1}\right|_r\right) + \cdots + w^n\left(\left.\frac{\partial}{\partial y^n}\right|_r\right)\right) \equiv ((\pi_1' \circ \psi)(r), \ldots, (\pi_n' \circ \psi)(r), w^1, \ldots, w^n),$$

where $\pi_i' : \mathbb{R}^n \to \mathbb{R}$ denotes the $i$th projection function of $\mathbb{R}^n$. We know that $(\coprod_{r \in V} T_r N, \tilde{\psi})$ is an admissible coordinate chart of $TN$. We have to prove that $\tilde{\psi} \circ F \circ \tilde{\varphi}^{-1} : \tilde{\varphi}((\coprod_{r \in V} T_r M)) \to \tilde{\psi}((\coprod_{r \in V} T_r N))$ is smooth. Here, for every $(x^1, \ldots, x^m, v^1, \ldots, v^m) \in \tilde{\varphi}(\coprod_{r \in U \cap V} T_r M)$, we have

$$\left( \tilde{\psi} \circ F \circ \tilde{\varphi}^{-1} \right) (x^1, \ldots, x^m, v^1, \ldots, v^m) = \tilde{\psi} (\tilde{\varphi}^{-1} (x^1, \ldots, x^m, v^1, \ldots, v^m))$$

$$= \left( \tilde{\psi} \circ F \right) \left( \varphi^{-1}(x^1, \ldots, x^m), v^1 \left( \frac{\partial}{\partial x^1} \Big|_{\varphi^{-1}(x^1, \ldots, x^m)} \right) + \cdots + v^m \left( \frac{\partial}{\partial x^m} \Big|_{\varphi^{-1}(x^1, \ldots, x^m)} \right) \right)$$

$$= \left( \tilde{\psi} \circ F \right) \left( \varphi^{-1}(x^1, \ldots, x^m), v^i \left( \frac{\partial}{\partial x^i} \Big|_{\varphi^{-1}(x^1, \ldots, x^m)} \right) \right)$$

$$= \tilde{\psi} \left( F \left( \varphi^{-1}(x^1, \ldots, x^m), v^i \left( \frac{\partial}{\partial x^i} \Big|_{\varphi^{-1}(x^1, \ldots, x^m)} \right) \right) \right)$$

$$= \tilde{\psi} \left( F(\varphi^{-1}(x^1, \ldots, x^m)), (dF_p) \left( v^i \left( \frac{\partial}{\partial x^i} \Big|_{\varphi^{-1}(x^1, \ldots, x^m)} \right) \right) \right)$$

$$= \tilde{\psi} \left( F(\varphi^{-1}(x^1, \ldots, x^m)), v^i \left( (dF_p) \left( \frac{\partial}{\partial x^i} \Big|_{\varphi^{-1}(x^1, \ldots, x^m)} \right) \right) \right)$$

$$= \tilde{\psi} \left( F(\varphi^{-1}(x^1, \ldots, x^m)), v^i \left( \left( \frac{\partial \left( \pi_j' \circ (\psi \circ F \circ \varphi^{-1}) \right)}{\partial x^i} (\varphi(\varphi^{-1}(x^1, \ldots, x^m))) \right) \left( \frac{\partial}{\partial y^j} \Big|_{F(\varphi^{-1}(x^1, \ldots, x^m))} \right) \right) \right)$$

$$= \tilde{\psi} \left( F(\varphi^{-1}(x^1, \ldots, x^m)), v^i \left( \left( \frac{\partial \left( \pi_j' \circ (\psi \circ F \circ \varphi^{-1}) \right)}{\partial x^i} (x^1, \ldots, x^m) \right) \left( \frac{\partial}{\partial y^j} \Big|_{F(\varphi^{-1}(x^1, \ldots, x^m))} \right) \right) \right)$$

$$= \tilde{\psi} \left( F(\varphi^{-1}(x^1, \ldots, x^m)), \left( v^i \left( \frac{\partial \left( \pi_j' \circ (\psi \circ F \circ \varphi^{-1}) \right)}{\partial x^i} (x^1, \ldots, x^m) \right) \right) \left( \frac{\partial}{\partial y^j} \Big|_{F(\varphi^{-1}(x^1, \ldots, x^m))} \right) \right)$$

$$= ((\pi_1' \circ \psi)(F(\varphi^{-1}(x^1, \ldots, x^m))), \ldots, (\pi_n' \circ \psi)(F(\varphi^{-1}(x^1, \ldots, x^m))),$$
$$v^i \left( \frac{\partial (\pi_1' \circ (\psi \circ F \circ \varphi^{-1}))}{\partial x^i} (x^1, \ldots, x^m) \right), \ldots, v^i \left( \frac{\partial (\pi_n' \circ (\psi \circ F \circ \varphi^{-1}))}{\partial x^i} (x^1, \ldots, x^m) \right) )$$

$$= ((\pi_1' \circ \psi \circ F \circ \varphi^{-1})(x^1, \ldots, x^m), \ldots, (\pi_n' \circ \psi \circ F \circ \varphi^{-1})(x^1, \ldots, x^m),$$
$$v^i \left( \frac{\partial (\pi_1' \circ \psi \circ F \circ \varphi^{-1})}{\partial x^i} (x^1, \ldots, x^m) \right), \ldots, v^i \left( \frac{\partial (\pi_n' \circ \psi \circ F \circ \varphi^{-1})}{\partial x^i} (x^1, \ldots, x^m) \right) ).$$

Thus,

$$\left( \tilde{\psi} \circ F \circ \tilde{\varphi}^{-1} \right) (x^1, \ldots, x^m, v^1, \ldots, v^m)$$
$$= ((\pi_1' \circ \psi \circ F \circ \varphi^{-1})(x^1, \ldots, x^m), \ldots, (\pi_n' \circ \psi \circ F \circ \varphi^{-1})(x^1, \ldots, x^m),$$
$$v^i \left( \frac{\partial (\pi_1' \circ \psi \circ F \circ \varphi^{-1})}{\partial x^i} (x^1, \ldots, x^m) \right), \ldots, v^i \left( \frac{\partial (\pi_n' \circ \psi \circ F \circ \varphi^{-1})}{\partial x^i} (x^1, \ldots, x^m) \right) )$$

Since $F : M \to N$ is smooth, $(U, \varphi)$ is an admissible coordinate chart of $M$ such that $p \in U$, $(V, \psi)$ is an admissible coordinate chart of $N$ such that $F(p) \in V$, $\psi \circ F \circ \varphi^{-1}$ is smooth. Since $\psi \circ F \circ \varphi^{-1}$ is smooth, each function

$(x^1, \ldots, x^m, v^1, \ldots, v^m) \rightarrow (\pi'_j \circ \psi \circ F \circ \varphi^{-1})(x^1, \ldots, x^m)$ is smooth, and each function

$$(x^1, \ldots, x^m, v^1, \ldots, v^m) \mapsto v^i \left( \frac{\partial \left( \pi'_j \circ \psi \circ F \circ \varphi^{-1} \right)}{\partial x^i} (x^1, \ldots, x^m) \right)$$

is smooth, and hence, $\tilde{\psi} \circ F \circ \tilde{\varphi}^{-1}$ is smooth. $\qquad \square$

## 5.4 Smooth Immersion

**Note 5.25** Let $M, N, P$ be smooth manifolds. Let $F : M \rightarrow N$, and $G : N \rightarrow P$ be smooth maps. Since $F : M \rightarrow N$ is smooth, by Theorem 5.24, $dF : TM \rightarrow TN$ is smooth. Similarly, $dG : TN \rightarrow TP$ is smooth. Since $dF : TM \rightarrow TN$ is smooth, and $dG : TN \rightarrow TP$ is smooth, $(dG) \circ (dF) : TM \rightarrow TP$ is smooth. Since $F : M \rightarrow N$ is smooth, and $G : N \rightarrow P$ is smooth, $G \circ F : M \rightarrow P$ is smooth, and hence, by Theorem 5.24, $d(G \circ F) : TM \rightarrow TP$ is smooth. We shall try to show that $d(G \circ F) = (dG) \circ (dF)$, that is, for every $(p, v) \in TM(= \coprod_{r \in M} T_r M)$, where $p \in M$, and $v \in T_p M$, $(d(G \circ F))(p, v) = ((dG) \circ (dF))(p, v)$.

$$\text{LHS} = (d(G \circ F))(p, v) = \left( (G \circ F)(p), \left( d(G \circ F)_p \right)(v) \right) = \left( G(F(p)), \left( d(G \circ F)_p \right)(v) \right)$$
$$= \left( G(F(p)), \left( (dG_{F(p)}) \circ (dF_p) \right)(v) \right) = \left( G(F(p)), (dG_{F(p)})((dF_p)(v)) \right),$$
$$\text{RHS} = ((dG) \circ (dF))(p, v) = (dG)((dF)(p, v)) = (dG)\left( F(p), (dF_p)(v) \right)$$
$$= \left( G(F(p)), (dG_{F(p)})((dF_p)(v)) \right).$$

Hence, LHS = RHS. Thus, we have shown that $d(G \circ F) = (dG) \circ (dF)$.

**Note 5.26** Let $M$ be a smooth manifold. Here, $\mathrm{Id}_M : M \rightarrow M$ is given by $\mathrm{Id}_M(x) = x$ for every $x$ in $M$. Clearly, $\mathrm{Id}_M$ is a smooth function. Hence, by Theorem 5.24, $d(\mathrm{Id}_M) : TM \rightarrow TM$ is smooth. Here, $\mathrm{Id}_{TM} : TM \rightarrow TM$ is a smooth function. We shall try to show that $d(\mathrm{Id}_M) = \mathrm{Id}_{TM}$, that is, for every $(p, v) \in TM(= \coprod_{r \in U} T_r M)$, where $p \in M$, and $v \in T_p M$, $(d(\mathrm{Id}_M))(p, v) = (\mathrm{Id}_{TM})(p, v)$.

$$\text{LHS} = (d(\mathrm{Id}_M))(p, v) = \left( \mathrm{Id}_M(p), \left( d(\mathrm{Id}_M)_p \right)(v) \right) = \left( p, \left( d(\mathrm{Id}_M)_p \right)(v) \right)$$
$$= \left( p, \left( d(\mathrm{Id}_M)_p \right)(v) \right) = (p, (\mathrm{Id}_{T_p M})(v)) = (p, v) = (\mathrm{Id}_{TM})(p, v) = \text{RHS}.$$

Thus, we have shown that $d(\mathrm{Id}_M) = \mathrm{Id}_{TM}$.

**Note 5.27** Let $M, N$ be smooth manifolds. Let $F : M \rightarrow N$ be a diffeomorphism. Since $F : M \rightarrow N$ is a diffeomorphism, $F : M \rightarrow N$ is smooth, and hence, $dF$ :

$TM \to TN$ is smooth. We want to prove that $dF : TM \to TN$ is a diffeomorphism, that is, $dF$ is 1–1 onto, and $(dF)^{-1}$ is smooth.

$dF : TM \to TN$ is 1–1: For this purpose, let $(dF)(p, v) = (dF)(q, w)$, where $p, q \in M$, and $v, w \in T_pM$. We have to show that $(p, v) = (q, w)$, that is, $p = q$, and $v = w$.

Here,    $(F(p), (dF_p)(v)) = (dF)(p, v) = (dF)(q, w) = (F(q), (dF_q)(w))$,    so $F(p) = F(q)$, and $(dF_p)(v) = (dF_q)(w)$. Since $F : M \to N$ is a diffeomorphism, $F$ is 1–1. Since $F$ is 1–1, and $F(p) = F(q)$, $p = q$. Since $p = q$, and $(dF_p)(v) = (dF_q)(w)$, $(dF_p)(v) = (dF_p)(w)$. Since $F : M \to N$ is a diffeomorphism, and $p \in M$, $dF_p : T_pM \to T_{F(p)}N$ is an isomorphism, and hence, $dF_p$ is 1–1. Since $dF_p$ is 1–1, and $(dF_p)(v) = (dF_p)(w)$, $v = w$. Thus, $dF : TM \to TN$ is 1–1.

$dF : TM \to TN$ is onto: For this purpose, let us take any $(q, w) \in TN$, where $q \in N$, and $w \in T_qN$.

Since $F : M \to N$ is a diffeomorphism, $F : M \to N$ is onto. Since $F : M \to N$ is onto, and $q \in N$, there exists $p \in M$ such that $F(p) = q$. Since $F(p) = q$, and $w \in T_qN$, $w \in T_{F(p)}N$. Since $F : M \to N$ is a diffeomorphism, and $p \in M$, $dF_p : T_pM \to T_{F(p)}N$ is an isomorphism, and hence, $T_pM \to T_{F(p)}N$ is onto. Since $dF_p : T_pM \to T_{F(p)}N$ is onto, and $w \in T_{F(p)}N$, there exists $v \in T_pM$ such that $(dF_p)(v) = w$. Since $p \in M$, and $v \in T_pM$, $(p, v) \in TM$. Also, $(dF)(p, v) = (F(p), (dF_p)(v)) = (q, (dF_p)(v)) = (q, w)$. This shows that $dF : TM \to TN$ is onto.

$(dF)^{-1} : TN \to TM$ is smooth: Since $F : M \to N$ is a diffeomorphism, $F^{-1} : N \to M$ is smooth, and hence, $d(F^{-1}) : TN \to TM$ is smooth.

Now, we shall try to show that $(dF)^{-1} = d(F^{-1})$, that is, for every $(q, w) \in TN$, where $q \in N$, and $w \in T_qN$, $(dF)^{-1}(q, w) = (d(F^{-1}))(q, w)$, that is, for every $(q, w) \in TN$, where $q \in N$, and $w \in T_qN$, $(q, w) = (dF)((d(F^{-1}))(q, w))$. For this purpose, let us take any $(q, w) \in TN$, where $q \in N$, and $w \in T_qN$. Now, since $dF : TM \to TN$ is onto, there exists $(p, v) \in TM$ such that $(F(p), (dF_p)(v)) = (dF)(p, v) = (q, w)$. It follows that $F(p) = q$, and $(dF_p)(v) = w$. Since $F : M \to N$ is a diffeomorphism, and $p \in M$, $dF_p : T_pM \to T_{F(p)}N$ is an isomorphism, and hence, $dF_p : T_pM \to T_{F(p)}N$ is 1–1 onto. Since $dF_p : T_pM \to T_{F(p)}N$ is 1–1 onto, and $(dF_p)(v) = w$, $v = (dF_p)^{-1}(w)$. Since $F : M \to N$ is a diffeomorphism, $F : M \to N$ is a 1–1 onto. Since $F : M \to N$ is a 1–1 onto, and $F(p) = q$, $p = F^{-1}(q)$. Now,

$$
\begin{aligned}
\text{RHS} &= (dF)((d(F^{-1}))(q, w)) = (dF)\left(F^{-1}(q), \left(d(F^{-1})_q\right)(w)\right) \\
&= (dF)\left(F^{-1}(q), \left(d(F^{-1})_q\right)(w)\right) \\
&= (dF)\left(F^{-1}(q), \left(d(F^{-1})_{F(p)}\right)(w)\right) \\
&= (dF)\left(F^{-1}(q), \left((dF_p)^{-1}\right)(w)\right) = (dF)\left(p, \left((dF_p)^{-1}\right)(w)\right) \\
&= (dF)(p, v) = (F(p), (dF_p)(v)) = (q, (dF_p)(v)) = (q, w) = \text{LHS}.
\end{aligned}
$$

Thus, $(dF)^{-1} = d(F^{-1})$. Since $(dF)^{-1} = d(F^{-1})$, and $d(F^{-1}) : TN \to TM$ is smooth, $(dF)^{-1} : TN \to TM$ is smooth.

This proves that if $F : M \to N$ is a diffeomorphism, then $dF : TM \to TN$ is a diffeomorphism.

**Note 5.28** Let $M, N$ be smooth manifolds. Let $F : M \to N$ be a diffeomorphism. From the Note 5.27, $dF : TM \to TN$ is a diffeomorphism. Since $dF : TM \to TN$ is a diffeomorphism, $dF : TM \to TN$ is 1–1 onto, and hence, $(dF)^{-1} : TN \to TM$. Since $F : M \to N$ is a diffeomorphism, $F^{-1} : N \to M$ is a diffeomorphism, and hence, by the Note 5.27, $d(F^{-1}) : TN \to TM$ is a diffeomorphism. Thus, $(dF)^{-1} : TN \to TM$, and $d(F^{-1}) : TN \to TM$.

Now, we shall try to show that $(dF)^{-1} = d(F^{-1})$, that is, for every $(q, w) \in TN$, where $q \in N$, and $w \in T_qN$, $(dF)^{-1}(q, w) = (d(F^{-1}))(q, w)$, that is, for every $(q, w) \in TN$, where $q \in N$, and $w \in T_qN$, $(q, w) = (dF)((d(F^{-1}))(q, w))$. For this purpose, let us take any $(q, w) \in TN$, where $q \in N$, and $w \in T_qN$. Now, since $dF : TM \to TN$ is onto, there exists $(p, v) \in TM$ such that $(F(p), (dF_p)(v)) = (dF)(p, v) = (q, w)$. It follows that $F(p) = q$, and $(dF_p)(v) = w$. Since $F : M \to N$ is a diffeomorphism, and $p \in M$, $dF_p : T_pM \to T_{F(p)}N$ is an isomorphism, and hence, $dF_p : T_pM \to T_{F(p)}N$ is 1–1 onto. Since $dF_p : T_pM \to T_{F(p)}N$ is 1–1 onto, and $(dF_p)(v) = w$, $v = (dF_p)^{-1}(w)$. Since $F : M \to N$ is a diffeomorphism, $F : M \to N$ is a 1–1 onto. Since $F : M \to N$ is a 1–1 onto, and $F(p) = q$, $p = F^{-1}(q)$. Now,

$$\begin{aligned}
\text{RHS} &= (dF)\left((d(F^{-1}))(q, w)\right) = (dF)\left(F^{-1}(q), \left(d(F^{-1})_q\right)(w)\right) \\
&= (dF)\left(F^{-1}(q), \left(d(F^{-1})_q\right)(w)\right) = (dF)\left(F^{-1}(q), \left(d(F^{-1})_{F(p)}\right)(w)\right) \\
&= (dF)\left(F^{-1}(q), \left((dF_p)^{-1}\right)(w)\right) = (dF)\left(p, \left((dF_p)^{-1}\right)(w)\right) \\
&= (dF)(p, v) = (F(p), (dF_p)(v)) = (q, (dF_p)(v)) = (q, w) = \text{LHS}.
\end{aligned}$$

Thus, we have shown that if $F : M \to N$ is a diffeomorphism, then $(dF)^{-1} = d(F^{-1})$.

**Definition** Let $t_0$ be a real number. Let $\varepsilon$ be any positive real number. Let $M$ be an $m$-dimensional smooth manifold. Let $\gamma : (t_0 - \varepsilon, t_0 + \varepsilon) \to M$ be a smooth map. Observe that the open interval $(t_0 - \varepsilon, t_0 + \varepsilon)$ is a one-dimensional smooth manifold. Since $\gamma : (t_0 - \varepsilon, t_0 + \varepsilon) \to M$ is a smooth map, and $t_0 \in (t_0 - \varepsilon, t_0 + \varepsilon)$, $d\gamma_{t_0} : T_{t_0}(t_0 - \varepsilon, t_0 + \varepsilon) \to T_{\gamma(t_0)}M$. The standard coordinate basis vector of $T_{t_0}\mathbb{R}$ is denoted by $\frac{d}{dt}\Big|_{t_0}$.

Since $T_{t_0}(t_0 - \varepsilon, t_0 + \varepsilon)$ is the same as $T_{t_0}\mathbb{R}$, and $\frac{d}{dt}\big|_{t_0}$ is the standard coordinate basis vector of $T_{t_0}\mathbb{R}$, $\frac{d}{dt}\big|_{t_0}$ is the standard coordinate basis vector of $T_{t_0}(t_0 - \varepsilon, t_0 + \varepsilon)$. Since $\frac{d}{dt}\big|_{t_0}$ is in $T_{t_0}(t_0 - \varepsilon, t_0 + \varepsilon)$, and $d\gamma_{t_0} : T_{t_0}(t_0 - \varepsilon, t_0 + \varepsilon) \rightarrow T_{\gamma(t_0)}M$, $(d\gamma_{t_0})(\frac{d}{dt}\big|_{t_0}) \in T_{\gamma(t_0)}M$. Here, $(d\gamma_{t_0})(\frac{d}{dt}\big|_{t_0})$ is denoted by $\gamma'(t_0)$, and is called the *velocity of $\gamma$ at $t_0$*. Since

$$\gamma'(t_0) = (d\gamma_{t_0})\left(\frac{d}{dt}\bigg|_{t_0}\right) \in T_{\gamma(t_0)}M, \text{ so } \gamma'(t_0) \in T_{\gamma(t_0)}M.$$

Since $\gamma'(t_0) \in T_{\gamma(t_0)}M$, $\gamma'(t_0)$ is a function from $C^\infty(M)$ to $\mathbb{R}$. Hence, for every $f \in C^\infty(M)$, we have

$$(f \circ \gamma)'(t_0) = \left(\frac{d}{dt}\bigg|_{t_0}\right)(f \circ \gamma) = \left((d\gamma_{t_0})\left(\frac{d}{dt}\bigg|_{t_0}\right)\right)(f) = (\gamma'(t_0))(f).$$

Thus, for every $f \in C^\infty(M)$, we have $(\gamma'(t_0))(f) = (f \circ \gamma)'(t_0)$.

Here, $\gamma(t_0) \in M$, and $M$ is an $m$-dimensional smooth manifold, so there exists an admissible coordinate chart $(U, \varphi)$ of $M$ such that $\gamma(t_0) \in U$. Let

$$\left(\frac{\partial}{\partial x^1}\bigg|_{\gamma(t_0)}, \ldots, \frac{\partial}{\partial x^m}\bigg|_{\gamma(t_0)}\right)$$

be the coordinate basis of $T_{\gamma(t_0)}M$ corresponding to $(U, \varphi)$, where

$$\frac{\partial}{\partial x^i}\bigg|_{\gamma(t_0)} \equiv \left(d(\varphi^{-1})_{\varphi(\gamma(t_0))}\right)\left(\frac{\partial}{\partial x^i}\bigg|_{\varphi(\gamma(t_0))}\right).$$

So the matrix representation of linear map $d\gamma_{t_0}$ is the following $m \times 1$ matrix:

$$\begin{bmatrix} \dfrac{d(\pi_1 \circ (\varphi \circ \gamma))}{dt}(t_0) \\[2mm] \dfrac{d(\pi_2 \circ (\varphi \circ \gamma))}{dt}(t_0) \\[2mm] \vdots \\[2mm] \dfrac{d(\pi_m \circ (\varphi \circ \gamma))}{dt}(t_0) \end{bmatrix},$$

where $\pi_i : \mathbb{R}^m \to \mathbb{R}$ denotes the $i$th projection function of $\mathbb{R}^m$. It follows that

$$\gamma'(t_0) = (d\gamma_{t_0})\left(\frac{d}{dt}\bigg|_{t_0}\right) = \left(\frac{d(\pi_1 \circ (\varphi \circ \gamma))}{dt}(t_0)\right)\left(\frac{\partial}{\partial x^1}\bigg|_{\gamma(t_0)}\right)$$

$$+\cdots+\left(\frac{d(\pi_m \circ (\varphi \circ \gamma))}{dt}(t_0)\right)\left(\frac{\partial}{\partial x^m}\bigg|_{\gamma(t_0)}\right)$$

$$= \left(\frac{d(\pi_i \circ (\varphi \circ \gamma))}{dt}(t_0)\right)\left(\frac{\partial}{\partial x^i}\bigg|_{\gamma(t_0)}\right).$$

Thus,

$$\gamma'(t_0) = \left(\frac{d(\pi_i \circ (\varphi \circ \gamma))}{dt}(t_0)\right)\left(\frac{\partial}{\partial x^i}\bigg|_{\gamma(t_0)}\right).$$

In the case of Euclidean space $\mathbb{R}^m$ for $M$, the above formula takes the form:

$$\gamma'(t_0) = \left(\frac{d(\pi_i \circ \gamma)}{dt}(t_0)\right)\left(\frac{\partial}{\partial x^i}\bigg|_{\gamma(t_0)}\right),$$

which is essentially the same as in the classical calculus.

**Theorem 5.29** *Let $M$ be an $m$-dimensional smooth manifold. Let $p \in M$. Let $v \in T_pM$. Then, there exist a positive real $\varepsilon$ and a smooth curve $\gamma : (-\varepsilon, \varepsilon) \to M$ such that $\gamma(0) = p$, and $\gamma'(0) = v$.*

*Proof* Since $p \in M$, and $M$ is an $m$-dimensional smooth manifold, there exists an admissible coordinate chart $(U, \varphi)$ of $M$ such that $p \in U$, and $\varphi(p) = 0$. For every $q \in U$, let

$$\left(\frac{\partial}{\partial x^1}\bigg|_q, \ldots, \frac{\partial}{\partial x^m}\bigg|_q\right)$$

be the coordinate basis of $T_qM$ corresponding to $(U, \varphi)$, where

$$\frac{\partial}{\partial x^i}\bigg|_q \equiv \left(d(\varphi^{-1})_{\varphi(q)}\right)\left(\frac{\partial}{\partial x^i}\bigg|_{\varphi(q)}\right).$$

Here,

$$\left(\frac{\partial}{\partial x^1}\bigg|_p, \ldots, \frac{\partial}{\partial x^m}\bigg|_p\right)$$

is the coordinate basis of $T_pM$, and $v \in T_pM$, there exist real numbers $v^1, \ldots, v^m$ such that

$$v = v^1 \left( \frac{\partial}{\partial x^1} \bigg|_p \right) + \cdots + v^m \left( \frac{\partial}{\partial x^m} \bigg|_p \right) = v^i \left( \frac{\partial}{\partial x^i} \bigg|_p \right).$$

Since $(U, \varphi)$ is an admissible coordinate chart of $M$, $\varphi$ is a mapping from $U$ onto $\varphi(U)$, where $\varphi(U)$ is an open neighborhood of $\varphi(p)(= 0)$ in $\mathbb{R}^m$. Since $t \mapsto (tv^1, \ldots, tv^m)$ is a continuous map from $\mathbb{R}$ to $\mathbb{R}^m$, and $\varphi(U)$ is an open neighborhood of $\varphi(p)(= (0, \ldots, 0) = (0v^1, \ldots, 0v^m))$ in $\mathbb{R}^m$, there exists $\varepsilon > 0$ such that for every $t \in (0 - \varepsilon, 0 + \varepsilon)$, we have $(tv^1, \ldots, tv^m) \in \varphi(U)$, and hence, for every $t \in (-\varepsilon, \varepsilon)$, we have $\varphi^{-1}(tv^1, \ldots, tv^m) \in U$. Now, let us define $\gamma : (-\varepsilon, \varepsilon) \to U(\subset M)$ as follows: For every $t \in (-\varepsilon, \varepsilon)$, $\gamma(t) \equiv \varphi^{-1}(tv^1, \ldots, tv^m)$. Here, $\gamma(0) = \varphi^{-1}(0v^1, \ldots, 0v^m) = \varphi^{-1}(0) = \varphi^{-1}(\varphi(p)) = p$. Thus, $\gamma(0) = p$. We shall try to show that $\gamma$ is smooth.

Since $(U, \varphi)$ is an admissible coordinate chart of $M$, $\varphi$ is a diffeomorphism from $U$ onto $\varphi(U)$, and hence, $\varphi^{-1}$ is smooth. Since $\varphi^{-1}$ is smooth, and $t \mapsto (tv^1, \ldots, tv^m)$ is smooth, their composite $t \mapsto \varphi^{-1}(tv^1, \ldots, tv^m)(= \gamma(t))$ is smooth, and hence, $\gamma$ is smooth. It follows that for every $t \in (-\varepsilon, \varepsilon)$ we have

$$(\pi_i \circ (\varphi \circ \gamma))(t) = \pi_i(\varphi(\gamma(t))) = \pi_i\big(\varphi\big(\varphi^{-1}(tv^1, \ldots, tv^m)\big)\big)$$
$$= \pi_i(tv^1, \ldots, tv^m) = tv^i = v^i t,$$

so

$$\frac{\mathrm{d}(\pi_i \circ (\varphi \circ \gamma))}{\mathrm{d}t}(0) = v^i.$$

Now, we shall try to show that $\gamma'(0) = v$. Here,

$$\gamma'(0) = \left( \frac{\mathrm{d}(\pi_i \circ (\varphi \circ \gamma))}{\mathrm{d}t}(0) \right) \left( \frac{\partial}{\partial x^i} \bigg|_{\gamma(0)} \right) = v^i \left( \frac{\partial}{\partial x^i} \bigg|_{\gamma(0)} \right) = v^i \left( \frac{\partial}{\partial x^i} \bigg|_p \right) = v.$$

Hence, $\gamma'(0) = v$.                                                                 □

**Note 5.30** Let $M$ be an $m$-dimensional smooth manifold, $N$ be an $n$-dimensional smooth manifold, and $F : M \to N$ be a smooth map. Let $\gamma : (t_0 - \varepsilon, t_0 + \varepsilon) \to M$ be a smooth map.

Since $\gamma : (t_0 - \varepsilon, t_0 + \varepsilon) \to M$ is a smooth map, and $F : M \to N$ is a smooth map, their composite $F \circ \gamma : (t_0 - \varepsilon, t_0 + \varepsilon) \to N$ is a smooth map, and hence, $(F \circ \gamma)'(t_0) \in T_{(F \circ \gamma)(t_0)}N = T_{F(\gamma(t_0))}N$. Since $\gamma : (t_0 - \varepsilon, t_0 + \varepsilon) \to M$ is a smooth map, $\gamma'(t_0) \in T_{\gamma(t_0)}M$. Since $\gamma : (t_0 - \varepsilon, t_0 + \varepsilon) \to M$, $\gamma(t_0) \in M$. Since $\gamma(t_0) \in M$, and $F : M \to N$ is smooth, $\mathrm{d}F_{\gamma(t_0)} : T_{\gamma(t_0)}M \to T_{F(\gamma(t_0))}N$. Since $\mathrm{d}F_{\gamma(t_0)} :

$T_{\gamma(t_0)}M \to T_{F(\gamma(t_0))}N$, and $\gamma'(t_0) \in T_{\gamma(t_0)}M$, $(dF_{\gamma(t_0)})(\gamma'(t_0)) \in T_{F(\gamma(t_0))}N$. Thus, $(dF_{\gamma(t_0)})(\gamma'(t_0))$ and $(F \circ \gamma)'(t_0)$ are elements of $T_{F(\gamma(t_0))}N$. We shall try to show that $(dF_{\gamma(t_0)})(\gamma'(t_0)) = (F \circ \gamma)'(t_0)$.

$$\text{RHS} = (F \circ \gamma)'(t_0) = \left(d(F \circ \gamma)_{t_0}\right)\left(\left.\frac{d}{dt}\right|_{t_0}\right) = \left((dF_{\gamma(t_0)}) \circ (d\gamma_{t_0})\right)\left(\left.\frac{d}{dt}\right|_{t_0}\right)$$

$$= (dF_{\gamma(t_0)})\left((d\gamma_{t_0})\left(\left.\frac{d}{dt}\right|_{t_0}\right)\right) = (dF_{\gamma(t_0)})(\gamma'(t_0)) = \text{LHS}.$$

Thus, we have shown that $(dF_{\gamma(t_0)})(\gamma'(t_0)) = (F \circ \gamma)'(t_0)$.

**Note 5.31** Let $M$ be an $m$-dimensional smooth manifold, $N$ be an $n$-dimensional smooth manifold, and $F : M \to N$ be a smooth map. Let $p \in M$. Let $v \in T_pM$. Let $\gamma : (-\varepsilon, \varepsilon) \to M$ be a smooth curve on $M$, such that $\gamma(0) = p$, and $\gamma'(0) = v$. (The existence of such a $\gamma$ has been established in Theorem 5.29.)

Since $F : M \to N$ is a smooth map, $dF_p : T_pM \to T_{F(p)}N$. Since $dF_p : T_pM \to T_{F(p)}N$, and $v \in T_pM$, $(dF_p)(v) \in T_{F(p)}N$. Here, $\gamma : (-\varepsilon, \varepsilon) \to M$ is smooth, and $F : M \to N$ is smooth, so their composite $F \circ \gamma : (-\varepsilon, \varepsilon) \to N$ is smooth, and hence, $(F \circ \gamma)'(0) \in T_{(F \circ \gamma)(0)}N = T_{F(\gamma(0))}N = T_{F(p)}N$. Thus, $(dF_p)(v)$ and $(F \circ \gamma)'(0)$ are members of $T_{F(p)}N$. We shall try to show that $(dF_p)(v) = (F \circ \gamma)'(0)$.

$$\text{RHS} = (F \circ \gamma)'(0) = (dF_{\gamma(0)})(\gamma'(0)) = (dF_p)(\gamma'(0)) = (dF_p)(v) = \text{LHS}.$$

This proves that $(dF_p)(v) = (F \circ \gamma)'(0)$. In short, we write $(dF_{\gamma(0)})(\gamma'(0)) = (F \circ \gamma)'(0)$.

**Definition** Let $M$ be an $m$-dimensional smooth manifold, $N$ be an $n$-dimensional smooth manifold, and $F : M \to N$ be a smooth map. Let $p \in M$.

It follows that $dF_p : T_pM \to T_{F(p)}N$. We know that $dF_p$ is a linear map from real linear space $T_pM$ to real linear space $T_{F(p)}N$. Here, $(dF_p)(T_pM)$ is a linear subspace of $T_{F(p)}N$. We know that $\dim T_pM = m$, and $\dim T_{F(p)}N = n$. Since $(dF_p)(T_pM)$ is a linear subspace of $T_{F(p)}N$, $\dim((dF_p)(T_pM)) \le \dim T_{F(p)}N = n$. Since $dF_p$ is a linear map from real linear space $T_pM$ to real linear space $T_{F(p)}N$, $\dim((dF_p)(T_pM)) \le \dim T_pM = m$. Since $\dim((dF_p)(T_pM)) \le m$, and $\dim((dF_p)(T_pM)) \le n$, $\dim((dF_p)(T_pM)) \le \min\{m, n\}$. Here, $\dim((dF_p)(T_pM))$ is called the *rank of $F$ at $p$*.

It follows that the rank of $F$ at $p$ is $r$ if and only if the matrix representation of linear map $dF_p$ has rank $r$. Clearly, the rank of $F$ at $p$ is less than or equal to $\min\{m, n\}$. Observe that the linear map $dF_p : T_pM \to T_{F(p)}N$ is onto if and only if $\dim((dF_p)(T_pM)) = \dim T_{F(p)}N (= n)$. Thus, the linear map $dF_p : T_pM \to T_{F(p)}N$ is onto if and only if the rank of $F$ at $p$ is $n$.

Observe that the linear map $dF_p : T_pM \to T_{F(p)}N$ is 1–1 if and only if $\dim((dF_p)(T_pM)) = \dim T_pM(= m)$. Thus, the linear map $dF_p : T_pM \to T_{F(p)}N$ is 1–1 if and only if the rank of $F$ at $p$ is $m$.

**Definition** Let $M$ be an $m$-dimensional smooth manifold, $N$ be an $n$-dimensional smooth manifold, and $F : M \to N$ be a smooth map. Let $k$ be a positive integer. If for every $p \in M$, the rank of $F$ at $p$ is $k$, then we write rank $F = k$, and say that $F$ *has constant rank k.* Let $M$ *be* an $m$-dimensional smooth manifold, $N$ be an $n$-dimensional smooth manifold, and $F : M \to N$ be a smooth map. Here, the linear map $dF_p : T_pM \to T_{F(p)}N$ is onto for every $p \in M$ if and only if the rank of $F$ at $p$ is $n$ for every $p \in M$. Thus, the linear map $dF_p : T_pM \to T_{F(p)}N$ is onto for every $p \in M$ if and only if rank $F = n$.

**Note 5.32** Let $M$ be an $m$-dimensional smooth manifold, $N$ be an $n$-dimensional smooth manifold, and $F : M \to N$ be a smooth map. Here, the linear map $dF_p : T_pM \to T_{F(p)}N$ is 1–1 for every $p \in M$ if and only if the rank of $F$ at $p$ is $m$ for every $p \in M$. Thus, the linear map $dF_p : T_pM \to T_{F(p)}N$ is 1–1 for every $p \in M$ if and only if rank $F = m$.

**Definition** Let $M$ be an $m$-dimensional smooth manifold, $N$ be an $n$-dimensional smooth manifold, and $F : M \to N$ be a smooth map. If for every $p \in M$, the linear map $dF_p : T_pM \to T_{F(p)}N$ is onto, then we say that $F$ is a *smooth submersion.* Thus, $F$ is a smooth submersion if and only if rank $F = n$.

**Definition** Let $M$ be an $m$-dimensional smooth manifold, $N$ be an $n$-dimensional smooth manifold, and $F : M \to N$ be a smooth map. If for every $p \in M$, the linear map $dF_p : T_pM \to T_{F(p)}N$ is 1–1, then we say that $F$ is a *smooth immersion.* Thus, $F$ is a smooth immersion if and only if rank $F = m$.

## 5.5  Inverse Function Theorem for Manifolds

**Note 5.33** Let $M(2 \times 3, \mathbb{R})$ be the collection of all $2 \times 3$ matrices with real entries, that is,

$$M(2 \times 3, \mathbb{R}) \equiv \left\{ \begin{bmatrix} a & b & c \\ d & e & f \end{bmatrix} : a, b, c, d, e, f \in \mathbb{R} \right\}.$$

We define addition of matrices as follows:

$$\begin{bmatrix} a & b & c \\ d & e & f \end{bmatrix} + \begin{bmatrix} a_1 & b_1 & c_1 \\ d_1 & e_1 & f_1 \end{bmatrix} \equiv \begin{bmatrix} a+a_1 & b+b_1 & c+c_1 \\ d+d_1 & e+e_1 & f+f_1 \end{bmatrix}.$$

Clearly, $M(2 \times 3, \mathbb{R})$ is a commutative group under addition as binary operation.

We define scalar multiplication of matrix as follows:

$$t\begin{bmatrix} a & b & c \\ d & e & f \end{bmatrix} \equiv \begin{bmatrix} ta & tb & tc \\ td & te & tf \end{bmatrix}.$$

Clearly, $M(2 \times 3, \mathbb{R})$ becomes a real linear space of dimension $(2 \times 3)$. Here, one of the basis of $M(2 \times 3, \mathbb{R})$ is

$$\left\{ \begin{bmatrix} 1 & 0 & 0 \\ 0 & 0 & 0 \end{bmatrix}, \begin{bmatrix} 0 & 1 & 0 \\ 0 & 0 & 0 \end{bmatrix}, \begin{bmatrix} 0 & 0 & 1 \\ 0 & 0 & 0 \end{bmatrix}, \begin{bmatrix} 0 & 0 & 0 \\ 1 & 0 & 0 \end{bmatrix}, \begin{bmatrix} 0 & 0 & 0 \\ 0 & 1 & 0 \end{bmatrix}, \begin{bmatrix} 0 & 0 & 0 \\ 0 & 0 & 1 \end{bmatrix} \right\}.$$

Clearly, the real linear space $\mathbb{R}^{2 \times 3}$ is isomorphic to the real linear space $M(2 \times 3, \mathbb{R})$. Since $\mathbb{R}^{2 \times 3}$ is isomorphic to $M(2 \times 3, \mathbb{R})$, we will not distinguish between

$$\begin{bmatrix} a & b & c \\ d & e & f \end{bmatrix} \quad \text{and} \quad (a, b, c, d, e, f).$$

Now, since $\mathbb{R}^{2 \times 3}$ is a $(2 \times 3)$-dimensional smooth manifold, $M(2 \times 3, \mathbb{R})$ is also a $(2 \times 3)$-dimensional smooth manifold. Let $M_2(2 \times 3, \mathbb{R})$ be the collection of all

$$\begin{bmatrix} a & b & c \\ d & e & f \end{bmatrix} \in M(2 \times 3, \mathbb{R})$$

such that

$$\text{rank} \begin{bmatrix} a & b & c \\ d & e & f \end{bmatrix} = 2.$$

If

$$\begin{bmatrix} a & b & c \\ d & e & f \end{bmatrix} \in M_2(2 \times 3, \mathbb{R}),$$

then

$$\text{rank} \begin{bmatrix} a & b & c \\ d & e & f \end{bmatrix} = 2,$$

that is, if

$$\begin{bmatrix} a & b & c \\ d & e & f \end{bmatrix} \in M_2(2 \times 3, \mathbb{R})$$

then

$$\left(\det\begin{bmatrix} b & c \\ e & f \end{bmatrix} \neq 0 \text{ or } \det\begin{bmatrix} a & c \\ d & f \end{bmatrix} \neq 0 \text{ or } \det\begin{bmatrix} a & b \\ d & e \end{bmatrix} \neq 0\right).$$

We want to show that $M_2(2 \times 3, \mathbb{R})$ is an open subset of $M(2 \times 3, \mathbb{R})$. For this purpose, let us fix any

$$\begin{bmatrix} a & b & c \\ d & e & f \end{bmatrix} \in M_2(2 \times 3, \mathbb{R}).$$

It follows that

$$\det\begin{bmatrix} b & c \\ e & f \end{bmatrix} \neq 0 \quad \text{or} \quad \det\begin{bmatrix} a & c \\ d & f \end{bmatrix} \neq 0 \quad \text{or} \quad \det\begin{bmatrix} a & b \\ d & e \end{bmatrix} \neq 0.$$

Case I: when

$$\det\begin{bmatrix} b & c \\ e & f \end{bmatrix} \neq 0$$

Here,

$$bf - ce = \det\begin{bmatrix} b & c \\ e & f \end{bmatrix} \neq 0.$$

Since there exists $\varepsilon > 0$, the open interval

$$\left(\det\begin{bmatrix} b & c \\ e & f \end{bmatrix} - \varepsilon, \det\begin{bmatrix} b & c \\ e & f \end{bmatrix} + \varepsilon\right)$$

does not contain 0. Let us define a function $f : \mathbb{R}^6 \to \mathbb{R}$ as follows: For every $(x_1, x_2, x_3, x_4, x_5, x_6)$ in $\mathbb{R}^6$

$$f(x_1, x_2, x_3, x_4, x_5, x_6) \equiv x_2 x_6 - x_5 x_3 = \det\begin{bmatrix} x_2 & x_3 \\ x_5 & x_6 \end{bmatrix}.$$

Clearly, $f$ is a continuous function. Also,

$$f(a, b, c, d, e, f) = \det\begin{bmatrix} b & c \\ e & f \end{bmatrix}.$$

Since $f$ is a continuous at $(a, b, c, d, e, f)$, there exists an open neighborhood $U$ of $(a, b, c, d, e, f)$ in $\mathbb{R}^6$ such that for every $(x_1, x_2, x_3, x_4, x_5, x_6) \in U$,

$$\det \begin{bmatrix} x_2 & x_3 \\ x_5 & x_6 \end{bmatrix} = f(x_1, x_2, x_3, x_4, x_5, x_6) \in (f(a,b,c,d,e,f) - \varepsilon, f(a,b,c,d,e,f) + \varepsilon)$$

$$= \left( \det \begin{bmatrix} b & c \\ e & f \end{bmatrix} - \varepsilon, \det \begin{bmatrix} b & c \\ e & f \end{bmatrix} + \varepsilon \right) \not\ni 0.$$

Thus, for every $(x_1, x_2, x_3, x_4, x_5, x_6) \in U$, we have

$$\det \begin{bmatrix} x_2 & x_3 \\ x_5 & x_6 \end{bmatrix} \neq 0,$$

and hence,

$$\text{rank} \begin{bmatrix} x_1 & x_2 & x_3 \\ x_4 & x_5 & x_6 \end{bmatrix} = 2$$

for every $(x_1, x_2, x_3, x_4, x_5, x_6) \in U$. Since $U$ is an open neighborhood of $(a, b, c, d, e, f)$ in $\mathbb{R}^6$,

$$\left\{ \begin{bmatrix} x_1 & x_2 & x_3 \\ x_4 & x_5 & x_6 \end{bmatrix} : (x_1, x_2, x_3, x_4, x_5, x_6) \in U \right\}$$

is an open neighborhood of

$$\begin{bmatrix} a & b & c \\ d & e & f \end{bmatrix} \text{ in } M(2 \times 3, \mathbb{R}).$$

Since

$$\text{rank} \begin{bmatrix} x_1 & x_2 & x_3 \\ x_4 & x_5 & x_6 \end{bmatrix} = 2$$

for every $(x_1, x_2, x_3, x_4, x_5, x_6) \in U$,

$$\left\{ \begin{bmatrix} x_1 & x_2 & x_3 \\ x_4 & x_5 & x_6 \end{bmatrix} : (x_1, x_2, x_3, x_4, x_5, x_6) \in U \right\} \subset M_2(2 \times 3, \mathbb{R}).$$

Since

$$\left\{ \begin{bmatrix} x_1 & x_2 & x_3 \\ x_4 & x_5 & x_6 \end{bmatrix} : (x_1, x_2, x_3, x_4, x_5, x_6) \in U \right\}$$

is an open neighborhood of

$$\begin{bmatrix} a & b & c \\ d & e & f \end{bmatrix} \text{ in } M(2 \times 3, \mathbb{R}),$$

and

$$\left\{ \begin{bmatrix} x_1 & x_2 & x_3 \\ x_4 & x_5 & x_6 \end{bmatrix} : (x_1, x_2, x_3, x_4, x_5, x_6) \in U \right\} \subset M_2(2 \times 3, \mathbb{R}),$$

$$\begin{bmatrix} a & b & c \\ d & e & f \end{bmatrix}$$

is an interior point of $M_2(2 \times 3, \mathbb{R})$.

Case II: when

$$\det \begin{bmatrix} a & c \\ d & f \end{bmatrix} \neq 0$$

This case is similar to the case I.

Case III: when

$$\det \begin{bmatrix} a & b \\ d & e \end{bmatrix} \neq 0$$

This case is similar to the case I. Hence, in all cases,

$$\begin{bmatrix} a & b & c \\ d & e & f \end{bmatrix}$$

is an interior point of $M_2(2 \times 3, \mathbb{R})$.


This proves that $M_2(2 \times 3, \mathbb{R})$ is an open subset of $M(2 \times 3, \mathbb{R})$. Since $M_2(2 \times 3, \mathbb{R})$ is an open subset of $M(2 \times 3, \mathbb{R})$, and $M(2 \times 3, \mathbb{R})$ is a $(2 \times 3)$-dimensional smooth manifold, $M_2(2 \times 3, \mathbb{R})$ is also a $(2 \times 3)$-dimensional smooth manifold.

Similarly, $M_2(3 \times 2, \mathbb{R})$ is a $(3 \times 2)$-dimensional smooth manifold. Its generalization to higher dimensions can also be given. Such smooth manifolds are known as the *manifolds of matrices of full rank*.

**Theorem 5.34** *Let M be an m-dimensional smooth manifold, N be an n-dimensional smooth manifold, and $F : M \rightarrow N$ be a smooth map. Let $p \in M$. If $dF_p : T_pM \rightarrow T_{F(p)}N$ is onto, then there exists an open neighborhood W of p such that for every $q \in W$, the linear map $dF_q : T_qM \rightarrow T_{F(q)}N$ is onto.*

*Proof* Here, $p \in M$, and $M$ is an $m$-dimensional smooth manifold, so there exists an admissible coordinate chart $(U, \varphi)$ of $M$ such that $p \in U$. For every $q \in U$, let

$$\left( \frac{\partial}{\partial x^1} \bigg|_q, \frac{\partial}{\partial x^2} \bigg|_q, \ldots, \frac{\partial}{\partial x^m} \bigg|_q \right)$$

be the coordinate basis of $T_qM$ corresponding to $(U, \varphi)$, where

$$\frac{\partial}{\partial x^i}\bigg|_q \equiv \left(\mathrm{d}(\varphi^{-1})_{\varphi(q)}\right)\left(\frac{\partial}{\partial x^i}\bigg|_{\varphi(q)}\right).$$

Here, $p \in M$, and $F : M \to N$, so $F(p) \in N$. Since $F(p) \in N$, and $N$ is an $n$-dimensional smooth manifold, there exists an admissible coordinate chart $(V, \psi)$ of $N$ such that $F(p) \in V$. For every $r \in V$, let

$$\left(\frac{\partial}{\partial y^1}\bigg|_r, \frac{\partial}{\partial y^2}\bigg|_r, \ldots, \frac{\partial}{\partial y^n}\bigg|_r\right)$$

be the coordinate basis of $T_r N$ corresponding to $(V, \psi)$, where

$$\frac{\partial}{\partial y^i}\bigg|_r \equiv \left(\mathrm{d}(\psi^{-1})_{\psi(r)}\right)\left(\frac{\partial}{\partial y^i}\bigg|_{\psi(r)}\right).$$

Here, the matrix representation of linear map $\mathrm{d}F_p$ is the following $n \times m$ matrix:

$$\left[\begin{array}{cc}
\dfrac{\partial(\pi_1 \circ (\psi \circ F \circ \varphi^{-1}))}{\partial x^1}(\varphi(p)) & \dfrac{\partial(\pi_1 \circ (\psi \circ F \circ \varphi^{-1}))}{\partial x^2}(\varphi(p)) \\[2ex]
\dfrac{\partial(\pi_2 \circ (\psi \circ F \circ \varphi^{-1}))}{\partial x^1}(\varphi(p)) & \dfrac{\partial(\pi_2 \circ (\psi \circ F \circ \varphi^{-1}))}{\partial x^2}(\varphi(p)) \\[2ex]
\cdots & \cdots \\[1ex]
\dfrac{\partial(\pi_n \circ (\psi \circ F \circ \varphi^{-1}))}{\partial x^1}(\varphi(p)) & \dfrac{\partial(\pi_n \circ (\psi \circ F \circ \varphi^{-1}))}{\partial x^2}(\varphi(p))
\end{array}\right.$$

$$\left.\begin{array}{c}
\cdots \quad \dfrac{\partial(\pi_1 \circ (\psi \circ F \circ \varphi^{-1}))}{\partial x^m}(\varphi(p)) \\[2ex]
\cdots \quad \dfrac{\partial(\pi_2 \circ (\psi \circ F \circ \varphi^{-1}))}{\partial x^m}(\varphi(p)) \\[1ex]
\cdots \quad \cdots \\[1ex]
\cdots \quad \dfrac{\partial(\pi_n \circ (\psi \circ F \circ \varphi^{-1}))}{\partial x^m}(\varphi(p))
\end{array}\right].$$

Since the linear map $\mathrm{d}F_p : T_p M \to T_{F(p)} N$ is onto, and $\dim T_{F(p)} N = n$, the rank of the matrix representation of linear map $\mathrm{d}F_p$ is $n$, that is,

$$\mathrm{rank} \left[\begin{array}{cc}
\dfrac{\partial(\pi_1 \circ (\psi \circ F \circ \varphi^{-1}))}{\partial x^1}(\varphi(p)) & \dfrac{\partial(\pi_1 \circ (\psi \circ F \circ \varphi^{-1}))}{\partial x^2}(\varphi(p)) \\[2ex]
\dfrac{\partial(\pi_2 \circ (\psi \circ F \circ \varphi^{-1}))}{\partial x^1}(\varphi(p)) & \dfrac{\partial(\pi_2 \circ (\psi \circ F \circ \varphi^{-1}))}{\partial x^2}(\varphi(p)) \\[2ex]
\cdots & \cdots \\[1ex]
\dfrac{\partial(\pi_n \circ (\psi \circ F \circ \varphi^{-1}))}{\partial x^1}(\varphi(p)) & \dfrac{\partial(\pi_n \circ (\psi \circ F \circ \varphi^{-1}))}{\partial x^2}(\varphi(p))
\end{array}\right.$$

$$\left.\begin{array}{c}
\cdots \quad \dfrac{\partial(\pi_1 \circ (\psi \circ F \circ \varphi^{-1}))}{\partial x^m}(\varphi(p)) \\[2ex]
\cdots \quad \dfrac{\partial(\pi_2 \circ (\psi \circ F \circ \varphi^{-1}))}{\partial x^m}(\varphi(p)) \\[1ex]
\cdots \quad \cdots \\[1ex]
\cdots \quad \dfrac{\partial(\pi_n \circ (\psi \circ F \circ \varphi^{-1}))}{\partial x^m}(\varphi(p))
\end{array}\right]$$

is $n$. Since the rank of $F$ at $p$ is less than or equal to $\min\{m, n\}$, and the rank of $F$ at $p$ is $n$, $n \leq \min\{m, n\} \leq m$, and hence, $n \leq m$. Here, $n \leq m$, and

$$
\text{rank}
\begin{bmatrix}
\dfrac{\partial(\pi_1 \circ (\psi \circ F \circ \varphi^{-1}))}{\partial x^1}(\varphi(p)) & \dfrac{\partial(\pi_1 \circ (\psi \circ F \circ \varphi^{-1}))}{\partial x^2}(\varphi(p)) \\[2mm]
\dfrac{\partial(\pi_2 \circ (\psi \circ F \circ \varphi^{-1}))}{\partial x^1}(\varphi(p)) & \dfrac{\partial(\pi_2 \circ (\psi \circ F \circ \varphi^{-1}))}{\partial x^2}(\varphi(p)) \\[2mm]
\cdots & \cdots \\[2mm]
\dfrac{\partial(\pi_n \circ (\psi \circ F \circ \varphi^{-1}))}{\partial x^1}(\varphi(p)) & \dfrac{\partial(\pi_n \circ (\psi \circ F \circ \varphi^{-1}))}{\partial x^2}(\varphi(p))
\end{bmatrix}
$$

$$
\begin{bmatrix}
\cdots & \dfrac{\partial(\pi_1 \circ (\psi \circ F \circ \varphi^{-1}))}{\partial x^m}(\varphi(p)) \\[2mm]
\cdots & \dfrac{\partial(\pi_2 \circ (\psi \circ F \circ \varphi^{-1}))}{\partial x^m}(\varphi(p)) \\[2mm]
\cdots & \cdots \\[2mm]
\cdots & \dfrac{\partial(\pi_n \circ (\psi \circ F \circ \varphi^{-1}))}{\partial x^m}(\varphi(p))
\end{bmatrix}
$$

is $n$, so

$$
\begin{bmatrix}
\dfrac{\partial(\pi_1 \circ (\psi \circ F \circ \varphi^{-1}))}{\partial x^1}(\varphi(p)) & \dfrac{\partial(\pi_1 \circ (\psi \circ F \circ \varphi^{-1}))}{\partial x^2}(\varphi(p)) \\[2mm]
\dfrac{\partial(\pi_2 \circ (\psi \circ F \circ \varphi^{-1}))}{\partial x^1}(\varphi(p)) & \dfrac{\partial(\pi_2 \circ (\psi \circ F \circ \varphi-1))}{\partial x^2}(\varphi(p)) \\[2mm]
\cdots & \cdots \\[2mm]
\dfrac{\partial(\pi_n \circ (\psi \circ F \circ \varphi^{-1}))}{\partial x^1}(\varphi(p)) & \dfrac{\partial(\pi_n \circ (\psi \circ F \circ \varphi^{-1}))}{\partial x^2}(\varphi(p))
\end{bmatrix}
$$

$$
\begin{bmatrix}
\cdots & \dfrac{\partial(\pi_1 \circ (\psi \circ F \circ \varphi^{-1}))}{\partial x^m}(\varphi(p)) \\[2mm]
\cdots & \dfrac{\partial(\pi_2 \circ (\psi \circ F \circ \varphi^{-1}))}{\partial x^m}(\varphi(p)) \\[2mm]
\cdots & \cdots \\[2mm]
\cdots & \dfrac{\partial(\pi_n \circ (\psi \circ F \circ \varphi^{-1}))}{\partial x^m}(\varphi(p))
\end{bmatrix}
$$

$\in M_n(n \times m, \mathbb{R}) \subset M(n \times m, \mathbb{R})$. Since $M_n(n \times m, \mathbb{R})$ is an open subset of $M(n \times m, \mathbb{R})$. It follows that $M_n(n \times m, \mathbb{R})$ is an open neighborhood of

$$
\begin{bmatrix}
\dfrac{\partial(\pi_1 \circ (\psi \circ F \circ \varphi^{-1}))}{\partial x^1}(\varphi(p)) & \dfrac{\partial(\pi_1 \circ (\psi \circ F \circ \varphi^{-1}))}{\partial x^2}(\varphi(p)) \\[2mm]
\dfrac{\partial(\pi_2 \circ (\psi \circ F \circ \varphi^{-1}))}{\partial x^1}(\varphi(p)) & \dfrac{\partial(\pi_2 \circ (\psi \circ F \circ \varphi^{-1}))}{\partial x^2}(\varphi(p)) \\[2mm]
\cdots & \cdots \\[2mm]
\dfrac{\partial(\pi_n \circ (\psi \circ F \circ \varphi^{-1}))}{\partial x^1}(\varphi(p)) & \dfrac{\partial(\pi_n \circ (\psi \circ F \circ \varphi^{-1}))}{\partial x^2}(\varphi(p))
\end{bmatrix}
$$

$$
\begin{bmatrix}
\cdots & \dfrac{\partial(\pi_1 \circ (\psi \circ F \circ \varphi^{-1}))}{\partial x^m}(\varphi(p)) \\[2mm]
\cdots & \dfrac{\partial(\pi_2 \circ (\psi \circ F \circ \varphi^{-1}))}{\partial x^m}(\varphi(p)) \\[2mm]
\cdots & \cdots \\[2mm]
\cdots & \dfrac{\partial(\pi_n \circ (\psi \circ F \circ \varphi^{-1}))}{\partial x^m}(\varphi(p))
\end{bmatrix}.
$$

Since $F : M \to N$ is a smooth map, $(U, \varphi)$ is an admissible coordinate chart of $M$ such that $p \in U$, and $(V, \psi)$ is an admissible coordinate chart of $N$ such that $F(p) \in V$, $\psi \circ F \circ \varphi^{-1}$ is smooth, and hence, each $\pi_i \circ (\psi \circ F \circ \varphi^{-1})$ is smooth. Since each $\pi_i \circ (\psi \circ F \circ \varphi^{-1})$ is smooth, and $\varphi$ is smooth, each

$$q \mapsto \frac{\partial(\pi_i \circ (\psi \circ F \circ \varphi^{-1}))}{\partial x^j}(\varphi(q))$$

is continuous, and hence,

$$q \mapsto \begin{bmatrix} \dfrac{\partial(\pi_1 \circ (\psi \circ F \circ \varphi^{-1}))}{\partial x^1}(\varphi(q)) & \dfrac{\partial(\pi_1 \circ (\psi \circ F \circ \varphi^{-1}))}{\partial x^2}(\varphi(q)) \\[2mm] \dfrac{\partial(\pi_2 \circ (\psi \circ F \circ \varphi^{-1}))}{\partial x^1}(\varphi(q)) & \dfrac{\partial(\pi_2 \circ (\psi \circ F \circ \varphi^{-1}))}{\partial x^2}(\varphi(q)) \\[2mm] \cdots & \cdots \\[2mm] \dfrac{\partial(\pi_n \circ (\psi \circ F \circ \varphi^{-1}))}{\partial x^1}(\varphi(q)) & \dfrac{\partial(\pi_n \circ (\psi \circ F \circ \varphi^{-1}))}{\partial x^2}(\varphi(q)) \end{bmatrix}$$

$$\begin{array}{c} \cdots \quad \dfrac{\partial(\pi_1 \circ (\psi \circ F \circ \varphi^{-1}))}{\partial x^m}(\varphi(q)) \\[2mm] \cdots \quad \dfrac{\partial(\pi_2 \circ (\psi \circ F \circ \varphi^{-1}))}{\partial x^m}(\varphi(q)) \\[2mm] \cdots \quad \cdots \\[2mm] \cdots \quad \dfrac{\partial(\pi_n \circ (\psi \circ F \circ \varphi^{-1}))}{\partial x^m}(\varphi(q)) \end{array}$$

is continuous. It follows that there exists an open neighborhood $W$ of $p$ in $M$ such that for every $q \in W$,

$$\begin{bmatrix} \dfrac{\partial(\pi_1 \circ (\psi \circ F \circ \varphi^{-1}))}{\partial x^1}(\varphi(q)) & \dfrac{\partial(\pi_1 \circ (\psi \circ F \circ \varphi^{-1}))}{\partial x^2}(\varphi(q)) \\[2mm] \dfrac{\partial(\pi_2 \circ (\psi \circ F \circ \varphi^{-1}))}{\partial x^1}(\varphi(q)) & \dfrac{\partial(\pi_2 \circ (\psi \circ F \circ \varphi^{-1}))}{\partial x^2}(\varphi(q)) \\[2mm] \cdots & \cdots \\[2mm] \dfrac{\partial(\pi_n \circ (\psi \circ F \circ \varphi^{-1}))}{\partial x^1}(\varphi(q)) & \dfrac{\partial(\pi_n \circ (\psi \circ F \circ \varphi^{-1}))}{\partial x^2}(\varphi(q)) \end{bmatrix}$$

$$\begin{array}{cc} \cdots \quad \dfrac{\partial(\pi_1 \circ (\psi \circ F \circ \varphi^{-1}))}{\partial x^m}(\varphi(q)) & \\[2mm] \cdots \quad \dfrac{\partial(\pi_2 \circ (\psi \circ F \circ \varphi^{-1}))}{\partial x^m}(\varphi(q)) & \in M_n(n \times m, \mathbb{R}), \\[2mm] \cdots \quad \cdots & \\[2mm] \cdots \quad \dfrac{\partial(\pi_n \circ (\psi \circ F \circ \varphi^{-1}))}{\partial x^m}(\varphi(q)) & \end{array}$$

and hence, for every $q \in W$, we have

$$\text{rank} \begin{bmatrix} \dfrac{\partial(\pi_1 \circ (\psi \circ F \circ \varphi^{-1}))}{\partial x^1}(\varphi(q)) & \dfrac{\partial(\pi_1 \circ (\psi \circ F \circ \varphi^{-1}))}{\partial x^2}(\varphi(q)) \\[2mm] \dfrac{\partial(\pi_2 \circ (\psi \circ F \circ \varphi^{-1}))}{\partial x^1}(\varphi(q)) & \dfrac{\partial(\pi_2 \circ (\psi \circ F \circ \varphi^{-1}))}{\partial x^2}(\varphi(q)) \\ \cdots & \\ \dfrac{\partial(\pi_n \circ (\psi \circ F \circ \varphi^{-1}))}{\partial x^1}(\varphi(q)) & \dfrac{\partial(\pi_n \circ (\psi \circ F \circ \varphi^{-1}))}{\partial x^2}(\varphi(q)) \end{bmatrix}$$

$$\begin{matrix} \cdots & \dfrac{\partial(\pi_1 \circ (\psi \circ F \circ \varphi^{-1}))}{\partial x^m}(\varphi(q)) \\[2mm] \cdots & \dfrac{\partial(\pi_2 \circ (\psi \circ F \circ \varphi^{-1}))}{\partial x^m}(\varphi(q)) \\ \cdots & \cdots \\ \cdots & \dfrac{\partial(\pi_n \circ (\psi \circ F \circ \varphi^{-1}))}{\partial x^m}(\varphi(q)) \end{matrix} \Bigg] = n.$$

It follows that for every $q \in W$, the linear map $dF_q : T_qM \to T_{F(q)}N$ is onto. $\qquad \square$

**Theorem 5.35** *Let $M$ be an $m$-dimensional smooth manifold, $N$ be an $n$-dimensional smooth manifold, and $F : M \to N$ be a smooth map. Let $p \in M$. If $dF_p : T_pM \to T_{F(p)}N$ is 1–1, then there exists an open neighborhood $W$ of $p$ such that for every $q \in W$, the linear map $dF_q : T_qM \to T_{F(q)}N$ is 1–1.*

*Proof* Here, $p \in M$, and $M$ is an $m$-dimensional smooth manifold, so there exists an admissible coordinate chart $(U, \varphi)$ of $M$ such that $p \in U$. For every $q \in U$,

$$\left( \frac{\partial}{\partial x^1}\bigg|_q, \frac{\partial}{\partial x^2}\bigg|_q, \ldots, \frac{\partial}{\partial x^m}\bigg|_q \right)$$

is the coordinate basis of $T_qM$ corresponding to $(U, \varphi)$, where

$$\frac{\partial}{\partial x^i}\bigg|_q \equiv \left( d(\varphi^{-1})_{\varphi(q)} \right) \left( \frac{\partial}{\partial x^i}\bigg|_{\varphi(q)} \right).$$

Here, $p \in M$, and $F : M \to N$, so $F(p) \in N$. Since $F(p) \in N$, and $N$ is an $n$-dimensional smooth manifold, there exists an admissible coordinate chart $(V, \psi)$ of $N$ such that $F(p) \in V$. For every $r \in V$,

$$\left( \frac{\partial}{\partial y^1}\bigg|_r, \frac{\partial}{\partial y^2}\bigg|_r, \ldots, \frac{\partial}{\partial y^n}\bigg|_r \right)$$

is the coordinate basis of $T_rN$ corresponding to $(V, \psi)$, where

$$\frac{\partial}{\partial y^i}\bigg|_r \equiv \left( d(\psi^{-1})_{\psi(r)} \right) \left( \frac{\partial}{\partial y^i}\bigg|_{\psi(r)} \right).$$

Here, the matrix representation of linear map $dF_p$ is the following $n \times m$ matrix:

$$
\begin{bmatrix}
\dfrac{\partial(\pi_1 \circ (\psi \circ F \circ \varphi^{-1}))}{\partial x^1}(\varphi(p)) & \dfrac{\partial(\pi_1 \circ (\psi \circ F \circ \varphi^{-1}))}{\partial x^2}(\varphi(p)) \\
\dfrac{\partial(\pi_2 \circ (\psi \circ F \circ \varphi^{-1}))}{\partial x^1}(\varphi(p)) & \dfrac{\partial(\pi_2 \circ (\psi \circ F \circ \varphi^{-1}))}{\partial x^2}(\varphi(p)) \\
\cdots & \cdots \\
\dfrac{\partial(\pi_n \circ (\psi \circ F \circ \varphi^{-1}))}{\partial x^1}(\varphi(p)) & \dfrac{\partial(\pi_n \circ (\psi \circ F \circ \varphi^{-1}))}{\partial x^2}(\varphi(p))
\end{bmatrix}
\begin{matrix}
\cdots & \dfrac{\partial(\pi_1 \circ (\psi \circ F \circ \varphi^{-1}))}{\partial x^m}(\varphi(p)) \\
\cdots & \dfrac{\partial(\pi_2 \circ (\psi \circ F \circ \varphi^{-1}))}{\partial x^m}(\varphi(p)) \\
\cdots & \cdots \\
\cdots & \dfrac{\partial(\pi_n \circ (\psi \circ F \circ \varphi^{-1}))}{\partial x^m}(\varphi(p))
\end{matrix}.
$$

Since the linear map $dF_p : T_pM \to T_{F(p)}N$ is 1–1, and $\dim T_pM = m$, the rank of the matrix representation of linear map $dF_p$ is $m$, that is,

$$
\mathrm{rank}
\begin{bmatrix}
\dfrac{\partial(\pi_1 \circ (\psi \circ F \circ \varphi^{-1}))}{\partial x^1}(\varphi(p)) & \dfrac{\partial(\pi_1 \circ (\psi \circ F \circ \varphi^{-1}))}{\partial x^2}(\varphi(p)) \\
\dfrac{\partial(\pi_2 \circ (\psi \circ F \circ \varphi^{-1}))}{\partial x^1}(\varphi(p)) & \dfrac{\partial(\pi_2 \circ (\psi \circ F \circ \varphi^{-1}))}{\partial x^2}(\varphi(p)) \\
\cdots & \cdots \\
\dfrac{\partial(\pi_n \circ (\psi \circ F \circ \varphi^{-1}))}{\partial x^1}(\varphi(p)) & \dfrac{\partial(\pi_n \circ (\psi \circ F \circ \varphi^{-1}))}{\partial x^2}(\varphi(p))
\end{bmatrix}
\begin{matrix}
\cdots & \dfrac{\partial(\pi_1 \circ (\psi \circ F \circ \varphi^{-1}))}{\partial x^m}(\varphi(p)) \\
\cdots & \dfrac{\partial(\pi_2 \circ (\psi \circ F \circ \varphi^{-1}))}{\partial x^m}(\varphi(p)) \\
\cdots & \cdots \\
\cdots & \dfrac{\partial(\pi_n \circ (\psi \circ F \circ \varphi^{-1}))}{\partial x^m}(\varphi(p))
\end{matrix}
$$

is $m$. Since the rank of $F$ at $p$ is less than or equal to $\min\{m, n\}$, and the rank of $F$ at $p$ is $m$, $m \le \min\{m, n\} \le n$, and hence, $m \le n$. Here, $m \le n$, and

$$
\mathrm{rank}
\begin{bmatrix}
\dfrac{\partial(\pi_1 \circ (\psi \circ F \circ \varphi^{-1}))}{\partial x^1}(\varphi(p)) & \dfrac{\partial(\pi_1 \circ (\psi \circ F \circ \varphi^{-1}))}{\partial x^2}(\varphi(p)) \\
\dfrac{\partial(\pi_2 \circ (\psi \circ F \circ \varphi^{-1}))}{\partial x^1}(\varphi(p)) & \dfrac{\partial(\pi_2 \circ (\psi \circ F \circ \varphi^{-1}))}{\partial x^2}(\varphi(p)) \\
\cdots & \\
\dfrac{\partial(\pi_n \circ (\psi \circ F \circ \varphi^{-1}))}{\partial x^1}(\varphi(p)) & \dfrac{\partial(\pi_n \circ (\psi \circ F \circ \varphi^{-1}))}{\partial x^2}(\varphi(p))
\end{bmatrix}
\begin{matrix}
\cdots & \dfrac{\partial(\pi_1 \circ (\psi \circ F \circ \varphi^{-1}))}{\partial x^m}(\varphi(p)) \\
\cdots & \dfrac{\partial(\pi_2 \circ (\psi \circ F \circ \varphi^{-1}))}{\partial x^m}(\varphi(p)) \\
\cdots & \cdots \\
\cdots & \dfrac{\partial(\pi_n \circ (\psi \circ F \circ \varphi^{-1}))}{\partial x^m}(\varphi(p))
\end{matrix}
$$

is $m$, so

$$
\begin{bmatrix}
\dfrac{\partial(\pi_1 \circ (\psi \circ F \circ \varphi^{-1}))}{\partial x^1}(\varphi(p)) & \dfrac{\partial(\pi_1 \circ (\psi \circ F \circ \varphi^{-1}))}{\partial x^2}(\varphi(p)) \\
\dfrac{\partial(\pi_2 \circ (\psi \circ F \circ \varphi^{-1}))}{\partial x^1}(\varphi(p)) & \dfrac{\partial(\pi_2 \circ (\psi \circ F \circ \varphi^{-1}))}{\partial x^2}(\varphi(p)) \\
\cdots & \cdots \\
\dfrac{\partial(\pi_n \circ (\psi \circ F \circ \varphi^{-1}))}{\partial x^1}(\varphi(p)) & \dfrac{\partial(\pi_n \circ (\psi \circ F \circ \varphi^{-1}))}{\partial x^2}(\varphi(p))
\end{bmatrix}
$$

$$
\begin{bmatrix}
\cdots & \dfrac{\partial(\pi_1 \circ (\psi \circ F \circ \varphi^{-1}))}{\partial x^m}(\varphi(p)) \\
\cdots & \dfrac{\partial(\pi_2 \circ (\psi \circ F \circ \varphi^{-1}))}{\partial x^m}(\varphi(p)) \\
\cdots & \cdots \\
\cdots & \dfrac{\partial(\pi_n \circ (\psi \circ F \circ \varphi^{-1}))}{\partial x^m}(\varphi(p))
\end{bmatrix}
$$

$\in \mathrm{M}_m(n \times m, \mathbb{R}) \subset \mathrm{M}(n \times m, \mathbb{R})$. Since $\mathrm{M}_m(n \times m, \mathbb{R})$ is an open subset of $\mathrm{M}(n \times m, \mathbb{R})$, it follows that $\mathrm{M}_m(n \times m, \mathbb{R})$ is an open neighborhood of

$$
\begin{bmatrix}
\dfrac{\partial(\pi_1 \circ (\psi \circ F \circ \varphi^{-1}))}{\partial x^1}(\varphi(p)) & \dfrac{\partial(\pi_1 \circ (\psi \circ F \circ \varphi^{-1}))}{\partial x^2}(\varphi(p)) \\
\dfrac{\partial(\pi_2 \circ (\psi \circ F \circ \varphi^{-1}))}{\partial x^1}(\varphi(p)) & \dfrac{\partial(\pi_2 \circ (\psi \circ F \circ \varphi^{-1}))}{\partial x^2}(\varphi(p)) \\
\cdots & \cdots \\
\dfrac{\partial(\pi_n \circ (\psi \circ F \circ \varphi^{-1}))}{\partial x^1}(\varphi(p)) & \dfrac{\partial(\pi_n \circ (\psi \circ F \circ \varphi^{-1}))}{\partial x^2}(\varphi(p))
\end{bmatrix}
$$

$$
\begin{bmatrix}
\cdots & \dfrac{\partial(\pi_1 \circ (\psi \circ F \circ \varphi^{-1}))}{\partial x^m}(\varphi(p)) \\
\cdots & \dfrac{\partial(\pi_2 \circ (\psi \circ F \circ \varphi^{-1}))}{\partial x^m}(\varphi(p)) \\
\cdots & \cdots \\
\cdots & \dfrac{\partial(\pi_n \circ (\psi \circ F \circ \varphi^{-1}))}{\partial x^m}(\varphi(p))
\end{bmatrix}.
$$

Since $F : M \to N$ is a smooth map, $(U, \varphi)$ is an admissible coordinate chart of $M$ such that $p \in U$, and $(V, \psi)$ is an admissible coordinate chart of $N$ such that $F(p) \in V$, $\psi \circ F \circ \varphi^{-1}$ is smooth, and hence, each $\pi_i \circ (\psi \circ F \circ \varphi^{-1})$ is smooth. Since each $\pi_i \circ (\psi \circ F \circ \varphi^{-1})$ is smooth, and $\varphi$ is smooth, each

$$q \mapsto \frac{\partial(\pi_i \circ (\psi \circ F \circ \varphi^{-1}))}{\partial x^j}(\varphi(q))$$

is continuous, and hence,

$$q \mapsto \begin{bmatrix} \frac{\partial(\pi_1 \circ (\psi \circ F \circ \varphi^{-1}))}{\partial x^1}(\varphi(q)) & \frac{\partial(\pi_1 \circ (\psi \circ F \circ \varphi^{-1}))}{\partial x^2}(\varphi(q)) \\ \frac{\partial(\pi_2 \circ (\psi \circ F \circ \varphi^{-1}))}{\partial x^1}(\varphi(q)) & \frac{\partial(\pi_2 \circ (\psi \circ F \circ \varphi^{-1}))}{\partial x^2}(\varphi(q)) \\ \ldots & \ldots \\ \frac{\partial(\pi_n \circ (\psi \circ F \circ \varphi^{-1}))}{\partial x^1}(\varphi(q)) & \frac{\partial(\pi_n \circ (\psi \circ F \circ \varphi^{-1}))}{\partial x^2}(\varphi(q)) \end{bmatrix}$$

$$\begin{array}{l} \ldots \quad \frac{\partial(\pi_1 \circ (\psi \circ F \circ \varphi^{-1}))}{\partial x^m}(\varphi(q)) \\ \ldots \quad \frac{\partial(\pi_2 \circ (\psi \circ F \circ \varphi^{-1}))}{\partial x^m}(\varphi(q)) \\ \ldots \quad \ldots \\ \ldots \quad \frac{\partial(\pi_n \circ (\psi \circ F \circ \varphi^{-1}))}{\partial x^m}(\varphi(q)) \end{array}$$

is continuous. It follows that there exists an open neighborhood $W$ of $p$ in $M$ such that for every $q \in W$,

$$\begin{bmatrix} \frac{\partial(\pi_1 \circ (\psi \circ F \circ \varphi^{-1}))}{1}(\varphi(q)) & \frac{\partial(\pi_1 \circ (\psi \circ F \circ \varphi^{-1}))}{\partial x^2}(\varphi(q)) \\ \frac{\partial(\pi_2 \circ (\psi \circ F \circ \varphi^{-1}))}{\partial x^1}(\varphi(q)) & \frac{\partial(\pi_2 \circ (\psi \circ F \circ \varphi^{-1}))}{\partial x^2}(\varphi(q)) \\ \ldots \\ \frac{\partial(\pi_n \circ (\psi \circ F \circ \varphi^{-1}))}{\partial x^1}(\varphi(q)) & \frac{\partial(\pi_n \circ (\psi \circ F \circ \varphi^{-1}))}{\partial x^2}(\varphi(q)) \end{bmatrix}$$

$$\begin{array}{l} \ldots \quad \frac{\partial(\pi_1 \circ (\psi \circ F \circ \varphi^{-1}))}{\partial x^m}(\varphi(q)) \\ \ldots \quad \frac{\partial(\pi_2 \circ (\psi \circ F \circ \varphi^{-1}))}{\partial x^m}(\varphi(q)) \\ \ldots \quad \ldots \\ \ldots \quad \frac{\partial(\pi_n \circ (\psi \circ F \circ \varphi^{-1}))}{\partial x^m}(\varphi(q)) \end{array} \in M_m(n, \mathbb{R}),$$

and hence, for every $q \in W$, we have

$$\text{rank} \begin{bmatrix} \dfrac{\partial(\pi_1 \circ (\psi \circ F \circ \varphi^{-1}))}{\partial x^1}(\varphi(q)) & \dfrac{\partial(\pi_1 \circ (\psi \circ F \circ \varphi^{-1}))}{\partial x^2}(\varphi(q)) \\[2mm] \dfrac{\partial(\pi_2 \circ (\psi \circ F \circ \varphi^{-1}))}{\partial x^1}(\varphi(q)) & \dfrac{\partial(\pi_2 \circ (\psi \circ F \circ \varphi^{-1}))}{\partial x^2}(\varphi(q)) \\[2mm] \cdots & \cdots \\[2mm] \dfrac{\partial(\pi_n \circ (\psi \circ F \circ \varphi^{-1}))}{\partial x^1}(\varphi(q)) & \dfrac{\partial(\pi_n \circ (\psi \circ F \circ \varphi^{-1}))}{\partial x^2}(\varphi(q)) \end{bmatrix}$$

$$\begin{bmatrix} \cdots & \dfrac{\partial(\pi_1 \circ (\psi \circ F \circ \varphi^{-1}))}{\partial x^m}(\varphi(q)) \\[2mm] \cdots & \dfrac{\partial(\pi_2 \circ (\psi \circ F \circ \varphi^{-1}))}{\partial x^m}(\varphi(q)) \\[2mm] \cdots & \cdots \\[2mm] \cdots & \dfrac{\partial(\pi_n \circ (\psi \circ F \circ \varphi^{-1}))}{\partial x^m}(\varphi(q)) \end{bmatrix} = m.$$

It follows that for every $q \in W$, the linear map $dF_q : T_q M \to T_{F(q)} N$ is 1–1. $\quad\square$

**Theorem 5.36** *Let $M$ be an $m$-dimensional smooth manifold, $N$ be an $m$-dimensional smooth manifold, and $F : M \to N$ be a smooth map. Let $p \in M$. If $dF_p :$ $T_p M \to T_{F(p)} N$ is invertible (i.e., 1–1 and onto); then, there exists a connected open neighborhood $U_0$ of $p$, and a connected open neighborhood $V_0$ of $F(p)$ such that $F|_{U_0} : U_0 \to V_0$ is a diffeomorphism.*

*Proof* Here, $p \in M$, and $M$ is an $m$-dimensional smooth manifold, so there exists an admissible coordinate chart $(U, \varphi)$ of $M$ such that $p \in U$. For every $q \in U$, let

$$\left( \left. \frac{\partial}{\partial x^1} \right|_q, \left. \frac{\partial}{\partial x^2} \right|_q, \ldots, \left. \frac{\partial}{\partial x^m} \right|_q \right)$$

be the coordinate basis of $T_q M$ corresponding to $(U, \varphi)$, where

$$\left. \frac{\partial}{\partial x^i} \right|_q \equiv \left( d(\varphi^{-1})_{\varphi(q)} \right) \left( \left. \frac{\partial}{\partial x^i} \right|_{\varphi(q)} \right).$$

Here, $p \in M$, and $F : M \to N$, so $F(p) \in N$. Since $F(p) \in N$, and $N$ is an $m$-dimensional smooth manifold, there exists an admissible coordinate chart $(V, \psi)$ of $N$ such that $F(p) \in V$, and $\psi(F(p)) = 0$. For every $r \in V$, let

$$\left( \left. \frac{\partial}{\partial y^1} \right|_r, \left. \frac{\partial}{\partial y^2} \right|_r, \ldots, \left. \frac{\partial}{\partial y^m} \right|_r \right)$$

be the coordinate basis of $T_r N$ corresponding to $(V, \psi)$, where

$$\frac{\partial}{\partial y^i}\bigg|_r \equiv \left(\mathrm{d}\big(\psi^{-1}\big)_{\psi(r)}\right)\left(\frac{\partial}{\partial y^i}\bigg|_{\psi(r)}\right).$$

Here, the matrix representation of linear map $\mathrm{d}F_p$ is the following $m \times m$ matrix:

$$\begin{bmatrix} \dfrac{\partial\big(\pi_1 \circ (\psi \circ F \circ \varphi^{-1})\big)}{\partial x^1}(\varphi(p)) & \dfrac{\partial\big(\pi_1 \circ (\psi \circ F \circ \varphi^{-1})\big)}{\partial x^2}(\varphi(p)) \\[2ex] \dfrac{\partial\big(\pi_2 \circ (\psi \circ F \circ \varphi^{-1})\big)}{\partial x^1}(\varphi(p)) & \dfrac{\partial\big(\pi_2 \circ (\psi \circ F \circ \varphi^{-1})\big)}{\partial x^2}(\varphi(p)) \\[2ex] \cdots & \cdots \\[1ex] \dfrac{\partial\big(\pi_m \circ (\psi \circ F \circ \varphi^{-1})\big)}{\partial x^1}(\varphi(p)) & \dfrac{\partial\big(\pi_m \circ (\psi \circ F \circ \varphi^{-1})\big)}{\partial x^2}(\varphi(p)) \end{bmatrix}$$

$$\begin{aligned} &\cdots \quad \dfrac{\partial\big(\pi_1 \circ (\psi \circ F \circ \varphi^{-1})\big)}{\partial x^m}(\varphi(p)) \\[2ex] &\cdots \quad \dfrac{\partial\big(\pi_2 \circ (\psi \circ F \circ \varphi^{-1})\big)}{\partial x^m}(\varphi(p)) \\[2ex] &\cdots \quad \cdots \\[1ex] &\cdots \quad \dfrac{\partial\big(\pi_m \circ (\psi \circ F \circ \varphi^{-1})\big)}{\partial x^m}(\varphi(p)) \end{aligned}\Bigg].$$

Since the linear map $\mathrm{d}F_p : T_pM \to T_{F(p)}N$ is 1–1 onto, and $\dim T_pM = m$,

$$\det\begin{bmatrix} \dfrac{\partial\big(\pi_1 \circ (\psi \circ F \circ \varphi^{-1})\big)}{\partial x^1}(\varphi(p)) & \dfrac{\partial\big(\pi_1 \circ (\psi \circ F \circ \varphi^{-1})\big)}{\partial x^2}(\varphi(p)) \\[2ex] \dfrac{\partial\big(\pi_2 \circ (\psi \circ F \circ \varphi^{-1})\big)}{\partial x^1}(\varphi(p)) & \dfrac{\partial\big(\pi_2 \circ (\psi \circ F \circ \varphi^{-1})\big)}{\partial x^2}(\varphi(p)) \\[2ex] \cdots & \cdots \\[1ex] \dfrac{\partial\big(\pi_n \circ (\psi \circ F \circ \varphi^{-1})\big)}{\partial x^1}(\varphi(p)) & \dfrac{\partial\big(\pi_n \circ (\psi \circ F \circ \varphi^{-1})\big)}{\partial x^2}(\varphi(p)) \end{bmatrix}$$

$$\begin{aligned} &\cdots \quad \dfrac{\partial\big(\pi_1 \circ (\psi \circ F \circ \varphi^{-1})\big)}{\partial x^m}(\varphi(p)) \\[2ex] &\cdots \quad \dfrac{\partial\big(\pi_2 \circ (\psi \circ F \circ \varphi^{-1})\big)}{\partial x^m}(\varphi(p)) \\[2ex] &\cdots \quad \cdots \\[1ex] &\cdots \quad \dfrac{\partial\big(\pi_n \circ (\psi \circ F \circ \varphi^{-1})\big)}{\partial x^m}(\varphi(p)) \end{aligned}\Bigg] \neq 0.$$

Since $F : M \to N$ is a smooth map, $(U, \varphi)$ is an admissible coordinate chart of $M$ such that $p \in U$, and $(V, \psi)$ is an admissible coordinate chart of $N$ such that $F(p) \in V$, $\psi \circ F \circ \varphi^{-1} : \varphi(U \cap F^{-1}(v)) \to \psi(V)$ is smooth. Also, $(\psi \circ F \circ \varphi^{-1})(\varphi(p)) = \psi(F(p)) = 0$. Clearly, $\varphi(U \cap F^{-1}(V))$ is an open neighborhood of

$\varphi(p)$ in $\mathbb{R}^m$. Further, $\psi(V)$ is an open neighborhood of 0 in $\mathbb{R}^m$. Now, by Theorem 4.1, there exists a connected open neighborhood $\hat{U}$ of $\varphi(p)$ such that

1. $\hat{U}$ is contained in $\varphi(U \cap F^{-1}(V))$,
2. $(\psi \circ F \circ \varphi^{-1})(\hat{U})$ is connected and is an open neighborhood of 0,
3. $(\psi \circ F \circ \varphi^{-1})$ has a smooth inverse on $(\psi \circ F \circ \varphi^{-1})(\hat{U})$.

Put $U_0 \equiv \varphi^{-1}(\hat{U})$, and $V_0 \equiv (F \circ \varphi^{-1})(\hat{U})$. We want to prove that $U_0$ is connected and is an open neighborhood of $p$ in $M$. Since $\hat{U}$ is connected, $\hat{U}$ is contained in $\varphi(U \cap F^{-1}(V))$, and $\varphi^{-1}$ is continuous, $\varphi^{-1}(\hat{U})(= U_0)$ is connected, and hence, $U_0$ is connected. Since $\hat{U}$ is open, $\hat{U}$ contained in $\varphi(U \cap F^{-1}(V))(\subset \varphi(U))$, and $\varphi(U)$ is open, $\hat{U}$ is open in $\varphi(U)$, and hence, $\varphi^{-1}(\hat{U})$ is open in $U$. Since $\varphi^{-1}(\hat{U})$ is open in $U$, and $U$ is open in $M$, $\varphi^{-1}(\hat{U})(= U_0)$ is open in $M$, and hence, $U_0$ is open in $M$. Since $\varphi(p) \in \hat{U}$, $p \in \varphi^{-1}(\hat{U})(= U_0)$, and hence, $p \in U_0$.

Since $\varphi^{-1}, F$ are continuous, $F \circ \varphi^{-1}$ is continuous. Since $F \circ \varphi^{-1}$ is continuous, $\hat{U}$ is connected, and $\hat{U}$ contained in $\varphi(U \cap F^{-1}(V))$, $(F \circ \varphi^{-1})(\hat{U})(= V_0)$ is connected, and hence, $V_0$ is connected. Since $(\psi \circ F \circ \varphi^{-1})$ has a smooth inverse on $(\psi \circ F \circ \varphi^{-1})(\hat{U})$, $\varphi \circ F^{-1} \circ \psi^{-1}$ is continuous. Since $\varphi \circ F^{-1} \circ \psi^{-1}$ is continuous, and $\psi$ is continuous, their composite $\varphi \circ F^{-1}$ is continuous. Since $\varphi \circ F^{-1}$ is continuous, and $\hat{U}$ is open, $(\varphi \circ F^{-1})^{-1}(\hat{U})(= (F \circ \varphi^{-1})(\hat{U}) = V_0)$ is open, and hence, $V_0$ is open. Since $\varphi(p) \in \hat{U}$, $V_0 = (F \circ \varphi^{-1})(\hat{U}) \ni (F \circ \varphi^{-1})(\varphi(p)) = F(p)$. Thus, $V_0$ is a connected open neighborhood of $F(p)$.

Since the inverse of $(\psi \circ F \circ \varphi^{-1})$ exists, $\psi \circ F \circ \varphi^{-1}$ is 1–1. Also, $\varphi, \psi^{-1}$ are 1–1. Since $\psi \circ F \circ \varphi^{-1}, \varphi, \psi^{-1}$ are 1–1, their composite $F$ is 1–1. It remains to be proved that $(F|_{U_0})^{-1} : V_0 \to U_0$ is smooth, that is, $(F^{-1}|_{V_0})$ is smooth. Since $\psi \circ F \circ \varphi^{-1}$ has a smooth inverse, $\varphi \circ F^{-1} \circ \psi^{-1}$ is smooth, and hence, $F^{-1}$ is smooth on $V_0$. $\quad\square$

**Note 5.37** The Theorem 5.36 is known as the *inverse function theorem for manifolds*.

**Definition** Let $M$ and $N$ be smooth manifolds. Let $F : M \to N$ be any function. If for every $p$ in $M$, there exists an open neighborhood $U$ of $p$ such that $F(U)$ is open in $N$, and the restriction $F|_U : U \to F(U)$ is a diffeomorphism, then we say that $F$ is a *local diffeomorphism*.

## 5.6 Shrinking Lemma

**Note 5.38** Let $M$ and $N$ be smooth manifolds. Let $F : M \to N$ be a local diffeomorphism. We shall try to show that $F$ is an open map. For this purpose, let us take any nonempty open subset $U$ of $M$.

We have to show that $F(U)$ is open in $N$, that is, each member of $F(U)$ is an interior point of $F(U)$, that is, for every $a$ in $U$, there exists an open neighborhood $V$ of $F(a)$ such that $V \subset F(U)$.

For this purpose, let us take any $a$ in $U$. Since $F : M \to N$ is a local diffeomorphism, and $a$ is in $M$, there exists an open neighborhood $U_1$ of $a$ such that $F(U_1)$ is open in $N$, and the restriction $F|_{U_1} : U_1 \to F(U_1)$ is a diffeomorphism. Since $U, U_1$ are open neighborhoods of $a$, $U \cap U_1$ is an open neighborhood of $a$, and hence, $U \cap U_1$ is open in $U_1$. Since $F|_{U_1} : U_1 \to F(U_1)$ is a diffeomorphism, $F|_{U_1} : U_1 \to F(U_1)$ is a homeomorphism. Since $F|_{U_1} : U_1 \to F(U_1)$ is a homeomorphism, and $U \cap U_1$ is open in $U_1$, $F|_{U_1}(U \cap U_1)(= F(U \cap U_1))$ is open in the open set $F(U_1)$, and hence, $F(U \cap U_1)$ is open in the open set $F(U_1)$. Since $F(U \cap U_1)$ is open in the open set $F(U_1)$, $F(U \cap U_1)$ is open in $N$. Since $a$ is in $U \cap U_1$, $F(a)$ is in $F(U \cap U_1)$. Thus, $F(U \cap U_1)$ is an open neighborhood of $F(a)$. Also, $F(U \cap U_1) \subset F(U)$.

**Note 5.39** Let $M, N, P$ be smooth manifolds. Let $F : M \to N$, and $G : N \to P$ be local diffeomorphisms. We shall try to show that their composite $G \circ F : M \to P$ is a local diffeomorphism.

Let us take any $a \in M$. We have to find an open neighborhood $U$ of $a$ such that $(G \circ F)(U)$ is open in $P$, and the restriction $(G \circ F)|_U : U \to (G \circ F)(U)$ is a diffeomorphism.

Since $F : M \to N$ is a local diffeomorphism, and $a \in M$, there exists an open neighborhood $U_1$ of $a$ such that $F(U_1)$ is open in $N$, and the restriction $F|_{U_1} : U_1 \to F(U_1)$ is a diffeomorphism. Since $F(a) \in N$, and $G : N \to P$ is a local diffeomorphism, there exists an open neighborhood $V$ of $F(a)$ such that $G(V)$ is open in $P$, and the restriction $G|_V : V \to G(V)$ is a diffeomorphism. Since $F(U_1)$ is an open neighborhood of $F(a)$, and $V$ is an open neighborhood of $F(a)$, $F(U_1) \cap V$ is an open neighborhood of $F(a)$. Here, $F(U_1) \cap V$ is an open neighborhood of $F(a)$ in $F(U_1)$, and $F|_{U_1} : U_1 \to F(U_1)$ is a diffeomorphism, $(F|_{U_1})^{-1}(F(U_1) \cap V)$ is an open neighborhood of $a$, and $F|_{(F|_{U_1})^{-1}(F(U_1) \cap V)} : (F|_{U_1})^{-1}(F(U_1) \cap V) \to F(U_1) \cap V$ is a diffeomorphism. Since $F(U_1) \cap V$ is an open neighborhood of $F(a)$ in $N$, and $G|_V : V \to G(V)$ is a diffeomorphism, $G(F(U_1) \cap V)$ is an open neighborhood of $G(F(a))(= (G \circ F)(a))$, and $G|_{(F(U_1) \cap V)} : F(U_1) \cap V \to G(F(U_1) \cap V)$ is a diffeomorphism. Since $F|_{(F|_{U_1})^{-1}(F(U_1) \cap V)} : (F|_{U_1})^{-1}(F(U_1) \cap V) \to F(U_1) \cap V$ is a diffeomorphism, and $G|_{(F(U_1) \cap V)} : F(U_1) \cap V \to G(F(U_1) \cap V)$ is a diffeomorphism, their composite $(G|_{(F(U_1) \cap V)}) \circ (F|_{(F|_{U_1})^{-1}(F(U_1) \cap V)}) : (F|_{U_1})^{-1}(F(U_1) \cap V) \to G(F(U_1) \cap V)$ is a diffeomorphism. Clearly, $(G|_{(F(U_1) \cap V)}) \circ (F|_{(F|_{U_1})^{-1}}(F(U_1) \cap V)) = (G \circ F)|_{(F|_{U_1})^{-1}(F(U_1) \cap V)}$. Thus, $(G \circ F)|_{(F|_{U_1})^{-1}(F(U_1) \cap V)} : (F|_{U_1})^{-1}(F(U_1) \cap V) \to G(F(U_1) \cap V)$ is a diffeomorphism, $(F|_{U_1})^{-1}(F(U_1) \cap V)$ is an open neighborhood of $a$, and

$$\left( (G \circ F)|_{\left( F|_{U_1} \right)^{-1}(F(U_1) \cap V)} \right) \left( \left( F|_{U_1} \right)^{-1}(F(U_1) \cap V) \right) \left( = (G \circ F) \left( \left( F|_{U_1} \right)^{-1}(F(U_1) \cap V) \right) \right.$$

$$= G \left( F \left( \left( F|_{U_1} \right)^{-1}(F(U_1) \cap V) \right) \right) = G(F(U_1) \cap V) \right)$$

is an open neighborhood of $(G \circ F)(a)$. It follows that $G \circ F : M \to P$ is a local diffeomorphism.

**Note 5.40** Let $M$, $N$ be smooth manifolds. Let $F : M \to N$ be a local diffeomorphism. We shall try to show that $F : M \to N$ is a local homeomorphism, that is, $F : M \to N$ is continuous, and for every $a \in M$, there exists an open neighborhood $U$ of $a$ such that $F(U)$ is an open neighborhood of $F(a)$ such that the restriction $F|_U : U \to F(U)$ is a homeomorphism.

First of all, we shall try to show that $F : M \to N$ is continuous. For this purpose, let us take any $a \in M$. Next, let us take any open neighborhood $V$ of $F(a)$. We have to find an open neighborhood $U$ of $a$ such that $F(U) \subset V$.

Since $a \in M$, and $F : M \to N$ is a local homeomorphism, there exists an open neighborhood $U_1$ of $a$ such that $F(U_1)$ is open in $N$, and the restriction $F|_{U_1} : U_1 \to F(U_1)$ is a diffeomorphism. Here, $F(U_1)$ is an open neighborhood of $F(a)$, and $V$ is an open neighborhood of $F(a)$, so $F(U_1) \cap V$ is an open neighborhood of $F(a)$, and hence, $F(U_1) \cap V$ is open in $F(U_1)$. Since $F(U_1) \cap V$ is open in $F(U_1)$, and $F|_{U_1} : U_1 \to F(U_1)$ is a diffeomorphism, $(F|_{U_1})^{-1}(F(U_1) \cap V)$ is open in $U_1$, and hence, $(F|_{U_1})^{-1}(F(U_1) \cap V)$ is open in $M$. Since $a \in U_1$, $(F|_{U_1})(a) = F(a) \in F(U_1) \cap V$, $a \in (F|_{U_1})^{-1}(F(U_1) \cap V)$, and hence, $(F|_{U_1})^{-1}(F(U_1) \cap V)$ is an open neighborhood of $a$.

It remains to be proved that $F((F|_{U_1})^{-1}(F(U_1) \cap V)) \subset V$. Here, $F|_{U_1} : U_1 \to F(U_1)$ is 1–1, so

$$F \left( \left( F|_{U_1} \right)^{-1}(F(U_1) \cap V) \right) = F(U_1) \cap V \subset V.$$

Thus, we have shown that $F : M \to N$ is continuous.

Now, we shall try to show that for every $a \in M$, there exists an open neighborhood $U$ of $a$, such that $F(U)$ is an open neighborhood of $F(a)$ and the restriction $F|_U : U \to F(U)$ is a homeomorphism.

For this purpose, let us fix any $a \in M$. Since $a \in M$, and $F : M \to N$ is a local diffeomorphism, there exists an open neighborhood $U$ of $a$ such that $F(U)$ is open in $N$, and the restriction $F|_U : U \to F(U)$ is a diffeomorphism. Since $F|_U : U \to F(U)$ is a diffeomorphism, $F|_U : U \to F(U)$ is a homeomorphism.

**Note 5.41** Let $M$, $N$ be smooth manifolds. Let $M_1$ be a nonempty open subset of $M$. Let $F : M \to N$ be a local diffeomorphism. We shall try to prove that $F|_{M_1} : M_1 \to N$ is a local diffeomorphism.

For this purpose, let us take any $a \in M_1$. We have to find an open neighborhood $U$ of $a$ in $M_1$ such that $(F|_{M_1})(U)$ is open in $N$, and the restriction $(F|_{M_1})|_U : U \to (F|_{M_1})(U)$ is a diffeomorphism.

Since $a \in M_1 \subset M$, $a \in M$. Since $a \in M$, and $F : M \to N$ is a local diffeomorphism, there exists an open neighborhood $U_1$ of $a$ such that $F(U_1)$ is open in $N$, and the restriction $F|_{U_1} : U_1 \to F(U_1)$ is a diffeomorphism. Since $M_1$ is an open neighborhood of $a$, $M_1 \cap U_1$ is an open neighborhood of $a$ in $U_1$. Since $M_1 \cap U_1$ is an open neighborhood of $a$ in $U_1$, and $F|_{U_1} : U_1 \to F(U_1)$ is a diffeomorphism, $(F|_{U_1})(M_1 \cap U_1)$ is open in $F(U_1)$, and hence, $(F|_{U_1})(M_1 \cap U_1)$ is open in $N$. Clearly, $(F|_{M_1})(M_1 \cap U_1)$ is open in $N$.

(Reason: Here, $(F|_{U_1})(M_1 \cap U_1)$ is open in $N$, and $(F|_{M_1})(M_1 \cap U_1) = (F|_{M_1 \cap U_1})(M_1 \cap U_1) = (F|_{U_1})(M_1 \cap U_1)$, so $(F|_{M_1})(M_1 \cap U_1)$ is open in $N$.)

Now, it remains to be showed that $(F|_{M_1})|_{(M_1 \cap U_1)} : (M_1 \cap U_1) \to (F|_{M_1})(M_1 \cap U_1)$ is a diffeomorphism, that is, $F|_{(M_1 \cap U_1)} : (M_1 \cap U_1) \to (F|_{M_1})(M_1 \cap U_1)$ is a diffeomorphism. Since $F|_{U_1} : U_1 \to F(U_1)$ is a diffeomorphism, its restriction $F|_{(M_1 \cap U_1)} : (M_1 \cap U_1) \to (F|_{M_1})(M_1 \cap U_1)$ is a diffeomorphism.

**Note 5.42** Let $M$ be an $m$-dimensional smooth manifold and $N$ be an $n$-dimensional smooth manifold. Let $F : M \to N$ be a diffeomorphism. We shall try to show that $F : M \to N$ is a local diffeomorphism. For this purpose, let us take any $a \in M$. We have to find an open neighborhood $U$ of $a$ in $M$ such that $F(U)$ is open in $N$, and the restriction $F|_U : U \to F(U)$ is a diffeomorphism.

Since $a \in M$, and $M$ is a smooth manifold, there exists an admissible coordinate chart $(U, \varphi)$ of $M$ such that $a \in U$. Since $F : M \to N$ is a diffeomorphism, $F : M \to N$ is 1-1. Since $F$ is 1-1, so $F|_U : U \to F(U)$ is 1-1 onto. Since $F : M \to N$ is a diffeomorphism, $F : M \to N$ is a homeomorphism. Since $F : M \to N$ is a homeomorphism, and $U$ is an open neighborhood of $a$, $F(U)$ is an open neighborhood of $F(a)$.

Now, it remains to be proved that:

1. $F|_U : U \to F(U)$ is smooth and
2. $(F|_U)^{-1} : F(U) \to U$ is smooth.

For 1: Since $F : M \to N$ is a diffeomorphism, $F : M \to N$ is smooth. Since $F : M \to N$ is smooth, $U$ is an open neighborhood of $a$, and $F(U)$ is an open neighborhood of $F(a)$, the restriction $F|_U : U \to F(U)$ is smooth.

For 2: Observe that $(F|_U)^{-1} = (F^{-1})|_{F(U)}$.

(Reason: Let $(y,x) \in$ LHS. So $(y,x) \in (F|_U)^{-1}$. Since $(y,x) \in (F|_U)^{-1}$, $(x,y) \in F|_U$. Since $(x,y) \in F|_U$, $x \in U$, and $(x,y) \in F$. Since $(x,y) \in F$, $(y,x) \in F^{-1}$. Since $(x,y) \in F$, $F(x) = y$. Since $x \in U$, $y = F(x) \in F(U)$. Since $(y,x) \in F^{-1}$, and $y \in F(U)$, $(y,x) \in (F^{-1})|_{F(U)} =$ RHS. Thus, LHS $\subset$ RHS. Next, let $(y,x) \in$ RHS. So $(y,x) \in (F^{-1})|_{F(U)}$. Since $(y,x) \in (F^{-1})|_{F(U)}$, $y \in F(U)$, and $(y,x) \in F^{-1}$. Since $y \in F(U)$, there exists $x_1 \in U$ such that $F(x_1) = y$. Since $(y,x) \in F^{-1}$, $(x,y) \in F$, and hence, $F(x) = y$. Since $F(x) = y$, and $F(x_1) = y$, $F(x) = F(x_1)$. Since $F(x) = F(x_1)$ and $F$ is 1–1, $x = x_1$. Since $x = x_1$, and $x_1 \in U$, $x \in U$. Since $x \in U$, and $(x,y) \in F$, $(x,y) \in F|_U$, and hence, $(y,x) \in (F|_U)^{-1} =$ LHS. Thus, RHS $\subset$ LHS. Hence, RHS $\subset$ LHS.)

Since $F : M \to N$ is a diffeomorphism, $F^{-1} : N \to M$ is smooth, and hence, $(F^{-1})|_{F(U)} : F(U) \to U$ is smooth. Since $(F^{-1})|_{F(U)} : F(U) \to U$ is smooth, and $(F|_U)^{-1} = (F^{-1})|_{F(U)}$, $(F|_U)^{-1} : F(U) \to U$ is smooth.

**Note 5.43** Let $M$ be an $m$-dimensional smooth manifold and $N$ be an $n$-dimensional smooth manifold. Let $F : M \to N$ be 1–1 onto. Let $F : M \to N$ be a local diffeomorphism. We shall try to show that $F : M \to N$ is a diffeomorphism.

Since $F : M \to N$ is a local diffeomorphism, by Note 5.40, $F : M \to N$ is a local homeomorphism, and hence, $F : M \to N$ is continuous. Since $F : M \to N$ is a local diffeomorphism, by Note 5.38, $F : M \to N$ is an open map. Since $F : M \to N$ is 1–1 onto, $F : M \to N$ is continuous, and $F : M \to N$ is an open map, $F : M \to N$ is a homeomorphism.

Now, we want to show that

1. $F : M \to N$ is smooth, and
2. $F^{-1} : N \to M$ is smooth.

For 1: Let us fix any $a \in M$. We have to find an admissible coordinate chart $(U, \varphi)$ of $M$ satisfying $a \in U$, and an admissible coordinate chart $(V, \psi)$ of $N$ satisfying $F(a) \in V$ such that $\psi \circ F \circ \varphi^{-1} : \varphi(U \cap F^{-1}(V)) \to \psi(V)$ is smooth.

Since $a \in M$, and $F : M \to N$ is a local diffeomorphism, there exists an open neighborhood $U_1$ of $a$ such that $F(U_1)$ is an open neighborhood of $F(a)$ in $N$, and the restriction $F|_{U_1} : U_1 \to F(U_1)$ is a diffeomorphism.

Since $a \in M$, and $M$ is an $m$-dimensional smooth manifold, there exists an admissible coordinate chart $(U_2, \varphi_2)$ of $M$ such that $a \in U_2$. Since $a \in U_2$, and $F : M \to N$, $F(a) \in N$. Since $F(a) \in N$, and $N$ is a smooth manifold, there exists an admissible coordinate chart $(V_2, \psi_2)$ of $N$ such that $F(a) \in V_2$. Since $U_1$ is an open neighborhood of $a$, and $U_2$ is an open neighborhood of $a$, $U_1 \cap U_2$ is an open neighborhood of $a$. Since $(U_2, \varphi_2)$ is an admissible coordinate chart of $M$ satisfying $a \in U_2$, and $U_1 \cap U_2$ is an open neighborhood of $a$, $(U_1 \cap U_2, \varphi_2|_{U_1 \cap U_2})$ is an admissible coordinate chart of $M$ satisfying $a \in U_1 \cap U_2$. Since $F|_{U_1} : U_1 \to F(U_1)$ is a diffeomorphism, the restriction $F|_{U_1 \cap U_2} : U_1 \cap U_2 \to F(U_1 \cap U_2)$ is a diffeomorphism. Since $F : M \to N$ is a homeomorphism, and $U_1 \cap U_2$ is an open neighborhood of $a$, $F(U_1 \cap U_2)$ is an open neighborhood of $F(a)$.

Since $(V_2, \psi_2)$ is an admissible coordinate chart of $N$ such that $V_2$ is an open neighborhood of $F(a)$, and $F(U_1 \cap U_2)$ is an open neighborhood of $F(a)$, $V_2 \cap F(U_1 \cap U_2)$ is an open neighborhood of $F(a)$. Since $(V_2, \psi_2)$ is an admissible coordinate chart $(V_2, \psi_2)$ of $N$, and $V_2 \cap F(U_1 \cap U_2)$ is an open neighborhood of $F(a)$, $(V_2 \cap F(U_1 \cap U_2), \psi_2|_{V_2 \cap F(U_1 \cap U_2)})$ is an admissible coordinate chart of $N$ such that $F(a) \in V_2 \cap F(U_1 \cap U_2)$.

Now, it remains to be showed that $(\psi_2|_{V_2 \cap F(U_1 \cap U_2)}) \circ F \circ (\varphi_2|_{U_1 \cap U_2})^{-1}$ is smooth. Since $F|_{U_1} : U_1 \to F(U_1)$ is a diffeomorphism, $F|_{U_1} : U_1 \to F(U_1)$ is smooth. Since $F|_{U_1} : U_1 \to F(U_1)$, $(U_1 \cap U_2, \varphi_2|_{U_1 \cap U_2})$ is an admissible coordinate chart of $M$ satisfying $a \in U_1 \cap U_2$, and $(V_2 \cap F(U_1 \cap U_2), \psi_2|_{V_2 \cap F(U_1 \cap U_2)})$ is an admissible coordinate chart of $N$ such that $F(a) \in V_2 \cap F(U_1 \cap U_2)$, $(\psi_2|_{V_2 \cap F(U_1 \cap U_2)}) \circ F|_{U_1} \circ (\varphi_2|_{U_1 \cap U_2})^{-1} (= (\psi_2|_{V_2 \cap F(U_1 \cap U_2)}) \circ F \circ (\varphi_2|_{U_1 \cap U_2})^{-1})$ is smooth, and hence, $(\psi_2|_{V_2 \cap F(U_1 \cap U_2)}) \circ F \circ (\varphi_2|_{U_1 \cap U_2})^{-1}$ is smooth.

Thus, we have shown that $F : M \to N$ is smooth.

For 2: Since $F : M \to N$ is 1–1 onto, $F^{-1} : N \to M$ is 1–1 onto. Now, since $F : M \to N$ is a local diffeomorphism, $F^{-1} : N \to M$ is a local diffeomorphism. Since $F^{-1} : N \to M$ is 1–1 onto and local diffeomorphism, as in 1, $F^{-1} : N \to M$ is a diffeomorphism.

**Theorem 5.44** *Let $M$ be an $m$-dimensional smooth manifold and $N$ be an $n$-dimensional smooth manifold. Let $F : M \to N$ be a local diffeomorphism. Then, $F$ is a smooth immersion and smooth submersion.*

*Proof* We have to prove that rank $F = m$, and rank $F = n$, that is, for every $p \in M$, the rank of $F$ at $p$ is $m$ and $m = n$. For this purpose, let us take any $p \in M$. Since $p \in M$, and $F : M \to N$ is a local diffeomorphism, there exists an open neighborhood $U$ of $p$ in $M$ such that $F(U)$ is open in $N$ and the restriction $F|_U : U \to F(U)$ is a diffeomorphism. Since $U$ is a nonempty open subset of $M$, and $M$ is an $m$-dimensional smooth manifold, $U$ is an $m$-dimensional smooth manifold, and hence, dim $T_p U = \dim U = m$. Since $F(U)$ is open in $N$, and $N$ is an $n$-dimensional smooth manifold, $F(U)$ is an $n$-dimensional smooth manifold, and hence dim $T_{F(p)}(F(U)) = \dim(F(U)) = n$. Since $F|_U : U \to F(U)$ is a diffeomorphism, and $p \in U$, the linear map $dF_p : T_p U \to T_{F(p)}(F(U))$ is an isomorphism. Since $dF_p : T_p U \to T_{F(p)}(F(U))$ is an isomorphism, $dF_p : T_p U \to T_{F(p)}(F(U))$ is 1–1, and hence, rank of $F$ at $p$ is dim $T_p U(= m)$.

Thus, for every $p \in M$, the rank of $F$ at $p$ is $m$. Since $dF_p : T_p U \to T_{F(p)}(F(U))$ is an isomorphism, $m = \dim T_p U = \dim T_{F(p)}(F(U)) = n$. $\square$

**Theorem 5.45** *Let $M$ be an $m$-dimensional smooth manifold and $N$ be an $n$-dimensional smooth manifold. Let $F : M \to N$ be a smooth immersion and smooth submersion map. Then, $F : M \to N$ is a local diffeomorphism.*

*Proof* For this purpose, let us take any $p \in M$. We have to find an open neighborhood $U$ of $p$ in $M$ such that $F(U)$ is open in $N$, and the restriction $F|_U: U \to F(U)$ is a diffeomorphism.

Since $p \in M$, and $F : M \to N$ is a smooth immersion, the linear map $dF_p : T_pM \to T_{F(p)}N$ is 1–1. Since $p \in M$, and $F : M \to N$ is a smooth submersion, the linear map $dF_p : T_pM \to T_{F(p)}N$ is onto. Since the linear map $dF_p : T_pM \to T_{F(p)}N$ is 1–1 onto, $m = \dim M = \dim T_pM = \dim(T_{F(p)}N) = n$.

Since $m = n$, and $N$ is an $n$-dimensional smooth manifold, $N$ is an $m$-dimensional smooth manifold. Since $M, N$ are $m$-dimensional smooth manifolds, $F : M \to N$ is a smooth map, and $dF_p : T_pM \to T_{F(p)}N$ is 1–1 onto, by Theorem 5.36, there exists an open neighborhood $U$ of $p$, and an open neighborhood $V$ of $F(p)$ such that $F|_U: U \to V$ is a diffeomorphism. Since $F|_U: U \to V$ is a diffeomorphism, $F|_U: U \to V$ is 1–1 onto, and hence, $F(U) = (F|_U)(U) = V$. Since $F(U) = V$, and $V$ is open, $F(U)$ is open. Since $F|_U: U \to V$ is a diffeomorphism, and $F(U) = V$, $F|_U: U \to F(U)$ is a diffeomorphism. Thus, $U$ is an open neighborhood of $p$, $F(U)$ is open in $N$, and $F|_U: U \to F(U)$ is a diffeomorphism. $\qquad \square$

**Theorem 5.46** *Let $M, N$ be $m$-dimensional smooth manifolds. Let $F : M \to N$ be a smooth immersion. Then, $F : M \to N$ is a local diffeomorphism.*

*Proof* We first try to show that $F : M \to N$ is a smooth submersion. For this purpose, let us take any $p \in M$. We have to show that the linear map $dF_p : T_pM \to T_{F(p)}N$ is onto.

Since $p \in M$, and $F : M \to N$ is a smooth immersion, the linear map $dF_p : T_pM \to T_{F(p)}N$ is 1–1. Since $\dim T_pM = \dim M = m = \dim N = \dim(T_{F(p)}N)$, $\dim T_pM = \dim(T_{F(p)}N)$. Since $\dim T_pM = \dim(T_{F(p)}N)$, and the linear map $dF_p : T_pM \to T_{F(p)}N$ is 1–1, $dF_p : T_pM \to T_{F(p)}N$ is onto. Since $F : M \to N$ is a smooth immersion and smooth submersion, by Theorem 5.45, $F : M \to N$ is a local diffeomorphism. $\qquad \square$

**Theorem 5.47** *Let $M, N$ be $m$-dimensional smooth manifolds. Let $F : M \to N$ be a smooth submersion. Then, $F : M \to N$ is a local diffeomorphism.*

*Proof* We first try to show that $F : M \to N$ is a smooth immersion. For this purpose, let us take any $p \in M$. We have to show that the linear map $dF_p : T_pM \to T_{F(p)}N$ is 1–1.

Since $p \in M$, and $F : M \to N$ is a smooth submersion, the linear map $dF_p : T_pM \to T_{F(p)}N$ is onto. Since $\dim T_pM = \dim M = m = \dim N = \dim(T_{F(p)}N)$, $\dim T_pM = \dim(T_{F(p)}N)$. Since $\dim T_pM = \dim(T_{F(p)}N)$, and the linear map $dF_p : T_pM \to T_{F(p)}N$ is onto, $dF_p : T_pM \to T_{F(p)}N$ is 1–1. Since $F : M \to N$ is a smooth submersion and smooth immersion, by Theorem 5.45, $F : M \to N$ is a local diffeomorphism. $\qquad \square$

**Note 5.48** Before going ahead, let us recall the constant rank Theorem 4.9:

Let $M$ be an $m$-dimensional smooth manifold, $N$ be an $n$-dimensional smooth manifold, $F : M \to N$ be a smooth map, and $r$ be the rank of $F$. Then, for every $p$ in

$M$, there exist admissible coordinate chart $(U, \varphi)$ in $M$ satisfying $p \in U$ and admissible coordinate chart $(V, \psi)$ in $N$ satisfying $F(p) \in V$, such that $F(U) \subset V$, and for every $(x_1, \ldots, x_r, x_{r+1}, \ldots, x_m)$ in $\varphi(U)$,

$$\left(\psi \circ F \circ \varphi^{-1}\right)(x_1, \ldots, x_r, x_{r+1}, \ldots, x_m) = \left(x_1, \ldots, x_r, \underbrace{0, \ldots, 0}_{n-r}\right).$$

**Theorem 5.49** *Let $M$ be a $m$-dimensional smooth manifold, $N$ be an $n$-dimensional smooth manifold, and $F : M \to N$ be smooth submersion (i.e., $F$ is of constant rank $n$). Then, for every $p$ in $M$, there exist admissible coordinate chart $(U, \varphi)$ in $M$ satisfying $p \in U$ and admissible coordinate chart $(V, \psi)$ in $N$ satisfying $F(p) \in V$ such that $F(U) \subset V$, and for every $(x_1, \ldots, x_n, x_{n+1}, \ldots, x_m)$ in $\varphi(U)$,*

$$\left(\psi \circ F \circ \varphi^{-1}\right)(x_1, \ldots, x_m) = (x_1, \ldots, x_n).$$

*Proof* Its proof is clear. □

**Theorem 5.50** *Let $M$ be a $m$-dimensional smooth manifold, $N$ be an $n$-dimensional smooth manifold, and $F : M \to N$ be smooth immersion (i.e., $F$ is of constant rank $m$).*

Then, for every $p$ in $M$, there exist admissible coordinate chart $(U, \varphi)$ in $M$ satisfying $p \in U$ and admissible coordinate chart $(V, \psi)$ in $N$ satisfying $F(p) \in V$ such that $F(U) \subset V$, and for every $(x_1, \ldots, x_m)$ in $\varphi(U)$,

$$\left(\psi \circ F \circ \varphi^{-1}\right)(x_1, \ldots, x_m) = \left(x_1, \ldots, x_m, \underbrace{0, \ldots, 0}_{n-m}\right).$$ □

*Proof* Its proof is clear.

**Theorem 5.51** *Let $M$ be an $m$-dimensional smooth manifold, $N$ be an $n$-dimensional smooth manifold, $F : M \to N$ be smooth, and $M$ be connected. Then, the following statements are equivalent:*

1. For every $p \in M$, there exist an admissible coordinate chart $(U, \varphi)$ in $M$ satisfying $p \in U$ and an admissible coordinate chart $(V, \psi)$ in $N$ satisfying $F(p) \in V$ such that $\psi \circ F \circ \varphi^{-1}$ behaves linearly, i.e., there exists a positive integer $r$ such that for every $(x_1, \ldots, x_r, x_{r+1}, \ldots, x_m)$ in $\varphi(U)$,

$$\left(\psi \circ F \circ \varphi^{-1}\right)(x_1, \ldots, x_r, x_{r+1}, \ldots, x_m) = \left(x_1, \ldots, x_r, \underbrace{0, \ldots, 0}_{n-r}\right).$$

2. $F$ has constant rank.

*Proof* $1 \Rightarrow 2$: Let us fix any $p \in M$. By the condition 1, there exist an admissible coordinate chart $(U, \varphi)$ in $M$ satisfying $p \in U$, an admissible coordinate chart $(V, \psi)$ in $N$ satisfying $F(p) \in V$, and a positive integer $r$ such that for every $(x_1, \ldots, x_r, x_{r+1}, \ldots, x_m)$ in $\varphi(U)$,

$$\left(\psi \circ F \circ \varphi^{-1}\right)(x_1, \ldots, x_r, x_{r+1}, \ldots, x_m) = \left(x_1, \ldots, x_r, \underbrace{0, \ldots, 0}_{n-r}\right).$$

For every $q \in U$,

$$\left(\left.\frac{\partial}{\partial x^1}\right|_q, \left.\frac{\partial}{\partial x^2}\right|_q, \ldots, \left.\frac{\partial}{\partial x^m}\right|_q\right)$$

is the coordinate basis of $T_q M$ corresponding to $(U, \varphi)$, where

$$\left.\frac{\partial}{\partial x^i}\right|_q \equiv \left(d(\varphi^{-1})_{\varphi(q)}\right)\left(\left.\frac{\partial}{\partial x^i}\right|_{\varphi(q)}\right).$$

For every $r \in V$,

$$\left(\left.\frac{\partial}{\partial y^1}\right|_r, \left.\frac{\partial}{\partial y^2}\right|_r, \ldots, \left.\frac{\partial}{\partial y^n}\right|_r\right)$$

is the coordinate basis of $T_r N$ corresponding to $(V, \psi)$, where

$$\left.\frac{\partial}{\partial y^i}\right|_r \equiv \left(d(\psi^{-1})_{\psi(r)}\right)\left(\left.\frac{\partial}{\partial y^i}\right|_{\psi(r)}\right).$$

We know that for every $q \in U$, the matrix representation of linear map $dF_q$ is the following $n \times m$ matrix:

$$\begin{bmatrix} \dfrac{\partial(\pi_1 \circ (\psi \circ F \circ \varphi^{-1}))}{\partial x^1}(\varphi(q)) & \dfrac{\partial(\pi_1 \circ (\psi \circ F \circ \varphi^{-1}))}{\partial x^2}(\varphi(q)) \\[2mm] \dfrac{\partial(\pi_2 \circ (\psi \circ F \circ \varphi^{-1}))}{\partial x^1}(\varphi(q)) & \dfrac{\partial(\pi_2 \circ (\psi \circ F \circ \varphi^{-1}))}{\partial x^2}(\varphi(q)) \\[2mm] \ldots & \ldots \\[2mm] \dfrac{\partial(\pi_n \circ (\psi \circ F \circ \varphi^{-1}))}{\partial x^1}(\varphi(q)) & \dfrac{\partial(\pi_n \circ (\psi \circ F \circ \varphi^{-1}))}{\partial x^2}(\varphi(q)) \end{bmatrix}$$

$$\begin{matrix} \ldots & \dfrac{\partial(\pi_1 \circ (\psi \circ F \circ \varphi^{-1}))}{\partial x^m}(\varphi(q)) \\[2mm] \ldots & \dfrac{\partial(\pi_2 \circ (\psi \circ F \circ \varphi^{-1}))}{\partial x^m}(\varphi(q)) \\[2mm] \ldots & \ldots \\[2mm] \ldots & \dfrac{\partial(\pi_n \circ (\psi \circ F \circ \varphi^{-1}))}{\partial x^m}(\varphi(q)) \end{matrix} \Bigg].$$

It follows that for every $q \in U$, the rank of the matrix representation of the linear map $dF_q$ is $r$. Since for every $q \in U$, the rank of $F$ at $q$ is equal to the rank of the matrix representation of the linear map $dF_q$, and the rank of the matrix representation of the linear map $dF_q$ is $r$, the rank of $F$ at $q$ is $r$.

Thus, we have shown that, corresponding to each $p \in M$, there exist a positive integer $r$ and an open neighborhood $U_{p,r}$ of $p$ such that for every $q \in U_{p,r}$, the rank of $F$ at $q$ is $r$. Observe that $\{\cup_{p \in M} U_{p,r} : r = 1, 2, \ldots, \min\{m, n\}\}$ is a finite partition of $M$ into open sets. Since $\{\cup_{p \in M} U_{p,r} : r = 1, 2, \ldots, \min\{m, n\}\}$ is a finite partition of $M$ into open sets, and $M$ is connected, $\{\cup_{p \in M} U_{p,r} : r = 1, 2, \ldots, \min\{m, n\}\}$ is a singleton set. It follows that for every $q \in M$, the rank of $F$ at $q$ is a constant. Thus, $F$ has constant rank.

$2 \Rightarrow 1$: Let $F$ has constant rank $r$. By the constant rank theorem, for every $p$ in $M$, there exist an admissible coordinate chart $(U, \varphi)$ in $M$ satisfying $p \in U$ and an admissible coordinate chart $(V, \psi)$ in $N$ satisfying $F(p) \in V$ such that $F(U) \subset V$, and for every $(x_1, \ldots, x_r, x_{r+1}, \ldots, x_m)$ in $\varphi(U)$,

$$\left(\psi \circ F \circ \varphi^{-1}\right)(x_1, \ldots, x_r, x_{r+1}, \ldots, x_m) = \left(x_1, \ldots, x_r, \underbrace{0, \ldots, 0}_{n-r}\right). \qquad \square$$

**Definition** Let $X$ be a topological space. If for every $x \in X$, there exist an open neighborhood $U_x$ of $x$ and a compact set $C$ such that $U_x \subset C$, then we say that $X$ is a *locally compact space*. Clearly, every compact space is locally compact.

**Note 5.52** Let $M$ be an $n$-dimensional smooth manifold. We shall try to show that $M$ is locally compact space. For this purpose, let us take any $p \in M$. By Lemma 4.49 of Chap. 4, there exists a countable collection $\{(U_1, \varphi_1), (U_2, \varphi_2), (U_3, \varphi_3), \ldots\}$ of admissible coordinate charts of $M$ such that

1. $\{U_1, U_2, U_3, \ldots\}$ is a basis of $M$,
2. each $\varphi_i(U_i)$ is an open ball in $\mathbb{R}^n$,
3. each closure $U_i^-$ of $U_i$ is a compact subset of M.

Since $\{U_1, U_2, U_3, \ldots\}$ is a basis of $M$, $\{U_1, U_2, U_3, \ldots\}$ is a cover of $M$, and hence, there exists a positive integer $k$ such that $p \in U_k$. By 3, $U_k^-$ is a compact subset of $M$. Further, $U_k \subset U_k^-$. Hence, $M$ is locally compact.

**Note 5.53** Let $X$ be a Hausdorff topological space. Then, we shall try to show that the following statements are equivalent:

1. $X$ is locally compact.
2. For every $x \in X$, there exists an open neighborhood $U_x$ of $x$ such that $U_x^-$ is compact.
3. There exists a basis $\mathcal{B}$ of $X$ such that for every $U \in \mathcal{B}$, $U^-$ is compact.

$3 \Rightarrow 2$: Let $\mathcal{B}$ be a basis of $X$ such that for every $U \in \mathcal{B}$, $U^-$ is compact. We have to prove 2. For this purpose, let us take any $x \in X$. Since $x \in X$, and $\mathcal{B}$ is a basis of $X$, there exists an open neighborhood $U_x$ of $x$ such that $U_x \in \mathcal{B}$, and $U_x^-$ is compact.

$2 \Rightarrow 1$: Let us take any $x \in X$. We have to find an open neighborhood $U_x$ of $x$ and a compact set $C$ such that $U_x \subset C$. Since $x \in X$, so by 2, there exists an open neighborhood $U_x$ of $x$ such that $U_x^-$ is compact. Also, $U_x \subset U_x^-$.

$1 \Rightarrow 3$: Let $\mathcal{B}$ be the collection of all open sets $U$ for which $U^-$ is compact. We shall try to show that $\mathcal{B}$ is a basis. For this purpose, let us take any open neighborhood $G$ of $x$. Since $x \in X$, and $X$ is locally compact, there exist an open neighborhood $U_x$ of $x$ and a compact set $C$ such that $U_x \subset C$. Since $U_x$ and $G$ are open neighborhoods of $x$, $U_x \cap G$ is an open neighborhood of $x$. Clearly, $U_x \cap G \subset C$. Since $C$ is compact, and $X$ is Hausdorff, $C$ is closed, and hence, $C^- = C$. Since $U_x \cap G \subset C$, $(U_x \cap G)^- \subset C^- = C$. Since $(U_x \cap G)^- \subset C$, $C$ is compact, and $(U_x \cap G)^-$ is closed, $(U_x \cap G)^-$ is compact. Since $U_x \cap G$ is open, and $(U_x \cap G)^-$ is compact, $U_x \cap G$ is in $\mathcal{B}$. Also, $U_x \cap G$ is an open neighborhood of $x$. This shows that $\mathcal{B}$ is a basis.

If $V \in \mathcal{B}$, then, by the definition of $\mathcal{B}$, $V^-$ is compact.

**Note 5.54** Let $X$ be a locally compact Hausdorff space. Let $G$ be a nonempty open subset of $X$. Then, we shall try to show that $G$ is locally compact Hausdorff. Since the topology of $X$ is Hausdorff, and $G$ has subspace topology of $X$, the topology of $G$ is Hausdorff.

Now, we want to show that $G$ is locally compact. For this purpose, let us take any $x$ in $G$. We have to find an open neighborhood $V_x$ of $x$ in $G$, and a compact subset $C_1$ of $G$ such that $V_x \subset C_1$.

Since $x$ is in $X$, and $X$ is locally compact, there exist an open neighborhood $U_x$ of $x$ in $X$ and a compact subset $C$ of $X$ such that $U_x \subset C$. Since $U_x$ is an open neighborhood of $x$ in $X$, $U_x \cap G$ is an open neighborhood of $x$ in $G$.

Now, we want to show that $(U_x \cap G)^- \cap G$ is compact in $G$. Since $C$ is compact, and $X$ is Hausdorff, $C$ is closed in $X$. Since $C$ is closed in $X$, and $U_x \cap G \subset C$, $(U_x \cap G)^- \subset C$. Since $(U_x \cap G)^- \subset C$, $C$ is compact, and $(U_x \cap G)^-$ is closed, $(U_x \cap G)^-$ is compact in $X$. Since $(U_x \cap G)^-$ is compact in $X$, and $G$ is open in $X$, $(U_x \cap G)^- \cap G$ is compact in $G$.

It remains to be showed that $U_x \cap G \subset (U_x \cap G)^- \cap G$. Since $U_x \cap G \subset (U_x \cap G)^-$, and $U_x \cap G \subset G$, $U_x \cap G \subset (U_x \cap G)^- \cap G$.

**Note 5.55** Let $X$ be a locally compact Hausdorff space. Let $F$ be a nonempty closed subset of $X$. Then, we shall try to show that $F$ is locally compact Hausdorff space. Since the topology of $X$ is Hausdorff, and $F$ has subspace topology of $X$, the topology of $F$ is Hausdorff.

Now, we want to show that $F$ is locally compact. For this purpose, let us take any $x$ in $F$. We have to find an open neighborhood $V_x$ of $x$ in $F$, and a compact subset $C_1$ of $F$ such that $V_x \subset C_1$.

Since $x$ is in $X$, and $X$ is locally compact, there exist an open neighborhood $U_x$ of $x$ in $X$ and a compact subset $C$ of $X$ such that $U_x \subset C$. Since $U_x$ is an open neighborhood of $x$ in $X$, $U_x \cap F$ is an open neighborhood of $x$ in $F$.

Now, we want to show that $(U_x \cap F)^- \cap F$ is compact in $F$. Since $U_x \subset C$, $U_x \cap F \subset C \cap F \subset C$. Since $C$ is compact, and $X$ is Hausdorff, $C$ is closed in $X$. Since $C$ is closed in $X$, and $U_x \cap F \subset C$, $(U_x \cap F)^- \subset C$. Since $(U_x \cap F)^-$ and $F$ are closed in $X$, $(U_x \cap F)^- \cap F$ is closed in $X$. Since $(U_x \cap F)^- \cap F \subset (U_x \cap F)^- \subset C$, $C$ is compact, and $(U_x \cap F)^- \cap F$ is closed, $(U_x \cap F)^- \cap F$ is compact in $X$. Since $(U_x \cap F)^- \cap F$ is compact in $X$, and $(U_x \cap F)^- \cap F \subset F$, $(U_x \cap F)^- \cap F$ is compact in $F$.

It remains to be showed that $U_x \cap F \subset (U_x \cap F)^- \cap F$. Since $U_x \cap F \subset (U_x \cap F)^-$, and $U_x \cap F \subset F$, $U_x \cap F \subset (U_x \cap F)^- \cap F$.

**Note 5.56** Let $X$ be a locally compact Hausdorff space. Let $x \in X$. Let $G$ be an open neighborhood of $x$. We shall try to show that there exists an open neighborhood $V_x$ of $x$ such that $V_x^- \subset G$, and $V_x^-$ is compact.

Since $X$ is a locally compact Hausdorff space, and $G$ is a nonempty open subset of $X$, by Note 5.54, $G$ is a locally compact Hausdorff space. Since $G$ is a locally compact Hausdorff space, and $x \in G$, there exist an open neighborhood $V_x$ of $x$ in $G$ and a compact subset $C$ of $G$ such that $V_x \subset C$. Since $V_x$ is open in $G$, and $G$ is open in $X$, $V_x$ is open in $X$. Since $V_x$ is open in $X$, and $x \in V_x$, $V_x$ is an open neighborhood of $x$ in $G$. Since $C$ is a compact subset of $G$, and $G \subset X$, $C$ is a compact subset of $X$. Since $C$ is a compact subset of $X$, and $X$ is a Hausdorff space, $C$ is a closed subset of $X$. Since $C$ is a closed subset of $X$, and $V_x \subset C$, $V_x^- \subset C \subset G$. Since $V_x^- \subset C$, $V_x^-$ is closed, and $C$ is a compact subset of $X$, $V_x^-$ is a compact subset of $X$.

This result is known as the *shrinking lemma*.

**Note 5.57** Let $X$ be a compact space. Let $\{F_n\}$ be a sequence of closed subsets of $X$. Let each $F_n$ be nonempty. Let $F_1 \supset F_2 \supset F_3 \supset \cdots$. We shall try to show that $\bigcap_{n=1}^{\infty} F_n$ is nonempty. If not, otherwise, let $\bigcap_{n=1}^{\infty} F_n = \emptyset$. We have to arrive at a contradiction.

Here, $X = \emptyset^c = (\bigcap_{n=1}^{\infty} F_n)^c = \bigcap_{n=1}^{\infty} ((F_n)^c)$. Further, since each $F_n$ is closed, each $(F_n)^c$ is open. Since each $(F_n)^c$ is open, and $\bigcup_{n=1}^{\infty} ((F_n)^c) = X$, $\{(F_n)^c\}$ is an open cover of $X$. Since $\{(F_n)^c\}$ is an open cover of $X$, and $X$ is compact, there exist positive integers $n_1 < n_2 < \cdots < n_k$ such that $\bigcup_{r=1}^{k} ((F_{n_r})^c) = X$.

Since $F_1 \supset F_2 \supset F_3 \supset \cdots$, $(F_1)^c \subset (F_2)^c \subset (F_3)^c \subset \cdots$. Now, since $n_1 < n_2 < \cdots < n_k$, $(F_{n_1})^c \subset (F_{n_2})^c \subset \cdots \subset (F_{n_k})^c$, and hence, $X = \bigcup_{r=1}^{k} ((F_{n_r})^c) = (F_{n_k})^c$. Since $(F_{n_k})^c = X$, $F_{n_k}$ is empty, which contradicts the assumption that each $F_n$ is nonempty.

**Theorem 5.58** *Let $X$ be a locally compact Hausdorff space. Let $\{U_n\}$ be a sequence of subsets of $X$. Let each $U_n$ be open and dense. Then, $\bigcap_{n=1}^{\infty} U_n$ is dense.*

*Proof* If not, otherwise, let $\bigcap_{n=1}^{\infty} U_n$ be not dense, that is, $(\bigcap_{n=1}^{\infty} U_n)^-$ is a proper subset of $X$. We have to arrive at a contradiction. Since $(\bigcap_{n=1}^{\infty} U_n)^-$ is a proper subset of $X$, there exists $a \in X$ such that $a \notin (\bigcap_{n=1}^{\infty} U_n)^-$. It follows that there exists an open neighborhood $V_a$ of $a$ such that $V_a \cap (\bigcap_{n=1}^{\infty} U_n) = \emptyset$.

Since $V_a$ is an open neighborhood of $a$, and $U_1$ is open, $V_a \cap U_1$ is open. Since $V_a$ is an open neighborhood of $a$, and $U_1$ is dense, $V_a \cap U_1$ is nonempty. Since $V_a \cap U_1$ is a nonempty open subset of $X$, and $X$ is a locally compact Hausdorff space, there exists a nonempty open set $W_1$ such that $W_1^- \subset V_a \cap U_1$, and $W_1^-$ is compact. Here, $W_1 \subset W_1^- \subset V_a \cap U_1$.

Since $W_1$ is a nonempty open subset of $X$, and $U_2$ is open and dense, $W_1 \cap U_2$ is a nonempty open subset of $X$. Since $W_1 \cap U_2$ is a nonempty open subset of $X$, and $X$ is a locally compact Hausdorff space, there exists a nonempty open set $W_2$ such that $W_2^- \subset W_1 \cap U_2$, and $W_2^-$ is compact. Clearly, $W_2^- \subset W_1 \subset W_1^-$. Also, $W_1^-, W_2^-$ are nonempty closed sets. Here, $W_2 \subset W_2^- \subset W_1 \cap U_2 \subset (V_a \cap U_1) \cap U_2 = V_a \cap (U_1 \cap U_2)$.

Since $W_2$ is a nonempty open subset of $X$, and $U_3$ is open and dense, $W_2 \cap U_3$ is a nonempty open subset of $X$. Since $W_2 \cap U_3$ is a nonempty open subset of $X$, and $X$ is a locally compact Hausdorff space, there exists a nonempty open set $W_3$ such that $W_3^- \subset W_2 \cap U_3$, and $W_3^-$ is compact. Clearly, $W_3^- \subset W_2 \subset W_2^-$. Also, $W_1^-, W_2^-, W_3^-$ are nonempty closed sets. Here, $W_3 \subset W_3^- \subset W_2 \cap U_3 \subset (V_a \cap (U_1 \cap U_2)) \cap U_3 = V_a \cap (U_1 \cap U_2 \cap U_3)$, etc.

Thus, we get a decreasing sequence $\{W_n^-\}$ of nonempty closed subsets of a compact set $W_1^-$. Hence, by Note 5.57, $\bigcap_{n=1}^{\infty} (W_n^-)$ is nonempty. Since $W_1^- \subset V_a \cap U_1, W_2^- \subset V_a \cap (U_1 \cap U_2), W_3^- \subset V_a \cap (U_1 \cap U_2 \cap U_3)$, etc., $\emptyset \subset \bigcap_{n=1}^{\infty} (W_n^-) \subset V_a \cap (\bigcap_{n=1}^{\infty} U_n) = \emptyset$. It follows that $\bigcap_{n=1}^{\infty} (W_n^-)$ is empty, a contradiction. $\square$

**Definition** Let $X$ be a topological space. Let $A$ be any subset of $X$. If $A^-$ contains no nonempty open set, then we say that $A$ is *nowhere dense*.

**Theorem 5.59** *Let $X$ be a locally compact Hausdorff space. Let $\{A_n\}$ be a sequence of nowhere dense subsets of $X$. Then, $\bigcup_{n=1}^{\infty} A_n$ has no interior point.*

*Proof* For every $n$, $A_n$ is nowhere dense, so $\emptyset = ((A_n)^-)^{\circ} = ((((A_n)^-)^c)^-)^c$, and hence, $(((A_n)^-)^c)^- = X$. It follows that $((A_n)^-)^c$ is an open dense set. So, by Theorem 5.58, $\bigcap_{n=1}^{\infty} ((A_n)^-)^c (= (\bigcup_{n=1}^{\infty} (A_n)^-)^c)$ is dense, and hence, $X = ((\bigcup_{n=1}^{\infty} (A_n)^-)^c)^-$. It follows that $\emptyset = (((\bigcup_{n=1}^{\infty} (A_n)^-)^c)^-)^c = (\bigcup_{n=1}^{\infty} (A_n)^-)^{\circ}$. $\square$

This theorem is known as *Baire category theorem*.

**Theorem 5.60** *Let $M$ be an $m$-dimensional smooth manifold, $N$ be an $n$-dimensional smooth manifold, and $F : M \to N$ be smooth. Let $F$ has constant rank $r$. Let $F : M \to N$ be onto. Then, $F$ is a smooth submersion.*

*Proof* We have to prove that rank $F = n$. If not, otherwise, let $r \equiv$ rank $F < n$. We have to arrive at a contradiction. Here, $r \equiv$ rank $F < n$, so $n - r$ is a positive integer. Let us take any $p \in M$.

By the constant rank theorem, there exist an admissible coordinate chart $(U_p, \varphi_p)$ in $M$ satisfying $p \in U_p$ and an admissible coordinate chart $(V_{F(p)}, \psi_{F(p)})$ in

$N$ satisfying $F(p) \in V_{F(p)}$, such that $F(U_p) \subset V_{F(p)}$, and for every $(x_1, \ldots, x_r, x_{r+1}, \ldots, x_m)$ in $\varphi_p(U_p)$,

$$\left(\psi_{F(p)} \circ F \circ \varphi_p^{-1}\right)(x_1, \ldots, x_r, x_{r+1}, \ldots, x_m) = \left(x_1, \ldots, x_r, \underbrace{0, \ldots, 0}_{(1\leq)n-r}\right)$$

$$\in \psi_{F(p)}\left(V_{F(p)}\right).$$

By Lemma 4.47, there exists an open neighborhood $W_p$ of $p$ such that $W_p \subset (W_p)^- \subset U_p$, and $(W_p)^-$ is compact. We shall try to show that $F((W_p)^-)$ is a nowhere dense subset of $N$.

Since $F : M \to N$ is smooth, $F : M \to N$ is continuous. Since $F : M \to N$ is continuous, and $(W_p)^-$ is compact, $F((W_p)^-)$ is a compact subset of $N$. Since $F((W_p)^-)$ is a compact subset of $N$, and the topology of $N$ is Hausdorff, $F((W_p)^-)$ is a closed subset of $N$. Since $F((W_p)^-)$ is a closed subset of $N$, $(F((W_p)^-))^- = F((W_p)^-)$. It remains to be showed that $F((W_p)^-)$ has no interior point.

Let us take any $q \in (W_p)^-$. Since $q \in (W_p)^- (\subset U_p)$, $q \in U_p$, and hence, $\varphi_p(q) \in \varphi_p(U_p)$. Put $\varphi_p(q) \equiv (y_1, \ldots, y_r, y_{r+1}, \ldots, y_m)$. So

$$\psi_{F(p)}(F(q)) = \left(\psi_{F(p)} \circ F \circ \varphi_p^{-1}\right)(\varphi_p(q))$$

$$= \left(\psi_{F(p)} \circ F \circ \varphi_p^{-1}\right)(y_1, \ldots, y_r, y_{r+1}, \ldots, y_m) \left(y_1, \ldots, y_r, \underbrace{0, \ldots, 0}_{(1\leq)n-r}\right)$$

$$\in \underbrace{\mathbb{R} \times \cdots \times \mathbb{R}}_{r} \times \underbrace{\{0\} \times \cdots \times \{0\}}_{(1\leq)n-r}.$$

Since $q \in U_p$, $F(q) \in F(U_p) \subset V_{F(p)}$. Since $F(q) \in V_{F(p)}$, and

$$F(q) \in \psi_{F(p)}^{-1}\left(\underbrace{\mathbb{R} \times \cdots \times \mathbb{R}}_{r} \times \underbrace{\{0\} \times \cdots \times \{0\}}_{(1\leq)n-r}\right),$$

$$F(q) \in V_{F(p)} \cap \psi_{F(p)}^{-1}\left(\underbrace{\mathbb{R} \times \cdots \times \mathbb{R}}_{r} \times \underbrace{\{0\} \times \cdots \times \{0\}}_{(1\leq)n-r}\right).$$

It follows that

$$F\left((W_p)^-\right) \subset V_{F(p)} \cap \psi_{F(p)}^{-1}\left(\underbrace{\mathbb{R} \times \cdots \times \mathbb{R}}_{r} \times \underbrace{\{0\} \times \cdots \times \{0\}}_{(1\leq)n-r}\right).$$

Clearly,

$$V_{F(p)} \cap \psi_{F(p)}^{-1}\left(\underbrace{\mathbb{R} \times \cdots \times \mathbb{R}}_{r} \times \underbrace{\{0\} \times \cdots \times \{0\}}_{(1\leq)n-r}\right)$$

has no interior point. (Reason: Here,

$$\psi_{F(p)}\left(V_{F(p)} \cap \psi_{F(p)}^{-1}\left(\underbrace{\mathbb{R} \times \cdots \times \mathbb{R}}_{r} \times \underbrace{\{0\} \times \cdots \times \{0\}}_{(1\leq)n-r}\right)\right)$$

$$= \psi_{F(p)}\left(V_{F(p)}\right) \cap \psi_{F(p)}\left(\psi_{F(p)}^{-1}\left(\underbrace{\mathbb{R} \times \cdots \times \mathbb{R}}_{r} \times \underbrace{\{0\} \times \cdots \times \{0\}}_{(1\leq)n-r}\right)\right)$$

$$\subset \psi_{F(p)}\left(V_{F(p)}\right) \cap \left(\underbrace{\mathbb{R} \times \cdots \times \mathbb{R}}_{r} \times \underbrace{\{0\} \times \cdots \times \{0\}}_{(1\leq)n-r}\right) \subset \left(\underbrace{\mathbb{R} \times \cdots \times \mathbb{R}}_{r} \times \underbrace{\{0\} \times \cdots \times \{0\}}_{(1\leq)n-r}\right),$$

and

$$\left(\underbrace{\mathbb{R} \times \cdots \times \mathbb{R}}_{r} \times \underbrace{\{0\} \times \cdots \times \{0\}}_{(1\leq)n-r}\right)$$

have no interior point, so

$$\psi_{F(p)}\left(V_{F(p)} \cap \psi_{F(p)}^{-1}\left(\underbrace{\mathbb{R} \times \cdots \times \mathbb{R}}_{r} \times \underbrace{\{0\} \times \cdots \times \{0\}}_{(1\leq)n-r}\right)\right)$$

has no interior point. Since $(V_{F(p)}, \psi_{F(p)})$ is an admissible coordinate chart of $N$, $\psi_{F(p)}$ is a homeomorphism from $V_{F(p)}$ onto $\psi_{F(p)}(V_{F(p)})$. Since $\psi_{F(p)}$ is a homeomorphism from $V_{F(p)}$ onto

$$\psi_{F(p)}(V_{F(p)}), \left( V_{F(p)} \cap \psi_{F(p)}^{-1} \left( \underbrace{\mathbb{R} \times \cdots \times \mathbb{R}}_{r} \times \underbrace{\{0\} \times \cdots \times \{0\}}_{(1\le)n-r} \right) \right) \subset V_{F(p)},$$

and

$$\psi_{F(p)} \left( V_{F(p)} \cap \psi_{F(p)}^{-1} \left( \underbrace{\mathbb{R} \times \cdots \times \mathbb{R}}_{r} \times \underbrace{\{0\} \times \cdots \times \{0\}}_{(1\le)n-r} \right) \right)$$

has no interior point,

$$V_{F(p)} \cap \psi_{F(p)}^{-1} \left( \underbrace{\mathbb{R} \times \cdots \times \mathbb{R}}_{r} \times \underbrace{\{0\} \times \cdots \times \{0\}}_{(1\le)n-r} \right)$$

has no interior point in $V_{F(p)}$. Since

$$V_{F(p)} \cap \psi_{F(p)}^{-1} \left( \underbrace{\mathbb{R} \times \cdots \times \mathbb{R}}_{r} \times \underbrace{\{0\} \times \cdots \times \{0\}}_{(1\le)n-r} \right)$$

has no interior point in $V_{F(p)}$, and $V_{F(p)}$ is open in $N$,

$$V_{F(p)} \cap \psi_{F(p)}^{-1} \left( \underbrace{\mathbb{R} \times \cdots \times \mathbb{R}}_{r} \times \underbrace{\{0\} \times \cdots \times \{0\}}_{(1\le)n-r} \right)$$

has no interior point in $N$.) Since

$$V_{F(p)} \cap \psi_{F(p)}^{-1} \left( \underbrace{\mathbb{R} \times \cdots \times \mathbb{R}}_{r} \times \underbrace{\{0\} \times \cdots \times \{0\}}_{(1\le)n-r} \right)$$

has no interior point in $N$, and

$$F((W_p)^-) \subset V_{F(p)} \cap \psi_{F(p)}^{-1} \left( \underbrace{\mathbb{R} \times \cdots \times \mathbb{R}}_{r} \times \underbrace{\{0\} \times \cdots \times \{0\}}_{(1\le)n-r} \right),$$

$F((W_p)^-)$ has no interior point. Thus, $F((W_p)^-)$ is a nowhere dense subset of $N$.

Here, $\{W_p : p \in M\}$ is an open cover of $M$. Since $M$ is a smooth manifold, the topology of $M$ is second countable. Since $M$ is a second countable space, and

$\{W_p : p \in M\}$ is an open cover of $M$, there exist $p_1, p_2, p_3, \ldots$ in $M$ such that $\{W_{p_n} : n \in \mathbb{N}\}$ is an open cover of $M$. Since $\{W_{p_n} : n \in \mathbb{N}\}$ is a cover of $M$, and $F : M \to N$ is onto, $\{F(W_{p_n}) : n \in \mathbb{N}\}$ is a cover of $N$. Since $W_{p_n} \subset (W_{p_n})^-$, $F(W_{p_n}) \subset F((W_{p_n})^-)$. Since, for every $n \in \mathbb{N}, F(W_{p_n}) \subset F((W_{p_n})^-)$, and $\{F((W_{p_n})^-) : n \in \mathbb{N}\}$ is a cover of $N$, $N = \bigcup_{n=1}^{\infty} F((W_{p_n})^-)$. Since $N$ is a smooth manifold, $N$ is a Hausdorff locally compact space. Since $N$ is a Hausdorff locally compact space, and for every $n \in \mathbb{N}, F((W_p)^-)$ is a nowhere dense subset of $N$, by Baire category theorem, $\bigcup_{n=1}^{\infty} F((W_{p_n})^-)(= N)$ has no interior point, and hence, $N$ has no interior point. This is a contradiction. $\qquad \square$

**Theorem 5.61** *Let $M$ be an $m$-dimensional smooth manifold, $N$ be an $n$-dimensional smooth manifold, and $F : M \to N$ be smooth. Let $F$ has constant rank $r$. Let $F : M \to N$ be 1–1. Then, $F$ is a smooth immersion.*

*Proof* We have to prove that rank $F = m$. If not, otherwise, let $r \equiv$ rank $F < m$. We have to arrive at a contradiction. Here, $r \equiv$ rank $F < m$, so $m - r$ is a positive integer. Let us take any $p \in M$.

By the constant rank theorem, there exist an admissible coordinate chart $(U, \varphi)$ in $M$ satisfying $p \in U$ and an admissible coordinate chart $(V, \psi)$ in $N$ satisfying $F(p) \in V$ such that $F(U) \subset V$, and for every $(x_1, \ldots, x_r, x_{r+1}, \ldots, x_m)$ in $\varphi(U)$,

$$\left(\psi \circ F \circ \varphi^{-1}\right)\left(x_1, \ldots, x_r, \underbrace{x_{r+1}, \ldots, x_m}_{(1 \le )m-r}\right) = \left(x_1, \ldots, x_r, \underbrace{0, \ldots, 0}_{(0 \le )n-r}\right) \in \psi(V).$$

Since $F : M \to N$ is 1–1, $\varphi^{-1}$ is 1–1, and $\psi$ is 1–1, their composite $\psi \circ F \circ \varphi^{-1}$ from $\varphi(U)$ to $V$ is 1–1. Put

$$\varphi(p) \equiv \left(a_1, \ldots, a_r, \underbrace{a_{r+1}, \ldots, a_m}_{(1 \le )m-r}\right).$$

Since

$$\left(a_1, \ldots, a_r, \underbrace{a_{r+1}, \ldots, a_m}_{(1 \le )m-r}\right) = \varphi(p) \in \varphi(U),$$

and $\varphi(U)$ is an open subset of $\mathbb{R}^m$, there exists $\varepsilon > 0$ such that

$$(a_1 - \varepsilon, a_1 + \varepsilon) \times \cdots \times (a_r - \varepsilon, a_r + \varepsilon) \times (a_{r+1} - \varepsilon, a_{r+1} + \varepsilon)$$
$$\times \cdots \times (a_m - \varepsilon, a_m + \varepsilon) \subset \varphi(U).$$

Since

$$\left(a_1,\ldots,a_r,\underbrace{a_{r+1},\ldots,a_m-\frac{\varepsilon}{2}}_{(1\le)m-r}\right), \left(a_1,\ldots,a_r,\underbrace{a_{r+1},\ldots,a_m+\frac{\varepsilon}{2}}_{(1\le)m-r}\right)$$

$$\in (a_1-\varepsilon,a_1+\varepsilon)\times\cdots\times(a_r-\varepsilon,a_r+\varepsilon)\times(a_{r+1}-\varepsilon,a_{r+1}+\varepsilon)$$
$$\times\cdots\times(a_m-\varepsilon,a_m+\varepsilon)\subset\varphi(U),$$

$$\left(\psi\circ F\circ\varphi^{-1}\right)\left(a_1,\ldots,a_r,\underbrace{a_{r+1},\ldots,a_m-\frac{\varepsilon}{2}}_{(1\le)m-r}\right)=\left(a_1,\ldots,a_r,\underbrace{0,\ldots,0}_{n-r}\right)$$

$$=\left(\psi\circ F\circ\varphi^{-1}\right)\left(a_1,\ldots,a_r,\underbrace{a_{r+1},\ldots,a_m+\frac{\varepsilon}{2}}_{(1\le)m-r}\right).$$

Since

$$\left(\psi\circ F\circ\varphi^{-1}\right)\left(a_1,\ldots,a_r,\underbrace{a_{r+1},\ldots,a_m-\frac{\varepsilon}{2}}_{(1\le)m-r}\right)$$

$$=\left(\psi\circ F\circ\varphi^{-1}\right)\left(a_1,\ldots,a_r,\underbrace{a_{r+1},\ldots,a_m+\frac{\varepsilon}{2}}_{(1\le)m-r}\right)$$

and $\psi\circ F\circ\varphi^{-1}$ is $1-1$,

$$\left(a_1,\ldots,a_r,\underbrace{a_{r+1},\ldots,a_m-\frac{\varepsilon}{2}}_{(1\le)m-r}\right)=\left(a_1,\ldots,a_r,\underbrace{a_{r+1},\ldots,a_m+\frac{\varepsilon}{2}}_{(1\le)m-r}\right),$$

which is a contradiction.                                                       $\square$

## 5.7 Global Rank Theorem

**Theorem 5.62** *Let M be an m-dimensional smooth manifold, N be an n-dimensional smooth manifold, and F : M → N be smooth. Let F has constant rank. Let F : M → N be 1–1 and onto. Then, F is a diffeomorphism.*

*Proof* Since $F : M \to N$ is onto, by Theorem 5.60, $F$ is a smooth submersion. Since $F : M \to N$ is 1–1, by Theorem 5.61, $F$ is a smooth immersion. Since $F : M \to N$ is submersion, and immersion, by Theorem 5.45, $F : M \to N$ is a local diffeomorphism. Since $F : M \to N$ is a local diffeomorphism, 1–1, and onto, by Note 5.43, $F : M \to N$ is a diffeomorphism.                                          □

**Note 5.63** Theorems 5.60, 5.61, and 5.62 together are known as the *global rank theorem.*

**Definition** Let $X, Y$ be topological spaces. Let $F : X \to Y$ be a mapping. If $F$ is 1–1 and continuous map, and $F$ is a homeomorphism from $X$ onto $F(X)$, where $F(X)$ has the subspace topology of $Y$, then we say that $F$ is a *topological embedding.*

**Note 5.64** Let $X, Y$ be topological spaces. Let $F : X \to Y$ be a mapping. Let $F$ be 1–1 and continuous. Let $F$ be an open mapping, that is, for every open subset $U$ of $X$, $F(U)$ is open in $Y$. We shall try to show that $F : X \to Y$ is a topological embedding.

Here, let us take any open subset $U$ of $X$. It remains to be proved that $F(U)$ is open in $F(X)$. Since $U$ is open in $X$, and $F$ is an open mapping, $F(U)$ is open in $Y$. Since $F(U)$ is open in $Y$, $F(U) \cap F(X)$ is open in $F(X)$. Since $F(U) \subset F(X)$, $F(U) \cap F(X) = F(U)$. Since $F(U) \cap F(X) = F(U)$, and $F(U) \cap F(X)$ is open in $F(X)$, $F(U)$ is open in $F(X)$.

**Note 5.65** Let $X, Y$ be topological spaces. Let $F : X \to Y$ be a mapping. Let $F$ be 1–1 and continuous. Let $F$ be a closed mapping, that is, for every closed subset $A$ of $X$, $F(A)$ is closed in $Y$. We shall try to show that $F : X \to Y$ is a topological embedding.

Here, let us take any open subset $U$ of $X$. It remains to be proved that $F(U)$ is open in $F(X)$. Since $U$ is open in $X$, the complement $U^c$ of $U$ is closed in $X$. Since $U^c$ is closed in $X$, and $F$ is a closed mapping, $F(U^c)$ is closed in $Y$. Since $F(U^c)$ is closed in $Y$, and $F(U^c) \subset F(X)$, $F(U^c) \cap F(X)(= F(U^c))$ is closed in $F(X)$, and hence, $F(U^c)$ is closed in $F(X)$. Since $F(U^c)$ is closed in $F(X)$, $F(X) - F(U^c)$ is open in $F(X)$. Since $F : X \to Y$ is 1–1, $F(X) - F(U^c) = F(X - U^c) = F(U)$. Since $F(X) - F(U^c) = F(U)$, and $F(X) - F(U^c)$ is open in $F(X)$, $F(U)$ is open in $F(X)$.

**Note 5.66** Let $X$ be a compact space and $Y$ be a Hausdorff space. Let $F : X \to Y$ be a continuous mapping and $F$ be 1–1. We shall try to show that $F : X \to Y$ is a topological embedding.

First of all, we shall try to show that $F : X \to Y$ is a closed map. For this purpose, let us take any closed subset $A$ of $X$. We have to show that $F(A)$ is closed in $Y$. Since $A$ is closed in the compact set $X$, $A$ is compact. Since $A$ is compact, and

$F : X \rightarrow Y$ is continuous, $F(A)$ is compact. Since $F(A)$ is a compact subset of $Y$, and $Y$ is Hausdorff, $F(A)$ is a closed subset of $Y$.

Thus, $F : X \rightarrow Y$ is a closed map. Since $F : X \rightarrow Y$ is 1–1, continuous, and closed, by the above note, $F : X \rightarrow Y$ is a topological embedding.

This result is known as *closed map lemma*.

**Definition** Let $X$, $Y$ be topological spaces. Let $F : X \rightarrow Y$ be a mapping. If for every compact subset $C$ of $Y$, $F^{-1}(C)$ is compact in $X$, then we say that $F : X \rightarrow Y$ is a *proper map*.

**Note 5.67** Let $X$ be a Hausdorff topological space and $Y$ be a Hausdorff locally compact space. Let $F : X \rightarrow Y$ be a continuous mapping. Let $F$ be a proper map. We shall try to show that $F$ is a closed mapping. For this purpose, let us take any closed subset $A$ of $X$. We have to show that $F(A)$ is closed in $Y$. If not, otherwise, let $F(A)$ be not closed. We have to arrive at a contradiction.

Since $F(A)$ is not closed, there exists $b \in Y$ such that $b$ is a limit point of $F(A)$, and $b \notin F(A)$. Since $b \in Y$, and $Y$ is a Hausdorff locally compact space, there exists an open neighborhood $U_b$ of $b$ such that $U_b^-$ is compact. Clearly, $b$ is a limiting point of $F(A) \cap (U_b^-)$.

(Reason: Let us take any open neighborhood $V_b$ of $b$. It follows that $V_b \cap U_b$ is an open neighborhood of $b$. Since $V_b \cap U_b$ is an open neighborhood of $b$, and $b$ is a limit point of $F(A)$, there exists a point $c \in V_b \cap U_b$, and $c \in F(A)$. Here, $c \in V_b \cap U_b$, so $c \in U_b(\subset (U_b^-))$, and hence, $c \in (U_b^-)$. Since $c \in (U_b^-)$, and $c \in F(A)$, $c \in F(A) \cap (U_b^-)$. Also, since $c \in V_b \cap U_b$, $c \in V_b$. Since $c \in F(A)$, and $b \notin F(A)$, $c \neq b$. This shows that $b$ is a limiting point of $F(A) \cap (U_b^-)$.)

Since $F : X \rightarrow Y$ is a continuous mapping, $F$ is a proper map, and $U_b^-$ is a compact subset of $Y$, $F^{-1}(U_b^-)$ is a compact subset of $X$. Since $F^{-1}(U_b^-)$ is a compact subset of $X$, and the topology of $X$ is Hausdorff, $F^{-1}(U_b^-)$ is a closed subset of $X$. Since $F^{-1}(U_b^-)$ is a closed subset of $X$, and $A$ is a closed subset of $X$, $F^{-1}(U_b^-) \cap A$ is a closed subset of $X$. Since $F^{-1}(U_b^-) \cap A$ is a closed subset of compact set $F^{-1}(U_b^-)$, $F^{-1}(U_b^-) \cap A$ is compact. Since $F^{-1}(U_b^-) \cap A$ is a compact subset of $X$, and $F : X \rightarrow Y$ is a continuous mapping, $F(F^{-1}(U_b^-) \cap A)$ is compact. Clearly, $F(F^{-1}(U_b^-) \cap A) = F(A) \cap (U_b^-)$.

(Reason: LHS $= F(F^{-1}(U_b^-) \cap A) \subset F(F^{-1}(U_b^-)) \cap F(A) \subset (U_b^-) \cap F(A) = F(A) \cap (U_b^-) =$ RHS. Next, let us take any $y \in F(A) \cap (U_b^-)$. We have to show that $y \in F(F^{-1}(U_b^-) \cap A)$. Since $y \in F(A) \cap (U_b^-)$, $y \in (U_b^-)$, and there exists $x \in A$ such that $F(x) = y$. Since $F(x) = y \in (U_b^-)$, $x \in F^{-1}(U_b^-)$. Since $x \in F^{-1}(U_b^-)$, and $x \in A$, $x \in F^{-1}(U_b^-) \cap A$. Therefore, $y = F(x) \in F(F^{-1}(U_b^-) \cap A)$.)

Since $F(F^{-1}(U_b^-) \cap A) = F(A) \cap (U_b^-)$, and $F(F^{-1}(U_b^-) \cap A)$ is compact, $F(A) \cap (U_b^-)$ is compact. Since $F(A) \cap (U_b^-)$ is a compact subset of $Y$, and the topology of $Y$ is Hausdorff, $F(A) \cap (U_b^-)$ is closed. Since $F(A) \cap (U_b^-)$ is closed, and $b$ is a limiting point of $F(A) \cap (U_b^-)$, $b \in F(A) \cap (U_b^-)(\subset F(A))$, and hence, $b \in F(A)$. This is a contradiction.

**Definition** Let $M$ be an $m$-dimensional smooth manifold, $N$ be an $n$-dimensional smooth manifold, and $F : M \rightarrow N$ be smooth. If $F$ is a smooth immersion, and $F : M \rightarrow N$ is a topological embedding, then we say that $F$ is a *smooth embedding of $M$ into $N$.*

**Note 5.68** Let $M$ be an $m$-dimensional smooth manifold, $N$ be an $n$-dimensional smooth manifold, $P$ be a $p$-dimensional smooth manifold, $F : M \rightarrow N$ be a smooth embedding, and $G : N \rightarrow P$ be a smooth embedding. We shall try to show that the composite map $G \circ F : M \rightarrow P$ is a smooth embedding, that is,

1. $G \circ F : M \rightarrow P$ is a smooth immersion, and
2. $G \circ F : M \rightarrow P$ is a topological embedding.

For 1: We have to show that $G \circ F : M \rightarrow P$ is a smooth immersion. For this purpose, let us take any $a \in M$. We have to show that the linear map $d(G \circ F)_a : T_aM \rightarrow T_{(G \circ F)(a)}P$ is 1–1.

Since $F : M \rightarrow N$ is a smooth embedding, $F : M \rightarrow N$ is a smooth immersion. Since $F : M \rightarrow N$ is a smooth immersion, and $a \in M$, the linear map $dF_a : T_aM \rightarrow T_{F(a)}N$ is 1–1. Since $G : N \rightarrow P$ is a smooth embedding, $G : N \rightarrow P$ is a smooth immersion. Since $G : N \rightarrow P$ is a smooth immersion, and $F(a) \in N$, the linear map $dG_{F(a)} : T_{F(a)}N \rightarrow T_{G(F(a))}P$ is 1–1. Since $dG_{F(a)} : T_{F(a)}N \rightarrow T_{G(F(a))}P$ is 1–1, and $dF_a : T_aM \rightarrow T_{F(a)}N$ is 1–1, their composite $(dG_{F(a)}) \circ (dF_a)(= d(G \circ F)_a)$ is 1–1, and hence, $d(G \circ F)_a$ is 1–1.

For 2: Here, we have to prove that $G \circ F : M \rightarrow P$ is a topological embedding, that is,

(a) $G \circ F : M \rightarrow P$ is 1–1.
(b) $G \circ F : M \rightarrow P$ is continuous.
(c) $G \circ F$ is a homeomorphism from $M$ onto $(G \circ F)(M)$, where the topology of $(G \circ F)(M)$ is the subspace topology of $P$.

For a: Since $F : M \rightarrow N$ is a smooth embedding, $F : M \rightarrow N$ is a topological embedding, and hence, $F : M \rightarrow N$ is 1–1. Similarly, $G : N \rightarrow P$ is 1–1. Since $F : M \rightarrow N$ is 1–1, and $G : N \rightarrow P$ is 1–1, their composite $G \circ F : M \rightarrow P$ is 1–1.
For b: Since $F : M \rightarrow N$ is a smooth embedding, $F : M \rightarrow N$ is a topological embedding, and hence, $F : M \rightarrow N$ is continuous. Similarly, $G : N \rightarrow P$ is continuous. Since $F : M \rightarrow N$ is continuous, and $G : N \rightarrow P$ is continuous, their composite $G \circ F : M \rightarrow P$ is continuous.
For c: Since $F : M \rightarrow N$ is a smooth embedding, $F : M \rightarrow N$ is a topological embedding, and hence, $F$ is a homeomorphism from $M$ onto $F(M)$, where $F(M)$ has the subspace topology of $N$. Since $G : N \rightarrow P$ is a smooth embedding, $G : N \rightarrow P$ is a topological embedding, and hence, $G$ is a homeomorphism from $N$ onto $G(N)$, where $G(N)$ has the subspace topology of $P$. Since $G$ is a homeomorphism from $N$ onto $G(N)$, the restriction $G|_{F(M)}$ is a homeomorphism from $F(M)$ onto $G(F(M))$, where the topology of $F(M)$ is the subspace topology

of $N$, and the topology of $G(F(M))$ is the subspace topology of $G(N)$. Since the topology of $G(F(M))$ is the subspace topology of $G(N)$, and the topology of $G$ $(N)$ is the subspace topology of $P$, the topology of $G(F(M))$ as the subspace topology of $G(N)$ and the topology of $G(F(M))$ as the subspace topology of $P$ are the same. It follows that $G|_{F(M)}$ is a homeomorphism from $F(M)$ onto $G(F(M))$, where the topology of $F(M)$ is the subspace topology of $N$, and the topology of $G(F(M))$ is the subspace topology of $P$. Since $F$ is a homeomorphism from $M$ onto $F(M)$, where $F(M)$ has the subspace topology of $N$, and $G|_{F(M)}$ is a homeomorphism from $F(M)$ onto $G(F(M))$, where the topology of $F$ $(M)$ is the subspace topology of $N$, and the topology of $G(F(M))$ is the subspace topology of $P$, the composite $(G|_{F(M)}) \circ F$ is a homeomorphism from $M$ onto $G(F(M))(= (G \circ F)(M))$, where the topology of $(G \circ F)(M)$ is the subspace topology of $P$. Further, since $(G|_{F(M)}) \circ F = G \circ F$, $G \circ F$ is a homeomorphism from $M$ onto $(G \circ F)(M)$, where the topology of $(G \circ F)(M)$ is the subspace topology of $P$.

**Note 5.69** Let $M$ be an $m$-dimensional smooth manifold, $N$ be an $n$-dimensional smooth manifold, and $F : M \to N$ be a 1–1 smooth immersion. Let $F$ be an open map. We shall try to show that $F : M \to N$ is a smooth embedding, that is, $F$ is a smooth immersion, and $F : M \to N$ is a topological embedding. Since $F : M \to N$ is a smooth immersion, it suffices to show that $F : M \to N$ is a topological embedding.

Since $F : M \to N$ is a smooth map, so $F : M \to N$ is continuous. Since $F$ is 1–1, continuous, and open, by Note 5.64, $F : M \to N$ is a topological embedding.

**Note 5.70** Let $M$ be an $m$-dimensional smooth manifold, $N$ be an $n$-dimensional smooth manifold, and $F : M \to N$ be a 1–1 smooth immersion. Let $F$ be a closed map. We shall try to show that $F : M \to N$ is a smooth embedding, that is, $F$ is a smooth immersion, and $F : M \to N$ is a topological embedding. Since $F : M \to N$ is a smooth immersion, it suffices to show that $F : M \to N$ is a topological embedding.

Since $F : M \to N$ is a smooth map, $F : M \to N$ is continuous. Since $F$ is 1–1, continuous, and closed, by Note 5.65, $F : M \to N$ is a topological embedding.

**Note 5.71** Let $M$ be an $m$-dimensional smooth manifold, $N$ be an $n$-dimensional smooth manifold, and $F : M \to N$ be a 1–1 smooth immersion. Let $F$ be a proper map. We shall try to show that $F : M \to N$ is a smooth embedding.

Since $M$ is an $m$-dimensional smooth manifold, $M$ is a Hausdorff locally compact space. Similarly, $N$ is a Hausdorff locally compact space. Since $F : M \to N$ is a smooth map, $F : M \to N$ is continuous. Further, since $F$ is a proper map, by Note 5.67, $F$ is a closed mapping.

Since $F : M \to N$ is a 1–1 smooth immersion, and $F$ is a closed map, by Note 5.70, $F : M \to N$ is a smooth embedding.

**Note 5.72** Let $M$ be an $m$-dimensional smooth manifold, $N$ be an $n$-dimensional smooth manifold, and $F : M \to N$ be a 1–1 smooth immersion. Let $M$ be compact. We shall try to show that $F : M \to N$ is a smooth embedding. We first try to show that $F : M \to N$ is proper map.

For this purpose, let us take any compact subset $C$ of $N$. We have to show that $F^{-1}(C)$ is compact in $M$. Since $C$ is compact in $N$, and the topology of $N$ is Hausdorff, $C$ is closed. Since $F : M \to N$ is smooth, $F : M \to N$ is continuous. Since $F : M \to N$ is continuous, and $C$ is a closed subset of $N$, $F^{-1}(C)$ is closed in $M$. Since $F^{-1}(C)$ is closed in $M$, and $M$ is compact, $F^{-1}(C)$ is compact in $M$.

This shows that $F : M \to N$ is proper map. Since $F : M \to N$ is a 1–1 smooth immersion, and $F : M \to N$ is proper map, by Note 5.71, $F : M \to N$ is a smooth embedding.

**Note 5.73** Let $M$ be an $m$-dimensional smooth manifold, $N$ be an $n$-dimensional smooth manifold, $F : M \to N$ be a 1–1 smooth immersion, and $m = n$. We shall try to show that $F : M \to N$ is a smooth embedding.

By Theorem 5.46, $F : M \to N$ is a local diffeomorphism, and hence, by Note 5.40, $F$ is an open map. Since $F : M \to N$ is a 1–1 smooth immersion, and $F$ is an open map, by Note 5.69, $F : M \to N$ is a smooth embedding.

**Theorem 5.74** *Let $M$ be an m-dimensional smooth manifold, $N$ be an n-dimensional smooth manifold, and $F : M \to N$ be smooth. Suppose that for every $p \in M$, there exists an open neighborhood $U$ of $p$ such that the restriction $F|_U : U \to N$ is a smooth embedding. Then, $F : M \to N$ is a smooth immersion.*

*Proof* We have to show that $F : M \to N$ is a smooth immersion, that is, rank $F = m$, that is, for every $p \in M$, the rank of $F$ at $p$ is $m$.

For this purpose, let us take any $p \in M$. By the given condition, there exists an open neighborhood $U$ of $p$ such that the restriction $F|_U : U \to N$ is a smooth embedding. Since $F|_U : U \to N$ is a smooth embedding, $F|_U : U \to N$ is a smooth immersion. Since $F|_U : U \to N$ is a smooth immersion, and $p \in U$, the linear map $d(F|_U)_p : T_p U \to T_{(F|_U)(p)} N$ is 1–1, that is, $d(F|_U)_p : T_p U \to T_{F(p)} N$ is 1–1. Since $U$ is an open neighborhood of $p$ in $M$, by Note 5.16, $T_p U$ and $T_p M$ are essentially the same. Since $T_p U$ and $T_p M$ are essentially the same, and $d(F|_U)_p : T_p U \to T_{F(p)} N$ is 1–1, $dF_p : T_p M \to T_{F(p)} N$ is 1–1. Since $dF_p : T_p M \to_{F(p)} N$ is 1–1, (rank of $F$ at $p$) $= \dim((dF_p)(T_p M)) = \dim(T_p M) = \dim M = m$. □

**Theorem 5.75** *Let $M$ be an m-dimensional smooth manifold, $N$ be an n-dimensional smooth manifold, and $F : M \to N$ be a smooth immersion. Let $p \in M$. Then, there exists an open neighborhood $U$ of $p$ such that the restriction $F|_U : U \to N$ is a smooth embedding.*

*Proof* By Theorem 5.50, there exist an admissible coordinate chart $(U_1, \varphi)$ in $M$ satisfying $p \in U_1$, an admissible coordinate chart $(V, \psi)$ in $N$ satisfying $F(p) \in V$ such that $F(U_1) \subset V$, and for every $(x_1, \ldots, x_m)$ in $\varphi(U_1)$,

$$\left(\psi \circ F \circ \varphi^{-1}\right)(x_1, \ldots, x_m) = \left(x_1, \ldots, x_m, \underbrace{0, \ldots, 0}_{n-m}\right).$$

Here, $(U_1, \varphi)$ is an admissible coordinate chart $(U_1, \varphi)$ in $M$ satisfying $p \in U_1$, so $U_1$ is an open neighborhood of $p$ in $M$. Since for every $(x_1, \ldots, x_m)$ in $\varphi(U_1)$,

$$\left(\psi \circ F \circ \varphi^{-1}\right)(x_1, \ldots, x_m) = \left(x_1, \ldots, x_m, \underbrace{0, \ldots, 0}_{n-m}\right),$$

and $F(U_1) \subset V$, $(\psi \circ F \circ \varphi^{-1}) : \varphi(U_1) \to \psi(V)$ and is 1–1. Since $(\psi \circ F \circ \varphi^{-1}), \varphi, \psi^{-1}$ are 1–1, their composite $\psi^{-1} \circ (\psi \circ F \circ \varphi^{-1}) \circ \varphi (= F|_{U_1})$ is a 1–1 mapping from $U_1$ to $V(\subset N)$. It follows that $F|_{U_1} : U_1 \to N$ is 1–1. Since $U_1$ is an open neighborhood of $p$ in $M$, and $M$ is an $m$-dimensional smooth manifold, there exists an open neighborhood $U$ of $p$ such that $U \subset U^- \subset U_1$, and $U^-$ is compact. Here, $U$ is a nonempty open subset of $M$ and $M$ is an $m$-dimensional smooth manifold, $U$ is also an $m$-dimensional smooth manifold. Now, since $F : M \to N$ is a smooth mapping, $F|_U : U \to N$ is a smooth map.

Now, it remains to be proved that $F|_U : U \to N$ is a smooth embedding, that is,

1. $F|_U : U \to N$ is a smooth immersion and
2. $F|_U : U \to N$ is a topological embedding.

For 1: Clearly, $F|_U : U \to N$ is a smooth immersion, that is, for every $q \in U$, the rank of $F|_U$ at $q$ is $m$.

(Reason: For this purpose, let us take any $q \in U$. Since $F : M \to N$ is a smooth immersion, and $q \in M$, the linear map $dF_q : T_q M \to T_{F(q)} N$ is 1–1. Since $U$ is an open neighborhood of $q$ in $M$, $T_q U$ and $T_q M$ are essentially the same. Since $T_q U$ and $T_q M$ are essentially the same, and $dF_q : T_q M \to T_{F(q)} N$ is 1–1, $d(F|_U)_q : T_q U \to T_{F(q)} N$ is 1–1. Since the linear map $d(F|_U)_q : T_q U \to T_{F(q)} N$ is 1–1, (rank of $F|_U$ at $q$) $= \dim((d(F|_U)_q)(T_q U)) = \dim(T_q U) = \dim(T_q M) = \dim M = m$. Hence, rank of $F|_U$ at $q$ is $m$.)

For 2: First of all, we shall try to show that $F|_{U^-} : U^- \to N$ is a topological embedding. Since $F|_{U_1} : U_1 \to N$ is 1–1, and $U^- \subset U_1$, $F|_{U^-} : U^- \to N$ is 1–1. Since $F : M \to N$ is smooth, $F : M \to N$ is continuous, and hence, the restriction $F|_{U^-} : U^- \to N$ is continuous. Further, since $U^-$ is compact, and $N$ is a Hausdorff space, by closed map Lemma 5.66, $F|_{U^-} : U^- \to N$ is a topological embedding. Since $F|_{U^-} : U^- \to N$ is a topological embedding, and $U \subset U^-$, the restriction $(F|_{U^-})|_U: U \to N$ is a topological embedding. Since $(F|_{U^-})|_U: U \to N$ is a topological embedding, and $(F|_{U^-})|_U = F|_U$, $F|_U: U \to N$ is a topological embedding. $\qquad\qquad\qquad\qquad\qquad\qquad\qquad\qquad\qquad\qquad\qquad\qquad\square$

**Theorem 5.76** *Let M be an m-dimensional smooth manifold, N be an n-dimensional smooth manifold, and $\pi: M \to N$ be a smooth submersion. Let $p \in M$. Then, there exist an open neighborhood $V^*$ of $\pi(p)$ in N and a smooth map $\sigma: V^* \to M$ such that $\pi \circ \sigma = \mathrm{Id}_{V^*}$, and $\sigma(\pi(p)) = p$.*

*Proof* Since $\pi: M \to N$ is a smooth submersion, and $p \in M$, by Theorem 5.49, there exist admissible coordinate chart $(U, \varphi)$ in $M$ satisfying $p \in U$, admissible coordinate chart $(V, \psi)$ in $N$ satisfying $\pi(p) \in V$ such that $\pi(U) \subset V$, and for every $(x_1, \ldots, x_n, x_{n+1}, \ldots, x_m)$ in $\varphi(U)$,

$$\left(\psi \circ \pi \circ \varphi^{-1}\right)(x_1, \ldots, x_n, x_{n+1}, \ldots, x_m) = (x_1, \ldots, x_n).$$

Here, $p \in U$, and $(U, \varphi)$ is an admissible coordinate chart in $M$, so $\varphi(p)$ is an element of the open subset $\varphi(U)$ of $\mathbb{R}^m$. Put $\varphi(p) \equiv (a_1, \ldots, a_n, a_{n+1}, \ldots, a_m)$. Here, $(a_1, \ldots, a_n, a_{n+1}, \ldots, a_m)$ is an element of the open subset $\varphi(U)$ of $\mathbb{R}^m$, so there exists $\varepsilon > 0$ such that

$$(a_1 - \varepsilon, a_1 + \varepsilon) \times \cdots \times (a_n - \varepsilon, a_n + \varepsilon) \times (a_{n+1} - \varepsilon, a_{n+1} + \varepsilon)$$
$$\times \cdots \times (a_m - \varepsilon, a_m + \varepsilon) \subset \varphi(U).$$

For every $(x_1, \ldots, x_n, x_{n+1}, \ldots, x_m) \in (a_1 - \varepsilon, a_1 + \varepsilon) \times \cdots \times (a_n - \varepsilon, a_n + \varepsilon) \times (a_{n+1} - \varepsilon, a_{n+1} + \varepsilon) \times \cdots \times (a_m - \varepsilon, a_m + \varepsilon)$, we have $(\psi \circ \pi \circ \varphi^{-1})(x_1, \ldots, x_n, x_{n+1}, \ldots, x_m) = (x_1, \ldots, x_n) \in \psi(V)$. It follows that

$$\psi\big(\pi\big(\varphi^{-1}((a_1 - \varepsilon, a_1 + \varepsilon) \times \cdots \times (a_n - \varepsilon, a_n + \varepsilon)$$
$$\times(a_{n+1} - \varepsilon, a_{n+1} + \varepsilon) \times \cdots \times (a_m - \varepsilon, a_m + \varepsilon))))$$
$$= \left(\psi \circ \pi \circ \varphi^{-1}\right)((a_1 - \varepsilon, a_1 + \varepsilon) \times \cdots \times (a_n - \varepsilon, a_n + \varepsilon)$$
$$\times(a_{n+1} - \varepsilon, a_{n+1} + \varepsilon) \times \cdots \times (a_m - \varepsilon, a_m + \varepsilon))$$
$$= (a_1 - \varepsilon, a_1 + \varepsilon) \times \cdots \times (a_n - \varepsilon, a_n + \varepsilon) \ni (a_1, \ldots, a_n).$$

Hence,

$$(a_1, \ldots, a_n)\big(= \left(\psi \circ \pi \circ \varphi^{-1}\right)(a_1, \ldots, a_n, a_{n+1}, \ldots, a_m)$$
$$= \left(\psi \circ \pi \circ \varphi^{-1}\right)(\varphi(p)) = \psi(\pi(p)))$$

is an interior point of

$$\psi\big(\pi\big(\varphi^{-1}((a_1 - \varepsilon, a_1 + \varepsilon) \times \cdots \times (a_n - \varepsilon, a_n + \varepsilon)$$
$$\times(a_{n+1} - \varepsilon, a_{n+1} + \varepsilon) \times \cdots \times (a_m - \varepsilon, a_m + \varepsilon)))).$$

Since $\psi(\pi(p))$ is an interior point of

$$\psi\big(\pi\big(\varphi^{-1}((a_1 - \varepsilon, a_1 + \varepsilon) \times \cdots \times (a_n - \varepsilon, a_n + \varepsilon) \\ \times (a_{n+1} - \varepsilon, a_{n+1} + \varepsilon) \times \cdots \times (a_m - \varepsilon, a_m + \varepsilon)))\big),$$

$$\psi\big(\pi\big(\varphi^{-1}((a_1 - \varepsilon, a_1 + \varepsilon) \times \cdots \times (a_n - \varepsilon, a_n + \varepsilon) \\ \times (a_{n+1} - \varepsilon, a_{n+1} + \varepsilon) \times \cdots \times (a_m - \varepsilon, a_m + \varepsilon))))$$

is an open neighborhood of $\psi(\pi(p))$. Since

$$\psi\big(\pi\big(\varphi^{-1}((a_1 - \varepsilon, a_1 + \varepsilon) \times \cdots \times (a_n - \varepsilon, a_n + \varepsilon) \\ \times (a_{n+1} - \varepsilon, a_{n+1} + \varepsilon) \times \cdots \times (a_m - \varepsilon, a_m + \varepsilon))))$$

is an open neighborhood of $\psi(\pi(p))$, and $\psi$ is a homeomorphism from $V$ onto $\psi(V)$,

$$\pi\big(\varphi^{-1}((a_1 - \varepsilon, a_1 + \varepsilon) \times \cdots \times (a_n - \varepsilon, a_n + \varepsilon) \\ \times (a_{n+1} - \varepsilon, a_{n+1} + \varepsilon) \times \cdots \times (a_m - \varepsilon, a_m + \varepsilon)))$$

is an open neighborhood of $\pi(p)$. Put

$$V^* \equiv \pi\big(\varphi^{-1}((a_1 - \varepsilon, a_1 + \varepsilon) \times \cdots \times (a_n - \varepsilon, a_n + \varepsilon) \\ \times (a_{n+1} - \varepsilon, a_{n+1} + \varepsilon) \times \cdots \times (a_m - \varepsilon, a_m + \varepsilon))).$$

Here, $V^*$ is an open neighborhood of $\pi(p)$. Let us define $\sigma : V^* \to M$ as follows: For every $q$ in $V^*$,

$$\sigma(q) \equiv \varphi^{-1}(\pi_1(\psi(q)), \ldots, \pi_n(\psi(q)), a_{n+1}, \ldots, a_m).$$

Since $\psi, \varphi^{-1}$, and all projection maps $\pi_i$ are smooth, $\sigma$ is smooth. Here, we want to prove that $\pi \circ \sigma = \mathrm{Id}_{V^*}$, that is, for every $q \in V^*$, $\pi(\sigma(q)) = q$.

$$\begin{aligned}
\mathrm{LHS} &= \pi(\sigma(q)) = \pi\big(\varphi^{-1}(\pi_1(\psi(q)), \ldots, \pi_n(\psi(q)), a_{n+1}, \ldots, a_m)\big) \\
&= (\pi \circ \varphi^{-1})(\pi_1(\psi(q)), \ldots, \pi_n(\psi(q)), a_{n+1}, \ldots, a_m) \\
&= \psi^{-1}\big((\psi \circ \pi \circ \varphi^{-1})(\pi_1(\psi(q)), \ldots, \pi_n(\psi(q)), a_{n+1}, \ldots, a_m)\big) \\
&= \psi^{-1}(\pi_1(\psi(q)), \ldots, \pi_n(\psi(q))) = \psi^{-1}(\psi(q)) = q = \mathrm{RHS}.
\end{aligned}$$

It remains to be proved that $\sigma(\pi(p)) = p$.

$$
\begin{aligned}
\text{LHS} &= \sigma(\pi(p)) \\
&= \varphi^{-1}\big(\pi_1\big((\psi \circ \pi \circ \varphi^{-1})(\varphi(p))\big), \ldots, \pi_n\big((\psi \circ \pi \circ \varphi^{-1})(\varphi(p))\big), a_{n+1}, \ldots, a_m\big) \\
&= \varphi^{-1}\big(\pi_1\big((\psi \circ \pi \circ \varphi^{-1})(a_1, \ldots, a_n, a_{n+1}, \ldots, a_m)\big), \ldots, \pi_n\big((\psi \circ \pi \circ \varphi^{-1}) \\
&\quad (a_1, \ldots, a_n, a_{n+1}, \ldots, a_m)\big), a_{n+1}, \ldots, a_m\big) \\
&= \varphi^{-1}\big(\pi_1(a_1, \ldots, a_n), \ldots, \pi_n(a_1, \ldots, a_n), a_{n+1}, \ldots, a_m\big) \\
&= \varphi^{-1}(a_1, \ldots, a_n, a_{n+1}, \ldots, a_m) = \varphi^{-1}(\varphi(p)) = p = \text{RHS.} \qquad \square
\end{aligned}
$$

**Theorem 5.77** *Let $M$ be an $m$-dimensional smooth manifold, $N$ be an $n$-dimensional smooth manifold, and $\pi : M \to N$ be a smooth map. If for every $p \in M$, there exist an open neighborhood $V$ of $\pi(p)$ in $N$ and a smooth map $\sigma : V \to M$ such that $\pi \circ \sigma = \mathrm{Id}_V$, and $\sigma(\pi(p)) = p$, then $\pi : M \to N$ is a smooth submersion.*

*Proof* We have to prove that $\pi : M \to N$ is a smooth submersion, that is, for every $p \in M$, the linear map $d\pi_p : T_p M \to T_{\pi(p)} N$ is onto. For this purpose, let us take any $p \in M$. By the given condition, there exist an open neighborhood $V$ of $\pi(p)$ in $N$ a smooth map $\sigma : V \to M$ such that $\pi \circ \sigma = \mathrm{Id}_V$, and $\sigma(\pi(p)) = p$. Since $V$ is an open neighborhood of $\pi(p)$ in $N$, $T_{\pi(p)}V$ and $T_{\pi(p)}N$ are essentially the same. Now, since $\pi, \sigma$ are smooth maps,

$$
\begin{aligned}
\mathrm{Id}_{T_{\pi(p)}N} = \mathrm{Id}_{T_{\pi(p)}V} &= \mathrm{d}(\mathrm{Id}_V)_{\pi(p)} = \mathrm{d}(\pi \circ \sigma)_{\pi(p)} \\
&= \big(d\pi_{\sigma(\pi(p))}\big) \circ \big(d\sigma_{\pi(p)}\big) = \big(d\pi_p\big) \circ \big(d\sigma_{\pi(p)}\big).
\end{aligned}
$$

Since

$$
\big(d\pi_p\big) \circ \big(d\sigma_{\pi(p)}\big) = \mathrm{Id}_{T_{\pi(p)}N}, \quad \text{so} \quad d\pi_p : T_p M \to T_{\pi(p)} N
$$

is onto. $\qquad \square$

**Note 5.78** If we combine the Theorems 5.76, and 5.77, then we get the result: Let $M$ be an $m$-dimensional smooth manifold, $N$ be an $n$-dimensional smooth manifold, and $\pi : M \to N$ be a smooth map.

$\pi : M \to N$ is a smooth submersion if and only if for every $p \in M$, there exist an open neighborhood $V$ of $\pi(p)$ in $N$ and a smooth map $\sigma : V \to M$ such that $\pi \circ \sigma = \mathrm{Id}_V$ and $\sigma(\pi(p)) = p$.

**Theorem 5.79** *Let $M$ be an $m$-dimensional smooth manifold, $N$ be an $n$-dimensional smooth manifold, and $\pi : M \to N$ be a smooth submersion. Then, $\pi : M \to N$ is an open map.*

*Proof* Let $G$ be a nonempty open subset of $M$. We have to show that $\pi(G)$ is open in $N$.

For this purpose, let us take any $\pi(p) \in \pi(G)$, where $p \in G$. Since $p \in M$, and $\pi : M \to N$ is a smooth submersion, by Theorem 5.76, there exist an open neighborhood $V$ of $\pi(p)$ in $N$ and a smooth map $\sigma : V \to M$ such that $\pi \circ \sigma = \mathrm{Id}_V$, and $\sigma(\pi(p)) = p$. Since $\sigma : V \to M$ is smooth, $\sigma : V \to M$ is continuous. Since $\sigma : V \to M$ is continuous, and $G$ is an open neighborhood of $p(= \sigma(\pi(p)))$ in $M$, $\sigma^{-1}(G)$ is an open neighborhood of $\pi(p)$ in $V$. Since $\sigma^{-1}(G)$ is an open neighborhood of $\pi(p)$ in $V$, and $V$ is open, $\sigma^{-1}(G)$ is an open neighborhood of $\pi(p)$ in $N$.

Now, it suffices to show that $\sigma^{-1}(G) \subset \pi(G)$. For this purpose, let us take any $q \in \sigma^{-1}(G)$. Since $q \in \sigma^{-1}(G) \subset V$, $q \in V$. Since $q \in V$, $q = \mathrm{Id}_V(q) = (\pi \circ \sigma)(q) = \pi(\sigma(q))$. Since $q \in \sigma^{-1}(G)$, $\sigma(q) \in G$. Since $\sigma(q) \in G$, $\pi(\sigma(q)) \in \pi(G)$. Since $\pi(\sigma(q)) \in \pi(G)$, and $\pi(\sigma(q)) = q$, $q \in \pi(G)$. Thus, we have shown that $\sigma^{-1}(G) \subset \pi(G)$.                    $\square$

**Definition** Let $X$ be a topological space and $Y$ be a nonempty set. Let $F : X \to Y$ be any onto map. Let $\mathcal{O}$ be the collection of all subsets $G$ of $Y$ such that $F^{-1}(G)$ is open in $X$. We shall try to show that $\mathcal{O}$ is a topology over $Y$.

1. Since $F^{-1}(\emptyset) = \emptyset$, and $\emptyset$ is open in $X$, $\emptyset \in \mathcal{O}$. Since $F^{-1}(Y) = X$, and $X$ is open in $X$, $Y \in \mathcal{O}$.
2. Let $G_1, G_2 \in \mathcal{O}$. We have to show that $G_1 \cap G_2 \in \mathcal{O}$, that is, $F^{-1}(G_1 \cap G_2)$ is open in $X$. Since $G_1 \in \mathcal{O}$, $F^{-1}(G_1)$ is open in $X$. Similarly, $F^{-1}(G_2)$ is open in $X$. It follows that $F^{-1}(G_1) \cap F^{-1}(G_2)(= F^{-1}(G_1 \cap G_2))$ is open in $X$, and hence, $F^{-1}(G_1 \cap G_2)$ is open in $X$.
3. Let $G_i \in \mathcal{O}$ for every $i \in I$. We have to show that $\cup_{i \in I} G_i \in \mathcal{O}$, that is, $F^{-1}(\cup_{i \in I} G_i)$ is open in $X$. Since, for every $i \in I$, $G_i \in \mathcal{O}$, $F^{-1}(G_i)$ is open in $X$. It follows that $\cup_{i \in I}(F^{-1}(G_i))(= F^{-1}(\cup_{i \in I} G_i))$ is open in $X$, and hence, $F^{-1}(\cup_{i \in I} G_i)$ is open in $X$.

Thus, we have shown that $\mathcal{O}$ is a topology over $Y$. The topology $\mathcal{O}$ is called the *quotient topology on $Y$ determined by $F$.*

**Definition** Let $X, Y$ be topological spaces. Let $F : X \to Y$ be any continuous, onto mapping. If the quotient topology on $Y$ determined by $F$ is the topology of $Y$, then we say that $F$ is a *quotient map.*

**Theorem 5.80** *Let $M$ be an $m$-dimensional smooth manifold, $N$ be an $n$-dimensional smooth manifold, and $\pi : M \to N$ be a smooth submersion. Let $\pi : M \to N$ be onto. Then, $\pi : M \to N$ is a quotient map.*

*Proof* For this purpose, let $G$ be any nonempty subset of $N$ such that $\pi^{-1}(G)$ is open in $M$. It suffices to show that $G$ is open in $N$.

Since $\pi : M \to N$ is a smooth submersion, by Theorem 5.79, $\pi : M \to N$ is an open map. Since $\pi : M \to N$ is an onto map, and $G$ is a subset of $N$, $\pi(\pi^{-1}(G)) = G$. Since $\pi : M \to N$ is an open map, and $\pi^{-1}(G)$ is open in $M$, $\pi(\pi^{-1}(G))(= G)$ is open in $N$, and hence, $G$ is open in $N$.                    $\square$

**Theorem 5.81** *Let $M$ be an $m$-dimensional smooth manifold, $N$ be an $n$-dimensional smooth manifold, and $P$ be a $p$-dimensional smooth manifold. Let $\pi : M \to N$ be a smooth submersion. Let $\pi : M \to N$ be onto. Let $F : N \to P$ be any map such that $F \circ \pi : M \to P$ is smooth. Then, $F : N \to P$ is smooth.*

*Proof* We have to prove that $F : N \to P$ is smooth, that is, for every $q$ in $N$, there exists an open neighborhood $V_q$ of $q$ in $N$ such that $F|_{V_q} : V_q \to P$ is smooth.

For this purpose, let us take any $q$ in $N$. Since $q$ is in $N$, and $\pi : M \to N$ is onto, there exists $p \in M$ such that $\pi(p) = q$. Since $\pi : M \to N$ is a smooth submersion, and $p \in M$, by Theorem 5.76, there exist an open neighborhood $V_q$ of $\pi(p)$ in $N$ and a smooth map $\sigma : V_q \to M$ such that $\pi \circ \sigma = \mathrm{Id}_{V_q}$, and $\sigma(\pi(p)) = p$. Since $\sigma : V_q \to M$ is smooth, and $F \circ \pi : M \to P$ is smooth, their composite $(F \circ \pi) \circ \sigma : V_q \to P$ is smooth. Here, $(F \circ \pi) \circ \sigma = F \circ (\pi \circ \sigma) = F \circ \mathrm{Id}_{V_q} = F|_{V_q}$. Since $F|_{V_q} = (F \circ \pi) \circ \sigma$, and $(F \circ \pi) \circ \sigma$ is smooth, $F|_{V_q}$ is smooth. $\qquad\square$

## 5.8 Properly Embedded

**Definition** Let $X, Y$ be any nonempty sets. Let $F : X \to Y$ be any mapping. Let $b \in Y$. The set $F^{-1}(b)$ is called the *fiber of $F$ over $b$*. Clearly, $X$ is partitioned into fibers of $F$.

**Theorem 5.82** *Let $M$ be an $m$-dimensional smooth manifold, $N$ be an $n$-dimensional smooth manifold, and $P$ be a $p$-dimensional smooth manifold. Let $\pi : M \to N$ be a smooth submersion. Let $\pi : M \to N$ be onto. Let $F : M \to P$ be any smooth map. Let $F$ be constant on fibers of $\pi$, that is, if $\pi(x) = \pi(y)$, then $F(x) = F(y)$. Then, there exists a unique smooth map $\hat{F} : N \to P$ such that $\hat{F} \circ \pi = F$.*

*Proof* **Existence**: We first try to show that $F \circ \pi^{-1}$ is a function. For this purpose, let $(x, y) \in F \circ \pi^{-1}$, and $(x, z) \in F \circ \pi^{-1}$. We have to show that $y = z$.

Since $(x, y) \in F \circ \pi^{-1}$, there exists $u$ such that $(x, u) \in \pi^{-1}$, and $(u, y) \in F$. Since $(x, u) \in \pi^{-1}$, $(u, x) \in \pi$. Since $(x, z) \in F \circ \pi^{-1}$, there exists $v$ such that $(x, v) \in \pi^{-1}$, and $(v, z) \in F$. Since $(x, v) \in \pi^{-1}$, $(v, x) \in \pi$. Since $(v, x) \in \pi$, $\pi(v) = x$. Similarly, $\pi(u) = x$. It follows that $\pi(u) = \pi(v)$. Since $\pi(u) = \pi(v)$, and $F$ is constant on fibers of $\pi$, $F(u) = F(v)$. Since $(v, z) \in F$, $F(v) = z$. Since $(u, y) \in F$, $F(u) = y$. Since $F(u) = y$, $F(v) = z$, and $F(u) = F(v)$, $y = z$. This proves that $F \circ \pi^{-1}$ is a function.

Since $\pi : M \to N$ is onto, and $F : M \to P$, $\mathrm{dom}\,(F \circ \pi^{-1}) = N$, and $\mathrm{ran}\,(F \circ \pi^{-1}) \subset P$. Since $F \circ \pi^{-1}$ is a function, $\mathrm{dom}\,(F \circ \pi^{-1}) = N$, and $\mathrm{ran}\,(F \circ \pi^{-1}) \subset P$, we can write $F \circ \pi^{-1} : N \to P$.

Put $\hat{F} \equiv F \circ \pi^{-1}$. Since $\hat{F} = F \circ \pi^{-1}$, and $F \circ \pi^{-1} : N \to P$, $\hat{F} : N \to P$. Now, $\hat{F} \circ \pi = (F \circ \pi^{-1}) \circ \pi = F \circ (\pi^{-1} \circ \pi)$. Clearly, $F \circ (\pi^{-1} \circ \pi) = F$.

(Reason: Let $(x, y) \in$ LHS $= F \circ (\pi^{-1} \circ \pi)$. So there exist $u$, $v$ such that $(x, u) \in \pi, (u, v) \in \pi^{-1}$, and $(v, y) \in F$. Since $(u, v) \in \pi^{-1}$, $(v, u) \in \pi$. It follows that $\pi(v) = u, \pi(x) = u$, and hence, $\pi(x) = \pi(v)$. Since $\pi(x) = \pi(v)$, and $F$ is constant on fibers of $\pi$, $F(x) = F(v)$. Since $(v, y) \in F$, $F(v) = y$. Since $F(v) = y$, and $F(x) = F(v)$, $F(x) = y$, and hence, $(x, y) \in F =$ RHS. Hence, LHS $\subset$ RHS. Next, let $(x, y) \in$ RHS $= F$. Since $(x, y) \in F$, and $F : M \to P$, $x \in M$, and $y \in P$. Since $x \in M$, and $\pi : M \to N$, there exists $u \in N$ such that $(x, u) \in \pi$. Since $(x, u) \in \pi$, $(u, x) \in \pi^{-1}$. Since $(x, u) \in \pi$, $(u, x) \in \pi^{-1}$, and $(x, y) \in F$, $(x, y) \in F \circ (\pi^{-1} \circ \pi) =$ LHS. Hence, RHS $\subset$ LHS. Since LHS $\subset$ RHS, and RHS $\subset$ LHS, LHS $=$ RHS.) Thus, $\hat{F} \circ \pi = F$.

Since $F \circ \pi^{-1} : N \to P$, and $\hat{F} = F \circ \pi^{-1}$, $\hat{F} : N \to P$. It remains to be showed that $\hat{F}$ is smooth. Since $\hat{F} \circ \pi = F$, and $F$ is a smooth map, $\hat{F} \circ \pi$ is smooth. Since $\hat{F} : N \to P$ is a map, and $\hat{F} \circ \pi$ is smooth, by Theorem 5.81, $\hat{F}$ is smooth.

This completes the proof of existence part.

**Uniqueness**: Let $F_1 : N \to P$ be a smooth mapping such that $F_1 \circ \pi = F$, and let $F_2 : N \to P$ be a smooth mapping such that $F_2 \circ \pi = F$. We have to show that $F_1 = F_2$.

Since $\pi : M \to N$ is onto, $\pi \circ \pi^{-1} = \mathrm{Id}_N$. Since $F_1 \circ \pi = F$, $F \circ \pi^{-1} = (F_1 \circ \pi) \circ \pi^{-1} = F_1 \circ (\pi \circ \pi^{-1}) = F_1 \circ \mathrm{Id}_N = F_1$. Thus, $F_1 = F \circ \pi^{-1}$. Similarly, $F_2 = F \circ \pi^{-1}$. It follows that $F_1 = F_2$. This proves the uniqueness part. $\square$

**Theorem 5.83** *Let $M$ be an $m$-dimensional smooth manifold, $N$ be an $n$-dimensional smooth manifold, and $P$ be a $p$-dimensional smooth manifold. Let $\pi_1 : M \to N$ be a smooth submersion onto map. Let $\pi_2 : M \to P$ be a smooth submersion onto map. Let $\pi_1$ be constant on fibers of $\pi_2$, and $\pi_2$ be constant on fibers of $\pi_1$. Then, there exists a unique diffeomorphism $F : N \to P$ such that $F \circ \pi_1 = \pi_2$.*

*Proof* **Existence**: We first try to show that $\pi_2 \circ \pi_1^{-1}$ is a function. For this purpose, let $(x, y) \in \pi_2 \circ \pi_1^{-1}$, and $(x, z) \in \pi_2 \circ \pi_1^{-1}$. We have to show that $y = z$.

Since $(x, y) \in \pi_2 \circ \pi_1^{-1}$, there exists $z$ such that $(x, z) \in \pi_1^{-1}$, and $(z, y) \in \pi_2$. Since $(x, z) \in \pi_1^{-1}$, $(z, x) \in \pi_1$. Since $(x, z) \in \pi_2 \circ \pi_1^{-1}$, there exists $w$ such that $(x, w) \in \pi_1^{-1}$, and $(w, z) \in \pi_2$. Since $(x, w) \in \pi_1^{-1}$, $(w, x) \in \pi_1$. Since $(w, z) \in \pi_2$, $\pi_2(w) = z$. Similarly, $\pi_2(z) = y$, $\pi_1(w) = x$, and $\pi_1(z) = x$. Since $\pi_1(w) = x = \pi_1(z)$, and $\pi_2$ is constant on fibers of $\pi_1$, $\pi_2(w) = \pi_2(z)$. Since $z = \pi_2(w) = \pi_2(z) = y$, $y = z$. This proves that $\pi_2 \circ \pi_1^{-1}$ is a function.

Since $\pi_1 : M \to N$ is onto, and $\pi_2 : M \to P$ is onto, dom $(\pi_2 \circ \pi_1^{-1}) = N$, and ran $(\pi_2 \circ \pi_1^{-1}) \subset P$. Since $\pi_2 \circ \pi_1^{-1}$ is a function, dom $(\pi_2 \circ \pi_1^{-1}) = N$, and ran $(\pi_2 \circ \pi_1^{-1}) \subset P$, we can write $\pi_2 \circ \pi_1^{-1} : N \to P$.

Put $F \equiv \pi_2 \circ \pi_1^{-1}$. Since $F = \pi_2 \circ \pi_1^{-1}$, and $\pi_2 \circ \pi_1^{-1} : N \to P$, $F : N \to P$. Now, $F \circ \pi_1 = (\pi_2 \circ \pi_1^{-1}) \circ \pi_1 = \pi_2 \circ (\pi_1^{-1} \circ \pi_1)$. Clearly, $\pi_2 \circ (\pi_1^{-1} \circ \pi_1) = \pi_2$.

(Reason: Let $(x, y) \in$ LHS $= \pi_2 \circ (\pi_1^{-1} \circ \pi_1)$. So there exist $u, v$ such that $(x, u) \in \pi_1, (u, v) \in \pi_1^{-1}$, and $(v, y) \in \pi_2$. Since $(u, v) \in \pi_1^{-1}$, $(v, u) \in \pi_1$. It follows that $\pi_1(v) = u$, and $\pi_1(x) = u$, and hence, $\pi_1(v) = \pi_1(x)$. Since $\pi_1(v) = \pi_1(x)$, and $\pi_2$ is constant on fibers of $\pi_1$, $\pi_2(v) = \pi_2(x)$. Since $(v, y) \in \pi_2$, $\pi_2(v) = y$. Since $\pi_2(v) = y$, and $\pi_2(v) = \pi_2(x)$, $\pi_2(x) = y$, and hence, $(x, y) \in \pi_2 = $ RHS. Hence, LHS $\subset$ RHS. Next, let $(x, y) \in$ RHS $= \pi_2$. Since $(x, y) \in \pi_2$ and $\pi_2 : M \to P$, $x \in M$, and $y \in P$. Since $x \in M$, and $\pi_1 : M \to N$, there exists $u \in N$ such that $(x, u) \in \pi_1$. Since $(x, u) \in \pi_1$, $(u, x) \in \pi_1^{-1}$. Since $(x, u) \in \pi_1$, $(u, x) \in \pi_1^{-1}$, and $(x, y) \in \pi_2$, $(x, y) \in \pi_2 \circ (\pi_1^{-1} \circ \pi_1) = $ LHS. Hence, RHS $\subset$ LHS. Since LHS $\subset$ RHS, and RHS $\subset$ LHS, LHS $=$ RHS.) Thus, $F \circ \pi_1 = \pi_2$.

Since $\pi_2 \circ \pi_1^{-1} : N \to P$, and $F = \pi_2 \circ \pi_1^{-1}$, $F : N \to P$. Now, we want to show that $F$ is 1–1, that is, $\pi_2 \circ \pi_1^{-1}$ is 1–1, that is, if $(x, z) \in \pi_2 \circ \pi_1^{-1}$, and $(y, z) \in \pi_2 \circ \pi_1^{-1}$, then $x = y$. For this purpose, let $(x, z) \in \pi_2 \circ \pi_1^{-1}$, and $(y, z) \in \pi_2 \circ \pi_1^{-1}$. We have to show that $x = y$.

Since $(x, z) \in \pi_2 \circ \pi_1^{-1}$, there exists $u$ such that $(x, u) \in \pi_1^{-1}$, and $(u, z) \in \pi_2$. Since $(x, u) \in \pi_1^{-1}$, $(u, x) \in \pi_1$. Since $(y, z) \in \pi_2 \circ \pi_1^{-1}$, there exists $v$ such that $(y, v) \in \pi_1^{-1}$, and $(v, z) \in \pi_2$. Since $(y, v) \in \pi_1^{-1}$, $(v, y) \in \pi_1$. Since $(v, y) \in \pi_1$, $\pi_1(v) = y$. Similarly, $\pi_1(u) = x$, $\pi_2(u) = z$, and $\pi_2(v) = z$. Since $\pi_2(u) = z = \pi_2(v)$, and $\pi_1$ is constant on fibers of $\pi_2$, $\pi_1(u) = \pi_1(v)$. Since $x = \pi_1(u) = \pi_1(v) = y$, $x = y$. Thus, we have shown that $F : N \to P$ is 1–1.

Now, we want to show that $F : N \to P$ is onto. For this purpose, let us take any $c \in P$. We have to find $y \in N$ such that $(y, c) \in F(= \pi_2 \circ \pi_1^{-1})$.

Since $\pi_2 : M \to P$ is onto, and $c \in P$, there exists $z \in M$ such that $(z, c) \in \pi_2$. Since $z \in M$, and $\pi_1 : M \to N$, there exists $y \in N$ such that $(z, y) \in \pi_1$, and hence, $(y, z) \in \pi_1^{-1}$. Since $(y, z) \in \pi_1^{-1}$, and $(z, c) \in \pi_2$, $(y, c) \in \pi_2 \circ \pi_1^{-1}$. Thus, we have shown that $F : N \to P$ is onto.

Now, we want to show that $F : N \to P$ is smooth. Since $F \circ \pi_1 = \pi_2$, and $\pi_2$ is smooth, $F \circ \pi_1$ is smooth. Since $\pi_1 : M \to N$ is a smooth submersion onto map, and $F \circ \pi_1$ is smooth, by Theorem 25, $F$ is smooth.

Lastly, we have to show that $F^{-1} : P \to N$ is smooth. Since $F \circ \pi_1 = \pi_2$, and $F$ is 1–1 onto, $F^{-1} \circ \pi_2 = \pi_1$. Since $F^{-1} \circ \pi_2 = \pi_1$, and $\pi_1$ is smooth, $F^{-1} \circ \pi_2$ is smooth. Since $\pi_2 : M \to P$ is a smooth submersion onto map, and $F^{-1} \circ \pi_2$ is smooth, by Theorem 5.81, $F^{-1}$ is smooth. Thus, we have shown that $F : N \to P$ is a diffeomorphism such that $F \circ \pi_1 = \pi_2$.

This completes the proof of existence part.

**Uniqueness**: Let $F_1 : N \to P$ be a smooth mapping such that $F_1 \circ \pi_1 = \pi_2$, and let $F_2 : N \to P$ be a smooth mapping such that $F_2 \circ \pi_1 = \pi_2$. We have to show that $F_1 = F_2$.

Since $\pi_1 : M \to N$ is onto, $\pi_1 \circ \pi_1^{-1} = \text{Id}_N$. Since $F_1 \circ \pi_1 = \pi_2$, $\pi_2 \circ \pi_1^{-1} = (F_1 \circ \pi_1) \circ \pi_1^{-1} = F_1 \circ (\pi_1 \circ \pi_1^{-1}) = F_1 \circ \text{Id}_N = F_1$. Thus, $F_1 = \pi_2 \circ \pi_1^{-1}$. Similarly, $F_2 = \pi_2 \circ \pi_1^{-1}$. It follows that $F_1 = F_2$. This proves the uniqueness part. $\square$

**Note 5.84** Before going ahead, let us recall the definition of open submanifold of $M$: Let $M$ be an $m$-dimensional smooth manifold, whose topology is $\mathcal{O}$, and differential structure is $\mathcal{A}$. Let $G$ be a nonempty open subset of $M$. Let $\mathcal{O}_1$ be the subspace topology over $G$ inherited from $M$, that is, $\mathcal{O}_1 = \{G_1 : G_1 \subset G, \text{ and } G_1 \in \mathcal{O}\}$.

Since $M$ is an $m$-dimensional smooth manifold, $M$ is a Hausdorff space. Since $M$ is a Hausdorff space, and $G$ is a subspace of $M$, $G$ with the subspace topology is a Hausdorff space. Since $M$ is an $m$-dimensional smooth manifold, $M$ is a second countable space. Since $M$ is a second countable space, and $G$ is a subspace of $M$, $G$ with the subspace topology is a second countable space. Put

$$\mathcal{A}_G \equiv \{(U, \varphi) : (U, \varphi) \in \mathcal{A}, \text{ and } U \subset G\}.$$

Now, we shall try to show that $\mathcal{A}_G$ is an atlas on $G$, that is,

1. $\cup\{U : (U, \varphi_U) \text{ is in } \mathcal{A}_G\} = G$,
2. all pairs of members in $\mathcal{A}_G$ are $C^\infty$ compatible.

For (1): By the definition of $\mathcal{A}_G$, it is clear that $\cup\{U : (U, \varphi) \text{ is in } \mathcal{A}_G\} \subset G$. So, it remains to be proved that $G \subset \cup\{U : (U, \varphi) \text{ is in } \mathcal{A}_G\}$. For this purpose, let us take any $p$ in $G$. Since $p$ is in $G$, and $G$ is a subset of $M$, $p$ is in $M$. Since $p$ is in $M$, and $M$ is an $m$-dimensional smooth manifold, there exists $(U, \varphi)$ in $\mathcal{A}$ such that $p$ is in $U$.

Since $(U, \varphi)$ is in $\mathcal{A}$, $U$ is an open subset of $M$. Since $U$ is open, and $G$ is open, $U \cap G$ is open. Thus, $U \cap G$ is an open neighborhood of $p$. We shall try to show that $(U \cap G, \varphi|_{(U \cap G)})$ is in $\mathcal{A}_G$, that is, $(U \cap G, \varphi|_{(U \cap G)}) \in \mathcal{A}$, and $U \cap G \subset G$. Since $U \cap G \subset G$, it remains to be showed that $(U \cap G, \varphi|_{(U \cap G)}) \in \mathcal{A}$.

Since $(U, \varphi)$ is in $\mathcal{A}$, $(U, \varphi_U)$ is an admissible coordinate chart of $M$. Now, since $U \cap G$ is a nonempty open subset of $U$, by the Lemma 1, $(U \cap G, \varphi|_{(U \cap G)})$ is an admissible coordinate chart of $M$, and hence, $(U \cap G, \varphi|_{(U \cap G)}) \in \mathcal{A}$.

For (2): Let us take any $(U, \varphi), (V, \psi) \in \mathcal{A}_G$. We have to show that $(U, \varphi)$ and $(V, \psi)$ are $C^\infty$ compatible.

Since $(U, \varphi) \in \mathcal{A}_G$, by the definition of $\mathcal{A}_G$, $(U, \varphi) \in \mathcal{A}$. Similarly, $(V, \psi) \in \mathcal{A}$. Since $(U, \varphi) \in \mathcal{A}$, $(V, \psi) \in \mathcal{A}$, and $\mathcal{A}$ is a differential structure, $(U, \varphi)$ and $(V, \psi)$ are $C^\infty$ compatible. Thus, we have shown that $\mathcal{A}_G$ is an atlas on $G$. Hence, $\mathcal{A}_G$ determines a unique smooth structure on $G$. Thus, the open subset $G$ of smooth manifold $M$ becomes an $m$-dimensional smooth manifold. Here, $G$ is called an *open submanifold of $M$*.

**Note 5.85** Let $M$ be an $m$-dimensional smooth manifold with smooth structure $\mathcal{A}$. Let $G$ be a nonempty open subset of $M$. In the open submanifold $G$ of $M$, the smooth structure of $G$ is determined by the atlas $\{(U, \varphi) : (U, \varphi) \in \mathcal{A}, \text{ and } U \subset G\}$ of $G$. Thus, $G$ becomes an $m$-dimensional smooth manifold.

We shall try to show that the inclusion map $\iota : G \to M$ (that is, for every $x$ in $G$, $\iota(x) \equiv x$) is a smooth embedding.

Since $\iota : G \to M$ is the identity mapping $\mathrm{Id}_G$, and $\mathrm{Id}_G$ is smooth, $\iota : G \to M$ is smooth.

Now, we want to show that $\iota : G \to M$ is a smooth immersion, that is, for every $p \in G$, the linear map $d\iota_p : T_p G \to T_{\iota(p)} M$ is 1–1. For this purpose, let us fix any $p \in G$. We have to show that the linear map $d\iota_p : T_p G \to T_{\iota(p)} M$ is 1–1. Clearly, $T_p G = T_p M$, and $T_{\iota(p)} M = T_p M$. So we have to show that the linear map $d\iota_p : T_p M \to T_p M$ is 1–1. Here, $\iota = \mathrm{Id}_G$, so we have to show that the linear map $d(\mathrm{Id}_G)_p : T_p M \to T_p M$ is 1–1.

Since $d(\mathrm{Id}_G)_p = \mathrm{Id}_{T_p G} = \mathrm{Id}_{T_p M}$, and $\mathrm{Id}_{T_p M} : T_p M \to T_p M$ is 1–1, $d(\mathrm{Id}_G)_p : T_p M \to T_p M$ is 1–1. Thus, we have shown that $\iota : G \to M$ is a smooth immersion. Also, clearly $\iota : G \to M$ is 1–1.

Since $G$ is open in $M$, $\iota : G \to M$ is an open map. Since $\iota : G \to M$ is a 1–1 immersion, and open, by Note 5.69, $\iota : G \to M$ is a smooth embedding.

**Conclusion**: Let $M$ be an $m$-dimensional smooth manifold. Let $G$ be a nonempty subset of $M$.

If $G$ is open in $M$, then there exists an $m$-dimensional smooth structure on $G$ such that the inclusion map $\iota : G \to M$ is a smooth embedding. Now, we want to prove its converse part:

Let $M$ be an $m$-dimensional smooth manifold. Let $G$ be a nonempty subset of $M$. Let $G$ be endowed with the subspace topology of $M$.

Suppose there exists an $m$-dimensional smooth structure $\mathcal{A}$ on $G$ such that the inclusion map $\iota : G \to M$ is a smooth embedding. We shall try to show that $G$ is open in $M$.

Since $\iota : G \to M$ is a smooth embedding, $\iota : G \to M$ is a smooth immersion. Since $G, M$ are $m$-dimensional smooth manifolds, and $\iota : G \to M$ is a smooth immersion, by Theorem 5.46, $\iota : G \to M$ is a local diffeomorphism. Since $\iota : G \to M$ is a local diffeomorphism, by Note 5.38, $\iota : G \to M$ is an open map. Since $\iota : G \to M$ is an open map, $G(= \iota(G))$ is an open subset of $M$.

**Definition** Let $M$ be an $m$-dimensional smooth manifold. Let $S$ be a nonempty subset of $M$. Let $S$ be endowed with the subspace topology of $M$. Let $k$ be a nonnegative integer.

If there exists an $(m - k)$-dimensional smooth structure $\mathcal{A}$ on $S$, such that the inclusion map $\iota : S \to M$ is a smooth embedding, then we say that $S$ is an *embedded submanifold of $M$ with codimension $k$*. Here, we also say that $M$ is an *ambient manifold for $S$*.

From the Note 5.85, we can write:

Let $M$ be an $m$-dimensional smooth manifold. Let $S$ be a nonempty subset of $M$. Let $S$ be endowed with the subspace topology of $M$. $S$ is open in $M$ if and only if $S$ is an embedded submanifold of $M$ with codimension 0. In this sense, we say that the open submanifolds of $M$ are exactly the embedded submanifolds of $M$ with codimension 0.

**Note 5.86** Let $M$ be an $m$-dimensional smooth manifold with smooth structure $\mathcal{A}$, $N$ be an $n$-dimensional smooth manifold with smooth structure $\mathcal{B}$, and $F : N \to M$ be a smooth embedding. Let $F(N)(\subset M)$ be endowed with the subspace topology of $M$.

Since $F : N \to M$ is a smooth embedding, $F : N \to M$ is a topological embedding, and hence, $F : N \to F(N)$ is a homeomorphism. Since $F : N \to F(N)$ is a homeomorphism, $N$ is homeomorphic to $F(N)$. Since $N$ is an $n$-dimensional smooth manifold, $N$ is an $n$-dimensional topological manifold. Since $N$ is an $n$-dimensional topological manifold, and $N$ is homeomorphic to $F(N)$, $F(N)$ is an $n$-dimensional topological manifold.

Let us take any $(U, \varphi) \in \mathcal{B}$. We shall try to show that $(F(U), \varphi \circ (F^{-1}))$ is a coordinate chart of $F(N)$. Clearly, $F(U)$ is open in $F(N)$. (Reason: Since $(U, \varphi) \in \mathcal{B}$, and $\mathcal{B}$ is a smooth structure on $N$, $U$ is open in $N$. Since $F : N \to F(N)$ is a homeomorphism, and $U$ is open in $N$, $F(U)$ is open in $F(N)$.) Here, $(U, \varphi) \in \mathcal{B}$, and $\mathcal{B}$ is a smooth structure on $N$, $\varphi(U)$ is open in $\mathbb{R}^n$.

Since $\varphi : U \to \varphi(U)$, $U \subset N$, and $F : N \to M$ is 1–1, $\varphi \circ (F^{-1})$ is a function whose domain is $F(U)$. Since $\varphi : U \to \varphi(U)$ is onto, $U \subset N$, and $F : N \to M$ is 1–1, the range of $\varphi \circ (F^{-1})$ is $\varphi(U)$. Since $\varphi : U \to \varphi(U)$ is 1–1, $U \subset N$, and $F : N \to M$ is 1–1, $\varphi \circ (F^{-1}) : F(U) \to \varphi(U)$ is 1–1.

Thus, we see that $\varphi \circ (F^{-1})$ is a 1–1 mapping from open subset $F(U)$ of $F(N)$ onto open subset $\varphi(U)$ of $\mathbb{R}^n$. Now, we shall try to show that $\varphi \circ (F^{-1}) : F(U) \to \varphi(U)$ is continuous.

Since $F : N \to M$ is a topological embedding, $F$ is a homeomorphism from $N$ onto $F(N)$, and hence, $F^{-1} : F(N) \to N$ is continuous and 1–1. Since $F^{-1} : F(N) \to N$ is continuous and 1–1, its restriction $(F^{-1})|_{F(U)} : F(U) \to U$ is continuous. Since $F : N \to M$ is 1–1, $(F^{-1})|_{F(U)} = (F|_U)^{-1}$. Since $(F^{-1})|_{F(U)} = (F|_U)^{-1}$, and $(F^{-1})|_{F(U)} : F(U) \to U$ is continuous, $(F|_U)^{-1} : F(U) \to U$ is continuous. Since $(F|_U)^{-1} : F(U) \to U$ is continuous, and $\varphi : U \to \varphi(U)$ is continuous, their composite $\varphi \circ ((F|_U)^{-1}) : F(U) \to \varphi(U)$ is continuous. Further, since $\varphi \circ ((F|_U)^{-1}) = \varphi \circ (F^{-1})$, $\varphi \circ (F^{-1}) : F(U) \to \varphi(U)$ is continuous.

Now, we shall try to show that $F \circ (\varphi^{-1}) : \varphi(U) \to F(U)$ is continuous. Since $(U, \varphi) \in \mathcal{B}$, and $\mathcal{B}$ is a smooth structure on $N$, $\varphi^{-1} : \varphi(U) \to U$ is continuous and 1–1. Since $F : N \to M$ is a smooth embedding, $F : N \to M$ is continuous. Since $F : N \to M$ is continuous, and $\varphi^{-1} : \varphi(U) \to U$ is continuous and 1–1, $F \circ (\varphi^{-1}) : \varphi(U) \to F(U)$ is continuous. Also, $(\varphi \circ (F^{-1}))^{-1} = F \circ (\varphi^{-1})$. It follows that $(\varphi \circ (F^{-1}))^{-1} : \varphi(U) \to F(U)$ is continuous.

Thus, we have shown that $(F(U), \varphi \circ (F^{-1}))$ is a coordinate chart of $F(N)$.

**Conclusion**: Let $M$ be an $m$-dimensional smooth manifold with smooth structure $\mathcal{A}$, $N$ be an $n$-dimensional smooth manifold with smooth structure $\mathcal{B}$, and $F : N \to M$ be a smooth embedding. Let $F(N)(\subset M)$ be endowed with the subspace topology of $M$. If $(U, \varphi) \in \mathcal{B}$, then $(F(U), \varphi \circ (F^{-1}))$ is a coordinate chart of $F(N)$.

**Note 5.87** Let $M$ be an $m$-dimensional smooth manifold with smooth structure $\mathcal{A}$, $N$ be an $n$-dimensional smooth manifold with smooth structure $\mathcal{B}$, and $F : N \to M$ be a smooth embedding. Let $F(N)(\subset M)$ be endowed with the subspace topology of $M$.

Since $M$ is an $m$-dimensional smooth manifold, $M$ is a Hausdorff space. Since $M$ is a Hausdorff space, and $F(N)$ is a subspace of $M$, $F(N)$ with the subspace topology is a Hausdorff space. Since $M$ is an $m$-dimensional smooth manifold, $M$ is a second countable space. Since $M$ is a second countable space, and $F(N)$ is a subspace of $M$, $F(N)$ with the subspace topology is a second countable space. Put

$$\mathcal{A}_F \equiv \left\{ \left( F(U), \varphi \circ \left( F^{-1} \right) \right) : (U, \varphi) \in \mathcal{B} \right\}.$$

From the Note 5.86, we see that $\mathcal{A}_F$ is a collection of coordinate charts of $F(N)$. Now, we shall try to show that $\mathcal{A}_F$ is an $n$-dimensional atlas on $F(N)$, that is,

1. $\cup \{ F(U) : (U, \varphi) \in \mathcal{B} \} = F(N)$,
2. all pairs of members in $\mathcal{A}_F$ are $C^\infty$ compatible.

For 1: Let us take any $(U, \varphi) \in \mathcal{B}$. Since $(U, \varphi) \in \mathcal{B}$, and $\mathcal{B}$ is a smooth structure of $N$, $U \subset N$. Since $U \subset N$, and $F : N \to M$, $F(U) \subset F(N)$. Hence, $\cup \{ F(U) : (U, \varphi) \in \mathcal{B} \} \subset F(N)$.

Now, it remains to be showed that $F(N) \subset \cup \{ F(U) : (U, \varphi) \in \mathcal{B} \}$. For this purpose, let us take any $F(p) \in F(N)$, where $p \in N$. Since $p \in N$, and $N$ is an $n$-dimensional smooth manifold with smooth structure $\mathcal{B}$, there exists $(V, \psi) \in \mathcal{B}$ such that $p \in V$. Since $p \in V$, $F(p) \in F(V) \subset \cup \{ F(U) : (U, \varphi) \in \mathcal{B} \}$. Hence, $F(N) \subset \cup \{ F(U) : (U, \varphi) \in \mathcal{B} \}$.

For 2: Let $\left( F(U), \varphi \circ \left( F^{-1} \right) \right) \in \mathcal{A}_F$, and $\left( F(V), \psi \circ \left( F^{-1} \right) \right) \in \mathcal{A}_F$, where $(U, \varphi) \in \mathcal{B}$, and $(V, \psi) \in \mathcal{B}$. Let $F(U) \cap F(V) \neq \emptyset$. We have to show that $\left( \psi \circ \left( F^{-1} \right) \right) \circ \left( \varphi \circ \left( F^{-1} \right) \right)^{-1}$ is smooth. Here,

$$\left( \psi \circ \left( F^{-1} \right) \right) \circ \left( \varphi \circ \left( F^{-1} \right) \right)^{-1} = \left( \psi \circ \left( F^{-1} \right) \right) \circ \left( F \circ \varphi^{-1} \right)$$
$$= \psi \circ \left( \left( F^{-1} \right) \circ F \right) \circ \varphi^{-1}$$
$$= \psi \circ \mathrm{Id}_N \circ \varphi^{-1} = \psi \circ \varphi^{-1}.$$

Since $F : N \to M$ is 1–1, $F(U \cap V) = F(U) \cap F(V) \neq \emptyset$, and hence, $U \cap V \neq \emptyset$. Since $U \cap V \neq \emptyset$, $(U, \varphi) \in \mathcal{B}$, $(V, \psi) \in \mathcal{B}$, and $\mathcal{B}$ is a smooth structure on $N$, $\psi \circ \varphi^{-1} (= \left( \psi \circ \left( F^{-1} \right) \right) \circ \left( \varphi \circ \left( F^{-1} \right) \right)^{-1})$ is smooth, and hence, $\left( \psi \circ \left( F^{-1} \right) \right) \circ \left( \varphi \circ \left( F^{-1} \right) \right)^{-1}$ is smooth.

Thus, we have shown that $\mathcal{A}_F$ is an $n$-dimensional atlas on $F(N)$, and hence, $F(N)$ is an $n$-dimensional smooth manifold, whose differential structure is determined by the atlas $\mathcal{A}_F$.

**Note 5.88** Let $M$ be an $m$-dimensional smooth manifold with smooth structure $\mathcal{A}$, $N$ be an $n$-dimensional smooth manifold with smooth structure $\mathcal{B}$, and $F : N \to M$ be a smooth embedding. Let $F(N)(\subset M)$ be endowed with the subspace topology of $M$.

From the Note 5.87, $F(N)$ is an $n$-dimensional smooth manifold, whose differential structure is determined by the atlas $\mathcal{A}_F \equiv \{(F(U), \varphi \circ (F^{-1})) : (U, \varphi) \in \mathcal{B}\}$.

Now, we want to show that $F : N \to F(N)$ is a diffeomorphism, that is,

1. $F : N \to F(N)$ is a homeomorphism,
2. $F : N \to F(N)$ is smooth, and $F^{-1} : F(N) \to N$ is smooth.

For 1: Since $F : N \to M$ is a smooth embedding, $F : N \to M$ is a homeomorphism.

For 2: Let us take any $p \in N$. Let $(U, \varphi) \in \mathcal{B}$, where $p \in U$. Let $(F(V), \psi \circ (F^{-1})) \in \mathcal{A}_F$, where $(V, \psi) \in \mathcal{B}$, and $F(p) \in F(V)$. We have to show that $(\psi \circ (F^{-1})) \circ F \circ \varphi^{-1}$ and $\varphi \circ F^{-1} \circ (\psi \circ (F^{-1}))^{-1}$ are smooth.

Here, $(\psi \circ (F^{-1})) \circ F \circ \varphi^{-1} = \psi \circ ((F^{-1}) \circ F) \circ \varphi^{-1} = \psi \circ \mathrm{Id}_N \circ \varphi^{-1} = \psi \circ \varphi^{-1}$, and $\varphi \circ F^{-1} \circ (\psi \circ (F^{-1}))^{-1} = \varphi \circ F^{-1} \circ (F \circ \psi^{-1}) = \varphi \circ (F^{-1} \circ F) \circ \psi^{-1} = \varphi \circ \mathrm{Id}_N \circ \psi^{-1} = \varphi \circ \psi^{-1}$. Since $F(p) \in F(V)$, there exists $q \in V$ such that $F(p) = F(q)$. Since $F(p) = F(q)$, and $F$ is 1–1, $p = q(\in V)$, and hence, $p \in U \cup V$. Since $p \in U \cup V$, $U \cup V \neq \emptyset$. Since $U \cup V \neq \emptyset$, $(U, \varphi) \in \mathcal{B}$, $(V, \psi) \in \mathcal{B}$, and $\mathcal{B}$ is a smooth structure on $N$, $\psi \circ \varphi^{-1}(= (\psi \circ (F^{-1})) \circ F \circ \varphi^{-1})$, and $\varphi \circ \psi^{-1}(= \varphi \circ F^{-1} \circ (\psi \circ (F^{-1}))^{-1})$ are smooth, and hence $(\psi \circ (F^{-1})) \circ F \circ \varphi^{-1}$, and $\varphi \circ F^{-1} \circ (\psi \circ (F^{-1}))^{-1}$ are smooth.

Thus, we have shown that $F : N \to F(N)$ is a diffeomorphism.

**Note 5.89** Let $M$ be an $m$-dimensional smooth manifold with smooth structure $\mathcal{A}$, $N$ be an $n$-dimensional smooth manifold with smooth structure $\mathcal{B}$, and $F : N \to M$ be a smooth embedding. Let $F(N)(\subset M)$ be endowed with the subspace topology of $M$.

From Note 5.87, $F(N)$ is an $n$-dimensional smooth manifold, whose differential structure is determined by the atlas $\mathcal{A}_F \equiv \{(F(U), \varphi \circ (F^{-1})) : (U, \varphi) \in \mathcal{B}\}$. Also, by Note 5.88, $F : N \to F(N)$ is a diffeomorphism.

Now, we want to show that the $n$-dimensional smooth manifold $F(N)$, with atlas $\mathcal{A}_F(= \{(F(U), \varphi \circ (F^{-1})) : (U, \varphi) \in \mathcal{B}\})$, is an embedded submanifold of $M$, that is, the inclusion map $\imath : F(N) \to M$ is a smooth embedding.

Clearly, $\imath = F \circ F^{-1}$. From the Note 5.88, $F : N \to F(N)$ is a diffeomorphism, so $F^{-1} : F(N) \to N$ is a diffeomorphism. Since $F^{-1}$ is a diffeomorphism, and $F : N \to M$ is a smooth embedding, their composition $F \circ F^{-1}(= \imath)$ is a smooth embedding, and hence, $\imath$ is a smooth embedding.

Thus, we have shown that $F(N)$ is an embedded submanifold of $M$ with codimension $m - n$.

**Note 5.90** Let $M$ be an $m$-dimensional smooth manifold with smooth structure $\mathcal{A}$, and $N$ be an $n$-dimensional smooth manifold with smooth structure $\mathcal{B}$. Here, the collection

$$\{(U \times V, (\varphi_U \times \psi_V)) : (U, \varphi_U) \in \mathcal{A}, (V, \psi_V) \in \mathcal{B}\}$$

is an atlas of the $(m + n)$-dimensional product manifold $M \times N$. Let us fix any $p \in N$.

Now, consider the mapping $F : M \to M \times N$ defined as follows: For every $x \in M$, $F(x) \equiv (x, p)$. We shall try to show that $F : M \to M \times N$ is a smooth embedding. For this purpose, it suffices to show that

1. $F$ is 1–1,
2. $F$ is a closed map, and
3. $F$ is a smooth immersion.

> For 1: It is clear that $F$ is 1–1.
>
> For 2: Let $A$ be a closed subset of $M$. We have to show that $F(A)$ is closed in $M \times N$. Here, $F(A) = A \times \{p\} = (M \times N) - ((A^c \times N) \cup (M \times \{p\}^c))$. Since $A$ is a closed subset of $M$, $A^c$ is open in $M$, and hence, $A^c \times N$ is open in $M \times N$. Since $N$ is an $n$-dimensional smooth manifold, the topology of $N$ is Hausdorff. Since the topology of $N$ is Hausdorff, and $p \in N$, $\{p\}$ is a closed subset of $N$, and hence, $\{p\}^c$ is open in $N$. It follows that $M \times \{p\}^c$ is open in $M \times N$. Since $M \times \{p\}^c$ is open in $M \times N$, and $A^c \times N$ is open in $M \times N$, $(A^c \times N) \cup (M \times \{p\}^c)$ is open in $M \times N$, and hence, $(M \times N) - ((A^c \times N) \cup (M \times \{p\}^c))(= F(A))$ is closed in $M \times N$. Thus, $F(A)$ is closed in $M \times N$.
>
> For 3: We have to show that $F : M \to M \times N$ is an immersion, that is, for every $q \in M$, the linear map $dF_q : T_q M \to T_{F(q)}(M \times N)$ is 1–1.
>
> For this purpose, let us fix any $q \in M$. We have to show that the linear map $dF_q : T_q M \to T_{F(q)}(M \times N)$ is 1–1. Since $q \in M$, and $M$ is an $m$-dimensional smooth manifold, there exists an admissible coordinate chart $(U, \varphi)$ of $M$ such that $q \in U$. Since $p \in N$, and $N$ is an $n$-dimensional smooth manifold, there exists an admissible coordinate chart $(V, \psi)$ of $N$ such that $p \in V$.
>
> For every $r \in U$,

$$\left( \left. \frac{\partial}{\partial x^1} \right|_r, \left. \frac{\partial}{\partial x^2} \right|_r, \cdots, \left. \frac{\partial}{\partial x^m} \right|_r \right)$$

is the coordinate basis of $T_r M$ corresponding to $(U, \varphi)$, where

$$\left. \frac{\partial}{\partial x^i} \right|_r \equiv \left( d(\varphi^{-1})_{\varphi(r)} \right) \left( \left. \frac{\partial}{\partial x^i} \right|_{\varphi(r)} \right).$$

Here, $(U \times V, (\varphi \times \psi))$ is an admissible coordinate chart of $M \times N$ such that $F(q) = (q,p) \in U \times V$. For every $(s,t) \in U \times V$,

$$\left( \left.\frac{\partial}{\partial y^1}\right|_{(s,t)}, \left.\frac{\partial}{\partial y^2}\right|_{(s,t)}, \cdots, \left.\frac{\partial}{\partial y^m}\right|_{(s,t)}, \left.\frac{\partial}{\partial y^{m+1}}\right|_{(s,t)}, \cdots, \left.\frac{\partial}{\partial y^{m+n}}\right|_{(s,t)} \right)$$

is the coordinate basis of $T_{(s,t)}(M \times N)$ corresponding to $(U \times V, (\varphi \times \psi))$, where

$$\left.\frac{\partial}{\partial y^i}\right|_{(s,t)} \equiv \left( d\left( (\varphi \times \psi)^{-1} \right)_{(\varphi \times \psi)(s,t)} \right) \left( \left.\frac{\partial}{\partial y^i}\right|_{(\varphi \times \psi)(s,t)} \right).$$

Observe that for every $u \in U$,

$$\begin{aligned}
\left( (\varphi \times \psi) \circ F \circ \varphi^{-1} \right)(\varphi(u)) &= ((\varphi \times \psi) \circ F)(u) \\
&= (\varphi \times \psi)(F(u)) = (\varphi \times \psi)(u,p) \\
&= (\varphi(u), \psi(p)).
\end{aligned}$$

Hence, the matrix representation of linear map $dF_q$ is the following $(m+n) \times m$ matrix:

$$\begin{bmatrix}
1 & 0 & 0 & \cdots & 0 \\
0 & 1 & 0 & \cdots & 0 \\
& & \vdots & & \\
0 & 0 & 0 & \cdots & 1 \\
0 & 0 & 0 & \cdots & 0 \\
0 & 0 & 0 & \cdots & 0 \\
& & \vdots & & \\
0 & 0 & 0 & \cdots & 0
\end{bmatrix}.$$

Clearly,

$$\text{rank} \begin{bmatrix}
1 & 0 & 0 & \cdots & 0 \\
0 & 1 & 0 & \cdots & 0 \\
& & \vdots & & \\
0 & 0 & 0 & \cdots & 1 \\
0 & 0 & 0 & \cdots & 0 \\
0 & 0 & 0 & \cdots & 0 \\
& & \vdots & & \\
0 & 0 & 0 & \cdots & 0
\end{bmatrix} = m,$$

Hence, $dF_q : T_q M \to T_{F(q)}(M \times N)$ is 1–1. Thus, $F$ is a smooth immersion.

Since $F : M \to M \times N$ is a smooth embedding, by Note 5.89, $F(N)(= M \times \{p\})$ is an embedded submanifold of $M \times N$ with codimension $(m + n) - n (= m)$, and $F(N)$ is diffeomorphic to $M$. Thus, we have shown that $M \times \{p\}$ is an embedded submanifold of $M \times N$ with codimension $m$, and $M \times \{p\}$ is diffeomorphic to $M$. In words, we say that a "slice" of a product manifold $M \times N$ is an embedded submanifold of $M \times N$, and each slice is diffeomorphic to $M$.

**Note 5.91** Let $M$ be an $m$-dimensional smooth manifold with smooth structure $\mathcal{A}$, and $N$ be an $n$-dimensional smooth manifold with smooth structure $\mathcal{B}$. Here, the collection

$$\{(U \times V, (\varphi \times \psi)) : (U, \varphi) \in \mathcal{A}, (V, \psi) \in \mathcal{B}\}$$

is an atlas of the $(m + n)$-dimensional product manifold $M \times N$.

Let $G$ be an open subset of $M$, and $f : G \to N$ be a smooth map. Put $\Gamma(f) \equiv \{(x, f(x)) : x \in G\}$ (called the *graph of f*). Now, consider the mapping $F : G \to M \times N$ defined as follows: For every $x \in G$, $F(x) \equiv (x, f(x))$. Clearly, $F(G) = \Gamma(f)$. We shall try to show that $F : G \to M \times N$ is a smooth embedding. For this purpose, it suffices to show that

1. $F$ is 1–1,
2. $F^{-1} : \Gamma(f) \to G$ is continuous,
3. $F : G \to M \times N$ is a smooth map, and
4. $F$ is a smooth immersion.

For 1: It is clear that $F$ is 1–1.

For 2: Clearly, $F^{-1} = \pi^1|_{\Gamma(f)}$, where $\pi^1 : M \times N \to M$ is defined by $\pi^1(x, y) \equiv x$, for every $(x, y) \in M \times N$. Since $\pi^1 : M \times N \to M$ is continuous, its restriction $\pi^1|_{\Gamma(f)}(= F^{-1})$ is continuous, and hence, $F^{-1} : \Gamma(f) \to G$ is continuous.

For 3: Let us take any $p \in G$. Since $p \in G$, and $G$ is an open submanifold of manifold $M$, there exists $(U, \varphi) \in \mathcal{A}$ satisfying $p \in U \subset G$. Here, $p \in G$, and $f : G \to N$, so $f(p) \in N$. Since $f(p) \in N$, and $N$ is an $n$-dimensional smooth manifold with smooth structure $\mathcal{B}$, there exists $(V, \psi) \in \mathcal{B}$ satisfying $f(p) \in V$. Thus, $(U \times V, (\varphi \times \psi))$ is an admissible coordinate chart of $(m + n)$-dimensional product manifold $M \times N$ satisfying $F(p) = (p, f(p)) \in U \times V$. We have to show that $(\varphi \times \psi) \circ F \circ \varphi^{-1}$ is smooth. Let us observe that, for every $q \in U$,
$((\varphi \times \psi) \circ F \circ \varphi^{-1})(\varphi(q)) = (\varphi \times \psi)(F(q)) = (\varphi \times \psi)(q, f(q)) = (\varphi(q), \psi(f(q))) = (\varphi(q), (\psi \circ f)(q)) = (\varphi(q), (\psi \circ f \circ \varphi^{-1})(\varphi(q)))$. Thus, for every $q \in U$,
$((\varphi \times \psi) \circ F \circ \varphi^{-1})(\varphi(q)) = (\varphi(q), (\psi \circ f \circ \varphi^{-1})(\varphi(q)))$. Now, since $f : G \to N$ is a smooth map, $\psi \circ f \circ \varphi^{-1}$ is smooth, and hence, $(\varphi \times \psi) \circ F \circ \varphi^{-1}$ is smooth.

For 4: We have to show that $F : G \to M \times N$ is an immersion, that is, for every $q \in G$, the linear map $dF_q : T_q M \to T_{F(q)}(M \times N)$ is 1–1.

For this purpose, let us fix any $q \in G$. We have to show that the linear map $dF_q : T_qM \to T_{F(q)}(M \times N)$ is 1–1. Since $q \in G(\subset M)$, and $M$ is an $m$-dimensional smooth manifold, there exists an admissible coordinate chart $(U, \varphi)$ of $M$ such that $q \in U$. Since $q \in U, f(q) \in N$. Since $f(q) \in N$, and $N$ is an $n$-dimensional smooth manifold, there exists an admissible coordinate chart $(V, \psi)$ of $N$ such that $f(q) \in V$. For every $r \in U$,

$$\left( \left.\frac{\partial}{\partial x^1}\right|_r, \left.\frac{\partial}{\partial x^2}\right|_r, \cdots, \left.\frac{\partial}{\partial x^m}\right|_r \right)$$

is the coordinate basis of $T_rM$ corresponding to $(U, \varphi)$, where

$$\left.\frac{\partial}{\partial x^i}\right|_r \equiv \left( d(\varphi^{-1})_{\varphi(r)} \right) \left( \left.\frac{\partial}{\partial x^i}\right|_{\varphi(r)} \right).$$

Here, $(U \times V, (\varphi \times \psi))$ is an admissible coordinate chart of $M \times N$ such that $F(q) = (q, f(q)) \in U \times V$. For every $(s, t) \in U \times V$,

$$\left( \left.\frac{\partial}{\partial y^1}\right|_{(s,t)}, \left.\frac{\partial}{\partial y^2}\right|_{(s,t)}, \cdots, \left.\frac{\partial}{\partial y^m}\right|_{(s,t)}, \left.\frac{\partial}{\partial y^{m+1}}\right|_{(s,t)}, \cdots, \left.\frac{\partial}{\partial y^{m+n}}\right|_{(s,t)} \right)$$

is the coordinate basis of $T_{(s,t)}(M \times N)$ corresponding to $(U \times V, (\varphi \times \psi))$, where

$$\left.\frac{\partial}{\partial y^i}\right|_{(s,t)} \equiv \left( d((\varphi \times \psi)^{-1})_{(\varphi \times \psi)(s,t)} \right) \left( \left.\frac{\partial}{\partial y^i}\right|_{(\varphi \times \psi)(s,t)} \right).$$

Observe that for every $u \in U$,

$$((\varphi \times \psi) \circ F \circ \varphi^{-1})(\varphi(u)) = ((\varphi \times \psi) \circ F)(u) = (\varphi \times \psi)(F(u))$$
$$= (\varphi \times \psi)(u, F(u)) = (\varphi(u), \psi(F(u)))$$
$$= (\varphi(u), (\psi \circ F \circ \varphi^{-1})(\varphi(u))).$$

Hence, the matrix representation of linear map $dF_q$ is the following $(m + n) \times m$ matrix:

$$\begin{bmatrix}
1 & 0 & & 0 \\
0 & 1 & & 0 \\
\vdots & \vdots & & \vdots \\
0 & 0 & & 1 \\
\frac{\partial(\pi_1 \circ (\psi \circ F \circ \varphi^{-1}))}{\partial x^1}(\varphi(q)) & \frac{\partial(\pi_1 \circ (\psi \circ F \circ \varphi^{-1}))}{\partial x^2}(\varphi(q)) & \cdots & \frac{\partial(\pi_1 \circ (\psi \circ F \circ \varphi^{-1}))}{\partial x^m}(\varphi(q)) \\
\frac{\partial(\pi_2 \circ (\psi \circ F \circ \varphi^{-1}))}{\partial x^1}(\varphi(q)) & \frac{\partial(\pi_2 \circ (\psi \circ F \circ \varphi^{-1}))}{\partial x^2}(\varphi(q)) & & \frac{\partial(\pi_2 \circ (\psi \circ F \circ \varphi^{-1}))}{\partial x^m}(\varphi(q)) \\
\vdots & \vdots & & \vdots \\
\frac{\partial(\pi_n \circ (\psi \circ F \circ \varphi^{-1}))}{\partial x^1}(\varphi(q)) & \frac{\partial(\pi_n \circ (\psi \circ F \circ \varphi^{-1}))}{\partial x^2}(\varphi(q)) & & \frac{\partial(\pi_n \circ (\psi \circ F \circ \varphi^{-1}))}{\partial x^m}(\varphi(q))
\end{bmatrix}.$$

Clearly,

$$\text{rank}\begin{bmatrix} 1 & 0 & & 0 \\ 0 & 1 & & 0 \\ \vdots & \vdots & & \vdots \\ 0 & 0 & & 1 \\ \dfrac{\partial(\pi_1 \circ (\psi \circ F \circ \varphi^{-1}))}{\partial x^1}(\varphi(q)) & \dfrac{\partial(\pi_1 \circ (\psi \circ F \circ \varphi^{-1}))}{\partial x^2}(\varphi(q)) & \cdots & \dfrac{\partial(\pi_1 \circ (\psi \circ F \circ \varphi^{-1}))}{\partial x^m}(\varphi(q)) \\ \dfrac{\partial(\pi_2 \circ (\psi \circ F \circ \varphi^{-1}))}{\partial x^1}(\varphi(q)) & \dfrac{\partial(\pi_2 \circ (\psi \circ F \circ \varphi^{-1}))}{\partial x^2}(\varphi(q)) & & \dfrac{\partial(\pi_2 \circ (\psi \circ F \circ \varphi^{-1}))}{\partial x^m}(\varphi(q)) \\ \vdots & \vdots & & \vdots \\ \dfrac{\partial(\pi_n \circ (\psi \circ F \circ \varphi^{-1}))}{\partial x^1}(\varphi(q)) & \dfrac{\partial(\pi_n \circ (\psi \circ F \circ \varphi^{-1}))}{\partial x^2}(\varphi(q)) & & \dfrac{\partial(\pi_n \circ (\psi \circ F \circ \varphi^{-1}))}{\partial x^m}(\varphi(q)) \end{bmatrix} = m,$$

Hence, $dF_q : T_q M \to T_{F(q)}(M \times N)$ is 1–1. Thus, $F$ is a smooth immersion.

Thus, we have shown that $F : G \to M \times N$ is a smooth embedding. Since $F : G \to M \times N$ is a smooth embedding, by Note 5.90, $F(G)(= \Gamma(f))$ is an embedded submanifold of $M \times N$ with codimension $(m + n) - m(= n)$, and $F(N)$ is diffeomorphic to $M$.

Thus, we have shown that $\Gamma(f)$ is an embedded submanifold of $M \times N$ with codimension $n$, and $\Gamma(f)$ is diffeomorphic to $G$.

**Definition** Let $M$ be an $m$-dimensional smooth manifold. Let $S$ be a nonempty subset of $M$. Let $S$ be endowed with the subspace topology of $M$. Let $k$ be a nonnegative integer. Let $S$ be an embedded submanifold of $M$ with codimension $k$. If the inclusion map $\imath : S \to M$ is a proper map, then we say that $S$ is *properly embedded*.

**Note 5.92** Let $M$ be an $m$-dimensional smooth manifold. Let $S$ be a nonempty subset of $M$. Let $S$ be properly embedded. We shall try to show that $S$ is a closed subset of $M$.

Since $M$ is an $m$-dimensional smooth manifold, $M$ is a Hausdorff locally compact space. Since $S$ is a nonempty subset of $M$, and the topology of $M$ is Hausdorff, the subspace topology of $S$ is Hausdorff. Since $S$ is properly embedded, $\imath : S \to M$ is a proper map. Since $S$ is a Hausdorff topological space, $M$ is a Hausdorff locally compact space, $\imath : S \to M$ is a continuous mapping, and $\imath$ is a proper map, by Note 5.67, $\imath : S \to M$ is a closed mapping, and hence, $\imath(S)(= S)$ is closed in $M$. Thus, $S$ is closed in $M$.

**Note 5.93** Let $M$ be an $m$-dimensional smooth manifold. Let $S$ be a nonempty closed subset of $M$. Let $S$ be an embedded submanifold of $M$. We shall try to show that $S$ is properly embedded, that is, the inclusion map $\imath : S \to M$ is a proper map, that is, for every compact subset $C$ of $M$, $\imath^{-1}(C)(= C \cap S)$ is compact in $S$.

For this purpose, let us take any compact subset $C$ of $M$. We have to show that $C \cap S$ is compact in $S$. Let $\{G_i\}_{i \in I}$ be any open cover of $C \cap S$. Here each $G_i$ is open in $S$, there exists an open subset $H_i$ of $M$ such that $G_i = H_i \cap S$. Since for each $i \in I$, $G_i = H_i \cap S$, and $S$ is a closed subset of $M$, $\{H_i\}_{i \in I} \cup \{S^c\}$ is an open cover of $C$ in $M$.

Since $\{H_i\}_{i \in I} \cup \{S^c\}$ is an open cover of $C$ in $M$, and $C$ is a compact subset of $M$, there exist $i_1, \ldots, i_k \in I$ such that $C \subset H_{i_1} \cup \cdots \cup H_{i_k} \cup S^c$, and hence, $C \cap S \subset (H_{i_1} \cup \cdots \cup H_{i_k} \cup S^c) \cap S = (H_{i_1} \cap S) \cup \cdots \cup (H_{i_k} \cap S) \cup (S^c \cap S) = (H_{i_1} \cap S) \cup \cdots \cup (H_{i_k} \cap S) = G_{i_1} \cup \cdots \cup G_{i_k}$. This shows that $C \cap S$ is compact in $S$.

Thus, we have shown that $S$ is properly embedded.

**Note 5.94** Let $M$ be an $m$-dimensional smooth manifold. Let $S$ be a nonempty compact subset of $M$. Let $S$ be an embedded submanifold of $M$. We shall try to show that $S$ is properly embedded.

Since $M$ is an $m$-dimensional smooth manifold, the topology of $M$ is Hausdorff. Since the topology of $M$ is Hausdorff, and $S$ is a nonempty compact subset of $M$, $S$ is closed in $M$. Since $S$ is closed in $M$, by the Note 5.92, $S$ is properly embedded.

**Note 5.95** Let $M$ be an $m$-dimensional smooth manifold with smooth structure $\mathcal{A}$, and $N$ be an $n$-dimensional smooth manifold with smooth structure $\mathcal{B}$. Here, the collection

$$\{(U \times V, (\varphi \times \psi)) : (U, \varphi) \in \mathcal{A}, (V, \psi) \in \mathcal{B}\}$$

is an atlas of the $(m + n)$-dimensional product manifold $M \times N$.

Let $f : M \to N$ be a smooth map. Here, the graph of $f$ is given by $\Gamma(f) \equiv \{(x, f(x)) : x \in M\}$. By Note 5.91, $\Gamma(f)$ is an embedded submanifold of $M \times N$ with codimension $n$, and $\Gamma(f)$ is diffeomorphic to $M$. Now, we want to show that $\Gamma(f)$ is a closed subset of $M \times N$, that is $(\Gamma(f))^c$ is open in $M \times N$.

For this purpose, let us take any point $(x, y) \in (\Gamma(f))^c$. Since $(x, y) \in (\Gamma(f))^c$, $(x, y) \notin \Gamma(f) = \{(x, f(x)) : x \in M\}$. It follows that $y \neq f(x)$. Here, $y, f(x)$ are distinct points of $N$, and in the topology of $N$, there exist an open neighborhood $V_1$ of $y$, and an open neighborhood $V$ of $f(x)$ such that $V_1$ and $V$ are disjoint. Since $f : M \to N$ is a smooth map, $f : M \to N$ is a continuous map. Since $f : M \to N$ is a continuous map, and $V$ is an open neighborhood of $f(x)$, there exists an open neighborhood $U$ of $x$ such that $f(U) \subset V$. Here, $U$ is an open neighborhood of $x$, and $V_1$ is an open neighborhood of $y$, $U \times V_1$ is an open neighborhood of $(x, y)$ in $M \times N$. Clearly, $U \times V_1$ is disjoint with $\Gamma(f)$. (Reason: If not, otherwise, let $(a, b) \in U \times V_1$, and $(a, b) \in \Gamma(f)$. Since $(a, b) \in \Gamma(f)$, $f(a) = b$. Since $(a, b) \in U \times V_1$, $a \in U$, and $b \in V_1$. Since $f(a) = b \in V_1$, $f(a) \in V_1$. Since $a \in U$, $f(a) \in f(U) \subset V$, and hence, $f(a) \in V$. Since $f(a) \in V_1$, and $f(a) \in V$, $V_1$ and $V$ are not disjoint, a contradiction.) Since $U \times V_1$ is disjoint with $\Gamma(f)$, the open neighborhood $U \times V_1$ of $(x, y)$ contained in $(\Gamma(f))^c$. Thus, $(x, y)$ is an interior point of $(\Gamma(f))^c$. Hence, $(\Gamma(f))^c$ is open in $M \times N$.

Thus, $\Gamma(f)$ is a closed subset of $M \times N$. Since $\Gamma(f)$ is an embedded submanifold of $M \times N$, and $\Gamma(f)$ is a closed subset of $M \times N$, by Note 5.93, $\Gamma(f)$ is properly embedded.

**Conclusion**: Let $M, N$ be smooth manifolds. Let $f : M \to N$ be a smooth map. Then, $\Gamma(f)$ is properly embedded in $M \times N$.

## 5.9 Regular Level Sets

**Definition** Let $M$ be a five-dimensional smooth manifold, and let $(U, \varphi)$ be an admissible coordinate chart of $M$. Any set among the following $10\left(= \binom{5}{2}\right)$ types:

$$\varphi^{-1}(\{(x_1, x_2, x_3, x_4, x_5) \in \varphi(U) : x_3 = \text{const}, x_4 = \text{const}, x_5 = \text{const}\}),$$
$$\varphi^{-1}(\{(x_1, x_2, x_3, x_4, x_5) \in \varphi(U) : x_2 = \text{const}, x_4 = \text{const}, x_5 = \text{const}\}),$$
$$\varphi^{-1}(\{(x_1, x_2, x_3, x_4, x_5) \in \varphi(U) : x_2 = \text{const}, x_3 = \text{const}, x_5 = \text{const}\}), \text{ etc.}$$

is called a two-*slice of U*. Observe that

$$\varphi^{-1}(\{(x_1, x_2, x_3, x_4, x_5) \in \varphi(U) : x_3 = \text{const}, x_4 = \text{const}, x_5 = \text{const}\})$$
$$\neq \varphi^{-1}(\{(x_1, x_2, x_3, x_4, x_5) \in \varphi(U) : x_2 = \text{const}, x_4 = \text{const}, x_5 = \text{const}\}).$$

(Reason: Since $\varphi(U)$ is a nonempty open subset of $\mathbb{R}^5$, $\{(x_1, x_2, x_3, x_4, x_5) \in \varphi$ $(U) : x_3 = \text{const}, x_4 = \text{const}, x_5 = \text{const}\} \neq \{(x_1, x_2, x_3, x_4, x_5) \in \varphi(U) : x_2 = \text{const}, x_4 = \text{const}, x_5 = \text{const}\}$. Now, since $\varphi$ is 1–1, $\varphi^{-1}(\{(x_1, x_2, x_3, x_4, x_5) \in \varphi(U) : x_3 = \text{const}, x_4 = \text{const}, x_5 = \text{const}\}) \neq \varphi^{-1}(\{(x_1, x_2, x_3, x_4, x_5) \in \varphi(U) : x_2 = \text{const}, x_4 = \text{const}, x_5 = \text{const}\})$.)

Also, if $(c_3, c_4, c_5) \neq (d_3, d_4, d_5)$, then $\{(x_1, x_2, x_3, x_4, x_5) \in \varphi(U) : x_3 = c_3, x_4 = c_4, x_5 = c_5\} \neq \{(x_1, x_2, x_3, x_4, x_5) \in \varphi(U) : x_3 = d_3, x_4 = d_4, x_5 = d_5\}$. Now, since $\varphi$ is 1–1, $\varphi^{-1}(\{(x_1, x_2, x_3, x_4, x_5) \in \varphi(U) : x_3 = c_3, x_4 = c_4, x_5 = c_5\}) \neq \varphi^{-1}(\{(x_1, x_2, x_3, x_4, x_5) \in \varphi(U) : x_3 = d_3, x_4 = d_4, x_5 = d_5\})$, etc.

**Definition** Let $M$ be a five-dimensional smooth manifold, and let $S$ be a nonempty subset of $M$. If for every $p$ in $S$, there exists an admissible coordinate chart $(U, \varphi)$ of $M$ such that $p \in U$, and $U \cap S(\subset U)$ is equal to a two-slice of $U$, then we say that $S$ *satisfies* local *two-slice condition* (see Fig. 5.2).

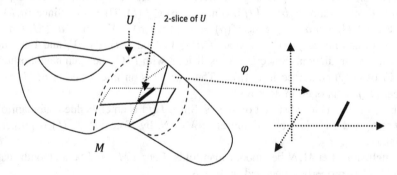

**Fig. 5.2** Admissible coordinate chart

**Note 5.96** Let $M$ be a five-dimensional smooth manifold. Let $S$ be a nonempty subset of $M$. Suppose that $S$ satisfies local two-slice condition. Let $S$ has the subspace topology inherited from $M$. We shall try to show that $S$ is a two-dimensional topological manifold.

Since $M$ is a smooth manifold, its topology is Hausdorff. Since the topology of $M$ is Hausdorff, and $S$ has the subspace topology inherited from $M$, the topology of $S$ is Hausdorff. Since $M$ is a smooth manifold, its topology is second countable. Since the topology of $M$ is second countable, and $S$ has the subspace topology inherited from $M$, the topology of $S$ is second countable.

Next, let us take any $p \in S$. Since $S$ satisfies local two-slice condition, there exists an admissible coordinate chart $(U, \varphi)$ of $M$ satisfying $p \in U$, and $U \cap S$ is a two-slice of $U$. Since $(U, \varphi)$ is an admissible coordinate chart of $M$, $U$ is open in $M$, and hence, $U \cap S$ is open in $S$. Since $(U, \varphi)$ is an admissible coordinate chart of five-dimensional smooth manifold $M$, $\varphi(U)$ is an open subset of $\mathbb{R}^5$. Since $U \cap S$ is a two-slice of $U$, for definiteness, let $U \cap S = \varphi^{-1}(\{(x_1, x_2, x_3, x_4, x_5) \in \varphi(U) : x_3 = a_3, x_4 = a_4, x_5 = a_5\})$, where $a_3, a_4, a_5$ are constants. It follows that $\varphi(U \cap S) = \{(x_1, x_2, x_3, x_4, x_5) \in \varphi(U) : x_3 = a_3, x_4 = a_4, x_5 = a_5\}$.

Let us define $\pi : \mathbb{R}^5 \to \mathbb{R}^2 \times \{a_3\} \times \{a_4\} \times \{a_5\}$ as follows: For every $(x_1, x_2, x_3, x_4, x_5) \in \mathbb{R}^5, \pi(x_1, x_2, x_3, x_4, x_5) \equiv (x_1, x_2, a_3, a_4, a_5)$. Clearly, $\pi$ is an open mapping. Since $\varphi(U)$ is an open subset of $\mathbb{R}^5$, and $\pi : \mathbb{R}^5 \to \mathbb{R}^2 \times \{a_3\} \times \{a_4\} \times \{a_5\}$ is open, $\pi(\varphi(U))(= \{\pi(x_1, x_2, x_3, x_4, x_5) : (x_1, x_2, x_3, x_4, x_5) \in \varphi(U)\} = \{(x_1, x_2, x_3, x_4, x_5) \in \varphi(U) : x_3 = a_3, x_4 = a_4, x_5 = a_5\} = \varphi(U \cap S))$ is open in $\mathbb{R}^2 \times \{a_3\} \times \{a_4\} \times \{a_5\}$, and hence, $\varphi(U \cap S)$ is open in $\mathbb{R}^2 \times \{a_3\} \times \{a_4\} \times \{a_5\}$. Let us define a function $j : \mathbb{R}^2 \times \{a_3\} \times \{a_4\} \times \{a_5\} \to \mathbb{R}^2$ as follows: For every $(x_1, x_2, a_3, a_4, a_5) \in \mathbb{R}^2 \times \{a_3\} \times \{a_4\} \times \{a_5\}, j(x_1, x_2, a_3, a_4, a_5) \equiv (x_1, x_2)$. Clearly, $j$ is a diffeomorphism. Since $j : \mathbb{R}^2 \times \{a_3\} \times \{a_4\} \times \{a_5\} \to \mathbb{R}^2$ is a diffeomorphism, $j : \mathbb{R}^2 \times \{a_3\} \times \{a_4\} \times \{a_5\} \to \mathbb{R}^2$ is a homeomorphism. Since $j : \mathbb{R}^2 \times \{a_3\} \times \{a_4\} \times \{a_5\} \to \mathbb{R}^2$ is a homeomorphism, and $\varphi(U \cap S)$ is open in $\mathbb{R}^2 \times \{a_3\} \times \{a_4\} \times \{a_5\}, j(\varphi(U \cap S))(= (j \circ \varphi)(U \cap S))$ is open in $\mathbb{R}^2$, and hence, $(j \circ \varphi)(U \cap S)$ is open in $\mathbb{R}^2$. Clearly, $(j \circ \varphi)(U \cap S) = \{(x_1, x_2) : (x_1, x_2, a_3, a_4, a_5) \in \varphi(U)\}$. (Reason: $(j \circ \varphi)(U \cap S) = j(\varphi(U \cap S)) = j(\{(x_1, x_2, x_3, x_4, x_5) \in \varphi(U) : x_3 = a_3, x_4 = a_4, x_5 = a_5\}) = j(\{(x_1, x_2, a_3, a_4, a_5) : (x_1, x_2, a_3, a_4, a_5) \in \varphi(U)\} = \{j(x_1, x_2, a_3, a_4, a_5) : (x_1, x_2, a_3, a_4, a_5) \in \varphi(U)\}) = \{(x_1, x_2) : (x_1, x_2, a_3, a_4, a_5) \in \varphi(U)\}$.)

Now, we shall try to show that $j \circ (\varphi|_{U \cap S})$ is a homeomorphism from $U \cap S$ onto $(j \circ \varphi)(U \cap S)$, that is,

1. $j \circ (\varphi|_{U \cap S})$ is 1–1
2. $(j \circ (\varphi|_{U \cap S}))(U \cap S) = (j \circ \varphi)(U \cap S)$,
3. $(j \circ (\varphi|_{U \cap S})) : U \cap S \to (j \circ \varphi)(U \cap S)$ is continuous,
4. $(j \circ (\varphi|_{U \cap S})) : U \cap S \to (j \circ \varphi)(U \cap S)$ is an open mapping.

For 1: Since $j$ is a homeomorphism, $j$ is 1–1. Since $(U, \varphi)$ is an admissible coordinate chart of $M$, $\varphi : U \to \varphi(U)$ is 1–1, and hence, $\varphi|_{U \cap S}$ is 1–1. Since $\varphi|_{U \cap S}$ is 1–1, and $j$ is 1–1, their composite $j \circ (\varphi|_{U \cap S})$ is 1–1.

For 2: Here,

$$\text{LHS} = \left( j \circ (\varphi|_{U \cap S}) \right)(U \cap S) = j\left( (\varphi|_{U \cap S})(U \cap S) \right)$$
$$= j(\varphi(U \cap S)) = (j \circ \varphi)(U \cap S) = \text{RHS}.$$

For 3: Since $j$ is a homeomorphism, $j$ is continuous. Since $(U, \varphi)$ is an admissible coordinate chart of $M$, $\varphi : U \to \varphi(U)$ is continuous, and hence, $\varphi|_{U \cap S}$ is continuous. Since $\varphi|_{U \cap S}$ is continuous, and $j$ is continuous, their composite $(j \circ (\varphi|_{U \cap S}))$ is continuous.

For 4: Since $j$ is a homeomorphism, $j$ is open. Since $(U, \varphi)$ is an admissible coordinate chart of $M$, $\varphi : U \to \varphi(U)$ is open, and hence, $\varphi|_{U \cap S}$ is open. Since $\varphi|_{U \cap S}$ is open, and $j$ is open, their composite $(j \circ (\varphi|_{U \cap S}))$ is open.

Thus, the ordered pair $(U \cap S, j \circ (\varphi|_{U \cap S}))$ is a coordinate chart of $S$. Also, $p \in U \cap S$. Hence, $S$ is a two-dimensional topological manifold. Put

$$\mathcal{A} \equiv \left\{ (U \cap S, j \circ (\varphi|_{U \cap S})) : (U, \varphi) \text{ is an admissible coordinate chart of } M \right\}.$$

We want to show that $\mathcal{A}$ is an atlas for $S$.

For this purpose, let us take any $(U \cap S, (j_1 \circ \varphi)|_{U \cap S}), (V \cap S, (j_2 \circ \psi)|_{V \cap S}) \in \mathcal{A}$, where $(U, \varphi), (V, \psi)$ are admissible coordinate charts of $M$, and $(U \cap S) \cap (V \cap S) \neq \emptyset$. We have to prove that

$$\left( j_2 \circ (\psi|_{V \cap S}) \right) \circ \left( j_1 \circ (\varphi|_{U \cap S}) \right)^{-1} : \left( j_1 \circ (\varphi|_{U \cap S}) \right)((U \cap S) \cap (V \cap S))$$
$$\to \left( j_2 \circ (\psi|_{V \cap S}) \right)((U \cap S) \cap (V \cap S))$$

is smooth, that is,

$$\left( j_2 \circ (\psi|_{V \cap S}) \right) \circ \left( j_1 \circ (\varphi|_{U \cap S}) \right)^{-1} : \left( j_1 \circ (\varphi|_{U \cap S}) \right)((U \cap V) \cap S)$$
$$\to \left( j_2 \circ (\psi|_{V \cap S}) \right)((U \cap V) \cap S)$$

is smooth, that is,

$$\left( j_2 \circ (\psi|_{V \cap S}) \right) \circ \left( j_1 \circ (\varphi|_{U \cap S}) \right)^{-1} : (j_1 \circ \varphi)((U \cap V) \cap S)$$
$$\to (j_2 \circ \psi)((U \cap V) \cap S)$$

is smooth, that is,

$$\left( j_2 \circ (\psi|_{V \cap S}) \right) \circ \left( (\varphi|_{U \cap S})^{-1} \circ (j_1)^{-1} \right) : (j_1 \circ \varphi)((U \cap V) \cap S)$$
$$\to (j_2 \circ \psi)((U \cap V) \cap S)$$

is smooth, that is,

$$\left(j_2 \circ \left((\psi|_{V\cap S}) \circ (\varphi|_{U\cap S})^{-1}\right) \circ (j_1)^{-1}\right) : (j_1 \circ \varphi)((U \cap V) \cap S)$$
$$\to (j_2 \circ \psi)((U \cap V) \cap S)$$

is smooth, that is,

$$\left(j_2 \circ \left((\psi \circ \varphi^{-1})|_{\varphi((U\cap V)\cap S)}\right) \circ (j_1)^{-1}\right) : (j_1 \circ \varphi)((U \cap V) \cap S)$$
$$\to (j_2 \circ \psi)((U \cap V) \cap S)$$

is smooth. Here, $\emptyset \neq (U\cap S)\cap(V\cap S) \subset U \cap V$, so $U\cap V \neq \emptyset$. Since $U\cap V \neq \emptyset$, and $(U,\varphi),(V,\psi)$ are admissible coordinate charts of $M$, $\psi \circ \varphi^{-1} : \varphi(U\cap V) \to \psi(U\cap V)$ is smooth, and hence, the restriction $(\psi \circ \varphi^{-1})|_{\varphi((U\cap V)\cap S)}$ is smooth. Since $j_2, (\psi \circ \varphi^{-1})|_{\varphi((U\cap V)\cap S)}, (j_1)^{-1}$ are smooth, their composite

$$\left(j_2 \circ \left((\psi \circ \varphi^{-1})|_{\varphi((U\cap V)\cap S)}\right) \circ (j_1)^{-1}\right) : (j_1 \circ \varphi)((U \cap V) \cap S)$$
$$\to (j_2 \circ \psi)((U \cap V) \cap S)$$

is smooth. Thus, we have shown that $\mathcal{A}$ is an atlas for $S$. Let $\iota : S \to M$ be the inclusion map of $S$.

We want to show that $\iota : S \to M$ is a smooth embedding, that is,

(a) $\iota : S \to M$ is smooth,
(b) $\iota : S \to M$ is a topological embedding, and
(c) $\iota : S \to M$ is a smooth immersion.

For a: Let us take any $p \in S$. So there exists $(U \cap S, j \circ (\varphi|_{U\cap S})) \in \mathcal{A}$, where $(U,\varphi)$ is an admissible coordinate chart of $M$, and $p \in U\cap S$. Since $p = \iota(p) \in M$, there exists an admissible coordinate chart $(V,\psi)$ of $M$ satisfying $p = \iota(p) \in V$. We have to show that

$$\psi \circ \iota \circ (j \circ (\varphi|_{U\cap S}))^{-1} \left(= \psi \circ (j \circ (\varphi|_{U\cap S}))^{-1} = \psi \circ \left((\varphi|_{U\cap S})^{-1} \circ j^{-1}\right)\right.$$
$$= \left(\psi \circ (\varphi|_{U\cap S})^{-1}\right) \circ j^{-1} = \left(\psi \circ (\varphi|_{U\cap S})^{-1}\right) \circ j^{-1}$$
$$\left.= (\psi \circ \varphi^{-1})|_{\varphi((U\cap V)\cap S)} \circ j^{-1}\right)$$

is smooth, that is, $(\psi \circ \varphi^{-1})|_{\varphi((U\cap V)\cap S)} \circ j^{-1}$ is smooth. Here, $U\cap V \neq \emptyset$. Since $U\cap V \neq \emptyset$, and $(U,\varphi),(V,\psi)$ are admissible coordinate charts of $M$, $\psi \circ \varphi^{-1} :$

$\varphi(U \cap V) \to \psi(U \cap V)$ is smooth, and hence, its restriction $(\psi \circ \varphi^{-1})\big|_{\varphi((U\cap V)\cap S)}$ is smooth. Since $(\psi \circ \varphi^{-1})\big|_{\varphi((U\cap V)\cap S)}$ is smooth, and $j^{-1}$ is smooth, their composite $(\psi \circ \varphi^{-1})\big|_{\varphi((U\cap V)\cap S)} \circ j^{-1}$ is smooth.

For b: Since $S$ has the subspace topology inherited from $M$, $\iota : S \to M$ is a topological embedding.

For c: Let us take any $p \in S$. We have to show that the linear map $d\iota_p : T_pS \to T_{\iota(p)}M$ is 1–1, that is, the linear map $d\iota_p : T_pS \to T_pM$ is 1–1.

Since $p \in S$, there exists $(U \cap S, j \circ (\varphi|_{U\cap S})) \in \mathcal{A}$, where $(U, \varphi)$ is an admissible coordinate chart of $M$, and $p \in U \cap S$. Here, $(U, \varphi)$ is an admissible coordinate chart of $M$ satisfying $\iota(p) \in U$. For every $q \in U \cap S$,

$$\left( \frac{\partial}{\partial x^1}\bigg|_q , \frac{\partial}{\partial x^2}\bigg|_q \right)$$

is the coordinate basis of $T_qS$ corresponding to $(U \cap S, j \circ (\varphi|_{U\cap S}))$, where

$$\frac{\partial}{\partial x^i}\bigg|_q \equiv \left( d\big( (j \circ (\varphi|_{U\cap S}))^{-1} \big)_{(j\circ(\varphi|_{U\cap S}))(q)} \right) \left( \frac{\partial}{\partial x^i}\bigg|_{(j\circ(\varphi|_{U\cap S}))(q)} \right).$$

For every $r \in U$,

$$\left( \frac{\partial}{\partial y^1}\bigg|_r , \frac{\partial}{\partial y^2}\bigg|_r , \frac{\partial}{\partial y^3}\bigg|_r , \frac{\partial}{\partial y^4}\bigg|_r , \frac{\partial}{\partial y^5}\bigg|_r \right)$$

is the coordinate basis of $T_rM$ corresponding to $(U, \varphi)$, where

$$\frac{\partial}{\partial y^i}\bigg|_r \equiv \left( d(\varphi^{-1})_{\varphi(r)} \right) \left( \frac{\partial}{\partial y^i}\bigg|_{\varphi(r)} \right).$$

Observe that for every $q \in U \cap S$,

$$\left( \varphi \circ \iota \circ (j \circ (\varphi|_{U\cap S}))^{-1} \right) ((j \circ (\varphi|_{U\cap S}))(q)) = \varphi(\iota(q)) = \varphi(q)$$
$$= (\varphi|_{U\cap S})(q) = j^{-1}((j \circ (\varphi|_{U\cap S}))(q)).$$

Also clearly, $j^{-1} : \mathbb{R}^2 \to \mathbb{R}^2 \times \{a_3\} \times \{a_4\} \times \{a_5\}$ is such that for every $(x^1, x^2)$ in $\mathbb{R}^2$, $j^{-1}(x^1, x^2) = (x_1, x_2, a_3, a_4, a_5)$. Hence, for every $q \in U \cap S$, the matrix representation of linear map $d\iota_q$ is the following $5 \times 2$ matrix:

$$\begin{bmatrix} 1 & 0 \\ 0 & 1 \\ 0 & 0 \\ 0 & 0 \\ 0 & 0 \end{bmatrix}.$$

Since the

$$\text{rank} \begin{bmatrix} 1 & 0 \\ 0 & 1 \\ 0 & 0 \\ 0 & 0 \\ 0 & 0 \end{bmatrix}$$

is 2, the linear map $d\iota_p : T_pS \to T_pM$ is 1–1. Thus, we have shown that $\iota : S \to M$ is a smooth embedding. Hence, $S$ is an embedded submanifold of $M$.

Since for every $q \in U \cap S$, the matrix representation of linear map $d\iota_q$ is the matrix

$$\begin{bmatrix} 1 & 0 \\ 0 & 1 \\ 0 & 0 \\ 0 & 0 \\ 0 & 0 \end{bmatrix},$$

$$(d\iota_q)\left(\left.\frac{\partial}{\partial x^1}\right|_q\right) = 1\left(\left.\frac{\partial}{\partial y^1}\right|_{\iota(q)}\right) + 0\left(\left.\frac{\partial}{\partial y^2}\right|_{\iota(q)}\right) + 0\left(\left.\frac{\partial}{\partial y^3}\right|_{\iota(q)}\right)$$

$$+ 0\left(\left.\frac{\partial}{\partial y^4}\right|_{\iota(q)}\right) + 0\left(\left.\frac{\partial}{\partial y^5}\right|_{\iota(q)}\right) = \left.\frac{\partial}{\partial y^1}\right|_{\iota(q)} = \left.\frac{\partial}{\partial y^1}\right|_q.$$

Thus, for every $q \in U \cap S$,

$$(d\iota_q)\left(\left.\frac{\partial}{\partial x^1}\right|_q\right) = \left.\frac{\partial}{\partial y^1}\right|_q.$$

Similarly, for every $q \in U \cap S$,

$$(d\iota_q)\left(\left.\frac{\partial}{\partial x^2}\right|_q\right) = \left.\frac{\partial}{\partial y^2}\right|_q.$$

Now, since

$$\left( (d\iota_q)\left( \frac{\partial}{\partial x^1}\bigg|_q \right), (d\iota_q)\left( \frac{\partial}{\partial x^2}\bigg|_q \right) \right)$$

is a basis of the real linear subspace $(d\iota_q)(T_qS)$ of $T_qM$,

$$\left( \frac{\partial}{\partial y^1}\bigg|_q, \frac{\partial}{\partial y^2}\bigg|_q \right)$$

is a basis of $(d\iota_q)(T_qS)$, while

$$\left( \frac{\partial}{\partial y^1}\bigg|_q, \frac{\partial}{\partial y^2}\bigg|_q, \frac{\partial}{\partial y^3}\bigg|_q, \frac{\partial}{\partial y^4}\bigg|_q, \frac{\partial}{\partial y^5}\bigg|_q \right)$$

is the coordinate basis of $T_qM$. Conclusion: For every $q \in U \cap S$, if

$$\left( \frac{\partial}{\partial y^1}\bigg|_q, \frac{\partial}{\partial y^2}\bigg|_q, \frac{\partial}{\partial y^3}\bigg|_q, \frac{\partial}{\partial y^4}\bigg|_q, \frac{\partial}{\partial y^5}\bigg|_q \right)$$

is the coordinate basis of $T_qM$, then

$$\left( \frac{\partial}{\partial y^1}\bigg|_q, \frac{\partial}{\partial y^2}\bigg|_q \right)$$

is a basis of $(d\iota_q)(T_qS)$.

**Note 5.97** As in Note 5.96, we can prove the following result:

Let $M$ be an $m$-dimensional smooth manifold. Let $S$ be a nonempty subset of $M$. Suppose that $S$ satisfies local $k$-slice condition. Let $S$ has the subspace topology inherited from $M$. Then, $S$ is a $k$-dimensional topological manifold. Also, there exists a smooth structure on $S$ such that $M$ becomes a $k$-dimensional embedded submanifold of $M$.

**Definition** Let $M$ be an $m$-dimensional smooth manifold, $N$ be an $n$-dimensional smooth manifold, and $\Phi : M \to N$ be a map. Any set of the form $\Phi^{-1}(c)$, where $c \in N$, is called a *level set of* $\Phi$.

**Note 5.98** Let $M$ be a three-dimensional smooth manifold, $N$ be a four-dimensional smooth manifold, and $\Phi : M \to N$ be a smooth map. Let 2 be the rank of $\Phi$. Let $c \in N$. We shall try to show that the level set $\Phi^{-1}(c)$ satisfies local two-slice condition.

By the constant rank Theorem 5.48, for every $p \in \Phi^{-1}(c)$ (that is, $\Phi(p) = c$) there exist admissible coordinate chart $(U, \varphi)$ in $M$ satisfying $p \in U$, and admissible coordinate chart $(V, \psi)$ in $N$ satisfying $\Phi(p) \in V$, such that $\Phi(U) \subset V$, and for every $(x_1, x_2, x_3)$ in $\varphi(U)$,

$$\left(\psi \circ \Phi \circ \varphi^{-1}\right)(x_1, x_2, x_3) = (x_1, x_2, 0, 0).$$

Now, we shall try to show that

$$\Phi^{-1}(c) \cap U = \varphi^{-1}(\{(x_1, x_2, x_3) \in \varphi(U) : x_1 = (\pi_1 \circ \psi)(c), x_2 = (\pi_2 \circ \psi)(c)\}),$$

that is,

$$\varphi\left(\Phi^{-1}(c) \cap U\right) = \{(x_1, x_2, x_3) \in \varphi(U) : x_1 = (\pi_1 \circ \psi)(c), x_2 = (\pi_2 \circ \psi)(c)\}.$$

Let $q \in \Phi^{-1}(c) \cap U$. It follows that $\Phi(q) = c$, and $q \in U$. Since $q \in U$, $\varphi(q) \in \varphi(U)$. Put $\varphi(q) \equiv (x_1, x_2, x_3)$. Hence, $(x_1, x_2, x_3) = \varphi(q) \in \varphi(U)$. Also,

$$
\begin{aligned}
(x_1, x_2, 0, 0) &= \left(\psi \circ \Phi \circ \varphi^{-1}\right)(x_1, x_2, x_3) \\
&= \left(\psi \circ \Phi \circ \varphi^{-1}\right)(\varphi(q)) = \psi(\Phi(q)) = \psi(c) \\
&= ((\pi_1 \circ \psi)(c), (\pi_2 \circ \psi)(c), (\pi_3 \circ \psi)(c), (\pi_4 \circ \psi)(c)).
\end{aligned}
$$

Hence, $x_1 = (\pi_1 \circ \psi)(c)$, and $x_2 = (\pi_2 \circ \psi)(c)$. Thus, $\Phi^{-1}(c) \cap U \subset \varphi^{-1}(\{(x_1, x_2, x_3) \in \varphi(U) : x_1 = (\pi_1 \circ \psi)(c), x_2 = (\pi_2 \circ \psi)(c)\})$. Next, let $q \in \varphi^{-1}(\{(x_1, x_2, x_3) \in \varphi(U) : x_1 = (\pi_1 \circ \psi)(c), x_2 = (\pi_2 \circ \psi)(c)\})$. We have to show that $q \in \Phi^{-1}(c) \cap U$, that is, $\Phi(q) = c$, and $q \in U$. Since $q \in \varphi^{-1}(\{(x_1, x_2, x_3) \in \varphi(U) : x_1 = (\pi_1 \circ \psi)(c), x_2 = (\pi_2 \circ \psi)(c)\}) \subset U$, $q \in U$.

Since $\varphi(q) \in \varphi(U)$, $((\pi_1 \circ \psi)(c), (\pi_2 \circ \psi)(c), (\pi_3 \circ \psi)(c), (\pi_4 \circ \psi)(c)) = \psi(c) = \psi(\Phi(q)) = (\psi \circ \Phi \circ \varphi^{-1})(\varphi(q)) = ((\pi_1 \circ \varphi)(q), (\pi_2 \circ \varphi)(q), 0, 0)$, and hence, $(\pi_3 \circ \psi)(c) = 0 = (\pi_4 \circ \psi)(c)$. Since $q \in \varphi^{-1}(\{(x_1, x_2, x_3) \in \varphi(U) : x_1 = (\pi_1 \circ \psi)(c), x_2 = (\pi_2 \circ \psi)(c)\})$, $\varphi(q) \in \{(x_1, x_2, x_3) \in \varphi(U) : x_1 = (\pi_1 \circ \psi)(c), x_2 = (\pi_2 \circ \psi)(c)\}$, and hence, $(\pi_1 \circ \varphi)(q) = (\pi_1 \circ \psi)(c)$, and $(\pi_2 \circ \varphi)(q) = (\pi_2 \circ \psi)(c)$. Since $(\pi_1 \circ \varphi)(q) = (\pi_1 \circ \psi)(c)$, $(\pi_2 \circ \varphi)(q) = (\pi_2 \circ \psi)(c)$, $(\pi_3 \circ \psi)(c) = 0 = (\pi_4 \circ \psi)(c)$, and $\psi(\Phi(q)) = ((\pi_1 \circ \varphi)(q), (\pi_2 \circ \varphi)(q), 0, 0)$, $\psi(\Phi(q)) = ((\pi_1 \circ \psi)(c), (\pi_2 \circ \psi)(c), 0, 0) = ((\pi_1 \circ \psi)(c), (\pi_2 \circ \psi)(c), (\pi_3 \circ \psi)(c), (\pi_4 \circ \psi)(c)) = \psi(c)$. Since $\psi(\Phi(q)) = \psi(c)$, $\Phi(q) = c$.

Thus, $\varphi(\Phi^{-1}(c) \cap U) = \{(x_1, x_2, x_3) \in \varphi(U) : x_1 = (\pi_1 \circ \psi)(c), x_2 = (\pi_2 \circ \psi)(c)\}$. Hence, the level set $\Phi^{-1}(c)$ satisfies the local 2-slice condition. Now, by Note 5.97, if $\Phi^{-1}(c)$ has the subspace topology inherited from $M$, then $\Phi^{-1}(c)$ is an embedded submanifold of $M$. Since $\Phi : M \to N$ is a smooth map, $\Phi : M \to N$ is continuous. Since $\Phi : M \to N$ is continuous, and $\{c\}$ is a closed subset of $N$, $\Phi^{-1}(c)$ is a closed subset of $M$. Since $\Phi^{-1}(c)$ is a closed subset of $M$, and $\Phi^{-1}(c)$ is an embedded submanifold of $M$, by Note 5.93, $\Phi^{-1}(c)$ is a properly embedded submanifold of $M$, and its codimension is 2.

**Note 5.99** The following result can be proved as in Note 5.98:

Let $M$ be an $m$-dimensional smooth manifold, $N$ be an $n$-dimensional smooth manifold, and $\Phi : M \to N$ be a smooth map. Let $r$ be the rank of $\Phi$. Let $c \in N$. Then, the level set $\Phi^{-1}(c)$ is a properly embedded submanifold of $M$, and its codimension is $r$.

**Note 5.100** Let $M$ be an $m$-dimensional smooth manifold, $N$ be an $n$-dimensional smooth manifold, and $\Phi : M \to N$ be a smooth submersion. Let $c \in N$. We shall try to show that the level set $\Phi^{-1}(c)$ is a properly embedded submanifold of $M$, and its codimension is $n$.

Since $\Phi : M \to N$ is a smooth submersion, $\Phi : M \to N$ is a smooth map, and rank $\Phi = n$. Since $\Phi : M \to N$ is a smooth map, and rank $\Phi = n$, by Note 5.99, the level set $\Phi^{-1}(c)$ is a properly embedded submanifold of $M$, and its codimension is $n$.

**Definition** Let $M$ be an $m$-dimensional smooth manifold, $N$ be an $n$-dimensional smooth manifold, and $\Phi : M \to N$ be a smooth map. Let $p \in M$.

If $d\Phi_p : T_pM \to T_{\Phi(p)}N$ is onto (that is, the rank of $\Phi$ at $p$ is $n$), then we say that $p$ is a *regular point of* $\Phi$. If $p$ is not a regular point of $\Phi$, then we say that $p$ is a *critical point of* $\Phi$.

**Note 5.101** Let $M$ be an $m$-dimensional smooth manifold, $N$ be an $n$-dimensional smooth manifold, and $\Phi : M \to N$ be a smooth map. Clearly, $\Phi$ is a smooth submersion if and only if every point of $M$ is a regular point of $\Phi$.

**Note 5.102** Let $M$ be an $m$-dimensional smooth manifold, $N$ be an $n$-dimensional smooth manifold, and $\Phi : M \to N$ be a smooth map. Let $m < n$. We shall try to show that every point of $M$ is a critical point of $\Phi$.

If not, otherwise, let there exists $p \in M$ such that $p$ is not a critical point of $\Phi$. Since $p$ is not a critical point of $\Phi$, $p$ is a regular point of $\Phi$, and hence, the linear map $d\Phi_p : T_pM \to T_{\Phi(p)}N$ is onto. Since the linear map $d\Phi_p : T_pM \to T_{\Phi(p)}N$ is onto, $n = \dim N = \dim T_{\Phi(p)}N \leq \dim T_pM = \dim M = m$. Thus, $m \not< n$, a contradiction.

Thus, every point of $M$ is a critical point of $\Phi$.

**Note 5.103** Let $M$ be an $m$-dimensional smooth manifold, $N$ be an $n$-dimensional smooth manifold, and $\Phi : M \to N$ be a smooth map. Let $G$ be the set of all regular points of $\Phi$. We shall try to show that $G$ is an open subset of $M$.

For this purpose, let us take any $p \in G$. Since $p \in G$, $p$ is a regular point of $\Phi$. Since $p$ is a regular point of $\Phi$, the linear map $d\Phi_p : T_pM \to T_{\Phi(p)}N$ is onto. Since the linear map $d\Phi_p : T_pM \to T_{\Phi(p)}N$ is onto, by Theorem 4.7, there exists an open neighborhood $W$ of $p$ such that for every $q \in W$, the linear map $dF_q : T_qM \to T_{F(q)}N$

is onto, and hence, the open neighborhood $W$ of $p$ is contained in $G$. It follows that $p$ is an interior point of $G$.

Thus, we have shown that $G$ is an open subset of $M$.

**Definition** Let $M$ be an $m$-dimensional smooth manifold, $N$ be an $n$-dimensional smooth manifold, and $\Phi : M \to N$ be a smooth map. Let $G$ be the set of all regular points of $\Phi$. Let $c \in N$.

If $\Phi^{-1}(c) \subset G$, then we say that $c$ is a *regular value of* $\Phi$, and $\Phi^{-1}(c)$ is a *regular value set*. If $c$ is not a regular value of $\Phi$, then we say that $c$ is a *critical value of* $\Phi$. If $c$ is a regular value of $\Phi$, then $\Phi^{-1}(c)$ is called a *regular level set*.

Clearly, if $\Phi^{-1}(c) = \emptyset$, then $c$ is a regular value of $\Phi$.

**Note 5.104** Let $M$ be an $m$-dimensional smooth manifold, $N$ be an $n$-dimensional smooth manifold, and $\Phi : M \to N$ be a smooth map. Let $c \in N$. Let $\Phi^{-1}(c)$ be a nonempty regular level set. We shall try to show that $\Phi^{-1}(c)$ is a properly embedded submanifold of $M$, and its codimension is $n$.

Let $G$ be the set of all regular points of $\Phi$. We know that $G$ is an open subset of $M$. Let $\mathcal{O}_1$ be the subspace topology over $G$ inherited from $M$, that is, $\mathcal{O}_1 = \{G_1 : G_1 \subset G, \text{ and } G_1 \in \mathcal{O}\}$. Let $\mathcal{A}$ be the differential structure of $M$. Put

$$\mathcal{A}_G \equiv \{(U, \varphi) : (U, \varphi) \in \mathcal{A}, \text{ and } U \subset G\}.$$

We know that $\mathcal{A}_G$ is an atlas on $G$. Thus, $G$ becomes an $m$-dimensional smooth manifold (called *open submanifold of M*). Also, $G$ is an embedded submanifold of $M$ with codimension 0. Since $\Phi : M \to N$ is a smooth map, and $G$ is an open submanifold of $M$, $\Phi|_G : G \to N$ is a smooth map.

Let us take any $p \in G$. We want to show that the linear map $d(\Phi|_G)_p : T_pG \to T_{(\Phi|_G)(p)}N$ is onto. Since $p \in G$, $p$ is a regular point of $M$, and hence, the linear map $d\Phi_p : T_pM \to T_{\Phi(p)}N$ is onto. Since $p \in G$, and $G$ is an open submanifold of $M$, $T_pM$ is equal to $T_pG$. Also, $(\Phi|_G)(p) = \Phi(p)$. It follows that the linear map $d(\Phi|_G)_p : T_pG \to T_{(\Phi|_G)(p)}N$ is onto.

Now, we want to show that $\Phi^{-1}(c)$ is an embedded submanifold of $M$. Since for every $p \in G$, the linear map $d(\Phi|_G)_p : T_pG \to T_{(\Phi|_G)(p)}N$ is onto, $\Phi|_G : G \to N$ is a smooth submersion. Since $\Phi|_G : G \to N$ is a smooth submersion, and $c \in N$, by Note 5.100, the level set $(\Phi|_G)^{-1}(c)$ is a properly embedded submanifold of $G$, and its codimension is $n$. Clearly, $(\Phi|_G)^{-1}(c) = \Phi^{-1}(c)$.

(Reason: Let $p \in (\Phi|_G)^{-1}(c)$. So $p \in G$, and $(\Phi|_G)(p) = c$. Since $p \in G$, $c = (\Phi|_G)(p) = \Phi(p)$, and hence, $p \in \Phi^{-1}(c)$. Thus, $(\Phi|_G)^{-1}(c) \subset \Phi^{-1}(c)$. Next, let $p \in \Phi^{-1}(c)$. Since $p \in \Phi^{-1}(c)$, $\Phi(p) = c$. Since $\Phi^{-1}(c)$ is a regular value set, $\Phi^{-1}(c) \subset G$. Since $p \in \Phi^{-1}(c) \subset G$, $p \in G$, and hence, $(\Phi|_G)(p) = \Phi(p) = c$. It follows that $p \in (\Phi|_G)^{-1}(c)$. Thus, $\Phi^{-1}(c) \subset (\Phi|_G)^{-1}(c)$.)

Since $(\Phi|_G)^{-1}(c) = \Phi^{-1}(c)$, and the level set $(\Phi|_G)^{-1}(c)$ is a properly embedded submanifold of $G$ with its codimension $n$, $\Phi^{-1}(c)$ is a properly embedded submanifold of $G$ with its codimension $n$. Since $\Phi^{-1}(c)$ is an embedded submanifold of $G$, the inclusion map $\iota_1 : \Phi^{-1}(c) \to G$ is a smooth embedding. Since $G$ is an embedded submanifold of $M$, the inclusion map $\iota_2 : G \to M$ is a smooth embedding. Since $\iota_1 : \Phi^{-1}(c) \to G$ is a smooth embedding, and $\iota_2 : G \to M$ is a smooth embedding, their composite $\iota_2 \circ \iota_1 : \Phi^{-1}(c) \to M$ is a smooth embedding. Since $\iota_2 \circ \iota_1 = \iota$, where $\iota$ denotes the inclusion map from $\Phi^{-1}(c)$ to $M$, and $\iota_2 \circ \iota_1 : \Phi^{-1}(c) \to M$ is a smooth embedding, the inclusion map $\iota : \Phi^{-1}(c) \to M$ is a smooth embedding. This shows that $\Phi^{-1}(c)$ is an embedded submanifold of $M$.

Now, we want to show that $\Phi^{-1}(c)$ is a properly embedded submanifold of $M$. For this purpose, let us take any compact set $K$ in $M$. We have to show that $(\iota_2 \circ \iota_1)^{-1}(K)$ is compact in $\Phi^{-1}(c)$. Here, $(\iota_2 \circ \iota_1)^{-1}(K) = ((\iota_1)^{-1} \circ (\iota_2)^{-1})(K) = (\iota_1)^{-1}(K \cap G) = (K \cap G) \cap \Phi^{-1}(c) = K \cap \Phi^{-1}(c)$. Since $K$ is compact in $M$, and $\Phi^{-1}(c)$ is closed in $M$, $K \cap \Phi^{-1}(c)$ is compact in $M$. Since $K \cap \Phi^{-1}(c)$ is compact in $M$, and $K \cap \Phi^{-1}(c) \subset \Phi^{-1}(c)$, $K \cap \Phi^{-1}(c)(= (\iota_2 \circ \iota_1)^{-1}(K))$ is compact in $\Phi^{-1}(c)$, and hence $(\iota_2 \circ \iota_1)^{-1}(K)$ is compact in $\Phi^{-1}(c)$.

Thus, $\Phi^{-1}(c)$ is a properly embedded submanifold of $M$ with codimension $n$.

## 5.10 Smooth Submanifolds

**Definition** Let $M$ be an $m$-dimensional smooth manifold and $S$ be a nonempty subset of $M$. If there exists a topology $\mathcal{O}$ over $S$ with respect to which $S$ becomes an $n$-dimensional topological manifold, and there exists a smooth structure on $S$ with respect to which the topological manifold $S$ becomes an $n$-dimensional smooth manifold such that the inclusion map $\iota : S \to M$ is smooth immersion, then we say that $S$ is an *immersed submanifold* (or *smooth submanifold*) *of $M$ with codimension* $m - n$.

**Note 5.105** Let $M$ be an $m$-dimensional smooth manifold and $S$ be a nonempty subset of $M$. Let $k \in \{0, 1, \ldots, m-1\}$. Let $S$ be an embedded submanifold of $M$ with codimension $k$. We shall try to show that $S$ is an immersed submanifold of $M$ with codimension $k$.

Put $\mathcal{O} \equiv \{G \cap S : G \text{ is open in } M\}$. We know that $\mathcal{O}$ (called the subspace topology of $S$) is a topology over $S$. Since $S$ is an embedded submanifold of $M$ with codimension $k$, $\mathcal{O}$ is a topology over $S$ with respect to which $S$ becomes a $(m-k)$-dimensional topological manifold. Also, there exists a smooth structure $\mathcal{A}$ on $S$, with respect to which the topological manifold $S$ becomes an $(m-k)$-dimensional smooth manifold such that the inclusion map $\iota : S \to M$ is a smooth embedding, and hence, $\iota : S \to M$ is a smooth immersion.

Hence, $S$ is an immersed submanifold of $M$ with codimension $k(= m - (m - k))$. This completes the proof. In short, every embedded submanifold is a smooth submanifold.

**Note 5.106** Let $M$ be an $m$-dimensional smooth manifold with smooth structure $\mathcal{A}$, $N$ be an $n$-dimensional smooth manifold with smooth structure $\mathcal{B}$, and $F : N \to M$ be a 1–1 smooth immersion. Put

$$\mathcal{O} \equiv \{F(V) : V \text{ is open in } N\}.$$

Clearly, $\mathcal{O}$ is a topology over $F(N)$, and $F$ is a homeomorphism from $N$ onto $F(N)$. Since $F$ is a homeomorphism from $N$ onto $F(N)$, $N$ is homeomorphic onto $F(N)$. Since $N$ is an $n$-dimensional smooth manifold, $N$ is an $n$-dimensional topological manifold. Since $N$ is an $n$-dimensional topological manifold, and $N$ is homeomorphic onto $F(N)$, $F(N)$ is an $n$-dimensional topological manifold. Since $N$ is an $n$-dimensional smooth manifold with smooth structure $\mathcal{B}$, and $F$ is a homeomorphism from $N$ onto $F(N)$, $\{(F(V), \psi \circ (F^{-1}|_{F(V)})) : (V, \psi) \in \mathcal{B}\}$ is a smooth structure on $F(N)$. Thus, $F(N)$ becomes an $n$-dimensional smooth manifold, and $F$ is a diffeomorphism from $N$ onto $F(N)$. It is clear that $\psi \circ (F^{-1}|_{F(V)}) = \psi \circ (F^{-1})$. It follows that the smooth structure on $F(N)$ is $\{(F(V), \psi \circ (F^{-1})) : (V, \psi) \in \mathcal{B}\}$.

Now, we want to prove that the inclusion map $\imath : F(N) \to M$ is a smooth map. Clearly, $\imath = F \circ F^{-1}$. Since $F : N \to F(N)$ is a diffeomorphism, $F^{-1} : F(N) \to N$ is a diffeomorphism. Since $F^{-1}$ is a diffeomorphism, and $F : N \to M$ is a 1–1 smooth immersion, their composition $F \circ F^{-1}(= \imath)$ is a 1–1 smooth immersion, and hence, $\imath$ is a smooth immersion. It follows that $F(N)$ is a smooth submanifold of $M$ with codimension $m - n$.

Conclusion: Let $M$ be an $m$-dimensional smooth manifold with smooth structure $\mathcal{A}$, $N$ be an $n$-dimensional smooth manifold with smooth structure $\mathcal{B}$, and $F : N \to M$ be a 1–1 smooth immersion. Then, $F(N)$ is a smooth submanifold of $M$ with codimension $m - n$.

**Note 5.107** Let $M$ be an $m$-dimensional smooth manifold and $S$ be a nonempty subset of $M$. Let $S$ be a smooth submanifold of $M$ with codimension 0. We shall try to show that $S$ is an embedded submanifold of $M$ with codimension 0.

Since $S$ is a smooth submanifold of $M$ with codimension 0, by its definition, there exists a topology $\mathcal{O}$ over $S$ with respect to which $S$ becomes an $m$-dimensional topological manifold, and there exists a smooth structure on $S$ with respect to which the topological manifold $S$ becomes an $m$-dimensional smooth manifold such that the inclusion map $\imath : S \to M$ is a smooth immersion. Here, $S$ is an $m$-dimensional smooth manifold, $M$ is an $m$-dimensional smooth manifold, and the inclusion map $\imath : S \to M$ is a 1–1 smooth immersion, so by Note 5.73, $\imath : S \to M$ is a smooth embedding. Since $\imath : S \to M$ is a smooth embedding, the inclusion map $\imath : S \to M$ is a topological embedding, and hence, $\imath$ is a homeomorphism from $S$ onto $\imath(S)(= S)$, where $\imath(S)$ has the subspace topology inherited from $M$. It follows that $\mathcal{O}$ is equal to the subspace topology of $S$ inherited from $M$. Since $S$ is endowed with the subspace topology inherited from $M$, and the inclusion map $\iota : S \to M$ is a smooth embedding, and the smooth structure on $S$ is $m$-dimensional, by the definition of embedded submanifold, $S$ is an embedded submanifold of $M$ with codimension 0.

**Note 5.108** Let $M$ be an $m$-dimensional smooth manifold and $S$ be a nonempty subset of $M$. Let $S$ be a smooth submanifold of $M$ with codimension $k$. Let the inclusion map $\iota : S \to M$ be proper.

We shall try to show that $S$ is an embedded submanifold of $M$ with codimension $k$.

Since $S$ is a smooth submanifold of $M$ with codimension $k$, by its definition, there exists a topology $\mathcal{O}$ over $S$ with respect to which $S$ becomes an $(m - k)$-dimensional topological manifold, and there exists a smooth structure on $S$ with respect to which the topological manifold $S$ becomes an $(m - k)$-dimensional smooth manifold such that the inclusion map $\iota : S \to M$ is a smooth immersion. Here, $S$ is an $(m - k)$-dimensional smooth manifold, $M$ is an $m$-dimensional smooth manifold, the inclusion map $\iota : S \to M$ is a 1–1 smooth immersion, and the inclusion map $\iota : S \to M$ is proper, so by Note 5.71, $\iota : S \to M$ is smooth embedding. Since $\iota : S \to M$ is a smooth embedding, the inclusion map $\iota : S \to M$ is a topological embedding, and hence, $\iota$ is a homeomorphism from $S$ onto $\iota(S)(= S)$, where $\iota(S)$ has the subspace topology inherited from $M$. It follows that $\mathcal{O}$ is equal to the subspace topology of $S$ inherited from $M$. Since $S$ is endowed with the subspace topology inherited from $M$, the inclusion map $\iota : S \to M$ is a smooth embedding, and the smooth structure on $S$ is $(m - k)$-dimensional, by the definition of embedded submanifold, $S$ is an embedded submanifold of $M$ with codimension $k$.

**Note 5.109** Let $M$ be an $m$-dimensional smooth manifold and $S$ be a nonempty subset of $M$. Let $S$ be a smooth submanifold of $M$ with codimension $k$. Let $S$ be compact. We shall try to show that $S$ is an embedded submanifold of $M$ with codimension $k$.

Since $S$ is a smooth submanifold of $M$ with codimension $k$, by its definition, there exists a topology $\mathcal{O}$ over $S$ with respect to which $S$ becomes an $(m - k)$-dimensional topological manifold, and there exists a smooth structure on $S$ with respect to which the topological manifold $S$ becomes an $(m - k)$-dimensional smooth manifold such that the inclusion map $\iota : S \to M$ is a smooth immersion. Here, $S$ is an $(m - k)$-dimensional smooth manifold, $M$ is an $m$-dimensional smooth manifold, the inclusion map $\iota : S \to M$ is a 1–1 smooth immersion, and $S$ is compact, so by Note 5.72, $\iota : S \to M$ is a smooth embedding. Since $\iota : S \to M$ is a smooth embedding, the inclusion map $\iota : S \to M$ is a topological embedding, and hence, $\iota$ is a homeomorphism from $S$ onto $\iota(S)(= S)$, where $\iota(S)$ has the subspace topology inherited from $M$. It follows that $\mathcal{O}$ is equal to the subspace topology of $S$ inherited from $M$. Since $S$ is endowed with the subspace topology inherited from $M$, the inclusion map $\iota : S \to M$ is a smooth embedding, and the smooth structure on $S$ is $(m - k)$-dimensional, by the definition of embedded submanifold, $S$ is an embedded submanifold of $M$ with codimension $k$.

**Note 5.110** Let $M$ be an $m$-dimensional smooth manifold and $S$ be a nonempty subset of $M$. Let $S$ be a smooth submanifold of $M$ with codimension $k$. Let $p \in S$.

We shall try to show that there exists an open neighborhood $U$ of $p$ in $S$ such that $U$ is an embedded submanifold of $M$ with codimension $k$.

Since $S$ is a smooth submanifold of $M$ with codimension $k$, there exists a topology $\mathcal{O}$ over $S$ with respect to which $S$ becomes a $(m - k)$-dimensional topological manifold, and there exists a smooth structure on $S$ with respect to which the topological manifold $S$ becomes a $(m - k)$-dimensional smooth manifold such that the inclusion map $\imath : S \rightarrow M$ is a smooth immersion. Since $\imath : S \rightarrow M$ is a smooth immersion, and $p \in S$, by Theorem 5.75, there exists an open neighborhood $U$ of $p$ in $S$ such that the restriction $\imath|_U : U \rightarrow M$ is a smooth embedding. Since $\imath|_U : U \rightarrow M$ is a smooth embedding, $\imath|_U : U \rightarrow M$ is a topological embedding, and hence, $\imath|_U$ is a homeomorphism from $U$ onto $(\imath|_U)(U)(= U)$, where $(\imath|_U)(U)$ has the subspace topology inherited from $M$. It follows that the topology over $U$ is the subspace topology inherited from $M$. Here, $S$ is a $(m - k)$-dimensional smooth manifold, and $U$ is an open neighborhood of $p$, so $U$ is a $(m - k)$-dimensional smooth manifold. Since $U$ is a $(m - k)$-dimensional smooth manifold, the topology over $U$ is the subspace topology inherited from $M$, and the inclusion map $\imath|_U : U \rightarrow M$ is a smooth embedding, $U$ is an embedded submanifold of $M$ with codimension $k$.

**Note 5.111** Let $M$ be an $m$-dimensional smooth manifold, $N$ be an $n$-dimensional smooth manifold, and $F : M \rightarrow N$ be a smooth map. Let $S$ be a nonempty subset of $M$. Let $S$ be a smooth submanifold of $M$. We shall try to show that the restriction $F|_S : S \rightarrow N$ is smooth.

Let the codimension of $S$ be $k$. Since $S$ is a smooth submanifold of $M$, there exists a topology $\mathcal{O}$ over $S$ with respect to which $S$ becomes an $(m - k)$-dimensional topological manifold, and there exists a smooth structure $\mathcal{A}$ on $S$ with respect to which the topological manifold $S$ becomes an $(m - k)$-dimensional smooth manifold such that the inclusion map $\imath : S \rightarrow M$ is a smooth immersion. Since $\imath : S \rightarrow M$ is a smooth immersion, $\imath : S \rightarrow M$ is smooth. Since $\imath : S \rightarrow M$ is smooth, and $F : M \rightarrow N$ is a smooth map, their composite $F \circ \imath : S \rightarrow N$ is smooth. Since $F \circ \imath : S \rightarrow N$ is smooth, and $F|_S = F \circ \imath$, $F|_S : S \rightarrow N$ is smooth.

**Note 5.112** Let $M$ be an $m$-dimensional smooth manifold, $N$ be an $n$-dimensional smooth manifold, and $S$ be a nonempty subset of $M$. Let $S$ be a smooth submanifold of $M$ with codimension $k$. Let $F : N \rightarrow M$ be a smooth map, and $F(N) \subset S$. Let $F : N \rightarrow S$ be continuous. We shall try to show $F : N \rightarrow S$ is smooth.

Let us fix any $p \in N$. Since $p \in N$, $F(p) \in F(N) \subset S$, and hence, $F(p) \in S$. Since $S$ is a smooth submanifold of $M$, there exists a topology $\mathcal{O}$ over $S$ with respect to which $S$ becomes an $(m - k)$-dimensional topological manifold, and there exists a smooth structure $\mathcal{A}$ on $S$ with respect to which the topological manifold $S$ becomes an $(m - k)$-dimensional smooth manifold such that the inclusion map $\imath : S \rightarrow M$ is a smooth immersion. Since the inclusion map $\imath : S \rightarrow M$ is a smooth immersion, by Theorem 5.46, $\imath : S \rightarrow M$ is a local diffeomorphism. Since $\imath : S \rightarrow M$ is a local diffeomorphism, and $F(p) \in S$, there exists an open neighborhood $U$ of

$F(p)$ in $S$ such that $\iota(U)$ is open in $M$, and the restriction $\iota|_U: U \to \iota(U)$ is a diffeomorphism, and hence, $(\iota|_U)^{-1} : \iota(U) \to U$ is smooth. Since $U$ is an open neighborhood of $F(p)$ in $S$, and $F : N \to S$ is continuous, there exists an open neighborhood $V$ of $p$ in $N$ such that $F(V) \subset U$. Since $V$ is open in $N$, $V$ is an open submanifold of $N$, and hence, $V$ is an embedded submanifold of $N$. Since $V$ is an embedded submanifold of $N$, $V$ is a smooth submanifold of $N$. Since $V$ is a smooth submanifold of $N$, and $F : N \to M$ is a smooth map, by Note 5.111, $F|_V: V \to M$ is a smooth map. Since $F|_V: V \to M$ is a smooth map, $(\iota|_U)^{-1} : \iota(U) \to U$ is smooth, and $(F|_V)(V) = F(V) \subset U = \iota(U)$, their composite $((\iota|_U)^{-1}) \circ (F|_V) : V \to U$ is smooth. Since $((\iota|_U)^{-1}) \circ (F|_V) : V \to U$ is smooth, and $((\iota|_U)^{-1}) \circ (F|_V) = F|_V$, $F|_V: V \to U$ is smooth. Since $F|_V: V \to U$ is smooth, and $V$ is an open neighborhood of $p$ in $N$, $F : N \to S$ is smooth at the point $p$. It follows that $F : N \to S$ is smooth.

**Note 5.113** Let $M$ be an $m$-dimensional smooth manifold, $N$ be an $n$-dimensional smooth manifold, and $S$ be a nonempty subset of $M$. Let $S$ be an embedded submanifold of $M$ with codimension $k$. Let $F : N \to M$ be a smooth map, and $F(N) \subset S$. Let $F : N \to S$ be continuous. We shall try to show $F : N \to S$ is smooth.

Since $S$ is an embedded submanifold of $M$ with codimension $k$, by Note 5.105, $S$ is a smooth submanifold of $M$ with codimension $k$. Since $F : N \to M$ is a smooth map, $F : N \to M$ is continuous. Since $S$ is an embedded submanifold of $M$, the topology over $S$ is the subspace topology inherited from $M$. Since $F : N \to M$ is continuous, $F(N) \subset S \subset M$, and the topology over $S$ is the subspace topology inherited from $M$, $F : N \to S$ is continuous. It follows, from Note 5.112, that $F : N \to S$ is smooth.

**Note 5.114** Let $M$ be an $m$-dimensional smooth manifold. Let $S$ be a nonempty subset of $M$. Suppose that $S$ satisfies local $k$-slice condition.

Since $S$ satisfies local $k$-slice condition, by Note 5.97, $S$ is an embedded submanifold of $M$. Since $S$ is an embedded submanifold of $M$, the topology over $S$ is the subspace topology inherited from $M$, and there exists a smooth structure $\mathcal{A}$ on $S$ with respect to which $S$ becomes a $k$-dimensional smooth manifold such that the inclusion map $\iota : S \to M$ is a smooth embedding.

Now, we want to show that such a differential structure $\mathcal{A}$ is unique. For this purpose, let $\mathcal{A}_1$ be a smooth structure on $S$ with respect to which $S$ becomes a $k$-dimensional smooth manifold such that the inclusion map $\iota : S \to M$ is a smooth embedding. Next, let $\mathcal{A}_2$ be a smooth structure on $S$ with respect to which $S$ becomes a $k$-dimensional smooth manifold such that the inclusion map $\iota : S \to M$ is a smooth embedding. We have to prove that $\mathcal{A}_1 = \mathcal{A}_2$.

Let $\mathcal{O}$ be the topology over $M$, and let $\mathcal{B}$ be the smooth structure of $M$. Since the inclusion map $\iota : S \to M$ is a smooth embedding from smooth manifold $(S, \mathcal{A}_1)$ to $(M, \mathcal{B})$, the inclusion map $\iota : S \to M$ is a smooth map from smooth manifold $(S, \mathcal{A}_1)$ to $(M, \mathcal{B})$. Since the inclusion map $\iota : S \to M$ is a smooth map from smooth manifold $(S, \mathcal{A}_1)$ to $(M, \mathcal{B})$, $(S, \mathcal{A}_2)$ is an embedded submanifold of $M$, and $\iota : S \to S$ is continuous, by Lemma 5.113, $\iota : (S, \mathcal{A}_1) \to (S, \mathcal{A}_2)$ is smooth. Similarly,

$\iota : (S, \mathcal{A}_2) \to (S, \mathcal{A}_1)$ is smooth. Since $\iota : (S, \mathcal{A}_1) \to (S, \mathcal{A}_2)$ is smooth, and $(\iota^{-1} =)\iota : (S, \mathcal{A}_2) \to (S, \mathcal{A}_1)$ is smooth, $\iota : (S, \mathcal{A}_1) \to (S, \mathcal{A}_2)$ is a diffeomorphism. Hence, $\mathcal{A}_1 = \mathcal{A}_2$.

**Note 5.115** Let $M$ be an $m$-dimensional smooth manifold. Let $S$ be a nonempty subset of $M$. Suppose that $S$ is a smooth submanifold of $M$.

Since $S$ is a smooth submanifold of $M$, there exists a topology $\mathcal{O}$ over $S$, and a smooth structure $\mathcal{A}$ on $S$ with respect to which $S$ becomes a $k$-dimensional smooth manifold such that the inclusion map $\iota : S \to M$ is a smooth immersion.

Now, we want to show that for fixed $\mathcal{O}$, the differential structure $\mathcal{A}$ is unique. Let us fix the topology $\mathcal{O}$ over $S$. Next, let $\mathcal{A}_1$ be a smooth structure on $S$ with respect to which $S$ becomes a $k$-dimensional smooth manifold such that the inclusion map $\iota : S \to M$ is a smooth immersion. Next, let $\mathcal{A}_2$ be a smooth structure on $S$ with respect to which $S$ becomes a $k$-dimensional smooth manifold such that the inclusion map $\iota : S \to M$ is a smooth immersion. We have to prove that $\mathcal{A}_1 = \mathcal{A}_2$.

Let $\mathcal{O}^*$ be the topology over $M$, and let $\mathcal{B}$ be the smooth structure of $M$. Since the inclusion map $\iota : S \to M$ is a smooth immersion from smooth manifold $(S, \mathcal{A}_1)$ to $(M, \mathcal{B})$, the inclusion map $\iota : S \to M$ is a smooth map from smooth manifold $(S, \mathcal{A}_1)$ to $(M, \mathcal{B})$. Since the inclusion map $\iota : S \to M$ is a smooth map from smooth manifold $(S, \mathcal{A}_1)$ to $(M, \mathcal{B})$, $(S, \mathcal{A}_2)$ is a smooth submanifold of $M$, and $\iota : S \to S$ is continuous, by Lemma 5.113, $\iota : (S, \mathcal{A}_1) \to (S, \mathcal{A}_2)$ is smooth. Similarly, $\iota : (S, \mathcal{A}_2) \to (S, \mathcal{A}_1)$ is smooth. Since $\iota : (S, \mathcal{A}_1) \to (S, \mathcal{A}_2)$ is smooth, and $(\iota^{-1} =)\iota : (S, \mathcal{A}_2) \to (S, \mathcal{A}_1)$ is smooth, $\iota : (S, \mathcal{A}_1) \to (S, \mathcal{A}_2)$ is a diffeomorphism. Hence, $\mathcal{A}_1 = \mathcal{A}_2$.

## 5.11 Tangent Space to a Submanifold

**Note 5.116** Let $M$ be an $m$-dimensional smooth manifold. Let $S$ be a nonempty subset of $M$. Let $S$ be an embedded submanifold of $M$.

Since $S$ is an embedded submanifold of $M$, the inclusion map $\iota : S \to M$ is a smooth embedding, and hence, $\iota : S \to M$ is a smooth immersion. Since $\iota : S \to M$ is a smooth immersion, for every $p \in S$, the linear map $d\iota_p : T_p S \to T_{\iota(p)} M$ is 1-1. Now, since $\iota(p) = p$, for every $p \in S$, the linear map $d\iota_p : T_p S \to T_p M$ is 1-1, and hence, the image set $(d\iota_p)(T_p S)$ is a subspace of the real linear space $T_p M$. Also, since $d\iota_p : T_p S \to T_p M$ is 1-1, $\dim((d\iota_p)(T_p S)) = \dim(T_p S) = \dim(S)$. That is why, we adopt the convention of not to distinguish between $(d\iota_p)(T_p S)$ and $T_p S$.

**Lemma 5.117** *Let $M$ be an $m$-dimensional smooth manifold. Let $S$ be a nonempty subset of $M$. Let $S$ be an embedded submanifold of $M$. Let $p \in S$. Let $v \in T_p M$. If there exists a smooth curve $\gamma : (-\varepsilon, \varepsilon) \to M$ whose image is contained in $S$, which is also smooth as a map into $S$, such that $\gamma(0) = p$, and $\gamma'(0) = v$, then $v \in T_p S$.*

*Proof* Let $\gamma : (-\varepsilon, \varepsilon) \to M$ be a smooth curve whose image is contained in $S$, which is also smooth as a map into $S$, such that $\gamma(0) = p$, and $\gamma'(0) = v$. We have to show that $v \in T_p S$.

Since $\gamma : (-\varepsilon, \varepsilon) \to S$ is a smooth map from one-dimensional smooth manifold $(-\varepsilon, \varepsilon)$ into smooth manifold $S$, and $0 \in (-\varepsilon, \varepsilon)$, $d\gamma_0 : T_0(-\varepsilon, \varepsilon) \to T_{\gamma(0)}S$. Now, since

$T_{\gamma(0)}S = T_p S$, $d\gamma_0 : T_0(-\varepsilon, \varepsilon) \to T_p S$, and hence, $v = \gamma'(0) = (d\gamma_0)(\frac{d}{dt}\big|_0) \in T_p S$.   $\square$

**Lemma 5.118** *Let $M$ be an $m$-dimensional smooth manifold. Let $S$ be a nonempty subset of $M$. Let $S$ be an embedded submanifold of $M$. Let $p \in S$. Let $v \in T_p M$. If $v \in T_p S$, then there exists a smooth curve $\gamma : (-\varepsilon, \varepsilon) \to M$ whose image is contained in $S$, which is also smooth as a map into $S$, such that $\gamma(0) = p$, and $\gamma'(0) = v$.*

*Proof* Let $v \in T_p S$. Here, $S$ is an embedded submanifold of $M$, so $S$ is a smooth manifold. Let $k$ be the dimension of smooth manifold $S$. Since $p \in S$, and $S$ is a smooth manifold, there exists an admissible coordinate chart $(U, \varphi)$ of $S$ such that $p \in U$, and $\varphi(p) = 0 \in \mathbb{R}^k$. For every $q \in U$, let

$$\left( \frac{\partial}{\partial x^1}\bigg|_q , \ldots, \frac{\partial}{\partial x^k}\bigg|_q \right)$$

be the coordinate basis of $T_q S$ corresponding to $(U, \varphi)$, where

$$\frac{\partial}{\partial x^i}\bigg|_q \equiv \left( d(\varphi^{-1})_{\varphi(q)} \right) \left( \frac{\partial}{\partial x^i}\bigg|_{\varphi(q)} \right).$$

Here,

$$\left( \frac{\partial}{\partial x^1}\bigg|_p , \ldots, \frac{\partial}{\partial x^k}\bigg|_p \right)$$

is the coordinate basis of $T_p S$, and $v \in T_p S$, so there exist real numbers $v^1, \ldots, v^k$ such that

$$v = v^1 \left( \frac{\partial}{\partial x^1}\bigg|_p \right) + \cdots + v^k \left( \frac{\partial}{\partial x^k}\bigg|_p \right) = v^i \left( \frac{\partial}{\partial x^i}\bigg|_p \right).$$

Since $(U, \varphi)$ is an admissible coordinate chart of $S$, $\varphi$ is a mapping from $U$ onto $\varphi(U)$, where $\varphi(U)$ is an open neighborhood of $\varphi(p)(= 0)$ in $\mathbb{R}^k$. Since $t \mapsto (tv^1, \ldots, tv^k)$ is a continuous map from $\mathbb{R}$ to $\mathbb{R}^k$, and $\varphi(U)$ is an open neighborhood of $\varphi(p)(= (0, \ldots, 0) = (0v^1, \ldots, 0v^k))$ in $\mathbb{R}^k$, there exists $\varepsilon > 0$ such that for every $t \in (0 - \varepsilon, 0 + \varepsilon)$, we have $(tv^1, \ldots, tv^k) \in \varphi(U)$, and hence, for every $t \in (-\varepsilon, \varepsilon)$, we have $\varphi^{-1}(tv^1, \ldots, tv^k) \in U$. Now, let us define $\gamma : (-\varepsilon, \varepsilon) \to U(\subset S \subset M)$ as follows: For every $t \in (-\varepsilon, \varepsilon), \gamma(t) \equiv \varphi^{-1}(tv^1, \ldots, tv^k)$.

Here, $(U, \varphi)$ is an admissible coordinate chart of $S$, so $U$ is open in $S$. Here, $\gamma(0) = \varphi^{-1}(0v^1, \ldots, 0v^k) = \varphi^{-1}(0) = \varphi^{-1}(\varphi(p)) = p$. Thus, $\gamma(0) = p$. We shall try to show that $\gamma$ as a function from $(-\varepsilon, \varepsilon)$ to smooth manifold $S$ is smooth.

Since $(U, \varphi)$ is an admissible coordinate chart of $S$, $\varphi$ is a diffeomorphism from $U$ onto $\varphi(U)$, and hence, $\varphi^{-1}$ is smooth. Since $\varphi^{-1}$ is smooth, and $t \mapsto (tv^1, \ldots, tv^k)$ is smooth, their composite $t \mapsto \varphi^{-1}(tv^1, \ldots, tv^k)(= \gamma(t))$ is smooth, and hence, $\gamma$ as a mapping from $(-\varepsilon, \varepsilon)$ to $S$ is smooth.

Now, we want to show that $\gamma$ is smooth as a map from smooth manifold $(-\varepsilon, \varepsilon)$ to smooth manifold $M$. For this purpose, let us take any $t_0 \in (-\varepsilon, \varepsilon)$. Here, $\gamma(t_0) = \varphi^{-1}(t_0 v^1, \ldots, t_0 v^k) \in U$.

Since $S$ is an embedded submanifold of $M$, there exists a smooth structure $\mathcal{A}$ on $S$ such that the inclusion map $\iota : S \to M$ is a smooth embedding (i.e., $\iota : S \to M$ is smooth, and smooth immersion.). Let $(W, \psi)$ be any admissible coordinate chart of $M$ such that $\gamma(t_0) \in W$. We have to show that $\psi \circ \gamma \circ (\mathrm{Id}_{(-\varepsilon, \varepsilon)})^{-1}$ is smooth.

Since $\iota : S \to M$ is smooth, $(U, \varphi)$ is an admissible coordinate chart of $S$ satisfying $\gamma(t_0) \in U$, and $(W, \psi)$ is any admissible coordinate chart of $M$ such that $(\iota(\gamma(t_0)) =) \gamma(t_0) \in W$, $\psi \circ \varphi^{-1}$ is smooth. Further, since $t \mapsto (tv^1, \ldots, tv^k)$ is smooth, $t \mapsto (\psi \circ \varphi^{-1})(tv^1, \ldots, tv^k)$ is smooth. For every $t \in (-\varepsilon, \varepsilon)$,

$$\left( \psi \circ \gamma \circ (\mathrm{Id}_{(-\varepsilon, \varepsilon)})^{-1} \right)(t) = (\psi \circ \gamma)\left( (\mathrm{Id}_{(-\varepsilon, \varepsilon)})^{-1}(t) \right) = (\psi \circ \gamma)(t) = \psi(\gamma(t))$$
$$= \psi(\varphi^{-1}(tv^1, \ldots, tv^k)) = (\psi \circ \varphi^{-1})(tv^1, \ldots, tv^k),$$

and $t \mapsto (\psi \circ \varphi^{-1})(tv^1, \ldots, tv^k)$ is smooth, so $\psi \circ \gamma \circ (\mathrm{Id}_{(-\varepsilon, \varepsilon)})^{-1}$ is smooth. For every $t \in (-\varepsilon, \varepsilon)$, we have

$$(\pi_i \circ (\varphi \circ \gamma))(t) = \pi_i(\varphi(\gamma(t))) = \pi_i(\varphi(\varphi^{-1}(tv^1, \ldots, tv^m)))$$
$$= \pi_i(tv^1, \ldots, tv^m) = tv^i = v^i t,$$

so

$$\frac{d(\pi_i \circ (\varphi \circ \gamma))}{dt}(0) = v^i.$$

Now, we shall try to show that $\gamma'(0) = v$. Here,

$$\gamma'(0) = \left( \frac{d(\pi_i \circ (\varphi \circ \gamma))}{dt}(0) \right) \left( \frac{\partial}{\partial x^i} \bigg|_{\gamma(0)} \right) = v^i \left( \frac{\partial}{\partial x^i} \bigg|_{\gamma(0)} \right) = v^i \left( \frac{\partial}{\partial x^i} \bigg|_p \right) = v.$$

Hence, $\gamma'(0) = v$. $\qquad\qquad\qquad\qquad\qquad\qquad\qquad\qquad\qquad\qquad\qquad\qquad\qquad\qquad\square$

**Lemma 5.119** *Let M be an m-dimensional smooth manifold. Let S be a nonempty subset of M. Let S be an embedded submanifold of M. Let $p \in S$. Let $v \in T_pS$. Let $f \in C^\infty(M)$ satisfying $f|_S = 0$. Then, $v(f) = 0$.*

*Proof* Since $f \in C^\infty(M)$, $f|_S \in C^\infty(S)$. (Reason: Since $f \in C^\infty(M)$, $f : M \to \mathbb{R}$ is smooth. Here, $S$ is an embedded submanifold of $M$, so the inclusion map $\iota : S \to M$ is smooth. Since $\iota : S \to M$ is smooth, and $f : M \to \mathbb{R}$ is smooth, their composite $(f|_S =) f \circ \iota : S \to \mathbb{R}$ is smooth, and hence, $f|_S \in C^\infty(S)$.) Since, by Note 5.116, we do not distinguish between $(d\iota_p)(T_pS)$ and $T_pS$, and $v \in T_pS$, $v \in (d\iota_p)(T_pS)$, and hence, there exists $w \in T_pS$ such that $(d\iota_p)(w) = v$. Now, $v(f) = ((d\iota_p)(w))(f) = w(f \circ \iota) = w(f|_S) = w(0) = 0$. □

**Lemma 5.120** *Let M be an m-dimensional smooth manifold. Let S be a nonempty subset of M. Let S be an embedded k-dimensional submanifold of M. Then, S satisfies the local k-slice condition.*

*Proof* Let us take any $p$ in $S$. Since $S$ is an embedded submanifold of $M$, there exists a smooth structure $\mathcal{A}$ on $S$ such that the inclusion map $\iota : S \to M$ is a smooth embedding (i.e., $\iota : S \to M$ is smooth, and smooth immersion). Since $\iota : S \to M$ is a smooth immersion, and $p \in S$, by Theorem 4.13, there exist admissible coordinate chart $(U, \varphi)$ in $S$ satisfying $p \in U$ and admissible coordinate chart $(V, \psi)$ in $M$ satisfying $(p =) \iota(p) \in V$ such that $\varphi(p) = 0$, $(\psi(p) =) \psi(\iota(p)) = 0$, and for every $(x_1, \ldots, x_k)$ in $\varphi(U)$,

$$\left(\psi \circ \varphi^{-1}\right)(x_1, \ldots, x_k) = \left(\psi \circ \iota \circ \varphi^{-1}\right)(x_1, \ldots, x_k) = \left(x_1, \ldots, x_k, \underbrace{0, \ldots, 0}_{m-k}\right).$$

Further, we may assume $\varphi(U) = C_\varepsilon^k(0)$, and $\psi(V) = C_\varepsilon^m(0)$ for some $\varepsilon > 0$. Since $(U, \varphi)$ is an admissible coordinate chart in $S$ satisfying $p \in U$, $U$ is an open neighborhood of $p$ in $S$, and hence, there exists an open neighborhood $W$ of $p$ in $M$ such that $W \cap S = U$. Since $W$ is an open neighborhood of $p$ in $M$, and $V$ is an open neighborhood of $p$ in $M$, their intersection $V \cap W$ is an open neighborhood of $p$ in $M$. Since $V \cap W (\subset V)$ is an open neighborhood of $p$ in $M$, and $(V, \psi)$ is an admissible coordinate chart in $M$ satisfying $p \in V$, $(V \cap W, \psi|_{V \cap W})$ is an admissible coordinate chart in $M$ satisfying $p \in V \cap W$. It suffices to show that

$$(V \cap W) \cap S = \left(\psi|_{V \cap W}\right)^{-1}\left(\left\{\left(x_1, \ldots, x_k, \underbrace{0, \ldots, 0}_{m-k}\right) : \left(x_1, \ldots, x_k, \underbrace{0, \ldots, 0}_{m-k}\right)\right.\right.$$
$$\left.\left. \in (\psi|_{V \cap W})(V \cap W)\right\}\right),$$

that is,

$$(\psi|_{V \cap W})((V \cap W) \cap S) = \left\{ \left( x_1, \ldots, x_k, \underbrace{0, \ldots, 0}_{m-k} \right) : \left( x_1, \ldots, x_k, \underbrace{0, \ldots, 0}_{m-k} \right) \right.$$
$$\left. \in (\psi|_{V \cap W})(V \cap W) \right\},$$

that is,

$$\psi((V \cap W) \cap S) = \left\{ \left( x_1, \ldots, x_k, \underbrace{0, \ldots, 0}_{m-k} \right) : \left( x_1, \ldots, x_k, \underbrace{0, \ldots, 0}_{m-k} \right) \in \psi(V \cap W) \right\},$$

that is,

$$\psi(U \cap V) = \left\{ \left( x_1, \ldots, x_k, \underbrace{0, \ldots, 0}_{m-k} \right) : \left( x_1, \ldots, x_k, \underbrace{0, \ldots, 0}_{m-k} \right) \in \psi(V \cap W) \right\}.$$

Let $\psi(q) \in$ LHS, where $q \in U \cap V$. Since $q \in U \cap V$, $q \in U$, and hence, $\varphi(q) \in \varphi(U)$. Put $\varphi(q) \equiv (y_1, \ldots, y_k)$. Thus, $(y_1, \ldots, y_k) \in \varphi(U)$, and hence,

$$\psi(q) = \psi\left(\varphi^{-1}(y_1, \ldots, y_k)\right) = (\psi \circ \varphi^{-1})(y_1, \ldots, y_k) = \left( y_1, \ldots, y_k, \underbrace{0, \ldots, 0}_{m-k} \right).$$

Since $q \in U \cap V$, $q \in V$. Since $q \in U$, and $W \cap S = U$, $q \in W$. Since $q \in W$, and $q \in V$, $q \in V \cap W$, and hence,

$$\left( y_1, \ldots, y_k, \underbrace{0, \ldots, 0}_{m-k} \right) = \psi(q) \in \psi(V \cap W).$$

Thus,

$$\text{LHS} = \psi(q) = \left( y_1, \ldots, y_k, \underbrace{0, \ldots, 0}_{m-k} \right) \in \left\{ \left( x_1, \ldots, x_k, \underbrace{0, \ldots, 0}_{m-k} \right) : \right.$$
$$\left. \left( x_1, \ldots, x_k, \underbrace{0, \ldots, 0}_{m-k} \right) \in \psi(V \cap W) \right\} = \text{RHS}.$$

Thus, LHS $\subset$ RHS. Next, let

$$\text{LHS} = \psi(q) \equiv \left( y_1, \ldots, y_k, \underbrace{0, \ldots, 0}_{m-k} \right) \in \text{RHS},$$

where $q \in V \cap W$. We have to show that $\psi(q) \in \psi(U \cap V)$. It suffices to show that $q \in U$. Here, $q \in V \cap W$, so

$$\left( y_1, \ldots, y_k, \underbrace{0, \ldots, 0}_{m-k} \right) = \psi(q) \in \psi(V) = C_\varepsilon^m(0)$$

$$= \underbrace{(-\varepsilon, \varepsilon) \times \cdots \times (-\varepsilon, \varepsilon)}_{k} \times \underbrace{(-\varepsilon, \varepsilon) \times \cdots \times (-\varepsilon, \varepsilon)}_{m-k},$$

and hence,

$$(y_1, \ldots, y_k) \in \underbrace{(-\varepsilon, \varepsilon) \times \cdots \times (-\varepsilon, \varepsilon)}_{k} = C_\varepsilon^k(0) = \varphi(U).$$

Since $(y_1, \ldots, y_k) \in \varphi(U)$, $\varphi^{-1}(y_1, \ldots, y_k) \in U$, and

$$\psi\big(\varphi^{-1}(y_1, \ldots, y_k)\big) = (\psi \circ \varphi^{-1})(y_1, \ldots, y_k) = \left( y_1, \ldots, y_k, \underbrace{0, \ldots, 0}_{m-k} \right) = \psi(q).$$

Now, since $\psi$ is 1–1, $q = \varphi^{-1}(y_1, \ldots, y_k) \in U$.                           $\square$

**Lemma 5.121** *Let M be an m-dimensional smooth manifold. Let S be a nonempty subset of M. Let S be an embedded k-dimensional submanifold of M. There exists a unique smooth structure over S with respect to which S is an embedded k-dimensional submanifold of M.*

*Proof* Since $S$ is an embedded submanifold of $M$, the topology over $S$ is the subspace topology inherited from $M$, and there exists a smooth structure $\mathcal{A}$ on $S$ with respect to which $S$ becomes a $k$-dimensional smooth manifold such that the inclusion map $\imath : S \to M$ is a smooth embedding.

By Lemma 5.120, $S$ satisfies the local $k$-slice condition. Since $S$ satisfies the local $k$-slice condition, by Note 5.114, the smooth structure $\mathcal{A}$ is unique.          $\square$

**Lemma 5.122** *Let M be an m-dimensional smooth manifold. Let S be a nonempty subset of M. Let S be an embedded k-dimensional submanifold of M. Let $p \in S$. Let $v \in T_pM$. Let $(f \in C^\infty(M)$ satisfying $f|_S = 0)$ implies $v(f) = 0$. Then, $v \in T_pS$.*

*Proof* By Lemma 5.121, there exists a unique smooth structure $\mathcal{A}$ over $S$ with respect to which $S$ is an embedded $k$-dimensional submanifold of $M$. Thus, the topology of $S$ is the subspace topology of $M$, and $(S, \mathcal{A})$ is a $k$-dimensional smooth manifold. Since $S$ is an embedded $k$-dimensional submanifold of $M$, by Lemma 5.120, $S$ satisfies the local $k$-slice condition. Since $S$ satisfies the local $k$-slice condition, and $p \in S$, there exists an admissible coordinate chart $(V, \psi)$ of $M$ such that $p \in V$, and $V \cap S (\subset V)$ is equal to a $k$-slice of $V$. For definiteness, let

$$V \cap S = \psi^{-1}(\{(x_1, \ldots, x_k, x_{k+1}, \ldots, x_m) \in \psi(V) : x_{k+1} = 0, \ldots, x_m = 0\}),$$

that is,

$$\psi(V \cap S) = \{(x_1, \ldots, x_k, x_{k+1}, \ldots, x_m) \in \psi(V) : x_{k+1} = 0, \ldots, x_m = 0\},$$

that is,

$$\psi(V \cap S) = \left\{ \left(x_1, \ldots, x_k, \underbrace{0, \ldots, 0}_{m-k}\right) : \left(x_1, \ldots, x_k, \underbrace{0, \ldots, 0}_{m-k}\right) \in \psi(V) \right\}.$$

Since $(V, \psi)$ is an admissible coordinate chart of $M$ such that $p \in V$, for every $q \in V$,

$$\left( \left.\frac{\partial}{\partial y^1}\right|_q, \cdots, \left.\frac{\partial}{\partial y^k}\right|_q, \left.\frac{\partial}{\partial y^{k+1}}\right|_q, \cdots \left.\frac{\partial}{\partial y^m}\right|_q \right)$$

is the coordinate basis of $T_q M$ corresponding to $(V, \psi)$, where

$$\left.\frac{\partial}{\partial y^i}\right|_q \equiv \left( d(\psi^{-1})_{\psi(q)} \right) \left( \left.\frac{\partial}{\partial y^i}\right|_{\psi(q)} \right).$$

By Note 5.96, there exists a smooth structure $\mathcal{B}$ on $S$ such that $S$ becomes a $k$-dimensional embedded submanifold of $M$, and

$$\left( \left.\frac{\partial}{\partial y^1}\right|_q, \cdots, \left.\frac{\partial}{\partial y^k}\right|_q \right)$$

is a basis of $(d\iota_q)(T_q S)$, where $\iota : S \to M$ denotes the inclusion map of $S$. Now, since $\mathcal{A}$ is unique, $\mathcal{A} = \mathcal{B}$. Since $p \in V$,

$$\left( \left.\frac{\partial}{\partial y^1}\right|_p, \cdots, \left.\frac{\partial}{\partial y^k}\right|_p, \left.\frac{\partial}{\partial y^{k+1}}\right|_p, \cdots \left.\frac{\partial}{\partial y^m}\right|_p \right)$$

is a basis of the real linear space $T_pM$. Now, since $v \in T_pM$, there exist real numbers $v^1, \ldots, v^k, v^{k+1}, \ldots, v^m$ such that

$$v = v^1 \left( \frac{\partial}{\partial y^1} \bigg|_p \right) + \cdots + v^k \left( \frac{\partial}{\partial y^k} \bigg|_p \right) + v^{k+1} \left( \frac{\partial}{\partial y^{k+1}} \bigg|_p \right) + \cdots + v^m \left( \frac{\partial}{\partial y^m} \bigg|_p \right).$$

We shall try to show that $v^{k+1} = 0$. Here, $\psi : V \to \psi(V)(\subset \mathbb{R}^m)$ is smooth, so the $(k + 1)$th projection map $\pi_{k+1} \circ \psi : V \to \mathbb{R}$ of $\psi$ is smooth. Here, $V$ is an open neighborhood of $p$ in $M$, and $M$ is a smooth manifold, there exists an open neighborhood $W$ of $p$ such that $p \in W \subset W^- \subset V$. By Lemma 4.58, there exists a smooth function $\chi : M \to [0, 1]$ such that

1. for every $x$ in $W^-, \chi(x) = 1$,
2. supp $\chi \subset V$.

If $x \in (M - (\text{supp } \chi)) \cap V$, then $x \in (M - (\text{supp } \chi))$, and hence, by the definition of support, $\chi(x) = 0$. It follows that if $x \in (M - (\text{supp } \chi)) \cap V$, then $(\chi(x))((\pi_{k+1} \circ \psi)(x)) = (0)((\pi_{k+1} \circ \psi)(x)) = 0$. This shows that the following function $f : M \to \mathbb{R}$ defined as follows: For every $x \in M$,

$$f(x) \equiv \begin{cases} (\chi(x))((\pi_{k+1} \circ \psi)(x)) & \text{if } \quad x \in V \\ 0 & \text{if } \quad M - (\text{supp } \chi), \end{cases}$$

is well defined. Since $V$ is open, $x \mapsto \chi(x)$ is smooth, and $x \mapsto (\pi_{k+1} \circ \psi)(x)$ is smooth, their product $x \mapsto (\chi(x))((\pi_{k+1} \circ \psi)(x))$ is smooth on $V$. Also, the constant function $x \mapsto 0$ is smooth on the open set $M - (\text{supp } \chi)$. Now, from the definition of $f : M \to \mathbb{R}, f$ is smooth, that is, $f \in C^\infty(M)$. From the definition of $f : M \to \mathbb{R}$, we can write: For every $x \in M$,

$$f(x) \equiv \begin{cases} (\chi(x))((\pi_{k+1} \circ \psi)(x)) & \text{if } \quad x \in V \\ 0 & \text{if } \quad x \notin V. \end{cases}$$

Clearly, $f|_S = 0$. (Reason: Let us take any $x \in S$. We have to show that $f(x) = 0$. For this purpose, it suffices to show that $(\pi_{k+1} \circ \psi)(x) = 0$ whenever $x \in V$. For this purpose, let $x \in V$. Since $x \in V$, and $x \in S$, $x \in V \cap S$. Since $x \in V \cap S$,

$$\psi(x) \in \psi(V \cap S) = \left\{ \left( x_1, \ldots, x_k, \underbrace{0, \ldots, 0}_{m-k} \right) : \left( x_1, \ldots, x_k, \underbrace{0, \ldots, 0}_{m-k} \right) \in \psi(V) \right\},$$

and hence, there exist real numbers $x_1, \ldots, x_k$, such that

$$\psi(x) = \left( x_1, \ldots, x_k, \underbrace{0, \ldots, 0}_{m-k} \right) \in \psi(V).$$

It follows that

$$(\pi_{k+1} \circ \psi)(x) = \pi_{k+1}(\psi(x)) = \pi_{k+1}\left(x_1, \ldots, x_k, \underbrace{0, \ldots, 0}_{m-k}\right) = 0.$$

Thus, $(\pi_{k+1} \circ \psi)(x) = 0.$) Since $f \in C^\infty(M)$, and $f|_S = 0$, by the assumption, $v(f) = 0$.

Since $W \subset V$, for every $x \in W$, we have $x \in V$, and $\chi(x) = 1$. Hence, for every $x \in W$, $f(x) = (\chi(x))((\pi_{k+1} \circ \psi)(x)) = (1)((\pi_{k+1} \circ \psi)(x)) = (\pi_{k+1} \circ \psi)(x)$. Thus, $f|_W = (\pi_{k+1} \circ \psi)|_W$. Now, since $W$ is an open neighborhood of $p$,

$$0 = 0(p) = (v(f))(p)$$

$$= \left(\left(v^1\left(\frac{\partial}{\partial y^1}\bigg|_p\right) + \cdots + v^k\left(\frac{\partial}{\partial y^k}\bigg|_p\right) + v^{k+1}\left(\frac{\partial}{\partial y^{k+1}}\bigg|_p\right)\right.\right.$$

$$+ \cdots + v^m\left(\frac{\partial}{\partial y^m}\bigg|_p\right)\right)(f)\right)(p) = \left(\left(v^1\left(\left(\frac{\partial}{\partial y^1}\bigg|_p\right)(f)\right)\right.\right.$$

$$+ \cdots + v^k\left(\left(\frac{\partial}{\partial y^k}\bigg|_p\right)(f)\right) + v^{k+1}\left(\left(\frac{\partial}{\partial y^{k+1}}\bigg|_p\right)(f)\right)$$

$$+ \cdots + v^m\left(\left(\frac{\partial}{\partial y^m}\bigg|_p\right)(f)\right)\right)(p) = \left(\left(v^1\left(\left(\frac{\partial}{\partial y^1}\bigg|_p\right)(f|_W)\right)\right.\right.$$

$$+ \cdots + v^k\left(\left(\frac{\partial}{\partial y^k}\bigg|_p\right)(f|_W)\right) + v^{k+1}\left(\left(\frac{\partial}{\partial y^{k+1}}\bigg|_p\right)(f|_W)\right)$$

$$+ \cdots + v^m\left(\left(\frac{\partial}{\partial y^m}\bigg|_p\right)(f|_W)\right)\right)(p) = \left(\left(v^1\left(\left(\frac{\partial}{\partial y^1}\bigg|_p\right)((\pi_{k+1} \circ \psi)|_W)\right)\right.\right.$$

$$+ \cdots + v^k\left(\left(\frac{\partial}{\partial y^k}\bigg|_p\right)((\pi_{k+1} \circ \psi)|_W)\right) + v^{k+1}\left(\left(\frac{\partial}{\partial y^{k+1}}\bigg|_p\right)((\pi_{k+1} \circ \psi)|_W)\right)$$

$$+ \cdots + v^m\left(\left(\frac{\partial}{\partial y^m}\bigg|_p\right)((\pi_{k+1} \circ \psi)|_W)\right)\right)(p).$$

Now, since

$$\left(\frac{\partial}{\partial y^i}\bigg|_p\right)((\pi_{k+1} \circ \psi)|_W) = \left(\left(d(\psi^{-1})_{\psi(p)}\right)\left(\frac{\partial}{\partial y^i}\bigg|_{\psi(p)}\right)\right)((\pi_{k+1} \circ \psi)|_W)$$

$$= \left(\frac{\partial}{\partial y^i}\bigg|_{\psi(p)}\right)(((\pi_{k+1} \circ \psi)|_W) \circ \psi^{-1}) = \left(\frac{\partial}{\partial y^i}\bigg|_{\psi(p)}\right)(\pi_{k+1})$$

$$= \begin{cases} 0 & \text{if } i \neq k+1 \\ 1 & \text{if } i = k+1, \end{cases}$$

$$0 = \left(\left(v^1(0) + \cdots + v^k(0) + v^{k+1}(1) + \cdots + v^m(0)\right)\right)(p) = v^{k+1}(1(p)) = v^{k+1} \cdot 1$$
$$= v^{k+1}.$$

Thus, $v^{k+1} = 0$.
  Similarly, $v^{k+2} = 0, \cdots, v^m = 0$. Hence,

$$v = v^1 \left(\frac{\partial}{\partial y^1}\bigg|_p\right) + \cdots + v^k \left(\frac{\partial}{\partial y^k}\bigg|_p\right) + 0 \left(\frac{\partial}{\partial y^{k+1}}\bigg|_p\right) + \cdots + 0 \left(\frac{\partial}{\partial y^m}\bigg|_p\right)$$

$$= v^1 \left(\frac{\partial}{\partial y^1}\bigg|_p\right) + \cdots + v^k \left(\frac{\partial}{\partial y^k}\bigg|_p\right) \in \text{span}\left(\frac{\partial}{\partial y^1}\bigg|_p, \cdots, \frac{\partial}{\partial y^k}\bigg|_p\right) = (d\iota_p)(T_pS).$$

Thus, $v \in (d\iota_p)(T_pS)$. Now, since we do not distinguish between $(d\iota_p)(T_pS)$ and $T_pS$, we can write $v \in T_pS$.                                                                                  □

# Chapter 6
# Sard's Theorem

The root of Sard's theorem lies in real analysis. It is one of the deep results in real analysis. How this result can be generalized into the realm of smooth manifold theory is only a later development. Its importance in manifold theory is for a definite reason. Upon applying Sard's theorem, Whitney was able to prove a startling property about smooth manifolds: For every smooth manifold, ambient space can be constructed, etc. For pedagogical reasons, we have given its proof in step-by-step manner using Taylor's inequality. Largely, this chapter is self-contained.

## 6.1 Measure Zero in Manifolds

**Lemma 6.1** *Let $A$ be any nonempty compact subset of $\mathbb{R}^4$. For every $k = 1, 2, 3, 4$, let $\mu_k$ denote the $k$-dimensional (Lebesgue) measure over $\mathbb{R}^k$.*

*If for every real $c$, $\mu_3(\{(y, z, w) : (c, y, z, w) \in A\}) = 0$, then $\mu_4(A) = 0$.*

*Proof* Since $A$ is compact in $\mathbb{R}^4$, by the Henie-Borel theorem, $A$ is bounded and closed. Since $A$ is bounded, there exist real numbers $a, b, a_2, b_2, a_3, b_3, a_4, b_4$ such that $a < b, a_2 < b_2, a_3 < b_3, a_4 < b_4$, and $A \subset [a, b] \times [a_2, b_2] \times [a_3, b_3] \times [a_4, b_4] \subset [a, b] \times \mathbb{R} \times \mathbb{R} \times \mathbb{R} = [a, b] \times \mathbb{R}^3$. Thus, $A \subset [a, b] \times \mathbb{R}^3$.

Let us fix any $\delta_1 > 0$. Put $\delta \equiv \frac{\delta_1}{2(b-a)} \ (> 0)$. It suffices to find a finite collection of four-dimensional rectangles that covers $A$, and sum of their $\mu_4$-measures is less than $\delta_1$.

Next let us fix any real $c \in [a, b]$.

Clearly, $\{(y, z, w) : (c, y, z, w) \in A\}$ is a compact subset of $\mathbb{R}^3$.

(Reason: Let us define a function $f : \mathbb{R}^4 \to \mathbb{R}^3$ as follows: For every $(x, y, z, w)$ in $\mathbb{R}^4$, $f(x, y, z, w) \equiv (y, z, w)$. Clearly, $f : \mathbb{R}^4 \to \mathbb{R}^3$ is continuous. It is clear that $\{c\} \times \mathbb{R}^3$ is a closed subset of $\mathbb{R}^4$. Since $\{c\} \times \mathbb{R}^3$ is a closed subset of $\mathbb{R}^4$, and $A$ is a closed subset of $\mathbb{R}^4$, their intersection $A \cap (\{c\} \times \mathbb{R}^3)$ is a closed subset of the compact set $A$, and hence, $A \cap (\{c\} \times \mathbb{R}^3)$ is compact. Since $A \cap (\{c\} \times \mathbb{R}^3)$ is

© Springer India 2014

R. Sinha, *Smooth Manifolds*, DOI 10.1007/978-81-322-2104-3_6

compact, and $f : \mathbb{R}^4 \to \mathbb{R}^3$ is continuous, the $f$-image set $f(A \cap (\{c\} \times \mathbb{R}^3))$ $(= \{f(c, y, z, w) : (c, y, z, w) \in A\} = \{(y, z, w) : (c, y, z, w) \in A\})$ is compact in $\mathbb{R}^3$, and hence, $\{(y, z, w) : (c, y, z, w) \in A\}$ is compact in $\mathbb{R}^3$.

Since the three-dimensional measure $\mu_3\{(y, z, w) : (c, y, z, w) \in A\} = 0$, and $\delta > 0$, by the definition of measure, there exists a countable collection $\{C_1, C_2, C_3, \ldots\}$ of open cubes in $\mathbb{R}^3$ such that $\{(y, z, w) : (c, y, z, w) \in A\} \subset C_1 \cup C_2 \cup C_3 \cup \cdots$, and $\mu_3(C_1) + \mu_3(C_2) + \mu_3(C_3) + \cdots < \delta$. Since $\{(y, z, w) : (c, y, z, w) \in A\}$ is a compact subset of $\mathbb{R}^3$, and $\{C_1, C_2, C_3, \ldots\}$ is an open cover of $\{(y, z, w) : (c, y, z, w) \in A\}$, there exists a positive integer $k_c$ such that $\{(y, z, w) : (c, y, z, w) \in A\} \subset C_1 \cup C_2 \cup \cdots \cup C_{k_c}$, and $\mu_3(C_1) + \mu_3(C_2) + \cdots + \mu_3(C_{k_c}) \le \mu_3(C_1) + \mu_3(C_2) + \mu_3(C_3) + \cdots < \delta$.

We claim that there exists a positive integer $n$ such that $A \cap ((c - \frac{1}{n}, c + \frac{1}{n}) \times \mathbb{R}^3) \subset (c - \frac{1}{n}, c + \frac{1}{n}) \times (C_1 \cup C_2 \cup \cdots \cup C_{k_c})$.

If not, otherwise, let for every positive integer $n$, $A \cap ((c - \frac{1}{n}, c + \frac{1}{n}) \times \mathbb{R}^3) \not\subset (c - \frac{1}{n}, c + \frac{1}{n}) \times (C_1 \cup C_2 \cup \cdots \cup C_{k_c})$. We have to arrive at a contradiction.

Thus, for every positive integer $n$, there exists $(x_n, y_n, z_n, w_n) \in A \cap ((c - \frac{1}{n}, c + \frac{1}{n}) \times \mathbb{R}^3)(\subset A)$ and $(x_n, y_n, z_n, w_n) \notin (c - \frac{1}{n}, c + \frac{1}{n}) \times (C_1 \cup C_2 \cup \cdots \cup C_{k_c})$. Since $\{(x_1, y_1, z_1, w_1), (x_2, y_2, z_2, w_2), (x_3, y_3, z_3, w_3), \ldots\}$ is a sequence in the compact set $A$, there exists a subsequence $\{(x_{n_1}, y_{n_1}, z_{n_1} w_{n_1}), (x_{n_2}, y_{n_2}, z_{n_2}, w_{n_2}), (x_{n_3}, y_{n_3}, z_{n_3}, w_{n_3}), \ldots\}$ of $\{(x_1, y_1, z_1, w_1), (x_2, y_2, z_2, w_2), (x_3, y_3, z_3, w_3), \ldots\}$ such that $\{(x_{n_1}, y_{n_1}, z_{n_1} w_{n_1}), (x_{n_2}, y_{n_2}, z_{n_2}, w_{n_2}), (x_{n_3}, y_{n_3}, z_{n_3}, w_{n_3}), \ldots\}$ is convergent, and hence, there exist real numbers $\alpha_1, \alpha_2, \alpha_3, \alpha_4$ such that $\lim_{i \to \infty}(x_{n_i}, y_{n_i}, z_{n_i} w_{n_i}) = (\alpha_1, \alpha_2, \alpha_3, \alpha_4)$. It follows that $\lim_{i \to \infty} x_{n_i} = \alpha_1$, $\lim_{i \to \infty} y_{n_i} = \alpha_2$, $\lim_{i \to \infty} z_{n_i} = \alpha_3$, $\lim_{i \to \infty} w_{n_i} = \alpha_4$. Since each $(x_n, y_n, z_n, w_n) \in A \cap ((c - \frac{1}{n}, c + \frac{1}{n}) \times \mathbb{R}^3)$, each $x_n \in (c - \frac{1}{n}, c + \frac{1}{n})$, and hence, $\lim_{n \to \infty} x_n = c$. Since $\lim_{n \to \infty} x_n = c$, $\alpha_1 = \lim_{i \to \infty} x_{n_i} = c$. Since $\alpha_1 = c$, and $\lim_{i \to \infty}(x_{n_i}, y_{n_i}, z_{n_i}, w_{n_i}) = (\alpha_1, \alpha_2, \alpha_3, \alpha_4)$, $\lim_{i \to \infty}(x_{n_i}, y_{n_i}, z_{n_i}, w_{n_i}) = (c, \alpha_2, \alpha_3, \alpha_4)$. Since $(x_n, y_n, z_n, w_n) \notin (c - \frac{1}{n}, c + \frac{1}{n}) \times (C_1 \cup C_2 \cup \cdots \cup C_{k_c})$, and $x_n \in (c - \frac{1}{n}, c + \frac{1}{n})$, $(y_n, z_n, w_n) \notin (C_1 \cup C_2 \cup \cdots \cup C_{k_c})$. Since $\lim_{i \to \infty}(x_{n_i}, y_{n_i}, z_{n_i} w_{n_i}) = (c, \alpha_2, \alpha_3, \alpha_4)$, $\lim_{i \to \infty}(y_{n_i}, z_{n_i} w_{n_i}) = (\alpha_2, \alpha_3, \alpha_4)$. Since each $(y_{n_i}, z_{n_i} w_{n_i})$ is in the closed set $(C_1 \cup C_2 \cup \cdots \cup C_{k_c})^c$, $(\alpha_2, \alpha_3, \alpha_4) = (\lim_{i \to \infty}(y_{n_i}, z_{n_i} w_{n_i})) \in (C_1 \cup C_2 \cup \cdots \cup C_{k_c})^c$, and hence, $(\alpha_2, \alpha_3, \alpha_4) \notin (C_1 \cup C_2 \cup \cdots \cup C_{k_c})(\supset \{(y, z, w) : (c, y, z, w) \in A\})$. It follows that $(\alpha_2, \alpha_3, \alpha_4) \notin \{(y, z, w) : (c, y, z, w) \in A\}$, and hence, $(c, \alpha_2, \alpha_3, \alpha_4) \notin A$. Since $\lim_{i \to \infty}(x_{n_i}, y_{n_i}, z_{n_i} w_{n_i}) = (c, \alpha_2, \alpha_3, \alpha_4)$, and each $(x_{n_i}, y_{n_i}, z_{n_i} w_{n_i})$ is in the closed set $A$, $(c, \alpha_2, \alpha_3, \alpha_4) = (\lim_{i \to \infty}(x_{n_i}, y_{n_i}, z_{n_i} w_{n_i})) \in A$, which is a contradiction. So our claim is true.

Thus, for every $c \in [a, b]$, there exist a positive integer $n_c$ and an open subset $G_c(\equiv C_1 \cup C_2 \cup \cdots \cup C_{k_c})$ of $\mathbb{R}^3$ such that $A \cap ((c - \frac{1}{n_c}, c + \frac{1}{n_c}) \times \mathbb{R}^3) \subset (c - \frac{1}{n_c}, c + \frac{1}{n_c}) \times G_c$. If we shrink $(c - \frac{1}{n_c}, c + \frac{1}{n_c})$ into another open interval $J$, that is, $J \subset (c - \frac{1}{n_c}, c + \frac{1}{n_c})$, then clearly $A \cap (J \times \mathbb{R}^3) \subset J \times G_c$. (Reason: Let $(x, y, z, w) \in \text{LHS} = A \cap (J \times \mathbb{R}^3)$. It follows that $x \in J$ and $(x, y, z, w) \in A$. Since $x \in J \subset (c - \frac{1}{n_c}, c + \frac{1}{n_c})$, $x \in (c - \frac{1}{n_c}, c + \frac{1}{n_c})$. Since $x \in (c - \frac{1}{n_c}, c + \frac{1}{n_c})$, and

$(x, y, z, w) \in A, \quad (x, y, z, w) \in A \cap ((c - \frac{1}{n_c}, c + \frac{1}{n_c}) \times \mathbb{R}^3) \subset (c - \frac{1}{n_c}, c + \frac{1}{n_c}) \times G_c,$
and hence, $(y, z, w) \in G_c$. Since $x \in J$, and $(y, z, w) \in G_c$, $(x, y, z, w) \in J \times G_c =$
RHS. Thus, LHS $\subset$ RHS).

Here, $\mu_3(G_{c_1}) = \mu_3(C_1 \cup \cdots \cup C_{k_{c_1}}) \leq \mu_3(C_1) + \cdots + \mu_3(C_{k_{c_1}}) < \delta.$ Thus,
$\mu_3(G_{c_1}) < \delta$. Similarly, $\mu_3(G_{c_1}) < \delta$, etc.

Since $[a, b]$ is compact, and $\{(c - \frac{1}{n_c}, c + \frac{1}{n_c}) : c \in [a, b]\}$ is an open cover of
$[a, b]$, there exist finite many $c_1, \ldots, c_l$ in $[a, b]$ such that $[a, b] \subset$
$(c_1 - \frac{1}{n_{c_1}}, c_1 + \frac{1}{n_{c_1}}) \cup \cdots \cup (c_l - \frac{1}{n_{c_l}}, c_l + \frac{1}{n_{c_l}})$. From the above observation, there
exist open intervals $J_1, \ldots, J_l$ such that $J_1 \subset (c_1 - \frac{1}{n_{c_1}}, c_1 + \frac{1}{n_{c_1}}), \ldots, J_l \subset$
$(c_l - \frac{1}{n_{c_l}}, c_l + \frac{1}{n_{c_l}})$, $[a, b] \subset J_1 \cup \cdots \cup J_l$, and $\mu_1(J_1) + \cdots + \mu_1(J_l) \leq 2(b - a)$.
Further, since $A \subset [a, b] \times \mathbb{R}^3$, so

$$A = A \cap ([a, b] \times \mathbb{R}^3) \subset A \cap ((J_1 \cup \cdots \cup J_l) \times \mathbb{R}^3) = A \cap ((J_1 \times \mathbb{R}^3) \cup \cdots \cup (J_l \times \mathbb{R}^3))$$
$$= (A \cap (J_1 \times \mathbb{R}^3)) \cup \cdots \cup (A \cap (J_l \times \mathbb{R}^3)) \subset (J_1 \times G_{c_1}) \cup \cdots \cup (J_l \times G_{c_l}).$$

Further, since $J_1 \times G_{c_1} = J_1 \times (C_1 \cup \cdots \cup C_{k_{c_1}}) = (J_1 \times C_1) \cup \cdots \cup (J_1 \times C_{k_{c_1}})$ and
$J_1 \times C_1, \ldots, J_1 \times C_{k_{c_1}}$ are rectangles in $\mathbb{R}^4$, $J_1 \times G_{c_1}$ is a finite union of rectangles.
Similarly, $J_1 \times G_{c_2}$ is a finite union of rectangles, etc. Now, since $A \subset$
$(J_1 \times G_{c_1}) \cup \cdots \cup (J_l \times G_{c_l})$, $A$ is covered by a finite collection of rectangles in
$\mathbb{R}^4$. Here, $\mu_4(J_1 \times G_{c_1}) + \cdots + \mu_4(J_l \times G_{c_l}) = \mu_1(J_1) \cdot \mu_3(G_{c_1}) + \cdots + \mu_1(J_l) \cdot \mu_3(G_{c_l})$
$< \mu_1(J_1) \cdot \delta + \cdots + \mu_1(J_l) \cdot \delta = (\mu_1(J_1) + \cdots + \mu_1(J_l))\delta \leq 2(b - a) \cdot \delta = 2(b - a) \cdot$
$\frac{\delta_1}{2(b-a)} = \delta_1$. $\qquad \square$

**Note 6.2** The Lemma 6.1 can be generalized as follows:

Let $A$ be any nonempty compact subset of $\mathbb{R}^n$. For every $k = 1, \ldots, n$, let $\mu_k$
denote the $k$-dimensional measure over $\mathbb{R}^k$.

If for every real $c$, $\mu_{n-1}(\{(x_2, \ldots, x_n) : (c, x_2, \ldots, x_n) \in A\}) = 0$, then $\mu_n(A) = 0$.

**Lemma 6.3** *Let $A$ be a nonempty compact subset of $\mathbb{R}$. Let $f : A \to \mathbb{R}$ be continuous. For every $k = 1, 2$, let $\mu_k$ denote the $k$-dimensional measure over $\mathbb{R}^k$.*

*Then, $\mu_2(\{(x, f(x)) : x \in A\}) = 0$.*

*In short, the graph of a continuous function has measure 0 in $\mathbb{R}^2$.*

*Proof* $A$ is a compact subset of $\mathbb{R}^1(= \mathbb{R})$. We have to prove that
$\mu_2(\{(x, f(x)) : x \in A\}) = 0$.

We want to apply the result of 6.2. Clearly, $\{(x, f(x)) : x \in A\}$ is a compact
subset of $\mathbb{R}^2$.

(Reason: First of all, we shall try to show that $\{(x, f(x)) : x \in A\}$ is a closed
subset of $\mathbb{R}^2$. For this purpose, let us take a convergent sequence $\{(x_n, f(x_n))\}$ in
$\{(x, f(x)) : x \in A\}$. Let $\lim_{n \to \infty}(x_n, f(x_n)) = (\alpha, \beta)$. We have to show that $(\alpha, \beta) \in$
$\{(x, f(x)) : x \in A\}$, that is, $\alpha \in A$ and $f(\alpha) = \beta$. Since $\lim_{n \to \infty}(x_n, f(x_n)) = (\alpha, \beta)$,
$\lim_{n \to \infty} x_n = \alpha$, and $\lim_{n \to \infty} f(x_n) = \beta$. Since $\{(x_n, f(x_n))\}$ is in $\{(x, f(x)) : x \in A\}$,
for each positive integer $n$, $x_n \in A$. Since $A$ is compact, $A$ is closed. Since $A$ is

closed, for each positive integer $n$, $x_n \in A$, and $\lim_{n\to\infty} x_n = \alpha$, $\alpha = (\lim_{n\to\infty} x_n) \in A$, and hence, $\alpha \in A$. Since $\{x_n\}$ is a convergent sequence in $A$, $\lim_{n\to\infty} x_n = \alpha \in A$, and $f : A \to \mathbb{R}$ is continuous, $\beta = \lim_{n\to\infty} f(x_n) = f(\alpha)$. Thus, we have shown that $\{(x, f(x)) : x \in A\}$ is a closed subset of $\mathbb{R}^2$. Clearly, $\{(x, f(x)) : x \in A\} \subset A \times f(A)$. Since $A$ is compact, and $f : A \to \mathbb{R}$ is continuous, $f(A)$ is compact. Since $A$ is compact, and $f(A)$ is compact, their Cartesian product $A \times f(A)$ is compact. Thus, $\{(x, f(x)) : x \in A\}$ is a closed subset of the compact set $A \times f(A)$, and hence, $\{(x, f(x)) : x \in A\}$ is compact.)

For every real $c$, $\{x_2 : (c, x_2) \in \{(x, f(x)) : x \in A\}\} = \emptyset$ or $\{f(c)\}$, so $\mu_1(\{x_2 : (c, x_2) \in \{(x, f(x)) : x \in A\}\}) = \mu_1(\emptyset)(= 0)$ or $\mu_1(\{f(c)\})(= 0)$, and hence, $\mu_1(\{x_2 : (c, x_2) \in \{(x, f(x)) : x \in A\}\}) = 0$. Now, we can apply Lemma 6.1. It follows that $\mu_2(\{(x, f(x)) : x \in A\}) = 0$. □

**Lemma 6.4** *Let $A$ be a nonempty compact subset of $\mathbb{R}^2$. Let $f : A \to \mathbb{R}$ be continuous. For every $k = 1, 2, 3$, let $\mu_k$ denote the k-dimensional measure over $\mathbb{R}^k$.*
*Then, $\mu_3(\{(x, y, f(x, y)) : (x, y) \in A\}) = 0$.*

*Proof* We have to prove that $\mu_3(\{(x, y, f(x, y)) : (x, y) \in A\}) = 0$ (see Fig. 6.1).

Clearly, $\{(x, y, f(x, y)) : (x, y) \in A\}$ is a compact subset of $\mathbb{R}^3$.

(Reason: First of all, we shall try to show that $\{(x, y, f(x, y)) : (x, y) \in A\}$ is a closed subset of $\mathbb{R}^3$. For this purpose, let us take any convergent sequence $\{(x_n, y_n, f(x_n, y_n))\}$ in $\{(x, y, f(x, y)) : (x, y) \in A\}$. Let $\lim_{n\to\infty}(x_n, y_n, f(x_n, y_n)) = (\alpha, \beta, \gamma)$. We have to show that $(\alpha, \beta, \gamma) \in \{(x, y, f(x, y)) : (x, y) \in A\}$, that is, $(\alpha, \beta) \in A$. Since $\lim_{n\to\infty}(x_n, y_n, f(x_n, y_n)) = (\alpha, \beta, \gamma)$, $\lim_{n\to\infty} x_n = \alpha$, $\lim_{n\to\infty} y_n = \beta$, and $\lim_{n\to\infty} f(x_n, y_n) = \gamma$. Here, $\{(x_n, y_n, f(x_n, y_n))\}$ is a sequence in $\{(x, y, f(x, y)) : (x, y) \in A\}$, so each $(x_n, y_n, f(x_n, y_n)) \in \{(x, y, f(x, y)) : (x, y) \in A\}$, and hence, each $(x_n, y_n) \in A$. Since $A$ is compact, $A$ is closed. Since $\lim_{n\to\infty} x_n = \alpha$, and $\lim_{n\to\infty} y_n = \beta$, $\lim_{n\to\infty}(x_n, y_n) = (\alpha, \beta)$. Since each $(x_n, y_n) = (\alpha, \beta)$, and $A$ is closed, $(\alpha, \beta) = (\lim_{n\to\infty}(x_n, y_n)) \in A$, and hence, $(\alpha, \beta) \in A$. Since each $(x_n, y_n) \in A$, $\lim_{n\to\infty}(x_n, y_n) = (\alpha, \beta) \in A$, and $f : A \to \mathbb{R}$ is continuous, $\gamma = \lim_{n\to\infty} f(x_n, y_n) = f(\alpha, \beta)$. Since $(\alpha, \beta) \in A$, $(\alpha, \beta, \gamma) = (\alpha, \beta, f(\alpha, \beta)) \in \{(x, y, f(x, y)) : (x, y) \in A\}$. Thus, we have shown that $\{(x, y, f(x, y)) : (x, y) \in A\}$ is closed. Here, $\{(x, y, f(x, y)) : (x, y) \in A\} \subset A \times f(A)$. Since $f : A \to \mathbb{R}$ is continuous, and $A$ is compact, $f(A)$ is compact. Since $A$ is compact, and $f(A)$ is compact,

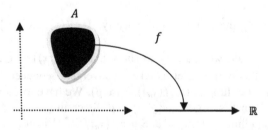

**Fig. 6.1** $\{(x, y, f(x, y)) : (x, y) \in A\}$ is a compact subset of $\mathbb{R}^3$

$A \times f(A)$ is compact in $\mathbb{R}^3$. Since $\{(x, y, f(x, y)) : (x, y) \in A\}$ is a closed subset of the compact set $A \times f(A)$, $\{(x, y, f(x, y)) : (x, y) \in A\}$ is compact.)

Now, by 6.2, it suffices to show that for every real $c$,

$$\mu_2(\{(s, t) : (c, s, t) \in \{(x, y, f(x, y)) : (x, y) \in A\}\}) = 0.$$

For this purpose, let us fix any real $c$. Since $A$ is a nonempty compact subset of $\mathbb{R}^2$, clearly $\{s : (c, s) \in A\}$ is a compact subset of $\mathbb{R}$. (See the proof of Lemma 6.1) Let us define a function $f_c : \{s : (c, s) \in A\} \to \mathbb{R}$ as follows: for every $s \in \{s : (c, s) \in A\}$, $f_c(s) \equiv f(c, s)$. Since $f_c$ is the composite of continuous function $s \mapsto (c, s)$, and continuous function $f : A \to \mathbb{R}$, $f_c : \{s : (c, s) \in A\} \to \mathbb{R}$ is a continuous function. Also, the domain of $f_c$, that is, $\{s : (c, s) \in A\}$ is compact. It follows, by Lemma 6.3, that $\mu_2(\{(x, f_c(x)) : x \in \{s : (c, s) \in A\}\}) = 0$. Now, it suffices to show that $\{(s, t) : (c, s, t) \in \{(x, y, f(x, y)) : (x, y) \in A\}\} = \{(x, f_c(x)) : x \in \{s : (c, s) \in A\}\}$.

Let us take any $(\eta, \varsigma) \in$ LHS. It follows that $(c, \eta, \varsigma) \in \{(x, y, f(x, y)) : (x, y) \in A\}$, and hence, $(c, \eta) \in A$, and $\varsigma = f(c, \eta) = f_c(\eta)$. Since $(c, \eta) \in A$, $\eta \in \{s : (c, s) \in A\}$, and hence, $(\eta, \varsigma) = (\eta, f_c(\eta)) \in$ RHS. Thus, LHS $\subset$ RHS. Next, let us take any $(\eta, \varsigma) \in$ RHS. It follows that there exists a real $x$ such that $(\eta, \varsigma) = (x, f_c(x))$. Since $(\eta, \varsigma) = (x, f_c(x))$, $\eta = x$, and $\varsigma = f_c(x) = f(c, x)$. Since $\varsigma = f(c, x)$, and $f : A \to \mathbb{R}$, $(c, x) \in A$, and hence, $(c, \eta, \varsigma) = (c, x, \varsigma) = (c, x, f(c, x)) \in \{(x, y, f(x, y)) : (x, y) \in A\}$. Since $(c, \eta, \varsigma) \in \{(x, y, f(x, y)) : (x, y) \in A\}$, $(\eta, \varsigma) \in$ LHS. Thus, RHS $\subset$ LHS. Since RHS $\subset$ LHS, and LHS $\subset$ RHS, LHS = RHS. $\square$

**Note 6.5** The Lemma 6.4 can be generalized as follows:

Let $n \in \{2, 3, 4, \ldots\}$. Let $A$ be a nonempty compact subset of $\mathbb{R}^{n-1}$. Let $f : A \to \mathbb{R}$ be continuous. For every $k = 1, \ldots, n$, let $\mu_k$ denote the $k$-dimensional measure over $\mathbb{R}^k$. Then, $\mu_n(\{(x, f(x)) : x \in A\}) = 0$.

**Lemma 6.6** *Let $X$ be a second countable, locally compact Hausdorff space. Then, there exists a sequence $\{K_n\}$ of compact subsets of $X$ such that*

1. *$\{K_n\}$ is a cover of $X$,*
2. *$K_1 \subset (K_2)^\circ \subset K_2 \subset (K_3)^\circ \subset K_3 \subset (K_4)^\circ \subset K_4 \subset \cdots$.*

*Proof* Since $X$ is a locally compact Hausdorff space, there exists a basis $\mathcal{B}$ of $X$ such that if $U \in \mathcal{B}$ then $U^-$ is compact. Since $\mathcal{B}$ is a basis of $X$, $\mathcal{B}$ is an open cover of $X$. Since $\mathcal{B}$ is an open cover of $X$, and $X$ is a second countable space, there exists a countable collection $\{U_1, U_2, U_3, \ldots\}$ of members of $\mathcal{B}$ which also covers $X$. Since $\{U_1, U_2, U_3, \ldots\}$ covers $X$, and for each positive integer $n$, $U_n \subset (U_n)^-$, $\{(U_1)^-, (U_2)^-, (U_3)^-, \ldots\}$ also covers $X$. Put $K_1 \equiv (U_1)^-$. Since $U_1 \in \mathcal{B}$, $(U_1)^- (= K_1)$ is compact, and hence, $K_1$ is compact. Since $K_1$ is compact, and $\{U_1, U_2, U_3, \ldots\}$ is an open cover of $K_1$, there exists a positive integer $n_1 > 1$ such that $K_1 \subset U_1 \cup \cdots \cup U_{n_1}$. Here, $(U_1 \cup \cdots \cup U_{n_1})^- = (U_1)^- \cup \cdots \cup (U_{n_1})^-$. Since $(U_1)^-, \ldots, (U_{n_1})^-$ are compact, $(U_1)^- \cup \cdots \cup (U_{n_1})^- (= (U_1 \cup \cdots \cup U_{n_1})^-)$ is compact, and hence, $(U_1 \cup \cdots \cup U_{n_1})^-$ is compact. Put $K_2 \equiv (U_1 \cup \cdots \cup U_{n_1})^-$.

Since $(U_1 \cup \cdots \cup U_{n_1})^- (= K_2)$ is compact, $K_2$ is compact. Since $U_1 \cup \cdots \cup U_{n_1}$ is an open set, and $(U_1 \cup \cdots \cup U_{n_1}) \subset (U_1 \cup \cdots \cup U_{n_1})^- = K_2$, $K_1 \subset (U_1 \cup \cdots \cup U_{n_1})$ $\subset (K_2)^{\circ} \subset K_2$. Thus, $K_1, K_2$ are compact sets satisfying $K_1 \subset (K_2)^{\circ} \subset K_2$. Since $K_2$ is compact, and $\{U_1, U_2, U_3, \ldots\}$ is an open cover of $K_2$, there exists a positive integer $n_2 > n_1$ such that $K_2 \subset U_1 \cup \cdots \cup U_{n_1} \cup \cdots \cup U_{n_2}$. Here, $(U_1 \cup \cdots \cup U_{n_2})^- = (U_1)^- \cup \cdots \cup (U_{n_2})^-$. Since $(U_1)^-, \ldots, (U_{n_2})^-$ are compact, $(U_1)^- \cup \cdots \cup (U_{n_2})^- (= (U_1 \cup \cdots \cup U_{n_2})^-)$ is compact, and hence, $(U_1 \cup \cdots \cup U_{n_2})^-$ is compact. Put $K_3 \equiv (U_1 \cup \cdots \cup U_{n_2})^-$. Since $(U_1 \cup \cdots \cup U_{n_2})^- (= K_3)$ is compact, $K_3$ is compact. Since $U_1 \cup \cdots \cup U_{n_2}$ is an open set, and $(U_1 \cup \cdots \cup U_{n_2}) \subset (U_1 \cup \cdots \cup U_{n_2})^- = K_3$, $K_2 \subset (U_1 \cup \cdots \cup U_{n_2}) \subset (K_3)^{\circ} \subset K_3$. Thus, $K_1, K_2, K_3$ are compact sets satisfying $K_1 \subset (K_2)^{\circ} \subset K_2 \subset (K_3)^{\circ} \subset K_3$, etc. This proves 2. Since $1 < n_1 < n_2 < n_3 < \cdots$, $U_1 \subset K_1, U_1 \cup \cdots \cup U_{n_1} \subset K_2, U_1 \cup \cdots \cup U_{n_2} \subset K_3$, etc., and $\{U_1, U_2, U_3, \ldots\}$ covers $X$, $\{K_1, K_2, K_3, \ldots\}$ is a cover of $X$. This proves 1.     $\square$

**Lemma 6.7** *Let $n \in \{2, 3, 4, \ldots\}$. Let $A$ be a nonempty open subset of $\mathbb{R}^{n-1}$. Let $f : A \to \mathbb{R}$ be continuous. For every $k = 1, \ldots, n$, let $\mu_k$ denote the $k$-dimensional measure over $\mathbb{R}^k$. Then, $\mu_n(\{(x, f(x)) : x \in A\}) = 0$.*

*Proof* Since $\mathbb{R}^{n-1}$ is a second countable space, and $A$ is a nonempty subset of $\mathbb{R}^{n-1}$, $A$ is a second countable space. Since $\mathbb{R}^{n-1}$ is locally compact Hausdorff space, and $A$ is a nonempty open subset of $\mathbb{R}^{n-1}$, by Note 5.54, $A$ is a locally compact Hausdorff. Since $A$ is a second countable, locally compact Hausdorff space, by Lemma 6.6, there exists a sequence $\{K_n\}$ of compact subsets of $X$ such that $\{K_n\}$ is a cover of $X$. Hence, by Note 6.5,

$$0 \le \mu_n(\{(x, f(x)) : x \in A\}) \le \mu_n\left(\left\{(x, f(x)) : x \in \bigcup_{n=1}^{\infty} K_n\right\}\right)$$

$$= \mu_n\left(\bigcup_{n=1}^{\infty} \{(x, f(x)) : x \in K_n\}\right) \le \sum_{n=1}^{\infty} \mu_n(\{(x, f(x)) : x \in K_n\})$$

$$= \sum_{n=1}^{\infty} \mu_n\left(\left\{\left(x, \left(f|_{K_n}\right)(x)\right) : x \in K_n\right\}\right) = \sum_{n=1}^{\infty} 0 = 0.$$

Hence, $\mu_n(\{(x, f(x)) : x \in A\}) = 0$.     $\square$

**Lemma 6.8** *Let $n \in \{2, 3, 4, \ldots\}$. Let $A$ be a nonempty closed subset of $\mathbb{R}^{n-1}$. Let $f : A \to \mathbb{R}$ be continuous. For every $k = 1, \ldots, n$, let $\mu_k$ denote the $k$-dimensional measure over $\mathbb{R}^k$. Then, $\mu_n(\{(x, f(x)) : x \in A\}) = 0$.*

*Proof* Since $\mathbb{R}^{n-1}$ is a second countable space, and $A$ is a nonempty subset of $\mathbb{R}^{n-1}$, $A$ is a second countable space. Since $\mathbb{R}^{n-1}$ is locally compact Hausdorff space, and $A$ is a nonempty closed subset of $\mathbb{R}^{n-1}$, by Note 5.55, $A$ is a locally compact Hausdorff. Since $A$ is a second countable, locally compact Hausdorff space, by Lemma 6.6, there exists a sequence $\{K_n\}$ of compact subsets of $X$ such that $\{K_n\}$ is a cover of $X$. Hence, by Note 6.5,

$$0 \leq \mu_n(\{(x,f(x)) : x \in A\}) \leq \mu_n\left(\left\{(x,f(x)) : x \in \bigcup_{n=1}^{\infty} K_n\right\}\right)$$

$$= \mu_n\left(\bigcup_{n=1}^{\infty}\{(x,f(x)) : x \in K_n\}\right) \leq \sum_{n=1}^{\infty} \mu_n(\{(x,f(x)) : x \in K_n\})$$

$$= \sum_{n=1}^{\infty} \mu_n\left(\left\{\left(x, \left(f|_{K_n}\right)(x)\right) : x \in K_n\right\}\right) = \sum_{n=1}^{\infty} 0 = 0.$$

Hence, $\mu_n(\{(x,f(x)) : x \in A\}) = 0.$ $\qquad\square$

**Definition** Let $V$ be any $n$-dimensional real linear space. Any subset of $V$ of the form $a + S$, where $S$ is a linear subspace of $V$, and $a \in V$, is called an *affine subspace of $V$ parallel to $S$ and passing through $a$*.

**Note 6.9** In real linear space $\mathbb{R}^3$, let $a + S$ be an affine subspace of $\mathbb{R}^3$ parallel to $S$ and passing through $a$. Let the dimension of the subspace $S$ be 2. We want to show that $\mu_3(a + S) = 0$. Since the dimension of the subspace $S$ is 2, there exist $(b_1, b_2, b_3), (c_1, c_2, c_3) \in \mathbb{R}^3$ such that $S = \{s(b_1, b_2, b_3) + t(c_1, c_2, c_3) : s, t \in \mathbb{R}\}$. Let $a \equiv (a_1, a_2, a_3)$. Hence,

$$a + S = (a_1, a_2, a_3) + \{s(b_1, b_2, b_3) + t(c_1, c_2, c_3) : s, t \in \mathbb{R}\}$$
$$= \{(a_1, a_2, a_3) + s(b_1, b_2, b_3) + t(c_1, c_2, c_3) : s, t \in \mathbb{R}\}$$
$$= \{(x, y, z) : x = b_1 s + c_1 t + a_1, y = b_2 s + c_2 t + a_2, z = b_3 s + c_3 t + a_3\}.$$

If $(x, y, z) \in a + S$, then there exist real numbers $s, t$ such that

$$\begin{cases} b_1 s + c_1 t = x - a_1 \\ b_2 s + c_2 t = y - a_2 \\ b_3 s + c_3 t = z - a_3. \end{cases}$$

From the first two equations, we have

$$\begin{vmatrix} b_1 & c_1 \\ b_2 & c_2 \end{vmatrix} s = \begin{vmatrix} x - a_1 & c_1 \\ y - a_2 & c_2 \end{vmatrix} \quad \text{and} \quad \begin{vmatrix} b_1 & c_1 \\ b_2 & c_2 \end{vmatrix} t = \begin{vmatrix} b_1 & x - a_1 \\ b_2 & y - a_2 \end{vmatrix}.$$

Now, from the third equation,

$$b_3 \begin{vmatrix} x - a_1 & c_1 \\ y - a_2 & c_2 \end{vmatrix} + c_3 \begin{vmatrix} b_1 & x - a_1 \\ b_2 & y - a_2 \end{vmatrix} = (z - a_3) \begin{vmatrix} b_1 & c_1 \\ b_2 & c_2 \end{vmatrix}.$$

Clearly, this takes the form $A_1 x + A_2 y + A_3 z = A_4$, and not all $A_1, A_2, A_3$ are zero. For definiteness, let $A_3 \neq 0$. It follows that

$$a + S = \left\{ \left( x, y, \frac{A_4}{A_1} - \frac{A_1}{A_3}x - \frac{A_2}{A_3}y \right) : (x, y) \in \mathbb{R}^2 \right\}.$$

Now, we can apply Lemma 6.8. It follows that

$$\mu_3(a + S) = \mu_3\left( \left( x, y, \frac{A_4}{A_1} - \frac{A_1}{A_3}x - \frac{A_2}{A_3}y \right) : (x, y) \in \mathbb{R}^2 \right) = 0. \qquad \square$$

We can easily generalize the above result as the following:

**Lemma 6.10** *Let $n \in \{2, 3, 4, \ldots\}$. Let $A$ be an affine subspace of $\mathbb{R}^n$ such that $A \neq \mathbb{R}^n$. For every $k = 1, \ldots, n$, let $\mu_k$ denote the $k$-dimensional measure over $\mathbb{R}^k$. Then, $\mu_n(A) = 0$.*

*Proof* Its proof is similar to the proof given in the Note 6.9. $\qquad \square$

**Definition** Let $M$ be an $m$-dimensional smooth manifold, and $N$ be an $n$-dimensional smooth manifold. Let $A$ be a nonempty subset of $M$. Let $F : A \rightarrow N$ be any map. If for every $p \in A$, there exist an open neighborhood $W$ of $p$ in $M$, and a function $\tilde{F} : W \rightarrow N$ such that for every $x \in W \cap A$, $\tilde{F}(x) = F(x)$, and $\tilde{F} : W \rightarrow N$ is a smooth function, then we say that $F : A \rightarrow N$ is *smooth on A.*

**Lemma 6.11** *Let $A$ be a nonempty subset of $\mathbb{R}^3$. For every $k = 1, 2, 3$, let $\mu_k$ denote the $k$-dimensional measure over $\mathbb{R}^k$. Let $\mu_3(A) = 0$. Let $F : A \rightarrow \mathbb{R}^3$ be smooth on $A$. Then, $\mu_3(F(A)) = 0$.*

*Proof* Let us take any $p \equiv (p_1, p_2, p_3) \in A$. Since $F : A \rightarrow \mathbb{R}^3$ is smooth on $A$, and $p \in A$, there exist an open neighborhood $W_p \equiv (p_1 - \varepsilon_p, p_1 + \varepsilon_p) \times (p_2 - \varepsilon_p, p_2 + \varepsilon_p) \times (p_3 - \varepsilon_p, p_3 + \varepsilon_p)$ of $p$ in $\mathbb{R}^3$, and a function $F_p : W_p \rightarrow \mathbb{R}^3$ such that for every $x \in W_p \cap A$, $F_p(x) = F(x)$, and $F_p : W_p \rightarrow \mathbb{R}^3$ is a smooth function. Since $A$ is a nonempty subset of $\mathbb{R}^3$, and $\mathbb{R}^3$ is a second countable space, $A$ is a second countable space. Since $A$ is a second countable space, and

$$\left\{ \left( p_1 - \frac{1}{2}\varepsilon_p, p_1 + \frac{1}{2}\varepsilon_p \right) \times \left( p_2 - \frac{1}{2}\varepsilon_p, p_2 + \frac{1}{2}\varepsilon_p \right) \right.$$
$$\left. \times \left( p_3 - \frac{1}{2}\varepsilon_p, p_3 + \frac{1}{2}\varepsilon_p \right) : p \equiv (p_1, p_2, p_3) \in A \right\}$$

is an open cover of $A$, there exists a countable collection of points

$$p^{(n)} \equiv \left( p_1^{(n)}, p_2^{(n)}, p_3^{(n)} \right) \in A (n = 1, 2, 3, \ldots)$$

such that

$$\left\{ \left( p_1^{(n)} - \frac{1}{2}\varepsilon_{p^{(n)}}, p_1^{(n)} + \frac{1}{2}\varepsilon_{p^{(n)}} \right) \times \left( p_2^{(n)} - \frac{1}{2}\varepsilon_{p^{(n)}}, p_2^{(n)} + \frac{1}{2}\varepsilon_{p^{(n)}} \right) \right.$$
$$\left. \times \left( p_3^{(n)} - \frac{1}{2}\varepsilon_{p^{(n)}}, p_3^{(n)} + \frac{1}{2}\varepsilon_{p^{(n)}} \right) : n = 1, 2, 3, \ldots \right\}$$

is an open cover of $A$. It follows that

$$A \subset \bigcup_{n=1}^{\infty} \left( \left[ p_1^{(n)} - \frac{1}{2}\varepsilon_{p^{(n)}}, p_1^{(n)} + \frac{1}{2}\varepsilon_{p^{(n)}} \right] \times \left[ p_2^{(n)} - \frac{1}{2}\varepsilon_{p^{(n)}}, p_2^{(n)} + \frac{1}{2}\varepsilon_{p^{(n)}} \right] \right.$$
$$\left. \times \left[ p_3^{(n)} - \frac{1}{2}\varepsilon_{p^{(n)}}, p_3^{(n)} + \frac{1}{2}\varepsilon_{p^{(n)}} \right] \right),$$

and hence,

$$A = A \cap \left( \bigcup_{n=1}^{\infty} \left( \left[ p_1^{(n)} - \frac{1}{2}\varepsilon_{p^{(n)}}, p_1^{(n)} + \frac{1}{2}\varepsilon_{p^{(n)}} \right] \times \left[ p_2^{(n)} - \frac{1}{2}\varepsilon_{p^{(n)}}, p_2^{(n)} + \frac{1}{2}\varepsilon_{p^{(n)}} \right] \right. \right.$$
$$\left. \left. \times \left[ p_3^{(n)} - \frac{1}{2}\varepsilon_{p^{(n)}}, p_3^{(n)} + \frac{1}{2}\varepsilon_{p^{(n)}} \right] \right) \right)$$
$$= \bigcup_{n=1}^{\infty} \left( A \cap \left( \left[ p_1^{(n)} - \frac{1}{2}\varepsilon_{p^{(n)}}, p_1^{(n)} + \frac{1}{2}\varepsilon_{p^{(n)}} \right] \times \left[ p_2^{(n)} - \frac{1}{2}\varepsilon_{p^{(n)}}, p_2^{(n)} + \frac{1}{2}\varepsilon_{p^{(n)}} \right] \right. \right.$$
$$\left. \left. \times \left[ p_3^{(n)} - \frac{1}{2}\varepsilon_{p^{(n)}}, p_3^{(n)} + \frac{1}{2}\varepsilon_{p^{(n)}} \right] \right) \right).$$

Hence,

$$F(A) = F \left( \bigcup_{n=1}^{\infty} \left( A \cap \left( \left[ p_1^{(n)} - \frac{1}{2}\varepsilon_{p^{(n)}}, p_1^{(n)} + \frac{1}{2}\varepsilon_{p^{(n)}} \right] \times \left[ p_2^{(n)} - \frac{1}{2}\varepsilon_{p^{(n)}}, p_2^{(n)} + \frac{1}{2}\varepsilon_{p^{(n)}} \right] \right. \right. \right.$$
$$\left. \left. \left. \times \left[ p_3^{(n)} - \frac{1}{2}\varepsilon_{p^{(n)}}, p_3^{(n)} + \frac{1}{2}\varepsilon_{p^{(n)}} \right] \right) \right) \right)$$
$$= \bigcup_{n=1}^{\infty} F \left( A \cap \left( \left[ p_1^{(n)} - \frac{1}{2}\varepsilon_{p^{(n)}}, p_1^{(n)} + \frac{1}{2}\varepsilon_{p^{(n)}} \right] \times \left[ p_2^{(n)} - \frac{1}{2}\varepsilon_{p^{(n)}}, p_2^{(n)} + \frac{1}{2}\varepsilon_{p^{(n)}} \right] \right. \right.$$
$$\left. \left. \times \left[ p_3^{(n)} - \frac{1}{2}\varepsilon_{p^{(n)}}, p_3^{(n)} + \frac{1}{2}\varepsilon_{p^{(n)}} \right] \right) \right).$$

It follows that

$$0 \le \mu_3(F(A)) = \mu_3 \left( \bigcup_{n=1}^{\infty} F\left( A \cap \left( \left[ p_1^{(n)} - \frac{1}{2}\varepsilon_{p^{(n)}}, p_1^{(n)} + \frac{1}{2}\varepsilon_{p^{(n)}} \right] \right. \right. \right.$$
$$\left. \times \left[ p_2^{(n)} - \frac{1}{2}\varepsilon_{p^{(n)}}, p_2^{(n)} + \frac{1}{2}\varepsilon_{p^{(n)}} \right] \times \left[ p_3^{(n)} - \frac{1}{2}\varepsilon_{p^{(n)}}, p_3^{(n)} + \frac{1}{2}\varepsilon_{p^{(n)}} \right] \right) \bigg) \bigg)$$
$$\le \sum_{n=1}^{\infty} \mu_3 \left( F\left( A \cap \left( \left[ p_1^{(n)} - \frac{1}{2}\varepsilon_{p^{(n)}}, p_1^{(n)} + \frac{1}{2}\varepsilon_{p^{(n)}} \right] \right. \right. \right.$$
$$\left. \times \left[ p_2^{(n)} - \frac{1}{2}\varepsilon_{p^{(n)}}, p_2^{(n)} + \frac{1}{2}\varepsilon_{p^{(n)}} \right] \times \left[ p_3^{(n)} - \frac{1}{2}\varepsilon_{p^{(n)}}, p_3^{(n)} + \frac{1}{2}\varepsilon_{p^{(n)}} \right] \right) \bigg) \bigg)$$
$$= \sum_{n=1}^{\infty} \mu_3 \left( F\left( A \cap \left( \left[ p_1^{(n)} - \frac{1}{2}\varepsilon_{p^{(n)}}, p_1^{(n)} + \frac{1}{2}\varepsilon_{p^{(n)}} \right] \right. \right. \right.$$
$$\left. \times \left[ p_2^{(n)} - \frac{1}{2}\varepsilon_{p^{(n)}}, p_2^{(n)} + \frac{1}{2}\varepsilon_{p^{(n)}} \right] \times \left[ p_3^{(n)} - \frac{1}{2}\varepsilon_{p^{(n)}}, p_3^{(n)} + \frac{1}{2}\varepsilon_{p^{(n)}} \right] \right) \bigg) \bigg)$$

Hence, it suffices to show that for every $n = 1, 2, 3, \ldots,$

$$\mu_3 \left( F_{p^{(n)}} \left( A \cap \left( \left[ p_1^{(n)} - \frac{1}{2}\varepsilon_{p^{(n)}}, p_1^{(n)} + \frac{1}{2}\varepsilon_{p^{(n)}} \right] \times \left[ p_2^{(n)} - \frac{1}{2}\varepsilon_{p^{(n)}}, p_2^{(n)} + \frac{1}{2}\varepsilon_{p^{(n)}} \right] \right. \right. \right.$$
$$\left. \times \left[ p_3^{(n)} - \frac{1}{2}\varepsilon_{p^{(n)}}, p_3^{(n)} + \frac{1}{2}\varepsilon_{p^{(n)}} \right] \right) \bigg) \bigg) = 0.$$

For this purpose, let us take any $\delta > 0$.
Since

$$F_{p^{(n)}} : \left( p_1^{(n)} - \varepsilon_{p^{(n)}}, p_1^{(n)} + \varepsilon_{p^{(n)}} \right) \times \left( p_2^{(n)} - \varepsilon_{p^{(n)}}, p_2^{(n)} + \varepsilon_{p^{(n)}} \right)$$
$$\times \left( p_3^{(n)} - \varepsilon_{p^{(n)}}, p_3^{(n)} + \varepsilon_{p^{(n)}} \right)$$
$$\to \mathbb{R}^3$$

is a smooth function, by a result similar to Theorem 3.33,

$$\left( F_{p^{(n)}} \right)' : \left( p_1^{(n)} - \varepsilon_{p^{(n)}}, p_1^{(n)} + \varepsilon_{p^{(n)}} \right) \times \left( p_2^{(n)} - \varepsilon_{p^{(n)}}, p_2^{(n)} + \varepsilon_{p^{(n)}} \right)$$
$$\times \left( p_3^{(n)} - \varepsilon_{p^{(n)}}, p_3^{(n)} + \varepsilon_{p^{(n)}} \right)$$
$$\to L\left( \mathbb{R}^3, \mathbb{R}^3 \right)$$

is continuous. Since

$$\left[ p_1^{(n)} - \frac{3}{4} \varepsilon_{p^{(n)}}, p_1^{(n)} + \frac{3}{4} \varepsilon_{p^{(n)}} \right] \times \left[ p_2^{(n)} - \frac{3}{4} \varepsilon_{p^{(n)}}, p_2^{(n)} + \frac{3}{4} \varepsilon_{p^{(n)}} \right]$$
$$\times \left[ p_3^{(n)} - \frac{3}{4} \varepsilon_{p^{(n)}}, p_3^{(n)} + \frac{3}{4} \varepsilon_{p^{(n)}} \right]$$

is a compact subset of the domain of the continuous function $(F_{p^{(n)}})'$, its image under $(F_{p^{(n)}})'$ is compact in the metric space $L(\mathbb{R}^3, \mathbb{R}^3)$ and hence is bounded. Thus, there exists a positive real number $C$ such that

$$\left( \sup \left\{ \left\| (F_{p^{(n)}})'(x) \right\| : x \in \left[ p_1^{(n)} - \frac{3}{4} \varepsilon_{p^{(n)}}, p_1^{(n)} + \frac{3}{4} \varepsilon_{p^{(n)}} \right] \times \left[ p_2^{(n)} - \frac{3}{4} \varepsilon_{p^{(n)}}, p_2^{(n)} + \frac{3}{4} \varepsilon_{p^{(n)}} \right] \right.$$
$$\left. \left. \times \left[ p_3^{(n)} - \frac{3}{4} \varepsilon_{p^{(n)}}, p_3^{(n)} + \frac{3}{4} \varepsilon_{p^{(n)}} \right] \right\} \right) \le C.$$

Since

$$F_{p^{(n)}} : \left( p_1^{(n)} - \frac{3}{4} \varepsilon_{p^{(n)}}, p_1^{(n)} + \frac{3}{4} \varepsilon_{p^{(n)}} \right) \times \left( p_2^{(n)} - \frac{3}{4} \varepsilon_{p^{(n)}}, p_2^{(n)} + \frac{3}{4} \varepsilon_{p^{(n)}} \right)$$
$$\times \left( p_3^{(n)} - \frac{3}{4} \varepsilon_{p^{(n)}}, p_3^{(n)} + \frac{3}{4} \varepsilon_{p^{(n)}} \right)$$
$$\to \mathbb{R}^3$$

is a smooth function, by a result similar to Theorem 3.28, for every $x, y$ in

$$\left( p_1^{(n)} - \frac{3}{4} \varepsilon_{p^{(n)}}, p_1^{(n)} + \frac{3}{4} \varepsilon_{p^{(n)}} \right) \times \left( p_2^{(n)} - \frac{3}{4} \varepsilon_{p^{(n)}}, p_2^{(n)} + \frac{3}{4} \varepsilon_{p^{(n)}} \right)$$
$$\times \left( p_3^{(n)} - \frac{3}{4} \varepsilon_{p^{(n)}}, p_3^{(n)} + \frac{3}{4} \varepsilon_{p^{(n)}} \right), \ \left| (F_{p^{(n)}})(y) - (F_{p^{(n)}})(x) \right| \le |y - x| C.$$

Since

$$A \cap \left( \left[ p_1^{(n)} - \frac{1}{2} \varepsilon_{p^{(n)}}, p_1^{(n)} + \frac{1}{2} \varepsilon_{p^{(n)}} \right] \times \left[ p_2^{(n)} - \frac{1}{2} \varepsilon_{p^{(n)}}, p_2^{(n)} + \frac{1}{2} \varepsilon_{p^{(n)}} \right] \right.$$
$$\left. \times \left[ p_3^{(n)} - \frac{1}{2} \varepsilon_{p^{(n)}}, p_3^{(n)} + \frac{1}{2} \varepsilon_{p^{(n)}} \right] \right) \subset A,$$

and

$$0 \le \mu_3 \left( A \cap \left( \left[ p_1^{(n)} - \frac{1}{2} \varepsilon_{p^{(n)}}, p_1^{(n)} + \frac{1}{2} \varepsilon_{p^{(n)}} \right] \times \left[ p_2^{(n)} - \frac{1}{2} \varepsilon_{p^{(n)}}, p_2^{(n)} + \frac{1}{2} \varepsilon_{p^{(n)}} \right] \right. \right.$$
$$\left. \left. \times \left[ p_3^{(n)} - \frac{1}{2} \varepsilon_{p^{(n)}}, p_3^{(n)} + \frac{1}{2} \varepsilon_{p^{(n)}} \right] \right) \right) \le \mu_3(A) = 0,$$

so

$$\mu_3\left(A \cap \left(\left[p_1^{(n)} - \frac{1}{2}\varepsilon_{p^{(n)}}, p_1^{(n)} + \frac{1}{2}\varepsilon_{p^{(n)}}\right] \times \left[p_2^{(n)} - \frac{1}{2}\varepsilon_{p^{(n)}}, p_2^{(n)} + \frac{1}{2}\varepsilon_{p^{(n)}}\right]\right.\right.$$
$$\left.\left.\times \left[p_3^{(n)} - \frac{1}{2}\varepsilon_{p^{(n)}}, p_3^{(n)} + \frac{1}{2}\varepsilon_{p^{(n)}}\right]\right)\right) = 0.$$

Since

$$\mu_3\left(A \cap \left(\left[p_1^{(n)} - \frac{1}{2}\varepsilon_{p^{(n)}}, p_1^{(n)} + \frac{1}{2}\varepsilon_{p^{(n)}}\right] \times \left[p_2^{(n)} - \frac{1}{2}\varepsilon_{p^{(n)}}, p_2^{(n)} + \frac{1}{2}\varepsilon_{p^{(n)}}\right]\right.\right.$$
$$\left.\left.\times \left[p_3^{(n)} - \frac{1}{2}\varepsilon_{p^{(n)}}, p_3^{(n)} + \frac{1}{2}\varepsilon_{p^{(n)}}\right]\right)\right) = 0,$$

there exists a countable collection of open cubes $B_{\delta_k}(k = 1, 2, 3, \ldots)$ of side length $\delta_k(> 0)$ such that

$$A \cap \left(\left[p_1^{(n)} - \frac{1}{2}\varepsilon_{p^{(n)}}, p_1^{(n)} + \frac{1}{2}\varepsilon_{p^{(n)}}\right] \times \left[p_2^{(n)} - \frac{1}{2}\varepsilon_{p^{(n)}}, p_2^{(n)} + \frac{1}{2}\varepsilon_{p^{(n)}}\right]\right.$$
$$\left.\times \left[p_3^{(n)} - \frac{1}{2}\varepsilon_{p^{(n)}}, p_3^{(n)} + \frac{1}{2}\varepsilon_{p^{(n)}}\right]\right) \subset \bigcup_{k=1}^{\infty} B_{\delta_k},$$

and

$$\sum_{k=1}^{\infty} \mu_3(B_{\delta_k}) < \frac{\delta}{\left(2\sqrt{3}C\right)^3}.$$

Here, for every $k = 1, 2, 3, \ldots,$

$$\begin{aligned}
\text{diam}\left((F_{p^{(n)}})(B_{\delta_k})\right) &= \sup\{|s - t| : s, t \in (F_{p^{(n)}})(B_{\delta_k})\} \\
&= \sup\{|(F_{p^{(n)}})(y) - (F_{p^{(n)}})(x)| : x, y \in B_{\delta_k}\} \\
&\leq \sup\{|y - x|C : x, y \in B_{\delta_k}\} \\
&= C \cdot \sup\{|y - x| : x, y \in B_{\delta_k}\} = C \cdot \text{diam } B_{\delta_k} = C \cdot \left(\sqrt{3}\delta_k\right),
\end{aligned}$$

so,

$$\begin{aligned}
\mu_3\left((F_{p^{(n)}})(B_{\delta_k})\right) &\leq \left(2 \cdot \text{diam }\left((F_{p^{(n)}})(B_{\delta_k})\right)\right)^3 \\
&\leq \left(2C \cdot \left(\sqrt{3}\delta_k\right)\right)^3 = \left(2\sqrt{3}C\right)^3 (\delta_k)^3 = \left(2\sqrt{3}C\right)^3 \cdot \mu_3(B_{\delta_k}).
\end{aligned}$$

It follows that

$$
0 \le \mu_3 \left( (F_{p^{(n)}}) \left( A \cap \left( \left[ p_1^{(n)} - \frac{1}{2}\varepsilon_{p^{(n)}}, p_1^{(n)} + \frac{1}{2}\varepsilon_{p^{(n)}} \right] \times \left[ p_2^{(n)} - \frac{1}{2}\varepsilon_{p^{(n)}}, p_2^{(n)} + \frac{1}{2}\varepsilon_{p^{(n)}} \right] \right. \right. \right.
$$
$$
\left. \left. \left. \times \left[ p_3^{(n)} - \frac{1}{2}\varepsilon_{p^{(n)}}, p_3^{(n)} + \frac{1}{2}\varepsilon_{p^{(n)}} \right] \right) \right) \right) \le \mu_3 \left( (F_{p^{(n)}}) \left( \bigcup_{k=1}^{\infty} B_{\delta_k} \right) \right)
$$
$$
= \mu_3 \left( \bigcup_{k=1}^{\infty} \left( (F_{p^{(n)}})(B_{\delta_k}) \right) \right) \le \sum_{k=1}^{\infty} \mu_3 \left( (F_{p^{(n)}})(B_{\delta_k}) \right) \le \sum_{k=1}^{\infty} \left( \left( 2\sqrt{3}C \right)^3 \cdot \mu_3(B_{\delta_k}) \right)
$$
$$
= \left( 2\sqrt{3}C \right)^3 \sum_{k=1}^{\infty} \mu_3(B_{\delta_k}) < \delta.
$$

It follows that

$$
\mu_3 \left( F_{p^{(n)}} \left( A \cap \left( \left[ p_1^{(n)} - \frac{1}{2}\varepsilon_{p^{(n)}}, p_1^{(n)} + \frac{1}{2}\varepsilon_{p^{(n)}} \right] \times \left[ p_2^{(n)} - \frac{1}{2}\varepsilon_{p^{(n)}}, p_2^{(n)} + \frac{1}{2}\varepsilon_{p^{(n)}} \right] \right. \right. \right.
$$
$$
\left. \left. \left. \times \left[ p_3^{(n)} - \frac{1}{2}\varepsilon_{p^{(n)}}, p_3^{(n)} + \frac{1}{2}\varepsilon_{p^{(n)}} \right] \right) \right) \right) = 0. \qquad \square
$$

We can easily generalize the above result as the following:

**Lemma 6.12** *Let A be a nonempty subset of $\mathbb{R}^n$. For every $k = 1, \ldots, n$, let $\mu_k$ denote the k-dimensional measure over $\mathbb{R}^k$. Let $\mu_n(A) = 0$. Let $F : A \to \mathbb{R}^n$ be smooth on A. Then, $\mu_n(F(A)) = 0$.*

*Proof* Its proof is similar to the proof given in Lemma 6.11. $\qquad \square$

**Definition** Let $M$ be an $n$-dimensional smooth manifold. Let $A$ be a nonempty subset of $M$. For every $k = 1, \ldots, n$, let $\mu_k$ denote the $k$-dimensional measure over $\mathbb{R}^k$. If for every admissible coordinate chart $(U, \varphi)$ for $M$, $\mu_n(\varphi(A \cap U)) = 0$, then we say that *A has measure zero in M*.

**Lemma 6.13** *Let M be an m-dimensional smooth manifold. Let A be a nonempty subset of M. For every $k = 1, \ldots, m$, let $\mu_k$ denote the k-dimensional measure over $\mathbb{R}^k$. Let $\{(U_i, \varphi_i) : i \in I\}$ be a collection of admissible coordinate charts for M such that $A \subset \bigcup_{i \in I} U_i$, and for every $i \in I$, $\mu_m(\varphi_i(A \cap U_i)) = 0$. Then, A has measure zero in M.*

*Proof* Let us take any admissible coordinate chart $(U, \varphi)$ for $M$. We have to prove that $\mu_m(\varphi(A \cap U)) = 0$. Since $M$ is an $m$-dimensional smooth manifold, $M$ is a second countable space. Since $M$ is a second countable space, and $A$ is a nonempty subset of $M$, $A$ is a second countable space. Since $A$ is a second countable space, and $\{U_i : i \in I\}$ is an open cover of $A$, there exists a countable subcollection $\{U_n : n = 1, 2, 3, \ldots\}$ of $\{U_i : i \in I\}$ such that $A \subset \bigcup_{n=1}^{\infty} U_n$, and hence, $A = A \cap (\bigcup_{n=1}^{\infty} U_n) = \bigcup_{n=1}^{\infty} (A \cap U_n)$. It follows that $\varphi(A \cap U) = \varphi((\bigcup_{n=1}^{\infty} (A \cap U_n)) \cap U) = \varphi(\bigcup_{n=1}^{\infty} (A \cap U_n \cap U)) = \bigcup_{n=1}^{\infty} (\varphi(A \cap U_n \cap U))$, and hence, $0 \le \mu_m(\varphi(A \cap$

$U)) = \mu_m(\bigcup_{n=1}^{\infty}(\varphi(A \cap U_n \cap U))) \le \sum_{n=1}^{\infty} \mu_m(\varphi(A \cap U_n \cap U))$. Now, it suffices to show that for each positive integer $n$, $\mu_m(\varphi(A \cap U_n \cap U)) = 0$.

If not, otherwise, let there exists a positive integer $n$ such that $0 < \mu_m(\varphi(A \cap U_n \cap U))$. We have to arrive at a contradiction. Since $0 < \mu_m(\varphi(A \cap U_n \cap U))$, $\varphi(A \cap U_n \cap U) \ne \emptyset$, and hence, $A \cap U_n \cap U \ne \emptyset$. It follows that $U_n \cap U \ne \emptyset$. Since $U_n \cap U \ne \emptyset$ and $(U_n, \varphi_n)$, $(U, \varphi)$ are admissible coordinate charts for $M$, $\varphi \circ ((\varphi_n)^{-1}) : \varphi_n(U_n \cap U) \to \varphi(U_n \cap U)$ is smooth. Here, $A \cap U_n \cap U \ne \emptyset$, so $\varphi_n(A \cap U_n \cap U)$ is a nonempty subset of $\varphi_n(U_n \cap U)(\subset \mathbb{R}^m)$. Since $\varphi_n(A \cap U_n \cap U)$ is a nonempty subset of $\varphi_n(U_n \cap U)$, and $\varphi \circ ((\varphi_n)^{-1}) : \varphi_n(U_n \cap U) \to \varphi(U_n \cap U)$ is smooth, the restriction $(\varphi \circ ((\varphi_n)^{-1}))|_{\varphi_n(A \cap U_n \cap U)}$ is smooth. Since $(U_n, \varphi_n) \in \{(U_i, \varphi_i) : i \in I\}$, $0 \le \mu_m(\varphi_n(A \cap U_n \cap U)) \le \mu_m(\varphi_n(A \cap U_n)) = 0$, and hence $\mu_m(\varphi_n(A \cap U_n \cap U)) = 0$. Now, we can apply Lemma 6.12. It follows that

$$\mu_m(\varphi(A \cap U_n \cap U)) = \mu_m\left(\left(\left(\varphi \circ \left((\varphi_n)^{-1}\right)\right)\Big|_{\varphi_n(A \cap U_n \cap U)}\right)(\varphi_n(A \cap U_n \cap U))\right)$$
$$= 0,$$

which is a contradiction.                                                                    □

**Lemma 6.14** *Let $M$ be an $m$-dimensional smooth manifold. Let $\{A_n\}$ be a collection of nonempty subsets of $M$. For every $k = 1, \ldots, m$, let $\mu_k$ denote the $k$-dimensional measure over $\mathbb{R}^k$. Let each $A_n$ has measure zero in $M$. Then, $\bigcup_{n=1}^{\infty} A_n$ has measure zero in $M$.*

*Proof* For this purpose, let us take any admissible coordinate chart $(U, \varphi)$ for $M$. We have to prove that $\mu_m(\varphi((\bigcup_{n=1}^{\infty} A_n) \cap U)) = 0$. Here, $0 \le \mu_m(\varphi((\bigcup_{n=1}^{\infty} A_n) \cap U)) = \mu_m(\varphi(\bigcup_{n=1}^{\infty}(A_n \cap U))) = \mu_m(\bigcup_{n=1}^{\infty}(\varphi(A_n \cap U))) \le \sum_{n=1}^{\infty} \mu_m(\varphi(A_n \cap U))$, so it suffices to show that for each positive integer $n$, $\mu_m(\varphi(A_n \cap U)) = 0$. If not, otherwise, let there exists a positive integer $n$ such that $0 < \mu_m(\varphi(A_n \cap U))$. We have to arrive at a contradiction. Since $A_n$ has measure zero in $M$, and $(U, \varphi)$ is an admissible coordinate chart for $M$, $\mu_m(\varphi(A_n \cap U)) = 0$, a contradiction.      □

**Lemma 6.15** *Let $M$ be an $m$-dimensional smooth manifold. Let $A$ be a nonempty subset of $M$. For every $k = 1, \ldots, m$, let $\mu_k$ denote the $k$-dimensional measure over $\mathbb{R}^k$. Let $A$ has measure zero in $M$. Then, $A^c(\equiv M - A)$ is a dense subset of $M$.*

*Proof* We have to show that $A^c$ is dense in $M$, that is, $(A^c)^- = M$, that is, $((A^c)^-)^c(= A^o) = \emptyset$. If not, otherwise, let $A^o \ne \emptyset$. We have to arrive at a contradiction. Since $A^o \ne \emptyset$, there exists a point $p \in A$ such that $p$ is an interior point of $A$. Since $p$ is an interior point of $A$, there exists an open neighborhood $G$ of $p$ in $M$ such that $p \in G \subset A$. Since $p \in A \subset M$, and $M$ is an $m$-dimensional smooth manifold, there exists an admissible coordinate chart $(U, \varphi)$ for $M$ such that $p \in U$. Clearly, $U \cap G(\subset U)$ is an open neighborhood of $p$. Since $U \cap G(\subset U)$ is an open neighborhood of $p$, and $(U, \varphi)$ is an admissible coordinate chart for $M$, $(U \cap G, \varphi|_{U \cap G})$ is an admissible coordinate chart for $M$. Here, $\varphi(p)$ is an interior point of the subset $(\varphi|_{U \cap G})(U \cap G)(= \varphi(U \cap G))$ of $\mathbb{R}^m$, and hence, $\mu_m(\varphi$

$(U \cap G)) > 0$. Since $(U \cap G, \varphi|_{U \cap G})$ is an admissible coordinate chart for $M$, and $A$ has measure zero in $M$,

$$0 = \mu_m\big((\varphi|_{U \cap G})(A \cap (U \cap G)))\big) = \mu_m(\varphi(A \cap (U \cap G)))$$
$$= \mu_m(\varphi(U \cap (A \cap G))) = \mu_m(\varphi(U \cap G)).$$

This is a contradiction.                                                          □

**Lemma 6.16** *Let $M$ be an $n$-dimensional smooth manifold, $N$ be an $n$-dimensional smooth manifold, $F : M \to N$ be a smooth map, and $A$ be a nonempty subset of $M$. Let $A$ has measure zero in $M$. Then, $F(A)$ has measure zero in $N$.*

*Proof* We have to prove that $F(A)$ has measure zero in $N$. For this purpose, let us take an admissible coordinate chart $(V, \psi)$ for $N$ such that $(F(A)) \cap V \neq \emptyset$. We have to show that $\mu_n(\psi((F(A)) \cap V)) = 0$, where $\mu_n$ denotes the $n$-dimensional measure over $\mathbb{R}^n$. Clearly, $F(A \cap F^{-1}(V)) = (F(A)) \cap V$.

(Reason:    LHS $= F(A \cap F^{-1}(V)) \subset (F(A)) \cap (F(F^{-1}(V))) \subset (F(A)) \cap V =$ RHS. So, LHS $\subset$ RHS. Next, let us take any $q \in$ RHS $= (F(A)) \cap V$. It follows that there exists $p \in A$ such that $F(p) = q \in V$. Since $F(p) \in V$, $p \in F^{-1}(V)$. Since $p \in F^{-1}(V)$, and $p \in A$, $p \in A \cap F^{-1}(V)$, and hence, $q = F(p) \in F(A \cap F^{-1}(V)) =$ LHS. Hence, RHS $\subset$ LHS).

Since $M$ is an $n$-dimensional smooth manifold, by Lemma 4.47, there exists a countable collection $\{(U_1, \varphi_1), (U_2, \varphi_2), (U_3, \varphi_3), \ldots\}$ of admissible coordinate charts for $M$ such that $\{U_1, U_2, U_3, \ldots\}$ is a basis of $M$.

Let us take any $p \in A \cap F^{-1}(V)$. Since $p \in A \cap F^{-1}(V)$, $p \in F^{-1}(V)$, and hence, $F(p) \in V$. Thus, $V$ is an open neighborhood of $F(p)$ in $N$. Since $F : M \to N$ is a smooth map, $F : M \to N$ is continuous. Since $F : M \to N$ is continuous, and $V$ is an open neighborhood of $F(p)$, $F^{-1}(V)$ is an open neighborhood of $p$ in $M$. Since $F^{-1}(V)$ is an open neighborhood of $p$ in $M$, and $\{U_1, U_2, U_3, \ldots\}$ is a basis of $M$, there exists a positive integer $k(p)$ such that $p \in U_{k(p)} \subset F^{-1}(V)$, and hence, $A \cap U_{k(p)} \subset A \cap F^{-1}(V)$. Since $p \in A \cap F^{-1}(V)$, $p \in A$. Since $p \in A$, and $p \in U_{k(p)}$, $p \in A \cap U_{k(p)}$. It follows that for every $p \in A \cap F^{-1}(V)$, there exists a positive integer $k(p)$ such that $p \in A \cap U_{k(p)} \subset A \cap F^{-1}(V)$, and hence, $A \cap F^{-1}(V) = \cup\{A \cap U_{k(p)} : p \in A \cap F^{-1}(V)\}$. Now, $(F(A)) \cap V = F(A \cap F^{-1}(V)) = F(\cup\{A \cap U_{k(p)} : p \in A \cap F^{-1}(V)\}) = \cup\{F(A \cap U_{k(p)}) : p \in A \cap F^{-1}(V)\}$, and hence $\psi((F(A)) \cap V) = \psi(\cup\{F(A \cap U_{k(p)}) : p \in A \cap F^{-1}(V)\}) = \cup\{\psi(F(A \cap U_{k(p)})) : p \in A \cap F^{-1}(V)\} = \cup\{(\psi \circ F)(A \cup U k(p)) : p \in A \cap F^{-1}(V)\}$. Thus, $\psi((F(A)) \cap V)$ is the union of countable many sets $(\psi \circ F)(A \cap U_{k(p)})$, where $p \in A \cap F^{-1}(V)$. Here, $\{k(p) : p \in A \cap F^{-1}(V)\} \subset \{1, 2, 3, \ldots\}$, so $\{k(p) : p \in A \cap F^{-1}(V)\}$ can be written as $\{k_1, k_2, k_3, \ldots\}$. It follows that

$$\big\{U_{k(p)} : p \in A \cap F^{-1}(V)\big\} = \{U_{k_1}, U_{k_2}, U_{k_3}, \ldots\},$$

and hence,

$$\cup\{(\psi \circ F)(A \cap U_{k(p)}) : p \in A \cap F^{-1}(V)\} = \bigcup_{j=1}^{\infty}((\psi \circ F)(A \cap U_{k_j})).$$

It follows that

$$\mu_n(\psi((F(A)) \cap V)) = \mu_n(\cup\{(\psi \circ F)(A \cap U_{k(p)}) : p \in A \cap F^{-1}(V)\})$$

$$= \mu_n\left(\bigcup_{j=1}^{\infty}((\psi \circ F)(A \cap U_{k_j}))\right) \le \sum_{j=1}^{\infty}\mu_n((\psi \circ F)(A \cap U_{k_j}))$$

$$= \sum_{j=1}^{\infty}\mu_n\left(\left(\psi \circ F \circ \left(\varphi_{k_j}\right)^{-1}\right)\left((\varphi_{k_j})(A \cap U_{k_j})\right)\right).$$

Now, it suffices to show that each

$$\mu_n\left(\left(\psi \circ F \circ \left(\varphi_{k_j}\right)^{-1}\right)\left((\varphi_{k_j})(A \cap U_{k_j})\right)\right) \text{ is 0.}$$

Since $F : M \to N$ is a smooth map, $\psi \circ F \circ (\varphi_{k_j})^{-1}$ is smooth. Also, since $A$ has measure zero in $M$, $\mu_n((\varphi_{k_j})(A \cap U_{k_j})) = 0$. Now, we can apply Lemma 6.12. It follows that

$$\mu_n\left(\left(\psi \circ F \circ \left(\varphi_{k_j}\right)^{-1}\right)\left((\varphi_{k_j})(A \cap U_{k_j})\right)\right) = 0. \qquad \square$$

**Note 6.17** In the following discussion, if $a \equiv (a^1, \ldots, a^n) \in \mathbb{R}^n$, and $\varepsilon > 0$, then

$$\left(a^1 - \frac{1}{2}\varepsilon, a^1 + \frac{1}{2}\varepsilon\right) \times \cdots \times \left(a^n - \frac{1}{2}\varepsilon, a^n + \frac{1}{2}\varepsilon\right)$$

will be denoted by $C_\varepsilon(a)$, and

$$\left[a^1 - \frac{1}{2}\varepsilon, a^1 + \frac{1}{2}\varepsilon\right] \times \cdots \times \left[a^n - \frac{1}{2}\varepsilon, a^n + \frac{1}{2}\varepsilon\right]$$

will be denoted by $C_\varepsilon[a]$. Here, $C_\varepsilon(a)$ will be called an *open cube of side length* $\varepsilon$, and $C_\varepsilon[a]$ will be called a *closed cube of side length* $\varepsilon$.

Let $G$ be a nonempty open subset of $\mathbb{R}^n$. Let $0 < \varepsilon$. We shall try to show that there exists a countable collection $\{U_n\}$ of open cubes such that $\bigcup_{n=1}^{\infty} U_n = G$, $\bigcup_{n=1}^{\infty} (U_n)^- = G$, and side length of each $U_n$ is strictly less than $\varepsilon$.

Let us take any $a \equiv (a^1, \ldots, a^n) \in G$. Since $a \in G$, and $G$ is open in $\mathbb{R}^n$, there exists a positive rational number $r$ such that $C_r(a) \subset G$, and $r < \varepsilon$. Now, since the collection $\{s : s \in \mathbb{R}^n$, and all coordinates of $s$ are rational numbers$\}$ is dense in $\mathbb{R}^n$, there exists $s \equiv (s^1, \ldots, s^n) \in \mathbb{R}^n$ such that $s \in C_{\frac{r}{2}}(a)$, and all coordinates of $s$ are rational numbers. Since $s \in C_{\frac{r}{4}}(a)$, for every $i = 1, \ldots, n$,

$$\left| a^i - s^i \right| = \left| s^i - a^i \right| < \frac{r}{4},$$

and hence, $a \in C_{\frac{r}{4}}(s)$. Now, clearly $C_{\frac{r}{4}}[s] \subset C_r(A)$. (Reason: Let us take any $x \equiv (x^1, \ldots, x^n) \in C_{\frac{r}{4}}[s]$. It follows that for every $i = 1, \ldots, n, |x^i - s^i| \leq \frac{r}{4}$. Now, for every $i = 1, \ldots, n$,

$$\left| x^i - a^i \right| \leq \left| x^i - s^i \right| + \left| s^i - a^i \right| \leq \frac{r}{4} + \left| s^i - a^i \right| < \frac{r}{4} + \frac{r}{4} = \frac{r}{2},$$

and hence, $x \in C_r(a)$.) Thus, for every $a \in G$, there exist a positive rational number $r$, and $s \in \mathbb{R}^n$ such that all coordinates of $s$ are rational numbers, and

$$a \in C_{\frac{r}{2}}(s) \subset \left( C_{\frac{r}{2}}(s) \right)^- = C_{\frac{r}{2}}[s] \subset C_r(A) \subset G.$$

Now, since the collection of all open cubes of the form $C_{\frac{r}{2}}(s)$, where $r$ is a positive rational number, and all coordinates of $s (\in \mathbb{R}^n)$ are rational numbers, is a countable set, there exists a countable collection $\{U_n\}$ of open cubes such that $\bigcup_{n=1}^{\infty} U_n = G, \bigcup_{n=1}^{\infty} (U_n)^- = G$, and side length of each $U_n$ is strictly less than $\varepsilon$. □

**Note 6.18** Let $G$ be a nonempty open subset of $\mathbb{R}^n$. Let $0 < \varepsilon$. We shall try to show that there exist a countable collection $\{U_n\}$ of open cubes, and a countable collection $\{V_n\}$ of open cubes such that $\bigcup_{n=1}^{\infty} U_n = G$, $\bigcup_{n=1}^{\infty} (U_n)^- = G$, $\bigcup_{n=1}^{\infty} V_n = G$, $U_n \subset (U_n)^- \subset V_n (n = 1, 2, 3, \ldots)$, and side length of each $V_n$ is strictly less than $\varepsilon$.

Let us take any $a \equiv (a^1, \ldots, a^n) \in G$. Since $a \in G$, and $G$ is open in $\mathbb{R}^n$, there exists a positive rational number $r$ such that $C_r(a) \subset G$, and $r < \varepsilon$. Now, since the collection $\{s : s \in \mathbb{R}^n$, and all coordinates of $s$ are rational numbers$\}$ is dense in $\mathbb{R}^n$, there exists $s \equiv (s^1, \ldots, s^n) \in \mathbb{R}^n$ such that $s \in C_{\frac{r}{4}}(a)$, and all coordinates of $s$ are rational numbers. Since $s \in C_{\frac{r}{4}}(a)$, for every $i = 1, \ldots, n$,

$$\left| a^i - s^i \right| = \left| s^i - a^i \right| < \frac{r}{8},$$

and hence, $a \in C_{\frac{r}{4}}(s)$. Now, clearly $C_{\frac{r}{4}}[s] \subset C_{\frac{r}{2}}(a)$. (Reason: Let us take any $x \equiv (x^1, \ldots, x^n) \in C_{\frac{r}{4}}[s]$. It follows that for every $i = 1, \ldots, n, |x^i - s^i| \leq \frac{r}{8}$. Now, for every $i = 1, \ldots, n$,

$$\left| x^i - a^i \right| \leq \left| x^i - s^i \right| + \left| s^i - a^i \right| \leq \frac{r}{8} + \left| s^i - a^i \right| < \frac{r}{8} + \frac{r}{8} = \frac{r}{4},$$

and hence, $x \in C_{\frac{r}{2}}(A)$.) Also, it is clear that $C_{\frac{r}{2}}[s] \subset C_r(A)$. (Reason: Let us take any $x \equiv (x^1, \ldots, x^n) \in C_{\frac{r}{2}}[s]$. It follows that for every $i = 1, \ldots, n, |x^i - s^i| \leq \frac{r}{4}$. Now, for every $i = 1, \ldots, n$,

$$\left|x^i - a^i\right| \le \left|x^i - s^i\right| + \left|s^i - a^i\right| \le \frac{r}{4} + \left|s^i - a^i\right| < \frac{r}{4} + \frac{r}{8} < \frac{r}{2},$$

and hence, $x \in C_r(A)$.) Thus, for every $a \in G$, there exist a positive rational number $r$, and $s \in \mathbb{R}^n$ such that all coordinates of $s$ are rational numbers, and

$$a \in C_{\frac{r}{4}}(s) \subset \left(C_{\frac{r}{4}}(s)\right)^- = C_{\frac{r}{4}}[s] \subset C_{\frac{r}{2}}[s] \subset C_r(A) \subset G.$$

Now, since the collection of all open cubes of the form $C_{\frac{r}{4}}(s)$, where $r$ is a positive rational number, and all coordinates of $s(\in \mathbb{R}^n)$ are rational numbers, is a countable set, there exist a countable collection $\{U_n\}$ of open cubes, and a countable collection $\{V_n\}$ of open cubes such that $\bigcup_{n=1}^{\infty} U_n = G$, $\bigcup_{n=1}^{\infty} (U_n)^- = G$, $\bigcup_{n=1}^{\infty} V_n = G$, $U_n \subset (U_n)^- \subset V_n (n = 1, 2, 3, \ldots)$, and side length of each $V_n$ is strictly less than $\varepsilon$. $\qquad\square$

## 6.2  Taylor's Inequality

**Note 6.19**  We want to prove the following result:

Let $p \in U \subset \mathbb{R}^2$. Let $U$ be an open set. Let $S \subset U$, and $S$ be star shaped with respect to point $p \equiv (p^1, p^2) \in S$. Let $f : U \to \mathbb{R}$ be $C^\infty$ on $U$. Let $x \equiv (x^1, x^2) \in S$. Then,

$$f(x) = f(p) + (x^1 - p^1)\left(\int_0^1 \left(\frac{\partial f}{\partial x^1}((1-t)p + tx)\right)dt\right)$$

$$+ (x^2 - p^2)\left(\int_0^1 \left(\frac{\partial f}{\partial x^2}((1-t)p + tx)\right)dt\right)$$

$$= f(p) + \frac{1}{1!}\left(\sum_{i \in \{1,2\}} (x^i - p^i)((D_i f)(p))\right)$$

$$+ \frac{1}{1!}\left(\sum_{(i,j) \in \{1,2\} \times \{1,2\}} (x^i - p^i)(x^j - p^j)\int_0^1 ((1-t)D_{ij}((1-t)p + tx))dt\right)$$

$$= f(p) + \frac{1}{1!}\left(\sum_{i \in \{1,2\}} (x^i - p^i)((D_i f)(p))\right)$$

$$+ \frac{1}{2!}\left(\sum_{(i,j) \in \{1,2\}^2} (x^i - p^i)(x^j - p^j)((D_{ij} f)(p))\right)$$

$$+ \frac{1}{3!}\left(\sum_{(i,j,k) \in \{1,2\}^3} (x^i - p^i)(x^j - p^j)(x^k - p^k)\int_0^1 ((1-t)^2 D_{ijk}((1-t)p + tx))dt\right)$$

$$= \cdots$$

*Proof* Let us take any $t$ satisfying $0 \leq t \leq 1$. Now, since $S$ is star shaped with respect to point $p$, and $x \in S$, $(1-t)p + tx$ is in $S$, and hence, $f((1-t)p + tx)$ is a real number. Here,

$$
\begin{aligned}
\frac{d}{dt} f((1-t)p + tx) &= \frac{d}{dt} f(p^1 + t(x^1 - p^1), p^2 + t(x^2 - p^2)) \\
&= \left( \frac{\partial f}{\partial x^1}((1-t)p + tx) \right) \left( \frac{d}{dt}(p^1 + t(x^1 - p^1)) \right) \\
&\quad + \left( \frac{\partial f}{\partial x^2}((1-t)p + tx) \right) \left( \frac{d}{dt}(p^2 + t(x^2 - p^2)) \right) \\
&= \left( \frac{\partial f}{\partial x^1}((1-t)p + tx) \right) (x^1 - p^1) \\
&\quad + \left( \frac{\partial f}{\partial x^2}((1-t)p + tx) \right) (x^2 - p^2),
\end{aligned}
$$

Hence,

$$
\begin{aligned}
(x^1 - p^1) &\int_0^1 \left( \frac{\partial f}{\partial x^1}((1-t)p + tx) \right) dt + (x^2 - p^2) \int_0^1 \left( \frac{\partial f}{\partial x^2}((1-t)p + tx) \right) dt \\
&= \int_0^1 \left( \left( \frac{\partial f}{\partial x^1}((1-t)p + tx) \right)(x^1 - p^1) + \left( \frac{\partial f}{\partial x^2}((1-t)p + tx) \right)(x^2 - p^2) \right) dt \\
&= f((1-t)p + tx)|_{t=0}^{t=1} = f(x) - f(p).
\end{aligned}
$$

Now,

$$
\begin{aligned}
f(x) = f(p) &+ (x^1 - p^1) \left( \int_0^1 \left( \frac{\partial f}{\partial x^1}((1-t)p + tx) \right) dt \right) \\
&+ (x^2 - p^2) \left( \int_0^1 \left( \frac{\partial f}{\partial x^2}((1-t)p + tx) \right) dt \right).
\end{aligned}
$$

On applying integration by parts, we get

$$
\begin{aligned}
f(x) = f(p) &+ (x^1 - p^1) \left( \int_0^1 ((D_1 f)((1-t)p + tx)) dt \right) \\
&+ (x^2 - p^2) \left( \int_0^1 ((D_2 f)((1-t)p + tx)) dt \right) \\
&= f(p) + (x^1 - p^1) \Big( (t((D_1 f)((1-t)p + tx)))|_{t=0}^{t=1} \\
&\quad - \int_0^1 t \Big( ((D_{11} f)((1-t)p + tx)) \frac{d}{dt}((1-t)p^1 + tx^1) \Big)
\end{aligned}
$$

$$+\left((D_{12}f)((1-t)p+tx))\frac{\mathrm{d}}{\mathrm{d}t}\left((1-t)p^2+tx^2\right)\right)\mathrm{d}t\right)$$

$$+\left(x^2-p^2\right)\left((t((D_2f)((1-t)p+tx)))|_{t=0}^{t=1}\mathrm{d}t\right.$$

$$-\int_0^1 t\left(((D_{21}f)((1-t)p+tx))\frac{\mathrm{d}}{\mathrm{d}t}\left((1-t)p^1+tx^1\right)\right.$$

$$+\left((D_{22}f)((1-t)p+tx))\frac{\mathrm{d}}{\mathrm{d}t}\left((1-t)p^2+tx^2\right)\right)\mathrm{d}t\right)$$

$$=f(p)+\left(x^1-p^1\right)((D_1f)(x)-\int_0^1 t\left(((D_{11}f)((1-t)p+tx))\left(x^1-p^1\right)\right.$$

$$+((D_{12}f)((1-t)p+tx))\left(x^2-p^2\right))\mathrm{d}t)$$

$$+\left(x^2-p^2\right)((D_2f)(x)-\int_0^1 t\left(((D_{21}f)((1-t)p+tx))\left(x^1-p^1\right)\right.$$

$$+((D_{22}f)((1-t)p+tx))\left(x^2-p^2\right))\mathrm{d}t)$$

$$=f(p)+\left(\left(x^1-p^1\right)((D_1f)(x))+\left(x^2-p^2\right)((D_2f)(x))\right)$$

$$-\left(x^1-p^1\right)^2\int_0^1 t(((D_{11}f)((1-t)p+tx)))\mathrm{d}t$$

$$-2\left(x^1-p^1\right)\left(x^2-p^2\right)\int_0^1 t((D_{12}f)((1-t)p+tx))\mathrm{d}t$$

$$-\left(x^2-p^2\right)^2\int_0^1 t(((D_{22}f)((1-t)p+tx)))\mathrm{d}t$$

$$=f(p)+\left(\left(x^1-p^1\right)\left((D_1f)(p)+\left(x^1-p^1\right)\int_0^1(D_{11}f)((1-t)p+tx)\mathrm{d}t\right.\right.$$

$$\left.+\left(x^2-p^2\right)\int_0^1(D_{12}f)((1-t)p+tx)\mathrm{d}t\right)$$

$$+\left(x^2-p^2\right)\left((D_2f)(p)+\left(x^1-p^1\right)\int_0^1(D_{21}f)((1-t)p+tx)\mathrm{d}t\right.$$

$$\left.\left.+\left(x^2-p^2\right)\int_0^1(D_{22}f)((1-t)p+tx)\mathrm{d}t\right)\right)$$

$$-\left(x^1-p^1\right)^2\int_0^1 t(((D_{11}f)((1-t)p+tx)))\mathrm{d}t$$

$$-2\left(x^1-p^1\right)\left(x^2-p^2\right)\int_0^1 t((D_{12}f)((1-t)p+tx))\mathrm{d}t$$

$$-\left(x^2-p^2\right)^2\int_0^1 t(((D_{22}f)((1-t)p+tx)))\mathrm{d}t$$

$$
\begin{aligned}
=\, & f(p) + \left( (x^1 - p^1)(D_1 f)(p) + (x^2 - p^2)(D_2 f)(p) \right) \\
& + (x^1 - p^1)^2 \int_0^1 ((D_{11}f)((1-t)p + tx) - t(((D_{11}f)((1-t)p + tx))))dt \\
& + 2(x^1 - p^1)(x^2 - p^2) \int_0^1 ((D_{12}f)((1-t)p + tx) - t(((D_{12}f)((1-t)p + tx)))dt \\
& + (x^2 - p^2)^2 \int_0^1 ((D_{22}f)((1-t)p + tx) - t(((D_{22}f)((1-t)p + tx))))dt \\
=\, & f(p) + \frac{1}{1!}\left( (x^1 - p^1)\frac{\partial f}{\partial x^1}(p) + (x^2 - p^2)\frac{\partial f}{\partial x^2}(p) \right) \\
& + \frac{1}{1!}\left( (x^1 - p^1)^2 \int_0^1 \left( (1-t)\frac{\partial^2 f}{\partial (x^1)^2}((1-t)p + tx) \right)dt \right. \\
& \qquad + 2(x^1 - p^1)(x^2 - p^2) \int_0^1 \left( (1-t)\frac{\partial^2 f}{\partial x^1 \partial x^2}((1-t)p + tx) \right)dt \\
& \qquad \left. + (x^2 - p^2)^2 \int_0^1 \left( (1-t)\frac{\partial^2 f}{\partial (x^2)^2}((1-t)p + tx) \right)dt \right) \\
=\, & f(p) + \frac{1}{1!}\left( \sum_{i \in \{1,2\}} (x^i - p^i)((D_i f)(p)) \right) \\
& + \frac{1}{1!}\left( \sum_{(i,j) \in \{1,2\} \times \{1,2\}} (x^i - p^i)(x^j - p^j) \int_0^1 ((1-t)D_{ij}((1-t)p + tx))dt \right).
\end{aligned}
$$

Similarly,

$$
\begin{aligned}
f(x) =\, & f(p) + \frac{1}{1!}\left( \sum_{i \in \{1,2\}} (x^i - p^i)((D_i f)(p)) \right) \\
& + \frac{1}{2!}\left( \sum_{(i,j) \in \{1,2\}^2} (x^i - p^i)(x^j - p^j)((D_{ij}f)(p)) \right) \\
& + \frac{1}{2!}\left( \sum_{(i,j,k) \in \{1,2\}^3} (x^i - p^i)(x^j - p^j)(x^k - p^k) \int_0^1 \left( (1-t)^2 D_{ijk}((1-t)p + tx) \right)dt \right) \\
=\, & \cdots
\end{aligned}
$$

This result can be generalized in $\mathbb{R}^n$, and is known as *Taylor's theorem*.

**Note 6.20** Let $p \in U \subset \mathbb{R}^n$. Let $U$ be an open set. Let $S \subset U$, and $S$ be star shaped with respect to point $p \equiv (p^1, \ldots, p^n) \in S$. Let $f : U \to \mathbb{R}$ be $C^\infty$ on $U$. Let $x \equiv (x^1, \ldots, x^n) \in S$. Let all the third-order partial derivatives of $f$ be bounded in absolute value by a constant $M$ on $S$. Then,

$$
\left| f(x) - \left\{ f(p) + \frac{1}{1!} \left( \sum_{i \in \{1,\ldots,n\}} (x^i - p^i)((D_i f)(p)) \right) \right.\right.
$$

$$
\left.\left. + \frac{1}{2!} \left( \sum_{(i,j) \in \{1,\ldots,n\}^2} (x^i - p^i)(x^j - p^j)((D_{ij}f)(p)) \right) \right\} \right|
$$

$$
= \left| \frac{1}{2!} \left( \sum_{(i,j,k) \in \{1,\ldots,n\}^3} (x^i - p^i)(x^j - p^j)(x^k - p^k) \int_0^1 \left((1-t)^2 D_{ijk}((1-t)p + tx)\right) dt \right) \right|
$$

$$
\leq \frac{1}{2!} \sum_{(i,j,k) \in \{1,\ldots,n\}^3} \left| (x^i - p^i)(x^j - p^j)(x^k - p^k) \int_0^1 \left((1-t)^2 D_{ijk}((1-t)p + tx)\right) dt \right|
$$

$$
= \frac{1}{2!} \sum_{(i,j,k) \in \{1,\ldots,n\}^3} |x^i - p^i||x^j - p^j||x^k - p^k| \left| \int_0^1 \left((1-t)^2 D_{ijk}((1-t)p + tx)\right) dt \right|
$$

$$
\leq \frac{1}{2!} \sum_{(i,j,k) \in \{1,\ldots,n\}^3} |x^i - p^i||x^j - p^j||x^k - p^k| \int_0^1 |1-t|^2 |D_{ijk}((1-t)p + tx)| dt
$$

$$
\leq \frac{1}{2!} \sum_{(i,j,k) \in \{1,\ldots,n\}^3} |x^i - p^i||x^j - p^j||x^k - p^k| \int_0^1 1 |D_{ijk}((1-t)p + tx)| dt
$$

$$
\leq \frac{1}{2!} \sum_{(i,j,k) \in \{1,\ldots,n\}^3} |x^i - p^i||x^j - p^j||x^k - p^k| \int_0^1 M dt
$$

$$
= \frac{1}{2!} \sum_{(i,j,k) \in \{1,\ldots,n\}^3} |x^i - p^i||x^j - p^j||x^k - p^k| M
$$

$$
\leq \frac{1}{2!} \sum_{(i,j,k) \in \{1,\ldots,n\}^3} |x - p||x - p||x - p| M
$$

$$
= \frac{1}{2!} \sum_{(i,j,k) \in \{1,\ldots,n\}^3} M|x - p|^3 = \frac{1}{3!} (n \times n \times n) \left( M|x - p|^3 \right)
$$

$$
= \frac{1}{2!} n^3 M|x - p|^3.
$$

Thus

$$\left| f(x) - \left\{ f(p) + \frac{1}{1!} \left( \sum_{i \in \{1,\dots,n\}} (x^i - p^i)((D_i f)(p)) \right) \right. \right.$$
$$\left. \left. + \frac{1}{2!} \left( \sum_{(i,j) \in \{1,\dots,n\}^2} (x^i - p^i)(x^j - p^j)((D_{ij} f)(p)) \right) \right\} \right|$$
$$\leq \frac{1}{2!} n^3 |x - p|^3 M,$$

etc.

**Conclusion**: Let $p \in U \subset \mathbb{R}^n$. Let $U$ be an open set. Let $S \subset U$, and $S$ be star shaped with respect to point $p \equiv (p^1, \dots, p^n) \in S$. Let $f : U \to \mathbb{R}$ be $C^\infty$ on $U$. Let $x \equiv (x^1, \dots, x^n) \in S$. Let all the third-order partial derivatives of $f$ be bounded in absolute value by a constant $M$ on $S$. Then,

$$\left| f(x) - \left\{ f(p) + \frac{1}{1!} \left( \sum_{i \in \{1,\dots,n\}} (x^i - p^i)((D_i f)(p)) \right) \right. \right.$$
$$\left. \left. + \frac{1}{2!} \left( \sum_{(i,j) \in \{1,\dots,n\}^2} (x^i - p^i)(x^j - p^j)((D_{ij} f)(p)) \right) \right\} \right|$$
$$\leq \frac{1}{2!} n^3 |x - p|^3 M.$$

In the following discussion, we shall use this inequality.

## 6.3 Sard's Theorem on $\mathbb{R}^n$

**Lemma 6.21** *Let $G$ be an open subset of $\mathbb{R}(= \mathbb{R}^1)$. Let $F : G \to \mathbb{R}^3$ be a smooth function. Let $F \equiv (F_1, F_2, F_3)$. For every $k = 1, 2, 3$, let $\mu_k$ denote the k-dimensional measure over $\mathbb{R}^k$. Then,*

$$\mu_3 \left( F \left( \left\{ x : x \in G, \text{and rank} \begin{bmatrix} (D_1 F_1)(x) \\ (D_1 F_2)(x) \\ (D_1 F_3)(x) \end{bmatrix} = 0 \right\} \right) \right) = 0.$$

*Proof* Put

$$C \equiv \left\{ x : x \in G, \text{and rank} \begin{bmatrix} (D_1 F_1)(x) \\ (D_1 F_2)(x) \\ (D_1 F_3)(x) \end{bmatrix} = 0 \right\}.$$

We have to show that $\mu_3(F(C)) = 0$. Here,

$$C = \left\{ x : x \in G, \text{ and rank} \begin{bmatrix} (D_1F_1)(x) \\ (D_1F_2)(x) \\ (D_1F_3)(x) \end{bmatrix} = 0 \right\}$$

$$= \{x : x \in G, \text{ and } (D_1F_1)(x) = 0, (D_1F_2)(x) = 0, (D_1F_3)(x) = 0\}$$

$$= \{x : x \in G, \text{ and } (D_1F_1)(x) = 0\} \cap \{x : x \in G, \text{ and } (D_1F_2)(x) = 0\}$$

$$\cap \{x : x \in G, \text{ and } (D_1F_3)(x) = 0\}$$

$$= (D_1F_1)^{-1}(\{0\}) \cap (D_1F_2)^{-1}(\{0\}) \cap (D_1F_3)^{-1}(\{0\}).$$

Since $F : G \to \mathbb{R}^3$ is a smooth function, $D_1F_1 : G \to \mathbb{R}$ is continuous. Since $D_1$ $F_1 : G \to \mathbb{R}$ is continuous, and $\{0\}$ is closed, $(D_1F_1)^{-1}(\{0\})$ is closed in $G$. Similarly, $(D_1F_2)^{-1}(\{0\}), (D_1F_3)^{-1}(\{0\})$ are closed in $G$. It follows that $(D_1F_1)^{-1}$ $(\{0\}) \cap (D_1F_2)^{-1}(\{0\}) \cap (D_1F_3)^{-1}(\{0\})(= C)$ is closed in $G$, and hence, $C$ is closed in $G$. Put

$$C_1 \equiv \{x : x \in G, \text{ and } 0 = (D_1F_1)(x) = (D_1F_2)(x) = (D_1F_3)(x)\}.$$

Clearly, $C_1 = C$, and $C_1$ is closed in $G$. Put

$$C_2 \equiv \{x : x \in G, \text{ and } 0 = (D_1F_1)(x) = (D_1F_2)(x) = (D_1F_3)(x) = (D_{11}F_1)(x)$$
$$= (D_{11}F_2)(x) = (D_{11}F_3)(x)\}.$$

Clearly, $C_2 \subset C_1$, and $C_2$ is closed in $G$. Similarly, we define $C_3 \equiv \{x : x \in G,$ and $0 = (D_1F_i)(x) = (D_{11}F_i)(x) = (D_{111}F_i)(x)$ for every $i = 1, 2, 3\}$, etc. Clearly, $\cdots \subset C_4 \subset C_3 \subset C_2 \subset C_1 = C$, and each $C_i$ is closed in $G$.

Now, we shall try to show that $\mu_3(F(C_1)) = 0$. Since $G$ is an open subset of $\mathbb{R}$, there exists a countable collection $\{U_n\}$ of open bounded intervals such that $\bigcup_{n=1}^{\infty} U_n = G$, and $\bigcup_{n=1}^{\infty}(U_n)^- = G$. Now, $F(C_1) = F(\{x : x \in G, \text{ and } 0 = (D_1F_1)(x) = (D_1F_2)$ $(x) = (D_1F_3)(x)\}) = F(\{x : x \in \bigcup_{n=1}^{\infty}(U_n)^-, \text{ and } 0 = (D_1F_1)(x) = (D_1F_2)(x) = (D_1F_3)(x)\}) = F(\bigcup_{n=1}^{\infty}\{x : x \in (U_n)^-, \text{ and } 0 = (D_1F_1)(x) = (D_1F_2)(x) = (D_1F_3)$ $(x)\}) = \bigcup_{n=1}^{\infty} F(\{x : x \in (U_n)^-, \text{ and } 0 = (D_1F_1)(x) = (D_1F_2)(x) = (D_1F_3)(x)\})$ $= \bigcup_{n=1}^{\infty} F(C_1 \cap ((U_n)^-))$, so $0 \leq \mu_3(F(C_1)) = \mu_3(\bigcup_{n=1}^{\infty} F(C_1 \cap ((U_n)^-))) \leq \sum_{n=1}^{\infty}$ $\mu_3(F(C_1 \cap ((U_n)^-)))$. Hence, it suffices to show that each $\mu_3(F(C_1 \cap ((U_n)^-))) = 0$. Let us first try to show that $\mu_3(F(C_1 \cap ((U_1)^-))) = 0$.

Since $D_{11}F_1 : G \to \mathbb{R}$ is continuous, and $(U_1)^-$ is compact, $(D_{11}F_1)((U_1)^-)$ is compact, and hence, $(D_{11}F_1)((U_1)^-)$ is bounded. Similarly, $(D_{11}F_2)((U_1)^-)$, $(D_{11}F_3)((U_1)^-)$ are bounded. It follows that $(D_{11}F_1)((U_1)^-) \cup (D_{11}F_2)((U_1)^-) \cup$ $(D_{11}F_3)((U_1)^-)$ is bounded, and hence, there exists a positive number $K$ such that $(D_{11}F_1)((U_1)^-) \cup (D_{11}F_2)((U_1)^-) \cup (D_{11}F_3)((U_1)^-)$ is contained in $[-K, K]$.

Case I: when $C_1 \cap ((U_1)^-) = \emptyset$. In this case,

$$\mu_3(F(C_1 \cap ((U_1)^-))) = \mu_3(F(\emptyset)) = \mu_3(\emptyset) = 0.$$

Thus, $\mu_3(F(C_1 \cap ((U_1)^-))) = 0$.

Case II: when $C_1 \cap ((U_1)^-) \neq \emptyset$. Now, let $R$ be the length of closed interval $(U_1)^-$. Let us fix any positive integer $N > 1$. Let us subdivide the closed interval $(U_1)^-$ into $N$ closed intervals $E_1, E_2, \ldots, E_N$ such that each $E_i$ is of length $\frac{R}{N}$. Thus, $(U_1)^- = E_1 \cup E_2 \cup \cdots \cup E_N$. It follows that $C_1 \cap ((U_1)^-) = C_1 \cap (E_1 \cup E_2 \cup \cdots \cup E_N) = (C_1 \cap E_1) \cup (C_1 \cap E_2) \cup \cdots \cup (C_1 \cap E_N) = \bigcup \{C_1 \cap E_k : C_1 \cap E_k \neq \emptyset\}$. Hence, $F(C_1 \cap ((U_1)^-)) = F(\bigcup \{C_1 \cap E_k : C_1 \cap E_k \neq \emptyset\}) = \bigcup \{F(C_1 \cap E_k) : C_1 \cap E_k \neq \emptyset\}$, and hence, $0 \leq \mu_3(F(C_1 \cap ((U_1)^-))) = \mu_3(\bigcup \{F(C_1 \cap E_k) : C_1 \cap E_k \neq \emptyset\}) \leq \sum \{\mu_3(F(C_1 \cap E_k)) : C_1 \cap E_k \neq \emptyset\}$. Now, it suffices to show that if $C_1 \cap E_k \neq \emptyset$, then $\mu_3(F(C_1 \cap E_k)) = 0$. For this purpose, let $C_1 \cap E_k \neq \emptyset$. Since $C_1 \cap E_k \neq \emptyset$, there exists $a \in C_1 \cap E_k$. Since $a \in C_1 \cap E_k$, $a \in E_k$.

Now, by Taylor's theorem, for every $x \in E_k$,

$$|F_1(x) - F_1(a)| = |F_1(x) - \{F_1(a) + 0\}|$$
$$= |F_1(x) - \{F_1(a) + (D_1 F_1)(a)\}| \leq \frac{1}{1!}\left(K\left(\frac{R}{N}\right)^2\right).$$

Thus, for every $x \in E_k$,

$$|F_1(x) - F_1(a)| \leq \frac{1}{1!}\left(K\left(\frac{R}{N}\right)^2\right).$$

Similarly, for every $x \in E_k$,

$$|F_2(x) - F_2(a)| \leq \frac{1}{1!}\left(K\left(\frac{R}{N}\right)^2\right) \quad \text{and} \quad |F_3(x) - F_3(a)| \leq \frac{1}{1!}\left(K\left(\frac{R}{N}\right)^2\right).$$

It follows that for every $x \in E_k$,

$$|F(x) - F(a)| = |(F_1(x), F_2(x), F_3(x)) - (F_1(a), F_2(a), F_3(a))|$$
$$= |(F_1(x) - F_1(a), F_2(x) - F_2(a), F_3(x) - F_3(a))|$$
$$\leq |F_1(x) - F_1(a)| + |F_2(x) - F_2(a)| + |F_3(x) - F_3(a)|$$
$$\leq \frac{1}{1!}\left(K\left(\frac{R}{N}\right)^2\right) + \frac{1}{1!}\left(K\left(\frac{R}{N}\right)^2\right) + \frac{1}{1!}\left(K\left(\frac{R}{N}\right)^2\right)$$
$$= 3\left(\frac{1}{1!}\left(K\left(\frac{R}{N}\right)^2\right)\right).$$

Hence, for every $x \in E_k$,

$$|F(x) - F(a)| \le 3\left(\frac{1}{1!}\left(K\left(\frac{R}{N}\right)^2\right)\right).$$

Thus,

$$F(C_1 \cap E_k) \subset F(E_k) \subset B_{3\left(\frac{1}{1!}\left(K\left(\frac{R}{N}\right)^2\right)\right)}(F(a)),$$

and hence,

$$0 \le \mu_3(F(C_1 \cap E_k)) \le \mu_3\left(B_{3\left(\frac{1}{1!}\left(K\left(\frac{R}{N}\right)^2\right)\right)}(F(a))\right) \le \left(2\left(3\left(\frac{1}{1!}\left(K\left(\frac{R}{N}\right)^2\right)\right)\right)\right)^3$$

$$\to 0 \text{ as } N \to \infty.$$

This shows that $\mu_3(F(C_1 \cap E_k)) = 0$. Hence, $\mu_3(F(C_1 \cap ((U_1)^-))) = 0$.

Thus, in all cases, $\mu_3(F(C_1 \cap ((U_1)^-))) = 0$. Similarly, $\mu_3(F(C_1 \cap ((U_2)^-))) = 0$, $\mu_3(F(C_1 \cap ((U_3)^-))) = 0$, etc. Thus, $\mu_3(F(C)) = \mu_3(F(C_1)) = 0$.  ∏

**Note 6.22** Similar to Lemma 6.21, we can prove the following result:

Let $G$ be an open subset of $\mathbb{R}(= \mathbb{R}^1)$. Let $F : G \to \mathbb{R}^n$ be a smooth function. Let $F \equiv (F_1, \ldots, F_n)$. For every $k = 1, \ldots, n$, let $\mu_k$ denote the $k$-dimensional measure over $\mathbb{R}^k$. Then,

$$\mu_n\left(F\left(\left\{x : x \in G, \text{ and } \text{rank}\begin{bmatrix}(D_1F_1)(x) \\ \vdots \\ (D_1F_n)(x)\end{bmatrix} = 0\right\}\right)\right) = 0.$$

**Lemma 6.23** *Let $m, n$ be positive integers satisfying $m \le n$. Let $G$ be a nonempty open subset of $\mathbb{R}^m$. Let $f : G \to \mathbb{R}^n$ be a smooth function, and $f \equiv (f_1, \ldots f_n)$. Let $\mu$ denote the Lebesgue measure over $\mathbb{R}^n$. Let $C$ be the collection of all critical points of $f$, that is,*

$$C \equiv \left\{x : x \in G, \text{ and rank}\begin{bmatrix}(D_1f_1)(x) & (D_mf_1)(x) \\ \vdots & \ddots \vdots \\ (D_1f_n)(x) & (D_mf_n)(x)\end{bmatrix} < n\right\}.$$

*Then, $\mu(f(C)) = 0$.*

*Proof* Since $G$ is a nonempty open subset of $\mathbb{R}^m$, there exists a countable collection $\{C_n\}$ of open cubes in $\mathbb{R}^m$ such that $\bigcup_{n=1}^{\infty} C_n = G$, $\bigcup_{n=1}^{\infty}(C_n)^- = G$, and for each $n = 1, 2, 3, \ldots$, the side length $R_n$ of $C_n$ is $< 1$. Now,

$$f(C) = f\left(\left\{x : x \in G, \text{ and rank} \begin{bmatrix} (D_1f_1)(x) & (D_mf_1)(x) \\ \vdots & \ddots \vdots \\ (D_1f_n)(x) & (D_mf_n)(x) \end{bmatrix} < n \right\}\right)$$

$$= f\left(\left\{x : x \in \bigcup_{n=1}^{\infty}(C_n)^-, \text{ and rank} \begin{bmatrix} (D_1f_1)(x) & (D_mf_1)(x) \\ \vdots & \ddots \vdots \\ (D_1f_n)(x) & (D_mf_n)(x) \end{bmatrix} < n \right\}\right)$$

$$= f\left(\bigcup_{n=1}^{\infty}\left\{x : x \in (C_n)^-, \text{ and rank} \begin{bmatrix} (D_1f_1)(x) & (D_mf_1)(x) \\ \vdots & \ddots \vdots \\ (D_1f_n)(x) & (D_mf_n)(x) \end{bmatrix} < n \right\}\right)$$

$$= \bigcup_{n=1}^{\infty} f\left(\left\{x : x \in (C_n)^-, \text{ and rank} \begin{bmatrix} (D_1f_1)(x) & (D_mf_1)(x) \\ \vdots & \ddots \vdots \\ (D_1f_n)(x) & (D_mf_n)(x) \end{bmatrix} < n \right\}\right)$$

$$= \bigcup_{n=1}^{\infty} f(C \cap (C_n)^-).$$

Since $f(C) = \bigcup_{n=1}^{\infty} f(C \cap (C_n)^-)$,

$$0 \le \mu(f(C)) = \mu\left(\bigcup_{n=1}^{\infty} f(C \cap (C_n)^-)\right) \le \sum_{n=1}^{\infty} \mu(f(C \cap (C_n)^-)).$$

Hence, it suffices to show that each $\mu(f(C \cap (C_n)^-)) = 0$.

We first try to show that $\mu(f(C \cap (C_1)^-)) = 0$. Since $f : G \to \mathbb{R}^n$ is a smooth function, each $(D_{ij}f_k) : G \to \mathbb{R}$ is a smooth function, and hence, each $(D_{ij}f_k) : G \to \mathbb{R}$ is continuous. Since $(D_{ij}f_k) : G \to \mathbb{R}$ is continuous, and $(C_1)^-$ is a compact subset of $G$, $(D_{ij}f_k)((C_1)^-)$ is compact, and hence, $(D_{ij}f_k)((C_1)^-)$ is bounded. Since $(D_{ij}f_k)((C_1)^-)$ is bounded, there exists a real $M_{ijk} > 0$ such that for every $z \in (C_1)^-$, $|(D_{ij}f_k)(z)| \le M_{ijk}$.

Put $M \equiv \max\{M_{ijk} : 1 \le i \le m, 1 \le j \le m, 1 \le k \le n\}$. Since $f : G \to \mathbb{R}$ is a smooth function, each $D_i f_j : G \to \mathbb{R}$ is a smooth function, and hence, each $D_i f_j : G \to \mathbb{R}$ is continuous. Since $D_i f_j : G \to \mathbb{R}$ is continuous, and $(C_1)^-$ is a compact subset of $G$, $(D_i f_j)((C_1)^-)$ is compact, and hence, $(D_i f_j)((C_1)^-)$ is bounded. Since $(D_i f_j)((C_1)^-)$ is bounded, there exists a real $M_{ij} > 0$ such that for every $z \in (C_1)^-$, $|(D_i f_j)(z)| \le M_{ij}$.

Put $M_0 \equiv \max\{M_{ij} : 1 \le i \le m, 1 \le j \le n\}$. Let us fix any positive integer $N > 1$. Let us subdivide the closed cube $(C_1)^-$ into $N^m$ closed cubes $E_1, E_2, \ldots, E_{N^m}$ such that each $E_i$ is of length $\frac{R_1}{N}$. Thus, $(C_1)^- = E_1 \cup E_2 \cup \cdots \cup E_{N^m}$. It follows that $C \cap (C_1)^- = C \cap (E_1 \cup E_2 \cup \cdots \cup E_{N^m}) = (C \cap E_1) \cup (C \cap E_2) \cup \cdots \cup (C \cap E_{N^m}) = \bigcup\{C \cap E_k : C \cap E_k \ne \emptyset, 1 \le k \le N^m\}$. Hence, $f(C \cap (C_1)^-) = f(\bigcup\{C \cap E_k : C \cap$

$E_k \neq \emptyset, 1 \leq k \leq N^m\}) = \bigcup\{f(C \cap E_k) : C \cap E_k \neq \emptyset, 1 \leq k \leq N^m\}$, and hence, $0 \leq \mu$ $(f(C \cap (C_1)^-)) = \mu(\bigcup\{f(C \cap E_k) : C \cap E_k \neq \emptyset, 1 \leq k \leq N^m\}) \leq \sum\{\mu(f(C \cap E_k))$ $: C \cap E_k \neq \emptyset, 1 \leq k \leq N^m\}$.

Let us take any positive integer $k$ satisfying $1 \leq k \leq N^m$, and $C \cap E_k \neq \emptyset$. Now, by Taylor's theorem, for every $x \equiv (x_1, \ldots, x_m) \in E_k$, and $y \equiv (y_1, \ldots, y_m) \in E_k$,

$$|f(y) - (f(x) + (f'(x))(y - x))|$$

$$= \left| \left( \begin{bmatrix} f_1(y) \\ \vdots \\ f_n(y) \end{bmatrix} - \left( \begin{bmatrix} f_1(x) \\ \vdots \\ f_n(x) \end{bmatrix} + \begin{bmatrix} (D_1 f_1)(x) & (D_m f_1)(x) \\ \vdots & \ddots \vdots \\ (D_1 f_n)(x) & (D_m f_n)(x) \end{bmatrix} \begin{bmatrix} y_1 - x_1 \\ \vdots \\ y_m - x_m \end{bmatrix} \right) \right) \right|$$

$$= \left| \left( \begin{bmatrix} f_1(y) \\ \vdots \\ f_n(y) \end{bmatrix} - \left( \begin{bmatrix} f_1(x) \\ \vdots \\ f_n(x) \end{bmatrix} + \begin{bmatrix} ((D_1 f_1)(x))(y_1 - x_1) + \cdots + ((D_m f_1)(x))(y_m - x_m) \\ \vdots \\ ((D_1 f_n)(x))(y_1 - x_1) + \cdots + ((D_m f_n)(x))(y_m - x_m) \end{bmatrix} \right) \right) \right|$$

$$= \left| \begin{bmatrix} f_1(y) - (f_1(x) + ((D_1 f_1)(x))(y_1 - x_1) + \cdots + ((D_m f_1)(x))(y_m - x_m)) \\ \vdots \\ f_n(y) - (f_n(x) + (((D_1 f_n)(x))(y_1 - x_1) + \cdots + ((D_m f_n)(x))(y_m - x_m))) \end{bmatrix} \right|$$

$$\leq |f_1(y) - (f_1(x) + ((D_1 f_1)(x))(y_1 - x_1) + \cdots + ((D_m f_1)(x))(y_m - x_m))| + \cdots$$
$$+ |f_n(y) - (f_n(x) + (((D_1 f_n)(x))(y_1 - x_1) + \cdots + ((D_m f_n)(x))(y_m - x_m)))|$$

$$\leq \underbrace{\frac{1}{1!} m^2 \left(\sqrt{m} \frac{R_1}{N}\right)^2 M + \cdots + \frac{1}{1!} m^2 \left(\sqrt{m} \frac{R_1}{N}\right)^2 M}_{n} = n \left( \frac{1}{1!} m^2 \left(\sqrt{m} \frac{R_1}{N}\right)^2 M \right)$$

$$= nm^3 M \frac{1}{N^2} (R_1)^2$$

$$< (nm^3 M) \frac{1}{N^2}.$$

Thus, for every $x, y \in E_k$,

$$|f(y) - (f(x) + (f'(x))(y - x))| < (nm^3 M) \frac{1}{N^2}.$$

Since $C \cap E_k \neq \emptyset$, there exists $a \in C \cap E_k \neq \emptyset$, and hence, for every $y \in C \cap E_k$,

$$|f(y) - (f(A) + (f'(A))(y - a))| < (nm^3 M) \frac{1}{N^2}.$$

Again, by Taylor's theorem, for every $x \equiv (x_1, \ldots, x_m), y \equiv (y_1, \ldots, y_m) \in E_k$,

$$|f(y) - f(x)| = \left| \begin{bmatrix} f_1(y) \\ \vdots \\ f_n(y) \end{bmatrix} - \begin{bmatrix} f_1(x) \\ \vdots \\ f_n(x) \end{bmatrix} \right| = \left| \begin{bmatrix} f_1(y) - f_1(x) \\ \vdots \\ f_n(y) - f_n(x) \end{bmatrix} \right|$$

$$\leq |f_1(y) - f_1(x)| + \cdots + |f_n(y) - f_n(x)|$$

$$\leq \underbrace{\frac{1}{0!} m^1 \left( \sqrt{m} \frac{R_1}{N} \right)^1 M_0 + \cdots + \frac{1}{0!} m^1 \left( \sqrt{m} \frac{R_1}{N} \right)^1 M_0}_{n}$$

$$= nm^{3/2} M_0 \frac{1}{N} R_1 < nm^{3/2} M_0 \frac{1}{N}.$$

Hence, for every $y \in C \cap E_k$, $|f(y) - f(a)| < nm^{3/2} M_0 \frac{1}{N}$. It follows that $f(C \cap E_k)$ is contained in an open ball of radius $nm^{3/2} M_0 \frac{1}{N}$ with center $f(a)$. Since $a \in C \cap E_k$, so $a \in C$, and hence,

$$\text{rank} \begin{bmatrix} (D_1 f_1)(a) & (D_m f_1)(a) \\ \vdots & \ddots \vdots \\ (D_1 f_n)(a) & (D_m f_n)(a) \end{bmatrix} < n.$$

So, the range space of $f'(a) : \mathbb{R}^m \to \mathbb{R}^n$ is contained in an $(n-1)$-dimensional subspace, say $V$, of $\mathbb{R}^n$. It follows that for every $y \in C \cap E_k$, $f(a) + (f'(a))(y - a)$ contained in $f(a) + V$. Since for every $y \in C \cap E_k$,

$$|f(y) - (f(a) + (f'(a))(y - a))| < (nm^3 M) \frac{1}{N^2},$$

so all points of $f(C \cap E_k)$ is at most $(nm^3 M) \frac{1}{N^2}$ distance "above" or "below" the affine hyperplane $f(a) + V$. Thus, $f(C \cap E_k)$ contained in a "box" whose height is

$$2 \left( (nm^3 M) \frac{1}{N^2} \right),$$

and $(n-1)$-dimensional base area is

$$\left( 2 \left( nm^{3/2} M_0 \frac{1}{N} \right) \right)^{n-1},$$

and hence,

$$\mu(f(C \cap E_k)) \leq \left( \left( 2 \left( nm^{3/2} M_0 \frac{1}{N} \right) \right)^{n-1} \right) \left( 2 \left( (nm^3 M) \frac{1}{N^2} \right) \right)$$

$$= \left( 2^n n^n m^{\frac{3(n+1)}{2}} (M_0)^{n-1} M \right) \frac{1}{N^{n+1}}.$$

It follows that

$$\mu(f(C \cap (C_1)^-)) \leq \sum \{\mu(f(C \cap E_k)) : C \cap E_k \neq \emptyset, 1 \leq k \leq N^m\}$$

$$\leq \left(2^n n^n m^{\frac{3(n+1)}{2}} (M_0)^{n-1} M\right) \frac{1}{N^{n+1}} N^m \rightarrow 0 \text{ as } N \rightarrow \infty,$$

because $m \leq n$. Thus, $\mu(f(C \cap (C_1)^-)) = 0$. Similarly, $\mu(f(C \cap (C_2)^-)) = 0$, etc. Thus, $\mu(f(C)) = 0$.                                                                     □

**Lemma 6.24** *Let $m, n$ be any positive integers. Let $G$ be a nonempty open subset of $\mathbb{R}^m$. Let $f : G \rightarrow \mathbb{R}^n$ be a smooth function, and $f \equiv (f_1, \ldots f_n)$. Let $C$ be the collection of all critical points of $f$, that is,*

$$C \equiv \left\{ x : x \in G, \text{ and rank} \begin{bmatrix} (D_1 f_1)(x) & (D_m f_1)(x) \\ \vdots & \ddots \vdots \\ (D_1 f_n)(x) & (D_m f_n)(x) \end{bmatrix} < n \right\}.$$

*Let for every $i = 1, 2, 3, \ldots,$*

$C_i \equiv \{x : x \in G, \text{ and all 1st, 2nd}, \ldots, i\text{th order partial derivatives of } f \text{ at } x \text{ are zero}\}.$

*Then,*

1. $\cdots \subset C_4 \subset C_3 \subset C_2 \subset C_1 \subset C \subset G,.$
2. *each $C_i$ and $C$ are closed in $G$.*

*Proof*

1. This is clear from the definitions of $C_i$ and $C$.
2. Since $f : G \rightarrow \mathbb{R}^n$ is smooth, each $D_j f_i : G \rightarrow \mathbb{R}$ is continuous. Since each $D_j f_i : G \rightarrow \mathbb{R}$ is continuous, and $\{0\}$ is closed in $\mathbb{R}$, each $(D_j f_i)^{-1}(\{0\})$ is closed in $G$. It follows that

$$\bigcap_{\substack{j \in \{1, \ldots, k\} \\ i \in \{1, \ldots, n\}}} (D_j f_i)^{-1}(\{0\}) \left(= \{x : x \in G, \text{ and } 0 = (D_j f_i)(x) \text{ for every } j \in \{1, \ldots, k\}, \right.$$

$$\left. \text{and for every } i \in \{1, \ldots, n\}\} = C_1\right)$$

is closed in $G$, and hence, $C_1$ is closed in $G$. Similarly, $C_2, C_3, \ldots$ are closed in $G$. Here,

$$C = \left\{ x : x \in G, \text{ and rank} \begin{bmatrix} (D_1f_1)(x) & (D_kf_1)(x) \\ \vdots & \ddots \vdots \\ (D_1f_n)(x) & (D_kf_n)(x) \end{bmatrix} < n \right\}$$

$$= \bigcap_{\substack{i_1,\ldots,i_n \in \{1,\ldots,k\}, \\ i_1,\ldots,i_n \text{ are distinct}}} \left\{ x : x \in G, \text{ and det} \begin{bmatrix} (D_{i_1}f_1)(x) & (D_{i_n}f_1)(x) \\ \vdots & \ddots \vdots \\ (D_{i_1}f_n)(x) & (D_{i_n}f_n)(x) \end{bmatrix} = 0 \right\}.$$

Now, since each $D_jf_i : G \to \mathbb{R}$ is continuous, each

$$\left\{ x : x \in G, \text{ and det} \begin{bmatrix} (D_{i_1}f_1)(x) & (D_{i_n}f_1)(x) \\ \vdots & \ddots \vdots \\ (D_{i_1}f_n)(x) & (D_{i_n}f_n)(x) \end{bmatrix} = 0 \right\}$$

is closed in $G$, and hence,

$$\bigcap_{\substack{i_1,\ldots,i_n \in \{1,\ldots,k\}, \\ i_1,\ldots,i_n \text{ are distinct}}} \left\{ x : x \in G, \text{ and det} \begin{bmatrix} (D_{i_1}f_1)(x) & (D_{i_n}f_1)(x) \\ \vdots & \ddots \vdots \\ (D_{i_1}f_n)(x) & (D_{i_n}f_n)(x) \end{bmatrix} = 0 \right\}$$

is closed in $G$. This shows that $C$ is closed in $G$. $\qquad\square$

**Lemma 6.25** *Let $m,n$ be any positive integers. For every $k = 1,\ldots,n$, let $\mu_k$ denote the $k$-dimensional measure over $\mathbb{R}^k$. Let $G$ be a nonempty open subset of $\mathbb{R}^m$. Let $f : G \to \mathbb{R}^n$ be a smooth function, and $f \equiv (f_1,\ldots,f_n)$. Let $C$ be the collection of all critical points of $f$, that is,*

$$C \equiv \left\{ x : x \in G, \text{ and rank} \begin{bmatrix} (D_1f_1)(x) & (D_mf_1)(x) \\ \vdots & \ddots \vdots \\ (D_1f_n)(x) & (D_mf_n)(x) \end{bmatrix} < n \right\}.$$

*Let, for every $i = 1, 2, 3, \ldots,$*

$C_i \equiv \{ x : x \in G, \text{ and all 1st, 2nd,}\ldots, i\text{th order partial derivatives of } f \text{ at } x \text{ are zero} \}.$

*Then,*

$$\mu_n\left(f\left(C_{[\frac{m}{n}]}\right)\right) = 0,$$

*where $[\frac{m}{n}]$ denotes the greatest integer less than or equal to $\frac{m}{n}$.*

*Proof* Since $G$ is a nonempty open subset of $\mathbb{R}^m$, there exists a countable collection $\{U_l\}$ of open cubes such that $\bigcup_{l=1}^{\infty} U_l = G$, $\bigcup_{l=1}^{\infty} (U_l)^- = G$, and side length of each $U_l$ is strictly less than 1. It follows that

$$C_{\left[\frac{m}{n}\right]} = C_{\left[\frac{m}{n}\right]} \cap G = C_{\left[\frac{m}{n}\right]} \cap \left( \bigcup_{l=1}^{\infty} (U_l)^- \right) = \bigcup_{l=1}^{\infty} \left( C_{\left[\frac{m}{n}\right]} \cap (U_l)^- \right),$$

and hence,

$$f\left(C_{\left[\frac{m}{n}\right]}\right) = f\left( \bigcup_{l=1}^{\infty} \left( C_{\left[\frac{m}{n}\right]} \cap (U_l)^- \right) \right) = \bigcup_{l=1}^{\infty} f\left( C_{\left[\frac{m}{n}\right]} \cap (U_l)^- \right).$$

Thus,

$$0 \le \mu_n\left( f\left( C_{\left[\frac{m}{n}\right]}\right) \right) = \mu_n\left( \bigcup_{l=1}^{\infty} f\left( C_{\left[\frac{m}{n}\right]} \cap (U_l)^- \right) \right) \le \sum_{l=1}^{\infty} \mu_n\left( f\left( C_{\left[\frac{m}{n}\right]} \cap (U_l)^- \right) \right),$$

and therefore, it suffices to show that each

$$\mu_n\left( f\left( C_{\left[\frac{m}{n}\right]} \cap (U_l)^- \right) \right) \text{ is zero.}$$

For this purpose, let us fix a positive integer $l$. We have to show that

$$\mu_n\left( f\left( C_{\left[\frac{m}{n}\right]} \cap (U_l)^- \right) \right) = 0.$$

Let us fix any positive integer $N > 1$. Let us subdivide the closed cube $(U_l)^-$ into $N^m$ closed cubes $E_1, E_2, \ldots, E_{N^m}$ such that each $E_i$ is of length $\frac{1}{N}$ (side length of $U_l$). For simplicity, let the side length of $U_l$ be 1. Thus, $(U_l)^- = E_1 \cup E_2 \cup \cdots \cup E_{N^m}$. It follows that

$$
\begin{aligned}
C_{\left[\frac{m}{n}\right]} \cap (U_l)^- &= C_{\left[\frac{m}{n}\right]} \cap (E_1 \cup E_2 \cup \cdots \cup E_{N^m}) \\
&= \left( C_{\left[\frac{m}{n}\right]} \cap E_1 \right) \cup \left( C_{\left[\frac{m}{n}\right]} \cap E_2 \right) \cup \cdots \cup \left( C_{\left[\frac{m}{n}\right]} \cap E_{N^m} \right) \\
&= \bigcup \left\{ C_{\left[\frac{m}{n}\right]} \cap E_k : C_{\left[\frac{m}{n}\right]} \cap E_k \ne \emptyset, 1 \le k \le N^m \right\}.
\end{aligned}
$$

Hence,

$$
\begin{aligned}
f\left( C_{\left[\frac{m}{n}\right]} \cap (U_l)^- \right) &= f\left( \bigcup \left\{ C_{\left[\frac{m}{n}\right]} \cap E_k : C_{\left[\frac{m}{n}\right]} \cap E_k \ne \emptyset, 1 \le k \le N^m \right\} \right) \\
&= \bigcup \left\{ f\left( C_{\left[\frac{m}{n}\right]} \cap E_k \right) : C_{\left[\frac{m}{n}\right]} \cap E_k \ne \emptyset, 1 \le k \le N^m \right\},
\end{aligned}
$$

and hence,

$$0 \le \mu_n\left(f\left(C_{[\frac{m}{n}]} \cap (U_l)^-\right)\right) = \mu_n\left(\bigcup\left\{f\left(C_{[\frac{m}{n}]} \cap E_k\right) : C_{[\frac{m}{n}]} \cap E_k \ne \emptyset, 1 \le k \le N^m\right\}\right)$$

$$\le \sum\left\{\mu_n\left(f\left(C_{[\frac{m}{n}]} \cap E_k\right)\right) : C_{[\frac{m}{n}]} \cap E_k \ne \emptyset, 1 \le k \le N^m\right\}.$$

Let us take any positive integer $k$ satisfying $1 \le k \le N^m$, and $C_{[\frac{m}{n}]} \cap E_k \ne \emptyset$. Let all the $([\frac{m}{n}] + 1)$ th order partial derivatives of $f$ be bounded in absolute value by a constant $M$ on the compact set $(U_l)^-$.

Now, by Taylor's theorem, for every $x \equiv (x_1, \ldots, x_m) \in C_{[\frac{m}{n}]} \cap E_k$, and $y \equiv (y_1, \ldots, y_m) \in C_{[\frac{m}{n}]} \cap E_k$,

$$\begin{aligned}
|f(x) - f(y)| &= |(f_1(x), \ldots, f_n(x)) - (f_1(y), \ldots, f_n(y))| \\
&= |(f_1(x) - f_1(y), \ldots, f_n(x) - f_n(y))| \\
&\le |f_1(x) - f_1(y)| + \cdots + |f_n(x) - f_n(y)| \\
&= \left|f_1(x) - \left\{f_1(y) + \frac{1}{1!}(0) + \frac{1}{2!}(0) + \cdots + \frac{1}{[\frac{m}{n}]!}(0)\right\}\right| + \cdots \\
&\quad + \left|f_n(x) - \left\{f_n(y) + \frac{1}{1!}(0) + \frac{1}{2!}(0) + \cdots + \frac{1}{[\frac{m}{n}]!}(0)\right\}\right| \\
&\le \underbrace{\frac{1}{[\frac{m}{n}]!}m^{[\frac{m}{n}]+1}|x - y|^{[\frac{m}{n}]+1}M + \cdots + \frac{1}{[\frac{m}{n}]!}m^{[\frac{m}{n}]+1}|x - y|^{[\frac{m}{n}]+1}M}_{n} \\
&= n\left(\frac{1}{[\frac{m}{n}]!}m^{[\frac{m}{n}]+1}|x - y|^{[\frac{m}{n}]+1}M\right) \\
&\le n\left(\frac{1}{[\frac{m}{n}]!}m^{[\frac{m}{n}]+1}\left(\sqrt{m}\frac{1}{N}\right)^{[\frac{m}{n}]+1}M\right).
\end{aligned}$$

It follows that $f(C_{[\frac{m}{n}]} \cap E_k)$ is contained in a closed ball of radius

$$n\left(\frac{1}{[\frac{m}{n}]!}m^{[\frac{m}{n}]+1}\left(\sqrt{m}\frac{1}{N}\right)^{[\frac{m}{n}]+1}M\right),$$

and hence,

$$\mu_n\left(f\left(C_{[\frac{m}{n}]} \cap E_k\right)\right) \le \left(2\left(n\left(\frac{1}{[\frac{m}{n}]!}m^{[\frac{m}{n}]+1}\left(\sqrt{m}\frac{1}{N}\right)^{[\frac{m}{n}]+1}M\right)\right)\right)^n.$$

Thus,

$$0 \le \mu_n\left(f\left(C_{\left[\frac{m}{n}\right]} \cap (U_l)^-\right)\right) \le N^m \left(\left(2\left(n\left(\frac{1}{\left[\frac{m}{n}\right]!} m^{\left[\frac{m}{n}\right]+1}\left(\sqrt{m}\frac{1}{N}\right)^{\left[\frac{m}{n}\right]+1} M\right)\right)\right)^n\right)$$

$\to 0$ as $N \to \infty$,

because

$$m < \left(\left[\frac{m}{n}\right]+1\right)n \quad \left(\text{that is}, \frac{m}{n} < \left[\frac{m}{n}\right]+1\right).$$

Thus,

$$\mu_n\left(f\left(C_{\left[\frac{m}{n}\right]} \cap (U_l)^-\right)\right) = 0. \qquad \square$$

**Lemma 6.26** *Let $m,n$ be any positive integers. For every $k = 1,\ldots,n$, let $\mu_k$ denote the k-dimensional measure over $\mathbb{R}^k$. Let $G$ be a nonempty open subset of $\mathbb{R}^m$. Let $f : G \to \mathbb{R}^n$ be a smooth function, and $f \equiv (f_1,\ldots,f_n)$. Let $C$ be the collection of all critical points of $f$, that is,*

$$C \equiv \left\{ x : x \in G, \text{and rank} \begin{bmatrix} (D_1f_1)(x) & (D_mf_1)(x) \\ \vdots & \ddots \vdots \\ (D_1f_n)(x) & (D_mf_n)(x) \end{bmatrix} < n \right\}.$$

*Then, $\mu_n(f(C)) = 0$.*

*Proof* (Induction on $m$): By Lemma 6.23, the theorem is true for $m = 1$.
   Case I: when $m \le n$. By Lemma 6.23, $\mu_n(f(C)) = 0$.
   Case II: when $n < m$. For $i = 1, 2, 3, \ldots$, put

$C_i \equiv \{x : x \in G, \text{and all partial derivatives of } f \text{ at } x \text{ upto } i\text{th order are zero}\},$

and $C_0 \equiv C$. By Lemma 6.24, $\cdots \subset C_4 \subset C_3 \subset C_2 \subset C_1 \subset C \subset G$. It follows that

$$C = (C - C_1) \cup (C_1 - C_2) \cup (C_2 - C_3) \cup \cdots \cup \left(C_{\left[\frac{m}{n}\right]-1} - C_{\left[\frac{m}{n}\right]}\right) \cup C_{\left[\frac{m}{n}\right]},$$

and hence,

$$f(C) = f\left((C - C_1) \cup (C_1 - C_2) \cup (C_2 - C_3) \cup \cdots \cup \left(C_{\left[\frac{m}{n}\right]-1} - C_{\left[\frac{m}{n}\right]}\right) \cup C_{\left[\frac{m}{n}\right]}\right)$$

$$= f(C - C_1) \cup f(C_1 - C_2) \cup \quad f(C_2 - C_3) \cup \cdots \cup f\left(C_{\left[\frac{m}{n}\right]-1} - C_{\left[\frac{m}{n}\right]}\right) \cup f\left(C_{\left[\frac{m}{n}\right]}\right).$$

It follows that

$$0 \leq \mu_n(f(C) = \mu_n(f(C - C_1) \cup f(C_1 - C_2) \cup f(C_2 - C_3)$$
$$\cup \cdots \cup f\left(C_{[\frac{m}{n}]-1} - C_{[\frac{m}{n}]}\right) \cup f\left(C_{[\frac{m}{n}]}\right))$$
$$\leq \mu_n(f(C - C_1)) + \mu_n(f(C_1 - C_2)) + \mu_n(f(C_2 - C_3))$$
$$+ \cdots + \mu_n\left(f\left(C_{[\frac{m}{n}]-1} - C_{[\frac{m}{n}]}\right)\right) + \mu_n\left(f\left(C_{[\frac{m}{n}]}\right)\right).$$

Since, by Lemma 6.25, $\mu_n(f(C_{[\frac{m}{n}]})) = 0$, it suffices to show that

1. $0 = \mu_n(f(C_1 - C_2)) = \mu_n(f(C_2 - C_3)) = \cdots = \mu_n(f(C_{[\frac{m}{n}]-1} - C_{[mn]}))$,
2. $\mu_n(f(C - C_1)) = 0$.

> For 1: For simplicity of discussion, here we shall assume 3 for $n$. Since $\mathbb{R}^m$ is a second countable space, there exists a countable basis $\mathcal{B}$ of $\mathbb{R}^m$. Since $C_2$ is closed in the open set $G$, $G - C_2$ is open in $\mathbb{R}^m$. Since $C_2 \subset C_1 \subset G$, $C_1 - C_2 \subset G - C_2 \subset G$.

Let us take any $a \in C_1 - C_2$. It follows that $a \in C_1$, and $a \notin C_2$. Since $a \in C_1$, for every $i \in \{1, \ldots, m\}$, and for every $j \in \{1, 2, 3\}$, $(D_i f_j)(a) = 0$. Further, since $a \notin C_2$, there exist $i_1, i_2 \in \{1, \ldots, m\}$, and $j_1 \in \{1, 2, 3\}$ such that $(D_{i_1 i_2} f_{j_1})(a) \neq 0$. For simplicity, let $(D_{11} f_1)(a) \neq 0$, that is, $(D_1(D_1 f_1))(a) \neq 0$. Observe that $D_1 f_1 : G \to \mathbb{R}$ is a smooth function.

Let us define a function $(F^1, F^2, \ldots, F^m) \equiv F : G - C_2 \to \mathbb{R}^m$ as follows: For every $x \equiv (x^1, x^2, \ldots, x^m) \in G - C_2$,

$$\left(F^1\left(x^1, x^2, \ldots, x^m\right), F^2\left(x^1, x^2, \ldots, x^m\right), \ldots, F^m\left(x^1, x^2, \ldots, x^m\right)\right) = F\left(x^1, x^2, \ldots, x^m\right)$$
$$\equiv \left((D_1 f_1)\left(x^1, \ldots, x^m\right), x^2, \ldots, x^m\right).$$

Clearly, $F(C_1 - C_2) \subset \{0\} \times \mathbb{R}^{m-1}$. (Reason: For every $x \equiv (x^1, x^2, \ldots, x^m) \in C_1 - C_2 (\subset G - C_2 \subset G)$, we have $x \in C_1$, and hence, $F(x) = ((D_1 f_1)(x), x^2, \ldots, x^m) = (0, x^2, \ldots, x^m) \in \{0\} \times \mathbb{R}^{m-1}$.) Since $(D_1 f_1) : G \to \mathbb{R}$ is a smooth function, $F : (G - C_2) \to \mathbb{R}^m$ is smooth. Here,

$$\det(F'(a)) = \det \begin{bmatrix} (D_1(D_1 f_1))(a) & (D_2(D_1 f_1))(a) & & (D_m(D_1 f_1))(a) \\ 0 & 1 & & 0 \\ \vdots & \vdots & \ddots & \vdots \\ 0 & 0 & & 1 \end{bmatrix}_{m \times m}$$
$$= (D_1(D_1 f_1))(a),$$

which is nonzero. So, by Theorem 4.1, there exists an open neighborhood $U_a$ of $a$ such that $U_a$ is contained in $G - C_2$, $F(U_a)$ is open in $\mathbb{R}^m$, and $F$ has a smooth inverse on $F(U_a)$. Thus, $F : U_a \to F(U_a)$ is a diffeomorphism, and $F^{-1} : F(U_a) \to$

$U_a$ is a diffeomorphism. Since $F^{-1} : F(U_a) \to U_a$ is a diffeomorphism, $U_a \subset G - C_2 \subset G$, and $f : G \to \mathbb{R}^3$ is smooth, their composite $f \circ (F^{-1}) : F(U_a) \to \mathbb{R}^3$ is a smooth function.

Since $U_a$ is an open neighborhood of $a$ in $\mathbb{R}^m$, there exists an open set $V_a \in \mathcal{B}$ such that $a \in V_a \subset (V_a)^- \subset U_a$, and $(V_a)^-$ is compact. Since $(V_a)^- \subset U_a \subset G - C_2$, $C_2 \cap (V_a)^- = \emptyset$.

Let us observe that $(C_1 - C_2) \cap (V_a)^-$ is compact.

(Reason: Since $C_1 \subset G$, $C_1$ is closed in $G$, and $G$ is open in $\mathbb{R}^m$, $C_1 \cup (G^c)$ is closed in $\mathbb{R}^m$. Since $C_1 \cup (G^c)$ is closed in $\mathbb{R}^m$, and $(V_a)^-(\subset U_a \subset G)$ is compact, their intersection $(C_1 \cup (G^c)) \cap ((V_a)^-)(= C_1 \cap ((V_a)^-))$ is compact, and hence, $C_1 \cap (V_a)^-$ is compact. Since $C_1 \cap (V_a)^-$ is compact, and $(C_1 - C_2) \cap (V_a)^- = (C_1 \cap (V_a)^-) - (C_2 \cap (V_a)^-) = (C_1 \cap (V_a)^-) - \emptyset = C_1 \cap (V_a)^-$, $(C_1 - C_2) \cap (V_a)^-$ is compact).

Since $f : G \to \mathbb{R}^3$ is a continuous function, and $(C_1 - C_2) \cap (V_a)^-$ is compact, $f((C_1 - C_2) \cap (V_a)^-)$ is compact. We shall try to show that $\mu_3(f((C_1 - C_2) \cap (V_a)^-)) = 0$. Now, since $f((C_1 - C_2) \cap (V_a)^-)$ is compact, by Lemma 6.1, for every real $c$, $\mu_2(\{(y^1, y^2) : (c, y^1, y^2) \in f((C_1 - C_2) \cap (V_a)^-)\}) = 0$. For this purpose, let us fix any real $c$. We have to show that $\mu_2(\{(y^1, y^2) : (c, y^1, y^2) \in f((C_1 - C_2) \cap (V_a)^-)\}) = 0$, that is, $\mu_2(\{(f_2(x^1, \ldots, x^m), f_3(x^1, \ldots, x^m)) : (x^1, \ldots, x^m) \in (C_1 - C_2) \cap (V_a)^-, \text{and} f_1(x^1, \ldots, x^m) = c\}) = 0$.

Let us take any $u \equiv (u^1, u^2, \ldots, u^m) \in F(U_a)$. Since $F^{-1} : F(U_a) \to U_a$ is a diffeomorphism, the rank of $(F^{-1})'(y)$ for every $y \in F(U_a)$ is $m$. Now, since $u \in F(U_a)$, the rank of $(F^{-1})'(u)$ is $m$. Since the rank of $(F^{-1})'(u)$ is $m$, and

$$((D_1 f_1, f_2, f_3) \circ (F^{-1}))'(u) = ((D_1 f_1, f_2, f_3)'((F^{-1})(u)))((F^{-1})'(u)),$$

$$\text{rank}(((D_1 f_1, f_2, f_3) \circ (F^{-1}))'(u)) = \text{rank}((D_1 f_1, f_2, f_3)'((F^{-1})(u))).$$

Since $u \in F(U_a)$, and $F$ is 1–1 over $U_a$, there exists a unique $x \equiv (x^1, x^2, \ldots, x^m) \in U_a$ such that $F(x) = u$, and hence, $F^{-1}(u) = x$. Also, since $((D_1 f_1)(x^1, x^2, \ldots, x^m), x^2, \ldots, x^m) = F(x^1, x^2, \ldots, x^m) = F(x) = u = (u^1, u^2, \ldots, u^m)$, $(D_1 f_1)(x) = (D_1 f_1)(x^1, x^2, \ldots, x^m) = u^1, x^2 = u^2, \ldots, x^m = u^m$. Now,

$$((D_1 f_1, f_2, f_3) \circ (F^{-1}))(u^1, u^2, \ldots, u^m) = (D_1 f_1, f_2, f_3)(F^{-1}(u^1, u^2, \ldots, u^m))$$
$$= (D_1 f_1, f_2, f_3)(F^{-1}(u)) = (D_1 f_1, f_2, f_3)(x)$$
$$= ((D_1 f_1)(x), f_2(x), f_3(x)) = (u^1, f_2(x), f_3(x))$$
$$= (u^1, f_2(F^{-1}(u)), f_3(F^{-1}(u)))$$
$$= (u^1, (f_2 \circ (F^{-1}))(u), (f_3 \circ (F^{-1}))(u))$$
$$= (u^1, (f_2 \circ (F^{-1}))(u^1, u^2, \ldots, u^m), (f_3 \circ (F^{-1}))(u^1, u^2, \ldots, u^m)).$$

Hence, for every $(u^1, u^2, \ldots, u^m) \in F(U_a)$,

$$((D_1f_1, f_2, f_3) \circ (F^{-1}))(u^1, u^2, \ldots, u^m)$$
$$= (u^1, (f_2 \circ (F^{-1}))(u^1, u^2, \ldots, u^m), (f_3 \circ (F^{-1}))(u^1, u^2, \ldots, u^m)).$$

It follows that for every $u \equiv (u^1, u^2, \ldots, u^m) \in F(U_a)$,

$$\text{rank}((D_1f_1, f_2, f_3)'((F^{-1})(u))) = \text{rank}\left(((D_1f_1, f_2, f_3) \circ (F^{-1}))'(u)\right)$$

$$= \text{rank} \begin{bmatrix} 1 & 0 & 0 \\ (D_1(f_2 \circ (F^{-1})))(u) & (D_2(f_2 \circ (F^{-1})))(u) & \cdots (D_m(f_2 \circ (F^{-1})))(u) \\ (D_1(f_3 \circ (F^{-1})))(u) & (D_2(f_3 \circ (F^{-1})))(u) & (D_m(f_3 \circ (F^{-1})))(u) \end{bmatrix}.$$

Now, let $x \equiv (x^1, x^2, \ldots, x^m) \in (C_1 - C_2) \cap U_a$. It follows that $x \in (C_1 - C_2) \cap U_a \subset (C_1 - C_2)$, and hence, $F(x) \in F((C_1 - C_2) \cap U_a) \subset F(U_a)$. Thus,

$$3 > 1 \geq \text{rank} \begin{bmatrix} (D_1(D_1f_1))(x) & (D_m(D_1f_1))(x) \\ 0 & \cdots 0 \\ 0 & 0 \end{bmatrix}_{3 \times m}$$

$$= \text{rank} \begin{bmatrix} (D_1(D_1f_1))(x) & (D_m(D_1f_1))(x) \\ (D_1f_2)(x) & \cdots (D_mf_2)(x) \\ (D_1f_3)(x) & (D_mf_3)(x) \end{bmatrix}_{3 \times m}$$

$$= \text{rank}((D_1f_1, f_2, f_3)'(x))$$

$$= \text{rank}((D_1f_1, f_2, f_3)'((F^{-1})(F(x))))$$

$$= \text{rank} \begin{bmatrix} 1 & 0 & 0 \\ (D_1(f_2 \circ (F^{-1})))(F(x)) & (D_2(f_2 \circ (F^{-1})))(F(x)) & \cdots (D_m(f_2 \circ (F^{-1})))(F(x)) \\ (D_1(f_3 \circ (F^{-1})))(F(x)) & (D_2(f_3 \circ (F^{-1})))(F(x)) & (D_m(f_3 \circ (F^{-1})))(F(x)) \end{bmatrix}.$$

It follows that

$$0 = \det \begin{bmatrix} 1 & 0 & 0 \\ (D_1(f_2 \circ (F^{-1})))(F(x)) & (D_2(f_2 \circ (F^{-1})))(F(x)) & (D_3(f_2 \circ (F^{-1})))(F(x)) \\ (D_1(f_3 \circ (F^{-1})))(F(x)) & (D_2(f_3 \circ (F^{-1})))(F(x)) & (D_3(f_3 \circ (F^{-1})))(F(x)) \end{bmatrix}$$

$$= \det \begin{bmatrix} (D_2(f_2 \circ (F^{-1})))(F(x)) & (D_3(f_2 \circ (F^{-1})))(F(x)) \\ (D_2(f_3 \circ (F^{-1})))(F(x)) & (D_3(f_3 \circ (F^{-1})))(F(x)) \end{bmatrix},$$

$$0 = \det \begin{bmatrix} 1 & 0 & 0 \\ (D_1(f_2 \circ (F^{-1})))(F(x)) & (D_2(f_2 \circ (F^{-1})))(F(x)) & (D_4(f_2 \circ (F^{-1})))(F(x)) \\ (D_1(f_3 \circ (F^{-1})))(F(x)) & (D_2(f_3 \circ (F^{-1})))(F(x)) & (D_4(f_3 \circ (F^{-1})))(F(x)) \end{bmatrix}$$

$$= \det \begin{bmatrix} (D_2(f_2 \circ (F^{-1})))(F(x)) & (D_4(f_2 \circ (F^{-1})))(F(x)) \\ (D_2(f_3 \circ (F^{-1})))(F(x)) & (D_4(f_3 \circ (F^{-1})))(F(x)) \end{bmatrix},$$

etc. Hence, for every $x \equiv (x^1, x^2, \ldots, x^m) \in (C_1 - C_2) \cap U_a$,

$$\text{rank} \begin{bmatrix} (D_2(f_2 \circ (F^{-1})))(F(x)) & (D_3(f_2 \circ (F^{-1})))(F(x)) \cdot & (D_m(f_2 \circ (F^{-1})))(F(x)) \\ (D_2(f_3 \circ (F^{-1})))(F(x)) & (D_3(f_3 \circ (F^{-1})))(F(x)) & \cdot (D_m(f_3 \circ (F^{-1})))(F(x)) \end{bmatrix} < 2.$$

Thus, for every $u \equiv (u^1, u^2, \ldots, u^m) \in F((C_1 - C_2) \cap U_a)$,

$$\text{rank} \begin{bmatrix} (D_2(f_2 \circ (F^{-1})))(u) & (D_3(f_2 \circ (F^{-1})))(u) \cdot & (D_m(f_2 \circ (F^{-1})))(u) \\ (D_2(f_3 \circ (F^{-1})))(u) & (D_3(f_3 \circ (F^{-1})))(u) & \cdot (D_m(f_3 \circ (F^{-1})))(u) \end{bmatrix} < 2.$$

Since $F(U_a)$ is open in $\mathbb{R}^m$, $\{(y^2, \ldots, y^m) : (0, y^2, \ldots, y^m) \in F(U_a)\}$ is open in $\mathbb{R}^{m-1}$.

Let us define a function $f_c : \{(y^2, \ldots, y^m) : (0, y^2, \ldots, y^m) \in F(U_a)\} \to \mathbb{R}^2$ as follows: For every $(y^2, \ldots, y^m)$ satisfying $(0, y^2, \ldots, y^m) \in F(U_a)$,

$$f_c(y^2, \ldots, y^m) = ((f_2 \circ (F^{-1}))(0, y^2, \ldots, y^m), (f_3 \circ (F^{-1}))(0, y^2, \ldots, y^m)).$$

Since $(y^2, \ldots, y^m) \mapsto (0, y^2, \ldots, y^m)$ is smooth, and $(f_2 \circ (F^{-1})) : F(U_a) \to \mathbb{R}$ is smooth, their composite $(y^2, \ldots, y^m) \mapsto (f_2 \circ (F^{-1}))(0, y^2, \ldots, y^m)$ is smooth. Similarly, $(y^2, \ldots, y^m) \mapsto (f_3 \circ (F^{-1}))(0, y^2, \ldots, y^m)$ is smooth. It follows that $f_c : \{(y^2, \ldots, y^m) : (0, y^2, \ldots, y^m) \in F(U_a)\} \to \mathbb{R}^2$ is smooth.

For every $u \equiv (u^2, \ldots, u^m)$ satisfying $(0, u^2, \ldots, u^m) \in F(U_a)$, we have $(D_1 (f_c)_1)(u^2, \ldots, u^m) = (D_2(f_2 \circ (F^{-1})))(0, u^2, \ldots, u^m)$, $(D_2(f_c)_1)(u^2, \ldots, u^m) = (D_3 (f_2 \circ (F^{-1})))(0, u^2, \ldots, u^m)$, etc., and hence,

$$(f_c)'(u) = (f_c)'(u^2, \ldots, u^m) = \begin{bmatrix} (D_1(f_c)_1)(u) & (D_2(f_c)_1)(u) \cdot & (D_{m-1}(f_c)_1)(u) \\ (D_1(f_c)_2)(u) & (D_2(f_c)_2)(u) & \cdot (D_{m-1}(f_c)_2)(u) \end{bmatrix}$$

$$= \begin{bmatrix} (D_2(f_2 \circ (F^{-1})))(0, u^2, \ldots, u^m) & (D_3(f_2 \circ (F^{-1})))(0, u^2, \ldots, u^m) \\ (D_2(f_3 \circ (F^{-1})))(0, u^2, \ldots, u^m) & (D_3(f_3 \circ (F^{-1})))(0, u^2, \ldots, u^m) \end{bmatrix}$$

$$\cdot \begin{matrix} (D_m(f_2 \circ (F^{-1})))(0, u^2, \ldots, u^m) \\ (D_m(f_3 \circ (F^{-1})))(0, u^2, \ldots, u^m) \end{matrix} \Bigg].$$

Clearly, every point of $\{(y^2, \ldots, y^m) : (0, y^2, \ldots, y^m) \in F((C_1 - C_2) \cap U_a)\}$ is a critical point of $f_c$. (Reason: Take any $u \equiv (u^2, \ldots, u^m) \in \{(y^2, \ldots, y^m) : (0, y^2, \ldots, y^m) \in F((C_1 - C_2) \cap U_a)\}$. It follows that $(0, y^2, \ldots, y^m) \in F((C_1 - C_2) \cap U_a) \subset F(U_a)$, and

$$\text{rank}((f_c)'(u)) = \text{rank} \begin{bmatrix} (D_2(f_2 \circ (F^{-1})))(0, u^2, \ldots, u^m) & (D_3(f_2 \circ (F^{-1})))(0, u^2, \ldots, u^m) \\ (D_2(f_3 \circ (F^{-1})))(0, u^2, \ldots, u^m) & (D_3(f_3 \circ (F^{-1})))(0, u^2, \ldots, u^m) \end{bmatrix}$$

$$\cdot \begin{matrix} (D_m(f_2 \circ (F^{-1})))(0, u^2, \ldots, u^m) \\ (D_m(f_3 \circ (F^{-1})))(0, u^2, \ldots, u^m) \end{matrix} \Bigg] < 2.$$

Now, by induction hypothesis, $\mu_2(f_c(\{(y^2,\ldots,y^m) : (0,y^2,\ldots,y^m) \in F((C_1 - C_2) \cap U_a)\})) = 0$. Now, since $(V_a)^- \subset U_a$,

$$0 \le \mu_2(\{((f_2 \circ (F^{-1}))(0,y^2,\ldots,y^m), (f_3 \circ (F^{-1}))(0,y^2,\ldots,y^m))$$
$$: (0,y^2,\ldots,y^m) \in F((C_1 - C_2) \cap (V_a)^-))\}$$
$$\le \mu_2(\{((f_2 \circ (F^{-1}))(0,y^2,\ldots,y^m), (f_3 \circ (F^{-1}))(0,y^2,\ldots,y^m))$$
$$: (0,y^2,\ldots,y^m) \in F((C_1 - C_2) \cap U_a))\}$$
$$= \mu_2((\{f_c(y^2,\ldots,y^m) : (0,y^2,\ldots,y^m) \in F((C_1 - C_2) \cap U_a)\})) = 0,$$

and hence,

$$\mu_2(\{((f_2 \circ (F^{-1}))(0,y^2,\ldots,y^m), (f_3 \circ (F^{-1}))(0,y^2,\ldots,y^m))$$
$$: (0,y^2,\ldots,y^m) \in F((C_1 - C_2) \cap (V_a)^-)\}) = 0.$$

It suffices to show that

$$\{(f_2(x^1,\ldots,x^m), f_3(x^1,\ldots,x^m)) : (x^1,\ldots,x^m)$$
$$\in (C_1 - C_2) \cap (V_a)^-, \text{ and } (D_1 f_1)(x^1,\ldots,x^m) = c\}$$
$$\subset \{((f_2 \circ (F^{-1}))(0,y^2,\ldots,y^m), (f_3 \circ (F^{-1}))(0,y^2,\ldots,y^m)) : (0,y^2,\ldots,y^m)$$
$$\in F((C_1 - C_2) \cap (V_a)^-)\}.$$

Let us take any $(f_2(x^1,\ldots,x^m), f_3(x^1,\ldots,x^m)) \in \text{LHS}$, where $(x^1,\ldots,x^m) \in (C_1 - C_2) \cap (V_a)^-$, and $f_1(x^1,\ldots,x^m) = c$. Now, since $(x^1,\ldots,x^m) \in (C_1 - C_2) \cap (V_a)^-$,

$$(0,x^2,\ldots,x^m) = ((D_1 f_1)(x^1,\ldots,x^m), x^2,\ldots,x^m) = F(x^1,\ldots,x^m)$$
$$\in F((C_1 - C_2) \cap (V_a)^-),$$

and hence, $F^{-1}(0,x^2,\ldots,x^m) = (x^1,\ldots,x^m)$. It follows that

$$(f_2(x^1,\ldots,x^m), f_3(x^1,\ldots,x^m)) = (f_2(F^{-1}(0,x^2,\ldots,x^m)), f_3(F^{-1}(0,x^2,\ldots,x^m)))$$
$$= ((f_2 \circ (F^{-1}))(0,x^2,\ldots,x^m), (f_3 \circ (F^{-1}))(0,x^2,\ldots,x^m)) \in \text{RHS}.$$

Thus, LHS $\subset$ RHS.

We have shown that for every $a \in C_1 - C_2$, there exists an open neighborhood $V_a \in \mathcal{B}$ of $a$ such that $\mu_3(f((C_1 - C_2) \cap (V_a)^-)) = 0$, and $a \in V_a \subset (V_a)^- \subset G$. Here, $\cup\{(C_1 - C_2) \cap (V_a)^- : a \in (C_1 - C_2)\} = (C_1 - C_2)$, so $f(C_1 - C_2) = f(\cup\{(C_1 - C_2) \cap (V_a)^- : a \in (C_1 - C_2)\}) = \cup\{f((C_1 - C_2) \cap (V_a)^-) : a \in (C_1 - C_2)\}$. Since each $V_a \in \mathcal{B}$, and $\mathcal{B}$ is a countable collection, $\{f((C_1 - C_2) \cap (V_a)^-) : a \in (C_1 - C_2)\}$ is a countable collection, and hence, $\mu_3(f(C_1 - C_2)) = \mu_3(\cup\{f((C_1 - C_2) \cap$

$(V_a)^-) : a \in (C_1 - C_2)\}) \leq \sum_{\text{countable}} \mu_3(f((C_1 - C_2) \cap (V_a)^-)) = \sum_{\text{countable}} 0 = 0.$
Thus, $\mu_3(f(C_1 - C_2)) = 0.$ Similarly,

$$0 = \mu_3(f(C_2 - C_3)) = \cdots = \mu_3\left(f\left(C_{[\frac{m}{n}]-1} - C_{[\frac{m}{n}]}\right)\right).$$

For 2: For simplicity of discussion, here, we shall assume 3 for $n$. Since $\mathbb{R}^m$ is a second countable space, there exists a countable basis $B$ of $\mathbb{R}^m$.

Since $C_1$ is closed in the open set $G$, $G - C_1$ is open in $\mathbb{R}^m$. Since $C_1 \subset C \subset G$, $C - C_1 \subset G - C_1 \subset G$. Now, since $f : G \to \mathbb{R}^3$ is a smooth function, $f|_{(G-C_1)} : (G - C_1) \to \mathbb{R}^3$ is smooth. Clearly, the collection of all critical points of $f|_{(G-C_1)}$ is $C - C_1$, that is,

$$C - C_1 = \left\{ x : x \in G - C_1, \text{and rank} \begin{bmatrix} (D_1f_1)(x) & (D_mf_1)(x) \\ (D_1f_2)(x) & \cdot (D_mf_2)(x) \\ (D_1f_3)(x) & (D_mf_3)(x) \end{bmatrix} < 3 \right\}.$$

Let us take any $a \in C - C_1$. Since $a \in C - C_1 \subset G$, $a \in G$. Now, since $a \notin C_1 = \{x : x \in G, \text{and all first order partial derivatives of } F \text{ at } x \text{ are zero}\}$, there exists $j \in \{1, 2, \ldots, m\}$, and $i \in \{1, 2, 3\}$ such that $(D_jf_i)(a) \neq 0$. For simplicity, let $i = j = 1$. Thus, $(D_1f_1)(a) \neq 0$. Let us define a function $(F^1, F^2, \ldots, F^m) \equiv F : (G - C_1) \to \mathbb{R}^m$ as follows: For every $x \equiv (x^1, x^2, \ldots, x^m) \in G - C_1$,

$$\left(F^1(x^1, x^2, \ldots, x^m), F^2(x^1, x^2, \ldots, x^m), \ldots, F^k(x^1, x^2, \ldots, x^m)\right) = F(x^1, x^2, \ldots, x^m)$$
$$\equiv (f_1(x^1, \ldots, x^m), x^2, \ldots, x^m).$$

Since $f : G \to \mathbb{R}^3$ is a smooth function, $f_1 : G \to \mathbb{R}$ is a smooth function, and hence, $F : (G - C_1) \to \mathbb{R}^k$ is smooth. Here,

$$\det(F'(A)) = \det \begin{bmatrix} (D_1f_1)(A) & (D_2f_1)(A) & (D_mf_1)(A) \\ 0 & 1 & 0 \\ \vdots & \vdots & \ddots \vdots \\ 0 & 0 & 1 \end{bmatrix}_{m \times m} = (D_1f_1)(A),$$

which is nonzero. So, by Theorem 4.1, there exists an open neighborhood $U_a$ of $a$ such that $U_a$ is contained in $(G - C_1)$, $F(U_a)$ is open in $\mathbb{R}^m$, and $F$ has a smooth inverse on $F(U_a)$. Thus, $F : U_a \to F(U_a)$ is a diffeomorphism, and $F^{-1} : F(U_a) \to U_a$ is a diffeomorphism. Since $F^{-1} : F(U_a) \to U_a$ is a diffeomorphism, $U_a \subset (G - C_1) \subset G$, and $f : G \to \mathbb{R}^3$ is smooth, their composite $f \circ (F^{-1}) : F(U_a) \to \mathbb{R}^3$ is a smooth function. Since $U_a$ is an open neighborhood of $a$ in $\mathbb{R}^m$, there exists an open set $V_a \in B$ such that $a \in V_a \subset (V_a)^- \subset U_a \subset (G - C_1)$, and $(V_a)^-$ is compact.

Let us observe that $(C - C_1) \cap (V_a)^-$ is compact.

(Reason: Since $C \subset G$, $C$ is closed in $G$, and $G$ is open in $\mathbb{R}^m$, $C \cup (G^c)$ is closed in $\mathbb{R}^m$. Since $C \cup (G^c)$ is closed in $\mathbb{R}^m$, and $(V_a)^- (\subset U_a \subset (G - C_1) \subset G)$ is compact, their intersection $(C \cup (G^c)) \cap ((V_a)^-) (= C \cap ((V_a)^-))$ is compact, and hence, $C \cap (V_a)^-$ is compact. Since $C \cap (V_a)^-$ is compact, and $(C - C_1) \cap (V_a)^- = (C \cap (V_a)^-) - (C_1 \cap (V_a)^-) = (C \cap (V_a)^-) - \emptyset = C \cap (V_a)^-$, $(C - C_1) \cap (V_a)^-$ is compact.) Since $f : G \to \mathbb{R}^3$ is a continuous function, and $(C - C_1) \cap (V_a)^- (\subset (V_a)^- \subset G)$ is compact, $f((C - C_1) \cap (V_a)^-)$ is compact. We shall try to show that $\mu_3(f((C - C_1) \cap (V_a)^-)) = 0$.

Now, since $f((C - C_1) \cap (V_a)^-)$ is compact, by Lemma 6.1, it suffices to show that for every real $c$, $\mu_2(\{(y^1, y^2) : (c, y^1, y^2) \in f((C - C_1) \cap (V_a)^-)\}) = 0$. For this purpose, let us fix any real $c$. We have to show that $\mu_2(\{(y^1, y^2) : (c, y^1, y^2) \in f((C - C_1) \cap (V_a)^-)\}) = 0$, that is, $\mu_2(\{(f_2(x^1, \ldots, x^m), f_3(x^1, \ldots, x^m)) : (x^1, \ldots, x^m) \in (C - C_1) \cap (V_a)^-, \text{and} f_1(x^1, \ldots, x^m) = c\}) = 0$.

Let us take any $u \equiv (u^1, u^2, \ldots, u^m) \in F(U_a)$. Since $F^{-1} : F(U_a) \to U_a$ is a diffeomorphism, the rank of $(F^{-1})'(y)$ for every $y \in F(U_a)$ is $m$. Now, since $u \in F(U_a)$, the rank of $(F^{-1})'(u)$ is $m$. Since the rank of $(F^{-1})'(u)$ is $m$, and

$$\left(f \circ (F^{-1})\right)'(u) = \left(f'((F^{-1})(u))\right)\left((F^{-1})'(u)\right),$$

$$\text{rank}\left(\left(f \circ (F^{-1})\right)'(u)\right) = \text{rank}(f'((F^{-1})(u))).$$

Since $u \in F(U_a)$, and $F$ is 1–1 over $U_a$, there exists a unique $x \equiv (x^1, x^2, \ldots, x^m) \in U_a$ such that $F(x) = u$, and hence, $F^{-1}(u) = x$. Also, since $(f_1(x^1, x^2, \ldots, x^m), x^2, \ldots, x^m) = F(x^1, x^2, \ldots, x^m) = F(x) = u = (u^1, u^2, \ldots, u^m)$, $f_1(x) = f_1(x^1, x^2, \ldots, x^m) = u^1, x^2 = u^2, \ldots, x^m = u^m$. Now, $(f \circ (F^{-1}))(u^1, u^2, \ldots, u^m) = f(F^{-1}(u^1, u^2, \ldots, u^m)) = f(F^{-1}(u)) = f(x) = (f_1(x), f_2(x), f_3(x)) = (u^1, f_2(x), f_3(x)) = (u^1, f_2(F^{-1}(u)), f_3(F^{-1}(u))) = (u^1, (f_2 \circ (F^{-1}))(u), (f_3 \circ (F^{-1}))(u)) = (u^1, (f_2 \circ (F^{-1}))(u^1, u^2, \ldots, u^m), (f_3 \circ (F^{-1}))(u^1, u^2, \ldots, u^m))$. Hence, for every $(u^1, u^2, \ldots, u^m) \in F(U_a)$,

$$\left(f \circ (F^{-1})\right)(u^1, u^2, \ldots, u^m) = \left(u^1, (f_2 \circ (F^{-1}))(u^1, u^2, \ldots, u^m),\right.$$
$$\left.(f_3 \circ (F^{-1}))(u^1, u^2, \ldots, u^m)\right).$$

It follows that for every $u \equiv (u^1, u^2, \ldots, u^m) \in F(U_a)$,

$$\text{rank}(f'((F^{-1})(u))) = \text{rank}\left(\left(f \circ (F^{-1})\right)'(u)\right)$$

$$= \text{rank} \begin{bmatrix} 1 & 0 & & 0 \\ (D_1(f_2 \circ (F^{-1})))(u) & (D_2(f_2 \circ (F^{-1})))(u) & \cdots & (D_m(f_2 \circ (F^{-1})))(u) \\ (D_1(f_3 \circ (F^{-1})))(u) & (D_2(f_3 \circ (F^{-1})))(u) & & (D_m(f_3 \circ (F^{-1})))(u) \end{bmatrix}$$

Now, let $x \equiv (x^1, x^2, \ldots, x^m) \in (C - C_1) \cap U_a$. It follows that $x \in (C - C_1) \cap U_a \subset (C - C_1)$, and hence, $F(x) \in F((C - C_1) \cap U_a) \subset F(U_a)$. Thus,

$$3 > \text{rank} \begin{bmatrix} (D_1 f_1)(x) & (D_m f_1)(x) \\ (D_1 f_2)(x) & \cdot \cdot (D_m f_2)(x) \\ (D_1 f_3)(x) & (D_m f_3)(x) \end{bmatrix}_{3 \times m} = \text{rank}(f'(x)) = \text{rank}\left(f'\left((F^{-1})(F(x))\right)\right)$$

$$= \text{rank} \begin{bmatrix} 1 & 0 & 0 \\ (D_1(f_2 \circ (F^{-1})))(F(x)) & (D_2(f_2 \circ (F^{-1})))(F(x)) & \cdot\cdot (D_m(f_2 \circ (F^{-1})))(F(x)) \\ (D_1(f_3 \circ (F^{-1})))(F(x)) & (D_2(f_3 \circ (F^{-1})))(F(x)) & (D_m(f_3 \circ (F^{-1})))(F(x)) \end{bmatrix}.$$

It follows that

$$0 = \det \begin{bmatrix} 1 & 0 & 0 \\ (D_1(f_2 \circ (F^{-1})))(F(x)) & (D_2(f_2 \circ (F^{-1})))(F(x)) & (D_3(f_2 \circ (F^{-1})))(F(x)) \\ (D_1(f_3 \circ (F^{-1})))(F(x)) & (D_2(f_3 \circ (F^{-1})))(F(x)) & (D_3(f_3 \circ (F^{-1})))(F(x)) \end{bmatrix}$$

$$= \det \begin{bmatrix} (D_2(f_2 \circ (F^{-1})))(F(x)) & (D_3(f_2 \circ (F^{-1})))(F(x)) \\ (D_2(f_3 \circ (F^{-1})))(F(x)) & (D_3(f_3 \circ (F^{-1})))(F(x)) \end{bmatrix},$$

$$0 = \det \begin{bmatrix} 1 & 0 & 0 \\ (D_1(f_2 \circ (F^{-1})))(F(x)) & (D_2(f_2 \circ (F^{-1})))(F(x)) & (D_4(f_2 \circ (F^{-1})))(F(x)) \\ (D_1(f_3 \circ (F^{-1})))(F(x)) & (D_2(f_3 \circ (F^{-1})))(F(x)) & (D_4(f_3 \circ (F^{-1})))(F(x)) \end{bmatrix}$$

$$= \det \begin{bmatrix} (D_2(f_2 \circ (F^{-1})))(F(x)) & (D_4(f_2 \circ (F^{-1})))(F(x)) \\ (D_2(f_3 \circ (F^{-1})))(F(x)) & (D_4(f_3 \circ (F^{-1})))(F(x)) \end{bmatrix},$$

etc. Hence, for every $x \equiv (x^1, x^2, \ldots, x^m) \in (C - C_1) \cap U_a$,

$$\text{rank} \begin{bmatrix} (D_2(f_2 \circ (F^{-1})))(F(x)) & (D_3(f_2 \circ (F^{-1})))(F(x)) & \cdot & (D_m(f_2 \circ (F^{-1})))(F(x)) \\ (D_2(f_3 \circ (F^{-1})))(F(x)) & (D_3(f_3 \circ (F^{-1})))(F(x)) & & (D_m(f_3 \circ (F^{-1})))(F(x)) \end{bmatrix} < 2.$$

Thus, for every $u \equiv (u^1, u^2, \ldots, u^m) \in F((C - C_1) \cap U_a)$,

$$\text{rank} \begin{bmatrix} (D_2(f_2 \circ (F^{-1})))(u) & (D_3(f_2 \circ (F^{-1})))(u) & \cdot & (D_m(f_2 \circ (F^{-1})))(u) \\ (D_2(f_3 \circ (F^{-1})))(u) & (D_3(f_3 \circ (F^{-1})))(u) & & (D_m(f_3 \circ (F^{-1})))(u) \end{bmatrix} < 2.$$

Since $F(U_a)$ is open in $\mathbb{R}^k$, $\{(y^2, \ldots, y^m) : (c, y^2, \ldots, y^m) \in F(U_a)\}$ is open in $\mathbb{R}^{m-1}$.

Let us define a function $f_c : \{(y^2, \ldots, y^m) : (c, y^2, \ldots, y^m) \in F(U_a)\} \to \mathbb{R}^2$ as follows: For every $(y^2, \ldots, y^m)$ satisfying $(c, y^2, \ldots, y^m) \in F(U_a)$,

$$f_c\big(y^2,\ldots,y^m\big) = \big((f_2 \circ (F^{-1}))(c,y^2,\ldots,y^m), (f_3 \circ (F^{-1}))(c,y^2,\ldots,y^m)\big).$$

Since $(y^2,\ldots,y^m) \mapsto (c,y^2,\ldots,y^m)$ is smooth, and $(f_2 \circ (F^{-1})) : F(U_a) \to \mathbb{R}$ is smooth, their composite $(y^2,\ldots,y^m) \mapsto (f_2 \circ (F^{-1}))(c,y^2,\ldots,y^m)$ is smooth. Similarly, $(y^2,\ldots,y^m) \mapsto (f_3 \circ (F^{-1}))(c,y^2,\ldots,y^m)$ is smooth. It follows that $f_c : \{(y^2,\ldots,y^m) : (c,y^2,\ldots,y^m) \in F(U_a)\} \to \mathbb{R}^2$ is smooth.

For every $u \equiv (u^2,\ldots,u^m)$ satisfying $(c,u^2,\ldots,u^m) \in F(U_a)$, $(D_1(f_c)_1)(u^2,\ldots,u^m) = (D_2(f_2 \circ (F^{-1})))(c,u^2,\ldots,u^m)$, $(D_2(f_c)_1)(u^2,\ldots,u^m) = (D_3(f_2 \circ (F^{-1})))(c,u^2,\ldots,u^m)$, etc., and hence,

$$(f_c)'(u) = (f_c)'(u^2,\ldots,u^m) = \begin{bmatrix} (D_1(f_c)_1)(u) & (D_2(f_c)_1)(u) \cdot & (D_{k-1}(f_c)_1)(u) \\ (D_1(f_c)_2)(u) & (D_2(f_c)_2)(u) & \cdot \ (D_{k-1}(f_c)_2)(u) \end{bmatrix}$$

$$= \begin{bmatrix} (D_2(f_2 \circ (F^{-1})))(c,u^2,\ldots,u^m) & (D_3(f_2 \circ (F^{-1})))(c,u^2,\ldots,u^m) \\ (D_2(f_3 \circ (F^{-1})))(c,u^2,\ldots,u^m) & (D_3(f_3 \circ (F^{-1})))(c,u^2,\ldots,u^m) \end{bmatrix}$$
$$\begin{matrix} \cdot & (D_k(f_2 \circ (F^{-1})))(c,u^2,\ldots,u^m) \\ \cdot & (D_k(f_3 \circ (F^{-1})))(c,u^2,\ldots,u^m) \end{matrix}\Bigg].$$

Clearly, every point of $\{(y^2,\ldots,y^m) : (c,y^2,\ldots,y^m) \in F(C \cap U_a)\}$ is a critical point of $f_c$. (Reason: Take any $u \equiv (u^2,\ldots,u^m) \in \{(y^2,\ldots,y^m) : (0,y^2,\ldots,y^m) \in F((C_1 - C_2) \cap U_a)\}$. It follows that $(0,y^2,\ldots,y^m) \in F((C - C_1) \cap U_a) \subset F(U_a)$, and

$$\text{rank}\big((f_c)'(u)\big)$$
$$= \text{rank} \begin{bmatrix} (D_2(f_2 \circ (F^{-1})))(c,u^2,\ldots,u^m) & (D_3(f_2 \circ (F^{-1})))(c,u^2,\ldots,u^m) \\ (D_2(f_3 \circ (F^{-1})))(c,u^2,\ldots,u^m) & (D_3(f_3 \circ (F^{-1})))(c,u^2,\ldots,u^m) \end{bmatrix}$$
$$\begin{matrix} \cdot & (D_k(f_2 \circ (F^{-1})))(c,u^2,\ldots,u^m) \\ \cdot & (D_k(f_3 \circ (F^{-1})))(c,u^2,\ldots,u^m) \end{matrix}\Bigg] < 2.)$$

Now, by induction hypothesis, $\mu_2(f_c(\{(y^2,\ldots,y^m) : (c,y^2,\ldots,y^m) \in F((C -C_1) \cap U_a)\})) = 0$. Since $(V_a)^- \subset U_a$,

$$0 \le \mu_2(\{((f_2 \circ (F^{-1}))(c,y^2,\ldots,y^m), (f_3 \circ (F^{-1}))(c,y^2,\ldots,y^m))$$
$$: (c,y^2,\ldots,y^m) \in F((C - C_1) \cap (V_a)^-)\})$$
$$\le \mu_2(\{((f_2 \circ (F^{-1}))(c,y^2,\ldots,y^m), (f_3 \circ (F^{-1}))(c,y^2,\ldots,y^m))$$
$$: (c,y^2,\ldots,y^m) \in F(C \cap U_a)\})$$
$$= \mu_2((\{f_c(y^2,\ldots,y^m) : (c,y^2,\ldots,y^m) \in F((C - C_1) \cap U_a)\})) = 0,$$

and hence,

$$\mu_2\big(\{\,((f_2\circ(F^{-1}))(c,y^2,\ldots,y^m),(f_3\circ(F^{-1}))(c,y^2,\ldots,y^m))$$
$$:(c,y^2,\ldots,y^m)\in F((C-C_1)\cap(V_a)^-)\}\big)=0.$$

It suffices to show that

$$\{(f_2(x^1,\ldots,x^m),f_3(x^1,\ldots,x^m)):(x^1,\ldots,x^m)\in(C-C_1)\cap(V_a)^-,$$
$$\text{and } f_1(x^1,\ldots,x^m)=c\}$$
$$\subset\{((f_2\circ(F^{-1}))(c,y^2,\ldots,y^m),(f_3\circ(F^{-1}))(c,y^2,\ldots,y^m))$$
$$:(c,y^2,\ldots,y^m)\in F((C-C_1)\cap(V_a)^-)\}.$$

Let us take any $(f_2(x^1,\ldots,x^m),f_3(x^1,\ldots,x^m))\in\text{LHS}$, where $(x^1,\ldots,x^m)\in(C-C_1)\cap(V_a)^-$, and $f_1(x^1,\ldots,x^m)=c$. Now, since $(x^1,\ldots,x^m)\in(C-C_1)\cap(V_a)^-$,

$$(c,x^2,\ldots,x^m)=(f_1(x^1,\ldots,x^m),x^2,\ldots,x^m)=F(x^1,\ldots,x^m)$$
$$\in F((C-C_1)\cap(V_a)^-),$$

and hence $F^{-1}(c,x^2,\ldots,x^m)=(x^1,\ldots,x^m)$. It follows that

$$(f_2(x^1,\ldots,x^m),f_3(x^1,\ldots,x^m))$$
$$=(f_2(F^{-1}(c,x^2,\ldots,x^m)),f_3(F^{-1}(c,x^2,\ldots,x^m)))$$
$$=((f_2\circ(F^{-1}))(c,x^2,\ldots,x^m),(f_3\circ(F^{-1}))(c,x^2,\ldots,x^m))\in\text{RHS}.$$

Hence, LHS $\subset$ RHS.

Thus, we have shown that for every $a\in(C-C_1)$, there exists an open neighborhood $V_a\in\mathcal{B}$ of $a$ such that $\mu_3(f((C-C_1)\cap(V_a)^-))=0$, and $a\in V_a\subset(V_a)^-\subset G$. Here, $\bigcup\{(C-C_1)\cap(V_a)^-:a\in(C-C_1)\}=(C-C_1)$, so $f((C-C_1))=f(\bigcup\{(C-C_1)\cap(V_a)^-:a\in(C-C_1)\})=\bigcup\{f((C-C_1)\cap(V_a)^-):a\in(C-C_1)\}$. Since each $V_a\in\mathcal{B}$, and $\mathcal{B}$ is a countable collection, $\{f((C-C_1)\cap(V_a)^-):a\in(C-C_1)\}$ is a countable collection, and hence,

$$\mu_3(f(C-C_1))=\mu_3\Big(\bigcup\{f((C-C_1)\cap(V_a)^-):a\in(C-C_1)\}\Big)$$
$$\le\underbrace{\sum}_{\text{countable}}\mu_3(f((C-C_1)\cap(V_a)^-))=\underbrace{\sum}_{\text{countable}}0=0.$$

Thus, $\mu_3(f(C-C_1))=0$.                                                          $\square$

## 6.4  Sard's Theorem on Manifolds

**Theorem 6.27** *Let $m, n$ be positive integers such that $m < n$. Let $M$ be an $m$-dimensional smooth manifold, and $N$ be an $n$-dimensional smooth manifold. Let $f : M \to N$ be a smooth mapping. Then, $f(M)$ has measure zero in $N$.*

*Proof* For simplicity, let $n = m + 1$. Let us define a mapping $F : M \times \mathbb{R} \to N$ as follows: For every $p \in M$, and $t \in \mathbb{R}$, $F(p, t) \equiv f(p)$. Clearly, $F$ is smooth. (Reason: We know that the mapping $(p, t) \mapsto p$ from product manifold $M \times \mathbb{R}$ to $M$ is smooth. Since $(p, t) \mapsto p$ from $M \times \mathbb{R}$ to $M$ is smooth, and $f : M \to N$ is smooth, their composite $(p, t) \mapsto f(p)(= F(p, t))$ from $M \times \mathbb{R}$ to $N$ is smooth, and hence, $F$ is smooth).

Observe that $\{(p, 0) : p \in M\}(\subset M \times \mathbb{R})$ is of measure zero in $M \times \mathbb{R}$. (Reason: Let $(U \times \mathbb{R}, (\varphi \times \mathrm{Id}_\mathbb{R}))$ be an admissible coordinate chart of the product manifold $M \times \mathbb{R}$, where $(U, \varphi)$ is an admissible coordinate chart of $M$, and $(\mathbb{R}, \mathrm{Id}_\mathbb{R})$ is an admissible coordinate chart of $\mathbb{R}$. We know that $M \times \mathbb{R}$ is an $(m + 1)$-dimensional smooth manifold. We have to show that $\mu_{m+1}((\varphi \times \mathrm{Id}_\mathbb{R})(\{(p, 0) : p \in M\} \cap (U \times \mathbb{R}))) = 0$, where $\mu_{m+1}$ denotes the $(m + 1)$-dimensional Lebesgue measure over $\mathbb{R}^{m+1}$. Here,

$$\mu_{m+1}((\varphi \times \mathrm{Id}_\mathbb{R})(\{(p, 0) : p \in M\} \cap (U \times \mathbb{R})))$$
$$= \mu_{m+1}((\varphi \times \mathrm{Id}_\mathbb{R})(\{(p, 0) : p \in U\}))\mu_{m+1}((\varphi(p) \times \mathrm{Id}_\mathbb{R}(0)) : p \in U)$$
$$= \mu_{m+1}((\varphi(p), 0) : p \in U) \leq \mu_{m+1}(\mathbb{R}^m \times \{0\}) = 0).$$

Since $F : M \times \mathbb{R} \to N$ is smooth, and $\{(p, 0) : p \in M\}(\subset M \times \mathbb{R})$ is of measure zero in $M \times \mathbb{R}$, by Lemma 6.16, $F(\{(p, 0) : p \in M\})(= \{F(p, 0) : p \in M\} = \{f(p) : p \in M\} = f(M))$ has measure zero in $N$, and hence, $f(M)$ has measure zero in $N$.                                                                                                                    □

**Theorem 6.28** *Let $M$ be an $m$-dimensional smooth manifold, and $N$ be an $n$-dimensional smooth manifold. Let $f : M \to N$ be a smooth mapping. Let $C$ be the collection of all critical points of $f$. Then, $f(C)$ has measure zero in $N$.*

*Proof* Let us take any admissible coordinate chart $(V, \psi)$ for $N$. We have to show that

$$\mu_n(\psi((f(C)) \cap V)) = 0,$$

where $\mu_n$ denotes the $n$-dimensional Lebesgue measure over $\mathbb{R}^n$. Since $M$ is an $m$-dimensional smooth manifold, by Lemma 4.49, there exists a countable collection $\{(U_1, \varphi_1), (U_2, \varphi_2), (U_3, \varphi_3), \ldots\}$ of admissible coordinate charts of $M$ such that $\{U_1, U_2, U_3, \ldots\}$ is a basis of $M$. Here, $C \subset M = U_1 \cup U_2 \cup U_3 \cup \cdots$, so

$$0 \leq \mu_n(\psi((f(C)) \cap V))$$
$$= \mu_n(\psi((f((C \cap U_1) \cup (C \cap U_2) \cup (C \cap U_3) \cup \cdots)) \cap V))$$
$$= \mu_n(\psi(((f(C \cap U_1)) \cup (f(C \cap U_2)) \cup (f(C \cap U_3)) \cup \cdots) \cap V))$$
$$= \mu_n(\psi(((f(C \cap U_1)) \cap V) \cup ((f(C \cap U_2)) \cap V) \cup ((f(C \cap U_3)) \cap V) \cup \cdots))$$
$$= \mu_n(\psi((f(C \cap U_1)) \cap V) \cup \psi((f(C \cap U_2)) \cap V) \cup \psi((f(C \cap U_3)) \cap V) \cup \cdots)$$
$$\leq \mu_n(\psi((f(C \cap U_1)) \cap V)) + \mu_n(\psi((f(C \cap U_2)) \cap V))$$
$$+ \mu_n(\psi((f(C \cap U_3)) \cap V)) + \cdots.$$

Since

$$0 \leq \mu_n(\psi((f(C)) \cap V))$$
$$\leq \mu_n(\psi((f(C \cap U_1)) \cap V))$$
$$+ \mu_n(\psi((f(C \cap U_2)) \cap V))$$
$$+ \mu_n(\psi((f(C \cap U_3)) \cap V)) + \cdots,$$

it suffices to show that for each $k = 1, 2, \ldots$,

$$\mu_n(\psi((f(C \cap U_k)) \cap V)) = 0.$$

Let us try to show that

$$\mu_n(\psi((f(C \cap U_1)) \cap V)) = 0.$$

Since $\varphi_1$ is a diffeomorphism from open set $U_1$ onto open set $\varphi_1(U_1)$, and $\psi$ is a diffeomorphism from open set $V$ onto the open set $\psi(V)$, the collection of all critical points of $f$ in $U_1$ is equal to $(\varphi_1)^{-1}$(the collection of all critical points of $(\psi \circ f \circ (\varphi_1)^{-1})$ in $\varphi_1(U_1)$). Since $f : M \to N$ is a smooth mapping, $\psi \circ f \circ (\varphi_1)^{-1}$ is smooth. Since $\psi \circ f \circ (\varphi_1)^{-1}$ is smooth, by Lemma 6.26, $0 = \mu_n((\psi \circ f \circ (\varphi_1)^{-1})$ (the collection of all critical points of $(\psi \circ f \circ (\varphi_1)^{-1})$ in $\varphi_1(U_1))) = \mu_n((\psi \circ f) ((\varphi_1)^{-1}$ (the collection of all critical points of $(\psi \circ f \circ (\varphi_1)^{-1})$ in $\varphi_1(U_1)))) = \mu_n ((\psi \circ f)(\text{collection of }$ all critical points of $f$ in $U_1)) = \mu_n(\psi(f(\text{collection of all }$ critical points of $f$ in $U_1))) = \mu_n(\psi((f(C \cap U_1)) \cap V))$. Thus, $\mu_n(\psi((f(C \cap U_1)) \cap V)) = 0$. Similarly,

$$\mu_n(\psi((f(C \cap U_2)) \cap V)) = 0, \mu_n(\psi((f(C \cap U_3)) \cap V)) = 0, \text{etc.}$$

$\square$

Theorem 6.28 is known as *Sard's theorem*.

# Chapter 7
# Whitney Embedding Theorem

This is the smallest chapter of this book, because it contains only two theorems which are due to Whitney. These theorems have three serious reasons to study. Firstly, in its proof the celebrated Sard's theorem got an application. Secondly, the statement of Whitney embedding theorem was contrary to the common belief that a smooth manifold may not have any ambient space. Thirdly, in its proof, Whitney used almost all tools of smooth manifolds developed at that time. Fortunately, in this chapter we have all the prerequisite for its proof in the special case of compact smooth manifolds. For the general case, which is more difficult, one can find its proof somewhere else.

## 7.1 Compact Whitney Embedding in $\mathbb{R}^N$

**Lemma 7.1** *Let $M$ be an $m$-dimensional smooth manifold. Let $M$ be compact. Then, there exists a positive integer $N$ such that $M$ is embedded in the Euclidean space $\mathbb{R}^N$.*

*Proof* Since $M$ is an open cover of $M$, by Lemma 4.52, there exists a countable collection $(U_1, \varphi_1), (U_2, \varphi_2), (U_3, \varphi_3), \ldots$ of admissible coordinate charts of $M$ such that

1. $U_1, U_2, U_3, \ldots$ covers $M$,
2. $U_1, U_2, U_3, \ldots$ is locally finite,
3. Each $\varphi_n(U_n)$ is equal to the open ball $B_3(0)$,
4. $\varphi_1^{-1}(B_1(0)), \varphi_2^{-1}(B_1(0)), \varphi_3^{-1}(B_1(0)), \ldots$ is an open cover of $M$.

Since $\varphi_1^{-1}(B_1(0)), \varphi_2^{-1}(B_1(0)), \varphi_3^{-1}(B_1(0)), \ldots$ is an open cover of $M$, and $M$ is compact, there exists a positive integer $k$ such that $\varphi_1^{-1}(B_1(0)), \ldots, \varphi_k^{-1}(B_1(0))$ is an open cover of $M$. For every $i = 1, \ldots, k$, $\varphi_i^{-1}$ is continuous, and $B_1[0]$ is compact, so $\varphi_i^{-1}(B_1[0])$ is compact. $\varphi_i^{-1}(B_1[0])$ is compact in $M$, and $M$ is Hausdorff, so $\varphi_i^{-1}(B_1[0])$ is a closed subset of $M$. Since $\varphi_i^{-1}(B_1[0])$ is a closed subset of $M$,

© Springer India 2014
R. Sinha, *Smooth Manifolds*, DOI 10.1007/978-81-322-2104-3_7

$\varphi_i^{-1}(B_2(0))$ is an open subset of $M$, and $\varphi_i^{-1}(B_1[0]) \subset \varphi_i^{-1}(B_2(0))$, by Lemma 4.58, there exists a smooth function $\psi_i : M \to [0, 1]$ such that

a. for every $x$ in $\varphi_i^{-1}(B_1[0])$, $\psi_i(x) = 1$,
b. supp $\psi_i \subset \varphi_i^{-1}(B_2(0))$.

Here, $\varphi_1 : U_1 \to \varphi_1(U_1)(\subset \mathbb{R}^m)$, so let $\varphi_1 \equiv (\varphi_{11}, \ldots, \varphi_{1m})$, where each $\varphi_{1i} : U_1 \to \mathbb{R}$. Similarly, let $\varphi_2 \equiv (\varphi_{21}, \ldots, \varphi_{2m})$, where each $\varphi_{2i} : U_2 \to \mathbb{R}$, etc.

Take any $p \in U_1 \cap (\varphi_1^{-1}(B_2[0]))^c$. Clearly, $\psi_1(p)\varphi_{11}(p) = 0$. (Reason: Since supp $\psi_1$ is contained in $\varphi_1^{-1}(B_2(0))$, $\psi_1(x) = 0$ for every $x \notin \varphi_1^{-1}(B_2(0))$. Since $p \in U_1 \cap (\varphi_1^{-1}(B_2[0]))^c = U_1 - (\varphi_1^{-1}(B_2[0])) \subset U_1 - (\varphi_1^{-1}(B_2(0)))$, $\psi_1(p) = 0$, and hence, $\psi_1(p)\varphi_{11}(p) = 0 \cdot \varphi_{11}(p) = 0$. This shows that the following definition of function is well-defined.

Let us define a function $\hat{\varphi}_{11} : M \to \mathbb{R}$ as follows: for every $p \in M$,

$$\hat{\varphi}_{11}(p) \equiv \begin{cases} \psi_1(p)\varphi_{11}(p) & \text{if } p \in U_1 \\ 0 & \text{if } p \in (\varphi_1^{-1}(B_2[0]))^c. \end{cases}$$

Now we want to show that $\hat{\varphi}_{11} : M \to \mathbb{R}$ is a smooth function. Since $(U_1, \varphi_1)$ is an admissible coordinate chart of $M$, $\varphi_1 : U_1 \to \varphi_1(U_1)$ is a diffeomorphism, and hence, $\varphi_1 : U_1 \to \varphi_1(U_1)$ is smooth. Since $(\varphi_{11}, \ldots, \varphi_{1m}) \equiv \varphi_1 : U_1 \to \varphi_1(U_1)(\subset \mathbb{R}^m)$ is smooth, $\varphi_{11} : U_1 \to \mathbb{R}$ is smooth. Since $\varphi_{11} : U_1 \to \mathbb{R}$ is smooth, $\psi_1 : M \to [0, 1]$ is smooth, and $U_1$ is an open subset of $M$, their product $p \mapsto \psi_1(p)\varphi_{11}(p)$ is smooth over open set $U_1$. Also the constant function $p \mapsto 0$ is smooth over the open set $(\varphi_1^{-1}(B_2[0]))^c$. Now since $\hat{\varphi}_{11}$ is well-defined, $\hat{\varphi}_{11} : M \to \mathbb{R}$ is smooth. Since the function $\hat{\varphi}_{11} : M \to \mathbb{R}$ is well-defined, for every $p \in M$,

$$\hat{\varphi}_{11}(p) \equiv \begin{cases} \psi_1(p)\varphi_{11}(p) & \text{if } p \in \varphi_1^{-1}(B_2[0]) \\ 0 & \text{if } p \in (\varphi_1^{-1}(B_2[0]))^c. \end{cases}$$

Similarly, the function $\hat{\varphi}_{12} : M \to \mathbb{R}$ defined as follows: for every $p \in M$,

$$\hat{\varphi}_{12}(p) \equiv \begin{cases} \psi_1(p)\varphi_{12}(p) & \text{if } p \in \varphi_1^{-1}(B_2[0]) \\ 0 & \text{if } p \in (\varphi_1^{-1}(B_2[0]))^c \end{cases}$$

is smooth, etc. Further the function $\hat{\varphi}_{21} : M \to \mathbb{R}$ is defined as follows: for every $p \in M$,

$$\hat{\varphi}_{21}(p) \equiv \begin{cases} \psi_2(p)\varphi_{21}(p) & \text{if } p \in \varphi_2^{-1}(B_2[0]) \\ 0 & \text{if } p \in (\varphi_2^{-1}(B_2[0]))^c \end{cases}$$

is smooth. Similarly, the function $\hat{\varphi}_{22} : M \to \mathbb{R}$ is defined as follows: for every $p \in M$,

$$\hat{\varphi}_{22}(p) \equiv \begin{cases} \psi_2(p)\varphi_{22}(p) & \text{if } p \in \varphi_2^{-1}(B_2[0]) \\ 0 & \text{if } p \in \left(\varphi_2^{-1}(B_2[0])\right)^c \end{cases}$$

is smooth, etc. Now let us define a function

$$F: M \to \left( \left( \underbrace{\mathbb{R} \times \cdots \times \mathbb{R}}_{m} \right) \times \cdots \times \underbrace{\left( \underbrace{\mathbb{R} \times \cdots \times \mathbb{R}}_{m} \right)}_{k} \right) \times \left( \underbrace{\mathbb{R} \times \cdots \times \mathbb{R}}_{k} \right) \left( = \mathbb{R}^{k(m+1)} \right)$$

as follows: for every $p \in M$,

$$F(p) \equiv ((\hat{\varphi}_{11}(p), \ldots, \hat{\varphi}_{1m}(p)), \ldots, (\hat{\varphi}_{k1}(p), \ldots, \hat{\varphi}_{km}(p)), (\psi_1(p), \ldots, \psi_k(p))).$$

Since each component function of $F$ is smooth, $F : M \to \mathbb{R}^N$ is smooth, where $N \equiv k(m + 1)$. We shall try to show that

1. $F : M \to \mathbb{R}^N$ is 1–1,
2. $F : M \to \mathbb{R}^N$ is a smooth immersion, that is, for every $p \in M$, the linear map $dF_p : T_pM \to T_{F(p)}\mathbb{R}^N(= \mathbb{R}^N)$ is 1–1.

For 1: Let $F(p) = F(q)$, where $p, q \in M$. We have to show that $p = q$. Here, $((\hat{\varphi}_{11}(p), \ldots, \hat{\varphi}_{1m}(p)), \ldots, (\hat{\varphi}_{k1}(p), \ldots, \hat{\varphi}_{km}(p)), (\psi_1(p), \ldots, \psi_k(p))) = F(p) = F(q) = ((\hat{\varphi}_{11}(q), \ldots, \hat{\varphi}_{1m}(q)), \ldots, (\hat{\varphi}_{k1}(q), \ldots, \hat{\varphi}_{km}(q)), (\psi_1(q), \ldots, \psi_k(q)))$, so $(\psi_1(p), \ldots, \psi_k(p)) = (\psi_1(q), \ldots, \psi_k(q))$, and for each $i = 1, \ldots, k$, $(\hat{\varphi}_{i1}(p), \ldots, \hat{\varphi}_{im}(p)) = (\hat{\varphi}_{i1}(q), \ldots, \hat{\varphi}_{im}(q))$. Since $p \in M = \varphi_1^{-1}(B_1(0)) \cup \cdots \cup \varphi_k^{-1}(B_1(0))$, there exists a positive integer $l$ such that $1 \leq l \leq k$, and $p \in \varphi_l^{-1}(B_1(0))$, and hence, $\psi_l(p) = 1$. Since $1 \leq l \leq k$, and $(\psi_1(p), \ldots, \psi_k(p)) = (\psi_1(q), \ldots, \psi_k(q))$, $1 = \psi_l(p) = \psi_l(q)$. Since $\psi_l(p) = \psi_l(q)$, and $\psi_l(p) = 1$, $\psi_l(q) = 1$. Since $\psi_l(p) \neq 0$, and supp $\psi_l$ is contained in $\varphi_l^{-1}(B_2(0))$, $p \in \varphi_l^{-1}(B_2(0))(\subset \varphi_l^{-1}(B_2[0]))$, and hence, $p \in \varphi_l^{-1}(B_2[0])$. Since $p \in \varphi_l^{-1}(B_2[0])$, $\hat{\varphi}_{l1}(p) = \psi_l(p)\varphi_{l1}(p) = 1\varphi_{l1}(p) = \varphi_{l1}(p), \ldots, \hat{\varphi}_{lm}(p) = \psi_l(p)\varphi_{lm}(p) = 1\varphi_{lm}(p) = \varphi_{lm}(p)$. Similarly, $\hat{\varphi}_{l1}(q) = \varphi_{l1}(q), \ldots, \hat{\varphi}_{lm}(q) = \varphi_{lm}(q)$. Since $\varphi_l(p) = (\varphi_{l1}(p), \ldots, \varphi_{lm}(p)) = (\hat{\varphi}_{l1}(p), \ldots, \hat{\varphi}_{lm}(p)) = (\hat{\varphi}_{l1}(q), \ldots, \hat{\varphi}_{lm}(q)) = (\varphi_{l1}(q), \ldots, \varphi_{lm}(q)) = \varphi_l(q)$, $\varphi_l(p) = \varphi_l(q)$. Since $\varphi_l(p) = \varphi_l(q)$, and $\varphi_l$ is 1–1, $p = q$.

For 2: Let us take any $p \in M$. We have to show that the linear map $dF_p : T_pM \to \mathbb{R}^N$ is 1–1. For this purpose, let $(dF_p)(v) = 0$, where $v \in T_pM$. We have to show that $v = 0$.

Since $p \in M = \varphi_1^{-1}(B_1(0)) \cup \cdots \cup \varphi_k^{-1}(B_1(0))$, there exists a positive integer $l$ such that $1 \leq l \leq k$, and $p \in \varphi_l^{-1}(B_1(0))$. It follows that for every $x \in \varphi_l^{-1}(B_1(0))(\subset \varphi_l^{-1}(B_2[0]))$, we have $\psi_l(x) = 1$, and $(\hat{\varphi}_{l1}(x), \ldots, \hat{\varphi}_{lm}(x)) = (\psi_l(x)\varphi_{l1}(x), \ldots, \psi_l(x)\varphi_{lm}(x)) = (1\varphi_{l1}(x), \ldots, 1\varphi_{lm}(x)) = \varphi_l(x)$. For simplicity, let $l = 1$. Thus, for every $x \in \varphi_1^{-1}(B_1(0))$, we have $(\hat{\varphi}_{11}(x), \ldots, \hat{\varphi}_{1m}(x)) = \varphi_1(x)$. Here, $0 = (dF_p)(v) = ((d(\hat{\varphi}_1)_p)(v), \ldots, (d(\hat{\varphi}_k)_p)(v), (d(\psi_1)_p)(v), \ldots, (d(\psi_k)_p)(v)) = ((d(\varphi_1)_p)(v), \ldots, (d(\hat{\varphi}_k)_p)(v), (d(\psi_1)_p)(v), \ldots, (d(\psi_k)_p)(v))$, so $(d(\varphi_1)_p)(v) = 0$. Since $\varphi_1$ is a diffeomorphism, $d(\varphi_1)_p$ is an isomorphism. Since $d(\varphi_1)_p$ is an isomorphism, and $(d(\varphi_1)_p)(v) = 0$, $v = 0$.

Since $F : M \to \mathbb{R}^N$ is 1–1, $F : M \to \mathbb{R}^N$ is a smooth immersion, and $M$ is compact, by Note 5.72, $F : M \to \mathbb{R}^N$ is a smooth embedding. □

**Note 7.2** The above Lemma 7.1 is known as the *compact Whitney embedding in* $\mathbb{R}^N$.

## 7.2 Compact Whitney Embedding in $\mathbb{R}^{2m+1}$

**Lemma 7.3** *Let $M$ be an m-dimensional smooth manifold. Let $M$ be compact. Then, there exists an embedding $F : M \to \mathbb{R}^{2m+1}$.*

*Proof* By Lemma 7.1, there exists a positive integer $N$ such that $M$ is embedded in the Euclidean space $\mathbb{R}^N$. Since $M$ is embedded in the Euclidean space $\mathbb{R}^N$, there exists a smooth embedding $f : M \to \mathbb{R}^N$. □

Case I when $N < 2m + 1$. Let us not distinguish between $(x_1, \ldots, x_N)(\in \mathbb{R}^N)$ and

$$\left( \underbrace{x_1, \ldots, x_N}_{N}, \underbrace{0, \ldots, 0}_{(2m+1)-N} \right) (\in \mathbb{R}^{2m+1}).$$

Thus, $\mathbb{R}^N \subset \mathbb{R}^{2m+1}$, $\mathbb{R}^N$ is an $N$-slice of $\mathbb{R}^{2m+1}$, and $\mathbb{R}^N$ satisfies local $N$-slice condition. It follows, by Note 5.97, that $\mathbb{R}^N$ is an $N$-dimensional embedded submanifold of $\mathbb{R}^{2m+1}$, and hence, there exists a smooth embedding $g : \mathbb{R}^N \to \mathbb{R}^{2m+1}$. Since $f : M \to \mathbb{R}^N$ is a smooth embedding, and $g : \mathbb{R}^N \to \mathbb{R}^{2m+1}$ is a smooth embedding, their composite map $g \circ f : M \to \mathbb{R}^{2m+1}$ is a smooth embedding.

Case II when $2m + 1 < N$. Since $M$ is an $m$-dimensional smooth manifold, and $\mathbb{R}$ is a 1-dimensional smooth manifold, the product $M \times M \times \mathbb{R}$ is a $(2m + 1)$-dimensional smooth manifold. Let us define a mapping $\alpha : M \times M \times \mathbb{R} \to \mathbb{R}^N$ as follows: for every $p, q \in M$, and $t \in \mathbb{R}$,

$$\alpha(p,q,t) \equiv t(f(p) - f(q)).$$

Clearly, $\alpha$ is smooth. (Reason: Since $(p,q,t) \mapsto p$ is smooth, and $f : M \to \mathbb{R}^N$ is smooth, their composite $(p,q,t) \mapsto f(p)$ is smooth. Similarly, $(p,q,t) \mapsto f(q)$ is smooth. Since $(p,q,t) \mapsto f(p)$ is smooth, and $(p,q,t) \mapsto f(q)$ is smooth, $(p,q,t) \mapsto (f(p), f(q))$ from $M \times M \times \mathbb{R}$ to $\mathbb{R}^N \times \mathbb{R}^N$ is smooth. Since $(p,q,t) \mapsto (f(p), f(q))$ from $M \times M \times \mathbb{R}$ to $\mathbb{R}^N \times \mathbb{R}^N$ is smooth, and $(x,y) \mapsto (x - y)$ from $\mathbb{R}^N \times \mathbb{R}^N$ to $\mathbb{R}^N$ is smooth, their composite $(p,q,t) \mapsto (f(p) - f(q))$ from $M \times M \times \mathbb{R}$ to $\mathbb{R}^N$ is smooth. Since $(p,q,t) \mapsto t$ from $M \times M \times \mathbb{R}$ to $\mathbb{R}$ is smooth, and $(p,q,t) \mapsto (f(p) - f(q))$ from $M \times M \times \mathbb{R}$ to $\mathbb{R}^N$ is smooth, $(p,q,t) \mapsto (t, (f(p) - f(q)))$ from $M \times M \times \mathbb{R}$ to $\mathbb{R} \times \mathbb{R}^N$ is smooth. Since $(p,q,t) \mapsto (t, (f(p) - f(q)))$ from $M \times M \times \mathbb{R}$ to $\mathbb{R} \times \mathbb{R}^N$ is smooth, and $(t,x) \mapsto tx$ from $\mathbb{R} \times \mathbb{R}^N$ to $\mathbb{R}^N$ is smooth, their composite $(p,q,t) \mapsto t(f(p) - f(q))$ $(= \alpha(p,q,t))$ from $M \times M \times \mathbb{R}$ to $\mathbb{R}^N$ is smooth, and hence, $\alpha : M \times M \times \mathbb{R} \to \mathbb{R}^N$ is smooth.) Since $\alpha : M \times M \times \mathbb{R} \to \mathbb{R}^N$ is smooth, and $\dim(M \times M \times \mathbb{R}) = 2m + 1 < N = \dim(\mathbb{R}^N)$, by Theorem 6.27, $\mu_N(\alpha(M \times M \times \mathbb{R})) = 0$, where $\mu_N$ denotes the $N$-dimensional Lebesgue measure over $\mathbb{R}^N$.

Let us define a mapping $\beta : TM \to \mathbb{R}^N$ as follows: for every $(p,v) \in TM(= \coprod_{r \in M} T_r M)$, where $p \in M$, and $v \in T_p M$,

$$\beta(p,v) = (df_p)(v).$$

Recall that $TM$, the tangent bundle of $M$, is a $2m$-dimensional smooth manifold. Clearly, $\alpha$ is smooth. (Reason: Since $f : M \to \mathbb{R}^N$ is smooth, by Theorem 5.24, $(p,v) \mapsto (f(p), (df_p)(v))$ from $TM$ to $T\mathbb{R}^N(= \coprod_{r \in \mathbb{R}^N}(T_r \mathbb{R}^N) = \coprod_{r \in \mathbb{R}^N} \mathbb{R}^N = \{(r,x) : r \in \mathbb{R}^N \text{ and } x \in \mathbb{R}^N\} = \mathbb{R}^N \times \mathbb{R}^N)$ is smooth. Since $(p,v) \mapsto (f(p), (df_p)(v))$ from $TM$ to $\mathbb{R}^N \times \mathbb{R}^N$ is smooth, and $(x,y) \mapsto y$ from $\mathbb{R}^N \times \mathbb{R}^N$ to $\mathbb{R}^N$ is smooth, their composite $(p,v) \mapsto (df_p)(v)(= \beta(p,v))$ from $TM$ to $\mathbb{R}^N$ is smooth, and hence, $\beta : TM \to \mathbb{R}^N$ is smooth.) Since $\beta : TM \to \mathbb{R}^N$ is smooth, and $\dim(TM) = 2m < 2m + 1 < N = \dim(\mathbb{R}^N)$, by Theorem 6.27, $\mu_N(\beta(TM)) = 0$. Since $\alpha(M \times M \times \mathbb{R}) \subset \mathbb{R}^N$, $\beta(TM) \subset \mathbb{R}^N$, $\mu_N(\alpha(M \times M \times \mathbb{R})) = 0$, and $\mu_N(\beta(TM)) = 0$,

$$0 \leq \mu_N((\alpha(M \times M \times \mathbb{R})) \cup (\beta(TM))) \leq \mu_N(\alpha(M \times M \times \mathbb{R})) + \mu_N(\beta(TM))$$
$$= 0 + 0 = 0,$$

and hence, $\mu_N((\alpha(M \times M \times \mathbb{R})) \cup (\beta(TM))) = 0$. This shows that $\mathbb{R}^N - ((\alpha(M \times M \times \mathbb{R})) \cup (\beta(TM)))$ is nonempty, and hence, there exists $a \in \mathbb{R}^N$ such that $a \notin \alpha(M \times M \times \mathbb{R})$, $a \notin \beta(TM)$, and $|a| \neq 0$.

Let $H_a(\{\equiv v : v \in \mathbb{R}^N, v \cdot a = 0\})$ denotes the orthogonal complement of $a$. We know that $H_a$ is a "hyperplane" of $\mathbb{R}^N$, so we shall not distinguish between $H_a$ and $\mathbb{R}^{N-1}$. Let us define a function $P_a : \mathbb{R}^N \to \mathbb{R}^{N-1}$ as follows: for every $v \in \mathbb{R}^N$,

$$P_a(v)\left(\equiv v - \frac{(v \cdot a)}{|a|^2} a\right)$$

is the orthogonal projection of $v$ on the hyperplane $H_a$. Clearly, $P_a : \mathbb{R}^N \to \mathbb{R}^{N-1}$ is linear, and smooth, and therefore, $d(P_a)_v = P_a$ for every $v$ in $\mathbb{R}^N$. Since $f : M \to \mathbb{R}^N$ is smooth, and $P_a : \mathbb{R}^N \to \mathbb{R}^{N-1}$ is smooth, their composite $(P_a \circ f) : M \to \mathbb{R}^{N-1}$ is smooth. Now we shall try to show that

1. $(P_a \circ f) : M \to \mathbb{R}^{N-1}$ is 1–1,
2. $(P_a \circ f) : M \to \mathbb{R}^{N-1}$ is an immersion, that is, for every $p \in M$, the linear map $d(P_a \circ f)_p : T_p M \to T_{(P_a \circ f)(p)}(\mathbb{R}^{N-1})(= \mathbb{R}^{N-1})$ is 1–1, that is, for every $p \in M$, $(d(P_a \circ f)_p)(w) = 0$ implies $w = 0$.

For 1: Let $(P_a \circ f)(p) = (P_a \circ f)(q)$, where $p, q \in M$. We have to show that $p = q$. We claim that $p = q$. If not, otherwise, let $p \neq q$. We have to arrive at a contradiction. Since $p \neq q$, and $f : M \to \mathbb{R}^N$ is a smooth embedding (and hence, $f$ is 1–1), $f(p) \neq f(q)$. Since $P_a(f(p)) = (P_a \circ f)(p) = (P_a \circ f)(q) = P_a(f(q))$, and $P_a : \mathbb{R}^N \to \mathbb{R}^{N-1}$ is linear,

$$(f(p) - f(q)) - \frac{(f(p) - f(q)) \cdot a}{|a|^2} a = P_a(f(p) - f(q)) = 0.$$

Since

$$(f(p) - f(q)) - \frac{(f(p) - f(q)) \cdot a}{|a|^2} a = 0,$$

and $f(p) \neq f(q)$, $(f(p) - f(q)) \cdot a \neq 0$, and hence,

$$a = \frac{|a|^2}{(f(p) - f(q)) \cdot a} (f(p) - f(q)) = \alpha\left(p, q, \frac{|a|^2}{(f(p) - f(q)) \cdot a}\right)$$
$$\in \alpha(M \times M \times \mathbb{R}).$$

Thus, $a \in \alpha(M \times M \times \mathbb{R})$, which is a contradiction. So our claim is true, that is, $p = q$. Thus, $(P_a \circ f) : M \to \mathbb{R}^{N-1}$ is 1–1.

For 2: Let us take any $p \in M$. Let $(d(P_a \circ f)_p)(w) = 0$ where $w \in T_p M$. We have to show that $w = 0$. If not, otherwise, let $w \neq 0$. We have to arrive at a contradiction. Since $f : M \to \mathbb{R}^N$ is a smooth embedding (and hence, $f : M \to \mathbb{R}^N$ is an immersion), and $p \in M$, the linear map $df_p : T_p M \to T_{f(p)}(\mathbb{R}^N)(= \mathbb{R}^N)$ is 1–1. Since the linear map $df_p : T_p M \to \mathbb{R}^N$ is 1–1, and $0 \neq w \in T_p M$, $(df_p)(w) \neq 0$. Here,

$$(df_p)(w) - \frac{((df_p)(w)) \cdot a}{|a|^2} a = P_a((df_p)(w)) = (P_a \circ (df_p))(w)$$

$$= \left( \left( d(P_a)_{F(p)} \right) \circ (df_p) \right)(w)$$

$$= \left( d(P_a \circ f)_p \right)(w) = 0, \text{ so } (df_p)(w)$$

$$= \frac{((df_p)(w)) \cdot a}{|a|^2} a.$$

Since

$$(df_p)(w) = \frac{((df_p)(w)) \cdot a}{|a|^2} a,$$

and $(df_p)(w) \neq 0$, $((df_p)(w)) \cdot a \neq 0$, and hence,

$$a = \frac{|a|^2}{((df_p)(w)) \cdot a}((df_p)(w)) = (df_p)\left( \frac{|a|^2}{((df_p)(w)) \cdot a} w \right)$$

$$= \beta \left( p, \frac{|a|^2}{((df_p)(w)) \cdot a} w \right) \in \beta(TM).$$

Thus, $a \in \beta(TM)$, which is a contradiction. Hence, $w = 0$.

Since $(P_a \circ f) : M \to \mathbb{R}^{N-1}$ is 1–1, $(P_a \circ f) : M \to \mathbb{R}^{N-1}$ is an immersion, and $M$ is compact, by Note 5.72, $(P_a \circ f) : M \to \mathbb{R}^{N-1}$ is a smooth embedding. Continuing this construction repeatedly for finite number of times, we get a map $F : M \to \mathbb{R}^{2m+1}$, which is an embedding. □

**Note 7.4** The Lemma 7.3 is known as the *compact Whitney embedding in* $\mathbb{R}^{2m+1}$.

# Bibliography

Apostol, T.M.: Mathematical Analysis, 2nd edn. Addison-Wesley, Reading (1974)
Boothby, W.M.: An Introduction to Differentiable Manifolds and Riemannian Geometry, 2nd edn. Academic Press, Orlando (1986)
Lee, J.M.: Introduction to Smooth Manifolds, 2nd edn. Springer, New York (2013)
Loring, W.Tu.: An Introduction to Manifolds, 2nd edn. Springer, New York (2012)
Rudin, W.: Principles of Mathematical Analysis, 3rd edn. McGraw-Hill, New York (1976)
Spivak, M.: Calculus on Manifolds. Westview Press, USA (1998)

© Springer India 2014                                                                                      485
R. Sinha, *Smooth Manifolds*, DOI 10.1007/978-81-322-2104-3

Printed in the United States
By Bookmasters